HETEROGENEOUS CATALYSIS for the SYNTHETIC CHEMIST

HETEROGENEOUS CATALYSIS for the SYNTHETIC CHEMIST

ROBERT L. AUGUSTINE

Department of Chemistry
Seton Hall University
South Orange, New Jersey

Marcel Dekker, Inc.   New York • Basel • Hong Kong

Library of Congress Cataloging-in-Publication Data

Augustine, Robert L.
    Heterogeneous catalysis for the synthetic chemist / Robert L.
Augustine.
        p. cm.
    Includes bibliographical references and index.
    ISBN 0-8247-9021-9 (hardcover : alk. paper)
    1. Heterogeneous catalysis. I. Title.
QD505.A94 1995
541.39'5–dc20
                                                        95-38138
                                                        CIP

The publisher offers discounts on this book when ordered in bulk quantities. For more information, write to Special Sales/Professional Marketing at the address below.

This book is printed on acid-free paper.

MARCEL DEKKER, INC.
270 Madison Avenue, New York, New York 10016

Current printing (last digit):
10 9 8 7 6 5 4 3 2 1

PRINTED IN THE UNITED STATES OF AMERICA

To
Marilyn

Foreword

Professor Augustine's stated intent in writing this book is to encourage the use of heterogeneous catalysts in organic synthesis. In three well-organized sections, he addresses basic concepts of catalysis and the influence of reaction variables, catalysts and catalyst preparation, and, replete with many examples, the scope of heterogeneously catalyzed reactions applicable to organic syntheses. I think he has succeeded admirably; this work should be of help to synthetic chemists using heterogeneous catalysts, neophyte or not.

Although heterogeneous catalysts promote a variety of reactions the discussion of hydrogenation is preponderant. This reflects reality. Hydrogenation has a broad scope and is a widely used reaction of great utility. Those involved in this area might find this section worth having for itself.

In a sense the title of this book does it an injustice. This fact-filled, heavily documented work should have a much larger audience than synthetic chemists. For instance, it should make an excellent text for a course on heterogeneous catalysis.

Paul N. Rylander

Preface

Catalyzed reactions are becoming more important in the synthesis of fine chemicals and pharmaceuticals. While most of these reactions are promoted by soluble, homogeneous catalysts, a strong case can be made for an increase in the extent to which heterogeneous catalysts should be used in these synthetic sequences. The development of heterogeneously catalyzed reactions, however, is held up by the fact that most synthetic chemists do not have the necessary basic understanding of the factors involved in heterogeneous catalysis. About the only exposure most chemists have to a heterogeneous catalyst is in the hydrogenation of an organic functional group. The usual procedure here is to locate a somewhat analogous reaction in the literature, obtain the catalyst from a commercial vendor and then attempt the reaction using the published conditions. Some attempts may then be made to improve the product yield by varying the temperature and/or pressure and possibly the solvent, but, for the most part, these attempts are based on a "let's try this and see what happens" approach.

This book was written to overcome this problem by providing the synthetic chemist with sufficient information to understand the effect that the different reaction variables can have on the outcome of a heterogeneously catalyzed reaction. While the complete coverage of these factors would require several volumes, it is felt that sufficient information is given here so a rational approach can be applied in selecting the reaction conditions needed to optimize the product yield. For those readers requiring more information, references to reviews and the original literature are provided.

Most of this information is given in the first two sections of the book while the third section covers those heterogeneously catalyzed reactions that can be used for the synthesis of more complex molecules. This later section may also be of interest to those catalytic chemists and chemical engineers concerned with the development of catalytic processes for use in the synthesis of fine chemicals.

While this book is based on a graduate chemistry course "Heterogeneous Catalysis" that is offered periodically at Seton Hall University and on the short courses ""Heterogeneous Catalysis" and "Selective Hydrogenation" that are given by the author, it is also built upon the concepts and data presented in the previous text, *Catalytic Hydrogenation, Techniques and Applications in Organic Chemistry*, which was published by the author some thirty years ago. One of the successful features of the old book was the 'recipes' provided for the hydrogenation of the various functional groups. In this present text, the catalytic hydrogenation of organic compounds is covered in even more detail, but the 'recipes' are not given. It was felt that while they were useful in the earlier stages of the development of hydrogenation technology, they were too specific and not the general procedures that were implied. The current book presents a more detailed discussion of the effects that the various reaction parameters can have on reaction selectivity so the reader can apply these concepts to the development of optimal reaction conditions for the preparation of specific compounds. The generally useful procedures are still given, but not emphasized as before.

Another departure from the previous text is the inclusion of chapters on other heterogeneously catalyzed reactions such as oxidations, solid acid and base catalyzed reactions and the like, as well as a more thorough coverage of the catalysts themselves. Not only are metal catalysts discussed but also the simple and complex oxides, the zeolites and clays, both as catalysts in their own right as well as being potential supports for the catalytically active metals. The factors involved in the preparation of the catalysts are also discussed in sufficient detail that the reader will understand the problems associated with catalysts having specific properties needed for obtaining optimal selectivity in a given reaction.

Grateful appreciation is given to my colleagues and students who have read portions of the manuscript and provided helpful suggestions for its improvement, especially Dr. Setrak, K. Tanielyan, Dr. Leslie Scott Szivos, Dr. Lisa K. Doyle, Dr. Sanjay V. Malhotra, Sr. Leo S. Posner, Dr. Steven Roberto and Dr. John R. Sowa. Special thanks and appreciation must go to my wife, Marilyn, for her patience, understanding and the encouragement to stay with this project to its completion.

Robert L. Augustine

Contents

HETEROGENEOUS CATALYSIS for the SYNTHETIC CHEMIST

1

Introduction

For the typical synthetic chemist the use of a heterogeneously catalyzed reaction usually involves the finding of a seemingly appropriate "recipe" to follow and the resignation to settle for whatever yield may be obtained. With other synthetically viable reactions a basic mechanistic understanding can frequently be used to modify the reaction conditions in order to optimize product formation. Such an approach is not readily applied to heterogeneously catalyzed processes because the type of mechanistic understanding needed by the synthetic chemist is not yet commonly available for this type of reaction. Further, the average synthetic chemist has little experience in dealing with heterogeneous systems so basic factors such as heat and mass transport, and the effect of the various reaction parameters on these factors are generally not understood.

At present the most important synthetic use of heterogeneous catalysis is for the hydrogenation and hydrogenolysis of various organic functional groups. Other reactions, such as catalytic oxidation and C–C bond formations using surface-promoted organometallic reactions, are used only infrequently. To develop these reactions as viable synthetic procedures, though, will require that all of the facets of heterogeneous catalysis be more fully understood by the practicing chemist.

This book is written with this goal in mind. It is based on the assumption that the reader will have had little, if any, experience with heterogeneous systems. The first part of the book is concerned with the basic concepts of heterogeneous catalysis and begins with a general discussion of catalysis, how catalysts function, the nature of the active sites on a catalyst surface and how they promote a reaction. This is followed by a treatment of the heat and mass transport aspects in heterogeneous systems and how these factors can be recognized. Next is a presentation of the reaction parameters typically found in heterogeneous catalysis (temperature, pressure, solvent, etc.) and a description of how each can affect the mass transport in a heterogeneous system. A discussion of the type of reactors used for heterogeneously catalyzed reactions follows. A basic treatment of reaction kinetics as applied to these systems concludes the first section. The second part of the book is devoted to the catalyst itself with chapters on metal catalysts, the materials used to support them, oxide catalysts and zeolites. The final section deals with heterogeneously catalyzed reactions of interest to the synthetic chemist.

This information is intended to provide a basic foundation for developing a working understanding of heterogeneous catalysis and heterogeneously

1

catalyzed reactions. Hopefully, this will encourage synthetic chemists to use such processes more frequently and, thus, increase the application of such systems to the preparation of a variety of materials.

The literature of heterogeneous catalysis is extensive so no attempt has been made to provide a complete literature review of any of the topics covered. Sufficient references are provided, though, to give an entree into these topics so that further information may be obtained as needed.

Of the numerous texts in this area, the classic book by Bond,[1] his more recent abbreviated version,[2] and those of Gasser[3] and Gates[4] provide a basic, general treatment of heterogeneous catalysis. Satterfield's[5] text gives a more practical treatment from an engineering perspective. While all of these texts provide some discussion of the kinetics of heterogeneously catalyzed processes Boudart and Djega-Mariadassou[6] have written the definitive treatment of this topic. The book by Anderson[7] provides an introduction into the use of metal catalysts with good discussions of preparation procedures and support materials. Those by Delannay,[8] Jones and McNicol,[9] Thomas and Lambert,[10] Anderson[11] and Anderson and Dawson[12,13] provide information on the techniques used for catalyst preparation and characterization. The procedures for specific reactions can be found in those books describing the catalytic hydrogenation and hydrogenolysis of organic functional groups.[14–22] Discussions of catalytic oxidations, although not exclusively heterogeneously catalyzed, have also been published.[23,24] While some reference will be made in the following sections to these general texts, most of the references will be from the original literature.

References

1. G. C. Bond, *Catalysis by Metals,* Academic Press, New York, 1962.
2. G. C. Bond, *Heterogeneous Catalysis, Principles and Applications*, Oxford University Press, Oxford, 1987.
3. R. P. H. Gasser, *An Introduction to Chemisorption and Catalysis by Metals*, Clarindon Press, Oxford, 1985.
4. B. C. Gates, *Catalytic Chemistry*, Wiley, New York, 1992.
5. C. N. Satterfield, *Heterogeneous Catalysis in Practice,* McGraw Hill, New York, 1980.
6. M. Boudart and G. Djega-Mariadassou, *Kinetics of Heterogeneous Catalytic Reactions,* Princeton University Press, Princeton, 1984.
7. J. R. Anderson, *Structure of Metallic Catalysts,* Academic Press, New York, 1975.
8. F. Delannay, *Characterization of Heterogeneous Catalysis*, (Chemical Industries Series, Vol. 15), Dekker, New York, 1984.
9. A. Jones and B. D. McNicol, *Temperature Programmed Reduction for Solid Material Characterization,* (Chemical Industries Series, Vol. 24), Dekker, NewYork, 1986.
10. J. M. Thomas and R. M. Lambert, *Characterization of Catalysts,* Wiley, New York, 1980.
11. R. B. Anderson, *Experimental Methods in Catalysis Research, Vol. I,* Academic Press, New York, 1968.
12. R. B. Anderson and P. T. Dawson, *Experimental Methods in Catalysis Research, Vol. II,* Academic Press, New York, 1976.

13 R. B. Anderson and P. T. Dawson, *Experimental Methods in Catalysis Research, Vol. III*, Academic Press, New York, 1976.

14. R. L. Augustine, *Catalytic Hydrogenation, Techniques and Applications in Organic Synthesis*, Dekker, New York, 1965.

15. P. N. Rylander, *Catalytic Hydrogenation over Platinum Metals*, Academic Press, New York, 1967.

16. P. N. Rylander, *Catalytic Hydrogenation in Organic Synthesis*, Academic Press, New York, 1979.

17. P. N. Rylander, *Hydrogenation Methods*, Academic Press, New York, 1985.

18. M. Freifelder, *Practical Catalytic Hydrogenation*, Wiley, New York, 1971.

19. P. N. Rylander, *Organic Synthesis with Noble Metal Catalysts*, Academic Press, New York, 1973.

20. A. P. G Kieboom and F. VanRantwijk, *Hydrogenation and Hydrogenolysis in Synthetic Organic Chemistry*, Delft University Press, 1977.

21. M. Bartok, *Stereochemistry of Heterogeneous Metal Catalysis*, Wiley, New York, 1985.

22. L. Cerveny (Ed.), *Catalytic Hydrogenation*, (Studies in Surface Science and Catalysis, Vol. 27), Elsevier, Amsterdam, 1986.

23. R. A. Sheldon and J. K. Kochi, *Metal Catalyzed Oxidations of Organic Compounds*, Academic Press, New York, 1981.

24. G. I. Golodets, *Heterogeneous Catalytic Reactions Involving Molecular Oxygen* (Studies in Surface Science and Catalysis, Vol. 15), Elsevier, Amsterdam, 1983.

SECTION ONE

Basic Concepts

2

Catalysis

A catalyst is a substance that increases the rate at which a chemical reaction approaches equilibrium without, itself, becoming permanently involved in the reaction. The key word in this definition is *permanently* since there is ample evidence showing that the catalyst and the reactants interact before a reaction can take place. The product of this interaction is a reactive intermediate from which the products are formed. This substrate:catalyst interaction can take place homogeneously with both the reactants and the catalyst in the same phase, usually the liquid, or it can occur at the interface between two phases. These heterogeneously catalyzed reactions generally utilize a solid catalyst with the interaction taking place at either the gas/solid or liquid/solid interface. Additional phase transport problems can arise when a gaseous reactant is also present in the liquid/solid system.

Since a catalyst merely increases the rate of a reaction it cannot be used to initiate a reaction that is thermodynamically unfavorable. The enthalpy of the reaction as well as other thermodynamic factors are a function of the nature of the reactants and the products only and, thus, cannot be modified by the presence of a catalyst. Kinetic factors, such as the reaction rate, activation energy, nature of the transition state, and so on, are the reaction characteristics that can be affected by a catalyst.

Effect on Activation Energies

To illustrate this point, consider the reactions of ethanol shown in Eqns. 2.1–3.

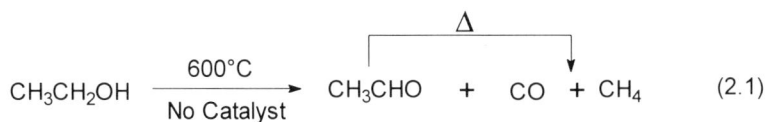

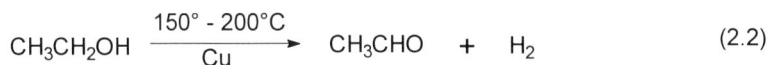

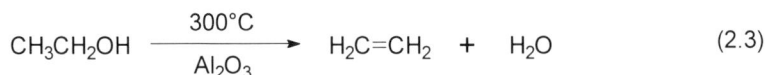

$$CH_3CH_2OH \xrightarrow[\text{No Catalyst}]{600°C} CH_3CHO \;\; + \;\; CO \; + \; CH_4 \qquad (2.1)$$

$$CH_3CH_2OH \xrightarrow[\text{Cu}]{150° - 200°C} CH_3CHO \;\; + \;\; H_2 \qquad (2.2)$$

$$CH_3CH_2OH \xrightarrow[\text{Al}_2O_3]{300°C} H_2C{=}CH_2 \;\; + \;\; H_2O \qquad (2.3)$$

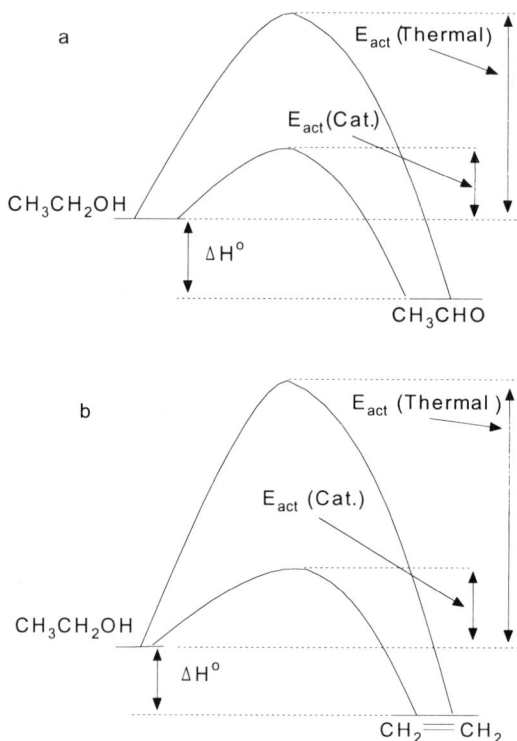

Fig. 2.1.   a) Energy curves for the thermal and copper catalyzed
           dehydrogenation of ethanol; b) energy curves for the
           thermal and Al₂O₃ catalyzed dehydration of ethanol.

The high temperatures required for the uncatalyzed dehydrogenation reaction indicate that the activation energy for this reaction is very high. In fact, the energy required is also sufficient to promote further decarbonylation of the product, acetaldehyde, to give methane and carbon monoxide as secondary products. As shown in Fig. 2.1a, though, the presence of a copper catalyst provides the reactant, ethanol, with a different, energetically more favored pathway for the formation of acetaldehyde. In this case the electron pair on the oxygen presumably interacts with the metal surface to give an intermediate with the carbinol C–H bond in proximity to the metal. Further reaction occurs to break this bond along with the O–H bond to form the aldehyde adsorbed on a dihydrido copper species (Eqn. 2.4). The aldehyde desorbs and reaction with oxygen removes the hydrogen to complete the reaction. The function of the copper catalyst in this case is to facilitate the breaking of the C–H bond and, thus, expedite the dehydrogenation.

(2.4)

The question remains, then, of why the dehydration products are formed only over alumina and not found in the thermal reaction. The dehydration reaction is thermodynamically viable but the activation energy for the uncatalyzed, thermal dehydration is greater than that for the thermal dehydrogenation. Alumina, however, because of its acidic character, promotes the dehydration to give the alkene in much the same way that this reaction proceeds when promoted by acids (Fig. 2.1b). The relative heights of the activation energies for the dehydration and dehydrogenation reactions (Fig. 2.1) are not relevant since the dehydrogenation is not a viable reaction over alumina and dehydration does not take place over copper. In most catalytic reactions the selectivity is generally determined by the nature of the catalyst being used with the other reaction parameters having a secondary importance.

The Catalyst Cycle

In a homogeneously catalyzed reaction the determination of the kinetic factors for the process is usually straightforward. In a solution, reactants and the soluble catalysts are uniformly distributed throughout the reaction medium and the reaction rate can be expressed as a function of the concentrations of these substances. A heterogeneously catalyzed process is more complex because the catalyst is not uniformly distributed throughout the reaction medium. Consider a two phase system, either vapor/solid or liquid/solid, with the solid phase the catalyst. In such a system several steps are needed to complete the catalytic cycle:

1. Transport of the reactants to the catalyst.

2. Interaction of the reactants with the catalyst (adsorption).

3. Reaction of adsorbed species to give the product(s).

4. Desorption of the product(s) from the catalyst.

5. Transport of the product(s) away from the catalyst.

The desired reaction takes place only in Step 3, but Steps 2 and 4 also involve chemical changes so any rate data obtained from such reactions includes all three steps. Any activation energy measurements also apply to the three step combination. The situation is further clouded by the fact that while Steps 1 and 5 merely involve a physical transport, either one can be the rate limiting step for the overall reaction process. Because of these complications the activation energies determined for heterogeneously catalyzed reactions are not the activation energies for the specific reactions in Step 3 but, rather, are apparent activation energies for the entire reaction sequence. Even with these limitations these apparent activation energies can be used to determine whether the rate limiting step is one of the chemical processes or merely a transport limitation. The rule of thumb is that if the apparent activation energy is greater than 10 kcal/mole (40 kJ/mole) then the chemical processes are rate limiting. If it is 3–4 kcal/mole (12–15 kJ/mole) or lower then the transport processes are assuming a greater degree of control over the reaction.[1] A more complete discussion of mass transport and its effect on vapor-and liquid-phase reactions is given in Chapter 5.

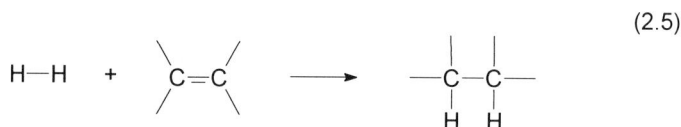

$$(2.5)$$

$$\text{H--H} \quad + \quad \underset{/}{\overset{\backslash}{\text{C}}} = \underset{\backslash}{\overset{/}{\text{C}}} \quad \longrightarrow \quad -\underset{|}{\overset{|}{\text{C}}}-\underset{\underset{\text{H}}{|}}{\overset{|}{\text{C}}}- $$

The general purpose of a catalyst is to provide a reaction pathway by which bonds may be weakened or broken. This is accomplished by the formation of reactive surface intermediates which, in turn, interact to give the reaction product(s). To illustrate this point, consider the addition of hydrogen to an alkene to give an alkane (Eqn. 2.5). This reaction involves the breaking of a H–H σ bond and a C=C π bond followed by the formation of two C–H σ bonds. While the overall process is exothermic there is no viable mechanistic pathway available by which this hydrogenation reaction can proceed in a concerted manner so that the energy needed for bond breaking comes from concomitant bond formation. Thus, an alternate route is needed to promote the bond breaking prior to the formation of the C–H bonds. This is provided by the catalyst.

Perhaps the oldest, viable organic reaction mechanism is that proposed by Horiuti and Polanyi[2] in 1934 for the hydrogenation of double bonds (Fig. 2.2). In this process the asterisk (*), indicates an interaction (adsorption) of the substrate with the catalyst surface. In Step 1 the H–H σ bond is broken and replaced by two, more reactive, M–H bonds. The π bond of the alkene is weakened by adsorption in Step 2. In Step 3 one of the activated hydrogen atoms reacts with the adsorbed alkene to give what is termed a "half-hydrogenated state" that is actually a surface organometallic species which then reacts in Step 4 with a second adsorbed hydrogen to give the alkane. This mechanism is widely accepted and used to explain all aspects of the alkene hydrogenation reaction but it suffers by not providing a better description of the adsorbed species and the nature of their interaction. Such matters can be described better by the application of basic organometallic chemistry concepts to the process.

$$H_2 \rightleftharpoons 2\ \underset{*}{H} \qquad\qquad (1)$$

$$\overset{\backslash}{\underset{/}{C}}=\overset{/}{\underset{\backslash}{C}} \rightleftharpoons \overset{\backslash}{\underset{/}{C}}\underset{*}{=}\overset{/}{\underset{\backslash}{C}} \qquad\qquad (2)$$

$$\underset{*}{\overset{\backslash}{\underset{/}{C}}}=\overset{/}{\underset{\backslash}{C}} + \underset{*}{\overset{H}{|}} \rightleftharpoons -\overset{|}{\underset{*}{C}}-\overset{|}{\underset{H}{C}}- \qquad\qquad (3)$$

$$-\overset{|}{\underset{*}{C}}-\overset{|}{\underset{H}{C}}- + \underset{*}{\overset{H}{|}} \longrightarrow -\overset{|}{\underset{H}{C}}-\overset{|}{\underset{H}{C}}- \qquad\qquad (4)$$

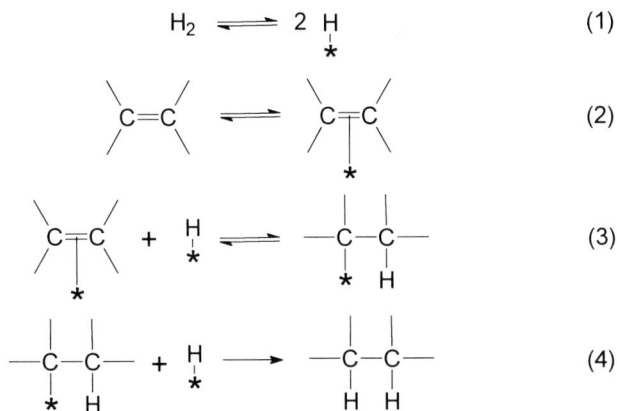

Fig. 2.2. Horiuti-Polanyi mechanism for alkene hydrogenation.

Adsorption

It has to be recognized that adsorption on a catalyst surface must be energetically favorable, have a relatively low activation energy and lead to the formation of reactive surface species. The reaction, preferably, should also be concerted with the energy needed for bond breaking resulting from concomitant bond formation. The weakening or breaking of the π bond of the alkene can be rationalized using the classic Chatt-Dewar-Duncanson[3,4] model for alkene complexation on a transition metal complex as depicted in Fig. 2.3a. In this model, π electrons are donated to an empty orbital of a metal atom on the catalyst surface. This is accompanied by back bonding from the filled d orbitals on the catalyst to the π^*

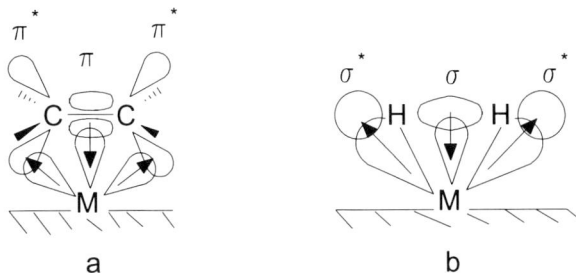

Fig. 2.3. a) Chatt-Dewar-Duncanson model for alkene adsorption on a metal active site;[3,4] b) analogous representation for hydrogen adsorption.

orbitals of the alkene. Both the donation of the bonding π electrons and the enhancement of the antibonding π^* orbitals serve to weaken the π bond of the alkene. The analogous procedure proposed for the adsorption of hydrogen is pictured in Fig. 2.3b. The latter case involves σ bond electron donation to the empty orbitals on a surface metal atom and back bonding to the σ^* orbitals. This results in the breaking of the H–H bond and the formation of a dihydrido species on the catalyst surface. This mode of adsorption is referred to as being dissociative since the adsorbing molecule is cleaved in the process. The alkene adsorption pictured in Fig. 2.3a is termed associative since the adsorbed species remains on the surface intact. The processes by which the adsorbed species interact is treated more fully in Chapter 4.

It should be recognized that while the adsorption of the reactants on the catalyst surface is a necessary feature in catalytic processes, adsorption, of itself, does not necessarily lead to a catalyzed reaction. For a reaction between adsorbed species to take place the adsorption of the reactants cannot be too strong nor too weak as illustrated by the volcano curve in Fig. 2.4. When the adsorption is too weak the amount of adsorbed species is too low to sustain the reaction. When strong adsorption occurs the substrate cannot leave the surface and the catalyst becomes poisoned for further reaction. For instance, hydrogen can be adsorbed on almost all metals and ethylene on most of them[5] yet only a few are capable of promoting the hydrogenation of ethylene. On metals such as Ti, V, Cr, Mo or W,

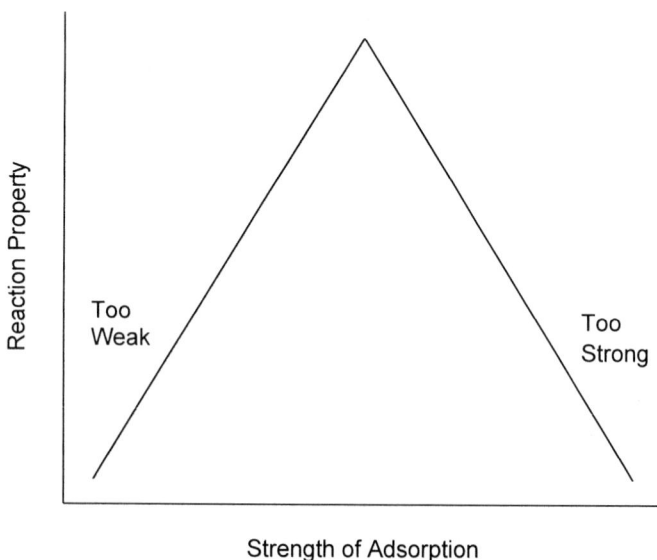

Fig. 2.4. Volcano curve relating strength of adsorption with reaction properties such as the rate of reaction.

ethylene adsorbs so strongly that the hydrogenation cannot be promoted. On other metals such as Mn, Au, or Ag hydrogen chemisorption is too weak so the formation of the active surface hydride cannot take place.

Physisorption and the BET Procedure

For practical purposes all adsorptions can be classified as one of two types. It can involve merely the van der Waals interaction between the substrate and the catalyst, a process that is termed physical adsorption or physisorption. Alternately, it can involve the formation of catalyst-substrate bonds as discussed above. This is termed chemical adsorption or chemisorption. While the latter is the basis for the chemistry of catalysts, physisorption is the basis for the BET procedure which is commonly used to measure the surface area of solids.[7-9]

In this technique a weighed sample of the material to be analyzed is placed in a tube of known volume and heated under vacuum to be degassed. The sample tube is then cooled in liquid nitrogen and a known amount of nitrogen gas is introduced into the cooled tube. After equilibration, the pressure is measured and the sequence repeated with successive pulses of nitrogen. Knowing the volume of the system, the temperature and the amount of nitrogen gas added with each pulse, the expected pressure in the absence of any adsorption can be calculated. From the difference between the calculated pressure and the observed pressure at each point the amount of nitrogen adsorbed can be determined. An idealized plot of observed pressure versus the volume adsorbed is shown in Fig. 2.5. The steep increase in the initial stages of the experiment shows the extent to which the first

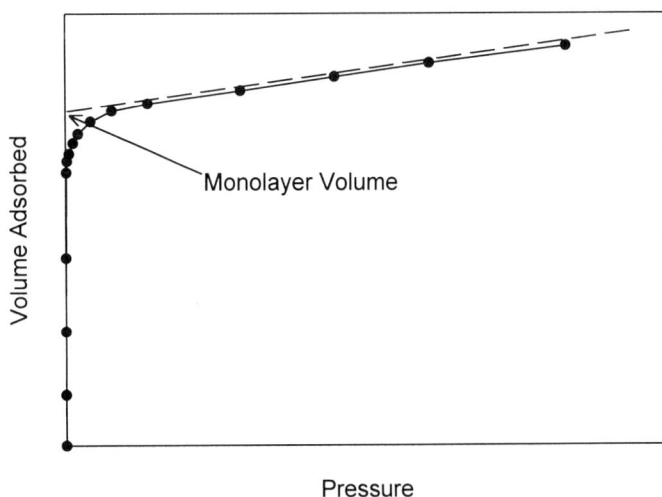

Fig. 2.5. Idealized BET plot for the determination of the surface area of catalysts and supports.

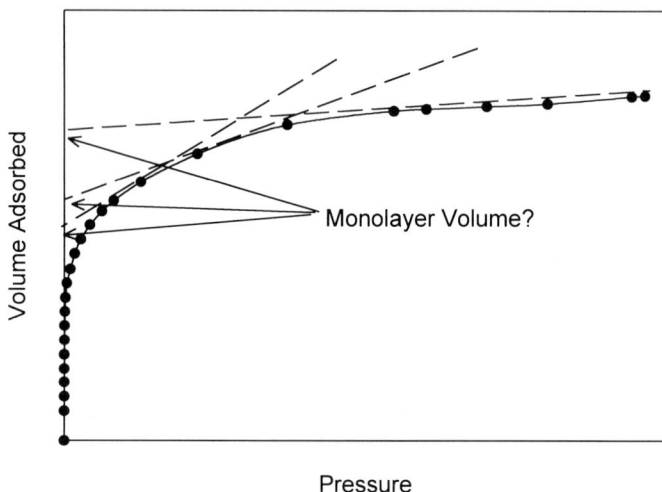

Fig. 2.6. One of the more typical types of curves observed in the determination of the surface area of catalysts and supports using the BET procedure.

few pulses of nitrogen are adsorbed. Unfortunately, with physisorption the adsorption process occurs beyond the formation of the first adsorption monolayer so additional layers of nitrogen molecules become stacked on top of the first resulting in a continuous adsorption curve. The volume of the monolayer adsorption should be determined by extrapolating the polylayer portion of the curve back to zero pressure as shown by the dotted line in Fig. 2.5.

The actual adsorption curves obtained are not as simple as that depicted in Fig. 2.5 but, rather, fall into one of several general patterns,[10,11] one of which is depicted in Fig. 2.6. With these curves extrapolation to zero pressure is not unambiguous so monolayer coverage is frequently determined by applying the BET equation in the form shown in Eqn. 2.6,[10] where V_{ads} is the volume of gas adsorbed at pressure P, P_0 is the saturation pressure, commonly 200–400 Torr, V_m is the volume of gas adsorbed at monolayer coverage and C is the BET constant that is related to the enthalpy of adsorption.

$$\frac{P}{V_{ads}(P_0 - P)} = \frac{1}{V_m C} + \frac{C-1}{V_m C} \times \frac{P}{P_0} \qquad (2.6)$$

By plotting the left side of Eqn. 2.6 against P/P_0 a straight line results with the relationships given in Eqns. 2.7–9. The BET surface area of the sample,

$$\text{Slope} = S = \frac{(C-1)}{V_m C} \tag{2.7}$$

$$\text{Intercept} = I = \frac{1}{V_m C} \tag{2.8}$$

$$V_m = \frac{1}{S+I} \tag{2.9}$$

expressed in m^2/g, is then calculated using Eqn. 2.10 where SA is the surface area, N is Avogadro's number, and A_m is the cross sectional area of the adsorbate molecule which, for nitrogen, is 16.2 $Å^2$.

$$SA = V_m \times N \times A_m \times 10^{-20} \tag{2.10}$$

By applying a vacuum to the system after adsorption is complete and measuring the rate of nitrogen desorption it is possible to obtain the pore volume of the material as well as the size distribution of the pores (under 50 nm) of the material.[8,9,11]

While these values are frequently reported with significant accuracy, what is usually needed is merely an evaluation of whether the material under investigation has a low surface area (less than 10 m^2/g), a moderate surface area (between 50 and 100 m^2/g), a high surface area (200–500 m^2/g) or a very high surface area (greater than 800 m^2/g). The pore distribution can be classed as predominantly microporous (small pore, less than 0.5 nm), mesoporous (moderate pore, 1.0 to 3.0 nm) or macroporous (large pore, greater than 5.0 nm).[12,13]

Chemisorption

Physisorption involves only a weak attraction between the substrate and the adsorbent but in chemisorption a chemical reaction takes place between the adsorbent and atoms on the catalyst surface. As a result, chemisorbed species are attached to the surface with chemical bonds and are more difficult to remove. If the adsorption of hydrogen on nickel is considered as an example, the reaction involves the breaking of an H–H bond and the formation of two Ni–H bonds on the surface. As shown in Fig. 2.3, this adsorption occurs by way of an initially adsorbed dihydrogen molecule. It proceeds via σ electron donation and back bonding to the σ^* orbitals of the hydrogen molecule with the final formation of the two surface M–H species.

The activation energies of most chemisorptions are very low, sometimes even zero.[14] The reason for this low activation energy is shown in Fig. 2.7[14] which illustrates the potential energy curves for both the physisorption (curve **P**)

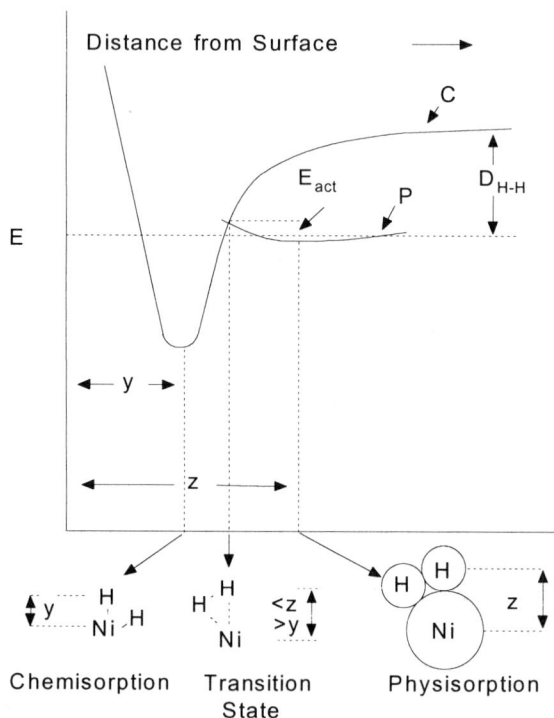

Fig. 2.7.   Energy curves for the chemisorption of hydrogen on nickel.
            (Redrawn from Ref. 14).

and chemisorption (curve **C**) of hydrogen on a nickel surface. The initial stage of
the reaction involves the physisorption of the hydrogen on the nickel with the
energy minimum of curve **P** corresponding to the sum of the van der Waals radii
of hydrogen and nickel (**z**). This physisorption process brings the hydrogen
molecule close enough to the nickel so the electron orbitals of the atoms on the
surface can begin to interact with the σ and σ^* molecular orbitals of the hydrogen.
As this interaction strengthens, the hydrogen molecule is attracted closer to the
nickel surface and the potential energy for the interaction increases slightly
following the physisorption energy curve (**P**) until it intersects with the
chemisorption curve (**C**). At this point attractive forces predominate resulting in a
shortening of the Ni–H distance along with the weakening and ultimate breaking
of the H–H bond.

The minimum of the chemisorption potential energy curve (**C**)
corresponds to the sum of the Ni and H atomic radii (**y**), a result of the formation
of the Ni–H bonds. Fig. 2.7 also shows the atomic arrangements at the various

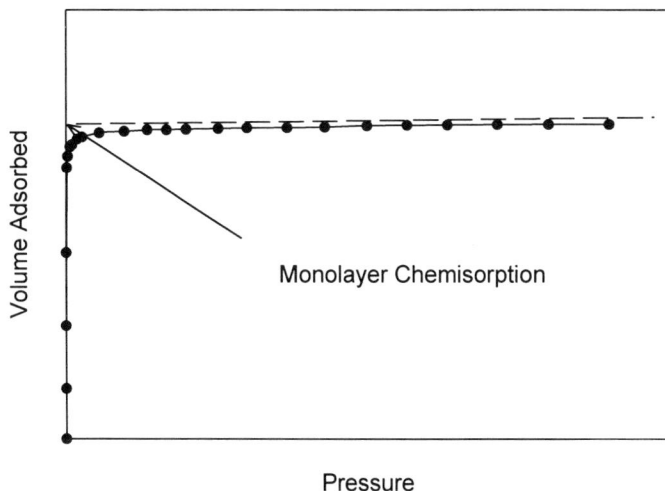

Fig. 2.8. Pressure-volume adsorbed relationship in the chemisorption of
          hydrogen on a metal catalyst as determined by the static
          method.

stages of this interaction as related to the appropriate positions on the potential
energy curves.

Metal Surface Areas by Selective Chemisorption

Since virtually all catalytically active metals chemisorb hydrogen, a measure of
the amount of hydrogen adsorbed on a metal is commonly used to determine the
metal surface area of catalysts. This is usually expressed as the catalyst dispersion
which is defined as the number of surface atoms divided by the total number of
metal atoms. Multiplying the dispersion by 100 gives a value referred to as the
percent exposed. The amount of hydrogen chemisorbed on a metal catalyst can be
determined in either a static mode or by the use of a pulse system The static mode
uses essentially the same approach as the BET method but is generally done at
ambient temperature since at this temperature physisorption of hydrogen is
virtually non-existent except as the precursor to chemisorption. The resulting
volume adsorbed-pressure curve (Fig. 2.8) is different from that of physisorption
in that only monolayer coverage is observed and the extrapolation to volume
adsorbed at zero pressure is straightforward.

In the pulse technique[15,16] a known amount of catalyst is placed in a flow
reactor under a stream of nitrogen or argon. Measured pulses of hydrogen are
then introduced into the carrier gas stream and passed through the catalyst bed.
The amount of hydrogen not adsorbed is measured by on-line gas chromatography
as depicted in Fig. 2.9. Subsequent pulses are introduced until no further

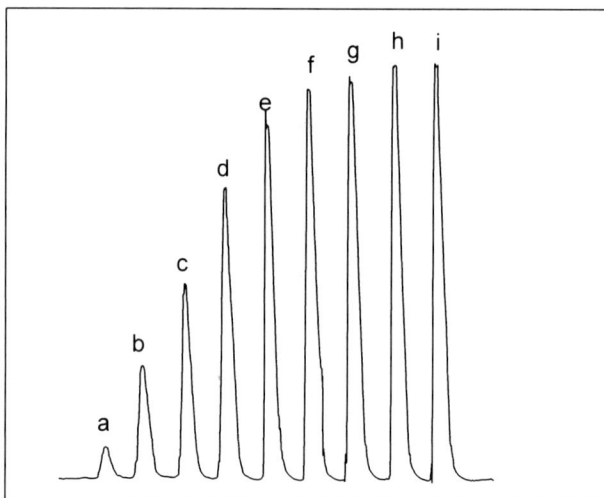

Fig. 2.9.   Gas chromatographic peaks showing the extent of
            hydrogen chemisorption as determined by the pulse
            technique.

adsorption takes place, indicated by the constant peak heights and areas of
subsequent peaks, as shown by peaks **h** and **i** in Fig. 2.9. The area under each of
these constant-sized peaks is related to the size of the hydrogen pulses introduced
into the reactor. The difference in area between these peaks and the smaller ones
(**a** through **g** in Fig. 2.9) determines the amount of hydrogen adsorbed.

Hydrogen chemisorption is the preferred method for determining metal
surface area. The process is simple and there is virtually no physisorption or
adsorption of the hydrogen on any oxide support. There are, however, some
potential problems. The pulse adsorption technique can give low values if some
of the hydrogen is only weakly adsorbed and is swept away by the carrier gas
stream between pulses. This problem can be minimized by using the slowest flow
rate consistent with good gas chromatographic analysis and spacing the pulses as
close together as possible while still having complete chromatographic separation
of the peaks. The use of small catalyst samples and shortened hydrogen pulses
can minimize the chromatographic problems.

The dissolution of hydrogen in palladium to form α- and β-phase
palladium hydrides has been assumed to present problems in determining the
surface area of palladium catalysts by hydrogen chemisorption, but under normal
chemisorption conditions this is probably not a factor. The α-phase Pd–H forms
initially by the migration of hydrogen atoms from the surface into the interstitial
volume of the palladium crystals. Hydrogen pressures near atmospheric are

required for this to occur. The β-phase Pd–H involves more extensive hydrogen dissolution resulting in a disruption of the palladium crystalline lattice. Higher pressures are needed for this to take place.[17-20] In a pulse chemisorption procedure the partial pressure of the hydrogen in the carrier gas stream is low. With 20 μl pulses in a 30 cc/min carrier gas stream, the partial pressure of hydrogen is less than 10 Torr, sufficient for chemisorption but not nearly enough for α- or β phase Pd–H formation. In the static system higher hydrogen pressures are involved but chemisorption measurements are generally terminated at pressures of about 200 Torr which, again, is probably too low for α- or β-phase Pd–H formation.

Another potential problem is the spillover of adsorbed hydrogen into the support material of the catalyst which is depicted in Fig. 2.10.[21,22] This occurs when the adsorbed hydrogen atoms migrate from the metal surface into the interstitial volume of the support material and are replaced on the metal by more hydrogen adsorbed from the atmosphere surrounding the catalyst. In some cases very large amounts of hydrogen can be spilled over into the support material but this requires exposure of the catalyst to at least one atmosphere of hydrogen for prolonged periods of time[22] and, in other cases, the use of higher temperatures.[21] Since neither of these conditions is present in the standard chemisorption procedures, the effect of hydrogen spillover on the determination of metal surface areas should be minimal.

The chemisorption of other gases has also been used to measure metal surface area. Oxygen has been used occasionally but has the disadvantage that it can oxidize some metals, particularly if higher temperatures are used. Silver and gold are the only metals for which oxygen chemisorption is routinely used. Elevated temperatures are required for gold.[23] Carbon monoxide is frequently used for palladium metal surface area determination. As discussed in the next section, though, problems can arise in the interpretation of CO chemisorption data

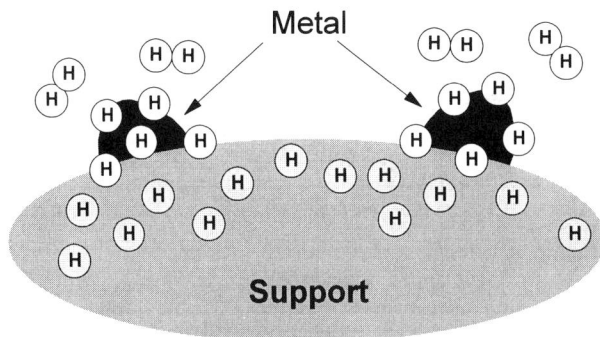

Fig. 2.10. Spillover of hydrogen from the surface of supported metal
particles into the interstitial volume of the support.

$$2\ Pt\ +\ H_2 \longrightarrow\ 2\ Pt\text{-}H \tag{2.11}$$

$$4\ Pt\text{-}H + 3O_2 \longrightarrow\ 4\ Pt\text{-}O\ +\ 2\ H_2O \tag{2.12}$$

$$2\ Pt\text{-}O + 3H_2 \longrightarrow\ 2\ Pt\text{-}H\ +\ 2\ H_2O \tag{2.13}$$

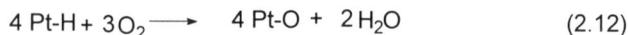

because of the many different surface species formed by its interaction with metals. Carbon monoxide should never be used to measure the surface areas of nickel catalysts because of the facile formation of the toxic $Ni(CO)_4$.

Another method for determining metal surface areas is the hydrogen-oxygen titration technique first described by Benson and Boudart.[24] In this procedure pulses of hydrogen of a known size are passed through the catalyst bed to fully cover the metal surface. These are followed by pulses of oxygen to react with the hydrogen to form water, with some also adsorbing onto the metal. Further pulses of hydrogen react first with the adsorbed oxygen and then adsorb onto the metal. The oxygen–hydrogen cycles are repeated and the water that is formed is absorbed either by the support material or an external dehydrating agent. The amount of each pulse of gas remaining after adsorption or reaction is determined by on-line gas chromatography using a procedure similar to that depicted by Fig. 2.9. The equations for this sequence are given in Eqns. 2.11–13 using platinum as an example but this procedure has also been applied to the determination of the metal surface areas of palladium[25] and rhodium[26] catalysts as well.

Chemisorption Stoichiometry

While all of these procedures can be used to measure the extent of selective chemisorption on metal surfaces there is a problem associated with the conversion of these data into a meaningful metal surface area.   Unlike the use of physisorption in the BET procedure, the metal surface area is not measured using the cross sectional area of the adsorbed gases but, rather, from the number of atoms or molecules adsorbed on the metal surface. The major difficulty lies in the determination of the stoichiometry of the adsorbed species with individual metal atoms on the catalyst surface. Generally a 1:1 substrate:surface metal atom ratio is used to measure the metal dispersion. However, infrared investigations of CO adsorbed on rhodium catalysts show the presence of three distinct surface species.[27,28] As depicted in Fig. 2.11 there is a *gem* dicarbonyl species, **I**, having infrared (ir) bands near 2100 and 2030 cm^{-1}, a linear monocarbonyl entity, **II**, with ir absorption at 2070 cm^{-1} and a bridged carbonyl, **III**, with a band at 1870 cm^{-1}. Similar carbonyls can be present on the surfaces of other metals. With all of these possibilities and no way of determining the relative amounts of each type present on the metal surface determining the stoichiometry of adsorption can be a significant problem.

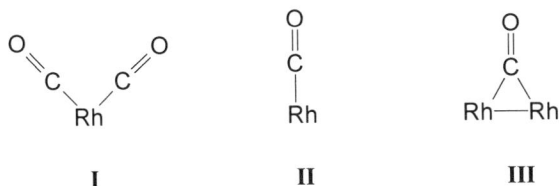

Fig. 2.11. Surface rhodium carbonyls.

A similar problem exists with hydrogen adsorption. As discussed in the next chapter, adsorption of hydrogen can take place on a single surface atom to give an MH_2 species or on two metal atoms to give M–H entities. Bridged adsorbed hydrogens are also possible. A number of attempts have been made to correlate hydrogen chemisorption data with metal particle size information which was obtained using transmission electron microscopic (TEM) analysis of the catalyst and/or X-ray line broadening (XRLB) calculations to determine average particle size. TEM is the most common comparison technique used. With this procedure particles as small as 0.3–0.5 nm can be observed when all of the experimental parameters are favorable. A lower resolution of 1.5–2.5 nm is more commonly the case. While particles of various sizes can be seen, the determination of reasonably exact particle size distributions can be a tedious task.[29-31] X-ray line broadening can be used to measure the average size of the metal particles but the lower limit with this technique is 4.0–5.0 nm. This procedure requires that both the particle and the support be crystalline.[32-34]

The difficulty with the use of these instrumental techniques as a means of determining the particle surface area is that while particle sizes or size distributions may be obtained, the shape of the metal particles is generally not easily discernible. Assumptions must be made concerning the shape before any surface area can be calculated. In one of these comparisons[35] a 2.5% Pt/SiO_2 catalyst was analyzed by TEM and XRLB as well as hydrogen chemisorption. The TEM data indicated an average particle size of 2.9 nm while XRLB gave values between 3.5nm and 5.0 nm depending on the reflection and particle shape assumptions made. Hydrogen chemisorption gave a dispersion of 0.24 using a 1:1 H:Pt stoichiometry. This corresponds to particle sizes between 2.9–4.5 nm depending on the particle shape used for the calculations. This general agreement, which was also observed in a number of similar studies, has led to the common acceptance of a 1:1 H:M stoichiometry in the determination of catalyst dispersion by hydrogen chemisorption. This stoichiometry is also used for both hydrogen and oxygen in the Boudart surface titration procedure.[24]

More detailed comparisons of TEM and chemisorption data have substantiated the 1:1 H:Pt stoichiometry for particles larger than 5.0nm but for smaller particles this relationship begins to fail. With smaller particles there is a higher ratio of the more coordinately unsaturated atoms present and these are capable of MH_2 formation which leads to an H:M stoichiometry greater than 1.[36-38] In hydrogen-oxygen titrations on platinum catalysts having small metal particles the H:Pt ratio increases while the O:Pt stoichiometry decreases from 1:1 to 0.5:1.[39] With gold catalysts the O:Au stoichiometry depends on the adsorption temperature. At 200°C it is approximately 0.5:1 while at 300°C it is about 1:1.[23]

Catalyst Complexity

Even though a 1:1 adsorption stoichiometry is commonly used and kinetic factors are determined from calculations based on the assumption that the catalyst surface can be considered uniform with all atoms having essentially the same reactivity, it is still generally recognized that the catalyst surface is not uniform, but, rather, is made up of various types of surface atoms distinguished by their differing environments and having different adsorption and reaction characteristics.[40] Dark field TEM data has been interpreted to show that the platinum particles on a Pt/SiO_2 catalyst have a very complex surface morphology[41] thus supporting the premise of a non-uniform metal surface.

Confirming data can be obtained from temperature programmed desorption (TPD) studies on supported catalysts.[42] In these experiments a catalyst sample is exposed to a substrate to produce an adsorbed surface species. The temperature of the catalyst is then raised at a fixed rate while a carrier gas flows through the catalyst bed and then into a gas chromatograph (GC). As the temperature rises the adsorbed material is desorbed and detected by the GC to give a TPD curve such as that shown in curve **I** of Fig. 2.12 which is typical of the thermal desorption of hydrogen adsorbed on a platinum catalyst.[43,44] The distinct peaks in the chromatogram strongly indicate the presence on the platinum surface of hydrogen adsorbed either in different ways or on different types of surface atoms. It was originally proposed that these desorption peaks corresponded to the desorption of surface species in which both monoatomic and diatomic forms of hydrogen were attached to one, two, or even three surface atoms.[44] No evidence was presented to correlate these suggested forms of adsorbed hydrogen with specific peak maxima in the TPD curve.

The possibility that these different desorption peaks were the result of hydrogen desorption from different types of surface sites was considered more recently[45] with further work supporting this conclusion.[46] The study by Tsuchiya, et al.[46] involved the TPD of hydrogen from platinum as shown in Fig. 2.12, but with several variations. First, the TPD was run to 200°C, stopped, and the reactor cooled to room temperature. A second TPD was then run on this catalyst without the introduction of more hydrogen. In this TPD, curve **II** in Fig. 2.12 was obtained. This is essentially a continuation of the previous run, picking up from where it was stopped. Further, exposure of the initial hydrogen-covered platinum catalyst to a pulse of propene produced propane by utilizing some of the adsorbed hydrogen for the reaction. When the platinum catalyst was heated to 200°C in a partial TPD and then cooled and exposed to propene, no propane

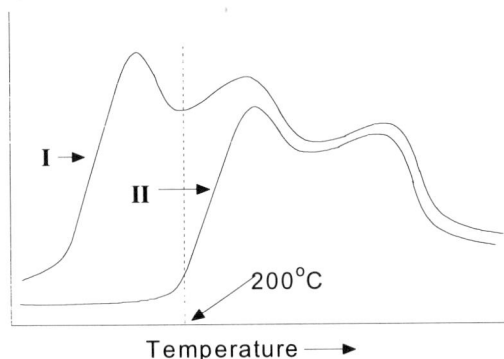

Fig. 2.12   Temperature programmed desorption (TPD) curves for the
             desorption of hydrogen from platinum catalysts. **I**) Full
             TPD curve; **II**) curve after partial desorption, cooling and
             then reheating. (Drawn using data from Ref. 44).

formation was observed. A separate hydrogen saturated catalyst was subjected to a partial TPD to 200°C, cooled, and then exposed to deuterium. The addition of propene gave predominantly propane-d_2. Reversal of the procedure with a partial TPD on a deuterated platinum catalyst, cooling, and exposure to hydrogen gave, on introduction of propene primarily propane-d_0.[46]

These results all suggest the presence on the platinum catalyst of different types of reactive sites. Those from which hydrogen is desorbed on heating to 200°C are apparently the ones responsible for alkene hydrogenation. The ability to restart a cooled, partial TPD and continuing with the same desorption mode as well as the general lack of scrambling observed in the hydrogen-deuterium experiments suggest that hydrogen does not migrate over the surface between the different types of sites under these conditions. The presence of the distinct peaks in these TPD curves supports the absence of surface migration. If surface migration occurred freely then there would be only a single, sharp peak in the TPD curve resulting from hydrogen migration to those sites from which it is more easily desorbed.

As discussed previously, the infrared spectra of CO adsorbed on supported rhodium catalysts show the presence of three different types of adsorption[27,28] as shown in Fig. 2.11. It has also been found that on ultradispersed Rh/SiO$_2$ catalysts which have very small rhodium particles only the *gem* dicarbonyl (**I**) bands were observed.[47] With rhodium catalysts having larger metal particles, bands associated with the linear form (**II**) appear and finally bridged carbonyls (**III**) are observed on even larger particles. These results are consistent with the presence on the rhodium catalysts of corner, edge and face atoms with the *gem* dicarbonyl (**I**) found on the corners, the linear (**II**) on edge atoms and bridging

(III) on faces. With the ultradispersed catalysts, the particles were thought to be two-dimensional rafts composed of about eight rhodium atoms.[47] Such particles would have only corner atoms. With somewhat larger three-dimensional particles the number of edge atoms would increase. With larger particles there would be an increase in the number of face atoms. A more complete discussion of the nature of the active sites on a catalyst surface is presented in Chapter 3.

References

1. H. C. Yao and P. H. Emmett, *J. Am. Chem. Soc.,* **81**, 4125 (1959).
2. I. Horiuti and M. Polanyi, *Trans. Faraday Soc.,* **30**, 1164 (1934).
3. M. J. S. Dewar, *Bull. soc. chim., France,* **18**, C71 (1951).
4. J. Chatt and L. A. Duncanson, *J. Chem. Soc.,* 2939 (1953).
5. G. C. Bond, *Heterogeneous Catalysis, Principles and Applications,* Oxford University Press, Oxford, 1987, pp 28-30.
6. S. Brunauer, P. H. Emmett and E. Teller, *J. Am. Chem. Soc.,* **60**, 309 (1938).
7. P. H. Emmett, *Adv. Catal.,* **1**, 65 (1948).
8. W. B. Innes, in *Experimental Methods in Catalysis Research, Vol. I,* (R. B. Anderson, Ed.) Academic Press, New York, 1968, p. 45.
9. K. S. W. Sing, in *Characterization of Catalysts,* (J. M. Thomas and R. M. Lambert, Eds.) Wiley, New York, 1980, p 12.
10. G. C. Bond, *Heterogeneous Catalysis, Principles and Applications,* Oxford University Press, Oxford, 1987, pp 18-21.
11. J. A. Moulijn, P. W. N. M. van Leeuwen and R. A. van Santen, *Stud. Surf. Sci. Catal.,* **79** (Catalysis), Elsevier, Amsterdam, 1993, p 419.
12. M. M. Dubinin, *Chem. Rev.,* **60**, 235 (1960).
13. IUPAC, *Pure Appl. Chem.,* **31**, 578 (1972).
14. G. C. Bond, *Heterogeneous Catalysis, Principles and Applications,* Oxford University Press, Oxford, 1987, pp 26-28.
15. J. Freel, *J. Catal.,* **25**, 139 (1972).
16. J. Sarkany and R.D. Gonzalez, *J. Catal.,* **76**, 75 (1982).
17. P. C. Aben, *J. Catal.,* **10**, 224 (1968).
18. M. Boudart and H. S. Hwang, *J. Catal.,* **39**, 44 (1975).
19. D. H. Everett and P. A. Sermon, *Zeit. Phys. Chem., Neue Folge,* **114**, 109 (1979).
20. A. Benedetti, G. Cocco, S. Enzo, F. Pina and L. Schiffini, *J. Chem. Phys.,* **78**, 120 (1981).
21. D. A. Dowden, *Catalysis (London),* **3**, 136 (1980).
22. R. L. Augustine, K. P. Kelly and R. W. Warner, *J. Chem. Soc., Faraday Trans. I,* **79**, 2639 (1983).
23. T. Fukashima, S. Galvano and G. Paravano, *J. Catal.,* **57**, 177 (1979).
24. J. E. Benson and M. Boudart, *J. Catal.,* **4**, 704 (1965).
25. J. E. Benson, H. S. Hwang and M. Boudart, *J. Catal.,* **30**, 146 (1973).
26. S. E. Wanke and N. A. Dougherty, *J. Catal.,* **24**, 367 (1972).
27. A. C. Yang and C. W. Garland, *J. Phys. Chem.,* **61**, 1504 (1957).
28. J. T. Yates, T. M. Duncan, S. D. Worley and R. W. Vaughn, *J. Chem. Phys.,* **70**, 1219 (1979).
29. F. Delannay, *Chem. Ind. (Dekker),* **15**, (Charact. Het. Catal.), 71 (1984).
30. A. Howie, in *Characterization of Catalysts,* (J. M. Thomas and R. M. Lambert, Eds.) Wiley, New York, 1980, p 89.
31. T. Baird, *Catalysis (London),* **5**, 172 (1982).

32. H. Klug and L. Alexander, *X-Ray Diffraction Procedures for Polycrystalline and Amorphous Materials, 2nd Ed.,* Wiley, New York 1974.
33. B. E. Warren and B. L. Averbach, *J. Appl. Phys.,* **21**, 595 (1950).
34. R. Canesan, H. Kucl, A. Saavedra and R. J. deAngelis, *J. Catal.,* **52**, 310, 320 (1978).
35. C. R. Adams, H. A. Benesi, R. M. Curtis and R. G. Meisenheimer, *J. Catal.,* **1**, 336 (1962).
36. J. Sinfelt and D. J. C. Yates, *J. Catal.,* **15**, 362 (1968).
37. M. Boudart and G. Djega-Mariadassou, *Kinetics of Heterogeneous Catalytic Reactions,* Princeton University Press, Princeton, 1984, p 23.
38. B. J. Kip, F. B. M. Duivenvoorden, D. C. Konigsberger and R. Prinz, *J. Catal.,* **105**, 26 1987.
39. G. R. Wilson and W. K. Hall, *J. Catal.,* **17**, 190 (1970).
40. G. D. Halsey, *Adv. Catal.,* **4**, 259 (1952).
41. O. L. Perez, D. Romeo and M. J. Yacaman, *J. Appl. Surf. Sci.,* **13**, 402 (1982).
42. R. J. Cvetanovic and Y. Amenomiya, *Adv. Catal.,* **17**, 103 (1967); *Catal. Revs.,* **6**, 21 (1972).
43. P. C. Aben, H. Van Der Eijk and J. M. Oelderik, *Int. Cong. Catal. (Proc.) 5th,* 717 (1972).
44. S. Tsuchiya, Y. Amenomiya and R. J. Cvetanovic, *J. Catal.,* **19**, 245 (1970).
45. K. J. Leary, J. N. Michaels and A. M. Stacy, *AIChE Journal,* **34**, 263 (1988).
46. S. Tsuchiya, N. Nakamura and M. Yoshioka, *Bull. Chem. Soc., Japan,* **51**, 981 (1978).
47. D. J. C. Yates, L. L. Murrell and E. B. Prestidge, *J. Catal.,* **57**, 41 (1979).

3

The Active Site

In order to understand the molecular processes taking place on a catalyst surface the nature of the sites promoting the reaction must be defined. While a catalyst surface is commonly considered to be of uniform composition for ease in interpreting kinetic data, it is generally accepted that the surface of a catalyst is composed of atoms having differing environments and numbers of neighboring atoms.[1] Considerable experimental data has been accumulated which indicates that these different types of surface atoms can have varying adsorption and reaction characteristics.[2-22] It might appear that this surface complexity could preclude the elucidation of any useful concepts concerning the interaction of substrates on a catalyst surface but a reasonable understanding of the mechanism of catalytic reactions has been developed, especially over the past several years.

Single Atom and Multiatom Sites

Since heterogeneously catalyzed reactions take place on the surface of the catalyst, the reaction rate is normally described as an *areal* turnover frequency (TOF) which is the number of molecules of product formed per unit time per surface atom or unit surface area of the catalyst. It would be expected, then, that varying the surface area, or dispersion, of the catalyst should cause a corresponding change in the areal TOF of a catalyzed reaction. With some reactions, such as those that involve the breaking of C–C bonds, alkane rearrangements, dehydrocyclizations and, presumably, other hydrogenolyses, this is observed. Other reactions, however, do not show the expected relationship between catalyst dispersion and areal TOF. With reactions such as hydrogenations, catalytic oxidations involving dioxygen, hydroformylations and similar processes, the areal TOF remains essentially constant when the catalyst dispersion is varied. Those reactions that show essentially no change in areal TOF with variations in catalyst dispersion have been called structure insensitive. Those where the areal TOF varies with change in catalyst dispersion are called structure sensitive.[23,24]

These terms are simply convenient labels for two broad classes of catalytic reactions and should not be considered an absolute characterization. While changes in areal TOF with variation in catalyst dispersion can be understood, the rationale presented to explain structure insensitivity is rather complex.[24] This is especially true in that a growing number of such reactions have been shown to have areal TOFs that vary with changing catalyst dispersion when they are run under appropriate reaction conditions.[25-27] It would seem, then, that a criterion

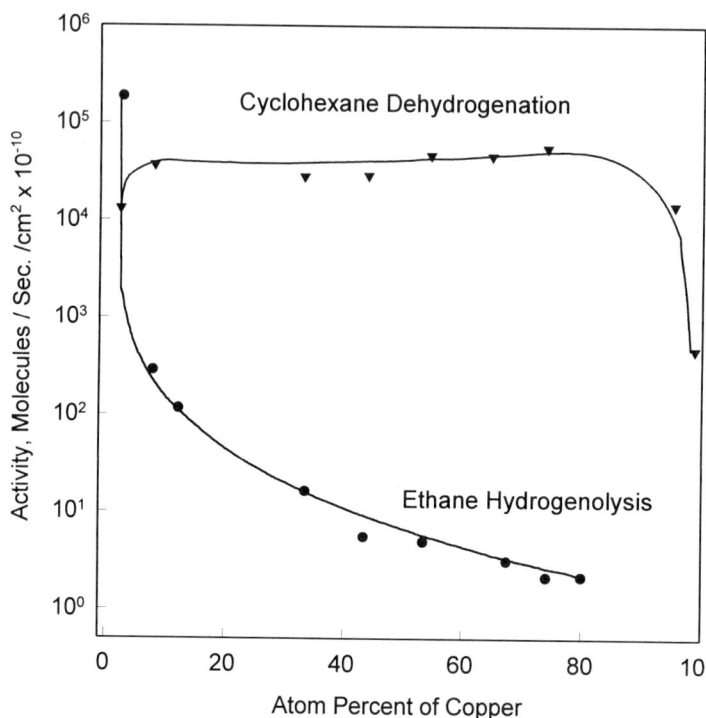

Fig. 3.1   Areal turnover frequencies for ethane hydrogenolyses and
cyclohexane dehydrogenations run over Ni/Cu catalysts
having increasing copper content. (Redrawn using data
from Ref. 28).

other than the relationship between reaction rate and catalyst dispersion is needed
to differentiate between these two classes of reactions.

In an interesting study, a classic structure sensitive reaction, ethane
hydrogenolysis, and a typical structure insensitive reaction, cyclohexane
dehydrogenation, were both run over a series of Ni/Cu catalysts having a wide
range of Ni:Cu ratios.[28] Nickel is the active species in promoting both of these
reactions while the copper is present only as a diluent of the nickel atoms on the
surface. A plot of the areal TOFs of these reactions versus the amount of copper
present in the catalyst (Fig. 3.1) shows that the rate of ethane hydrogenolysis
decreased rapidly as the amount of copper in the catalyst increased. The rate of
cyclohexane dehydrogenation, on the other hand, remained constant until the
amount of copper present in the catalyst was quite high. These results indicate
that ethane hydrogenolysis is promoted by ensembles of nickel atoms and that
these ensembles must have a minimum critical size to catalyze this reaction. The

$$
\begin{array}{l}
\triangle \qquad\qquad \text{Hydrogenolysis} \\
\qquad \overset{H_2}{\diagdown} \overset{}{Re_3} \\
\qquad\qquad\qquad\qquad\qquad\searrow \qquad CH_3CH_2CH_3 \\
\qquad\qquad\qquad\qquad\nearrow \\
\qquad \overset{H_2}{\diagup}\, Re_1 \\
H_2C{=}CHCH_3 \qquad \text{Hydrogenation}
\end{array}
\qquad (3.1)
$$

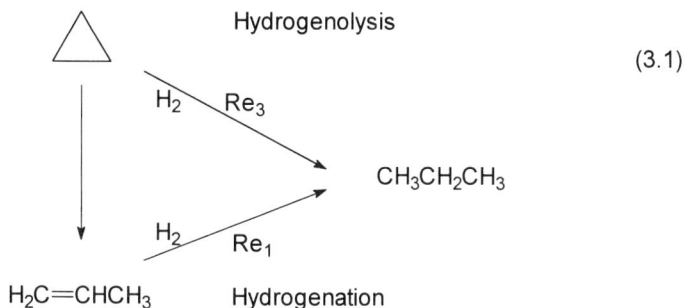

increase in copper content dilutes the nickel on the surface and decreases the number of such ensembles that are available for this reaction. The active sites on which the dehydrogenation takes place are not influenced by the presence of copper until a considerable amount has been added so it is logical to assume that these sites are single atoms and that their number decreases only when the surface coverage by copper is extensive.

Similar conclusions have been drawn from a study of the hydrogenolysis and isomerization of cyclopropane over catalysts made up of supported rhenium clusters of differing nuclearity.[29] The hydrogenolysis of cyclopropane was found to take place only over those catalysts having at least a Re_3 nuclearity while the hydrogenation of propene, an isomerization product of cyclopropane, took place on Re_1 species (Eqn. 3.1).

These results indicate that, perhaps, a better definition of structure sensitive reactions would be those that occur over ensembles of surface atoms while structure insensitive reactions are those that are promoted by single atom active sites.

It is necessary, however, to define more clearly the nature of the specific active sites, be they ensembles or single atoms, that are responsible for promoting specific reactions. A number of approaches to this problem have been made over the years with varying degrees of success.

Metal Single Crystals as Model Catalysts

Some of the more extensive studies in this area have involved the use of well-characterized metal single crystals as catalysts for a variety of reactions.[11-22] The use of single crystals is advantageous in that the exact arrangement of atoms on their surfaces can be determined by low energy electron diffraction (LEED) analysis under high vacuum conditions.[30] After the LEED analysis is completed, the single crystal is moved from the vacuum chamber through a pressure lock into a compartment in which the desired reaction can be run at pressures up to atmospheric with product analysis accomplished by gas chromatography and/or mass spectrometry.[13,18-22]

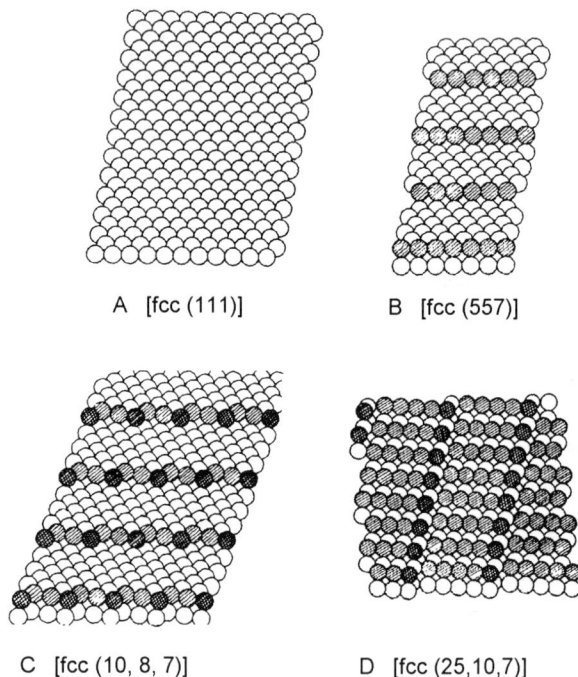

A [fcc (111)]   B [fcc (557)]

C [fcc (10, 8, 7)]   D [fcc (25,10,7)]

Fig. 3.2.   Surface structure of platinum single crystal catalysts with the
            corresponding Miller index notations given in brackets.
            (Reproduced, with permission, from Ref. 14.)

Reaction data obtained from one single crystal is difficult to evaluate in
terms of specific site activity.   But, when a reaction is run over several single
crystals such as those pictured in Fig. 3.2[14] active site identification for specific
reactions is more easily accomplished. In this series the crystals have varying
numbers of corner, edge and face atoms on their surfaces. Running the same
reaction over all of the crystals in this series and then comparing reaction rates
and/or selectivities with changes in the relative amounts of specific types of atoms
gives an indication of the type of site on which the reaction takes place.   For
instance, when cyclohexane was passed at 250°–300°C over the series of platinum
single crystal catalysts shown in Fig. 3.2 the products formed were benzene,
which was produced by C–H bond breaking, and n-hexane and smaller alkanes,
which were formed by C–C hydrogenolysis (Eqn. 3.2).[14] The extent of benzene
formation over these single crystal catalysts increased in the order A < B < C < D.
As this is also the order of increasing corner atom densities, C–H bond breaking

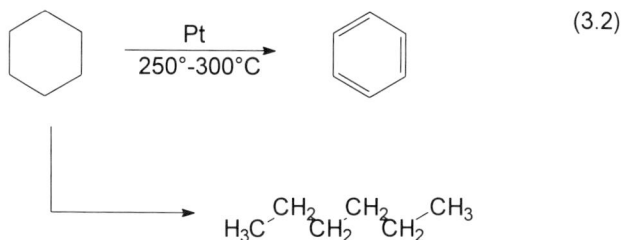

$$(3.2)$$

must be taking place on corner atoms.   By the principle of microscopic reversibility[31] it can be concluded that C–H bond formation or catalytic hydrogenation of an alkene must also occur on these same corner sites.   Since cyclohexane dehydrogenation is a structure insensitive reaction, these corners must be considered single atom active sites and not part of a reactive ensemble.

In this same series of reactions it was also found that C–C bond cleavage took place most readily over the 111 face catalyst (A in Fig. 3.2) and to a considerably less extent on the others. This reaction, which is structure sensitive, obviously takes place over ensembles of face atoms. It was also found[17,18] that the dehydrocyclization of hexane to benzene, another structure sensitive reaction, took place four times faster on the 111 face of a platinum crystal than on one cleaved to expose the 100 face. The hydrogenolysis of ethane, on the other hand, was found to be significantly faster on Ni (100) than on Ni (111).[22] Thus, it is not sufficient merely to define a site as an ensemble of face atoms; the orientation of the atoms must also be specified.

A series of structure sensitive reactions was studied using a combination of single crystals and supported metal catalysts.[19-22]   The amount of reaction observed using the 111 face crystal was compared with the results obtained using a series of supported catalysts having differing particle sizes. Reaction data from those supported catalysts having the largest particles most closely resembled the results obtained with the 111 single crystal catalyst. This is not surprising since the larger metal particles have a higher proportion of their surfaces composed of face atoms, probably in the more closely packed 111 orientation.

While these results show a relationship between single crystals and supported metals when used for structure sensitive reactions, similar relations for those reactions promoted by a single atom site may not be as straightforward since there are more different types of single atom sites possible on a metal surface than there are groups of face atoms.

Over the past several years a number of instrumental techniques have been applied to the characterization of catalysts.[32,33]   Ultraviolet,[34] infrared,[35,36] and raman[37] spectroscopic data have provided information concerning the nature of substrates adsorbed on a catalyst surface as well as the composition and structure of many oxide catalysts. Extended X-ray absorption fine structure (EXAFS) is

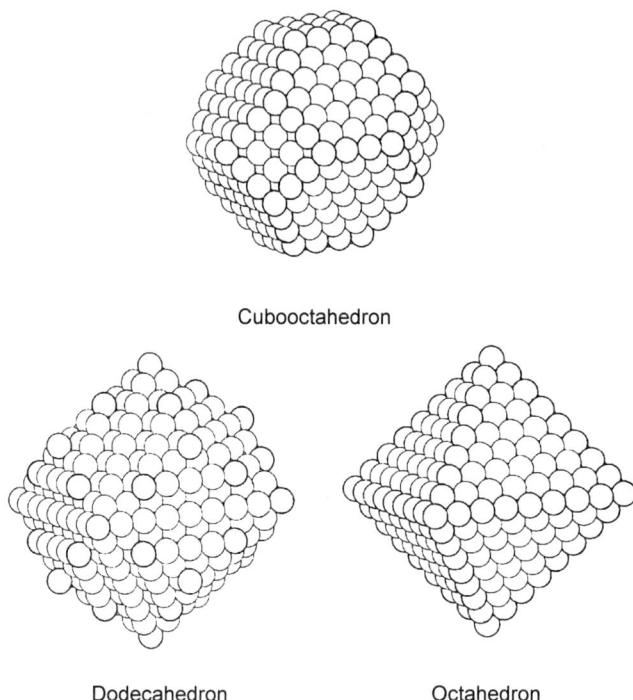

Cubooctahedron

Dodecahedron   Octahedron

Fig. 3.3.   Some idealized crystal shapes possible with an fcc crystal
           lattice. (Reproduced, with permission, from Ref. 47).

used to provide the average coordination number of atoms of specific elements in
a catalyst, the nature of the atoms surrounding these target atoms, and the average
distance separating the target atom from its neighbors.[38-40] Electron microscopy
can be used to detect and quantify very small metal particles on a support matrix.
With special techniques information concerning particle morphology may be
obtained.[41-45] However, none of these instrumental methods can, as yet, be
applied to the characterization of the different types of atoms present on the
surface of the small metal particles found in dispersed metal catalysts.   Other
approaches have to be used.

Ideal Crystal Models

Supported metal particles were compared to the idealized crystal shapes available
in various crystalline arrangements. Some of the shapes derived from fcc crystals
(the crystalline orientation of most catalytically active metals) are depicted in Fig.
3.3.[46] The various types of surface atoms were differentiated by the number of

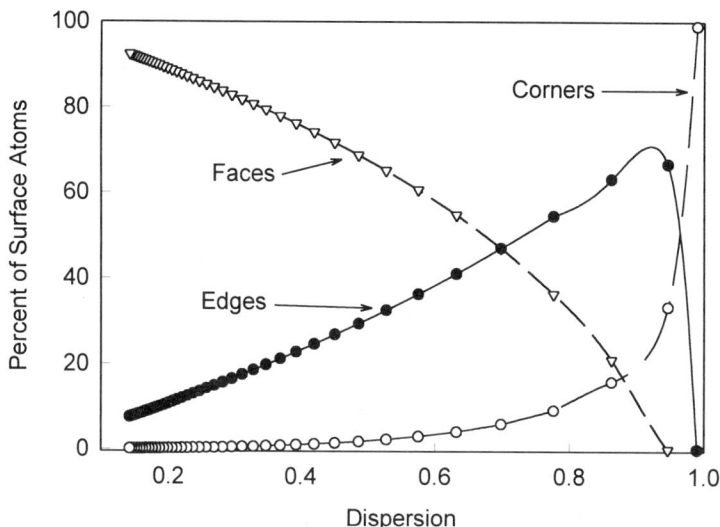

Fig. 3.4.   Relationship between the different types of surface atoms
and the catalyst dispersion as calculated for cubooctahedral
crystals. (Drawn using data from Ref. 46).

their nearest neighbors. The amounts of each of the different types of atoms
present on the surface of the different crystalline shapes were calculated for
crystals of varying sizes. The change in the relative amounts of each type with
increasing particle size or decreasing dispersion, was then plotted as shown in Fig.
3.4 for the cubooctahedron. Regardless of the particle shape, the percent of corner
atoms on the surface decreased rapidly with decreasing dispersion while the
percentage of face atoms increased. The edge atom component of the surface
went through a maximum at a size that depended on the shape of the particle
under consideration.[46,47]

       The basic concept was to run a given reaction over a series of supported
metal catalysts that were identical except for the size of the metal particles and
then to relate changes in activity and/or selectivity with one of the trends shown in
Fig. 3.4. The use of such information to evaluate specific reactions, however,
relies on the basic assumption that the metal crystallites present in supported
metal catalysts are well defined crystalline shapes and that all of the particles in
the catalyst series have, at least, similar shapes. This is, at best, a rather tenuous
assumption.

       While this work did show that a number of different types of single atom
sites can be present on the surface of dispersed metal catalysts, the possibility was

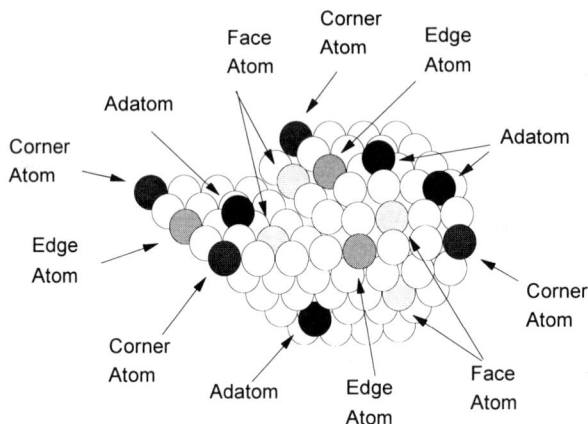

Fig. 3.5. Different types of surface atoms on an irregular metal particle.

not considered of particles having irregular shapes such as that pictured in Fig. 3.5. A weak beam electron microscopic investigation has shown that the platinum particles present on a supported catalyst have a very complex surface structure,[48] a factor that negates the practical application of this approach to the determination of the types of reactions taking place on specific types of surface atoms.

Homogeneous Catalyst Models

Some proposals have been made that attempted to relate observed catalyst activity with the mechanisms of similar reactions taking place on soluble, monatomic homogeneous catalysts so the nature of the active site could be related to the geometry of the homogeneous complex.[49,50] In this approach, the sites on dispersed metal catalysts were considered to be related to octahedral species from which increasing numbers of ligands were removed. These sites were designated ^3M, ^2M, and ^1M depending on the degree of coordinative unsaturation and represent, respectively, corner, edge and face atoms as shown in Fig. 3.6. Reaction of hydrogen on these sites can give the $^3MH_2$, ^3MH, $^2MH_2$, ^2MH and ^1MH species. The $^3MH_2$ and $^2MH_2$ moieties result from the adsorption of a dihydrogen molecule on a single surface atom. The ^3MH, ^2MH and ^1MH species come from the dissociative adsorption of hydrogen on two surface atoms. Of these, the $^2MH_2$ and ^1MH entities are fully saturated so they cannot be expected to be involved in further reactions. The $^3MH_2$, ^3MH and ^2MH, however, do have coordination sites available for the adsorption of an additional substrate, such as a double bond, so they can be expected to take part in a hydrogenation reaction.[50,51]

The complex, $(\Phi_3P)_3RhHCO$, (1) which promotes double bond hydrogenation and isomerization by the mechanism shown in Scheme 3.1,[52] is

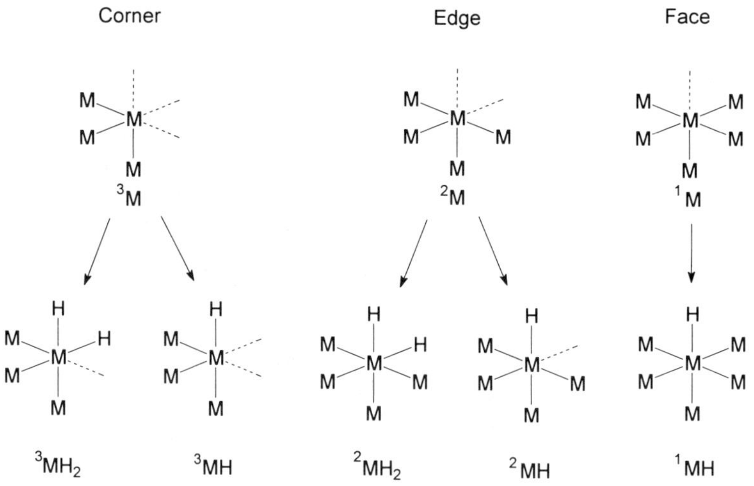

Fig. 3.6   Proposed monatomic surface sites based on homogeneous catalyst
analogs.

Scheme 3.1

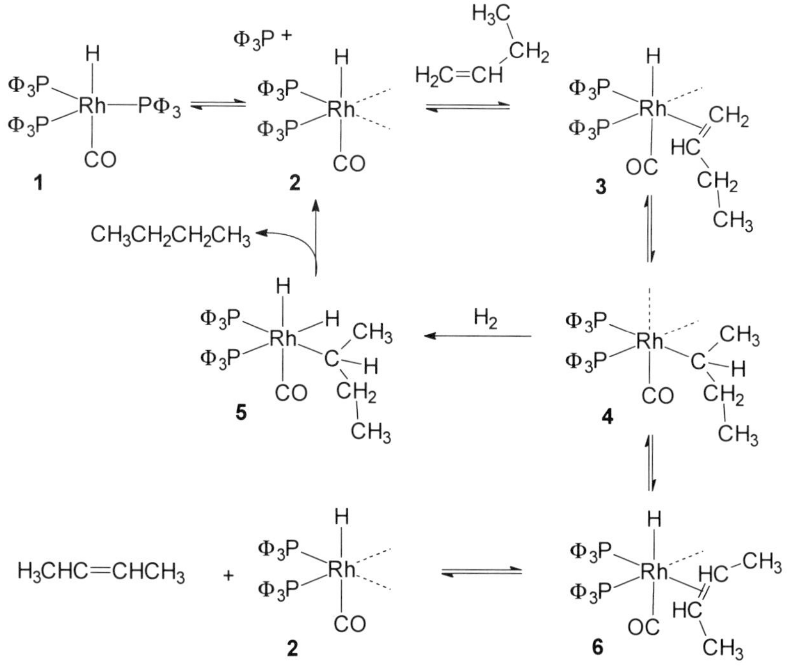

considered the analog of a ^{3}MH site.[50,51] Alkene hydrogenation over **1** proceeds by way of the five-coordinate hydride, **3**, having an adsorbed olefin. Hydrogen insertion gives the metalalkyl, **4**. Addition of dihydrogen leads to the dihydrido metalalkyl, **5** which, on reductive elimination gives the alkane and regenerates the active monohydride, **2**. When insufficient hydrogen is available to react with **4**, β-elimination can take place to regenerate the initial hydrido metal alkene, **3**, or the isomeric species, **6**. Desorption of the olefin from **3** regenerates the starting alkene while from **6** the isomeric alkene is obtained. Alkene hydrogenation over the ^{3}MH (**7**) sites is considered to take place by an analogous process that is illustrated in Scheme 3.2.[50,51] Alkene adsorption followed by hydrogen insertion gives the metalalkyl or half hydrogenated state, **9**, which has sufficient coordinate unsaturation to accommodate both atoms of a dihydrogen molecule to produce the surface dihydrido species, **10**. Reductive elimination gives the alkane and regenerates the 3MH site, **7**. Double bond isomerization can occur when hydrogen availability is insufficient. β-Elimination on **9** can take place to give the isomeric adsorbed alkenes, **8** and **11**, which can desorb the isomeric alkenes to regenerate the active site, **7**.

It should be noted that this process is simply a more detailed description of the Horiuti-Polanyi[53] mechanism shown in Fig. 2.2. If the amount of hydrogen

Scheme 3.2

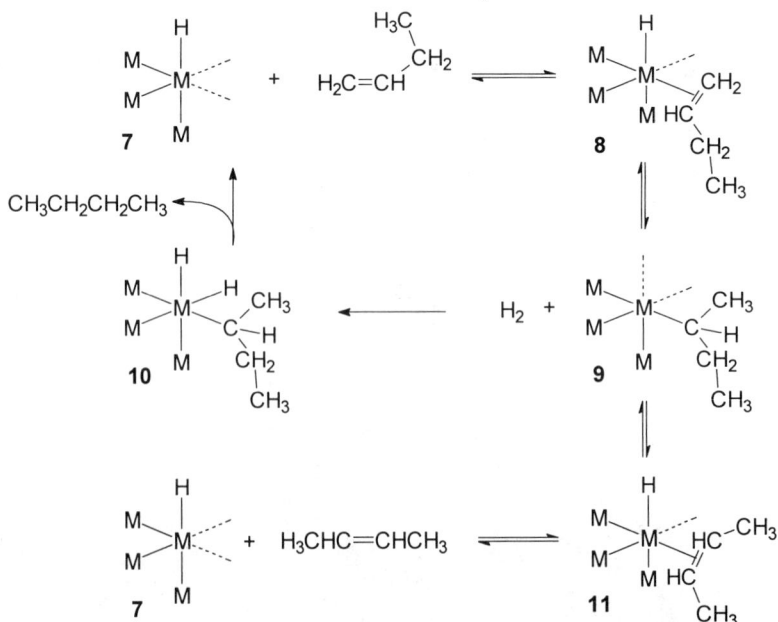

available to the catalyst is limited, the addition of adsorbed hydrogen to the metalalkyl (half-hydrogenated state), **9,** will be decreased and double bond isomerization will be favored. Under high hydrogen availability conditions the final step in the saturation cycle takes place readily and double bond isomerization is minimized.

The homogeneous entity comparable to the $^3MH_2$ site is the Wilkinson catalyst, $(\Phi_3P)_3RhCl$ **(12)**,[54] which promotes alkene hydrogenation by the mechanism shown in Scheme 3.3.[54] Here dihydrogen reacts with the active unsaturated species, **13,** to give the pentacoordinate dihydride, **14.** Alkene adsorption gives **15** which undergoes hydrogen insertion to give the hydrido metalalkyl, **16.** Reductive elimination then takes place rapidly to give the alkane and the active catalyst, **13.** The corresponding reaction sequence for a ^3M site **(17)** is shown in Scheme 3.4 with the $^3MH_2$ species, **18,** adsorbing a double bond to give **19** which can undergo hydrogen insertion to form the hydrido species, **20.** This, on reductive elimination, gives the alkane and regenerates the ^3M site.[51,55]

In contrast to hydrogenations taking place over **1** or a ^3MH site, in an alkene hydrogenation over $(\Phi_3P)_3RhCl$ **(12)** or a $^3MH_2$ site, the two hydrogens are added to the double bond almost simultaneously so, while a metalalkyl intermediate, **15** or **19,** is formed it has a very short lifetime before reductive elimination occurs. When **12** is used as a hydrogenation catalyst, double bond isomerization is not observed during an alkene hydrogenation.[54] By analogy, isomerization should not take place on $^3MH_2$ sites either. It might appear that the

Scheme 3.3

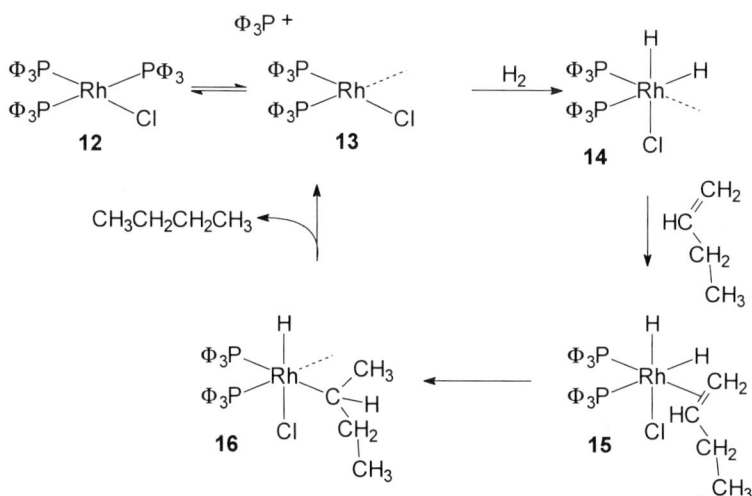

Scheme 3.4

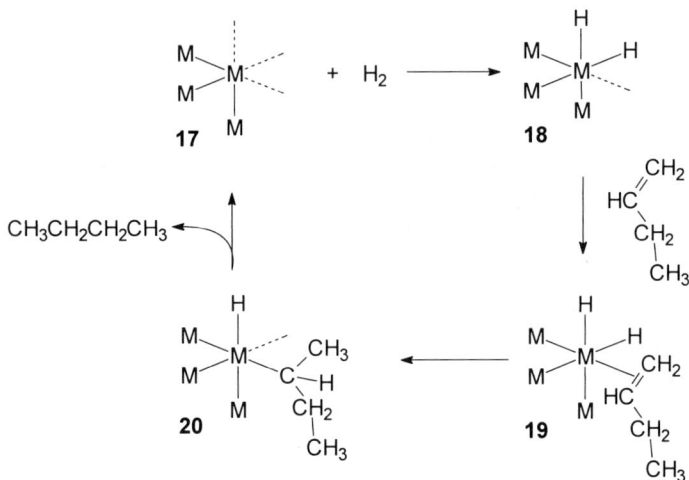

reaction sequence proposed for these sites is different from the Horiuti-Polanyi mechanism but this is not the case. Instead, it is merely a special instance in which the second hydrogen transfer step takes place very rapidly after the first. With the two hydrogens adsorbed on the same atom that coordinates the alkene, it would be difficult to rationalize anything different. The sequence pictured in Scheme 3.4 differs from that shown in Scheme 3.2 in the source of the two hydrogen atoms that are added to the double bond. In the ^{3}MH cycle (Scheme 3.2) each hydrogen atom comes from a different dihydrogen molecule while in the ^{3}MH$_2$ cycle (Scheme 3.4) both hydrogen atoms come from the same dihydrogen.

While there are no appropriate homogeneous analogs for the ^{2}MH site, **21**, a reaction sequence similar to that suggested for the ^{3}MH sites can be proposed (Scheme 3.5). The primary difference here is the lack of sufficient unsaturation on the metalalkyl, **23**, for the adsorption of both hydrogens of a dihydrogen molecule as in the ^{3}MH sequence (Scheme 3.2). Since the neighboring surface atoms are probably saturated with hydrogen, the possibility of hydrogen adsorption utilizing adjacent metal atoms is also unrealistic so the only course of action for the metalalkyl is β-elimination to give either **22** or **24** with possible double bond isomerization.

Steady State Studies

Experimental data has been sought to substantiate some of the ideas developed from these models. One of these[56] utilized an interesting probe molecule, (+)apopinene, **25**, (Eqn. 3.3) which has a number of special characteristics. The

Scheme 3.5

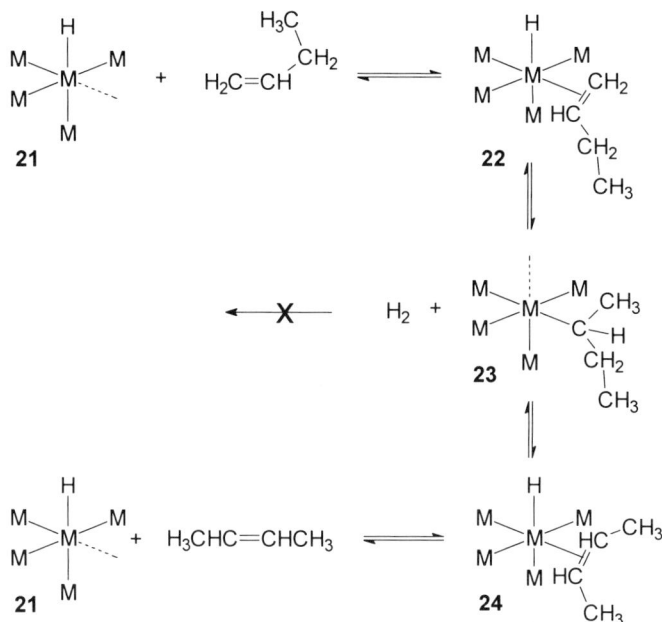

gem dimethyl groups on the one bridge effectively limit adsorption of the double bond to the opposite side of the molecule. (+)Apopinene is chiral but the double bond isomer, **26**, is its mirror image so hydrogen addition to either enantiomer will take place at the same rate. Since the saturated product has a plane of symmetry and, thus, is optically inactive, either saturation or isomerization will result in a loss of optical activity.

The rate of hydrogen addition to **25** and/or **26**, k_a, was determined by measuring the hydrogen consumption in the classic manner and the rate of isomerization, k_i, was obtained from corrected optical rotation measurements.[56] The hydrogenation/isomerization of **25** was run over series of catalysts of varying dispersions. Typical k_i/k_a versus dispersion plots obtained using a series of supported platinum catalysts are shown in Fig. 3.7.[57] Initially it was suggested that saturation took place over corner atoms,[58] but when a comparison was made to the predicted edge and corner atom compositions of octahedral crystals using the van Hardtefeld[46] models (the dashed lines in Fig. 3.7) it was proposed that both saturation and isomerization took place on edge atoms.[57]

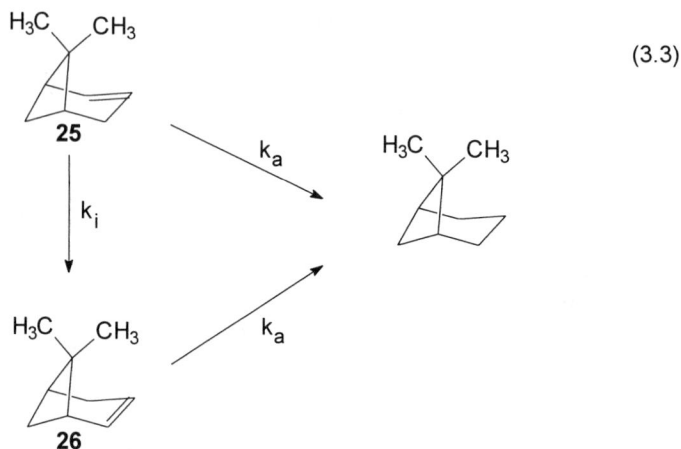

(3.3)

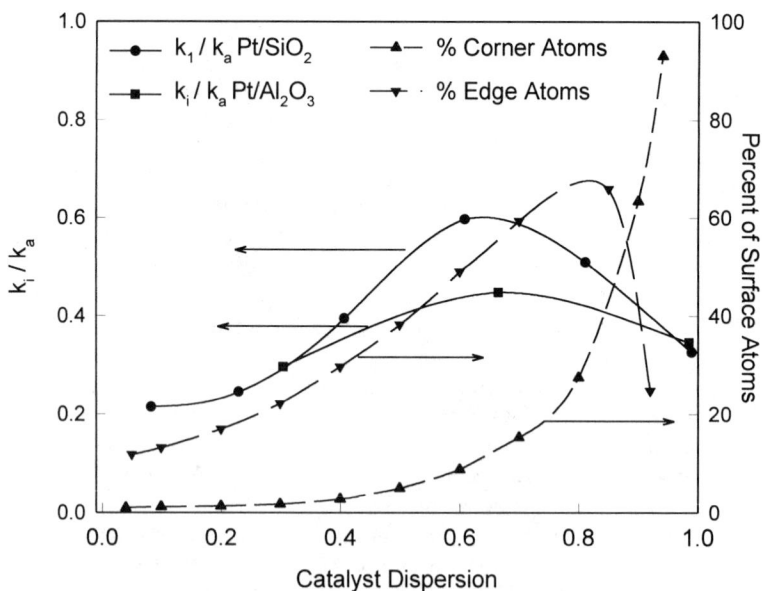

Fig. 3.7.   Relationship between the k_i/k_a ratio and catalyst dispersion in the apopinene hydrogenation over Pt/SiO$_2$ and Pt/Al$_2$O$_3$ catalysts. (Solid Lines) Redrawn from data taken from Ref. 57. Percent of corner and edge atoms on idealized octahedral crystals of varying sizes. (Dashed Lines) (Drawn using data from Ref. 46).

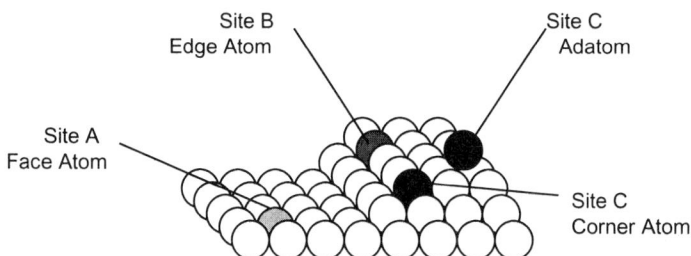

Fig. 3.8. Different types of reactive sites on a palladium catalyst.

In another study, 1-butene was deuterated and the resulting butane and isomeric butenes were subjected to microwave spectroscopic analysis to determine the position of the deuterium atoms in each material. On the basis of these findings mechanistic proposals were presented for reactions considered to take place over each of the three different types of sites shown in Fig. 3.8.[59-62]

The corners and adatoms (C sites) are those on which saturation and, possibly, isomerization can take place. Edge atoms (B sites) promote isomerization while the only reaction considered possible for face atoms (A sites) is the promotion of sigmatropic isomerizations, a process that has subsequently been shown not to occur on metal surfaces.[63] These sites, then, do not appear to take part in an alkene hydrogenation. For comparison, the reactions proposed for Type C sites are essentially those discussed previously for the ^{3}M (Scheme 3.4) and ^{3}MH sites (Scheme 3.2). Type B sites are the same as the ^{2}MH described in Scheme 3.5 and the Type A sites are equivalent to ^{1}M sites.

Single Turnover Reaction Sequence

In each of these studies the site:product relationship was determined on the basis of product distribution data obtained from standard, steady state catalytic reactions. While this approach can provide evidence for the type of site(s) responsible for the formation of certain products it cannot give any indication of the number of such sites that are present on the catalyst surface. Since the activities of the various types of sites are different, it is possible that a small number of very active sites could dominate product formation. In order to relate the extent of product formation with the number of specific types of sites present an experimental arrangement is needed which obviates these site activity differences. One way of doing this is to use the catalyst surface as a stoichiometric reagent so that each site reacts only once. In this way there will always be a 1:1 site:product molecule ratio regardless of the rates at which the different types of sites react.

This has been accomplished with the single turnover (STO) procedure shown in Fig. 3.9.[51,64-66] In this reaction sequence, a sample of the catalyst is

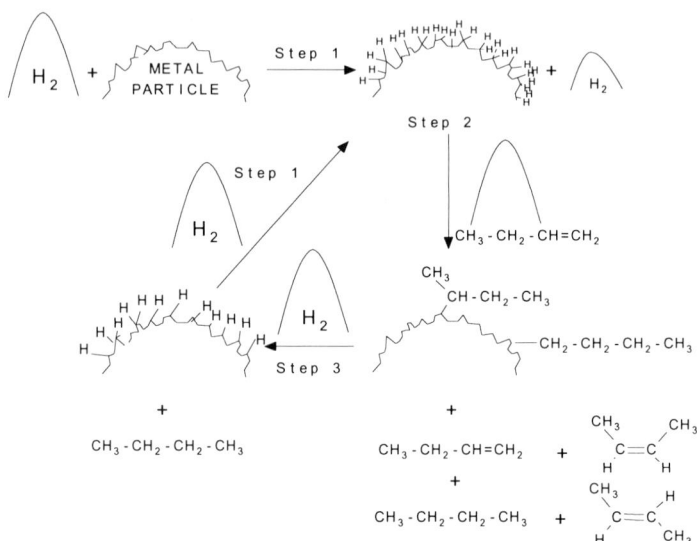

Fig. 3.9. Single turnover reaction sequence. (Reproduced, with permission, from Ref. 66).

placed in a small flow reactor and purged with a stream of purified helium. A pulse of hydrogen, of sufficient size to saturate all of the surface atoms on the catalyst, is then passed through the catalyst bed (Step 1). The excess hydrogen is removed in the carrier gas stream so the only hydrogen available for further reaction is that adsorbed on the catalyst surface. A pulse of 1-butene is then passed through the catalyst to react with the adsorbed hydrogen (Step 2). The reaction mixture is analyzed by on-stream gas chromatography which separates the unreacted 1-butene from the saturation product, butane, and the isomerization products, *cis* and *trans* 2-butene. Some hydrocarbon remains on the catalyst as the metalalkyl and is removed as butane by a second hydrogen pulse completing the STO reaction sequence (Step 3).

   Since it has been shown that hydrogen migration across the catalyst surface is unlikely,[67] it follows that in this STO procedure each site reacts only once and, thus, there is a 1:1 relationship between the number of specific sites present and the number of molecules of each product formed over these sites.[51] The number of butane molecules produced by the initial reaction of butene with the hydrogen covered catalyst corresponds to the number of "direct saturation sites". It has been proposed that these sites give butane by the $^3MH_2$ reaction cycle shown in Scheme 3.4 and, thus, they have been labeled ^3M sites.[51] The formation of butane by reaction of the second pulse of hydrogen with the metalalkyl (Step 3) occurs, presumably, by way of the ^3MH reaction sequence shown in Scheme 3.2. These two-step saturation sites are labeled ^3MH.

Isomerization apparently takes place over sites labeled ^{2}M by the sequence shown in Scheme 3.5.[51]

When the time between the introduction of the first hydrogen pulse and the 1-butene pulse was lengthened those hydrogen atoms that are weakly held on the surface are removed in the carrier gas stream and the amount of butane that was formed decreased. Since no other changes in product composition were observed, this weakly bound hydrogen appears to be found only on some of the direct saturation sites. Thus, there are two types of these sites, those having strongly adsorbed hydrogen and those with the hydrogen weakly adsorbed. These sites have been labeled ^{3}M$_I$ and ^{3}M$_R$ respectively.[51]

The STO procedure can be used to measure the amounts of four different types of sites involved in the hydrogenation of a double bond, the direct saturation sites, ^{3}M$_I$ and ^{3}M$_R$, two-step saturation sites, ^{3}MH, and isomerization sites, ^{2}M. In addition, there are sites that only adsorb hydrogen but do not take part in alkene hydrogenation. These are labeled ^{1}M and the number of such sites present is measured by the relationships shown in Eqns. 3.4 and 3.5.[51] They are based on the mechanisms shown in Schemes 3.2, 3.4 and 3.5 in which the ^{3}M$_I$ and ^{3}M$_R$ sites initially adsorb two hydrogen atoms and all of the other sites only one.

$$H_{Adsorbed} = (2 \times [^3M]) + [^3MH] + [^2M] + [^1M] \qquad (3.4)$$

$$[^1M] = H_{Ads.} - \{(2 \times [^3M]) + [^3MH] + [^2M]\} \qquad (3.5)$$

To be certain that all of the sites are reacting in this sequence it is essential that the time the reactant pulse is in contact with the catalyst is sufficient for the diffusion of reactant molecules throughout the catalyst bed. To see if this is occurring, the pulse size and carrier gas flow rates are changed and the product composition analyzed. If the product formation remains constant with changes in the amount of time the reactant is in contact with the catalyst, it can be concluded that the reaction has proceeded to completion under the conditions used. Application of this procedure to platinum, palladium, and rhodium catalysts on a number of supports[65] has shown that the STO reaction sequence can be used to determine the densities of the saturation sites, ^{3}M$_I$, ^{3}M$_R$ and ^{3}MH, on all of these catalysts. However, the extent of isomerization varies with changes in pulse size and carrier gas flow rate so valid isomerization site (^{2}M) densities cannot be obtained. Carbon supported catalysts cannot be characterized using this procedure because the carbon absorbs the hydrocarbon pulses thus making the product analysis impossible.

Site–Reactivity Relationships

Being able to measure the amounts of the saturation sites on the surface of a dispersed metal catalyst provides a means of further defining the specific activity of these sites. This has been accomplished by using a series of STO characterized catalysts having varying reactive site densities to promote the same reaction. The trends in the extent of product formation were then compared to the changes in specific site densities on the catalysts as was done using the platinum single crystals shown in Fig. 3.2.[13] Series of supported catalysts having the same metal load and differing site densities are easily prepared by the reduction of the catalyst precursor in flowing hydrogen at different temperatures.[68]

A series of ten Pt/CPG (controlled pore glass) catalysts were used for the dehydrogenation of cyclohexane to benzene at 300°C. The plot of the amount of benzene formed versus the total number of direct and two step saturation sites is shown in Fig. 3.10a.[66,69] The direct relationship between these factors up to about 95% benzene formation shows that the saturation sites (^3M and ^3MH) which form C–H bonds on alkene hydrogenation are also those on which the C–H bonds are broken to produce benzene. The rest of the curve shows almost complete conversion over those catalysts with the highest saturation site densities. Figs. 3.10b and 3.10c show plots of the extent of benzene formation with changes in isomerization site (^2M) densities and catalyst dispersions, respectively. Neither of these show any significant relationship, thus, confirming that this reaction takes place over the STO characterized saturation sites on the platinum surface.

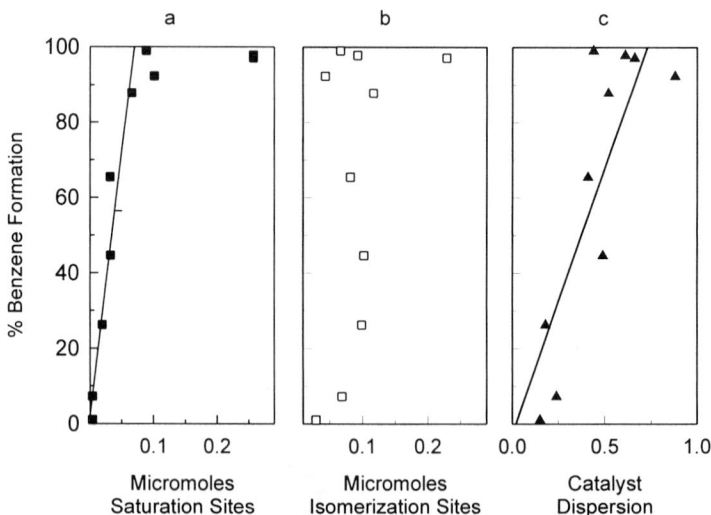

Fig. 3.10. The extent of benzene formation by cyclohexane
dehydrogenation run over a series of STO characterized
Pt/CPG catalysts compared with: a) the number of STO
saturation sites ($^3M + {}^3MH$); b) the number of
isomerization sites (^2M); c) the catalyst dispersion.
(Redrawn using data from Ref. 69).

As discussed previously, this reaction was also run under these same conditions over the series of specifically cleaved platinum single crystals shown in Fig. 3.2.[13] The results of these experiments show that it was the corner atoms on these crystals that promoted C–H bond breaking. Thus, the saturation sites on the dispersed metal catalysts are also corner atoms. Since this saturation site description agrees with that proposed on the basis of the butene deuteration described previously,[59-62] it is likely that the isomerization sites, ^2M, are edge atoms and the hydrogenation inactive sites, ^1M, are face atoms. A similar approach can be used to determine the nature of the active sites responsible for promoting almost any type of reaction.[65,70]

Site Specific Turnover Frequencies

It was mentioned previously that the rate of a heterogeneously catalyzed reaction is expressed as a turnover frequency (TOF) which is the number of times an active site reacts per unit time. Since active site concentrations have not been available, in most cases the TOF is expressed as the number of molecules formed per unit time per surface atom or unit surface area. The ability to use the STO procedure to measure active site densities also provides a means of determining specific site TOFs. It is apparent that the total number of molecules formed in a catalytic reaction per unit time is the sum of the production from each active site. Thus, the reaction TOF can be expressed as the sum of the products of the specific site TOF and the specific site densities as shown in Eqn. 3.6.[71]

$$\text{Rate} = (A \times [^3M_I]) + (B \times [^3M_R]) + (C \times [^3MH]) \qquad (3.6)$$

In this equation the **Rate** is the molar TOF of the reaction, moles of product formed/mole of metal catalyst/unit time. The terms in [ ] are the STO measured site densities given in moles of site/mole of metal. The specific site TOFs, **A**, **B** and **C**, have units of moles of product/mole of site/unit time. Of these factors, the site densities are available from an STO characterization of the catalyst and the **Rate** is determined for the specific reaction run over the STO characterized catalyst. When a series of at least three STO characterized catalysts is used for the same reaction, run under the same conditions, the specific site TOFs can be calculated from the simultaneous equations expressed as in Eqn. 3.6. When this approach was used in the hydrogenation of cyclohexene over a series of seven Pt/CPG catalysts specific site TOF values for the ^3M_I, ^3M_R and ^3MH sites were found to be 2.1, 18.2 and 5.2 moles of product/mole of site/second, respectively.[71] Not surprisingly, that site with the weakly held hydrogen was the most active and that on which the hydrogen was strongly held was the least active.

While each of the different types of saturation sites has a different reactivity in hydrogenation reactions, in the oxidation of CO[65] or iso-propanol[70] each of these sites has essentially the same activity.

The presence on the surface of a dispersed metal catalyst of at least three distinct corner sites having different activities is, however, not compatible with the octahedral models of the 3M sites shown in Fig. 3.4 and used in Schemes 3.2 and 3.4 to develop analogies with specific homogeneous catalysts. A more detailed description of these corner atom sites and others present on the surface of metal catalysts is presented in the next chapter in conjunction with a discussion of the surface electronic orbitals of such species.

Oxide Catalysts

Siegel[72] has proposed octahedral analogs for the different types of oxide catalyst sites which are similar to the ^3M, ^2M and ^1M sites described for metal catalysts, but the identification of active sites on oxide catalysts has not been investigated as extensively as with metal catalysts. Those identifications that have been published have generally been based on some type of spectral analysis of an adsorbed species. These analyses, however, usually do not provide the complete information needed to accurately describe these sites.[35] Investigations of supported oxides have primarily been concerned with changes in oxidation state as well as the nature of the adsorbed species. These are complex systems and definite conclusions regarding the nature of the surface adsorption sites has not been forthcoming.

Data for simple oxides is more easily interpreted but most simple oxides are relatively inactive as catalysts. Ultraviolet spectral analysis of alkaline earth oxides indicate the presence of three kinds of uncoordinated ions corresponding to corner, edge and face atoms.[73-77] similar to those proposed by Siegel.[72] Infrared data from hydrogen chemisorption on ZnO shows the presence of two species, one of which is active in alkene hydrogenations.[78,79] TPD data showed that this active species desorbed in two stages, indicating the presence of two similar species having similar spectral properties but differing in their ease of desorption.[80] An infrared study of oxygen adsorbed on α Fe_2O_3 (hematite) showed two complex groups of spectral bands. One, ranging from $1350-1250 cm^{-1}$ is ascribed to the presence of O_2^- and the other group, $1100-900$ cm^{-1}, to the presence of O_2^{-2}. TPD determinations showed that the O_2^- species desorbed near $150°C$ while the O_2^{-2} was not desorbed below $300°C$. Heating at higher temperatures was found to drive off the lattice oxygen.[81]

References

1. G. D. Halsey, *Adv. Catal.,* 4, 259 (1952).
2. A. S. Al-Ammar, S. J.Thompson and G. Webb, *J. Chem. Soc., Chem. Commun.,* 323 (1977).
3. G. F. Berndt, S. J. Thompson and G. Webb, *J. Chem. Soc., Faraday Trans. I,* **79**, 195 (1983).
4. L. Gonzalez-Tejuca, K. Aika, S. Namba and J. Turkevich, *J. Phys. Chem.,* **81,** 1399 (1977).
5. T. Harada, A. Tai, M. Yamamoto, H. Ozaki and Y. Izumi, *Stud. Surf. Sci. Catal.,* **7** (New Horiz. Catal., Pt.A), 364 (1981).
6. R. Lopez, *Coll. Czech. Chem. Commun.,* **48**, 2269 (1983).

7. N. Y. Topsoe and H. Topsoe, *J. Catal.,* **84**, 386 (1983).
8. H. Topsoe and B. S. Clausen, *Appl. Catal.,* **25**, 273 (1986).
9. W. M. H. Sachtler, *Ultramicroscopy,* **20**, 135 (1986).
10. P. A. Sermon, G. Georigiades, M. S. W. Vong, M. A. Martin-Luengo and P. N. Reyes, *Proc. R. Soc. Lond. A,* **410**, 353 (1987).
11. G. A. Somorjai, *Catal. Revs.,* **7**, 87 (1972); **18**, 173 (1978).
12. G. A. Somorjai, *Accounts Chem. Res.,* **9**, 248 (1976).
13. G. A. Somorjai and S. M. Davis, *J. Catal.,* **65**, 78 (1980).
14. R. K. Herz, W. D. Gillespie and G. A. Somorjai, *J. Catal.,* **67**, 371 (1981).
15. W. D. Gillespie, R. K. Herz, E. E. Petersen and G. A. Somorjai, *J. Catal.,* **70**, 147 (1981).
16. S. M. Davis, F. Zaera and G. A. Somorjai, *J. Am. Chem. Soc.,* **104**, 7453 (1982).
17. G. A. Somorjai and S. M. Davis, *ChemTech.,* **13,** 508 (1983).
18. G. A. Somorjai and S. M. Davis, *Plat. Met. Rev.,* **27**, 54 (1984).
19. D. W. Goodman, *Chem. Ind. (Dekker),* **22**, (Catal. Org. React.), 171 (1985).
20. D. W. Goodman, *Accounts Chem. Res.,* **17**, 194 (1984).
21. D. W. Goodman, *Ann. Rev. Phys. Chem.,* **37**, 425 (1986).
22. D. W. Goodman, *Surf. Sci.,* **123**, L679 (1982).
23. M. Boudart, *Adv. Catal.,* **20**, 153 (1969).
24. R. Burch, *Catalysis (London),* **7**, 149 (1985).
25. R. Z. C. van Meerten, A. H. G. M. Beaumont, P. F. M. T. van Nisselrooiz and J. W. E. Coenen, *Surf. Sci.,* **135**, 565 (1983).
26. P. H. Otero-Schipper, W. A. Wachter, J. B. Butt, R. L. Burwell, Jr., and J. B. Cohen, *J. Catal.,* **50**, 494 (1977).
27. P. H. Otero-Schipper, W. A. Wachter, J. B. Butt, R. L. Burwell, Jr., and J. B. Cohen, *J. Catal.,* **53**, 414 (1978).
28. J. H. Sinfelt, J. L. Carter and D. J. C. Yates, *J. Catal.,* **24**, 283 (1972).
29. P. S. Kirlin and B. C. Gates, *Nature,* **325**, 38 (1987).
30. H. E. Farnsworth, in *Experimental Methods in Catalysis Research,* **Vol. I,** (R. B. Anderson, Ed.) Academic Press, NY, 1968, p 268.
31. D. G. Miller, *Chem. Rev.,* **60**, 15 (1960).
32. B. Imelik and J. C. Vedrine, *Catalyst Charactrization,* Plenum Press, New York, 1994.
33. J. A. Moulijin, P. W. N. M. van Leewuen and R. A. van Santen, *Stud. Surf. Sci. Catal.,* **79** (Catalysis), 363 (1993).
34. E. Garbowski and H. Praliaud, in *Catalyst Charactrization,* (B. Imelik and J. C. Vedrine Eds.) Plenum Press, New York, 1994, p 61.
35. A. Zecchina, E. Garrone and E. Guglielminotti, *Catalysis (London),* 6, 90 (1983).
36. G. Coudurier and F. Lefebvre, in *Catalyst Charactrization,* (B. Imelik and J. C. Vedrine Eds.) Plenum Press, New York, 1994. p 11.
37. E. Garbowski and G. Coudurier, in *Catalyst Charactrization,* (B. Imelik and J. C. Vedrine Eds.) Plenum Press, New York, 1994. p 45.
38. G. N. Greaves, *Catal. Today,* **2**, 581 (1988).
39. J. Evans, *Catalysis (London),* **8**, 1 (1989).
40. B. Morawech, in *Catalyst Charactrization,* (B. Imelik and J. C. Vedrine Eds.) Plenum Press, New York, 1994. p 377.
41. T. Baird, *Catalysis (London),* **5**, 172 (1982).
42. M. Jose-Yacaman and M. Avalos-Borja, *Catal. Revs.,* **34**, 55 (1992).

43. A. K. Datke and D. J. Smith, *Catal. Revs.*, **34**, 129 (1992).

44. S. B. Rice and S. A. Bradley, *Catal. Today*, **21**, 71 (1994).

45. P. Gallezot and C. Leclerq, in *Catalyst Charactrization*, (B. Imelik and J. C. Vedrine Eds.) Plenum Press, New York, 1994. p 509.

46. R. van Hardefeld and F. Hartog, *Surf. Sci.*, **15**, 189 (1969).

47. R. van Hardefeld and F. Hartog, *Adv. Catal.*, **22**, 75 (1979).

48. O. L. Perez, D. Romeu and M. J. Yacaman, *Appl. Surf. Sci.*, **13**, 402 (1982).

49. R. L. Augustine and J. F. Van Peppen, *Ann. N. Y. Acad. Sci.*, **172**, 244 (1970).

50. S. Siegel, J. Outlaw, Jr. and N. Garti, *J. Catal.*, **52**, 102 (1978).

51. R. L. Augustine and R. W. Warner, *J. Catal.*, **80**, 358 (1983).

52. C. O'Connor and G. Wilkinson, *J. Chem. Soc., A*, 2665 (l968).

53. I. Horiuti and M. Polanyi, *Trans. Faraday Soc.*, **30**, 1164 (1934).

54. J. A. Osborne, F. H. Jardine, J. F. Young and G. Wilkinson, *J. Chem.Soc., A*, 1711 (l966).

55. R. L. Augustine and P.J. O'Hagan, *Chem. Ind., (Dekker)* **40** (Catal. Org. React.), 111 (1989).

56. G. V. Smith, O. Zahraa, A. Molnar, M. M. Khan, B. Rihter and W. E. Brower, *J. Catal.*, **83**, 238 (1983).

57. F. Notheisz, M. Bartok, D. Ostgard and G. V. Smith, *J. Catal.*, **101**, 213 (1986).

58. G. V. Smith, A. Molnar, M. M. Khan, D. Ostgard and N. Yoshida, *J. Catal.*, **98**, 502 (1986).

59. M. J. Ledoux, *Nouv. J. Chim.*, **2**, 9 (1978).

60. M. J. Ledoux, F. G. Gault, A. Bouchy and G. Roussy, *J. Chem. Soc., Faraday Trans. II.*, **74**, 652 (1978).

61. M. J. Ledoux and F. G. Gault, *J. Catal.*, **60**, 15 (1979).

62. M. J. Ledoux, *J. Catal.*, **70**, 375 (1981).

63. G. V. Smith, Private communication.

64. R. L. Augustine and R. W. Warner, *J. Org. Chem.*, **46**, 2614 (1981).

65. R. L. Augustine, D. R. Baum, K. G. High, L. S. Szivos and S. T. O'Leary, *J. Catal.*, **127**, 675 (1991).

66. R. L. Augustine, *Catal. Today*, **12**, 139 (1992).

67. S. Tsuchiya, N. Nakamura and M. Yoshioka, *Bull. Chem. Soc., Japan*, **51**, 981 (1978).

68. R. L. Augustine, K. P. Kelly and Y.-M. Lay, *Appl. Catal.*, **19**, 87 (1985).

69. R. L. Augustine and M. M. Thompson, *J. Org. Chem.*, **52**, 1911 (1987).

70. R. L. Augustine and L. Doyle, *Chem. Ind. (Dekker)* **53** (Catal. Org. React.), 279 (1994).

71. R. L. Augustine, M. M. Thompson and M. A. Doran, *J. Chem. Soc., Chem. Commun.*, 1173 (1987).

72. S. Siegel, *J. Catal*, **30**, 139 (1973).

73. R. L. Nelson and J. W. Hale *Discuss. Faraday Soc.*, **52**, 77 (1971).

74. A. Zecchina, M. G. Lofthouse and F. S. Stone, *J. Chem. Soc., Faraday Trans. I*, **71**, 1476 (1975).

75. A. Zecchina and F. S. Stone, *J. Chem. Soc., Faraday Trans. I*, **72**, 2364 (1976).

76. E. Garrone, A. Zecchina and F. S. Stone, *Phils. Mag. B*, **42**, 683 (1980).

77. S. Coluccia, A. J. Tench and R. L. Segall, *J. Chem. Soc., Faraday Trans. I*, **75**, 176 (1979).

78. R. P. Eischens, W. A. Pliskin and M. J. D. Low, *J. Catal.*, **1**, 180 (1962).
79. A. L. Dent and R. J. Kokes, *J. Phys. Chem.*, **73**, 3772 (1969).
80. G. L. Griffith and J. T. Yates, Jr., *J. Catal.*, **73**, 396 (1982).
81. F. Al-Mashta, J. Sheppard, V. Lorenzelli and G. Busca, *J. Chem. Soc., Faraday Trans. I*, 78, 979 (1982).

Surface Frontier Molecular Orbitals

As mentioned in Chapter 3, the octahedral models used to describe the active sites on metal surfaces are not compatible with the presence of three different types of saturation sites on a catalyst surface so another model must be developed. On consideration of the fcc crystal structure, which is that of most catalytically active metals, it can be seen that the bulk atoms in these metals are bound to twelve nearest neighbor atoms using the lobes of the t_{2g} d orbitals. The octahedrally oriented e_g orbitals are directed toward but not bonded to the next nearest neighbors in the crystal lattice as shown in Fig. 4.1.[1] This atomic orientation precludes the presence of any octahedral arrangement involving M–M bonds.

Surface Electron Orbitals

With this orbital arrangement and the assumption of localized surface orbitals, the orientation of those orbitals extending from atoms in a 111 and a 100 plane are depicted in Fig. 4.2.[2] The angular orientation of these orbitals from the 111 face would seem to preclude the type of adsorption shown in Fig. 2.3 which takes place on a single surface atom but should permit an adsorption process utilizing several surface atoms. This will be discussed later. Single atom adsorption could take place on 100 face atoms if the perpendicular e_g orbital were empty so it could accept electrons from the substrate. The t_{2g} orbitals are properly oriented for backbonding as depicted in Fig. 2.3. If this were to occur, though, nothing further

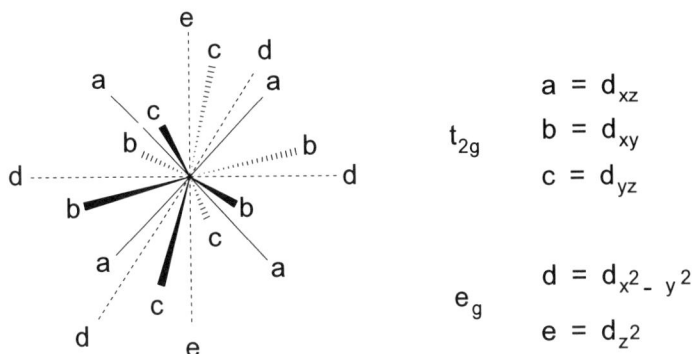

$$t_{2g} \quad \begin{aligned} a &= d_{xz} \\ b &= d_{xy} \\ c &= d_{yz} \end{aligned}$$

$$e_g \quad \begin{aligned} d &= d_{x^2 - y^2} \\ e &= d_{z^2} \end{aligned}$$

Fig. 4.1. Orbital orientations of the d electrons.

Top View   Side View

100
Face

t_{2g} e_g

111
Face

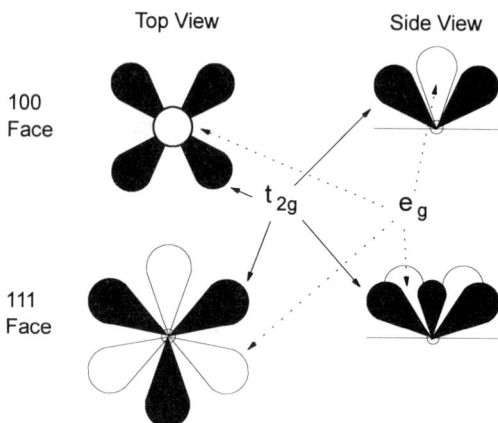

Fig. 4.2. Localized surface orbital orientations for atoms in a 100 and 111 plane.

could take place since there is no means available for adsorbing a second reactant on these atoms if they react as single atom active sites. These face atoms can either promote a reaction via adsorption on a multiatom ensemble of atoms or the single atoms could adsorb one reactant but do nothing further.

This localization of the surface orbitals has been supported by EHMO calculations on three and four layer slabs of various metals in 111 and 100 orientations.[3-6] While these calculations can be applied to a large number of atoms in either the 111 or 100 planes of a metal, they cannot easily be used to determine the electron energies for a single atom placed at the corner or edge of such an arrangement.

Surface Sites on an fcc Metal

One way of calculating the electron energy levels of the various types of surface sites is to consider each type of site as a "surface complex" made up of the central, active atom surrounded by the neighboring "ligand" atoms. In an fcc metal, a bulk metal atom is surrounded by twelve near neighbor ligands as shown in both top and side views from the 111 and 100 orientations in Fig. 4.3 This is the parent species and each of the possible surface complexes can be formed by the removal of the appropriate ligands from the parent.

The surface complex made up of an atom in a 111 plane is formed by removing ligands 10, 11 and 12 from the parent. An atom in a 100 plane is produced by removing atoms 4, 9, 10 and 12. A tetrahedral corner site is composed of the central atom and ligands"1, 2, and 3, while an octahedral corner has ligands 2, 3, 6 and 7. Table 4.1 lists all of the surface arrangements possible

Table 4.1

Types of surface atoms on FCC metals[7]

Site Description[a]	Number of Nearest Neighbors	Ligands	Designation
111 (face)	9	1, 2, 3, 4, 5, 6, 7, 8, 9,	A
100 (face)	8	1, 2, 3, 5, 6, 7, 8, 11	B
111-100 (wide edge)	7	1, 2, 3, 5, 6, 7, 8	C
111-111 (wide edge)	7	1, 2, 3, 4, 5, 6, 7	D
111-111 (narrow edge)	6	2, 3, 5, 6, 7, 8	E
111-100 (narrow edge)	5	2, 4, 5, 6, 7	F
111-111-100 (cubooctahedral corner)	6	1, 2, 3, 6, 7, 8	G
111-111-100-100 (corner)	5	1, 2, 3, 7, 8	H
111-111-111 (wider corner)	5	1, 2, 3, 6, 7	I
111-111-111 (wide Corner)	4	1, 3, 7, 8	J
111-111-111-111 (octahedral corner)	4	2, 3, 6, 7	K
111-111-100 (narrow corner)	3	1, 3, 8	L
111-111-111 (tetrahedral corner)	3	1, 2, 3	M

[a] As depicted in Figs. 4.4-4.6.

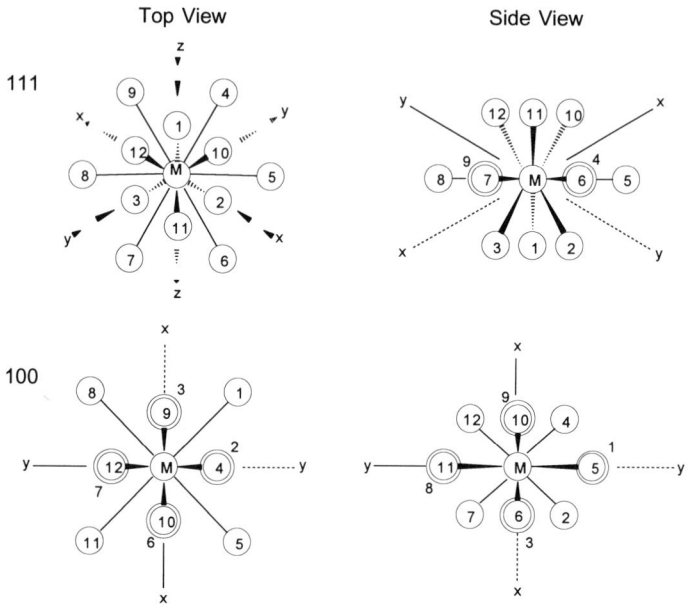

Fig. 4.3.   Top and side views from the 111 and 100 planes of the twelve
            coordinate complex representing a metal atom in an fcc crystalline
            arrangement surrounded by its twelve nearest neighbor atoms.

with an fcc crystal structure with the type of site defined and the ligand numbers
associated with it.[7,8] In this list are atoms in the two low Miller index planes, 111
and 100, types A and B. The 110 plane is considered for reaction purposes to be
the ends of a staggered series of 111 planes. These sites are also depicted in Fig.
4.4.

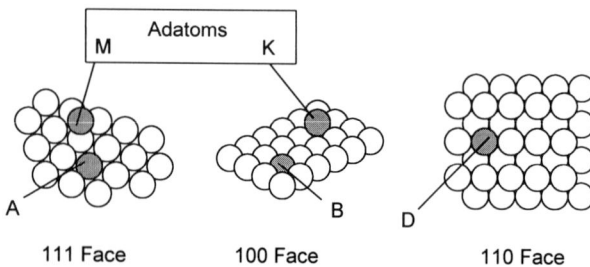

Fig. 4.4.   Depiction of the atomic orientations in the low Miller
            index planes of fcc metals and the adatoms present on
            the 111 and 100 planes.

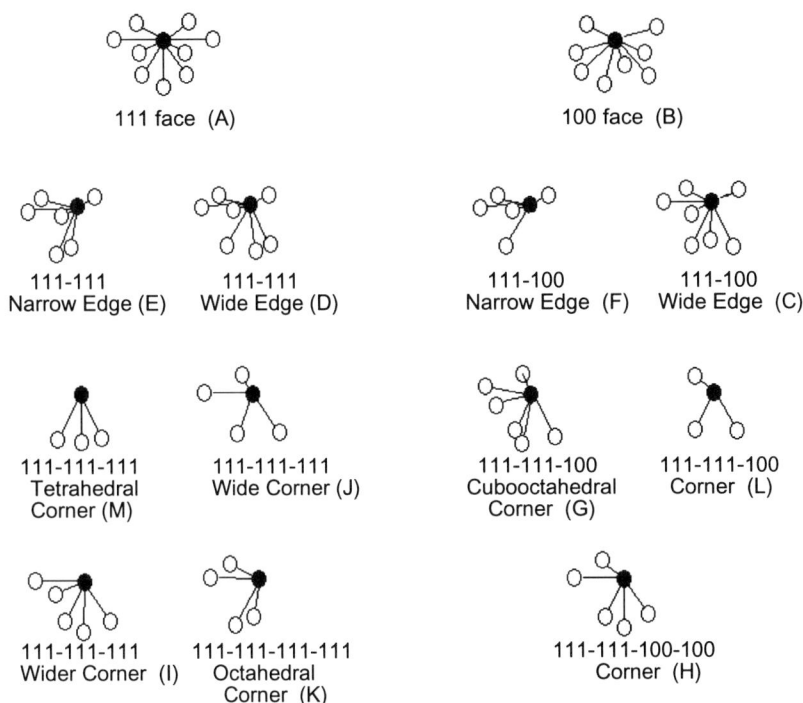

111 face (A)

100 face (B)

111-111
Narrow Edge (E)

111-111
Wide Edge (D)

111-100
Narrow Edge (F)

111-100
Wide Edge (C)

111-111-111
Tetrahedral
Corner (M)

111-111-111
Wide Corner (J)

111-111-100
Cubooctahedral
Corner (G)

111-111-100
Corner (L)

111-111-111
Wider Corner (I)

111-111-111-111
Octahedral
Corner (K)

111-111-100-100
Corner (H)

Fig. 4.5.   Structural representations of the various types of atoms present
on the surface of an fcc metal.

The 111 plane can intersect another plane at either 60° or 120° so there are
two 111–111 edge sites, labeled D and E, and two 111–100 edge sites, labeled C
and F. The 90° orientation required for the intersection of two 100 planes is not
possible. The seven different corner atoms range from the cubooctahedral, F,
with six neighboring atoms to the tetrahedral corner, M, having only three. Here,
too, the 60° and 120° plane intersections show up in the corner atom
arrangements. Structures of these sites are depicted in Fig. 4.5.[7,8]

Other views of some of these sites are illustrated with the crystal shapes
shown in Figs. 4.4 and 4.6. It can be seen in the cubic crystal that the atoms on
the corner are not bonded along the edges of the cube. Instead these corner atoms
are bonded to three nearest neighbors and are, in reality, tetrahedral corners (M) as
shown more clearly on the tetrahedral crystal. The face atoms on the cube are in a
100 plane (B). What appear to be edge atoms, though, are the surface sites shown
in Fig. 4.5 as the 111–111–100–100 corner sites (H).

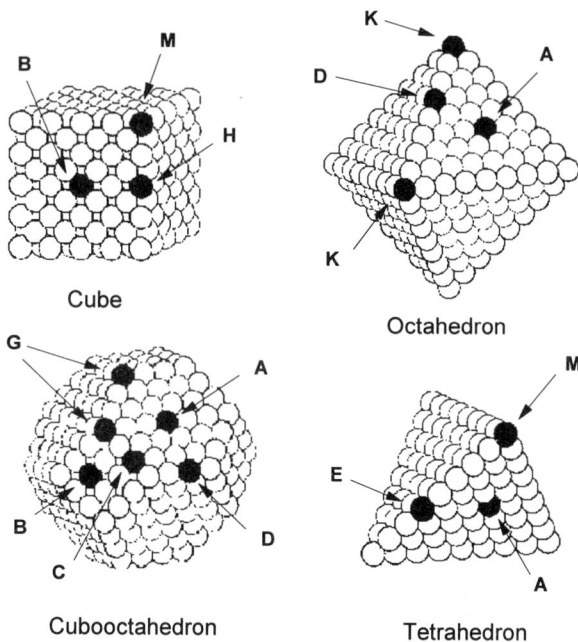

Fig. 4.6.   Idealized fcc crystal structures and the types of sites present on
them.

All of the face atoms on the tetrahedron and octahedron are in a 111 plane
(A). The tetrahedron has the 111–111 narrow edge (E) while the octahedron has
the wide edge (D). The octahedron, obviously, has octahedral corners (K) and the
cubooctahedron has cubooctahedral corners (G). Both 111 (A) and 100 (B) face
atoms are present on the cubooctahedron along with the 111–111 wide edge (D)
and the 111–100 wide edge (C).

If the metal catalyst particles were present only in the form of these
idealized crystals, then the number of active corner atoms present would be very
low. However, STO evaluations of dispersed metal catalysts have shown that
these active atoms are present in rather large amounts, at times as high as 30%–
35% of the total metal atoms present. Such high surface concentrations of the
highly unsaturated atoms can only be accounted for by the presence of the
irregular particle shapes that were observed using dark field TEM imaging
techniques.[9] Additional active sites are probably present as adatoms on the 111
(M) and 100 (K) planes as shown in Fig. 4.4.

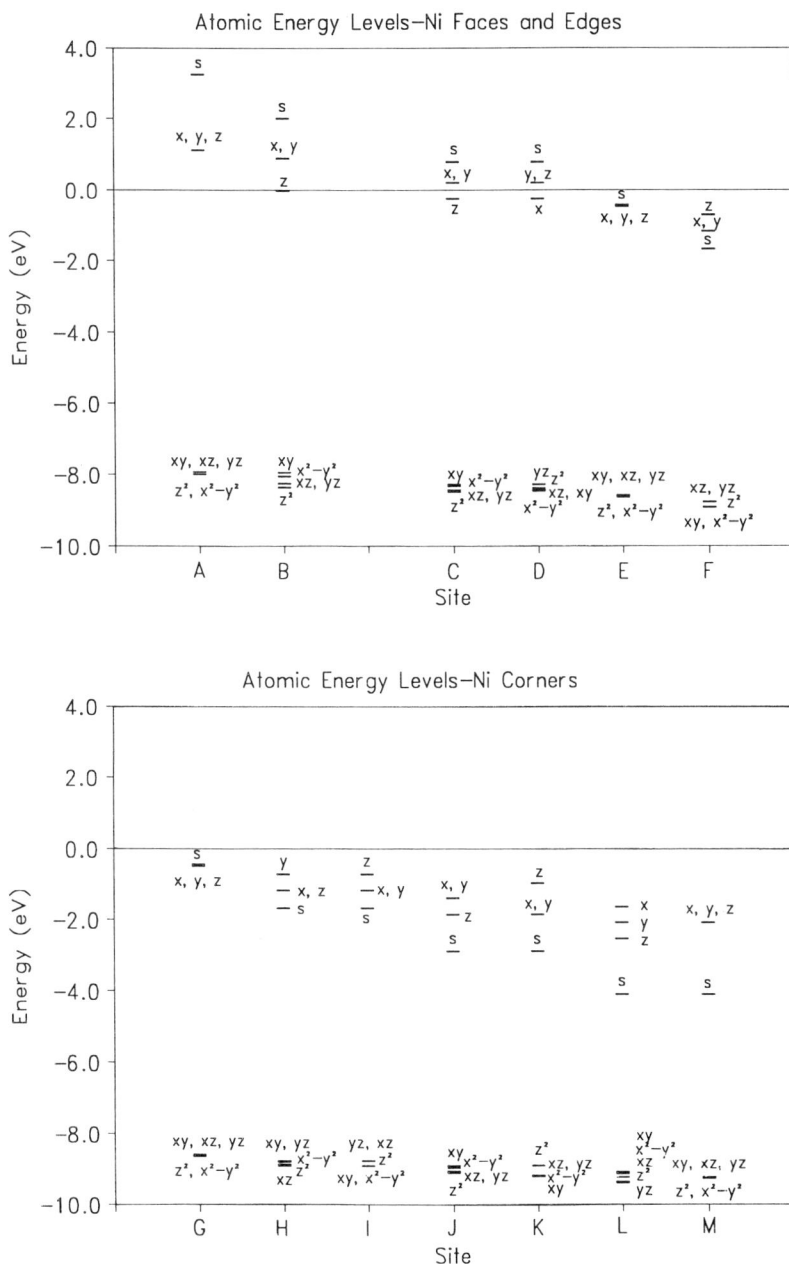

Fig. 4.7   Electron energy levels for the various types of Ni surface atoms.
Horizontal equivalence indicates orbital degeneracy. Vertical
orientation depicts relative energy values.

Angular Overlap Model

If, to a first approximation, the sites described above can be considered as "surface complexes", then methods for calculating electron energy levels of normal complexes could be applied to determining the electron energy levels of these sites as well. The Angular Overlap Model (AOM)[10,11] is particularly suited for this task. This is a simple approximation of the full molecular orbital treatment that contains all of the important characteristics of M–L interactions. AOM assumes that the energy values for the entire complex can be considered to be the sum of the energies of each single M–L interaction. These M–L energies are proportional to the square of the overlap integral between orbitals on the M and the L as determined by the angle between them in the coordination sphere. In addition to the angular component of the overlap integral there is also a radial component that is a function of the specific M and L and the distance between them.

The angular component of the overlap integral for these surface complexes is determined by using the values of the spherical coordinates of each of the ligands to solve the equations developed by Schäffer[12] for this purpose. To maintain a continuity between the various types of sites, the x, y and z Cartesian coordinates were set coincident with the p_x, p_y and p_z orbitals of the central atom. In this way the axes required for measuring the spherical coordinates of the ligand atoms were fixed in a constant orientation for all of the surface complexes. The radial components were determined using EHMO calculations on the appropriate M–M diatomic species using published Huckel parameters[5-6,13,14] for the surface atoms and the interatomic distance of the bulk metal. The s, p and d orbital energies for the surface sites on a Ni catalyst are shown in Fig. 4.7. These values and comparable data for other metals can be used in Frontier Molecular Orbital (FMO) considerations of adsorptions and catalytic reactions taking place on specific sites.[7,8]

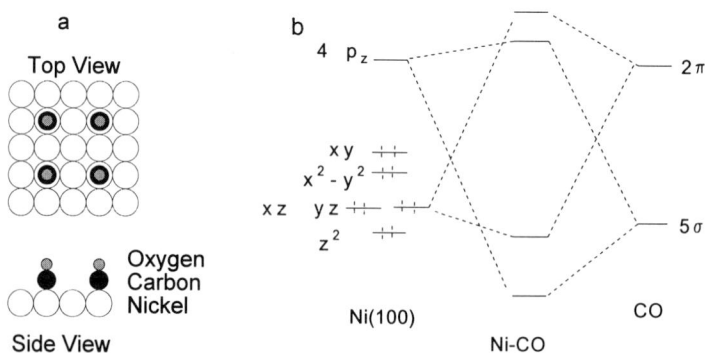

Fig. 4.8. a) Representation of the mode of CO adsorption on Ni 100 atoms;[5] b) orbital interactions in this adsorption.

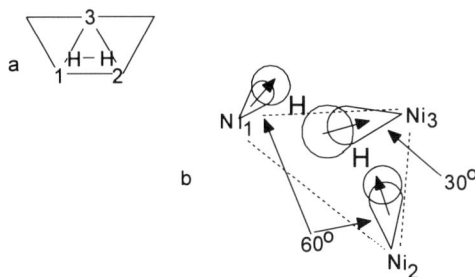

Fig. 4.9. a) Calculated H_2 adsorption orientation on a Ni 111 face;[3]
b) FMO depiction of the electron overlap for this
adsorption.

Frontier Molecular Orbital Considerations

The adsorption of CO on the 100 face of nickel, for instance, has been calculated to take place on top of the atom as shown in Fig. 4.8.[5] This is the type of adsorption discussed previously for 100 face atoms using only the d orbitals. However, since nickel is a d_{10} element the d orbitals on this site are filled, so this simplified adsorption scheme does not apply. The 100 site energy levels shown in Fig. 4.8b indicate, though, that the p_z is the LUMO and is ideally situated for accepting the 2σ electrons from the CO. The HOMO of this site is the d_{xy} that is directed toward the neighboring atoms in the 100 plane so it is not available for backbonding. Neither is the next lower orbital, the d_{x2-y2} which is also oriented in the 100 plane. Backbonding can only take place with the d_{xz} or d_{yz} orbitals.

Hydrogen adsorption on the 111 face of nickel has been calculated to take place in the orientation shown in Fig. 4.9a[3] which suggests the need for a three atom site for this interaction. As Fig. 4.9b shows, the orientation of the LUMO p_x orbital of Ni_3 at an angle of 30° above the plane situates it properly for overlap with the H–H σ bonding electrons. The HOMO d_{xz} or d_{yz} orbitals on Ni_1 and Ni_2 project 60° above the plane and can overlap with the σ* orbitals on hydrogen to provide the needed backbonding. After breaking the H–H bond, two Ni–H bonds are formed on two of these Ni atoms.[7,8]

Specific site reactivity data[15] indicated that the platinum catalyzed carbon monoxide oxidation takes place on the corner atoms of the platinum particles. Fig. 4.10a shows the electron energies for a platinum tetrahedral corner. Fig. 4.10b is an FMO projection of a mechanism for this reaction as it would, presumably, take place on a tetrahedral corner of a platinum catalyst. On this site the LUMO is the symmetrical 6s orbital and the HOMO is one of the degenerate d_{xy}, d_{xz} and d_{yz} orbitals. Oxygen adsorption can take place by electron donation to the LUMO 6s and backbonding by the d_{xz} orbitals. At this point the LUMO is a degenerate 6p orbital and the HOMO is either the d_{xy} or d_{yz}. Carbon monoxide adsorption takes place using the LUMO p_x with backbonding by the HOMO d_{xy} to give the surface

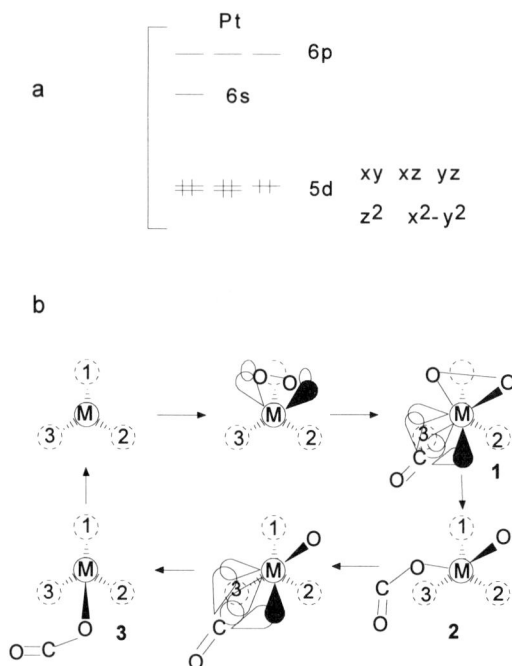

Fig. 4.10. a) Electron energy levels for a Pt atom on a tetrahedral
           corner; b) FMO representation of the oxidation of CO
           over this atom.

species, **1**, with both reactants adsorbed. Oxygen insertion into the Pt–C bond
gives the oxocarbonate, **2**, which loses carbon dioxide and then adsorbs a second
molecule of carbon monoxide in the same way as the first. Another oxygen
insertion gives the carbonate, **3**, which, on loss of carbon dioxide, regenerates the
active site and completes the catalytic cycle.

The mechanism for alkene hydrogenation was discussed in Schemes 3.2
and 3.4 using the octahedral complexes as models for the ^{3}MH and ^{3}MH$_2$ sites.
Since such an octahedral orientation does not exist in an fcc metal, an FMO
approach is needed to illustrate these reactions as they would take place on
specific corner atoms. While there is nothing in the electron energy values for
these corner sites that would suggest which one might be a ^{3}MH site or a ^{3}M
site as delineated in the STO reaction sequence,[15,16] it can reasonably be assumed
that a site that adsorbs only one hydrogen atom does so in conjunction with two
other atoms as shown in Fig. 4.9 Thus, some corner atoms that are adjacent to
111 faces as depicted in Fig. 4.11, can adsorb hydrogen through a three atom
ensemble and, thus, become the two-step saturation sites,^{3}MH. Other surface
atoms, however, particularly adatoms that do not have any other atoms in the

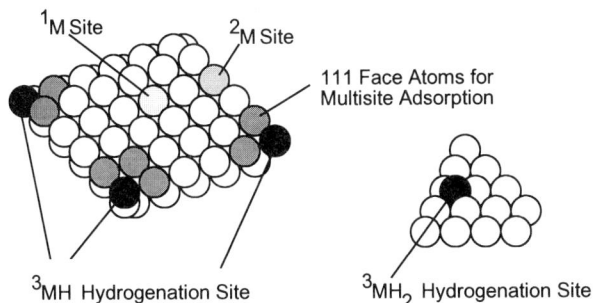

Fig. 4.11. A representation of the different types of STO characterized
sites on a metal catalyst particle.

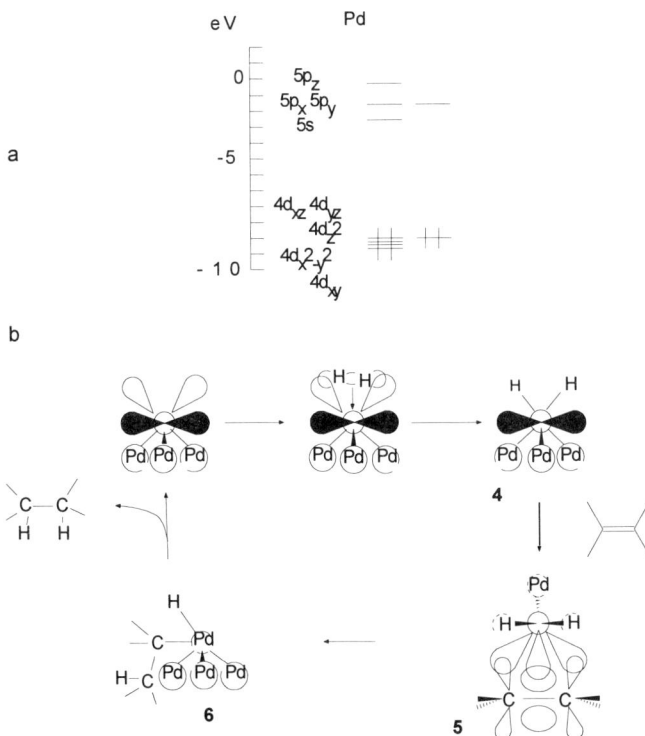

Fig. 4.12. a) Electron energy levels for a Pd atom on an octahedral
corner; b) FMO representation of the hydrogenation of an
alkene over this atom.

parallel orientation needed for multiatom adsorption, could adsorb both atoms of a dihydrogen molecule and react as a $^3MH_2$ site. The edge atoms, then, are the ^2M sites and the ^1M sites are the face atoms as depicted in Fig. 4.11.

The FMO treatment of an alkene hydrogenation taking place on a palladium octahedral corner is shown in Fig. 4.12. On this site the LUMO is the 5s orbital and the HOMO the degenerate d_{xz} and d_{yz}. Hydrogen adsorption utilizing these orbitals will give the dihydride, having the side view **4**, while blocking additional adsorption from the top of the site. Alkene adsorption, which is viewed from the top in **5**, can then take place using the LUMO p_x with backbonding from the d_{xy}, the only filled orbital having the proper orientation to do so. Hydrogen insertion followed by reductive elimination from the hydrido metalalkyl, side view in **6**, gives the alkane, regenerates the active site and completes the catalytic cycle.

While the FMO treatment depicted in Fig. 4.12 fully describes the surface interactions taking place when a double bond is hydrogenated, this approach is more complex than is needed for most purposes. The reaction sequences shown in Schemes 3.2, 3.4 and 3.5, which are based on an octahedral species however, are incorrect. Scheme 4.1 shows the reaction sequence for alkene hydrogenation over a ^3MH site that is analogous to Scheme 3.2 but uses a more descriptive surface site model. Instead of the depiction of the ^3MH site as a corner atom as

Scheme 4.1

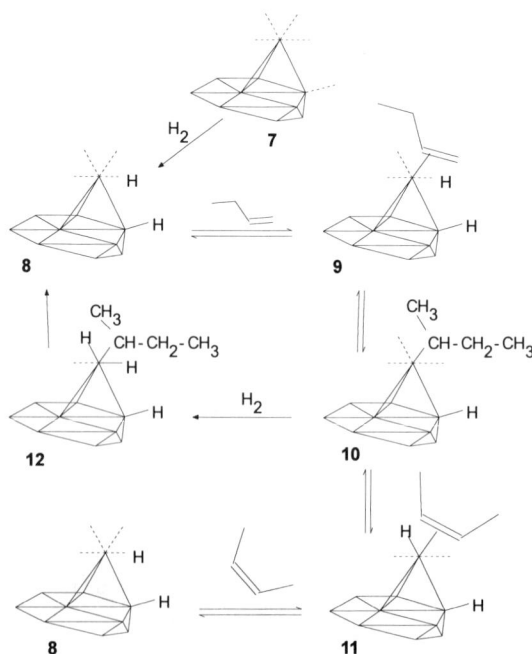

Scheme 4.2

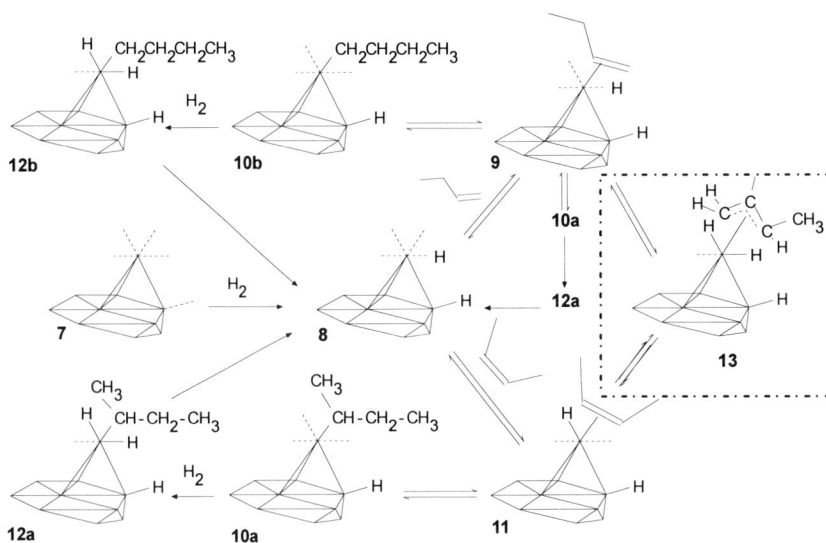

shown in Fig. 4.11, the site here is an adatom located on the edge of a metal crystallite, **7**. Ensemble adsorption of dihydrogen gives the ^{3}MH site, **8**, with the second hydrogen atom located on the adjacent edge atom. Alkene adsorption followed by hydrogen insertion gives the adsorbed metalalkyl, **10**. β-Elimination of this adsorbed species can regenerate the initially adsorbed alkene, **9**, or the isomeric species, **11**. Dihydrogen addition to **10** gives the MH$_2$ species, **12**. Reductive elimination then produces the alkane and regenerates the ^{3}MH site, **8**.

This, however, is still a simplified approach in that it does not take into consideration the formation of the primary metalalkyl nor the possible intermediacy of an adsorbed π-allyl species as occurs over palladium and, to some extent, nickel catalysts.[17]

The reaction sequence outlined in Scheme 4.2 is more complete, showing the primary metalalkyl, **10b**, as well as the adsorbed π-allyl, **13**. With catalysts other than palladium or nickel this species is not present so the boxed portion of this reaction sequence would not take place.

The corresponding reaction path for the hydrogenation of an alkene on an adatom, **14**, is shown in Scheme 4.3. Here the ^{3}MH$_2$ site, **15**, can adsorb the alkene to give the adsorbed species, **16** or **17**, which are then saturated using the two hydrogen atoms on the active site. The adsorbed π-allyl, **20**, may also take part with a palladium catalyst. The formation of this species provides the pathway for the facile double bond isomerization observed over palladium catalysts.[17]

Scheme 4.3

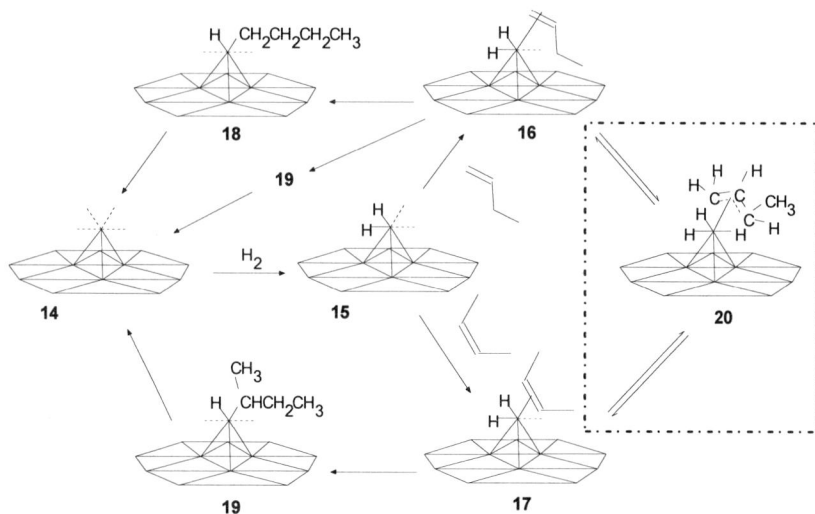

The pathway for double bond isomerization taking place over edge atoms, **21**, which are converted to ^{2}MH sites, **22**, is depicted in Scheme 4.4.

Scheme 4.4

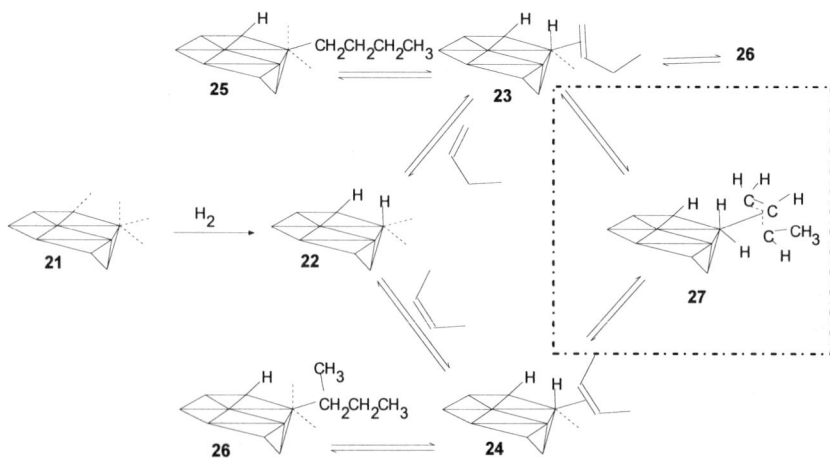

References

1. J. B. Goodenough, *Magnetism and the Chemical Bond*, Interscience, New York, 1963.
2. G. C. Bond, *Disc. Faraday Soc.*, **41**, 200 (1966).
3. J.-Y. Saillard and R. Hoffmann, *J. Am. Chem. Soc.*, **106**, 2006 (1984).
4. C. Zheng, Y. Apeloig and R. Hoffmann, *J. Am. Chem. Soc.*, **110**, 749 (1988).
5. S.-S. Sung and R. Hoffmann, *J. Am. Chem. Soc.*, **107**, 578 (1985).
6. J. Silvestre and R. Hoffman, *Langmuir*, **1**, 621 (1985).
7. R. L. Augustine, K. M. Lahanas and F. Cole, *Stud. Surf. Sci. Catal.*, **75** (New Frontiers in Catalysis), 1568 (1993).
8. R. L. Augustine and K. M. Lahanas, *Chem. Ind. (Dekker)*, **53** (Catal. Org. React.), 279 (1994).
9. O. L. Perez, D. Romeo and M. J. Yacaman, *Appl. Surf. Sci.*, **13**, 402 (1982).
10. E. Larsen and G. N. LaMer, *J. Chem. Ed.*, **51**, 633 (1974).
11. R. S. Drago, *Physical Methods in Inorganic Chemistry*, Saunders, NY, 1977, pp. 396-406.
12. C. E. Schäffer, *Pure and Applied Chem.*, **24**, 361 (1970).
13. D. L. DuBois and R. Hoffmann, *Nouv. J. Chim.*, **1**, 479 (1977).
14. S. Komiya, T. A. Albright, R. Hoffmann and J. K.Kochi, *J. Am. Chem. Soc.*, **99**, 8440 (1977).
15. R. L. Augustine, D. R. Baum, K. G. High, L. S. Szivos and S. T. O'Leary, *J. Catal.*, **127**, 675 (1991).
16. R. L. Augustine and R. W. Warner, *J. Catal.*, **80**, (1983) 358.
17. R. L. Augustine, F. Yaghmaie and J. F. Van Peppen, *J. Org. Chem.*, **49**, 1865 (1984).

5

Reaction Variables

Catalytic reactions can take place in either the liquid or vapor phase. Liquid phase reactions can be run in either a continuous manner or as a batch process while vapor phase reactions are run only in a continuous mode. In a batch reaction the catalyst, reactants, and other components of the reaction mixture are placed in an appropriate reaction vessel, the reaction is run and the products removed from the vessel and separated from the catalyst. In a continuous system the reactants are passed through the catalyst and the products removed at the same rate as the reactants are added. The applicability of vapor phase processes is limited by the volatility and thermal stability of the reactants and products so such processes are not commonly involved in the preparation of even moderately complex molecules. Because of this, primary attention will be placed here on liquid phase processes with vapor phase systems of secondary importance. A discussion of the different types of reactors used for each of these processes is found in the following chapter. The present discussion is concerned with the effect that the different reaction parameters can have on the outcome of a catalytic reaction.

Of all the reaction variables involved in a heterogeneously catalyzed reaction, the most important is the nature of the catalyst to be used. Factors associated with catalyst preparation and selection will be discussed in Sections II and III. The relative importance of the other reaction parameters will depend on a number of factors. Reactions that run in a continuous or flow system have different requirements from those run in a batch mode. Generally, parameters such as the quantity of catalyst, the size of the catalyst particles, the temperature of the system, the concentration of the substrate(s), and, when gaseous reactants are used, the reaction pressure, are important variables in heterogeneously catalyzed reactions. In flow reactions the catalyst:substrate contact time can frequently have a significant impact on the outcome of the reaction. In liquid phase batch processes catalyst agitation can also play an important role. The one constant parameter in almost all liquid phase reactions is the presence of a solvent, the nature of which is an important factor in heterogeneously catalyzed liquid phase reactions.

Solvent Effects

The most obvious reason for the selection of a particular solvent is to dissolve, at least partially, the substrate on which the reaction is being run. Complete solution before reaction is not necessary. All that is required is that some of the substrate

Scheme 5.1

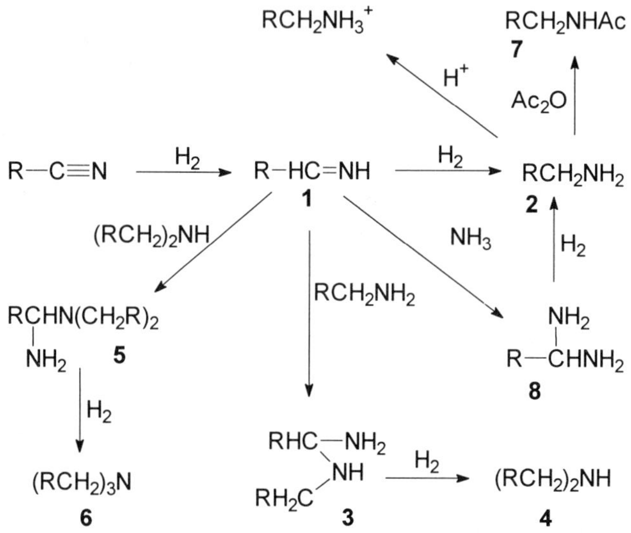

is in solution. It is essential, though, that all of the product remain in solution throughout the reaction since, otherwise, it would precipitate on the catalyst and block the active sites from further reaction.

The amount of solvent to use can be of importance in developing a successful catalytic process. As outlined in Scheme 5.1, the hydrogenation of a nitrile first produces an aldimine, **1**, which is then hydrogenated further to give the primary amine, **2**. The by-products of these hydrogenations are the secondary, **4**, and tertiary amines, **6**, produced by condensation of the product amine, **2** or **4**, with the imine to give an aminal, **3** or **5**, which is subsequently hydrogenolyzed. With a dilute solution of the nitrile this condensation is minimized and primary amine production increases.   A similar effect is also observed in the hydrogenation of nitro groups. As illustrated in Eqn. 5.1 the intermediate is the nitroso compound, **9**, which can condense with either the product aniline, **11**, or intermediate hydroxylamine, **10**.[1] Diluting the reaction mixture can minimize the extent of such condensations.

It is not always necessary to use a large quantity of solvent to have a low concentration of dissolved substrate. Using a solvent in which the reactant is only partially dissolved will have the same effect.   The nitroimidazole, **12**, was selectively hydrogenated only in concentrated ammonia, a solvent in which the concentration of dissolved reactant was low enough that the amount of the intermediate nitroso compound in solution was minimized and condensation with

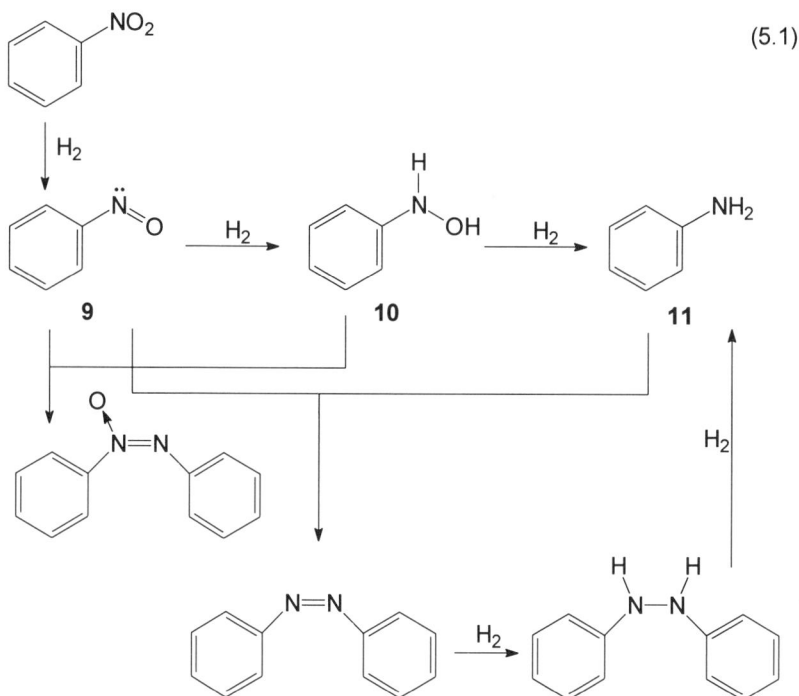

(5.1)

the product amine was effectively prevented. In other solvents the solubility of the imidazole was too high and dimeric product formation was predominant.[2]

Beyond solubility characteristics other factors are also important in selecting an appropriate solvent.[3] Sometimes a reactive material such as acetic anhydride or ammonia is needed to trap an intermediate or to react with a product molecule to prevent further reaction. In the nitrile hydrogenation described in Scheme 5.1, primary amine formation is enhanced by the addition of acetic anhydride to the reaction mixture. This material reacts with the initially formed primary amine, **2**, to form the amide, **7** which prevents the amine from condensing with the imine intermediate.[4] Hydrogenations of nitriles over nickel catalysts are frequently run in an ammoniacal solution. The ammonia, being present in excess, reacts preferentially with the imine, **1**, to give the aminal, **8**, which is further hydrogenolyzed to the primary amine, **2**.[5] Acidic media can be used to protonate product amines and, thus, keep them from reacting further.[6]

The catalytic reactions of polar groups is influenced by the polarity of the solvent. The hydrogenation of $\Delta^{1,9}$-octalone-2 (**13**) produces both the *cis* (**14**) and *trans* (**15**) β decalones (Eqn. 5.3). As shown by the data in Table 5.1, in aprotic

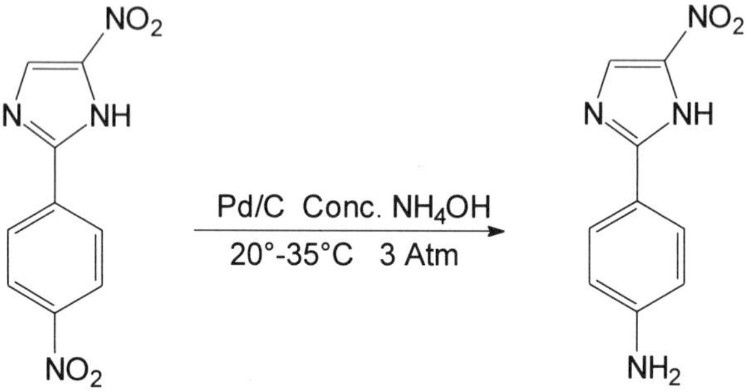

(5.2)

12

solvents the amount of the *cis* product, **14**, obtained increases with increasing solvent dielectric constant while with hydroxylic solvents the reverse is true.[7,8]

These observations have been rationalized by assuming that the *cis* product is formed primarily by the 1–4 addition of hydrogen to the enone while 1–2 addition to the double bond gives nearly an equal mixture of the *cis* and *trans* products. An increase in the dielectric constant will increase the polarity of the enone[9,10] and, thus, the affinity toward 1–4 addition, as is observed with the aprotic solvents.[11] In protic media, though, the hydroxy groups on the solvent molecules hydrogen bond with the carbonyl oxygen which modifies the way it adsorbs on the catalyst.[11] This hinders the 1–4 addition of hydrogen to the adsorbed species. The smaller the alcohol the more effective this interaction and the less 1–4 addition is favored. In methanol, hydrogen bonding with the carbonyl oxygen occurs readily and 1–2 addition of hydrogen to the double bond is preferred. In *tert*-butanol, hydrogen bonding is expected to be minimal but since this solvent is polar, 1–4 addition predominates.[7,8] These reactions are discussed in more detail in Chapter 15.

(5.3)

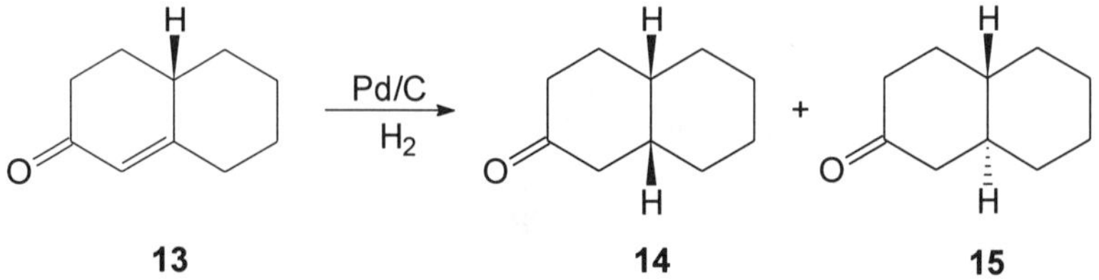

13   **14**   **15**

Table 5.1

Relationship between solvent dielectric constant and product
stereochemistry in the hydrogenation of β-octalone (**13**) over Pd
catalysts.[7,8]

	Solvent	Dielectric Constant	% cis β Decalone (**14**)
Aprotic	Dimethyl Formamide	38.0	79
	Acetone	21.0	63
	Ethyl Acetate	6.0	57
	Diethyl Ether	4.3	58
	Dioxane	2.2	52
	Cyclohexane	2.1	51
	n-Hexane	1.9	48
Protic	Methanol	33.6	41
	iso-Propanol	26.0	49
	Ethanol	25.1	55
	n-Propanol	21.8	68
	sec-Butanol	18.7	70
	n-Butanol	17.8	71
	tert-Butanol	10.9	91

The effect that a solvent exerts on the rate of a heterogeneously catalyzed reaction depends on the reaction substrate, the catalytically active species and the nature of the catalyst support.[12] Because of this complexity, attempts to derive a general equation relating reaction rate with some property or properties of the solvent have been unsuccessful.[13,14] The primary reason for this failure is that the solvents are not present merely as diluents in which the reactants are dissolved, they are also capable of interacting with the catalyst surface so they can also modify the extent to which different types of substrates are adsorbed.

In the hydrogenation of methylenecyclohexane, **16**, over a palladium catalyst, 1-methylcyclohexene, **17**, is also produced by double bond isomerization (Eqn. 5.4).[15] In an ethanol solution both of these isomers are hydrogenated over palladium at essentially the same rate but, as depicted in Fig. 5.1, when a 1:1 ethanol:benzene solvent is used the rate of hydrogenation of **17** is decreased but the hydrogenation of **16** is not affected initially, but drops off significantly after only a short reaction time. At this point analysis of the reaction mixture showed that the only olefin present was **17** which, as shown by the bottom curve in Fig. 5.1, was hydrogenated very slowly under these conditions. These results indicate that benzene evidently competes with **17** for adsorption sites on the catalyst but that **16** is more readily adsorbed than is benzene.[14,15]

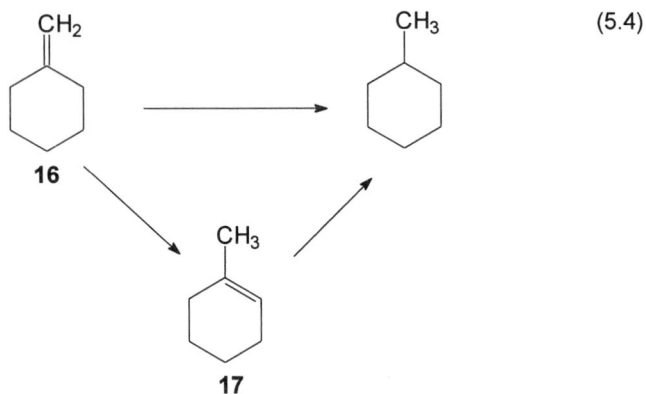

$$\text{(5.4)}$$

Benzene is not the only solvent that can be adsorbed on a catalyst surface. Any material having unshared electron pairs can be expected to interact with the catalyst to some extent. The STO procedure[16] described in Chapter 3 provides a simple means of evaluating the extent of a solvent:catalyst interaction and its

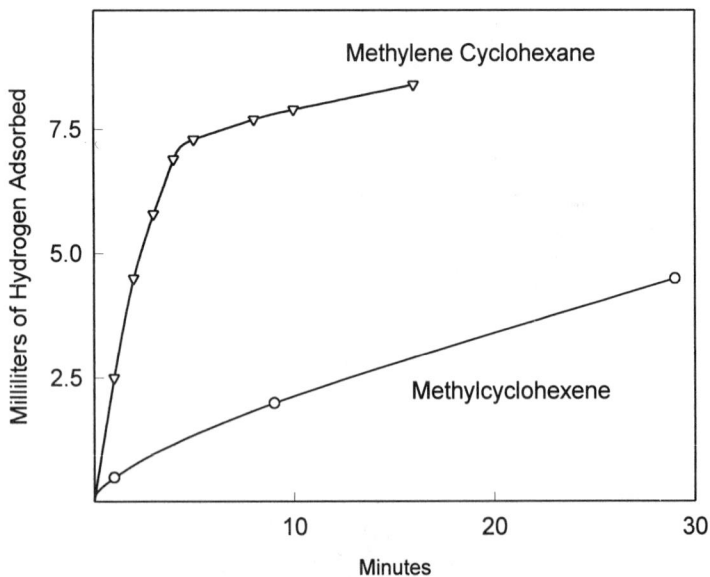

Fig. 5.1.   Hydrogen uptake in the hydrogenation of methylenecyclohexane and 1-methylcyclohexene over Pd/C in a 1:1
ethanol:benzene solvent. (Redrawn using data from Ref. 15).

Table 5.2

Effect of solvent on the amount of saturation and isomerization observed with the single turnover (STO) procedure run over Pt/CPG catalysts.[17]

Solvent	% Direct Saturation	% Two-Step Saturation	% Double Bond Isomerization
None	41	20	11
Pentane	39	17	10
Ethyl Acetate	38	19	7
Tetrahydrofuran	20	12	18
Methanol	17	10	43

influence on the reaction characteristics of the various types of sites present on a dispersed metal catalyst. This was accomplished by first using the STO procedure to measure the amount of butane formed by direct and two-step saturation for a given catalyst. The catalyst was then exposed to the vapors of various solvents and the extent of saturation was measured again.[17]

Table 5.2 lists the results obtained using a supported platinum catalyst. These data show that in the presence of pentane or ethyl acetate the amount of saturation observed was within experimental error of that found with the solvent free system indicating that little or no interaction with the catalyst was taking place with these solvents. This was expected with pentane but not with ethyl acetate since the ester has both carbonyl and ether oxygens and it should adsorb on the platinum surface through the unshared electron pairs on these atoms. The amount of STO saturation[16] decreased significantly in the presence of tetrahydrofuran (THF) and, especially, with ethanol showing that adsorption of these solvents does take place on the metal surface. Similar effects were also observed with supported palladium catalysts.[17] The decrease in saturation product indicates that this adsorption is occurring on the saturation sites of the catalyst that, as discussed previously, are the more coordinately unsaturated corner atoms and adatoms on the metal surface.

As the extent of saturation decreased, the amount of isomerization increased, showing that adsorption of the solvent molecules on these corner atoms and adatoms results in a decrease in the surface coordination. This effectively converts those saturation sites that are capable of adsorbing two hydrogen atoms and an alkene into the less unsaturated isomerization sites that can only hold one hydrogen and the alkene molecule as shown in Fig. 5.2 for ethanol adsorption on an adatom.[17] This latter species is analogous to the ^2M isomerization sites described in Scheme 4.4.

Since these reactions were run in the vapor phase it was considered possible that the degree of solvent:catalyst interaction may not be the same as that found in a liquid phase reaction. While the solvent:catalyst interaction should

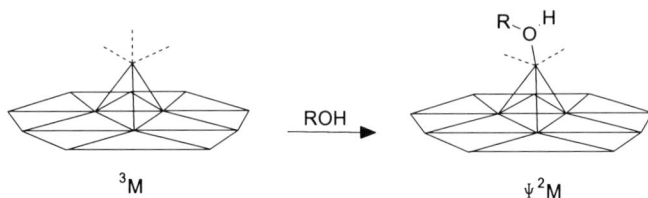

Fig. 5.2.   Alcohol interaction with ^3M saturation sites to produce
ψ ^2M isomerization sites.

have an influence on the overall rate of the reaction it will not necessarily be the same for all catalysts. It was not expected that the extent of interaction would be the same for all types of active sites and, as discussed in Chapter 3, different catalysts will have different ratios of the various types of surface sites.

Fig. 5.3 shows the relationship between the nature of the solvent and the reaction rates for the hydrogenation of 4-methyl-1-cyclohexene run over several Pt/SiO_2 catalysts. It can be seen that while there is a general decline in rate on

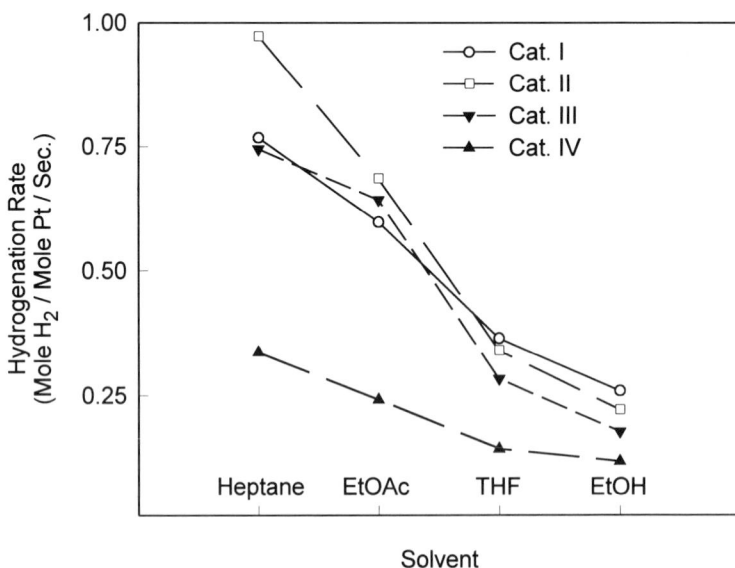

Fig. 5.3.   Solvent effect on the rate of hydrogenation of 4-methyl-1-
cyclohexene over Pt/SiO_2 catalysts. (Redrawn using data from
Ref. 18).

Table 5.3

Solvent effect on specific site turnover frequencies (TOF) for the
hydrogenation of 4-methyl-1-cyclohexene run over Pt/SiO$_2$ catalysts.[18]

Solvent	^3M_I TOF	^3M_R TOF	^{3}MH TOF
Heptane	1.25	11.00	3.25
Ethyl Acetate	0.50	11.00	2.00
Tetrahydrofuran	0.10	5.00	2.00
Ethanol	0.10	2.00	2.0

progressing from heptane through ethyl acetate to THF and ethanol for all of the catalysts, there is no direct parallel observed between any of the catalysts.[18]

Since the overall turnover frequency (TOF) for a hydrogenation reaction is the sum of the TOFs for each type of saturation site as described in Eqn. 3.6, it should be possible to determine the extent of solvent interaction with different types of sites by calculating the specific site TOFs for the same reaction run in different solvents. These data are listed in Table 5.3 for the hydrogenation of 4-methyl-1-cyclohexene over a series of STO characterized Pt/SiO$_2$ catalysts.[18]

The TOFs obtained in heptane are the "true" site rates with essentially no solvent:catalyst interaction taking place. In ethyl acetate the ^3M_I and ^{3}MH site TOFs decreased but the more active ^3M_R TOF remained the same. It is not clear whether this decrease in site activity was the result of the blocking of a number of the sites by the solvent molecules that would decrease the site density or if it was simply that an adsorbed solvent molecule lowered the activity of the site. Most likely, it is a combination of both factors. In any event there was a 60% loss of activity by the ^3M_I sites and a 40% loss by the ^{3}MH sites in ethyl acetate. In THF there was a stronger interaction resulting in a 92% loss in ^3M_I site activity, a 55% loss by the ^3M_R sites and a 40% loss by the ^{3}MH. In ethanol, a solvent even more strongly adsorbed on the catalyst, there was a 92% decrease in ^3M_I activity, an 82% loss in ^3M_R and a 40% loss in ^{3}MH.[18] These data provide a good example of the importance of the competitive adsorption of the solvent in a heterogeneously catalyzed reaction.

The nature of the solvent can also have a significant impact on which functional group will be selectively hydrogenated from a mixture of substrates.[19] It has been reported that polar solvents tend to increase the adsorption of non-polar substrates on a catalyst surface while non-polar solvents facilitate the adsorption of polar substrates. In the hydrogenation of a mixture of cyclohexene and acetone over a nickel catalyst, the use of a polar solvent favored the selective hydrogenation of cyclohexene while with a non-polar solvent acetone hydrogenation was preferred.[20] Evidently the polar substrate, acetone, is more favorably solvated in a polar solvent and, thus, is not as readily adsorbed on the catalyst surface as is the non-polar substrate, cyclohexene. In a non-polar medium the reverse is true. A similar solvent influenced selectivity was observed in the selective hydrogenation of 1-hexene and 2-methyl-3-hydroxypropene over a Pt/SiO$_2$ catalyst.[21] In methanol, 1-hexene hydrogenation was favored by about a 3:1 ratio while in cyclohexane, the unsaturated alcohol was selectively hydrogenated in a 9:1 ratio.

Care has to be taken in designing solvent systems for such selective reactions because the catalyst support material can also play a role in determining the nature of the solvent present in the neighborhood of the catalyst particles. Consider the hydrogenation of a mixture of cyclohexene and cyclohexanone over supported ruthenium catalysts.[12] When this mixture was hydrogenated over Ru/SiO$_2$ in water a slight selectivity for cyclohexene hydrogenation was observed as shown in Fig. 5.4a. In cyclohexane containing 5% water a selectivity toward cyclohexanone hydrogenation was observed (Fig. 5.4b). With Ru/C or ruthenium on silanized silica in water containing 5% cyclohexane, cyclohexene hydrogenation was favored (Fig. 5.4c and d). In water, cyclohexanone is preferentially solvated so the hydrogenation of cyclohexene is slightly favored. In the mixed solvents, though, the hydrophilicity of the support determined the nature of the solvent found around the catalyst particles. In water:cyclohexane mixtures the hydrophilic silica was surrounded by water in bulk cyclohexane. Since the water would contain more cyclohexanone than cyclohexene, cyclohexanone hydrogenation was favored. With the hydrophobic carbon or silanized silica the support was surrounded by cyclohexane in bulk water. Here the cyclohexene concentration in the solvent surrounding the catalyst was higher than cyclohexanone so cyclohexene hydrogenation was preferred.[12]

This is different from the situation described previously in that here there is a solvent mixture that partitions the substrates with cyclohexanone present primarily in the water and the cyclohexene in the cyclohexane. The substrate found in the solvent layer which surrounds the active component of the catalyst is the one that is preferentially hydrogenated.

Although specific examples will be discussed in later sections, some generalizations concerning solvent selection are appropriate at this time. Halogenated hydrocarbons, which can be hydrogenolyzed to produce halogen acids and solvents such as pyridine or benzene that are strongly adsorbed on the catalyst should be avoided unless their modifying effects are desired. Materials with hydrogenatable groups such as acetone, acetonitrile or benzene should not be used because partial hydrogenation of the solvent can invalidate reaction data. Alcohols are the most commonly used solvents but they have been shown to modify the catalyst surface.[15,17,18] Useful alternatives that apparently do not interact as much with the catalyst include ethers and esters. If solubility is not a problem, saturated hydrocarbons should be considered since they do not adsorb on the catalyst under moderate reaction conditions and, thus, do not influence the adsorption of the substrate molecules. Dioxane should never be used with a Raney nickel catalyst at elevated temperatures as explosions can occur.[22] Acetic acid is commonly used in conjunction with PtO$_2$ because it neutralizes a basic

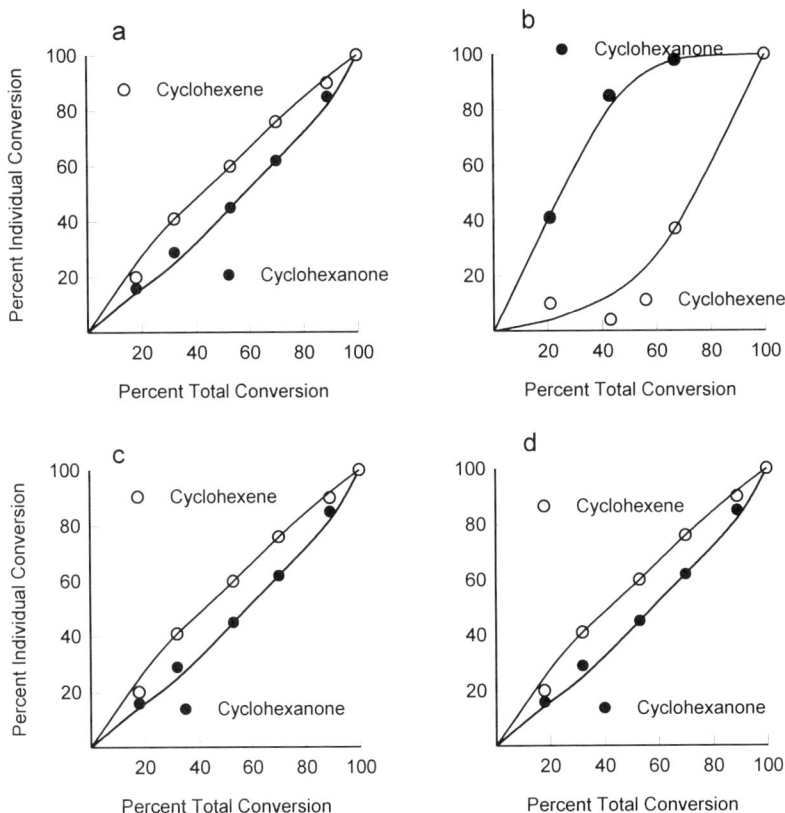

Fig. 5.4.   Selectivity in the hydrogenation of a mixture of cyclohexene
and cyclohexanone over: a) Ru/SiO$_2$ in water; b) Ru/SiO$_2$ in
cyclohexane/5% water; c) Ru/C in water/5% cyclohexane; d)
Ru/SiO$_2$-TMS in water/5% cyclohexane. (Redrawn, using
data from Ref. 12).

impurity present in the oxide.[23]   Acetic acid can, however, inhibit other noble
metal catalysts so its use as a solvent is not generally advisable.   Acetic acid
should not be used with base metal or oxide catalysts as catalyst dissolution can
occur.

Solvent quantity is dictated by factors such as the need for dilution to
minimize unwanted side reactions, the cost of separating and recovering the
solvent, and the solubility of reactants and products.   While it may be
economically preferable to hydrogenate liquid substrates without a solvent,
control of some mass transport steps may be limited and attaining high reaction

selectivity made more difficult. Because the amount of solvent in the reactor is usually large with respect to the amount of catalyst, slight impurities in the solvent can exert a marked influence on catalyst activity and selectivity. It is best to use as pure a solvent as possible in any catalytic process. In large scale, commercial processes this may not be feasible so developmental work on such reactions should be done using the grade of solvent that will be used in the final process.

Mass Transport

Previous chapters discussed the nature of the active sites present on the surface of a heterogeneous catalyst and the ways in which the reactants can be adsorbed on these sites and their interaction once adsorbed. Before this can take place, however, the reactants must first come in contact with the catalyst surface and this factor is probably the least understood by the typical synthetic chemist. Almost all synthetic reactions are run in a homogeneous, single phase, system. In such cases the concentrations of the reactants are uniformly distributed throughout the reaction medium and even though the reactant concentrations decrease with time the probability of reactant molecules interacting is the same for all volume segments of the solution.

In a heterogeneously catalyzed reaction, though, the reactants must come in contact with the active sites on catalyst particles that are not uniformly distributed throughout the reaction medium. In this case the extent of the reaction and, in many cases, the nature of the products can be determined not by the rate at which the reaction takes place on the active sites, but by the transport of the reactants to the active sites.

Fig. 5.5 illustrates the factors associated with reactant transport in a heterogeneously catalyzed liquid phase hydrogenation of an alkene. The three phases, gaseous hydrogen, liquid solvent containing the dissolved hydrogen and olefin, and the solid catalyst, are separated by gas/liquid and liquid/solid interfacial barriers. It is the need for the reactants to cross these barriers that is the primary distinction between heterogeneous and homogeneous reactions.

A hydrogen molecule must go through the gas/liquid interface to dissolve in the liquid phase. It must then migrate through the solution and pass through the liquid/solid barrier before it can reach the catalyst particle where it has to find an active site on which to adsorb and be activated. When the active sites are on the surface of the catalyst particle this is relatively straightforward. Most generally, though, the catalyst is composed of small metal particles supported on porous materials so the hydrogen must migrate through the pores of the support before reaching an active site for adsorption.

The alkene, being in solution, does not have to pass through a gas/liquid barrier but these molecules still have to move through the solution to the catalyst, go through the liquid/solid interface and migrate through the catalyst particle to reach the active sites. Reaction can then take place between the adsorbed hydrogen and alkene to produce the alkane. This then desorbs, migrates out of the

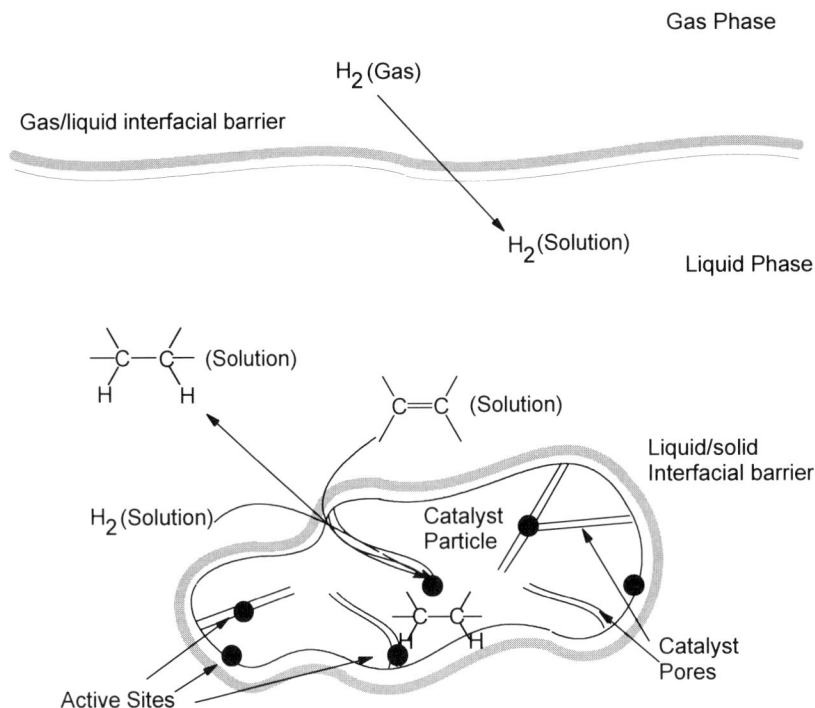

Fig. 5.5. Mass transport steps in the liquid phase hydrogenation of an alkene.

catalyst particle, passes through the solid/liquid interface, and, finally, goes into solution.

In addition to these mass transport steps, heat conduction can also be important in heterogeneously catalyzed processes. For exothermic reactions the heat generated at the catalytic site must be dissipated away from the catalyst and into the reaction medium while heat must be supplied to the active sites for endothermic reactions. In liquid phase processes heat transport is generally not a significant factor since the liquid tends to equalize the temperature throughout the reaction medium and, thus, facilitate temperature control. In vapor phase processes, however, heat transport can be a significant problem.

Of all of these steps, only the adsorption of the reactants, the reaction between the adsorbed species, and the desorption of the product are chemical reactions. The rest are physical processes that can strongly influence not only the apparent rate of the reaction but also the nature of the products obtained. To fully utilize heterogeneous catalysis as a viable synthetic procedure, then, it is important to recognize not only how these different transport steps can influence the outcome of the reaction but also how to determine when a reaction is limited by one or more of these transport steps and how the various reaction parameters can influence both the chemical reaction as well as the mass transport steps of the process.

Three Phase Batch Reactions

Since most heterogeneously catalyzed, synthetically useful reactions are run in a liquid phase batch reactor, the reaction variables involved in such systems will be discussed first. Table 5.4 lists those reaction variables that can affect the observed

Table 5.4

Effect of reaction variables on the different stages of a three phase catalytic process.[24]

Reaction Parameter	Primary Influence	Secondary Influence	Little or No Influence
Agitation	Gas/Liquid Mass Transport	Liquid/Solid Mass Transport	Chemical Reaction
Pressure	Gas/Liquid Mass Transport		Liquid/Solid Mass Transport (Liquid Reagent)
	Liquid/Solid Mass Transport (Gas Reagent)		
	Chemical Reaction		
Amount of Catalyst	Liquid/Solid Mass Transport		Gas/Liquid Mass Transport
Size of Catalyst Particles	Liquid/Solid Mass Transport		Gas/Liquid Mass Transport
Temperature	Chemical Reaction	Gas/Liquid Mass Transport	
		Liquid/Solid Mass Transport	
Concentration of the Liquid Reagent	Liquid/Solid Mass Transport (Liquid Reagent)		Gas/Liquid Mass Transport
	Chemical Reaction		Liquid/Solid Mass Transport (Gas Reagent)

rate of a reaction, such as a liquid phase hydrogenation, and the extent to which they influence the various mass transport and chemical steps of the process.[24]

Quantity of Catalyst

A number of reviews have been published that provide quantitative discussions of the effect that these reaction parameters exert on the different steps of a three phase reaction.[24-27] For simplification, however, a catalyzed reaction can be considered to be characterized by a kinetic rate factor, k_r, and a mass transport factor, k_m. The relationship of these factors and the catalyst quantity, x, to the reaction rate is given by Eqn. 5.5 and its inverse, Eqn. 5.6.[28]

$$r = \frac{k_r k_m x}{k_r x - k_m} \tag{5.5}$$

$$\frac{1}{r} = \frac{1}{k_m} + \frac{1}{k_r x} \tag{5.6}$$

Fig. 5.6 shows a plot of Eqn 5.5 which represents a reaction run with varying quantities of catalyst at a fixed pressure, temperature and agitation rate. When mass transport is an insignificant factor in the reaction, k_m approaches zero and the rate, r, equals $k_r x$, which is the asymptote of the curve at low catalyst quantities. With larger amounts of catalyst, the curve approaches the second asymptote, $r = k_m$, for a reaction controlled by mass transport effects. There is

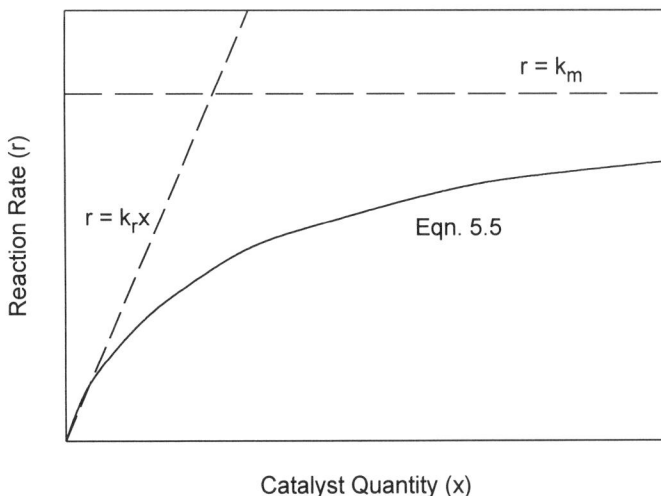

Fig. 5.6. Graph of Equation 5.5 showing the asymptotes.

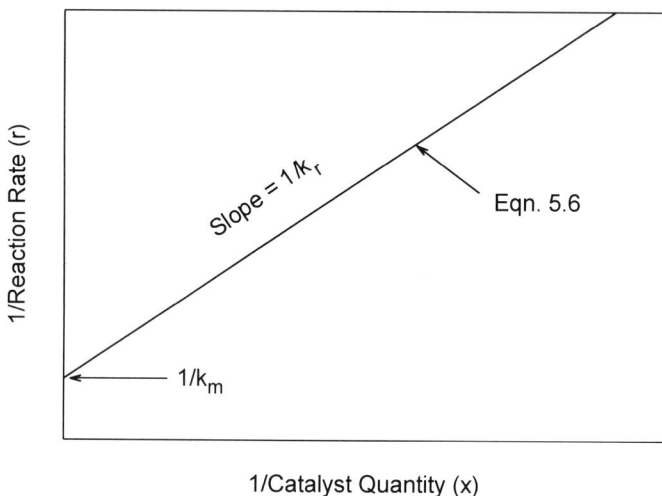

Fig. 5.7.   Graph of Equation 5.6.

usually some contribution of both k_r and k_m in all heterogeneously catalyzed reactions. For practical purposes measuring these asymptotic values is difficult but they can be estimated from the plot of the reciprocal, Eqn. 5.6, shown in Fig. 5.7.[28] In a plot of the reciprocal of the catalyst weight versus the reciprocal of the rate, the y-intercept is the reciprocal of k_m, which is the rate of the reaction if it were completely controlled by mass transport effects. The slope of the line in this simplified approach is the reciprocal of k_r, the rate of the reaction when mass transport limitations are absent; for reactions run in the so-called kinetic regime.[28]. However, in the more rigid treatment of these factors the slope of this line is also proportional to the sum of the liquid/solid mass transport and the chemical reaction-pore diffusion limitation on the reaction. At present, there is no way of separating these factors but, as discussed later, other data can be used to differentiate them.[24]

As an example, Fig. 5.8 shows a plot of the reciprocal of the rate versus the reciprocal of the catalyst weight for the hydrogenation of nitrobenzene run over varying quantities of a Pd/C catalyst at 30°C and one atmosphere pressure at a fixed rate of agitation.[24] The y-intercept (0.143×10^3 min/g-mole) is small compared to the reciprocal of the rate of the hydrogenation run over the smallest amount of catalyst (the right hand point, 1.43×10^3 min/g-mole). These numbers indicate that the gas/liquid mass transport had little effect on the hydrogenation run with this small amount of catalyst. For the run with the highest amount of catalyst (the left hand point) the intercept is about 30% of the reciprocal rate (0.45×10^3 min/g-mole) which indicates that some mass transport limitation was

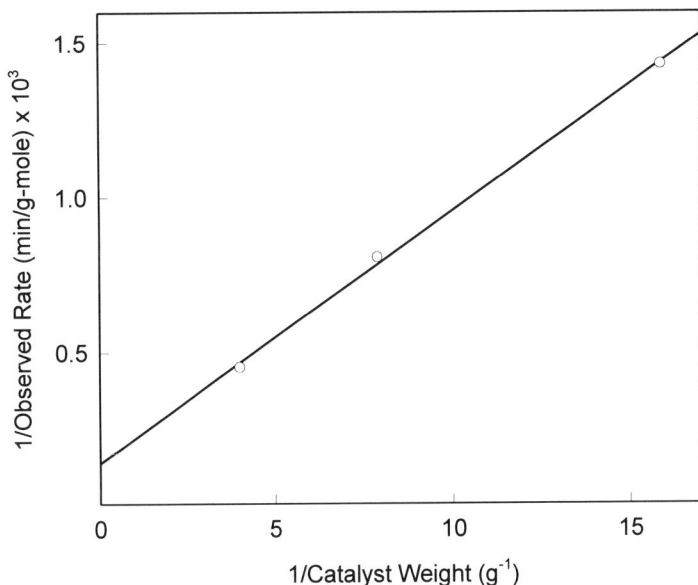

Fig. 5.8.   Effect of catalyst quantity on the rate of nitrobenzene
            hydrogenation over Pd/C in ethanol solution. (Redrawn using
            data from Ref. 24).

present in this later reaction. In Fig. 5.9 is a similar plot for the hydrogenation of
thiophene over different amounts of a cobalt catalyst.[24]   Here, the intercept
(0.57 x 10^3 min/g-mole) is about 70% of the reciprocal rate for the reaction run
over the largest amount of catalyst (0.8 x 10^3 min/g-mole), showing that a
considerable degree of mass transport control was present in that reaction. Even
with the lowest amount of catalyst the reciprocal rate (1.69 x 10^3 min/g-mole) was
only about three times larger than the intercept indicating that mass transport was
important there as well.[24]

It is important to recognize that any change in the quantity of catalyst used
will affect the observed rate even when the chemical reactions are limiting.
However, when the observed rate is normalized to a unit weight of catalyst, the
normalized rate will not change with changes in the amount of catalyst used if the
reaction is not gas/liquid or liquid/solid mass transport limited. This provides a
more simple procedure for determining whether a reaction is being run under
mass transfer limitations. It is only necessary to run the reaction first under the
commonly used conditions and then to repeat it using about one-half of the
catalyst. If the normalized rates are within ±5% of each other then mass transport
limitation is minimal. It is better to use a smaller amount of catalyst for the
second reaction. If a larger amount is used the increase in quantity could be

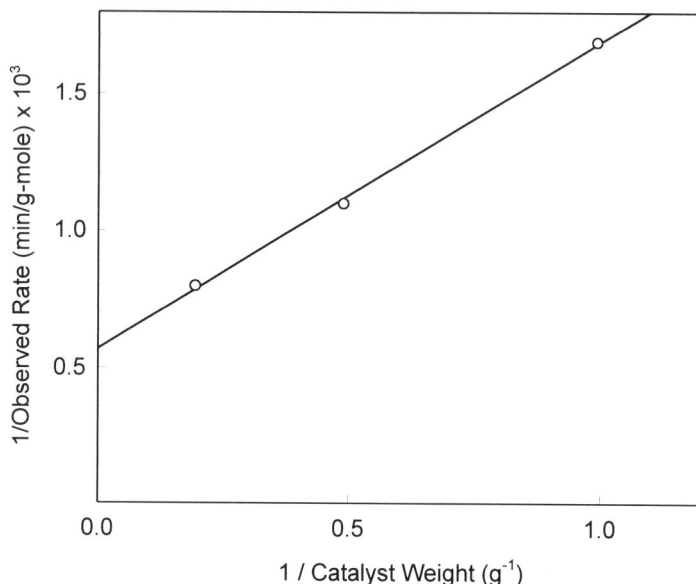

Fig. 5.9.   Effect of catalyst quantity on the rate of thiophene
            hydrogenation over a modified cobalt catalyst in a
            hydrocarbon solvent. (Redrawn using data from Ref. 24).

enough to place mass transport limitations on the reaction and, thus, not provide
an unambiguous evaluation of the standard reaction conditions.

    For a given rate of agitation and, therefore, a set quantity of hydrogen
available in solution, varying the quantity of catalyst is also a reasonable way of
modifying the hydrogen availability to the catalyst. A large quantity of catalyst
represents a large liquid/solid interfacial area. A given concentration of dissolved
hydrogen is spread over this entire area so the concentration per unit area is
smaller and the driving force needed to go through the interface is lower. With a
small quantity of catalyst the interfacial area is smaller. The concentration per
unit area is larger so more dissolved substrate will proceed through the interface
to adsorb on the active sites of the catalyst. Thus, for reactions requiring a high
hydrogen availability, small amounts of catalyst should be used while those
reactions needing a low hydrogen availability should use larger quantities of
catalyst. These factors hold when comparing different weights of the same
catalyst. With different catalysts the concentrations of the active sites on each or,
at the very least, their dispersions, should also be considered. Catalysts having
high STO saturation site densities[16] (Chapter 3) or high dispersions are more
active than are those with lower active site densities or dispersions. These factors

must also be considered in determining the amount of catalyst to use when high or low hydrogen availability is desired.

Since hydrogen solubility varies from solvent to solvent it might be thought that the choice of solvent would be a factor in determining the extent of hydrogen availability to the catalyst. In general, though, hydrogen solubility is of secondary importance when compared to variations in agitation and catalyst quantity.[24]

These considerations apply most readily to the gaseous reactant and not to the organic species dissolved in the solvent. Generally this material is present in much greater amounts than the quantity of a gaseous reactant such as hydrogen. Thus, transport of the organic substrate is seldom rate limiting except, possibly, near the end of the reaction when the concentration of this material is low.[29] For selective hydrogenations it may be advantageous to stop the reaction before the theoretical uptake of hydrogen to obtain a high selectivity. With the proper selection of reaction parameters a 90–95% conversion with optimum selectivity should be attainable.

Because different batches of catalyst can differ in activity it is essential in testing for mass transport limitations that samples of the same batch of catalyst be used.

Agitation

Agitation of a heterogeneous reaction mixture is important to keep the catalyst particles suspended as uniformly as possible in the reaction medium. This is generally accomplished by some form of stirring but shaking or rocking the reaction vessel can also be used for the same purpose. In most cases a vigorous stirring is sufficient to provide optimum catalyst suspension. In those cases where stirring proves to be inadequate, shaking the reactor can sometimes be more effective. In neither case should the agitation be so severe that catalyst particles become stuck to the side of the reactor, away from the reaction mixture and, thus, unavailable for promoting the reaction.

When a gaseous reactant is used, agitation is of primary importance in facilitating the transport of the gas molecules through the gas/liquid interface. The greater the rate of agitation, the larger the liquid surface area exposed to the gas and the more gas molecules that can pass through the interface into solution. Thus, the extent of agitation can be used as a convenient way of controlling the amount of the gaseous reagent that is dissolved and, thereby, available to the catalyst. Varying the rate of agitation is used as a common test for the extent to which mass transport controls a catalytic reaction. Increasing the agitation affects the mass transport factor, k_m, but not the kinetic factor, k_r so with a sufficiently high rate of agitation mass transport effects can be minimized.

To determine if mass transport through the gas/liquid interface is the limiting step in a reaction several experiments should be run at different agitation rates with all other parameters kept constant. When changes in agitation result in different observed rates as in the section A of Fig. 5.10, it can be concluded that gas/liquid mass transport is limiting the reaction. Increasing the agitation until the

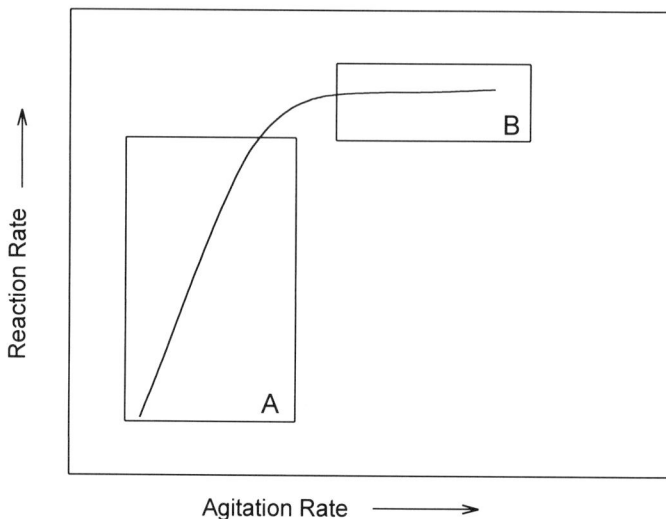

Fig. 5.10.   Idealized representation of the effect of agitation rate on the
rate of a heterogeneously catalyzed three phase reaction.

observed rate remains essentially constant as in section B in Fig. 5.10, provides
conditions under which this limitation is no longer present. Such conditions are
commonly referred to as the kinetic regime of the reaction.

The hydrogenation of nitrobenzene proceeds through several intermediate
compounds that can condense to give products other than aniline (Eqn. 5.1). To
optimize aniline formation and minimize the coupling side reactions the amount
of hydrogen made available to the catalyst should be as high as possible as is the
case when high agitation rates are used.[1] For some selective hydrogenations, such
as those discussed in Section III, a low hydrogen availability is needed to
minimize over hydrogenation. For such reactions it may be necessary to use only
enough agitation to keep the catalyst suspended but yet maintain a minimum
gas/liquid contact area. Failure to agitate the catalyst at all could produce a
situation in which the hydrogen availability to the catalyst is so low that
anomalous results could be obtained. It is best, from a practical standpoint, to
avoid such situations.

Pore Diffusion and the Size of the Catalyst Particles

One aspect of mass transport that is not easily measured is the effect on the
reaction rate caused by the diffusion of the reactants through the pores of the
catalyst particle to reach an active site and the migration of product molecules
back through the pores to the liquid/solid interface and into the solution.[24] As
discussed in Chapter 13 most metal catalysts are prepared by impregnating a

support material with a metal salt and then reducing the salt to give the supported metal. The supports usually have a high surface area because of an extensive pore structure. Problems in pore diffusion are related to the size of the catalyst particles, the size of the pores in the support materials and the location of the catalytically active material within these pores. The catalyst particles referred to in the present context are the species placed in the reactor and not the metal crystallites present on the support.

The smaller gaseous molecules will migrate within the pores of the support material faster than the larger organic substrates. If a high hydrogen availability is required for a reaction, a catalyst having active sites on or near the surface of the support is needed and not one with the catalytically active species found deep within the pores. It is best in these cases to use small catalyst particles made up of support materials having uniformly distributed wide pores that provide a ready access of the substrates to the active sites on the metal crystallites present in the catalyst. If this is not possible, consideration should be given to the use of preparation procedures that will deposit the catalytically active metal particulates on or near the surface of the support and not in the pores or to use as the support a low surface area material such as α Al_2O_3, which has a low pore volume.

In cases where a partially hydrogenated product is desired it is particularly important to have the active sites on or near the surface of the catalyst granules because with catalysts having active species within the pores further reaction can take place as the partially saturated product migrates out of the pores past other active sites.

With larger catalyst pellets, liquid/solid mass transport and diffusion through the pores of the catalyst to reach an active site can easily become rate limiting. A simple test to see if such limitation is present is to first run the reaction using the standard catalyst particles and then to repeat the reaction using the same conditions and weight of catalyst but with the catalyst particles crushed to a smaller size. A change in observed rate indicates the presence of liquid/solid mass transport and/or pore diffusion limitations on the reaction with the larger sized catalyst particles.

As discussed later, this factor is more commonly of concern in flow reaction systems. The catalysts generally used with batch reactors are fine powders with which these diffusion limitations are minimal.

Pressure

Increasing the pressure of the gaseous reactant not only increases the amount present in the gas phase but also increases gas/liquid transport and the solubility of the gas in the liquid phase. This, in turn, facilitates liquid/solid transport of this species. All of these factors increase the availability of the gaseous reagent to the catalyst. Fig. 5.11 shows a typical plot for the relationship between hydrogen pressure and the reaction rate at a fixed catalyst quantity and agitation rate.[28] At lower values an increase in pressure promotes an increase in rate but above a given value further increases in pressure have little or no effect on the rate. In the

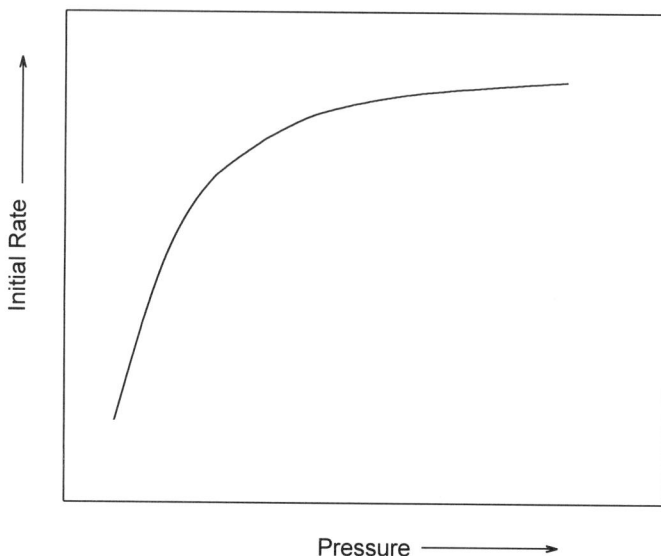

Fig. 5.11. Effect of hydrogen pressure on initial rate of reaction.

lower pressure regime, mass transport control is evident while at higher pressures the rate of the reaction is controlled by the chemical processes and further increases in the hydrogen pressure have little effect. In the past it was not uncommon to use hydrogen pressures of between 135–270 atmospheres (2000–4000 psig) to run a catalytic hydrogenation possibly because this relationship was not recognized and it was thought that the higher pressures would facilitate the reaction. However, with few exceptions it is seldom necessary to use pressures in excess of 65-100 atmospheres (1000–1500 psig) and many hydrogenations are run at atmospheric or only slightly elevated pressures. In most cases if a reaction will not run under a pressure of 100 atmospheres it is a good idea to reexamine the reaction system with particular attention to the catalyst and solvent used, and to explore other possibilities that might effect the reaction at a lower pressure. As a general rule it is best to run a reaction at the lowest pressure consistent with obtaining a reasonable reaction rate.

Temperature

Since the reactant adsorption, interaction of adsorbed species and product desorption are the only chemical processes in the sequence of steps shown in Fig. 5.5, they are most affected by changes in the reaction temperature. Mass transport is also influenced by changes in reaction temperature, but to a much smaller degree.[28] A significant change in reaction rate when the temperature is changed indicates that mass transport limitations are minimal or non-existent. As a general rule an apparent activation energy for a catalytic reaction of about 10 kcal/mole (40 kJ/mole) or greater indicates that the rate is controlled by one or more of these chemical reaction steps.   If the apparent activation energy is less than 3–4 kcal/mole (10–15 kJ/mole) mass transport limitation is indicated.[30]   Changes in the reaction temperature should be made with care, especially if reaction selectivity is desired.   Higher temperatures can cause a lessening of reaction selectivity, the production of unwanted side reactions, and a shortening of the life of the catalyst. As with the reaction pressure, a reaction should be run at as low a temperature as is consistent with maintaining an acceptable reaction rate.

Three Phase Continuous Reactions

While most laboratory scale reactions are run in a batch mode there are a large number of commercially important processes that are run in a continuous manner. Since some of the continuous reactors described in Chapter 6 are potentially useful for laboratory scale processes and many lab scale reactions will have to be adapted to large scale continuous flow systems, a knowledge of the effect that reaction variables can have on heterogeneously catalyzed reactions run in a continuous manner is important.

In the three phase flow reactors the catalyst is suspended in a mobile liquid phase while the gas/liquid interface is maximized by stirring or bubbling the gas through the liquid.   The most common reactor used for such systems is the continuous stirred tank reactor (CSTR) which is described in Chapter 6. In such systems all of the parameters that were described previously will have about the same effect on the reaction as they do in a batch reactor.   However, since the liquid phase is mobile, the catalyst:substrate contact time or liquid flow rate can also be an important reaction variable. In such systems the liquid flow must be such that there is a reasonable chance of having a sufficient number of substrate molecules come in contact with the catalyst particles. If this contact time is too long, secondary reactions can occur and reaction selectivity may decrease. The slower the flow rate, the more the reaction will approximate one run in the batch mode.   If the contact time is too short a low conversion to product will be observed. A low conversion can frequently be increased by repeatedly cycling the reaction liquid through the reactor until the desired conversion is obtained. Selectivity can also be enhanced by this cycling process if a product separation step is included before the reaction mixture is re-introduced into the reactor. The ability to optimize the catalyst:substrate interaction time is a particular advantage of flow systems. The use of such a technique in laboratory scale reactions could also be beneficial.

Two Phase Liquid Reactions

In liquid phase reactions that do not involve a gaseous reactant there are fewer transport steps than in the three phase systems. As depicted in Fig. 5.12, the dissolved substrates migrate to the catalyst particle, pass through the liquid/solid interface, and migrate through the pores of the catalyst particle to reach an active site on which they are adsorbed and interact. The product then desorbs and

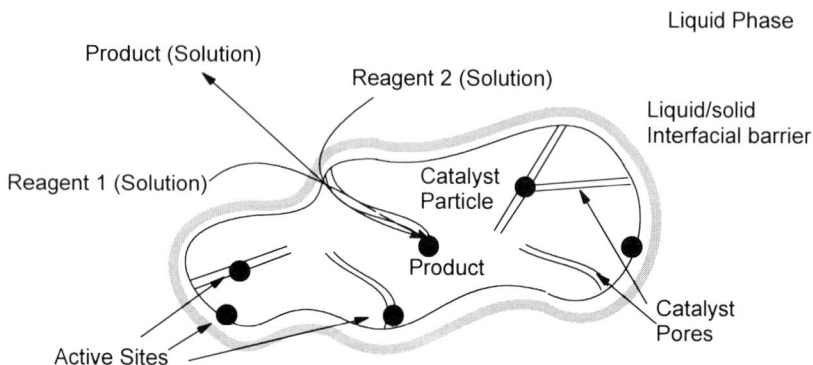

Fig. 5.12.   Mass transport steps in liquid phase reactions catalyzed by a
              solid catalyst.

migrates out of the catalyst particle, through the solid/liquid interface and into
solution.   As in any catalytic process, only the adsorption, reaction, and
desorption steps are chemical processes so temperature changes will influence
only these steps. The same activation energy criteria for mass transport limitation
apply here as in other systems.

The ways in which reaction parameters affect a two phase batch reaction
are similar to those considered above for the three phase systems. Since there is
no gas phase, agitation only serves to keep the catalyst suspended making it more
accessible to the dissolved reactants so it only has a secondary effect on mass
transfer processes. Substrate concentration and catalyst quantity are the two most
important reaction variables in such reactions since both have an influence on the
rate of migration of the reactants through the liquid/solid interface.   Also of
significant importance are the factors involved in minimizing pore diffusion
factors; the size of the catalyst particles and their pore structure.

The absence of the gas phase also simplifies the running of the two phase
flow reactions.   In these systems all that is needed is to pass the solution
containing the reactants through a fixed bed of catalyst held in a reactor. As in all
liquid phase reactions, the presence of the solvent facilitates temperature control
and minimizes temperature gradients within the reactor.   The most significant
mass transport effects in such systems are the migration through the liquid/solid
interface and pore diffusion through the catalyst particles.   Increasing the
concentration of the reactants will increase the rate of interfacial migration. If
reaction selectivity is no problem then the substrate concentration can be as high
as convenient for the catalyst quantity and flow rate used. Since the reactants are
passed through a bed of catalyst, a reasonable flow rate can only be obtained when
larger catalyst particles are used. In this type of reaction pore diffusion can have a
significant effect on the rate of reaction.

For a given catalyst bed and reactant concentration these mass transport effects can be minimized by increasing the reactant:catalyst contact time. If the reaction rate or selectivity is unchanged when the contact time is increased by lowering the flow rate, the reaction can be considered to be free of transport limitations. Another test for mass transport limitation can be made by varying the substrate:catalyst ratio by changing the amount of catalyst in the reactor. This is not as easy to do as in a liquid phase batch system because the flow characteristics of the continuous reactor depend on the quantity and shape of the catalyst particles used. These factors should remain constant in order to obtain a viable rate evaluation. There are three methods that can be used to test for transport limitation in this way. In one, the loading of the catalytically active species on the support is changed. The second involves the making of catalyst particles with various blends of the catalyst and an inert material.[31] In the third, the catalyst particles are diluted with particles of an inert material.

The first of these methods is the most likely to give ambiguous results. Modifying the catalyst load can lead to the formation of significantly different catalytic species.[32] Changes in the reaction rate that are observed when such modifications are made could be due to the presence of differing numbers of active sites and not to changes in mass transport. The second approach is the most widely used, particularly in vapor phase reactions, but the third is not only a viable alternative but is also easier to accomplish. The inert material used in either case should have the same geometrical and physical characteristics as the catalyst. The catalyst support material is a good choice. A problem with both techniques is the requirement that the dilution be homogeneous, so extreme care must be taken in mixing the catalyst and inert material to assure such uniformity. In either option the catalyst is diluted until the reaction rate or conversion per unit weight of catalyst no longer changes at a given flow rate.

Vapor Phase Reactions

As in the two phase liquid reactions there are fewer mass transport steps in vapor phase reactions than in three phase processes. These steps are shown in Fig. 5.13. The gaseous reactants must pass through the gas/solid interface to reach the catalyst particle. They then migrate through the particle to become adsorbed on the active sites. After reaction the product desorbs, migrates back through the particle to the solid/vapor interface which it passes through to enter the vapor phase in the reactor.

This type of reaction is analogous to the two phase liquid flow process described above but since there is no liquid present in these vapor phase reactions, the control of the temperature is more difficult and care must be taken to minimize any temperature gradients in the catalyst bed. To test for the presence of such gradients, the temperature should be measured at several points within the catalyst bed. If the temperature is not uniform throughout the system, the mode of heating or, possibly, the reactor design may need modification. As in liquid phase reactions, higher temperatures can result in catalyst deactivation and lower

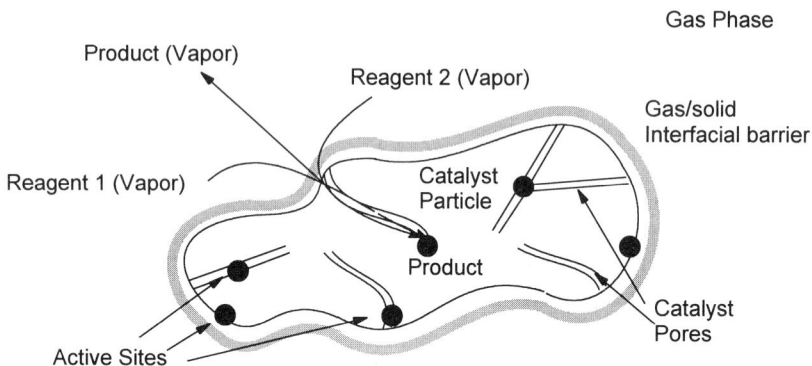

Fig. 5.13. Mass transport steps in vapor phase reactions catalyzed by a
          solid catalyst.

reaction selectivities so the temperature of the reaction should be kept only as
high as needed to keep both reagents and products in the vapor phase and
maintain a reasonable reaction rate.

While concentration gradients along the direction of flow (axial gradients)
are to be expected in any continuous reaction, gradients perpendicular to the flow
(radial gradients) can result in lower catalyst activity and should be minimized.
Radial concentration gradients can be caused by improper reactant flow
characteristics through the reactor or by mass transport limitation. In liquid phase
reactions radial gradients are usually minimal but this is not the case in vapor
phase reactions. If improper reactant flow characteristics are the cause of any
concentration gradients, a modification of the reactor is required.

The extent of mass transport control in the reaction is a function of the gas
pressure and flow rate as well as the quantity and shape of the catalyst. As
described for the two phase liquid flow reactions, the possibility of mass transport
limitation can be determined by examining the change in product formation for a
given flow rate produced by varying the substrate:catalyst contact time or the
catalyst:substrate ratio.

In vapor phase reactions the reactant ratios can be varied over a wide range
thus presenting another factor to be considered in selecting optimum conditions
for a specific reaction. In addition, the catalyst:reactant ratio can be modified
further by diluting the reactant flow with an inert gas. Although this is sometimes
advantageous, dilution lowers the partial pressure of the reactants and this affects
their mass transport and adsorption characteristics. It should be suspected that
reactions using dilute reactant gas streams are mass transport limited unless
shown otherwise. Mass transport limitation can be said to be absent when there is
no change in the observed rate of reaction with changes in reactant dilution.
When relatively inactive catalysts are used, dilution can sometimes be useful. As

Table 5.5

Common tests for the determination of mass transfer limitations of
heterogeneously catalyzed reactions.

Limitation	Parameter	Test
General Mass Transport	Temperature	If apparent activation energy is > 10 kcal/mole reaction is not mass transport limited.
Gas/Liquid Mass Transport	Agitation	Increase agitation until the rate of the reaction remains essentially constant.
	Quantity of Catalyst	Decrease the catalyst quantity until no change is observed in the normalized rate of the reaction.
Liquid/Solid Mass Transport	Quantity of Catalyst (Batch System)	Decrease the catalyst quantity until no change is observed in the normalized reaction rate.
	Quantity of Catalyst (Flow System)	Dilute catalyst bed with inert particles of support until no change is observed in the degree of conversion at a given flow rate.
Pore Diffusion	Catalyst Particle Size (Batch System)	Crush catalyst and compare rate with that obtained with uncrushed catalysts.
	Catalyst:Substrate Contact Time Flow System	Decrease flow rate until the degree of conversion becomes constant.
	Catalyst Particle Size	Crush catalyst and compare rate with that obtained with uncrushed catalysts.
Temperature Gradient (Vapor Phase)	Temperature	Measure the temperature at several points in the catalyst bed to assure uniform heating.

with solvents, impurities in the inert carrier gas should be carefully monitored.
Table 5.5 summarizes the various tests used to determine the degree to which
mass transfer limitation steps may have influenced a given reaction.

Effect of Mass Transport on Reaction Selectivity

The manner in which mass transport can affect reaction selectivity depends on the
kind of reactions involved.[24] Three general types of selectivity have been
defined[33] and mass transport effects each in a different way.

   Type I selectivity is that occurring when two simultaneous reactions are
taking place.[33] An example of this type is the hydrogenation of the C4 acetylenes

$$(5.7)$$

$$H_3C-CH_2-C\equiv CH \longrightarrow H_3C-CH_2CH_2-CH_3$$

$$H_2C=CH-HC=CH_2 \longrightarrow H_3C-CH_2CH_2-CH_3$$

that are produced as impurities in the preparation of butadiene (Eqn. 5.7). On the appropriate catalyst the rate of hydrogenation of the triple bonds will be high compared to diene hydrogenation. In this instance the more rapid reaction will be influenced more by mass transport limitations, so running the reaction under those conditions in which hydrogen transport is the limiting factor will lower the apparent selectivity of the catalyst. The same effect will be noted if the active sites on the catalyst or the quantity of catalyst is increased. The more active the catalyst, the greater the effect that mass transport has on the reaction selectivity.

$$(5.8)$$

$$H_3C-CH_2-CH_2-C\overset{O}{\underset{H}{\diagdown}}$$

$$H_3C-HC{:}CH-C\overset{O}{\underset{H}{\diagdown}}$$

$$H_3C-HC{:}CHCH_2-OH$$

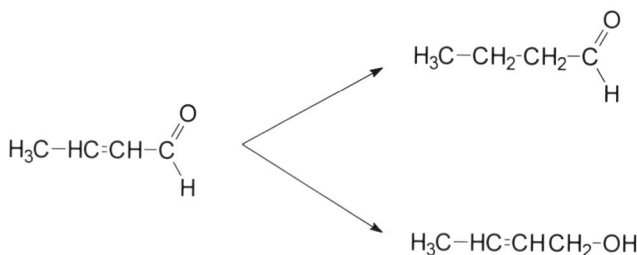

Type II selectivity involves the differentiation between two parallel reactions in which different products are formed by separate paths from the same starting material.[33] This type of selectivity is encountered in the hydrogenation of crotonaldehyde to either butyraldehyde or 2 buten-1-ol (Eqn. 5.8). When both reactions are of the same kinetic order changes in mass transport will influence them both to the same extent and there will be no effect on reaction selectivity. When the reactions are of different kinetic orders, that one with the higher order will be more affected by mass transport limitation.

$$(5.9)$$

$$R-C\equiv C-R \longrightarrow \overset{R}{\underset{H}{\diagup}}C=C\overset{R}{\underset{H}{\diagdown}} \longrightarrow R-CH_2-CH_2-R$$

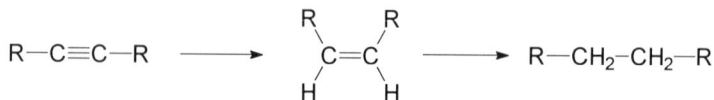

Type III selectivity is that found with serial reactions such as the hydrogenation of an acetylene first to an alkene and then to the alkane (Eqn. 5.9).[33] In such reactions the transport of the organic substrate through the liquid/solid interface can have a significant impact on the reaction selectivity. When the amount of acetylene present at the catalyst surface is lessened there is more of an opportunity for the alkene to be hydrogenated further to the alkane. Decreasing the amount of organic substrate will decrease reaction selectivity, but a decrease in hydrogen availability will increase the selectivity since there will be less hydrogen available for reaction with the alkene. Using a catalyst having the active species present primarily on the surface will also improve the selectivity of such reactions.

References

1. J. R. Kosak, *Chem. Ind. (Dekker)*, **5**, (Catal. Org. React.), 461 (1981).
2. W. H. Jones, S. H. Pines and M. Sletzinger, *Ann. N. Y. Acad. Sci.*, **214**, 150 (1973).
3. P. N. Rylander, *Catalysis in Organic Synthesis* (W. H. Jones, Ed.) Academic Press, New York, 1980, p. 155.
4. F. E. Gould, G. S. Johnson and A. F. Ferris, *J. Org. Chem.*, **25**, 1658 (1960).
5. C. F. Winans and H. Adkins, *J. Am. Chem. Soc.*, **54**, 306 (1932).
6. M. Freifelder and Y. H. Ng, *J. Pharm. Soc.*, **54**, 1204 (1965).
7. R. L. Augustine, D. C. Migliorini, R. E. Foscante, C. S. Sodano and M. J. Sisbarro, *J. Org. Chem.*, **34,** 1075 (1969).
8. R. L. Augustine, *Adv. Catal.*, **25**, 56 (1976).
9. E. M. Kosower, *J. Am. Chem. Soc.*, **80**, 3253 (1958).
10. E. M. Kosower and D. C. Remy, *Tetrahedron*, **5**, 281 (1959).
11. F. J. McQuillin, W. O. Ord and P. L. Simpson, *J. Chem. Soc.*, 5996 (1963).
12. P. G. J. Koopman, H. M. A. Buurmans, A. P. G. Kieboom and H. van Bekkum, *Recl. Trav. Chim. Pays-Bas*, **100**, 156 (1981).
13. L. Cerveny, J. Cervena and V. Ruzicka, *Coll. Czech. Chem. Commun.*, **37**, 1946 (1972).
14. L. Cerveny, A. Prochazha, M. Zenezny and V. Ruzicka, *Coll. Czech. Chem. Commun.*, **38**, 3134 (1973).
15. R. L. Augustine, V. Nocito, J. F. Van Peppen, R. W. Warner and F. Yaghmaie, *Catalysis in Organic Synthesis*, (W. H. Jones, Ed.) Academic Press, New York, 1980, p. 173.
16. R. L. Augustine and R. W. Warner, *J. Catal.*, **80**, (1983) 358.
17. R. L. Augustine, R. W. Warner and M. J. Melnick, *J. Org. Chem.*, **49,** 4853 (1984).
18. R. L. Augustine and P. Techasauvapak, *J. Mol. Catal.*, **87**, 95 (1994).
19. L. Cerveny and V. Ruzicka, *Catal. Rev.*, **24**, 503 (1982).
20. J. P. Wauquier and J. C. Jungers, *Bull. soc. chim.*, *France*, 1280 (1957).
21. L. Cerveny, R. Junova and V. Ruzicka, *Coll. Czech. Chem. Commun.*, **44**, 2328 (1979).
22. R. Mozingo, *Org. Synthesis*, **Coll. Vol. III,** 181 (1955).
23. C. W. Keenan, B. W. Geisemann and H. A. Smith, *J. Am. Chem. Soc.*, **76**, 229 (1954).
24. G. W. Roberts, *Catalysis in Organic Synthesis*, P. N. Rylander and H. Greenfield, Eds.) Academic Press, New York, 1976, p. 1.
25. C. N. Satterfield, *Mass Transfer in Heterogeneous Catalysis*, M.I.T. Press, London, 1970.

26.   G. J. K. Acres and B. J. Cooper, *J. Appl. Chem. Biotechnol.*, **22**, 769 (1972).
27.   A. J. Bird and D. J. Thompson, *Catalysis in Organic Synthesis,* (W. H. Jones, Ed.), Academic Press, New York, 1980, p. 61.
28.   W. R. Alcorn and T. J. Sullivan, *Chem. Ind. (Dekker)*, **18** (Catal. Org. React.), 221 (1984).
29.   Ref. 25, pp 107-123.
30.   H. C. Yao and P. H. Emmett, *J. Am. Chem., Soc.,* **81,** 4125 (1959).
31.   R. M. Koros and E. S. Nowak, *Chem. Eng. Sci.,* **22**, 470 (1967).
32.   R. L. Augustine, D. R. Baum, K. G. High, L. S. Szivos and S. T. O'Leary, *J. Catal.,* **127**, 675 (1991).
33.   A. Wheeler, *Adv. Catal.,* **3**, 249 (1951).

6

Catalytic Reactors

While the proper selection of reaction variables is essential in developing a viable heterogeneously catalyzed process, it has to be recognized that the type of reactor in which the reaction is run can also be important. A properly designed reactor will provide an environment in which mass and temperature transport problems are minimized so control over the different facets of a catalytic reaction is more easily attained. Catalytic reactions can be run in either a batch mode or as a continuous process. Obviously, each of these requires a different type of operating system. Batch reactors are favored for the production of smaller amounts of a number of different compounds while continuous reactors are generally dedicated to the preparation of a single material. Batch reactors also find use in the development of new catalytic processes, especially during the initial phases of such work. The following is a brief description of the different types of reactors commonly used for heterogeneously catalyzed reactions. More complete information can be found in those review articles concerned with reactors used for gas-liquid-solid and gas-solid multiphase reactions.[1-6]

Batch Reactors

Most synthetically useful heterogeneously catalyzed reactions involve the hydrogenation of functional groups, a reaction that is typically run as a batch process. As depicted by the schematic in Fig. 6.1,[1] the apparatus in which these, and other three phase reactions are run, must be capable of containing the gas

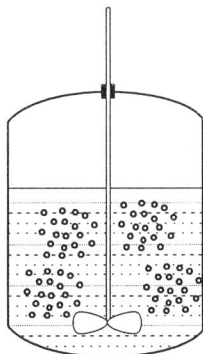

Fig. 6.1.   A schematic representation of a batch reactor for use in a heterogeneously catalyzed reaction. (Reproduced, with permission, from Ref. 1.)

while providing a means for the gaseous and liquid reactants to come in contact with the solid catalyst.[1-3] There are three general classes of reactors used for these reactions: those needed for high pressures of the gas; those used for low pressures; and those used for reactions run at atmospheric or sub-atmospheric pressure.

Stirred Tank Reactors

The most commonly used high pressure reactor is the stirred autoclave, also referred to as a stirred tank reactor (STR). These reactors are commercially available in sizes ranging from 50 ml to 500 gallons. A typical small scale reactor

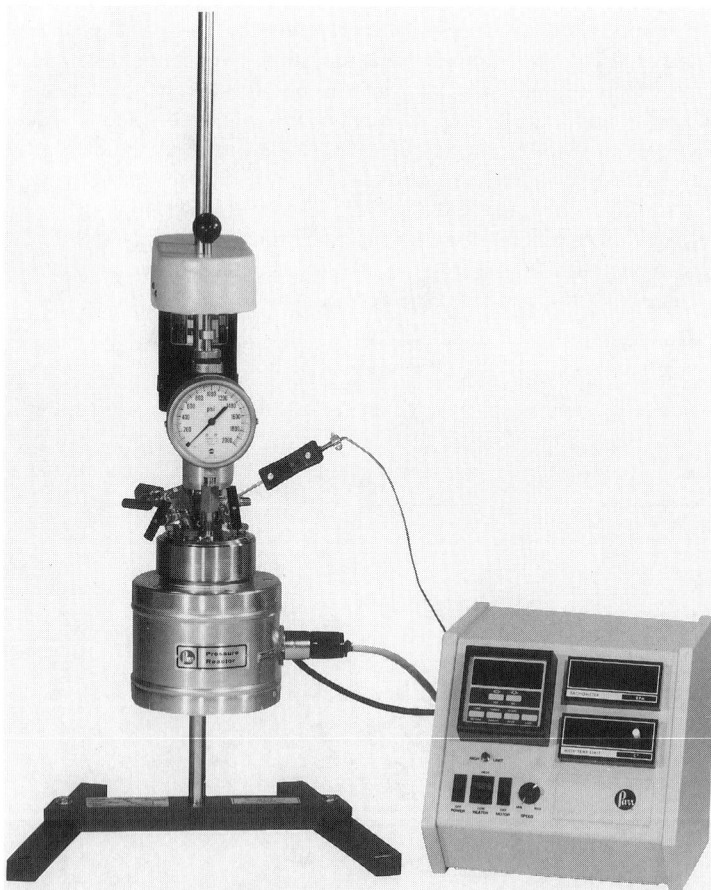

Fig. 6.2.   A small scale stirred autoclave with temperature controller.
(Courtesy Parr Instrument Co.)

is shown in Fig. 6.2 along with its temperature controller. These systems are available in a variety of metals for specific purposes but stainless steel is most commonly used. The bottom of the reactor, or tank, holds the reaction mixture while the reactor head is usually fitted on the outside of the reactor with gas inlets and outlets, a pressure gauge, some type of stirring mechanism and a rupture disk for the release of pressure in the event of a rapid pressure rise in the reactor. On the inside is generally found a cooling coil through which an external fluid is passed to either cool the reaction mixture or assist in controlling an exothermic reaction. Also present are vapor phase and liquid phase sampling tubes to facilitate reaction monitoring and a well in which a thermal sensor is placed for control of the reaction temperature.

With apparatus designed for maximum operating pressures of 200–350 atmospheres the reactor head is generally bolted to the tank with the high pressure seal made with a compressible gasket. Other versions of autoclaves have quick opening closures that can be used to pressure maxima around 100–125 atmospheres.

The essential feature of these reactors, though, is the stirring devise used to keep the solid catalyst uniformly suspended in the liquid reaction medium to maximize the gas/liquid contact area and to promote interfacial mass transport. Good agitation is also needed for optimum temperature control of the reaction. A commonly used stirrer is the six blade Rushton turbine[7] shown in Fig. 6.3. This

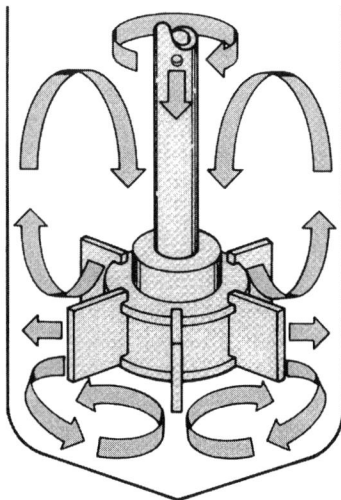

Fig 6.3.   Reactor agitation as provided by a six blade Rushton turbine.[7] Gas dispersion is enhanced by holes in the hollow shaft. (Courtesy Autoclave Engineers.)

arrangement provides both axial and radial fluid flow within the reactor. When used with a hollow shaft, gas dispersion is enhanced by drawing the gas through a hole in the upper part of the shaft and down through an exit at the bottom of the shaft, near the turbine blades. Ideal mixing of fine catalyst suspensions is generally obtained with this stirring arrangement, particularly with the smaller STRs. With larger reactors, baffles on the tank wall may be needed to maximize the agitation of the reaction mixture.

An alternate agitator design is shown in Fig. 6.4. This pitched blade turbine, sometimes referred to as an "axial flow impeller", is specially suited for high speed liquid/solid applications where tank baffles may be impractical. The fluid can be directed up or down depending on the direction of the pitch of the blades.

In practice these reactors are first loaded with the catalyst and non-gaseous reactant either as a liquid or in solution. For optimum agitation the liquid level should not exceed two-thirds the volume of the reactor. Because of the need to keep the catalyst suspended as uniformly as possible in the reaction mixture, the catalytic material is usually in the form of a fine powder. Since dry catalysts of this type can ignite the solvent vapor if added to the reactor after the liquid, the common procedure is to put the catalyst in the reactor first and then wet it carefully with the solvent, liquid reactant, or solution. If some catalyst has to be added to a reactor containing a volatile liquid it should be poured in as a slurry made with the reaction solvent.

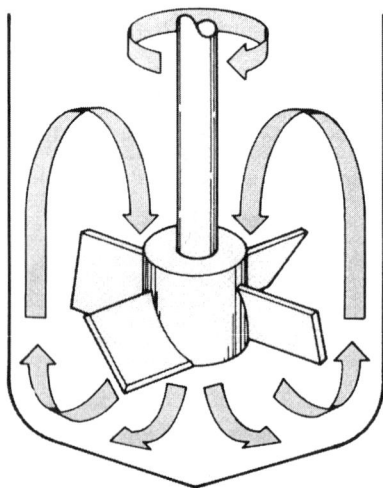

Fig. 6.4.   Reactor agitation using a pitched blade turbine.
(Courtesy Autoclave Engineers.)

Fig. 6.5. Parr hydrogenator commonly used for low pressure
          hydrogenation reactions. (Courtesy Parr Instrument Co.)

After the reactor is sealed and pressurized with the reactant gas, stirring is
begun and the temperature raised to the desired level. After stabilization of the
temperature, the progress of the reaction can be monitored by the pressure drop
within the reactor. If needed, gas can be added periodically to the reactor. Vapor
phase and/or liquid phase samples can also be withdrawn from time to time to
determine the extent of product formation. If the reaction rate data is to have any
significance the size of the withdrawn samples should be very small with respect
to the volume of the reaction mixture.

After the reaction is completed the agitation is continued while the reactor
is cooled. The pressure is then carefully released, the reactor opened and the
reaction mixture removed.

These reactors can also be used for catalytic reactions that do not involve a
gaseous reagent. In such cases it is best to displace the air in the apparatus with
an inert gas before initiating the reaction.

Parr Hydrogenator

The most common low pressure reactor is the Parr hydrogenator shown in Fig. 6.5. This system is composed of a reaction bottle that is connected to a gas reservoir that has a capacity of about four liters. Two versions of the apparatus are commercially available: a standard model that takes 250 mL and 500 mL reaction bottles and that can be used to pressures up to about 60 psig; and a larger size that uses 1 liter and 2 liter bottles that have about a 40 psig pressure limit. The bottles are connected to the gas reservoir by a tube inserted through a rubber stopper in the mouth of the bottle. The common black rubber stoppers may contain catalyst poisons that can be extracted into the reaction mixture by some solvents. The use of neoprene stoppers will minimize this problem.

Special reactor bottles having working volumes of 1 mL to 1000 mL are also available. As depicted in Fig. 6.6, these bottles have threaded Teflon adapters to hold the hydrogen inlet tube so catalyst contamination by the reactor is virtually eliminated. The smaller bottles, which have working volumes between 1 mL and 20 mL, use a spacer to secure them into the shaker assembly (Fig. 6.6a). Those with capacities of 20 mL to 200 mL (Fig. 6.6b) fit directly into the shaker assembly of the smaller Parr apparatus while 500 mL and 1000 mL vessels of this type (Fig. 6.6c) are available for the larger unit.

Gas-liquid mass transport limitation can be a problem with reactions run in the larger reactor bottles using the standard shaking speed for agitation. This can

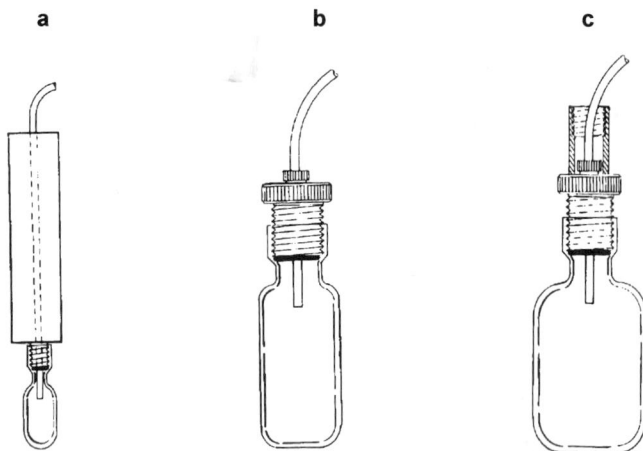

Fig. 6.6.   Threaded reactor bottles for the Parr Hydrogenator. a) Small scale bottle with threaded Nylon spacer; b) bottle for standard reactor; and c) bottle for large scale reactor, each with threaded Teflon adapters. (Courtesy Ace Glass, Inc.)

sometimes be overcome by using smaller reactors particularly with a small quantity of catalyst. Increasing the shaking rate can also be beneficial but the agitation should not be so severe that catalyst particles are isolated on the upper walls of the reactor bottles.

As with any catalytic reactor, the catalyst is first placed in the reaction bottle and the liquid substrate or solution carefully added to it. The bottle is placed in a protective shield, placed in the rocking cradle and connected to the gas reservoir. The bottle is alternately pressurized from the reservoir, isolated from the reservoir, and vented, several times to remove the air in the reaction system. The reaction bottle is then shaken by means of a motor driven eccentric. If needed the reactor can be heated by a controlled heating jacket.

The progress of the reaction is monitored by the drop in the pressure of the system. With the standard apparatus having the reactor open to the reservoir, a 0.1 mole of gas uptake corresponds to approximately an 8 psig pressure drop. To detect smaller quantities of gas uptake, the reactor can be isolated from the reservoir so the gauge reads only the pressure of the gas in the reactor bottle. With a half full 500 mL reactor bottle a 0.1 mole gas uptake corresponds to about a 150 psig pressure drop.[8] Approximate pressure drops for other volumes and quantities of gas consumed are easily calculated using the ideal gas law.

Atmospheric Pressure Apparatus

There are a number of devices described in the literature for running three phase reactions under an atmosphere of the gaseous reactant.[9-17] The most common are adaptations of the sloping manifold apparatus[9,17] shown in Fig. 6.7. The catalyst and reaction liquid are placed in a standard taper round bottom flask to which is added a magnetic stirring bar. The flask is then attached to the apparatus and the gas burette is filled with a liquid such as aqueous copper sulfate. After isolating the gas burette the apparatus is alternately evacuated and filled with the reaction gas several times to remove the air from the system. After the last filling, the liquid in the burette is also displaced by the gas. After closing the gas inlet to the apparatus, the internal pressure is equalized to atmospheric and the liquid volume in the burette is noted. The reaction is initiated by stirring the reaction flask with a magnetic stirrer. Gas uptake will result in a decrease in the internal pressure of the reactor as measured on the mercury manometer. The leveling bulb is then raised to increase the volume of the liquid in the burette which displaces the gas and raises the internal pressure to atmospheric. The liquid level is noted and the difference recorded as the amount of gas reacted. This process is continued until the reaction is completed.

Several simplifying modifications have been added to this basic system.[17] In one, a stopcock was placed at the bottom of the burette through which the liquid could be "titrated" into the burette from a raised leveling bulb. The other was the attachment to the manifold of a shorter U-tube which could be isolated from the system by a stopcock. Filling this with water or a similar liquid provides

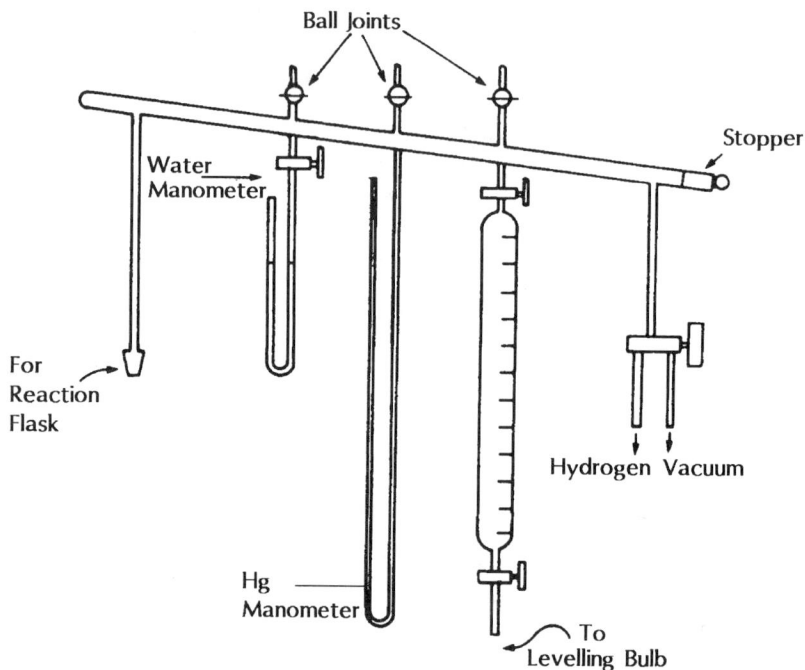

Fig. 6.7. Sloping manifold atmospheric pressure hydrogenator.
(Reproduced, with permission, from Ref. 9.)

a sensitive gauge for discerning small pressure changes within the reactor. With these modifications the uptake of small quantities of gas can be measured.

While these reactors are commonly used under ambient conditions, temperature control of the reaction can be accomplished by using jacketed reaction flasks such as that pictured in Fig. 6.8[18] with a thermostatted liquid pumped through the jacket.

Magnetic stirring is frequently used in such systems but this often leads to gas/liquid mass transport limitation of the reaction. To overcome this, other means of agitation can be used. In such cases it may be necessary to connect the reactor flask to the manifold by a length of flexible tubing. When this is needed it is best to use rubber or polypropylene tubing rather than Tygon since oxygen bleeding through the latter can cause problems in interpreting the reaction data.

Automatic Systems

The monitoring of the gas uptake in these batch reactors can be a monotonous process so a number of different types of automated systems have been developed.[17,19-24] Most of these are one-of-a-kind, and were built and used in a

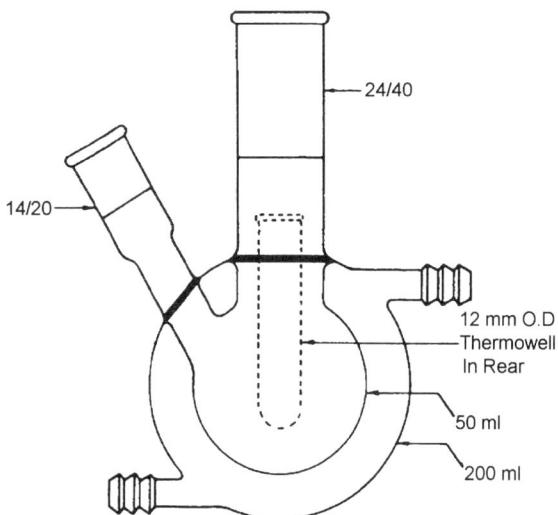

Fig. 6.8.   Jacketed reaction flask for use with the Sloping Manifold
            hydrogenator. (Reproduced, with permission, from Ref. 18.)

single laboratory. The descriptions of some of these units have been published
with the kind of automation dependent on the type of apparatus being used. These
fall into two general categories. In one the hydrogen pressure is kept constant by
changing the volume of the gas in the system as in the atmospheric pressure
apparatus described above. In the other type, the volume of the system remains
constant and the amount of gas consumed in the reaction is determined by the
corresponding pressure change. The STRs and Parr hydrogenator are examples of
this type of system.

     The use of a constant pressure apparatus provides for a more facile
interpretation of the reaction data since there is no change in the pressure of the
gas present in the reaction system as the reaction proceeds. In constant volume
apparatus, the amount of gas available for further reaction steadily decreases with
increasing reactant conversion, making the evaluation of the reaction rate data
more complex. Automation of these constant volume systems is not a difficult
procedure since all that changes is the internal pressure of the reactor and this can
be recorded by means of an appropriately situated pressure transducer.

     Automation of the constant pressure apparatus is not as simple. In these
systems the hydrogen pressure in the reactor is kept constant by changing the
volume of the system as the gas is consumed. This can be accomplished manually
by raising a leveling bulb filled with a liquid and attached to the bottom of a gas
burette by a flexible tube.[9] Automation of such systems has been achieved by the
use of a motor-driven device to raise the bulb with the motor being regulated by

impulses from some type of pressure sensing device in the system.[17,23] The motor drive can be electronically calibrated to provide automated volume/time data. Another technique involves a motor-driven gas syringe to change the volume of the system.[19,20] This later technique can be used at elevated pressures as well as atmospheric. The capacity of all of these systems is determined by the size of the burette or syringe, and it is necessary to change the size of the gas holding device to vary the scale of the reactor. They suffer from a further disadvantage in that they, generally, have both mechanical and electrical features that are difficult to design, build and maintain.

A more general system involves filling a reservoir tank with hydrogen to a pressure higher than that used in the reactor. Hydrogen is then fed into the reactor through a pressure regulator so the pressure in the reactor is held constant. The gas consumption is measured by the pressure drop of the reservoir.

Another automated reactor system is that shown in Fig. 6.9.[24] It has a solenoid valve as its only "moving" component and it is fully computerized for data acquisition and evaluation. It can be used at essentially any pressure and for any size reaction without modification of the system. The basic operation of this apparatus is simple. As the hydrogenation proceeds a pressure drop occurs inside the reactor, **1**.   The reactor pressure is monitored by a sensitive pressure transducer, **2**, which is interfaced with a microcomputer, **4**. When the computer detects this pressure drop, a signal is sent to a timing device, **5**, which then opens the solenoid valve, **6**, for a prescribed length of time, permitting a known volume of hydrogen to be introduced into the reactor. The number of hydrogen pulses introduced and the time at which each was introduced into the reactor are recorded by the computer for future data evaluation.

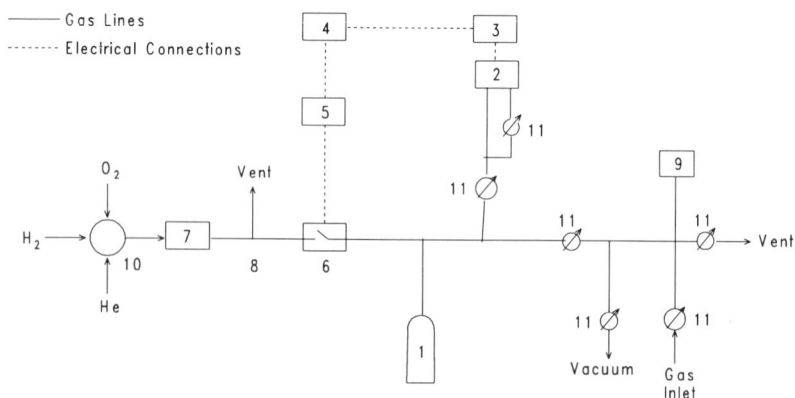

Fig. 6.9   Schematic representation of a computerized constant volume - constant pressure apparatus. (Reproduced, with permission, from Ref. 24.)

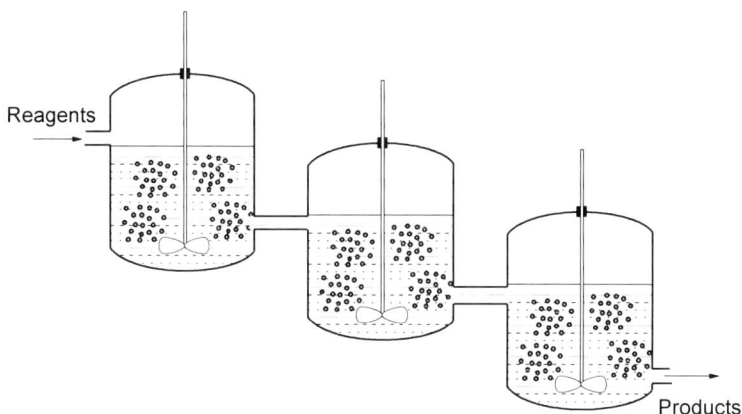

Fig. 6.10. Continuous stirred tank reactors (CSTR) in series.
(Reproduced, with permission, from Ref. 1.)

The size of the hydrogen pulse is determined by the time interval during which the valve, **6**, is held open and the pressure of the incoming gas as regulated by **7**. Calibration is effected by introducing pulses into a bubble meter or by displacing liquid in a burette. This apparatus can be used for any reaction involving a gaseous reactant. Further, by venting to the atmosphere through valve **8** and monitoring for a pressure increase in the reactor, this apparatus can also be used for reactions in which a gas is evolved. In such reactions the pulse valve is opened to release a known volume of gas from the reactor.

Catalyst agitation in the reactor can by accomplished by stirring, shaking or any other means. The reaction can be run at atmospheric pressure as well as sub-atmospheric or elevated pressures as well. The only criteria to meet for the higher pressure reactions is the need to use a reactor and pressure transducer compatible with the elevated pressures and to have the pressure of the gas before the regulator, **7**, higher than the reaction pressure. The reactor vessels shown in Fig. 6.6 are particularly suited for use at low pressures with this apparatus.

Continuous Reactors

In continuous or flow reactors the reactant fluid moves continually through the catalyst so at least a semblance of a steady state condition can be established inside the reactor. There are two general types of continuous reactors; those in which the catalyst particles are packed into a fixed bed through which the reactant fluid is passed and those in which the reactants pass through a slurry of the catalyst particles.

An example of the latter type is the continuous stirred tank reactor (CSTR), which is essentially an STR into which the reactants are added continuously with the partially reacted material removed at the same rate.[1,3] In

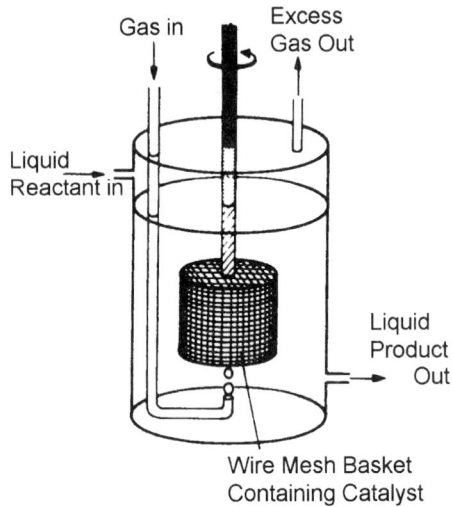

Fig. 6.11. Representation of a continuous stirred tank reactor
using a spinning basket to hold the catalyst.
(Reproduced, with permission, from Ref. 25.)

some systems of this type the catalyst is present as a fine powder which is kept in suspension by the same type of mixing described for the STR. With essentially perfect mixing, the effluent has the same composition as the contents of the reactor. Since it is possible in such a system that some reactant molecules can move rapidly through the reactor, the product yield will generally be lower than in a batch process. It is common to use a series of these CSTRs with the reactant stream from one reactor passing into another as depicted in Fig. 6.10.[1] In such an arrangement there is a stepwise increase in product formation on progressing from one reactant to another. A series of several reactors is generally needed to obtain high conversion.

Another problem encountered with this type of system is the loss of some catalyst along with the effluent so a catalyst recycling system is also usually included in the design of the system. One way of eliminating this problem is to place the catalyst pellets in a basket that is attached to the stirrer shaft as shown in Fig. 6.11.[25] A more detailed picture of a spinning basket reactor is given in Fig. 6.12. In this type of arrangement the baskets are placed on the stirring shaft at right angles to each other with propellers mounted above and below the baskets to direct the flow of the reactant fluid through the catalyst filled spinning baskets.[26]

A variation of this system is the fixed basket apparatus shown in Fig. 6.13. Here the gas and liquid reactants are circulated by means of the impellers, placed above and below the catalyst bed, through the catalyst that is held in the circular basket. This type of reactor, however, is usually used for vapor phase reactions.[27]

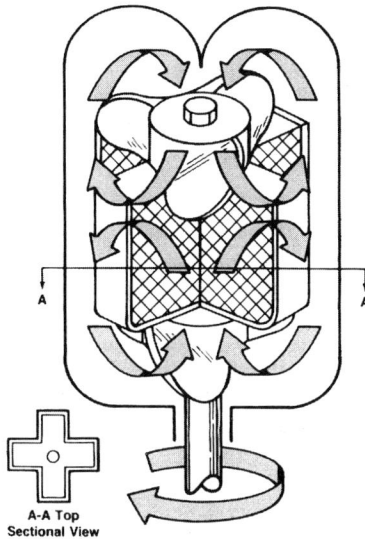

Fig. 6.12. Carberry spinning basket reactor.[26] (Courtesy of Autoclave Engineers.)

The spinning basket and fixed basket reactors can also be used in batch reactions.

Another continuous reaction system is the fluidized bed reactor shown in Fig. 6.14.[1] In this arrangement the reactant fluid is passed upwards through a bed

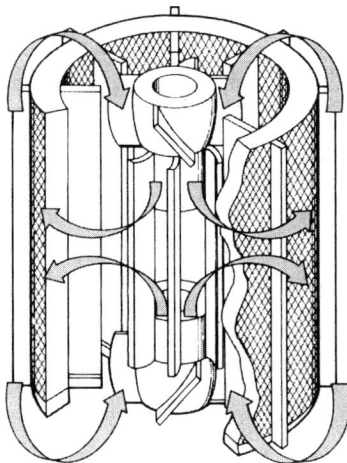

Fig. 6.13. Fixed basket reactor. (Courtesy Autoclave Engineers.)

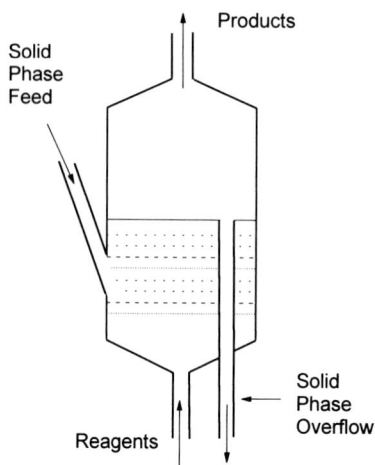

Fig. 6.14.   Schematic representation of a fluidized bed reactor.
(Reproduced, with permission, from Ref. 1.)

of fine catalyst particles at such a rate that the catalyst particles are placed in constant motion and the bed appears to be in a state of ebolition.   When the reactants are gases the fluidized bed facilitates temperature control.   With this reactor, however, solid is lost and must be continually replaced.   When the reactant is a liquid, this type of arrangement is sometimes referred to as a slurry phase reactor.

When the catalyst is in a fixed bed, catalyst attrition can be minimized by having the catalyst present as course granules or pellets.   With these, however, pore diffusion can present problems.   The most common fixed bed reactor used for three phase reactions is the trickle bed reactor shown in Fig. 6.15.[25,28]   In these reactors the liquid reagent is passed down through the catalyst bed while the gas is introduced either in the same direction or counter-currently to the liquid flow with the product collected from the bottom of the reactor.   With too fast a liquid flow, channeling through the catalyst bed can take place along with incomplete wetting of the catalyst, which results in hot spots in the bed.

Another flow system is the tubular reactor shown in Fig. 6.16.   Here the reactants are passed in a continuous stream into the inlet tube, A, through a bed of catalyst pellets surrounding the thermocouple well, E, and the products exit by way of the port, B.   The reaction mixture passes over a series of baffles, D, to attain the proper temperature before reaching the catalyst bed.   The entire system is encased in the jacket, C, and sealed with the end joints, E.   With such apparatus no attempt is made to induce mixing of the elements of the reacting fluid at different points along the direction of flow.   The assumption is made for these reactors that no element of the reacting fluid overtakes any other on passage

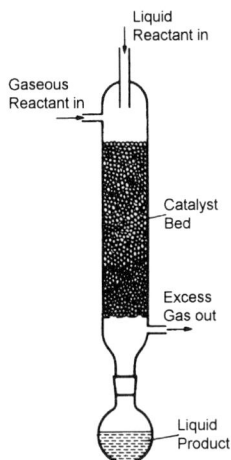

Fig. 6.15.   Trickle bed reactor. (Reproduced, with permission,
            from Ref. 25.)

through the catalyst bed. The fluid moves through the reactor like a plug so they
are also referred to as plug-flow reactors. They can be used for either vapor phase
or liquid phase reactions.

A useful small scale liquid phase fixed bed reactor is easily made by
replacing the analytical column of an HPLC system with one containing the
catalyst. The liquid reactants are pumped through the catalyst by an HPLC pump,
and the reactor can be heated using any LC column heater. When fine catalyst

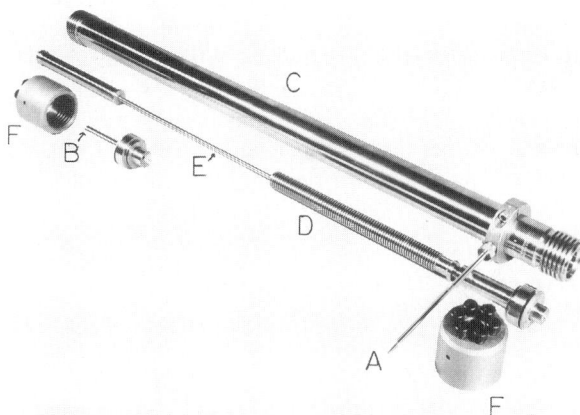

Fig. 6.16. Small scale flow reactor. (Courtesy Parr Instrument Co.)

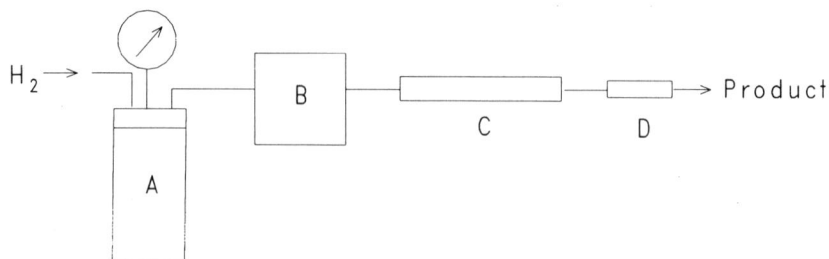

Fig. 6.17. Outline of a liquid phase continuous selective hydrogenation
apparatus. (Reproduced, with permission, From Ref. 24.)

particles are used in such a system they can become so tightly packed that very
high pressures are required to force the reactants through the catalyst bed. To
avoid this problem larger catalyst particles should be used. At times mixing the
finer catalyst particles with small silica beads can also be helpful.

A modification of this system, which is depicted in Fig. 6.17, has been
used for the selective hydrogenation of a number of substrates.[24] This is
composed of a stainless steel high pressure bomb, A, which is used as the solvent
reservoir, an HPLC pump, B, a stainless steel reactor tube with temperature
control capabilities, C, to hold the catalyst and flow restrictor back pressure
regulators, D, at the end of the flow system. This is a liquid phase flow system in
which the only hydrogen available for reaction is that which is dissolved in the
solvent in the reservoir.

With this system the reactant and solvent are placed in the high pressure
reservoir and this is filled with hydrogen to a pressure sufficient so the amount of
hydrogen dissolved in the solvent is that required for the desired selective
hydrogenation of the substrate. The mixture of reactant, solvent and dissolved
hydrogen is pumped through the reactor and samples of the product stream
analyzed by gas chromatography. The pressure in the reactor is held sufficiently
high to prevent dissolution of the hydrogen by the back pressure regulator-flow
restrictor. The amount of hydrogen in the solution can be varied by changing the
hydrogen pressure in the reservoir. Maximum selectivity is obtained when the
pressure is regulated to provide a near stoichiometric amount of hydrogen for the
reaction.[24]

Membrane Reactors

Membrane reactors are another type of system used for continuous reactions.
They are unique in that they combine reaction and separation in a single operation
with the membrane not only selectively removing some of the reactants or
products from the reacting area but also frequently acting as the catalyst or
support for the catalyst used in the reaction.[4-6,29,30] The operation of such
reactors is most easily described using a palladium membrane reactor for a

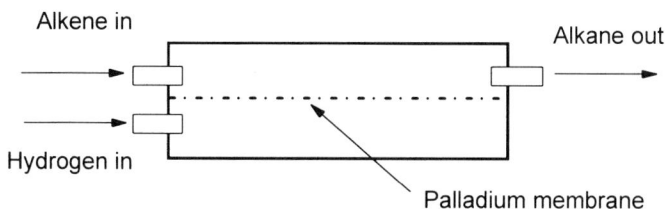

Fig. 6.18. Simplified drawing of a palladium membrane reactor used for the hydrogenation of an alkene.

hydrogenation reaction. A simplified picture of such a system is given in Fig. 6.18. Here a reactant, such as cyclohexene, is passed through the reactor on one side of the membrane with hydrogen entering the system on the other side. Diffusion through the palladium places adsorbed hydrogen atoms on the palladium surface on the other side of the membrane where it is available for saturating the alkene. This type of reactor can be used for both liquid phase and vapor phase reactions.[31,32]

References

1. K. G. Denbigh and J. C. R. Turner, *Chemical Reactor Theory, An Introduction*, Cambridge University Press, Cambridge, 1984.
2. J. F. Jenck, *Stud. Surf. Sci. Catal.*, **59** (Heterog. Catal. Fine Chem., II), 1 (1991).
3. P. L. Mills, P. A. Ramachandran and R. V. Chaudhari, *Rev. Chem. Eng.*, **8**, 1 (1992).
4. J. Shu, B. P. A. Grandjean, A. Van Nest and S. Kaliaguine, *Can. J. Chem. Eng.*, **69**, 1036 (1991).
5. T. T. Tsotsis and R. G. Minet, *Chem. Ind. (Dekker)*, **51** (Computer Aided Design of Catalysts), 471 (1993).
6. G. Saracco and V. Specchia, *Catal. Revs.*, **36**, 305 (1994).
7. J. H. Rushton, *Chem. Ing. Prog.*, **46**, 395 (1950).
8. R. L. Augustine, *Catalytic Hydrogenation*, Dekker, New York, 1965, p. 8.
9. Ref. 8, p. 11.
10. L. M. Joskel, *Ind. Eng. Chem., Anal. Ed.*, **15**, 590 (1943).
11. R. V. Savacool and G. E. Ullgot, *Anal. Chem.*, **24**, 714 (1952).
12. C. A. Brown and H. C. Brown, *J. Org. Chem.*, **31**, 3989 (1966).
13. D. C. Bishop and D. Hartley, *Chem. Ind. (London)*, 680 (1971).
14. W. F. Massler, R. E. Burnett, J. D. Morrison, P. A. Chang and C. M. Wheeler, *J. Chem. Educ.*, **52**, 202 (1975).
15. K. D. Cowan and E. J. Eisenbraun, *Chem. Ind. (London)*, 221 (1976).
16. K. D. Cowan, D. L. Bymaster, H. Hall and E. J. Eisenbraun, *Chem. Ind. (London)*, 105 (1986).
17. R. L. Augustine, G. W. Ewing and H. A. Ashworth, *Chem. Ind. (Dekker)*, **5** (Catal. Org. React.), 441 (1981).

18. Ref. 8, p. 20.
19. W. K. Rohwedder, *Rev. Sci. Instrum.*, **37**, 1734 (1966).
20. H. B. Tinker, J. H. Craddock and F. E. Paulik, *Rev. Sci. Instrum.*, **39**, 590 (1968).
21. G. W. H. A. Mansveld, A. P. G. Kieboom, W. T. M. DeGroot and H. VanBekkem, *Anal. Chem.*, **42**, 813 (1970).
22. J. J. Szakasits, *Anal. Chem.*, **42**, 1708 (1970).
23. M. E. Cain and G. Knight, *Chem. Ind. (London)*, 1125 (1971).
24. R. L. Augustine, S. K. Tanielyan and G. Wolosh, *Chem. Ind. (Dekker)*, **53** (Catal. Org. React.), 547 (1994).
25. G. C. Bond, *Heterogeneous Catalysis*, Clarendon Press, Oxford, 1974, p. 58; Second Edition, 1987, p 75.
26. J. J. Carberry, *Ind. Eng. Chem.*, **56**, 39 (1964).
27. J. A. Mahoney, *J. Catal.*, **32**, 247 (1974).
28. K. M. Ng and C. F. Chu, *Chem. Ing. Prog.*, **83**, 55 (1987).
29. J. N. Armor, *Chemtech*, **22**, 557 (1992).
30. V. M. Gryaznov, *Platinum Met. Rev.*, **36**, 70 (1992).
31. T. S. Farris and J. N. Armor, *Appl. Catal.*, A, **96**, 25 (1993).
32. J. N. Armor and T. S. Farris, *Stud. Surf. Sci. Catal.*, **75** (New Frontiers in Catalysis, Pt. B), 1363 (1993).

7

Reaction Kinetics

In order to use a heterogeneously catalyzed process more effectively some knowledge of the mechanism by which the adsorbed species interact is needed. A common method for obtaining this information, though indirectly, is by a detailed kinetic study of the reaction. While a complete discussion of heterogeneous kinetics is beyond the scope of this book, a brief presentation is needed. A more thorough coverage can be found elsewhere.[1-10]

Langmuir-Hinshelwood Kinetics

The first step in any catalyzed reaction is the activation of the substrate by adsorption on the catalyst as represented by Eqn. 7.1, where M is an active site on the catalyst, S is a substrate molecule, and P is the product. In such situations the rate of the reaction will depend on the amount of S adsorbed on the catalyst, that is, on the concentration of $M-S$ as shown by Eqn. 7.2.

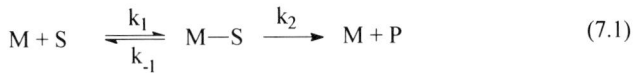

$$M + S \underset{k_{-1}}{\overset{k_1}{\rightleftharpoons}} M-S \xrightarrow{k_2} M + P \qquad (7.1)$$

$$\text{Rate} = R = k_2[M\text{-}S] \qquad (7.2)$$

While it is generally recognized that, as described in Chapter 3, a catalyst surface is composed of a number of different types of surface sites with varying adsorption and reaction characteristics, this model is not easily used for kinetic analyses. Instead, it is assumed that every surface atom on the catalyst is capable of adsorbing a substrate molecule, that they all do so with equal energy, and that there can be only one substrate molecule adsorbed on each surface atom. The amount of the substrate, S, added to the reaction mixture is known, but the amount of S adsorbed on the catalyst as $M-S$ is not. However, the total number of surface atoms, M_0, can be determined by procedures such as those discussed in Chapter 2 so the number of atoms on which a substrate is adsorbed can be written as in Eqn. 7.3.

$$[M_0] = [M] + [M - S] \qquad (7.3)$$

The equilibrium constant, **K**, for the adsorption equilibrium in Eqn. 7.1 can be written as Eqn. 7.4.

$$K = \frac{k_1}{k_{-1}} = \frac{[M - S]}{[M][S]} \tag{7.4}$$

Combining Eqns. 7.3 and 7.4 gives Eqn. 7.5.

$$K = \frac{[M - S]}{[M][S]} = \frac{[M - S]}{([M_o] - [M - S])[S]} = \frac{[M - S]}{[M_o][S] - [M - S][S]} \tag{7.5}$$

Rearranging gives Eqn. 7.6.

$$[M - S] = \frac{K[M_o][S]}{1 + K[S]} \tag{7.6}$$

With a unit surface area and having [**M-S**] defined as Θ, Eqn. 7.6 becomes Eqn. 7.7 and Eqn. 7.2 becomes Eqn. 7.8.

$$\Theta = \frac{K[S]}{1 + K[S]} \tag{7.7}$$

$$R = k\Theta = \frac{k\,K[S]}{1 + K[S]} \tag{7.8}$$

In this case **R** is generally expressed as an areal turnover frequency (TOF), the number of molecules reacted per unit time per unit surface area. If the catalytic process follows this model the rate, **R**, will be related to [S] as shown in Fig. 7.1. The rate is first order at low values of [S] and decreases to zero order as [S] increases. By analogy, [**M–S**] is expected to be small when **S** is weakly adsorbed and the reaction should be first order. When **S** is strongly adsorbed a larger value for [**M–S**] should result and the reaction will be zero order.

Unimolecular reactions of this type, though, are generally of little synthetic interest. Most synthetically useful catalytic reactions involve an interaction between two adsorbed species to give the product. In such cases Eqn. 7.1 must be expanded to Eqn. 7.9.

$$M + A \underset{}{\overset{K_A}{\rightleftharpoons}} M-A \qquad (7.9)$$

$$M + B \underset{}{\overset{K_B}{\rightleftharpoons}} M-B$$

$$M-A \ + \ M-B \ \xrightarrow{k} \ P + M + M$$

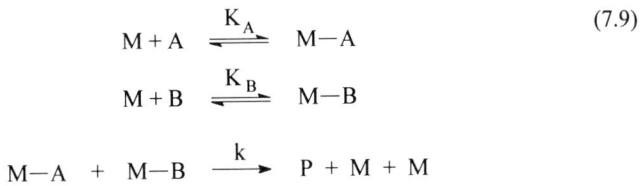

The rate of the reaction, **R**, is then given by Eqn. 7.10.

$$R = k[M - A][M - B] \qquad (7.10)$$

Following the line of reasoning used for the unimolecular process Eqns. 7.11 and 7.12 can be developed,

$$[M - A] = \Theta_A = \frac{K_A[A]}{1 + K_A[A]} \qquad (7.11)$$

$$[M - B] = \Theta_B = \frac{K_B[B]}{1 + K_B[B]} \qquad (7.12)$$

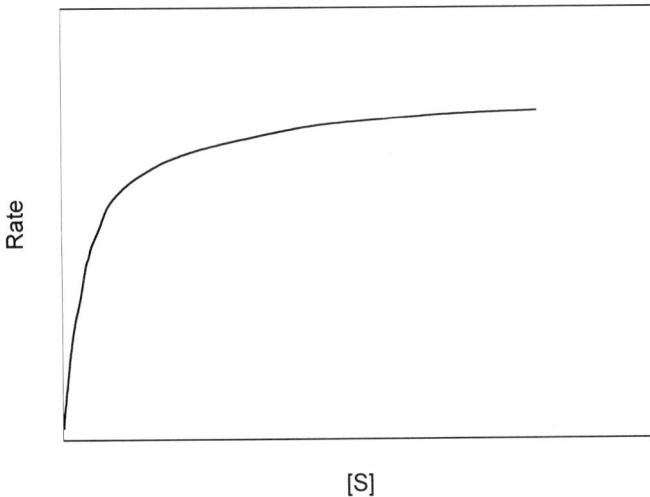

Fig. 7.1. Relationship between reaction rate and substrate concentration in a heterogeneously catalyzed reaction.

and Eqn. 7.10 becomes Eqn. 7.13.

$$R = k \ [M - A][M - B] = \frac{k \ K_A[A] \ K_B[B]}{\langle 1 + K_A[A] \rangle \langle 1 + K_B[B] \rangle} \qquad (7.13)$$

In this instance it is assumed that either a molecule of **A** or a molecule of **B** is adsorbed on each surface atom and that the reaction takes place by the interaction of species adsorbed on adjacent atoms. Reactions of this type in which the rate limiting step involves the reaction of these chemisorbed species are said to follow Langmuir-Hinshelwood kinetics.

It is apparent from Eqn. 7.13 that even with the simplifying assumptions, this rate equation is complex and can be even more so if **A** and **B** are not adsorbed equally. The kinetic equations for each possibility are then derived and compared with the experimental data with the conclusion that the kinetic equation that best fits the experimental facts will define the reaction mechanism. This may not always be the case, however, since different mechanisms may lead to kinetic equations having the same mathematical form but with constants that have different meanings depending on the assumptions made. Further, some of the more complex mechanisms can give kinetic expressions with so many adjustable constants that they can easily be made to fit the experimental data but still prove little about the actual mechanism.

To simplify the process the formulas for a number of kinetic models have been published in a several texts.[7-10] In one of these[10] mathematical equations were developed for five different common types of reactions. With the various mechanisms and rate controlling steps, though, there are six different formulations for one type and ten for each of the others. As stated above, some of these kinetic equations have the same mathematical form but the constants have different meanings so a great deal of judgment is needed in assigning a precise mechanism to such reactions. This reliance on mathematical derivations makes the application of this approach difficult for individuals whose primary interest is not reaction kinetics. While the use of graphical procedures to interpret kinetic data for organic reactions is not uncommon, this approach is not usually used in heterogeneous catalysis even though the common mathematical expressions can be rewritten in linear form.

Given the discussion in Chapter 3 concerning the nature of the active sites on the surface of a catalyst along with the basic organometallic concepts used to derive the reaction mechanisms that were presumed to take place on them, it would appear that the assumptions used in deriving these kinetic expressions are not valid. Certainly, the catalyst surface is non-uniform and each type of site has different adsorption and reaction characteristics. Also, with a single atom active site established for many reactions, the assumption that each reactant must be adsorbed on different sites cannot be correct. Both reactants must be adsorbed on the same atom in order for a reaction to take place. Granted, Langmuir-

Hinshelwood kinetics have played an important role in developing an understanding of heterogeneously catalyzed reactions and that they have been useful in solving reactor and process design problems but there is no reason why these basic concepts cannot be modified to give a more accurate assessment of the actual processes taking place on a catalyst surface.

Michaelis-Menten Kinetics

Interestingly, a fully appropriate model was developed at the same time as the Langmuir model using a similar basic approach. This is the Michaelis-Menten equation which has proved to be so useful in the interpretation of enzyme kinetics and, thereby, understanding the mechanisms of enzyme reactions. Another advantage in using this model is the fact that a graphical presentation of the data is commonly used to obtain the reaction kinetic parameters. Some basic concepts and applications will be presented here but a more complete discussion can be found in a number of texts.[11-13]

As with the Langmuir treatment discussed above, the development of the Michaelis-Menten equation also begins with the rapid and reversible formation of a reactive complex, **E–S**, between the enzyme, **E**, or active site and the substrate, **S**. As shown in Eqn. 7.14, this complex then reacts to give the product, **P**, regenerating the enzyme for further reaction.

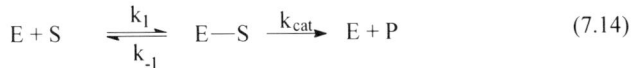

$$E + S \; \underset{k_{-1}}{\overset{k_1}{\rightleftharpoons}} \; E\!-\!S \; \xrightarrow{k_{cat}} \; E + P \qquad (7.14)$$

The first order rate constant for this reaction is k_{cat} while the rate of the reaction, v, is given by Eqn. 7.15.

$$v = k_{cat}[E - S] \qquad (7.15)$$

The stability of the **E–S** complex is determined by the magnitude of the dissociation constant, K_s, as defined by Eqn. 7.16.

$$K_S = \frac{k_{-1}}{k_1} = \frac{[E][S]}{[E - S]} \qquad (7.16)$$

It is at this point that the Langmuir and Michaelis-Menten approaches diverge. The development of the Langmuir isotherm (Eqn. 7.7) was based on the *adsorption* coefficient of the catalyst-substrate complex (Eqn. 7.4) while the Michaelis-Menten equation is based on the *dissociation* constant of this complex (Eqn. 7.16).

The amount of unreacted enzyme present, E, is equal to the total enzyme, E_o, minus that complexed with the substrate. Combining Eqns. 7.16 and 7.17 gives Eqn. 7.18.

$$[E] = [E_o] - [E - S] \qquad (7.17)$$

$$[E - S] = \frac{[E_o][S]}{K_S + [S]} \qquad (7.18)$$

The rate of the reaction is then expressed as Eqn. 7.19.

$$v = \frac{k_{cat}[E_o][S]}{K_S + [S]} \qquad (7.19)$$

Under zero order conditions for S, K_S is much smaller than $[S]$ leading to Eqn. 7.20,

$$V_{max} = k_{cat}[E_o] \qquad (7.20)$$

so a more general form for Eqn. 7.19 can be written as Eqn. 7.21.

$$v = \frac{V_{max}}{K_M + [S]} \qquad (7.21)$$

Here V_{max} is the maximum rate at which the reaction can take place when running under zero order reaction conditions with no diffusional constraints. Under these conditions it is related to k_{cat} by Eqn. 7.20. V_{max} represents the maximum number of substrate molecules that can be converted to product per active site per unit time; it is the maximum TOF for the reaction. K_M is known as the Michaelis Constant. In simple systems it is equal to K_S, the true dissociation constant of the $E-S$ complex. For most purposes, though, K_M must be considered as an apparent dissociation constant that may be treated as the overall dissociation constant of all enzyme bound species. When there is only one $E-S$ complex and all adsorption steps are fast, k_{cat} is the first order rate constant for the conversion of $E-S$ into product. More commonly k_{cat} is a first order rate constant that is associated with the reactions of all $E-S$ complexes in the system.

As in heterogeneously catalyzed processes, the rate of enzyme reactions usually follow saturation kinetics with respect to the concentration of S as shown in Fig. 7.1. This curve is redrawn in Fig. 7.2 to show the relationship to V_{max} and K_M which is the substrate concentration when $v = 1/2\ V_{max}$. At low values of $[S]$

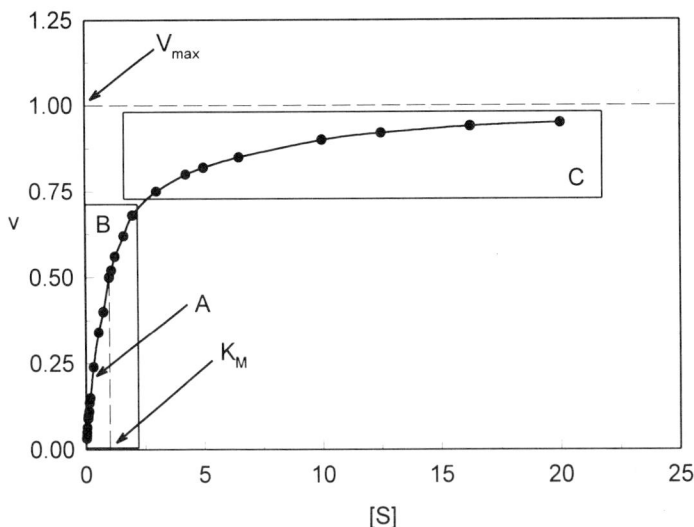

Fig. 7.2.   Relationship between the reaction rate and the substrate
concentration in an enzyme reaction showing the
relationship to V_{max} and K_M.

the concentration of S is much smaller than K_M and the rate of reaction is first
order in S following Eqn. 7.22.

$$v = \frac{k_{cat}\,[E_o][S]}{K_M} = \frac{V_{max}\,[S]}{K_M} \qquad (7.22)$$

When the concentration of S is large with respect to K_M the reaction becomes
zero order in S and the rate expression is given by Eqn. 7.23.

$$v = k_{cat}[E_o] = V_{max} \qquad (7.23)$$

The curve in Fig. 7.2 is a rectangular hyperbola with asymptotes at
$S = -K_M$ and $v = V_{max}$.   Since it is difficult to draw such curves accurately and
even more difficult to estimate their asymptotes, various forms of graphical
representations of the reaction data have been used to obtain values for V_{max}, K_M
and other kinetic parameters of enzyme reactions.

Before discussing these graphic techniques it would be helpful to reiterate
the reasons for using these enzyme analogies for heterogeneously catalyzed
reactions. The most important of these is to incorporate into the kinetic analyses

of heterogeneously catalyzed processes the concept of the single atom active site on which all of the reactants adsorb and interact. Secondly, the commonly used graphical approach can provide a more simple means of determining the relative strengths of adsorption for each of the reactants, the maximum TOF which can be expected for the reaction and other pertinent kinetic parameters. This procedure can also provide an indication of the reaction mechanism as well as the types of inhibition, or activation which may be operating.

While almost all synthetically useful catalytic reactions involve two reactants it will be best to first illustrate this graphical approach using a unimolecular process and the way it can be affected by inhibitors. After this some different types of bisubstrate reactions will be discussed. More detailed analyses of the kinetics of multisubstrate systems have been published in texts[11-13] and shorter articles.[14,15]

To maintain continuity, the terms K_M, V_{max}, v and k_{cat} will be used as defined above but in all mechanistic illustrations the enzyme, E, will be replaced by the metal site, M.

Transforming the Michaelis-Menten equation into a linear form facilitates the graphical analyses of the data. One of the more widely used methods for accomplishing this is the double reciprocal or Lineweaver-Burk plot.[11,12] Inverting both sides of Eqn. 7.19 and substituting Eqn. 7.20 gives Eqn. 7.24, a Lineweaver-Burk double reciprocal, Eqn. 7.24.

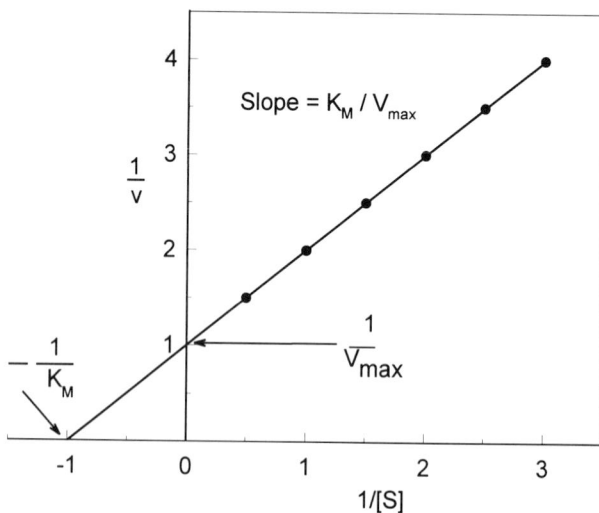

Fig. 7.3. Double reciprocal (Lineweaver-Burk) plot for reactions run using optimal [S] values for the B portion of the curve in Fig. 7.2.

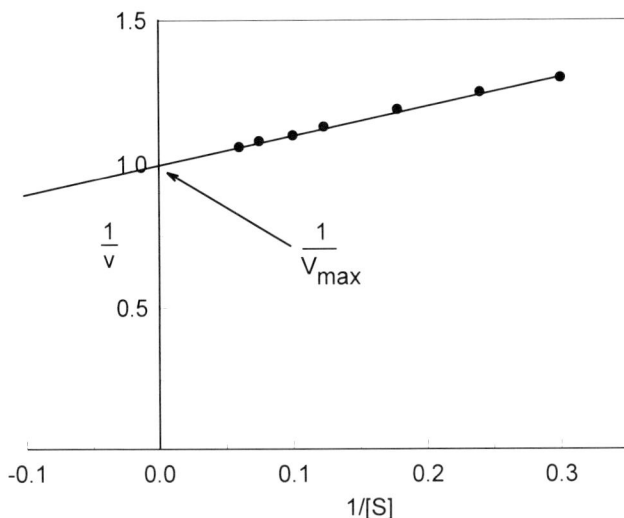

Fig. 7.4. Double reciprocal plot for reactions run using high substrate
          concentrations.

$$\frac{1}{v} = \frac{K_M}{V_{max}} \frac{1}{[S]} + \frac{1}{V_{max}} \qquad (7.24)$$

Plotting **1/v** against **1/[S]** gives a line such as that shown in Fig. 7.3. Here
the y-axis intercept is equal to **1/V$_{max}$**, the x-axis intercept to **-1/K$_M$**,and the slope
to **K$_M$/V$_{max}$** so the pertinent kinetic data are readily obtained.

For maximum reliability in such plots, the substrate concentrations should
be chosen from the region near **K$_M$** with the actual values selected to give
somewhat evenly spaced reciprocals.[16] The plot shown in Fig. 7.3 was obtained
using substrate concentrations in the middle or **B** range as illustrated in Fig. 7.2.
With high substrate concentrations, such as those in the **C** range in Fig. 7.2, the
resulting Lineweaver-Burk plot will be nearly horizontal (Fig. 7.4). While
1/V$_{max}$ can be measured in this plot the slope will be close to zero making an
accurate determination of **K$_M$** difficult. If the substrate concentrations are low
with respect to **K$_M$** (A range in Fig. 7.2) the Lineweaver-Burk double reciprocal
plot will intercept the axes too close to the origin to give any reasonably accurate
values for either **V$_{max}$** or **K$_M$** as illustrated in Fig. 7.5.

Inhibited Reactions

While plots such as that shown in Fig. 7.3 are relatively easy to construct and
interpret, the simplified reactions to which they are applied are seldom

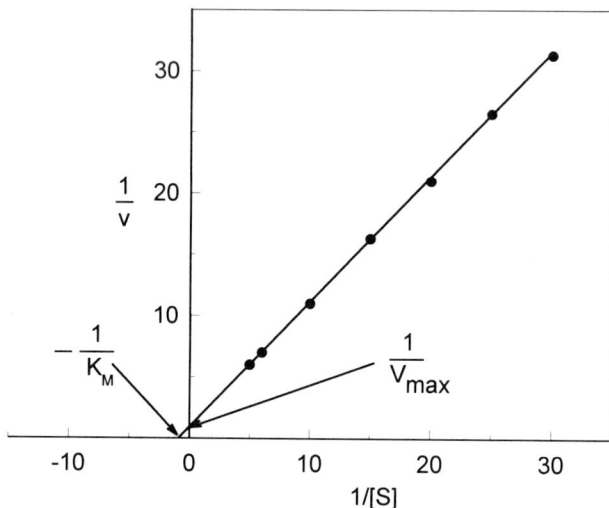

Fig. 7.5. Double reciprocal plot for reactions run using low substrate
concentrations.

encountered in synthetically useful processes. A degree of complexity can be
added to this unimolecular process, though, by investigating the inhibition of such
reactions. The different types of inhibition can be differentiated by the types of
double reciprocal plots obtained. Inhibition can be classed as being competitive,
non-competitive, or uncompetitive with mixed type inhibition also a possibility.
The procedure used to distinguish between these different types of inhibition
patterns as well as to obtain the appropriate kinetic parameters is to first run the
reaction using several different substrate concentrations in the absence of the
inhibitor, **I**, and then to repeat the process at several concentrations of **I** to obtain a
family of rate curves with varying [**S**] and [**I**].

Competitive Inhibition

Competitive inhibition results when an inhibitor, **I**, binds reversibly to the active
site and prevents the substrate, **S**, from binding, and vice versa. Both **I** and **S** are
mutually exclusive on the active site.[17] The equilibria describing this inhibition
pattern are shown in Eqn. 7.25.

Since some of the active sites are blocked by the inhibitor a higher
concentration of **S** will be required to give the same initial rate as observed in the
absence of **I**. At an infinitely high [**S**] all of the sites can be made to react with **S**
so the V_{max} of the reaction will not be affected by **I**.

The double reciprocal equation for the reaction sequence shown in Eqn.
7.25 is given by Eqn. 7.26.

$$M + S \; \underset{K_S}{\rightleftharpoons} \; M{-}S \; \xrightarrow{k_{cat}} \; M + P \qquad (7.25)$$

$$+$$

$$I$$

$$K_I \; \Big\updownarrow$$

$$M{-}I$$

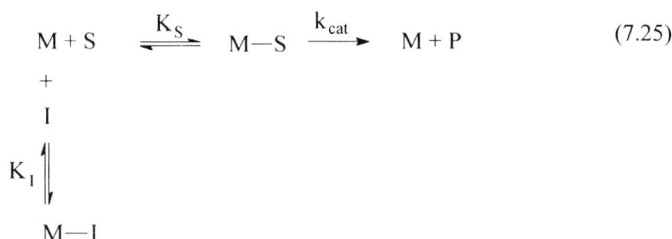

$$\frac{1}{v} = \frac{K_M}{V_{max}} \left(1 + \frac{[I]}{K_I}\right)\frac{1}{[S]} + \frac{1}{V_{max}} \qquad (7.26)$$

This is similar to Eqn. 7.24, the double reciprocal equation for the simple unimolecular reaction. The difference is that in Eqn. 7.26, K_M is modified and should be considered to be $K_{M(apparent)}$ which is defined in Eqn. 7.27.

$$K_{M(app)} = K_M\left(1 + \frac{[I]}{K_I}\right) \qquad (7.27)$$

Fig. 7.6 shows the double reciprocal plots for reactions run in the presence of several concentrations of I. The series of lines intersecting the y-axis at the same point is the diagnostic pattern for the competitive inhibition of a reaction. As in the uninhibited reaction the y-axis intercept is $1/V_{max}$. This is the same for all values of [I] but the x-axis intercepts are the different values of $-1/K_{M(app)}$ for each concentration of I. The slopes of the lines are proportional to $K_{M(app)}$ which, as shown in Eqn. 7.27, also contains the term, K_I, the dissociation constant for the M–I complex. These slopes follow Eqn. 7.28,

$$\text{Slope} = \frac{K_{M(app)}}{V_{max}} = \frac{K_M}{V_{max}}\left(1 + \frac{[I]}{K_I}\right) \qquad (7.28)$$

which can be rewritten in linear form as Eqn. 7.29.

$$\text{Slope} = \frac{K_M}{V_{max}\,K_I}\,[I] + \frac{k_M}{V_{max}} \qquad (7.29)$$

A replot of the slope of each line in Fig. 7.6 versus the concentration of I gives the line shown in Fig. 7.7a which has a slope of $K_M/V_{max}K_I$, a y-axis

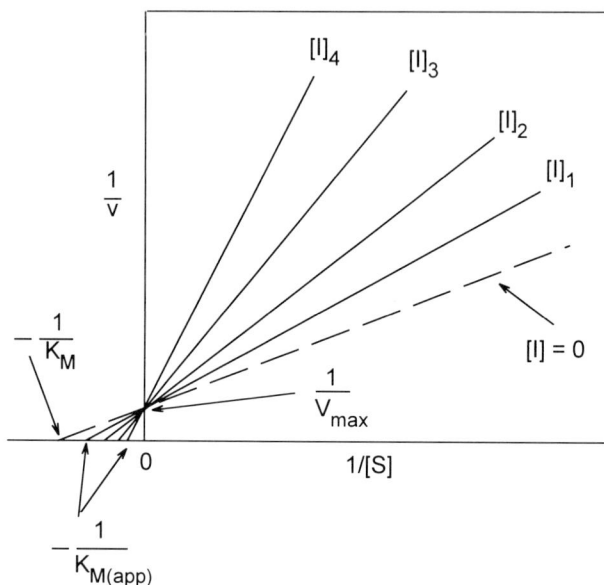

Fig. 7.6.   Double reciprocal plots for reactions run at different fixed
concentrations of a competitive inhibitor, I.

intercept of K_M/V_{max} and an x-axis intercept at $-K_I$. The y-axis intercept is also
the slope of the line at $[I] = 0$.

The $K_{M(app)}$ found at the x-axis intercepts in Fig. 7.6 can also be written
as a linear function of $[I]$ as in Eqn. 7.30.

$$K_{M(app)} = \frac{K_M}{K_I} [I] + K_M \qquad (7.30)$$

The replot of $K_{M(app)}$ versus $[I]$, shown in Fig. 7.7b, has a y-axis intercept of
K_M, an x-axis intercept of $-K_I$ and a slope of K_M/K_I. From these three plots
(Figs. 7.6, 7.7a and 7.7b) the dissociation constants for both the M-S and M-I
complexes are obtained along with a determination of the maximum rate the
reaction can attained under zero order conditions with no diffusion control. If the
total number of active sites, M_0, is known, the first order rate constant for the
formation of the product, k_{cat}, can be calculated using Eqn. 7.31.

$$k_{cat} = \frac{V_{max}}{[M_0]} \qquad (7.31)$$

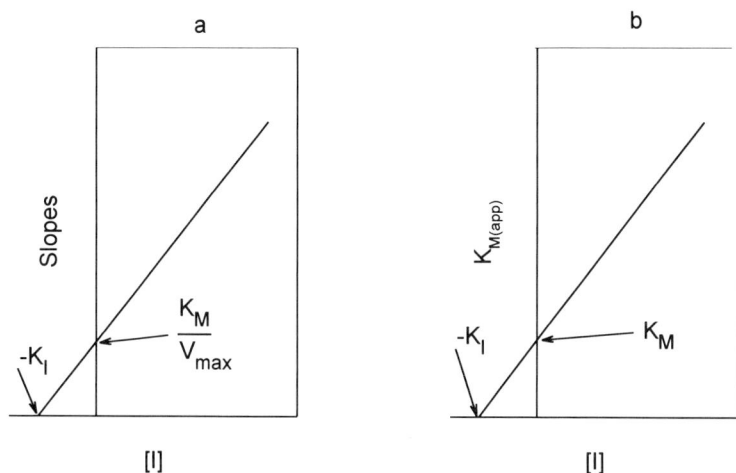

Fig. 7.7.   Replots of the data taken from Fig. 7.6. a) Slope versus [**I**];
b) $V_{max(apparent)}$ versus [**I**].

Non-Competitive Inhibition

A classic non-competitive inhibitor has no effect on substrate binding and vice-versa. Inhibitor and substrate molecules adsorb independently on different sites but, while the inhibitor does not affect the adsorption of the substrate it does inhibit the further reaction of the adsorbed species.[18] On a metal catalyst this type of inhibition could arise when the substrate is adsorbed on a corner atom and the inhibitor on face atoms near the corner. The resulting steric or electronic factors could prevent further reaction of the adsorbed substrate.

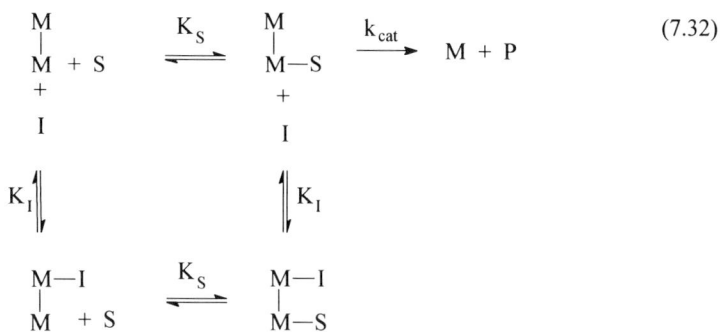

$$\text{(7.32)}$$

The equations showing non-competitive inhibition are given in Eqn. 7.32. According to this series of equilibria, at any concentration of **I** some of the active sites will have both an adsorbed **S** and a neighboring **I**, an arrangement which is non-reactive. In this case the V_{max} of the reaction will decrease.

The reciprocal form of the rate equation for Eqn. 7.32 is given in Eqn. 7.33 with both the slope and the y-axis intercept modified by $V_{max(apparent)}$ which is defined in Eqn. 7.34.

$$\frac{1}{v} = \frac{K_M}{V_{max}}\left(1 + \frac{[I]}{K_I}\right)\frac{1}{[S]} + \frac{1}{V_{max}}\left(1 + \frac{[I]}{K_I}\right) \tag{7.33}$$

$$\frac{1}{V_{max(app)}} = \frac{\left(1 + \dfrac{[I]}{K_I}\right)}{V_{max}} \tag{7.34}$$

Plots of **1/v** versus **1/[S]** at various concentrations of **I** are shown in Fig. 7.8. This type of plot with all of the lines intersecting the x-axis at the same point, **$-1/K_M$**, is characteristic of reactions run with non-competitive inhibition. The

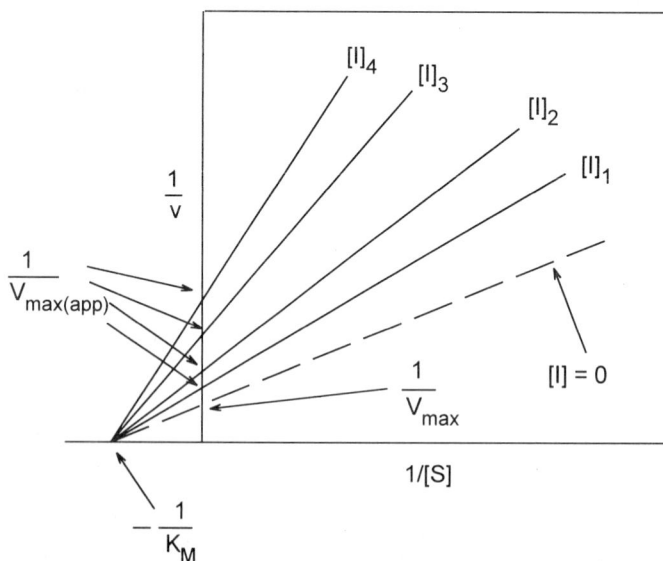

Fig. 7.8.   Double reciprocal plots for reactions run with different concentrations of a non-competitive inhibitor, **I**.

slopes of these lines follow Eqn. 7.28 and has the linear form shown by Eqn. 7.29.

The replot of the slopes versus $[I]$ has an x-axis intercept of $-K_I$. The y-axis intercept is K_M/V_{max} which is the slope of the reaction plot at $[I] = 0$. The slope of the replot is equal to the slope of the uninhibited reaction multiplied by $1/K_I$.

The y-axis intercepts of the reciprocal plots in Fig. 7.8 are also linear functions of $[I]$ as shown in Eqn. 7.35.

$$\frac{1}{V_{max(app)}} = \frac{1}{V_{max} \, K_I} [I] + \frac{1}{V_{max}} \qquad (7.35)$$

The replot of the y-axis intercepts in Fig. 7.8 versus $[I]$ has an x-axis intercept of $-K_I$, a y-axis intercept of $1/V_{max}$ and a slope of $1/V_{max} \times 1/K_I$.

Uncompetitive Inhibition

Uncompetitive inhibition results when both the substrate and the inhibitor adsorb on the same site and the resulting complex is unreactive.[19]   The equilibria associated with this type of inhibition is shown in Eqn. 7.36.

$$M + S \; \underset{}{\overset{K_S}{\rightleftharpoons}} \; M\!-\!S \; \overset{k_{cat}}{\longrightarrow} \; M + P \qquad (7.36)$$

$$+$$
$$I$$
$$K_I \updownarrow$$
$$M \overset{S}{\underset{I}{<}}$$

The reciprocal form of the rate expression for this equation is given by Eqn. 7.37.

$$\frac{1}{v} = \frac{K_M}{V_{max}} \frac{1}{[S]} + \frac{1}{V_{max}} \left(1 + \frac{[I]}{K_I} \right) \qquad (7.37)$$

The plots of $1/v$ versus $1/[S]$ at various concentrations of I are shown in Fig. 7.9. The slopes of the lines are not modified by either $[I]$ or K_I so the lines for the plots associated with different concentrations of I are parallel to the line from the uninhibited reaction.   The y-axis intercepts are $1/V_{max(app)}$ and the x-axis intercepts are $1/K_{M(app)}$.   The $1/V_{max(app)}$ can be rewritten in linear form as in Eqn. 7.35.   The replot of $1/V_{max(app)}$ versus $[I]$ has a slope of $1/V_{max} \times 1/K_I$, a y-axis intercept of $1/V_{max}$ and an x-axis intercept of $-K_I$. The linear equation for

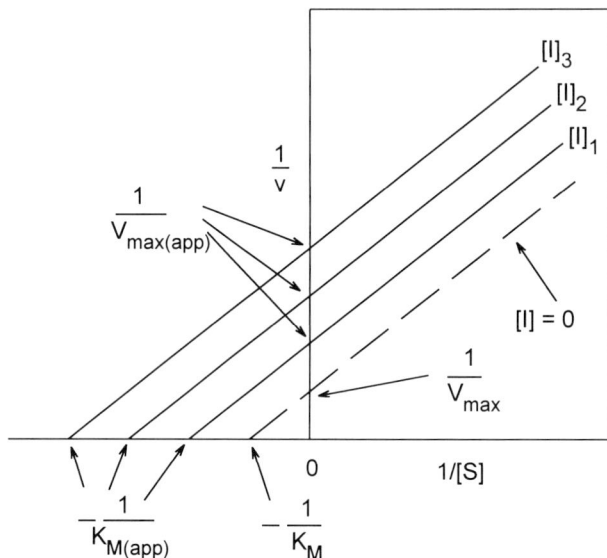

Fig. 7.9.   Double reciprocal plots for reactions run in the presence of different fixed concentrations of an uncompetitive inhibitor, **I**.

$1/K_{M(app)}$, the x-axis intercepts in Fig. 7.9, is given in Eqn. 7.30.   The replot of $1/K_{M(app)}$ versus **[I]** also has a slope of $1/K_I \times 1/V_{max}$ and an x-axis intercept of $-K_I$.   The y-axis intercept, though, is $1/K_M$.   Thus, values for K_M, K_I and V_{max} can be obtained directly with k_{cat} determined by using Eqn. 7.31.

From the above discussion it can be seen that a competitive inhibitor has no effect on V_{max} but does influence $K_{M(app)}$.   Non-competitive inhibition, however, results in a decrease in V_{max} but no change in K_M while uncompetitive inhibition produces a decrease in both V_{max} and K_M.

Mixed Type Inhibition

Not all inhibitions fall cleanly into one of these three patterns.   Frequently a mixed type inhibition is encountered in which V_{max} and K_M are affected along with the slopes of the reciprocal plots.   There are a variety of mixed type inhibition patterns and these are thoroughly covered elsewhere[20] so only the simplest of these, that one in which **M–I** has a lower affinity than **M** for **S** and the **MSI** complex is non-productive, will be discussed here.   The equilibria for this type of inhibition are shown in Eqn. 7.38 with $\alpha > 1$.

This system can be considered a mixture of competitive and non-competitive inhibition.   The reciprocal form of the rate equation for Eqn. 7.38 is given by Eqn. 7.39.

$$(7.38)$$

$$M + S \; \underset{}{\overset{K_S}{\rightleftharpoons}} \; M{-}S \; \xrightarrow{k_{cat}} \; M + P$$

$$+ \qquad\qquad +$$
$$I \qquad\qquad I$$

$$K_I \Big\updownarrow \qquad\qquad \Big\updownarrow K_I$$

$$I{-}M \; + \; S \; \overset{K_S}{\rightleftharpoons} \; M\!\!\overset{I}{\underset{S}{\diagdown}}$$

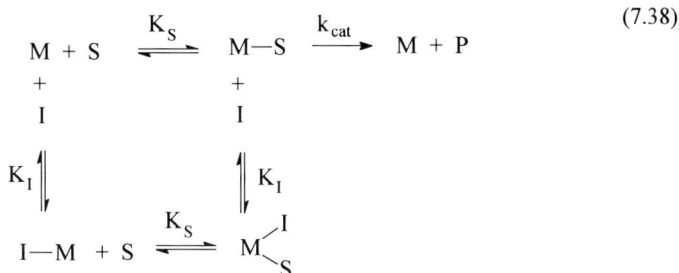

$$\frac{1}{v} = \frac{K_M}{V_{max}}\left(1 + \frac{[I]}{K_I}\right)\frac{1}{[S]} + \frac{1}{V_{max}}\left(1 + \frac{[I]}{\alpha K_I}\right) \qquad (7.39)$$

The plots of $1/v$ versus $1/[S]$ at various concentrations of **I** are shown in Fig. 7.10. These plots form a series of intersecting lines having differing x-axis and y-axis intercepts and slopes. As in the previously discussed inhibitions, K_I, K_M and V_{max} are all associated with these slopes and axis intercepts but an additional factor is present here, the parameter, α, which is a measure of the relative strengths of **M–I** and **M–S** adsorption.

As in the simpler inhibition plots the slopes of the lines follow Eqn. 7.28, but the x-axis intercepts are more complex following Eqn. 7.40,

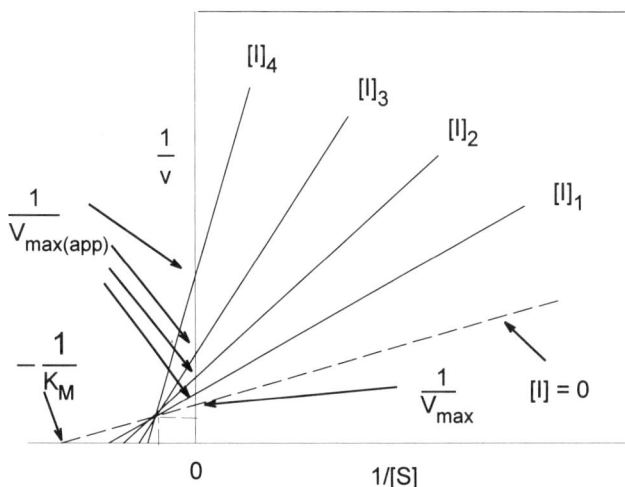

Fig. 7.10. Double reciprocal plots for reactions run in the presence of different fixed concentrations of a mixed type inhibitor, **I**, with $\alpha > 1$.

$$\frac{\left(1 + \dfrac{[I]}{\alpha K_I}\right)}{K_M\left(1 + \dfrac{[I]}{K_I}\right)} = \frac{\left(1 + \dfrac{[I]}{\alpha K_I}\right)}{K_{M(app)}} \tag{7.40}$$

while the y-axis intercepts are given by Eqn. 7.41.

$$\frac{\left(1 + \dfrac{[I]}{\alpha K_I}\right)}{V_{max}} \tag{7.41}$$

The intersection point of the lines corresponds on the x-axis to $-1/\alpha K_I$ and on the y-axis to Eqn. 7.42.

$$\frac{1}{v} = \frac{1}{V_{max}}\left(\frac{\alpha - 1}{\alpha}\right) \tag{7.42}$$

The linear form of the equations for the slopes of the lines in Fig. 7.10 is given in Eqn. 7.29. Replotting the slopes versus $[I]$ for these lines gives a plot with an x-axis intercept of $-K_I$ and a y-axis intercept of K_M/V_{max}. The y-axis intercepts in Fig. 7.10 can be rewritten in linear form as Eqn. 7.43.

$$\frac{1}{V_{max(app)}} = \frac{1}{\alpha K_I V_{max}}[I] + \frac{1}{V_{max}} \tag{7.43}$$

Replotting these intercepts versus $[I]$ gives a plot that has a y-axis intercept of $1/V_{max}$ and an x-axis intercept of $-\alpha K_I$ so all of the pertinent parameters for Eqn. 7.39 can be obtained.

Bisubstrate Reactions

Most synthetically useful reactions involve the interaction of two substrates on a catalytically active site to give the reaction product or products. Such systems are more complex than the simple unimolecular reaction but resemble somewhat the inhibited unimolecular processes just described. There are, however, some significant differences in that the two adsorbed substrates must interact to form products and not give a non-productive species. Thus, for two reactants, **A** and **B**, it is necessary to look at the initial rate data for a series of reactions that are run keeping the concentration of **A** constant while varying the concentration of **B** as well as a reaction series in which [**B**] is kept constant while [**A**] changes. The K_A, which is the dissociation constant for **A** at a fixed value of [**B**] may not be the true

K_A, or $K_{M(A)}$, but a $K_{A(apparent)}$ that changes as [B] varies. The same can be said concerning the effect of [A] on K_B. The V_{max} found at a saturating concentration of A may not be the same when [B] is saturating. The true $K_{M(A)}$ is found when [B] is saturating, that is when the reaction is zero order in B. The true $K_{M(B)}$ is found when [A] is saturating and the true V_{max} when the reaction is zero order in both A and B.[21]

Typically, the kinetics of bisubstrate reactions are studied by measuring the initial reaction rates over a range of concentrations of one substrate, A, while holding the concentration of the second substrate, B, constant and doing this for several fixed values of [B]. If specific concentrations of A are used for all of the reaction series, these same rates can also be used to examine changes in reaction rate when [B] is varied at fixed concentrations of A for several values of [A].

There are three general patterns observed with bisubstrate reactions: random, sequential and ping-pong.

Random Bisubstrate Reactions

If two reactants, A and B, bind to the catalytically active site randomly and if the binding of one changes the dissociation constant of the other by a factor, α, the system can be described by Eqn. 7.44.[22] This reaction sequence is similar to the non-competitive inhibition described by Eqn. 7.32 but has the **MAB** complex as the active species rather than the non-productive entity present during inhibition.

$$ (7.44) $$

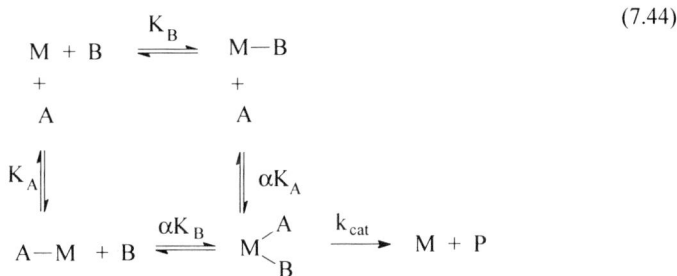

Since the rate of the reaction depends on the concentrations of both A and B the rate equation can be written in two forms to show the rate dependency on each. Eqn. 7.45 is the reciprocal rate expression when [B] is held constant and [A] varies. The rate expression for those reactions run with [A] constant and [B] changing is symmetrical to Eqn. 7.45. This is also true for all of the other equations and plots described below.

$$ \frac{1}{v} = \frac{\alpha K_A}{V_{max}}\left(1 + \frac{K_B}{[B]}\right)\frac{1}{[A]} + \frac{1}{V_{max}}\left(1 + \frac{\alpha K_B}{[B]}\right) \qquad (7.45) $$

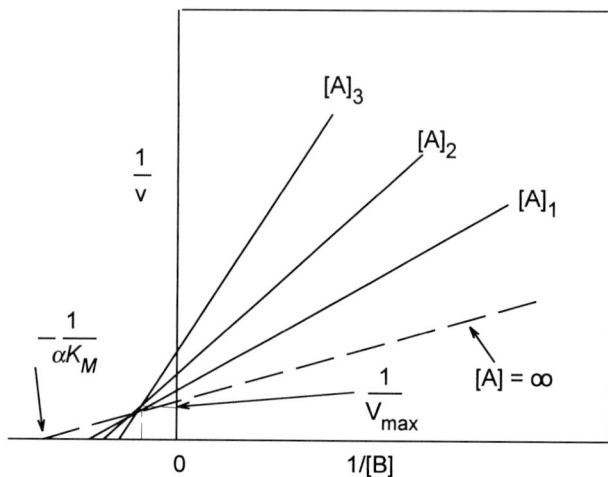

Fig. 7.11. Double reciprocal plots for reactions run at different fixed
concentrations of **A** for a random bisubstrate reaction with
$\alpha < 1$.

In Fig. 7.11 are shown the plots of **1/v** versus **1/[B]** at constant **[A]** for
several values of **[A]**. These plots intersect at a point above the x-axis and with
slopes having Eqn. 7.46.

$$\text{Slope} = \frac{\alpha K_B}{V_{max}} \left(1 + \frac{K_A}{[A]} \right) \qquad (7.46)$$

The lines intersect the y-axis at:

$$\frac{1}{V_{max}} \left(1 + \frac{\alpha K_A}{[A]} \right) \qquad (7.47)$$

The x-axis intercepts for Fig. 7.11 are shown in Eqn. 7.48.

$$-\Delta K_B = - \frac{\left(1 + \dfrac{\alpha K_A}{[A]} \right)}{\alpha K_B \left(1 + \dfrac{K_A}{[A]} \right)} \qquad (7.48)$$

The plot in Fig. 7.11 for the reaction run under zero order conditions in **A**, [A] is saturating, has a y-axis intercept at $1/V_{max}$ and an x-axis intercept at $-1/\alpha K_B$. The point of intersection of the lines in both plots corresponds on the x-axis to $-1/K_B$ and on the y-axis to:

$$\frac{1}{V_{max}} (1 - \alpha) \tag{7.49}$$

This is the case when α is less than one, that is, when the binding of one reactant increases the binding affinity of the other. When α equals one the binding of one reactant has no effect on the other and the lines intersect on the x-axis at $-1/K_B$. If the binding of one reactant decreases the binding of the other, α is greater than one and the lines of these plots will intersect below the x-axis. The intersection still corresponds on the x-axis to $-1/K_B$.

The value for V_{max} is found from the y-axis intercepts of the replots of $1/V_{max(app)}$ versus $1/[A]$ using the linear Eqn. 7.50. This replot gives value for αK_A from the x-axis intercept.

$$\frac{1}{V_{max(app)}} = \frac{\alpha K_A}{V_{max}} \frac{1}{[A]} + \frac{1}{V_{max}} \tag{7.50}$$

The linear equation for the slopes of the lines in Fig. 7.11 is given by Eqn. 7.51.

$$\text{Slope} = \frac{\alpha K_A K_B}{V_{max}} \frac{1}{[B]} + \frac{\alpha K_A}{V_{max}} \tag{7.51}$$

The x-axis intercepts of the replot of the slopes versus $1/[A]$ give values for K_A. While this value, and that for αK_A, can be used to calculate the value of α, this parameter can also be obtained from replots of $1/\Delta K_B$ (Eqn. 7.48) versus $1/[A]$ using the linear Eqn. 7.52.

$$\frac{1}{\Delta K_B} = \frac{K_A}{K_B(1 - \alpha)} \frac{1}{[A]} + \frac{1}{K_B(1 - \alpha)} \tag{7.52}$$

As mentioned above, the equations and plots for reactions run holding [A] constant with varying [B] are symmetrical to those which were obtained from reactions in which [B] was kept constant and [A] changed. All of the kinetic parameters can be obtained from either set of data.

Ordered Bisubstrate Reactions

The equilibria shown in Eqn. 7.53 describes the system in which the reactants **A** and **B** combine with the catalyst in the specific order, **A** adsorbing before **B**.[23] Such cases can arise when the adsorption of **A** causes the electronic changes in the active site which are necessary for **B** to adsorb. Without this modification of the electronic nature of the site **B** could not be adsorbed.

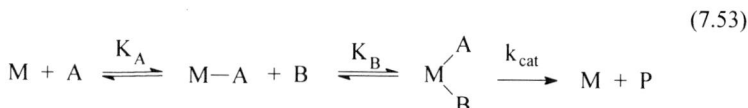

(7.53)

$$M + A \xrightleftharpoons{K_A} M{-}A + B \xrightleftharpoons{K_B} M \begin{smallmatrix} A \\ \\ B \end{smallmatrix} \xrightarrow{k_{cat}} M + P$$

The reciprocal rate expression for reactions run with varying concentrations of **B** at a constant concentration of **A** is given by Eqn. 7.54.

$$\frac{1}{v} = \frac{K_B}{V_{max}} \left(1 + \frac{K_A}{[A]} \right) \frac{1}{[B]} + \frac{1}{V_{max}} \tag{7.54}$$

with the 1/v versus 1/[B] plots depicted in Fig. 7.12. Here the lines for all of the plots intersect at the y-axis at the point $1/V_{max}$. The limiting plot, that run under zero order conditions in **A**, intersects the x-axis at $-1/K_B$ while the other plots intersect the x-axis at:

$$-\frac{1}{K_{B(app)}} = -\frac{1}{K_B \left(1 + \frac{K_A}{[A]} \right)} \tag{7.55}$$

from which the linear Eqn. 7.56 can be derived.

$$K_{B(app)} = K_A K_B \frac{1}{[A]} + K_B \tag{7.56}$$

The replot of $K_{B(app)}$ versus 1/[A] has a slope of $K_A K_B$, a y-axis intercept at K_B and an x-axis intercept at $-1/K_A$.

The linear expression for the slopes of the plots in Fig. 7.12 is given by Eqn. 7.57.

$$\text{Slope} = \frac{K_A K_B}{V_{max}} \frac{1}{[A]} + \frac{K_B}{V_{max}} \tag{7.57}$$

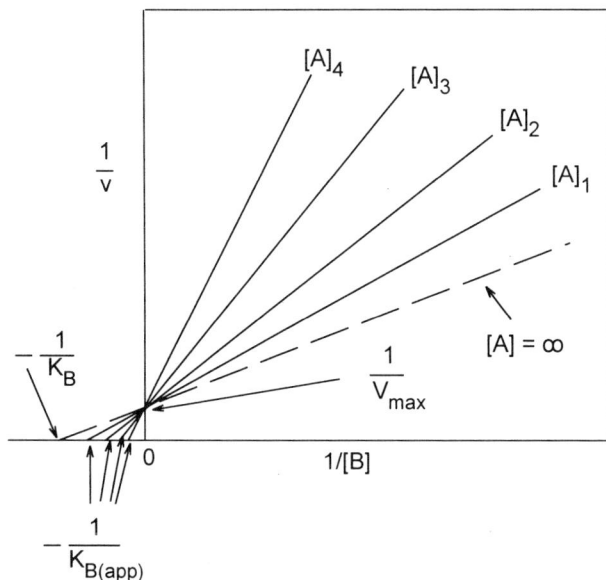

Fig. 7.12.   Plots of **1/v** versus **1/[B]** for an ordered bisubstrate
reaction where **A** adsorbs before **B**.

The replot of slope versus **1/[A]** has a slope of $K_A K_B / V_{max}$, a y-axis intercept at
K_B / V_{max} and an x-axis intercept of $-1/K_A$.

   In contrast to what is observed with a random bisubstrate reaction, with an
ordered bisubstrate process the rate data obtained on varying the concentration of
A at constant concentrations of **B** do not give reciprocal plots which are
symmetrical to Fig. 7.12.   Comparing Eqn. 7.54 with Eqn. 7.58, the double
reciprocal equation for reactions in which **[A]** is varied at constant **[B]**, shows the
differences between the two systems.

$$\frac{1}{v} = \frac{K_A}{V_{max}} \left(\frac{K_B}{[B]} \right) \frac{1}{[A]} + \frac{1}{V_{max}} \left(1 + \frac{K_B}{[B]} \right) \qquad (7.58)$$

   These differences are further illustrated by comparing Fig. 7.12 with Fig.
7.13, the plot of **1/v** versus **1/[A]**.   In Fig. 7.13 the lines intersect at a point above
the x-axis but not on the y-axis as in Fig. 7.12.   The intercept correlates on the y-
axis to $1/V_{max}$ and on the x-axis to $-1/K_A$.   When the reaction is run under zero
order conditions in **B**, the slope of the plot is zero and the y-intercept is $1/V_{max}$.

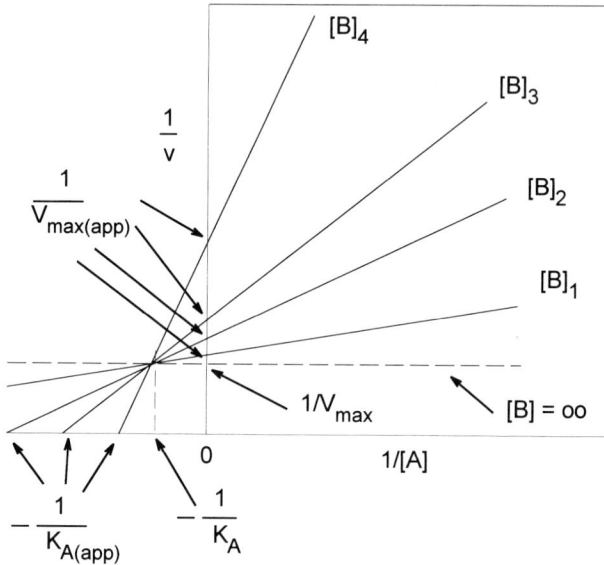

Fig. 7.13. Plots of **1/v** versus **1/[B]** for an ordered bisubstrate reaction
when **A** adsorbs before **B**.

The plots in Fig. 7.13 intercept the y-axis at:

$$\frac{1}{V_{max(app)}} = \frac{\left(1 + \dfrac{K_B}{[B]}\right)}{V_{max}} = \frac{K_B}{V_{max}}\frac{1}{[B]} + \frac{1}{V_{max}} \tag{7.59}$$

and the x-axis at:

$$-\frac{1}{K_{A(app)}} = -\frac{\left(1 + \dfrac{[B]}{K_B}\right)}{K_A} = -\left(\frac{1}{K_A K_B}[B] + \frac{1}{K_A}\right) \tag{7.60}$$

A replot of **1/V$_{max(app)}$** versus **1/[B]** has a slope of **K$_B$/V$_{max}$**. The replot of **1/K$_{A(app)}$** versus **[B]** has a slope of **1/K$_A$K$_B$**, a y-axis intercept at **1/K$_A$** and an x-axis intercept at **-K$_B$**.

The slopes of the lines in Fig. 7.13 follow Eqn. 7.61.

$$\text{Slope} = \frac{K_A K_B}{V_{max}} \frac{1}{[B]} \tag{7.61}$$

A plot of the slopes versus $1/[B]$ has a slope of $K_A K_B/V_{max}$ and an intercept at the origin. The obtaining of such a plot verifies that the reaction is ordered with **A** adding to the active site before **B**. K_A is the ratio of the slope of the slope replot and the slope of the y-axis intercept plot.

Ping-Pong Bisubstrate Reactions

A ping-pong mechanism is one in which a molecule of one product is released between the addition of the two reactants.[24]   The equilibria for this type of reaction is shown in Eqn. 7.62.

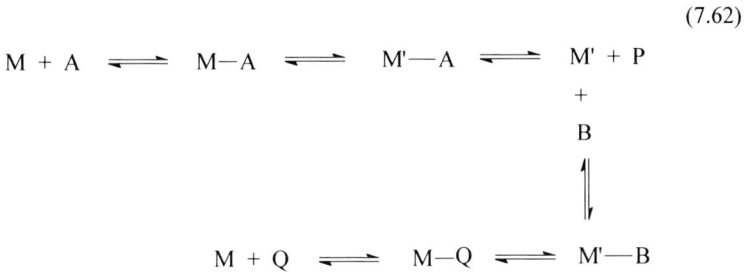

$$(7.62)$$

$$M + A \rightleftharpoons M{-}A \rightleftharpoons M'{-}A \rightleftharpoons M' + P$$

$$+$$

$$B$$

$$\big\updownarrow$$

$$M + Q \rightleftharpoons M{-}Q \rightleftharpoons M'{-}B$$

In this sequence the metal site, **M**, adsorbs a substituent, **A**, to give the **M–A** complex. This is changed to another complex, **M'–P**, which dissociates to give one of the products, **P**, and a modified site, **M'**. **M'** then reacts with the second substrate, **B**, to give **M'–B** which converts to **M–Q** then desorbs the second product, **Q**, and regenerates the initial active site, **M**.

The reciprocal form of the rate expressions for this reaction sequence run, with [A] changing and [B] constant, is shown in Eqn. 7.63. The reverse gives an equation symmetrical in **A** and **B**.

$$\frac{1}{v} = \frac{K_{M(A)}}{V_{max}} \frac{1}{[A]} + \frac{1}{V_{max}} \left(1 + \frac{K_{M(B)}}{[B]} \right) \tag{7.63}$$

The plots of $1/v$ versus $1/[A]$ at several concentrations of **B** are shown in Fig. 7.14. These plots are characterized by a series of parallel lines having a slope of $K_{M(A)}/V_{max}$.   The y-axis intercepts are given by Eqn. 7.64 and the x-axis intercepts by Eqn. 7.65.

$$\frac{1}{V_{max(app)}} = \frac{\left(1 + \dfrac{K_{M(B)}}{[B]}\right)}{V_{max}} = \frac{K_{M(B)}}{V_{max}}\frac{1}{[B]} + \frac{1}{V_{max}} \tag{7.64}$$

$$-\frac{1}{K_{M(A)(app)}} = -\frac{\left(1 + \dfrac{K_{M(B)}}{[B]}\right)}{K_{M(A)}} = -\left(\frac{K_{M(B)}}{K_{M(A)}}\frac{1}{[B]} + \frac{1}{K_{M(A)}}\right) \tag{7.65}$$

The replot of the y-axis intercepts versus $1/[A]$ has a slope of $K_{M(B)}/V_{max}$, a y-axis intercept of $1/V_{max}$ and an x-axis intercept of $-1/K_{M(B)}$. Replotting the x-axis intercepts from Fig. 7.14 versus $1/[B]$ gives a line with a slope of $K_{M(B)}/K_{M(A)}$, a y-axis intercept of $1/K_{M(A)}$ and an x-axis intercept of $-1/K_{M(B)}$.

As with the random bisubstrate reactions, the plots and replots for data obtained when $[B]$ is varied and $[A]$ is constant are symmetrical to those obtained when $[A]$ is varied and $[B]$ is held constant. All of the pertinent information can

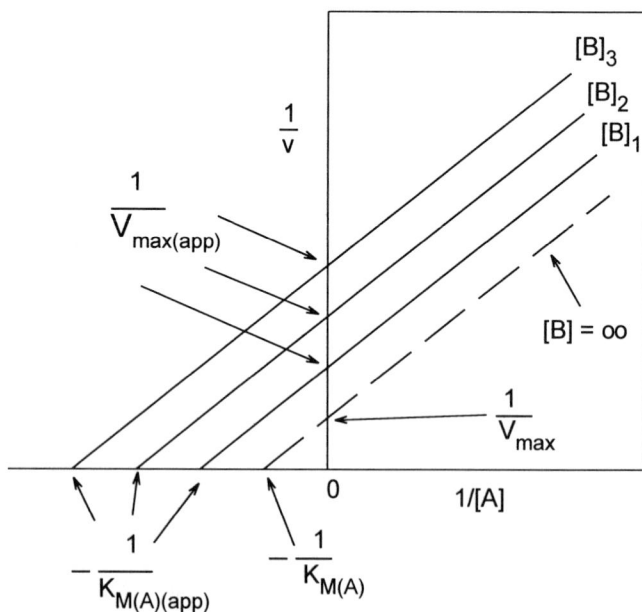

Fig. 7.14. Double reciprocal plots for ping-pong reactions run at several fixed concentrations of **B**.

Scheme 7.1

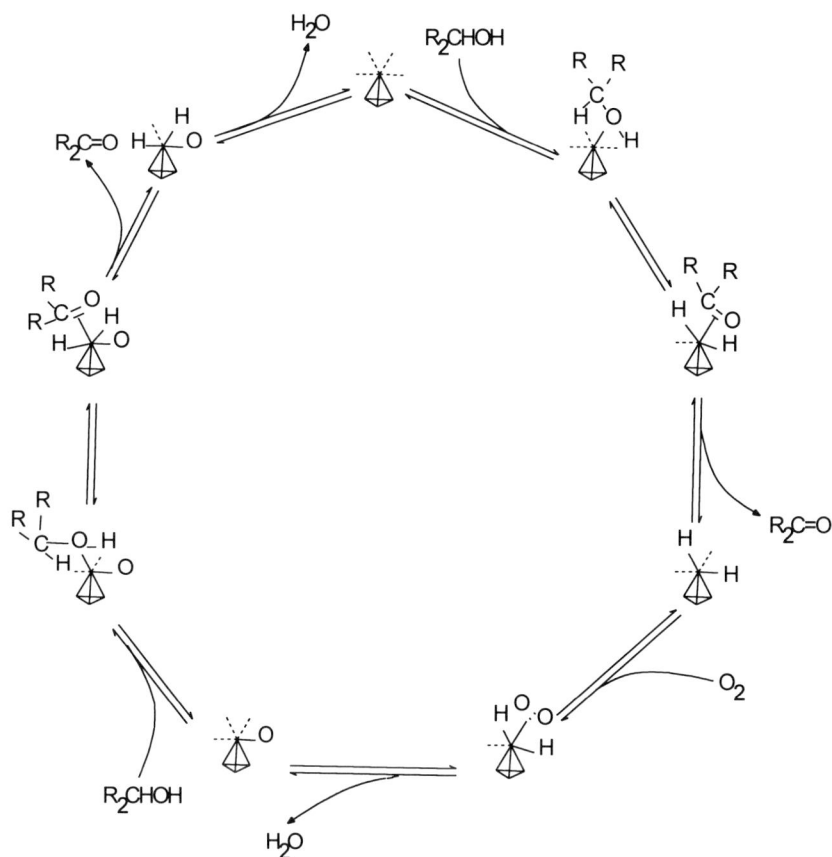

be obtained from one set of plots so it is not necessary to use both sets except as a check on the results.

Further Applications

The previous discussion was intended to show the nature of the Michaelis-Menten approach to the examination of the kinetics of heterogeneously catalyzed reactions. It is not, by any means, a complete presentation. A similar approach can be used to investigate the effect of inhibitors or activators in bisubstrate reactions with the results used to indicate, among other things, which reactant is being inhibited or activated and the mode of that action.[11] While this approach has not been generally applied to the determination of the mechanism of heterogeneously catalyzed processes it does have some advantages over the

Table 7.1

Rates of 2-propanol oxidation using a Pt/SiO_2 catalyst at various partial pressures of O_2.

	Rate (moles oxygen consumed/minute/mole Pt)		
[2-propanol]	760 mm oxygen	300 mm oxygen	235 mm oxygen
0.0 M	0.0	0.0	0.0
0.044 M	-	-	0.49 (± .02)
0.058 M	0.78 (± .04)	0.70 (± .03)	-
0.066 M	0.94 (± .05)	0.82 (± .04)	0.64 (± .03)
0.075 M	-	-	0.76 (± .04)
0.088 M	1.23 (± .06)	1.03 (± .05)	0.72 (± .04)
0.11 M	-	1.07 (± .05)	0.84 (± .04)
0.13 M	1.45 (± .07)	1.21 (± .06)	-
0.18 M	1.77 (± .09)	-	-

classic Langmuir-Hinshelwood kinetics, particularly for individuals who are not well versed in the details of this later method.

An example of the use of Michaelis-Menten kinetics in one investigation can, perhaps, illustrate the utility of this method. The platinum catalyzed oxidation of secondary alcohols to ketones using oxygen as the oxidant has been the subject of many investigations.[25] The reported mechanistic data suggest that the reaction proceeds by way of an initial dehydrogenation followed by the oxidation of the adsorbed hydrogen by the oxygen reactant as illustrated in Scheme 7.1. This reaction appears to be in the ping-pong category just discussed. The double reciprocal rate expression for this reaction is given by Eqn. 7.66 where $K_{M(alc)}$ and $K_{M(O_2)}$ are, respectively, the dissociation constants for the alcohol and oxygen adsorbed on the catalyst.

$$\frac{1}{v} = \frac{K_{M(Alc)}}{V_{max}} \frac{1}{[Alc]} + \frac{1}{V_{max}} \left(1 + \frac{K_{M(O_2)}}{[O_2]} \right) \tag{7.66}$$

The rates of 2-propanol oxidation were measured for reactions run over an 8% Pt/SiO_2 catalyst in water at 20°C. The initial rates of oxygen uptake (about 10% conversion) were determined at three fixed partial pressures of oxygen over a range of alcohol concentrations. The resulting data are listed in Table 7.1.

The plots of $1/v$ versus $1/[Alc]$ for the three partial pressures of oxygen are shown in Fig. 7.15. The three parallel lines are indicative of a ping-pong

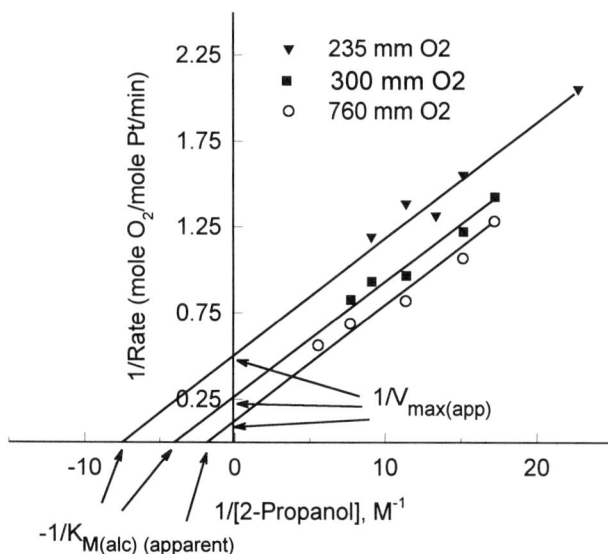

Fig. 7.15.   Double reciprocal plot of 2-propanol oxidations run under various partial pressures of oxygen.

mechanism and this supports the mechanistic proposal in Scheme 7.1. These lines have slopes of $K_{M(Alc)}/V_{max}$, y-axis intercepts of $1/V_{max(app)}$ and x-axis intercepts of $-1/_{K_{M(Alc)(app)}}$. Eqns. 7.67 and 7.68 give the linear forms for these factors.

$$\frac{1}{V_{max(app)}} = \frac{\left(1 + \dfrac{K_{M(O_2)}}{[O_2]}\right)}{V_{max}} = \frac{K_{M(O_2)}}{V_{max}}\frac{1}{[O_2]} + \frac{1}{V_{max}} \tag{7.67}$$

$$\frac{1}{K_{M(Alc)(app)}} = \frac{\left(1 + \dfrac{K_{M(O_2)}}{[O_2]}\right)}{K_{M(Alc)}} = \frac{K_{M(O_2)}}{K_{M(Alc)}}\frac{1}{[O_2]} + \frac{1}{K_{M(Alc)}} \tag{7.68}$$

The replot of $1/V_{max(app)}$ versus $1/[O_2]$ is shown in Fig. 7.16 and for $1/K_{M(Alc)(app)}$ versus $1/[O_2]$ in Fig. 7.17. In Fig. 7.16 the y-axis intercept gives a value of 27 min^{-1}(mole site)$^{-1}$ for V_{max} while the x-axis intercept shows $K_{M(O_2)}$ to be 0.0057M. The y-axis intercept in Fig. 7.17 gives a value for $K_{M(Alc)}$ of

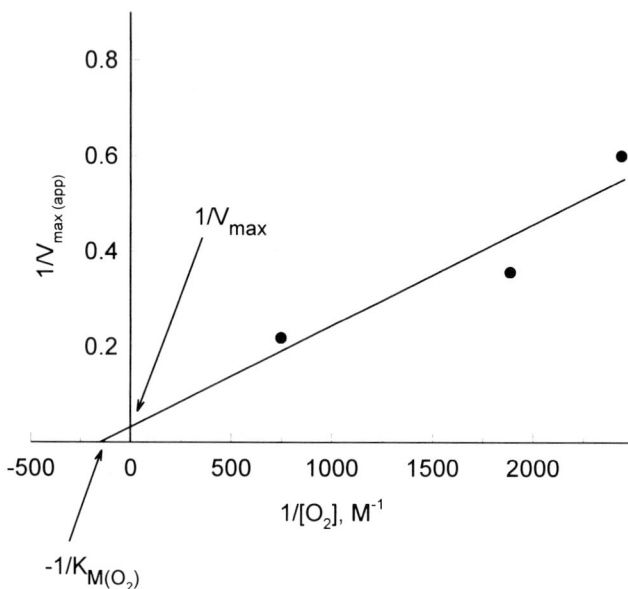

Fig. 7.16.   Replot of the y-axis intercepts in Fig. 7.15 versus $1/[O_2]$.

1.0M and from the x-axis intercept a value of 0.0043M for $K_{M(O_2)}$. One reason for this discrepancy in the value of $K_{M(O_2)}$ is that only three partial pressures of oxygen were used so the replots had rather poor correlation coefficients.

These K_M values show that oxygen is bound to the platinum catalyst two-hundred times more strongly than 2-propanol and that platinum is more easily saturated with oxygen than alcohol during the oxidation. This is one reason why the rate data were obtained at only three partial pressures of oxygen. The rates obtained at one atmosphere of oxygen were almost zero order in oxygen so the oxygen pressures that could practically be used for this study were limited.

The value for V_{max}, 27^{-1}(mole site)$^{-1}$, is the maximum rate for this reaction when it is run under saturation conditions for both substrates. In contrast to the single type of active site found in most enzymes, there are a number of different types of sites present on the surface of the platinum catalyst used for these oxidations. It was shown in a parallel study that 2-propanol oxidation takes place over the coordinately unsaturated corner atoms, that is, the single turnover (STO) characterized ^3M_I, ^3M_R and ^3MH sites (see Chapter 2). It was also shown that the specific site turnover frequencies (TOF) for these sites are 5.5, 7.9 and 5.0 moles O_2 uptake/mole site/minute respectively.

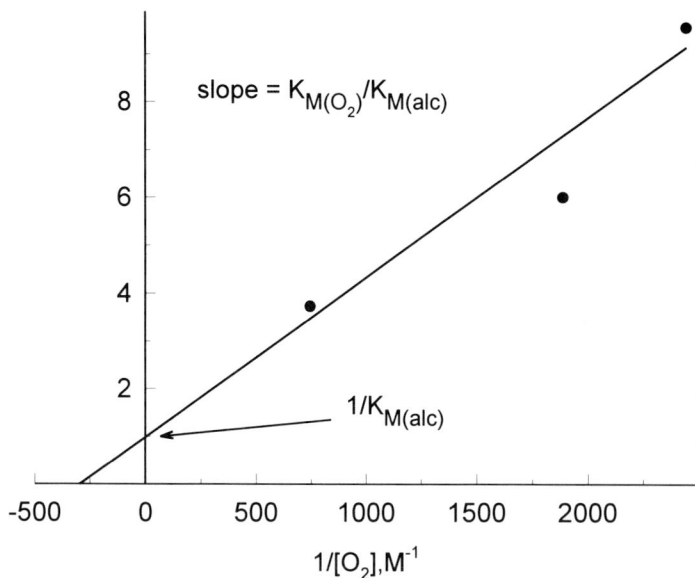

Fig. 7.17. Replot of the x-axis intercepts in Fig. 7.15 versus $1/[O_2]$.

The active sites on the 8% Pt/SiO_2 catalyst used for the kinetic investigation were composed of 44% ^3M_I and 28% each of ^3M_R and ^3MH sites. Combining these data gives an overall TOF for the reaction of about 6 $min^{-1}(moles\ site)^{-1}$. Considering that V_{max} is the maximum rate when the reaction is run under zero order conditions in both O_2 and alcohol and that in the present case it was near zero order in oxygen but first order in 2-propanol, this discrepancy between the TOF and V_{max} is not unreasonable.

References

1. G. C. Bond, *Catalysis by Metals,* Academic Press, New York, 1962.
2. G. C. Bond, *Heterogeneous Catalysis, Principles and Applications,* Oxford University Press, Oxford, 1987.
3. R. P. H. Gasser, *An Introduction to Chemisorption and Catalysis by Metals,* Clarindon Press, Oxford, 1985.
4. C. N. Satterfield, *Heterogeneous Catalysis in Practice,* McGraw Hill, New York, 1980.
5. M. Boudart and G. Djega-Mariadassou, *Kinetics of Heterogeneous Catalytic Reactions,* Princeton University Press, Princeton, 1984.
6. I. Langmuir, *Trans. Faraday Soc.,* **17,** 621 (1921).
7. K. J. Laidler, *Chemical Kinetics, 2nd Ed.,* McGraw Hill, New York, 1965.
8. J. M. Smith, *Chemical Engineering Kinetics, 2nd Ed.,* McGraw Hill, New York, 1970.

9. O. A. Hougen and K. M. Watson, *Chemical Process Principles, Part III: Kinetics and Catalysis,* Wiley, New York, 1947.
10. J. M. Thomas and W. J. Thomas, *Introduction to the Principles of Heterogeneous Catalysis,* Academic Press, New York, 1967.
11. I. H. Segal, *Enzyme Kinetics,* Wiley, New York, 1975.
12. A. Fersht, *Enzyme Structure and Mechanism, 2nd Ed.,* Freeman, New York, 1985.
13. A. Cornish-Bowden, *Principles of Enzyme Kinetics,* Butterworths, London, 1976.
14. W. W. Cleland, *The Enzymes,* **2**, 1 (1970).
15. K. Dalziel, *The Enzymes,* **10**, 2 (1975).
16. Ref. 11, pp 46-48.
17. Ref. 11, pp 100-125.
18. Ref. 11, pp 125-136.
19. Ref. 11, pp 136-143.
20. Ref. 11, pp 170 ff.
21. Ref. 11, p 273.
22. Ref. 11, pp 273-320.
23. Ref. 11, pp320 ff.
24. Ref. 11, pp 606 ff.
25. R. L. Augustine and L. K. Doyle, *J. Catal.,* **141**, 58 (1993).

SECTION TWO

The Catalyst

8

The Catalyst

Of all the reaction parameters involved in a heterogeneously catalyzed process, the selection of the catalyst is probably the most critical in determining the outcome of a particular reaction. The chapters in this section provide the reader with the information needed to make the selection of the catalyst more efficient. Most of the catalysts used for synthetic reactions are composed of the catalytically active species dispersed on a supposedly inert support material. The nature of the support can influence both catalyst stability and activity. It can also define the procedure best applicable for the preparation of a specific catalyst. Because of this, the commonly used supports are discussed in Chapter 9 to provide the background needed in the discussion of supported catalysts presented in the following chapters.

Heterogeneous catalysts can be either oxides or metals. The various types of oxide catalysts are outlined in Fig. 8.1 and discussed in detail in Chapter 10. These oxide catalysts are used primarily as either solid acids or bases or as promoters for oxidation reactions.

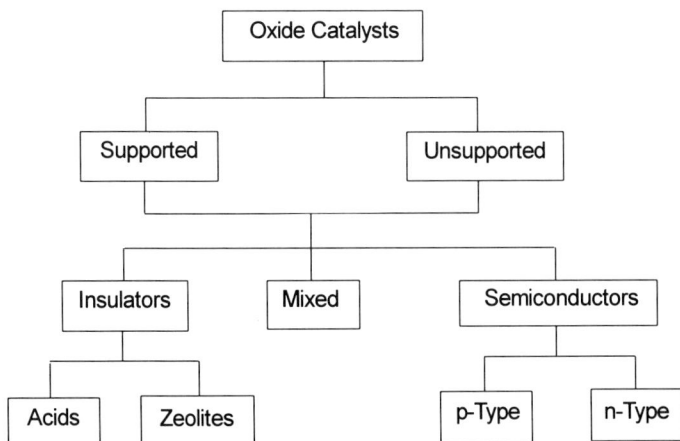

Fig. 8.1   Forms of oxide catalysts.

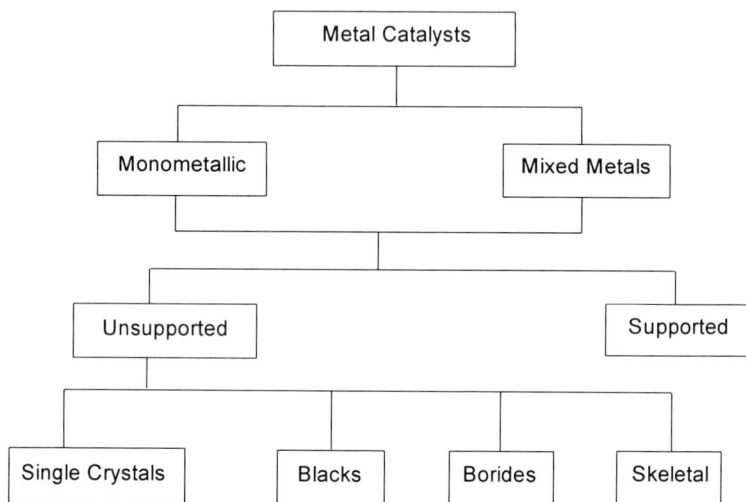

Fig. 8.2.   Forms of metal catalysts.

Most synthetically useful catalytic processes are run over metal catalysts which, as outlined in Fig. 8.2, can be composed of a single metallic component or a mixture of metals. Either of these types can be supported or unsupported. Metal catalysts are used primarily for hydrogenations, hydrogenolyses, isomerizations and oxidative dehydrogenations. They are rather easily prepared in a pure form and can be characterized without too much difficulty. Because of this, metal catalysts are generally preferred for basic research. Such materials have been used to obtain almost all of the fundamental information on which the various theories of catalysis have been derived. A general discussion of catalytically active metals and the factors influencing their activity is presented in Chapter 11 while Chapter 12 deals with the preparation and properties of the various types of unsupported or bulk metal catalysts. The preparation and properties of the supported metal catalysts are presented in Chapter 13.

While the literature concerned with catalyst preparation is extensive, there are several monographs that deserve special mention. These are the proceedings of the five conferences held since 1976 which were devoted to the factors involved in the preparation of catalysts, both oxides and metals.[1-5]

References

1.   *Studies in Surface Science and Catalysis*, **1**, (Preparation of Catalysts) (B. Delmon, P. Jacobs and G. Poncelet, Eds.) Elsiever, Amsterdam, 1976.

2. *Studies in Surface Science and Catalysis*, **3**, (Preparation of Catalysts, II) (B. Delmon, P. Grange, P. Jacobs and G. Poncelet, Eds.) Elsiever, Amsterdam, 1979.
3. *Studies in Surface Science and Catalysis*, **16**, (Preparation of Catalysts, III) (G. Poncelet, P. Grange and P. A. Jacobs, Eds.) Elsiever, Amsterdam, 1983.
4. *Studies in Surface Science and Catalysis*, **31**, (Preparation of Catalysts, IV) (B. Delmon, P. Grange, P. A. Jacobs and G. Poncelet, Eds.) Elsiever, Amsterdam, 1987.
5. *Studies in Surface Science and Catalysis*, **63**, (Preparation of Catalysts, V) (G. Poncelet, P. A. Jacobs, P. Grange and B. Delmon, Eds.) Elsiever, Amsterdam, 1991.

Catalyst Supports

The active sites on a heterogeneous catalyst are found on its surface. The most efficient catalysts usually have a large catalytically active surface area exposed to the reaction medium. One way of maximizing the active surface of a catalyst is for it to be present as a very fine powder. However, heating powdered catalysts usually results in sintering or the agglomeration of the small particulates into larger, less efficient, entities. This process is facilitated with powdered materials since, as shown in Fig. 9.1a, the individual particulates are in close contact with their neighbors. Heat promotes the coalescence into increasingly larger entities and a corresponding loss of active surface area.[1] The extent of agglomeration is a

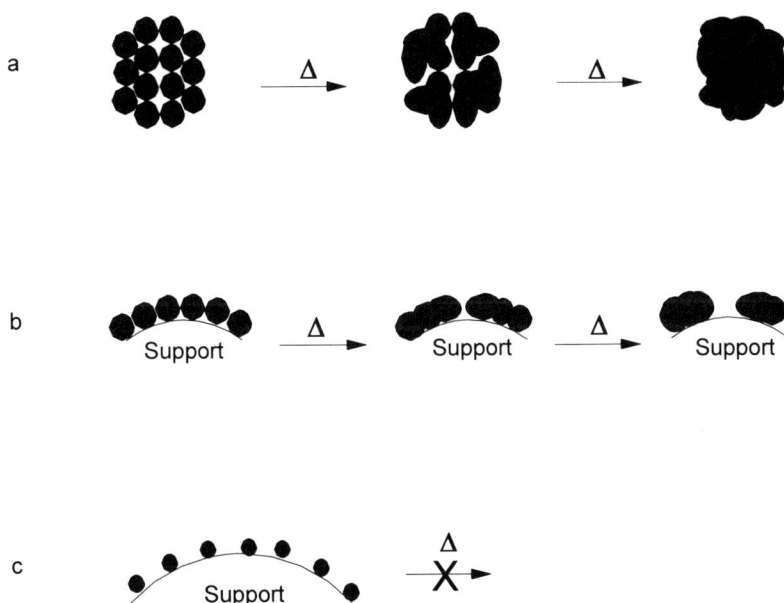

Fig. 9.1. Effect of catalyst load and the presence of a support on the sintering of a catalyst. a) Unsupported powdered catalyst; b) high load supported catalyst; c) low load supported catalyst.

function of the particular catalyst, the temperature and the time the catalyst is exposed to the heat. For most catalytically active metal powders sintering can be observed at temperatures as low as 100°C but oxide catalysts can generally be used at higher temperatures without a significant loss of active surface area.

If the catalytically active particulates can be kept separate from each other thermal agglomeration can be eliminated or, at least, minimized. The most effective way of accomplishing this is to attach the catalyst species to the surface of a thermostable support material as shown in Fig. 9.1c. For comparison, a powdered platinum black catalyst can sinter at temperatures as low as 100°C but a supported catalyst, such as Pt/Al_2O_3, does not usually sinter at temperatures under 450–500°C,

The environment of the dispersed catalyst particulates in these supported catalysts is influenced by a number of factors, primarily the relative amount of catalytically active material present, the nature of the support and the strength of the interaction between the support and the catalyst particles. Metal catalysts having a low metal load are usually composed of very small crystallites widely scattered over the surface of the support (Fig. 9.1c). Such crystallites generally have a high dispersion so the metal is efficiently used. As the metal load increases the crystallites are closer together and, also, frequently larger in size (Fig. 9.1b). The further apart the crystallites, the more resistant they are to thermal sintering. Obviously, if the catalyst crystallites are in contact with each other, thermal coalescence can be expected to take place with a resulting loss of active surface area. However, as shown in Fig. 9.1b, the presence of the support can limit the size of the sintered particles.

A second sintering mechanism, which applies primarily to metals, involves the interaction of the metal with some species in the reaction medium to form a surface intermediate that is sufficiently volatile that it can migrate from one particle to another. Since smaller metal particles have a higher percentage of their component atoms on the surface than larger ones, the formation of a volatile surface species will be favored on the smaller crystallites and the result will be the growth of the larger particulates and the loss of the smaller ones. Halide ions frequently promote sintering by this mechanism so every effort should be made to remove them from the catalyst when metal halides are used in the preparation procedure. A strongly reducing atmosphere can also facilitate metal sintering at high temperatures[2] but this can be minimized by running the reduction at a lower temperature, if possible.

The objective in the preparation of supported catalysts is to have the catalytically active crystallites separated. When this is the case the only way sintering can occur is if the catalyst particulates migrate across the support surface or if there is a vapor phase sintering promoter present in the reaction medium. The lower the catalyst load and the higher the support surface area, the less likely that sintering will take place. The migration ability of the catalytically active species depends primarily on the strength with which it is bonded to the support. If there is a weak interaction the catalyst particles can move across the support surface and sintering may occur at temperatures only slightly higher than those needed to coalesce the unsupported material. On the other hand, if the interaction is very strong it approaches chemical compound formation on the support surface with a possible loss in catalytic activity. With most supported catalysts the support interaction is, ideally, somewhere between these extremes. A more complete discussion of the catalyst support interaction is found at the end of this chapter.

While the low metal load, highly dispersed catalysts provide a maximum surface area per unit weight of metal, those catalysts having larger catalyst particulates spread over most of the support surface provide a maximum metal surface area per volume of catalyst.[1] Generally the more expensive precious metal catalysts have low metal loads and are highly dispersed while catalysts containing the less expensive base metals have higher metal loadings, usually 20–40% or higher. The relationship between the catalyst load and crystallite size or catalyst dispersion is dependent on a number of factors, of which one of the more important is the surface area of the support.

The support can also be used to absorb small quantities of materials that are toxic to the catalyst and, thus, prevent them from interacting with the catalytically active surface. Activated carbon supports are particularly effective in this regard.

There are a number of physical characteristics of the support that are important for the proper performance of the supported catalyst.[3] These are hardness, density, pore volume, pore size, pore distribution, particle size and particle shape. The surface area is directly related to pore size, distribution and volume. Support particles must be sufficiently hard to withstand the abrasion associated with the type of process in which they are to be used so they will not be broken down into smaller particles and thus change the operating conditions of the system. In flow processes using a fixed bed reactor the catalyst particles must be able to withstand the pressure and vibration induced by the flow of the reaction medium. The breakdown of the catalyst into finer particles can have a significant impact on the flow characteristics of the reactor. Reactions run in a fixed bed reactor usually require larger supported catalyst pellets to keep the pressure drop over the catalyst bed at an acceptable level. Reactions run in a batch mode using a stirred autoclave generally use catalysts composed of small support particles so the catalyst can be efficiently suspended in the reaction medium by the action of the stirrer. The density of the catalyst must be such as to provide the optimum suspension in the reaction medium. In a stirred reactor, particle breakdown can change the suspension characteristics of the catalyst and also increase the difficulty of subsequent catalyst separation from the reaction medium.

Generally, one considers that the maximum surface area for the support or catalyst is the best possible arrangement but this is not always true. First it must be recognized that the surface area is not only associated with the external surface of the particle but also with the surface of all the pores within the particles. A particle having small diameter pores will have a higher surface area than one having larger diameter pores with both having the same total pore volume. For vapor phase reactions involving small molecules, catalyst particles having a large number of smaller pores and higher surface areas are preferred since the multitude of pores provides access of the smaller reactant molecules to the catalytically active sites present throughout the catalyst particles. Reactions of synthetic interest, however, generally involve larger molecules and are usually run in a liquid medium. For such reactions, catalysts having the smaller pores are inefficient since the diffusion restraints would severely hinder the reaction. Catalysts for these reactions, then, should have larger pores with the catalytically active sites located near the surface of the support particles so diffusion of the reactants to the active sites will not be a significant factor in the reaction.

Since all of these characteristics are related it is necessary to balance one against the other to optimize the performance of the catalyst. Highly porous catalysts will have a low density and be relatively easy to break down into smaller particles. As the hardness and abrasion resistance of the particle increases, though, the porous nature decreases so some balance must be sought to optimize the catalyst for a specific application.[3]

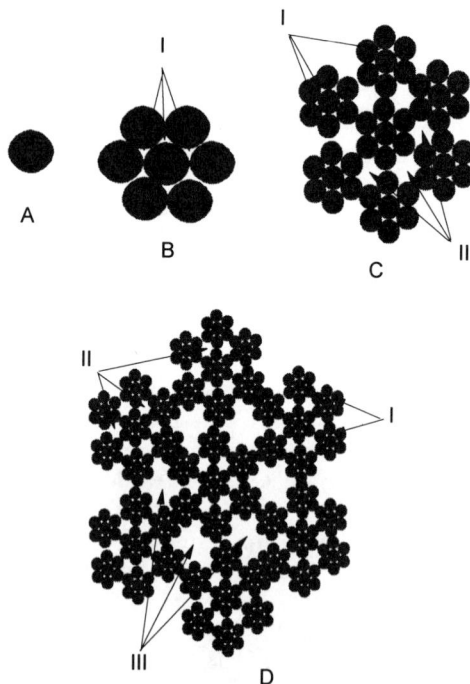

Fig. 9.2.   Pore development in the agglomeration of the basic unit, A, into progressively larger particles.

It is obvious from this discussion that control of pore size and distribution is important in developing the optimum catalyst for a specific process. The most common perception of the pores in a catalyst particle is that they are essentially holes of various diameters drilled into the particle at various angles or, perhaps, they resemble various sized tunnels randomly oriented throughout the particle. While such representations may be helpful in discussing the effects of pore diffusion and similar catalytic factors, it does not provide any basis for the development of methods that can produce supports or supported catalysts having specific pore characteristics.

For this purpose it is best to understand the development of the porous structure of a given solid particle. All solids are built up from individual units that can be molecules, a unit crystal or, even, a small densely packed arrangement of molecules or unit crystals. This basic building block is denoted as A in Fig. 9.2. The coalescence of several of these units gives a larger system, B, with small interces between the basic units. These are designated as Type I pores. Coalescence of several B units forms a larger system, C, in which larger, Type II, pores are also present. Further agglomeration of C units gives D in which the Type III pores are even larger. [3]

Pore size and distribution can be regulated by controlling the size of the initial building block, A, and the size of the final particle, such as D in Fig. 9.2. The larger the A particle is, the larger the Type I pores are. If A is large enough and the final particle size is held to B, then a narrow pore size distribution will result. The problem is to have a sufficiently large initial species, A, so that B will be reasonably large to be useful as a support. This is sometimes accomplished using a colloidal particle, such as silica gel, as the initial building block. Such colloidal species can have small interces as well but they are of atomic scale and relatively unimportant in catalysis. A more common system is represented by C in which there is a bimodal distribution of pores, that is, small Type I and larger Type II.

Not only does the size of the basic building block effect the final pore sizes but the shape of the particle is also important. Large, irregularly shaped particles will pack loosely on agglomeration with the production of large pores. If small particles are also present they may pack into the voids thus reducing the porosity of the material considerably.[4] If the basic building block, A, or any of the other larger systems do not coalesce in a uniform, close packed manner, the pore size will be dependent on the extent of linear "polymerization" and branching as depicted in Fig. 9.3.[5]

Another method of controlling pore size distribution is to incorporate an additive into the system during the preparation of the support and then to remove it selectively after the support particles have been formed. Common materials used for this purpose are pieces of cellulose, cotton, or other carbonaceous material that can be removed by ignition to leave behind voids corresponding to the shape of the added material.[4]

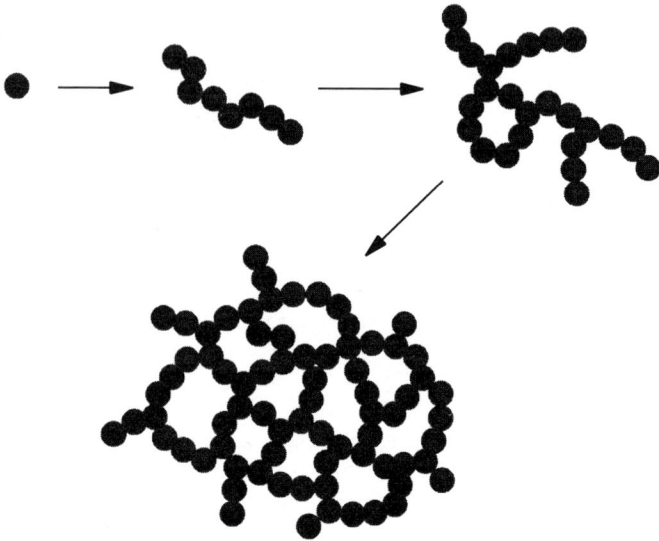

Fig. 9.3.   Pore formation in a solid formed by the linear polymerization of
the basic component.

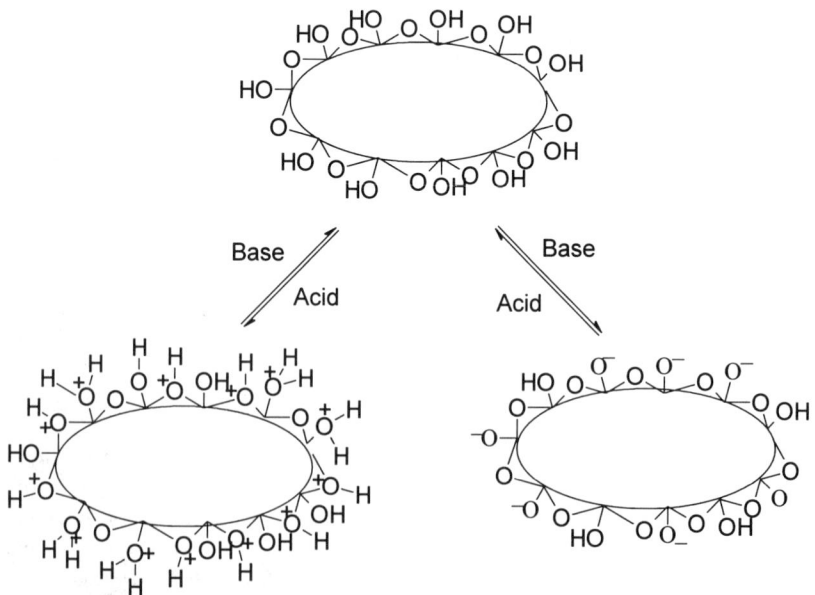

Fig. 9.4.   Formation of charged surfaces on an oxide support.

Table 9.1

Isoelectric points of various oxides.[6]

Type	Oxide	IEP	Adsorbents
Acidic	Sb_2O_3 WO_3 SiO_2	<0.4 <0.5 1.0 - 2.0	Cations
Amphoteric	MnO_2 SnO_2 TiO_2 $\gamma\, Fe_2O_3$ ZrO_2 CeO_2 $\alpha,\, \gamma\, Al_2O_3$	3.9 - 4.5 ~5.5 ~6 6.5 - 6.9 ~6.7 ~6.75 7.0-9.0	Cations or Anions
Basic	Y_2O_3 $\alpha\, Fe_2O_3$ ZnO La_2O_3 MgO	~8.9 8.4 - 9.0 8.7 - 9.7 ~10.4 12.1 - 12.7	Anions

Another critical factor in the preparation of supported catalysts is the process used to incorporate the catalytically active species or its precurser onto the support. As discussed in more detail in subsequent chapters, this can be accomplished in several ways but generally there is some interaction between the functional groups on the surface of the support and the material from which the catalytically active species is eventually derived.

The adsorption character of a support material is governed by the nature of its surface functionality. Most of the common support materials are inorganic oxides that have various forms of **M–O–M**, oxide, and **M–OH**, hydroxy groups on their surfaces. The most notable exception is carbon which has carboxylic acids and phenolic hydroxy groups on the surface. The interaction with the catalyst precurser, then, involves these surface functionalities usually by way of a ligand or ion exchange.

The surface oxygen containing groups of an oxide can react with either acid or base so the nature of the charge on the oxide surface depends on the pH of the solution around the particle. As depicted in Fig. 9.4[6] an acidic medium protonates the surface oxygens to give a positively charged surface. In base a negatively charged surface is produced. Somewhere between these extremes there is a pH value at which the overall charge on the particle is zero. This value is characteristic of the oxide and is referred to as its zero point of charge (ZPC) or its isoelectric point (IEP). The more acidic oxides, such as silica, have low IEPs indicating a facile production of an anionic surface. Basic oxides, such as magnesia, have high IEPs and are easily protonated to give a cationic surface. Oxides with intermediate IEPs are amphoteric and can have a positive or a negative surface depending on the reaction medium. Table 9.1 lists the IEP values for several oxides.[6]

When the surface has a negative charge, cationic species are attracted to it and become adsorbed. A positive surface interacts with negative species. However, information other than the IEP is frequently needed to maximize the adsorption of the catalyst precurser. Electrophoresis provides a method of measuring the velocity of a charged particle in an electric field. The zeta (ζ) potential is a measure of this electrophoretic velocity with respect to the potential applied to the particles. As an example, a plot of the zeta potential versus pH for aluminum hydroxide is shown in Fig. 9.5.[6] The IEP of this material corresponds to the pH at zero potential. This plot shows that aluminum hydroxide is amphoteric with an IEP of about 7.5. It can adsorb either anionic or cationic species depending on the pH of the adsorbing solution. Anionic precursors are best adsorbed at a pH lower than 4 and cationic precursors at a pH greater than 9.

The types of solids that have been used as catalyst supports are many and varied. Some are listed in Table 9.1.[6] The most common, however, are alumina, silica, and carbon, with titania assuming some importance over the past several years.

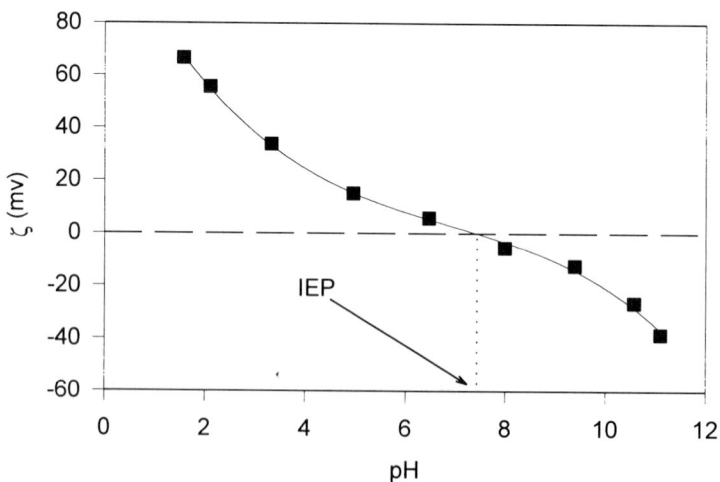

Fig. 9.5.   Effect of pH on the zeta potential (ζ) of Al(OH)$_3$.
(Redrawn using data from Ref. 6.)

Alumina

There are a number of materials that have been referred to as aluminas.[7-9] These species differ from each other in their chemical composition and crystalline structure.   There are the trihydroxides, $Al(OH)_3$, of which the two crystalline forms, Gibbsite and Bayerite, are the most common. Loss of a molecule of water leads to the oxyhydroxides, $AlO(OH)$, boehmite, pseudo (ψ) boehmite and diaspore which differ from each other in their crystalline arrangement. Further dehydration leads to the transition aluminas that have the generic formula, $Al_2O_3 \cdot xH_2O$ with $0 < x < 1$.   Complete dehydration gives corundum or α alumina. The relationship and approximate temperatures of transition between these species are shown in Fig. 9.6.[7-9] These materials can have surface areas that range from 0.1 to 400 m^2/g, pore volumes between 0.1 and 1.5 cm^3/g and average pore sizes between 2 nm and 170 μm. It is little wonder why "alumina" has been used so extensively as a catalyst support since one can prepare an alumina having the surface characteristics needed for almost any process.

The trihydroxides are made by treating solutions of aluminum salts with base or by the hydrolysis of aluminum alkyls.[9] They are usually voluminous, amorphous gels which, on drying, give highly porous materials with high surface areas. Heating to 120°C causes the loss of physically adsorbed water. This frequently results in a destruction of the pore structure caused by the rupture of the pore walls by the heated water. If the hydroxide is suspended in a low

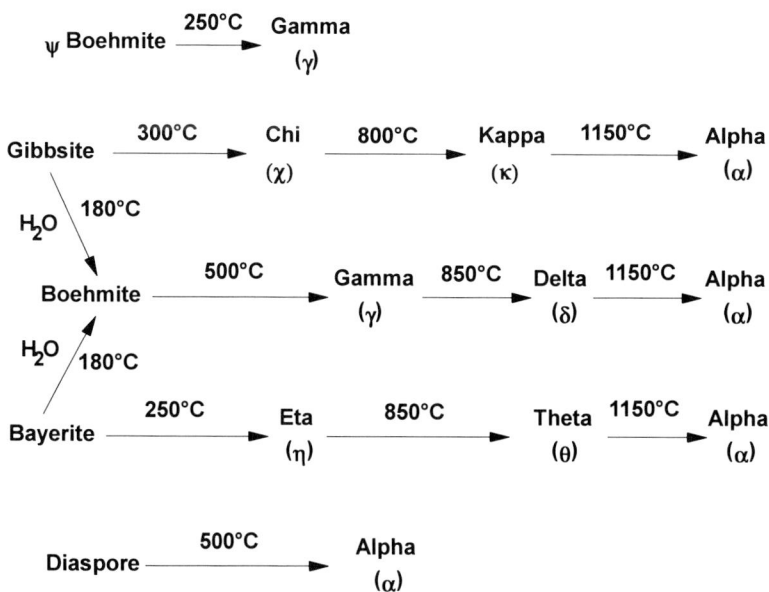

Fig. 9.6.   Relationship between the various "aluminas".

molecular weight alcohol, the water in the hydroxide is replaced by the alcohol. Evaporation of the alcohol does not cause the rupture of the pore walls so the porosity and surface area of the dried species are higher than those obtained by drying the aqueous material. Further heating results in the loss of water formed by conversion of adjacent surface hydroxy groups to a surface oxide (Eqn. 9.1). This begins at about 300°C and continues until almost no surface hydroxy groups are present, a process that requires heating to near 1200°C.

(9.1)

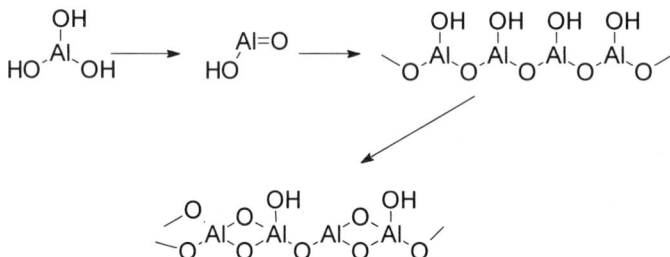

The most common catalyst supports are the transitional aluminas, particularly γ alumina, which is best prepared by heating ψ boehmite.[9] This process gives a γ alumina having a surface area between 150–300 m^2/g, a pore volume between 0.5–1 cm^3/g and a large number of pores in the 3–12 nm range. The γ aluminas prepared from other sources have significantly lower surface areas and pore volumes.[9] In contrast, α alumina, the most dehydrated form of alumina, is essentially non-porous with surface areas between 0.1 and 5 m^2/g.[8]

The hydroxy oxides, especially ψ boehmite, have excellent coalescing properties so they are particularly good for forming large support pellets and granules. They are also used as "wash coats" to provide porous surfaces for catalyst adhesion to non-porous materials such as metals and ceramics.

The data presented in Fig. 9.5 shows the IEP of Gibbsite, $Al(OH)_3$, to be near 7.5.[6] The relationships between the zeta potential and pH for γ alumina[10] and α alumina[11] are shown in Fig. 9.7a and b, respectively. Comparison of Fig. 9.7 with Fig. 9.5 shows that the IEP of the aluminas is influenced by their degree of surface hydroxylation. As the number of surface hydroxy groups decreases the IEP increases as do the pH values for optimum cation and anion interaction. On going from Gibbsite to γ alumina the extent of surface hydroxylation decreases significantly, going from essentially fully hydroxylated to a species having about 8–12 OH groups per nm^2.[12] The IEP changes from about 7.5 for Gibbsite to about 8.75 for γ alumina. Anion interaction with γ alumina can take place at pH values as high as 7 but cation interactions require a pH of 10 or greater. The almost fully dehydrated α alumina has an even higher IEP, about 9.5. It can interact with anions at a pH near 8 but cationic interactions need a pH of 10.5 or greater.

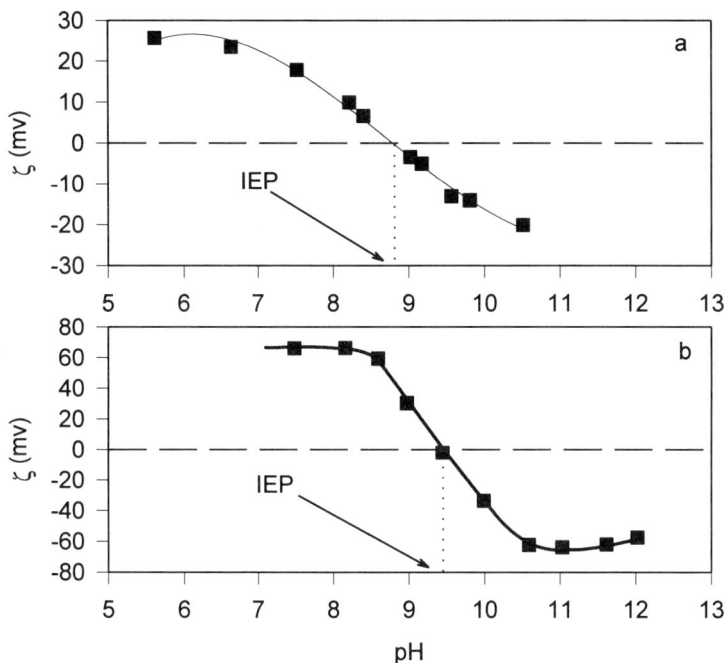

Fig. 9.7.   Effect of pH on the zeta potential (ζ) of a) γ alumina[10]
and b) α alumina. [Redrawn using data from Ref. 10 (a)
and Ref. 11 (b).]

The IEPs of aluminas, however, can be changed by the addition of modifiers. The addition of cations such as Li^+ and Na^+ results in an increase in the IEP while the presence of an anion such as F^- significantly decreases the IEP. Thus, adsorption of Li^+ or Na^+ enlarges the pH range in which negative species can be adsorbed on the surface.[13] Fortunately, only a relatively small amount of these cations is needed since higher concentrations could lead to catalyst deactivation. The addition of a small amount of F^- as the modifier extends the pH range for adsorption of positive species and increases the extent of adsorption of such moieties at pH values lower than normally needed.[13] Another way of changing the IEP of alumina without the introduction of a modifier that could also have an effect on the resulting catalyst is to increase the temperature of the alumina suspension used to interact with the catalyst precurser. Increasing the temperature of the alumina suspension to 50°C results in an increase of the IEP to facilitate the adsorption of negatively charged species. Cooling the suspension to 10°C extends the pH range for positive ion adsorption.[13]

Because of its amorphous nature, alumina can react with and dissolve in strongly acidic or basic reaction media.

Silica

Silica is produced by the reaction of an alkali metasilicate with an acid. This is commonly done by bubbling carbon dioxide through a dilute solution of sodium metasilicate (water glass) or by the addition of dilute acid to a pH of about 7.[14,15] Typically the solution will gel and the resulting material is washed free of the Na^+ and then dried. The silanol groups condense to form polymeric siloxanes giving an amorphous material made up primarily of nearly spherical particles composed of non-ordered SiO_4 tetrahedra.[16] Heating the gel to about 100°C removes only the physically adsorbed water giving the material commonly referred to as silica gel. Further heating to 200°C results in some dehydration of the surface hydroxy groups but their number still approximates one per surface silicon atom. Exposure to hot water rehydrates the surface producing gem dihydroxy silicon atoms (Eqn. 9.2). Heating to 500°C decreases the surface hydroxide concentration to about 20–30% of that present on the 200°C heated material. Exposure to temperatures of 1000°C reduces the surface hydroxides to about half of that found on heating to 500°C.[16]

(9.2)

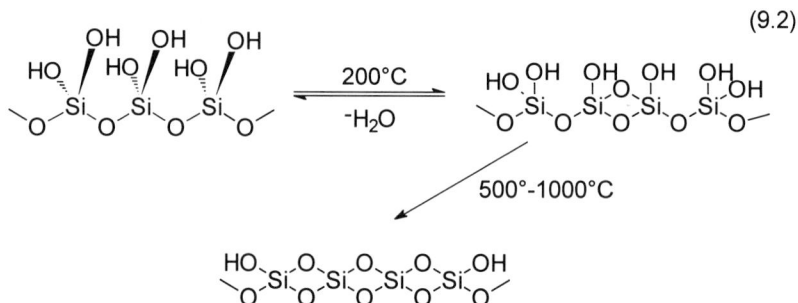

As with alumina, pore disruption caused by the evaporation of the water at temperatures near 120°C can take place on drying the gel. If the gel is first taken up in an alcohol to replace the internal water, the pore shrinkage is minimized. Evaporation of the aqueous material gives gels with surface areas of about 200–800 m^2/g while those obtained from alcoholic suspensions have surface areas of about 500–800 m^2/g. Rapid drying of the gels promotes an increase in the pore volume and average pore diameter with a broadened pore size distribution. This effect is enhanced when critical phase drying is used.[15]

A colloidal silica with 10–500 nm particles can be prepared by the hydrolysis of tetraethoxysilane with aqueous ammonia. The particle morphology appears to be independent of the chemical properties of the silane but depends on the nature of the reaction medium. Large, irregularly shaped particles are produced at lower temperatures using solutions having a high ammonia concentration. Smaller, spherical particles result when the tetraethoxysilane concentration is low.[17]

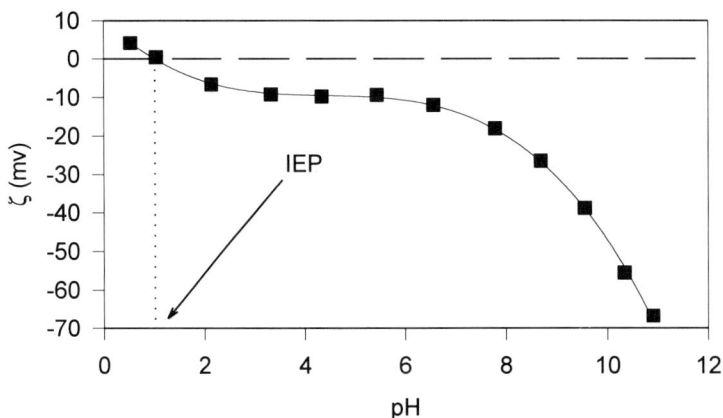

Fig. 9.8.   Effect of pH on the zeta potential (ζ) of SiO_2. (Redrawn using
data from Ref. 6.)

A finely divided non-porous silica is produced by the flame hydrolysis of $SiCl_4$. The resulting material, known as Cabosil or Aerosil is produced as 5–40 nm particles with surface areas between 50 and 400 m^2/g. This method of preparation produces a very pure form of silica that is frequently favored for basic research.[16]

Another common form of silica is diatomaceous earth, also known as kieselguhr.[14,16] This is a naturally occurring material composed of fossilized diatoms. It contains 70–90% SiO_2, has 0.2–0.7 μm pores and a surface area of about 15–40 m^2/g. It is a common support for nickel catalysts used in large scale reactions.

The IEP for silica is about pH 1, indicating the acidic nature of this material. The zeta potential versus pH curve for silica is given in Fig. 9.8.[6] This curve shows that while the surface becomes negative at low pH values the surface charge is significant only at a pH above 7. Thus, silica can only adsorb cations and only when in a solution having a pH greater than 7.

Silica is more resistant to acidic media than is alumina but it is adversely affected by a strongly alkaline environment.[14]

Titania

The titania used for catalysis is commonly prepared by the aqueous hydrolysis of titanium salts or by the flame hydrolysis of $TiCl_4$.[14-18] It is found in two crystalline forms, rutile and anatase, with the form produced depending on the temperature and preparation procedure used. Anatase is more stable at the temperatures normally used for catalytic processes and, thus, is the more common support. The anatase prepared by flame hydrolysis of $TiCl_4$ has surface areas near 40–80 m^2/g and mean pore diameters of about 50 nm. That prepared by aqueous hydrolytic procedures generally has a somewhat lower surface area. Rutile usually has a lower surface area than does anatase.

The titania obtained by the acid hydrolysis of tetrabutoxytitanium has surface areas of 190–200 m^2/g and average pore diameters of 15–20 nm.[19] Even higher surface areas can be obtained using supercritical drying.[20]

There are about 4–5 surface hydroxy groups per square nanometer on anatase and these are of two different types.[18] One type is basic in nature and can be exchanged by F$^-$. This type of site is responsible for the adsorption of acidic species onto the surface and initiates the adsorption of base sensitive compounds such as MoO$_2$(acac)$_2$. The second type of surface hydroxide is acidic. It interacts strongly with base and initiates the adsorption of acid sensitive compounds such as Fe(acac)$_3$.[21] This amphoteric nature of titania is illustrated by the zeta potential versus pH curve shown in Fig. 9.9.[6] Cationic catalyst precursers are adsorbed at pH values greater than 7–8 and anionic species from solutions having a pH less than about 5.

Titania is different from silica and alumina in two significant ways. While silica and alumina are insulator oxides, titania is a semiconductor, a factor that makes it useful for heterogeneous photocatalytic reactions. Secondly, the surface Ti^{+4} species are readily reduced to Ti^{+3} in the presence of appropriate reducing agents. One such agent is the hydrogen adsorbed on the catalytically active metal crystallites on the titania surface. This hydrogen can spillover into the support and reduce the Ti^{+4} species adjacent to the metal. This reduction causes a significant change in the character of the support. The effect this has on the catalytic activity of the metal is discussed later in this chapter as a strong metal-support interaction (SMSI).

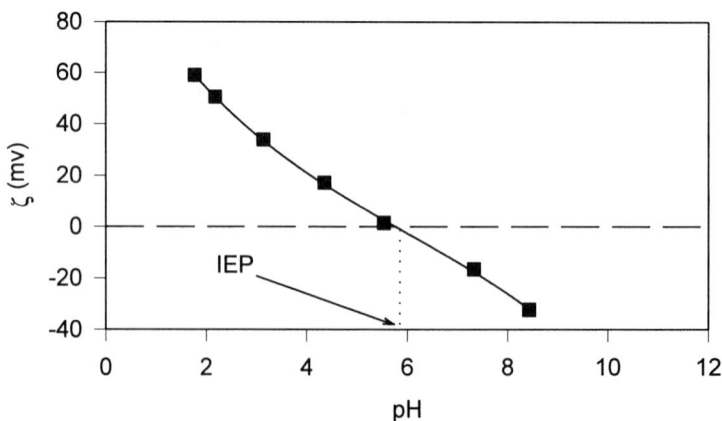

Fig. 9.9.   Effect of pH on the zeta potential (ζ) of TiO$_2$.
(Redrawn using data from Ref. 6.)

Carbon

The common carbon containing catalyst supports are carbon black, charcoal and graphite. Blacks are produced by the thermal dehydrogenation of hydrocarbons, primarily methane. They have surface areas in the 70–250 m^2/g range with no well developed pore structure.[22] Charcoals are produced by the pyrolysis of natural or synthetic organic polymers, a process accomplished by heating the material to 700–900°C in the absence of air. These materials are then activated. The so-called physical activation requires treatment with carbon dioxide and/or steam at 400–600°C. In this process some of the carbon is burned away to give a pore structure to the charcoal. The degree to which this is developed depends on the raw material used and the temperature and time of activation. A second activation procedure, chemical activation, involves the impregnation of the raw material with the activating agent prior to pyrolysis. The most common activating agents are zinc chloride or phosphoric acid. In this case the activation takes place during carbonization with most of the activating agent still incorporated into the carbon structure. Washing out this material produces the pore structure of the charcoal. The surface areas of the charcoals also depends on the type of material being pyrolyzed. Some of the most common are listed in Table 9.2.[22] The pore size distributions in these charcoals varies widely.

Table 9.2

Surface areas of different types of carbon supports.[22]

Carbon	Surface Area (m^2/g)
Graphite, natural	0.1 - 20
Graphite, synthetic	0.3 - 25
Graphitized carbon blacks	20 - 100
Carbon blacks	70 - 250
Activated carbons	
Wood charcoal	300 - 900
Peat charcoal	350 - 1000
Coal carbon	300 - 1000
Nutshell charcoals	700 - 1500

During the carbonization process some graphite formation also takes place. Further heating to 1000°C can increase the amount of graphite and produces the graphitized charcoals. The extent of graphitization depends on the nature of the raw material. For simplicity, these materials can be divided into two broad categories; those which are composed primarily of cellulose and those containing mainly lignin. Since cellulose does not melt on carbonization it is more difficult to convert to graphite. The cellulosic materials give low graphite charcoals.[22]   Materials, such as nutshells, which contain more lignin than cellulose, give a more graphitic carbon on carbonization and activation.[23]   The more graphite present, the more stable the carbon is to oxidation.

The functional groups present in charcoal are phenols, carboxylic acids, quinones, ketones and lactones. They are essentially acidic supports. The nature and extent of the functionalities on the charcoal particle surface are a function of the material used in the carbonization and the type and duration of the activation procedure. In addition, treatment of these charcoals with oxidizing agents such as nitric acid or hydrogen peroxide increases the number of acid species present.[24] A similar treatment will also functionalize the non-porous carbon blacks.[25] Because of this it is difficult to draw any general conclusions concerning the adsorption capabilities of these charcoals other than to say that being acidic they will most readily adsorb cationic species.

A common charcoal is Norit, frequently used as a decolorizing agent in the synthetic laboratory. A plot of the zeta potential of Norit versus pH is shown in Fig. 9.10.[26] This charcoal has an IEP at about pH 4.5 and can adsorb cations from solutions having a pH between 6–8.

In contrast to carbon black and charcoal, graphite is a highly crystalline material composed of stacked planes of aromatic rings 0.335 nm apart (Fig.

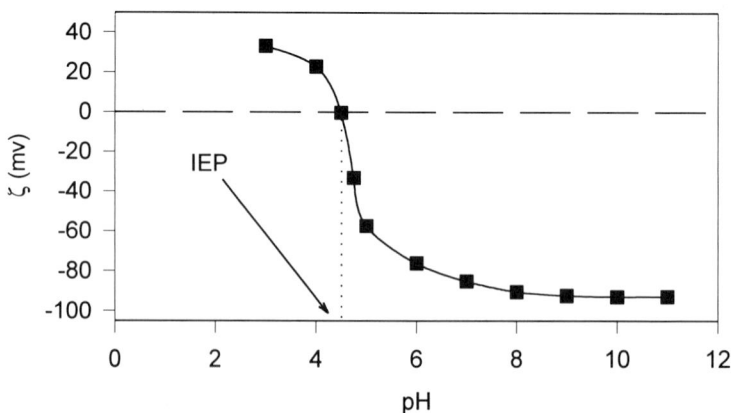

Fig. 9.10.   Effect of pH on the zeta potential (ζ) of Norit carbon.
(Redrawn using data from Ref. 26.)

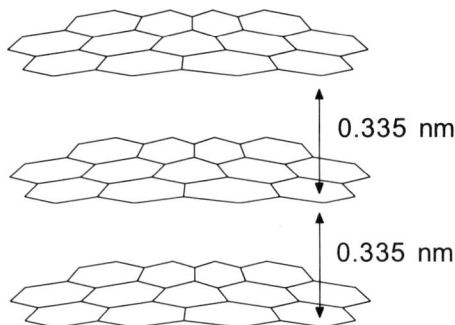

Fig. 9.11. Planar structure of graphite.

9.11).[22] Except for this interplanar area, graphite particles are essentially non-porous and thus, graphite is not commonly used as a catalyst support. If, however, the graphite is subjected to strong oxidizing agents such as nitric acid, hydrogen peroxide or sodium hypochlorite, the edges of these planes are oxidized to give acidic functional groups. These modified graphites can then be used to adsorb cations from neutral to moderately basic media.[25]

The electronic character of the stacked aromatic systems makes it possible for atoms or molecules to slip between these layers, either accepting or donating electrons to bond with the carbon system, a process referred to as intercalation. Platinum metal graphite intercalates have been prepared by reducing platinum halide graphite intercalates. These catalysts are particularly useful for the selective hydrogenation of acetylenes to *cis* alkenes (Eqn. 9.3).[22]

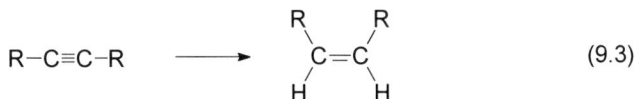

$$R-C{\equiv}C-R \longrightarrow \overset{\displaystyle R}{\underset{\displaystyle H}{}}C{=}C\overset{\displaystyle R}{\underset{\displaystyle H}{}} \qquad (9.3)$$

Catalyst-Support Interactions

Initially, the support material was thought to be inert and to serve simply as a vehicle for keeping the catalytically active species separated and, thus, minimize sintering. This is accomplished because catalyst crystallites are attached to the support material by some type of chemical bond so they are not free to migrate across the surface and agglomerate or coalesce with other crystallites to form larger particles.

Since the catalyst particles are anchored to the support through some form of bonding the support can, potentially, also influence the activity of the

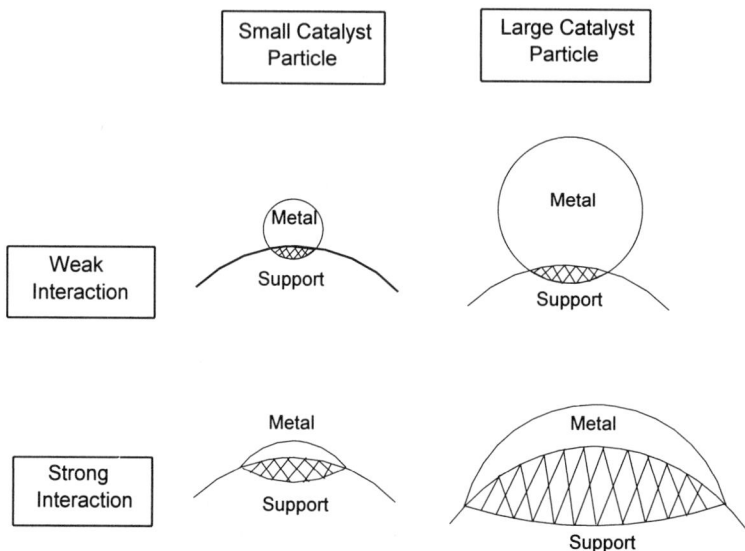

Fig. 9.12. Effect of catalyst particle size and the strength of the metal
support interaction on the metal:support contact area.

catalyst.[27-29] This support effect can be envisioned as taking place in two distinct ways. In one, the support could modify the electronic character of the catalyst particles and this could affect the adsorption and reaction characteristics of the catalytically active sites. Another possibility is that the extent of attraction between the support and the catalyst could influence the shape or geometry of the catalyst particles. The electronic effect could change the activity of the sites on the metal surface while the geometric effect would modify the number of active sites present. Although there are a large number of publications concerned with catalyst support interactions, there is, at present, no viable way of adequately determining if an observed catalytic change is brought about by an electronic modification of the catalytic sites or by a modification of the number of sites present on the catalyst particle. Most of the work in this area has been concerned with the effect of the support on the activities of dispersed metal catalysts so the rest of this discussion will center specifically on metal-support interactions.

Obviously, the extent of any of these metal-support interactions will be influenced by the size of the metal particles. In particular, the electronic influence of the support will be more pronounced with small metal crystallites than with large particles since there are fewer metal atoms over which to spread the electronic influence of the support. The geometric effect that can also be considered a measure of the "wetability" of the support by the metal can be envisioned as taking place with large metal particles as well as smaller ones. As depicted in Fig. 9.12, this effect can be more pronounced with larger particles than with smaller ones. Small catalyst particles having a weak interaction with the support have only a small interfacial area between the two solids (hatched area in Fig. 9.12). As the particle gets larger, this contact area with the support also grows, but not extensively. When the interaction is stronger, the metal can spread out over the support. In this case, the larger metal particles have a much greater contact area with the support. Since a larger contact area can also lead to enhanced electronic interactions, Fig. 9.12 also serves to illustrate the interrelationship between the geometric and electronic influence of the support on the catalyst.

Metal-support interactions can be defined as being weak, medium or strong.[28] Non-reducible metal oxides such as silica, alumina, and magnesia as well as carbon or graphite are considered to exert only a weak influence on the metal and, thus, exhibit only a weak metal-support interaction (WMSI).[28] Zeolites (Chapters 10 and 13) exert a medium metal-support interaction (MMSI)[28] while metals supported on reducible oxides when reduced at high temperatures exhibit a strong metal-support interaction (SMSI).[27-32]

The data on the extent of WMSI with differing support materials is difficult to interpret. An electron spectroscopy for chemical analysis (ESCA)[33] study indicated that there was a strong interaction between nickel and an alumina support.[34] Further, with a low nickel content, the nickel was present only as nickel metal while at high metal contents some nickel oxide was also present. In contrast, there appeared to be little interaction between the nickel and a silica support with the nickel present primarily as nickel oxide regardless of the nickel content of the catalyst.[34] Other data show, however, that the support material has a strong influence on the ease of reduction of nickel oxide with NiO/SiO_2 the most easily reduced and NiO/Al_2O_3 somewhat more difficult.[35] Since the metal-support interaction (MSI) is a measure of how strongly the metal particles are attached to the support, it was reasoned that there should be a correlation between the strength of this interaction and the temperature required to sinter the catalyst. Heating supported nickel catalysts in a vacuum showed that the interaction with the support decreased in the order $C > SiO_2 > Al_2O_3$.[36] These findings contradict the conclusions from the ESCA studies on nickel supported on silica and alumina mentioned above.[34]

As might be expected MSI depends not only on the support but also on the metal. Sintering supported silver catalysts in a vacuum indicated that MSI decreased in the order $SiO_2 > Al_2O_3 > C$, while with platinum catalysts it was $Al_2O_3 > SiO_2 > C$.[37] To further complicate the picture, sintering studies run in a hydrogen atmosphere gave an MSI order for platinum supported catalysts of $Al_2O_3 > C > SiO_2$.[38]

Support effects have also been reported for a number of reactions. A Ni/SiO_2 catalyst that was considered on the basis of metal particle size and X-ray photoelectron spectroscopy (XPS)[39-41] data to have a weaker MSI than a Ni/SiO_2–MgO catalyst, had a turnover frequency for the hydrogenation of carbon

Table 9.3

Effect of pretreatment on hydrogen adsorption on a Pt/TiO_2 catalyst.[32]

Sample	Sequential treatments on one sample	H/M
2% Pt/TiO_2	H_2, 175°C	0.93
	H_2, 500°C	0.05
	H_2, 175°C	0.05
	1) O_2, 400°C	0.89
	2) H_2, 175°C	
	H_2, 500°C	0.03

monoxide that was thirty to forty times larger than that observed with the Ni/SiO_2–MgO.[42] Ru/SiO_2 was also shown to be more active for this reaction than Ru/Al_2O_3,[43,44] while in the hydrogenation of carbon dioxide Rh/Al_2O_3 had a higher activity than Rh/SiO_2.[45] The water-gas shift reaction was promoted more readily by Pt/Al_2O_3 than by Pt/SiO_2,[46] and Pd/Al_2O_3 was two orders of magnitude more effective than Pd/SiO_2 for the catalytic reactions of neopentane.[47]

These results are a small fraction of the published data concerning metal-support interactions. While, as discussed later, there are some logical conclusions that can be drawn concerning the nature of the strong metal-support interactions there are too many factors involved to be able to sort through the mass of data concerned with the interaction of non-reducible or non-reduced supports and the metal particles on them.

One obvious problem concerns the comparison of catalysts having varying dispersions or metal particle sizes since, as discussed above, the MSI are particle size dependent. Less obvious factors are the metal precurser salts and the methods used to prepare the catalysts. It is known that trace amounts of some of the ions that may be present in the catalyst preparation can influence catalyst activity. Such ions may or may not be present in the metal salts used to prepare the catalyst and the amount of such material left on the catalyst can vary with different preparation procedures. It is also possible that one support may be more able than another to absorb such ions to keep them away from the catalytically active sites on the metal.[28] It is best to keep these matters in mind when evaluating any MSI data.

Strong Metal-Support Interactions

A significant change in catalytic activity occurs when a metal-supported on a reducible oxide such as titania is heated in hydrogen at relatively high temperatures. For one thing these high temperature reduced (HTR) materials show a dramatic decrease in hydrogen and carbon monoxide chemisorption, a factor which has become diagnostic of catalysts exhibiting strong metal-support interactions (SMSI).[28-32]   An example of how SMSI can affect hydrogen chemisorption is shown by the data in Table 9.3. Reducing a 2% Pt/TiO_2 catalyst at 175°C [low temperature reduction (LTR)] gave a normal, highly dispersed catalyst with an H/M ratio of 0.93. Heating this same catalyst at 500°C under hydrogen put it into the SMSI state with a decrease in the H/M ratio to 0.05. Further low temperature treatment with hydrogen had no effect on the hydrogen chemisorption but the initial high H/M ratio could be restored by first heating the SMSI catalyst in oxygen and followed by an LTR treatment.

Similar results were obtained on HTR treatment of platinum, palladium, rhodium, ruthenium or iridium supported on the readily reducible oxides, V_2O_3[31], Ta_2O_5[31], Nb_2O_5[31,48,49], ZrO_2,[50], CeO_2[51] and Nd_2O_3[52] in addition to TiO_2. With the more easily reducible SnO_2 low temperature reduction was sufficient to put the catalyst in the SMSI state.[53]   Other oxides such as Sc_2O_3[31], HfO_2[31], MgO[31], ZrO_2[31] and Y_2O_3[31] do not exhibit this effect. There are some data showing that La_2O_3[52] and ZrO_2[31] do not exhibit SMSI but other results indicate that these supports may do so under certain conditions.[55] There are some reports that very high temperature reduction of Al_2O_3 supported catalysts will put them into the SMSI state.[56-58]

It is obvious from these findings that a reduction of the support material is taking place using the metal activated hydrogen at high temperatures.   The presence of such reduction has been verified spectroscopically.[59] At first it was thought that this effect was the result of an electronic interaction between the catalytic metal particles and the cations in the reduced support but more extensive work has shown that it is more than that. Transmission electron microscopy (TEM)[59] and extended x-ray absorption fine structure (EXAFS)[60] data has shown that in the SMSI state small platinum particles on reduced titania acquire an almost two dimensional structure because of an enhanced attraction between the platinum and the reduced titanium oxide, TiO_x, in the support.[61] Further work has shown that this attraction is sufficiently strong that the TiO_x migrates over the surface of the metal particles as depicted in Fig. 9.13. This encapsulation is responsible, at least in part, for the observed decrease in chemisorption capabilities.[61-63]

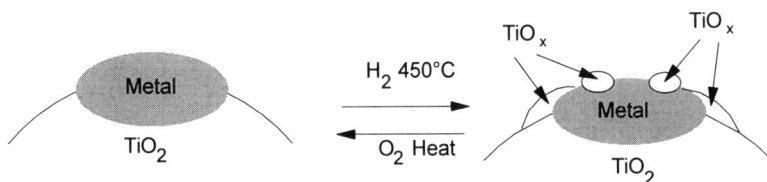

Fig. 9.13.   Migration of the reduced species, TiO_x, over a metal catalyst in the SMSI state.

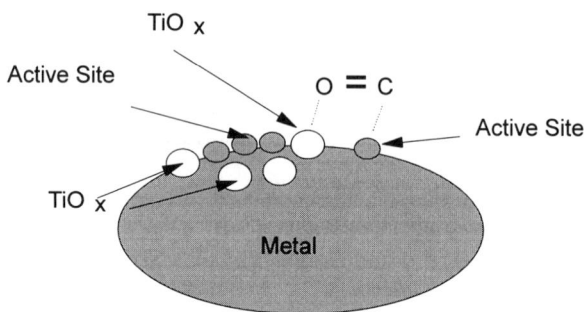

Fig. 9.14. Proposed mode of carbon monoxide adsorption on a
metal catalyst in the SMSI state.

As discussed earlier, oxide supported metals have the particles anchored to
the support through **M–O–S** bonding. On reduction this is converted, at least in
part, to **M–S** bonding. The tendency of a supported metal to enter the SMSI state
is determined by the strength of this **M–S** bond, a factor involved in the migration
of the reduced support over the metal. This, in turn, is dependent not only on the
ease of support reduction, a factor that controls the amount of reduced support in
the vicinity of the metal, but also on the nature of the metal. It has been found, for
instance, that platinum enters the SMSI state more easily than does rhodium and
that platinum is also more stable in the SMSI state than is rhodium.[62] These data
indicate that the Pt–Ti bond is stronger than the Rh–Ti bond. Titania supported
nickel catalysts having large nickel crystallites were found to require more severe
conditions for inducing SMSI than were needed for Ni/TiO_2 catalysts having
smaller nickel particulates.[64]

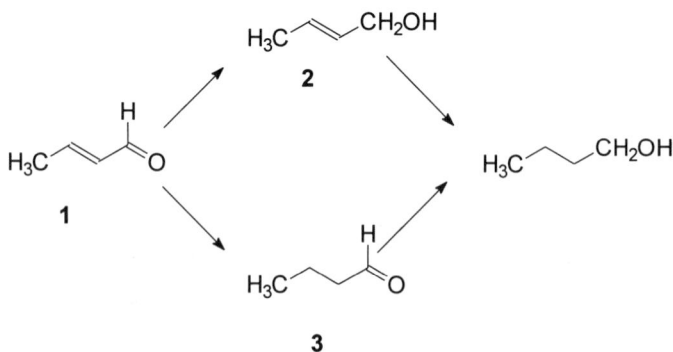

$$(9.4)$$

Even though the chemisorption of both hydrogen and carbon monoxide are reduced on catalysts in the SMSI state, these catalysts show enhanced activity for the hydrogenation of carbon monoxide.[65] It has been proposed that this increase in activity is brought about by a cooperative adsorption of the carbon monoxide molecule at the interface between the platinum and the TiO_x overlayer with the carbon atom adsorbed on the platinum and the oxygen on the oxygen deficient TiO_x as depicted in Fig. 9.14. This enhanced carbonyl hydrogenation has also been observed in the vapor phase hydrogenation of acetone where the Pt/TiO_2 catalysts in the SMSI state had turnover frequencies five hundred times greater than those observed with Pt/SiO_2 catalysts.[66] The selective hydrogenation of crotonaldehyde, **1**, to crotonalcohol, **2**, (Eqn. 9.4) was also enhanced over HTR Pt/TiO_2 catalysts. LTR Pt/TiO_2, Pt/SiO_2 and Pt/Al_2O_3 catalysts gave only butryaldehyde, **3**. With the HTR Pt/TiO_2 crotonalcohol, **2**, was produced in 37% selectivity at low conversions.[67]

References

1.   J. W. Geus, *Stud. Surf. Sci. Catal.*, **16**, (Prep. Catal., III), 1 (1983).
2.   S. P. S. Andrew, *Stud. Surf. Sci. Catal.*, **1**, (Prep. Catal.), 429 (1976).
3.   A. B. Stiles, *Catalyst Supports and Supported Catalysts*, (A. B. Stiles, Ed.) Butterworths, Boston, 1987, p 87.
4.   D. L. Trimm and A. Stanislaus, *App. Catal.*, **21**, 215 (1986).
5.   J. R. Anderson, *Structure of Metallic Catalysts*, Academic Press, New York, 1975, p 41.
6.   J. P. Brunelle, *Stud. Surf. Sci. Catal.*, **3**, (Prep. Catal., II), 211 (1979).
7.   R. Poisson, J.-P. Brunelle and P. Nortier, *Catalyst Supports and Supported Catalysts*, (A. B. Stiles, Ed.) Butterworths, Boston, 1987, p 11.
8.   Ref. 5, p 46.
9.   R. K. Oberlander, *Appl. Ind. Catal.*, **3**, 63 (1984).
10. M. Escudey and F. Gil-Llambias, *J. Colloid Interface Sci.*, **107**, 272 (1985).
11. R. J. Hunter, *Zeta Potential in Colloid Science*, Academic Press, New York, 1981, p 233.
12. Ref. 5, p 49.
13. L. Vordonis, A. Akratopulu, P. G. Koutsoukos and A. Lycorghiotis, *Stud. Surf. Sci. Catal.*, **31**, (Prep. Catal., IV), 309 (1987).
14. A. B. Stiles, *Catalyst Supports and Supported Catalysts*, (A. B. Stiles, Ed.) Butterworths, Boston, 1987, p 57.
15. M. E. Winyall, *Appl. Ind. Catal.*, **3**, 43 (1984).
16. Ref. 5, p 39.
17. A. van Blaaderen and A. P. M. Kentgens, *J. Non-Cryst. Solids*, **149**, 161 (1992).
18. Ref. 5, p56.
19. M. Schneider and A. Baiker, *J. Mater. Chem.*, **2**, 587 (1992).
20. S. Utamapanya, K. J. Klabunde and J. R. Schlup, *Chem. Mater.*, **3**, 175 (1991).
21. J. A. R. van Veen, *Z. Physik. Chem.*, **162**, 215 (1989).
22. A. J. Bird, *Catalyst Supports and Supported Catalysts*, (A. B. Stiles, Ed.) Butterworths, Boston, 1987, p 107.

23.   R. Rudriguez-Reinoso and M. Molina-Sabio, *Carbon*, **30**, 1111 (1992).
24.   R. Burmeister, B. Despeyroux, K. Deller, K. Siebold and P. Albers, *Stud. Surf. Sci. Catal.*, **78** (Heterog. Catal. Fine Chem.), 361 (1993).
25.   D. Richard and P. Gallezot, *Stud. Surf. Sci. Catal.*, **31** (Prep. Catal., IV), 71 (1987).
26.   A. Gupta and K. Ofori-Ansah, *Proc. Australas. Inst. Min. Metall*, **289**, 239 (1984).
27.   G. C. Bond, *Stud. Surf. Sci. Catal.*, **11** (Met. Support-Met. Addit. Eff. Catal.), 1 (1982).
28.   G. C. Bond and R. Burch, *Catalysis (London)*, **6**, 27 (1983).
29.   S. J. Tauster, S. C. Fung, R. T. K. Baker and J. A. Horsley, *Science*, **211**, 1121 (1981).
30.   S. J. Tauster, *Accounts Chem. Res.*, **20**, 389 (1987).
31.   S. J. Tauster and S. C. Fung, *J. Catal.*, **55**, 29 (1978).
32.   S. J. Tauster, S. C. Fung and R. L. Garten, *J. Am. Chem. Soc.*, **100**, 170 (1978).
33.   J. S. Brinen, *Accounts Chem. Res.*, **9**, 86 (1976).
34.   M. Wu, R. Chin and D. M. Hercules, *Spectroscopy Lett.*, **11**, 615 (1978).
35.   H. Hattori, H. Imai and K. Tanabe, *Appl. Catal.*, **4**, 87 (1982).
36.   M. Arai, T. Ishikawa and Y. Nishiyama, *J. Phys. Chem.*, **86**, 577 (1982).
37.   M. Arai, T. Ishikawa, T. Nakayama and Y. Nishiyama, *J. Colloid Interface Sci.*, **97**, 254 (1984).
38.   R. T. K. Baker, E. B. Prestridge and R. L. Garten, *J. Catal.*, **56**, 390 (1979).
39.   M. W. Roberts, *Chem. Soc. Revs*, **6**, 373 (1977).
40.   C. D. Batich, P. H. Holloway and M. A. Kosinski, *Chemtech*, **16**, 494 (1986).
41.   J. Grimblot, L. Gengembre and A. D'Huysser, *J. Electron Spectrosc. Relat. Phenom.*, **52**, 485 (1990).
42.   P. J. Reucroft, H. Parekh, P. Ganesan, S. N. Russell and B. H. Davis, *Appl. Catal.*, **3**, 65 (1982).
43.   E. Zaglli and J. L. Falconer, *J. Catal.*, **69**, 1 (1981).
44.   F. Stoop, A. M. G. Verbiest and K. van der Weile, *Appl. Catal.*, **25**, 51 (1986).
45.   F. Solymosi, A. Erdöhelyi and T. Bánsági, *J. Catal.*, **68**, 371 (1981).
46.   D. C. Grenoble, M. J. Estadt and D. F. Ollis, *J. Catal.*, **67**, 90 (1981).
47.   W. Juszczyk, Z. Karpinski, L. Ratajczykowa, Z. Stanasiuk, J. Zielinski, L. L. Sheu and W. M. H. Sachtler, *J. Catal.*, **120**, 68 (1989).
48.   G. Sankar, S. Vasudevan and C. N. R. Rao, *J. Phys. Chem.*, **92**, 1878 (1988).
49.   K. Kunimori, K. Ito, K. Iwai and T. Uchijima, *Chem. Lett.*, 573 (1986).
50.   C. Dall'Agnol, A. Gervasini, F. Morazzoni, F. Pinna, G. Strukul and L. Zanderighi, *J. Catal.*, **96**, 106 (1985).
51.   D. Kalakkad, A. K. Datye and H. Robota, *Appl. Catal., B*, **1**, 191 (1992).
52.   S. Bernal, F. J. Botana, R. García, M. L. López and J. M. Rodríguez-Izquierdo, *Inorg. Chim. Acta*, **140**, 49 (1987).
53.   P. A. Sermon, V. A. Self and E. P. S. Barrett, *J. Mol. Catal.*, **65**, 377 (1991).
54.   P. Turlier and G. A. Martin, *React. Kinet. Catal. Lett.*, **25**, 1 (1984).

55. R. F. Hicks, Q.-J. Yen and A. T. Bell. *Appl. Surf. Sci.,* **19**, 315 (1984).
56. T. Ren-Yuan, W. Rong-An and L. Li-Wu, *Appl. Catal.,* **10**, 163 (1984).
57. O. Okada, M. Ipponmatsu, M. Kawai, K. Aika and T. Onishi, *Chem. Lett.,* 1041 (1984).
58. S. I. Abasov, V. Yu. Borvkov and V. B. Kazansky, *Catal. Lett.,* **15**, 269 (1992).
59. T. Baird, *Catalysis (London),* **5**, 172 (1982).
60. J. Evans, *Catalysis (London),* **8**, 1 (1989).
61. J. H. A. Martens, R. Prins, H. Zandbergen and D. C. Koningsberger, *J. Phys. Chem.,* **92**, 1903 (1988).
62. J. B. F. Anderson, R. Burch and J. A. Cairns, *Appl. Catal.,* **25**, 173 (1986).
63. B. Viswanathan, T. Lakshmi and U. D. Mary, *Indian J. Technol.,* **30**, 99 (1992).
64. E. I. Ko, S. Winston and C. Woo, *J. Chem. Soc., Chem. Commun.,* 740 (1982).
65. M. E. Levin, M. Salmeron, A. T. Bell and G. A. Somorjai, *J. Catal.,* **106**, 401 (1987).
66. B. Sen and M. A. Vannice, *J. Catal.,* **113**, 52 (1988).
67. M. A. Vannice and B. Sen, *J. Catal.,* **115**, 65 (1989).

Oxide Catalysts

Historically, oxide catalysts have been used primarily for vapor phase reactions in the petroleum and petrochemical industries. Recent work, however, has shown that these catalysts can also be effective in promoting a number of synthetically useful reactions. While simple oxides show activity for some oxidations they are more commonly used as solid acids or bases. Complex oxides can act as acids or bases as well as oxidation catalysts. Complex oxides can range in composition from the simple, amorphous, binary oxides to the more complex ternary and quaternary systems. The use of zeolites and clays can impart shape selectivity to a number of reactions, a feature that makes these systems particularly appealing for use in synthesis.

Oxide catalysts fall into two general categories. They are either electrical insulators or they can act as semiconductors. Insulator oxides are those in which the cationic material has a single valence so they have stoichiometric M:O ratios. The simple oxides, MgO, Al_2O_3 and SiO_2 and the more complex zeolites, which are aluminosilicates, fall into this category. These materials are not effective as oxidation catalysts and find most use as solid acids[1] or bases.[2,3]

Semiconductor oxides are most commonly used in oxidations. They are materials in which the metallic species is relatively easily cycled between two valence states. This can be two different positive oxidation states as in Fe_2O_3, V_2O_5, TiO_2, CuO or NiO, or the interconversion between the positive ion and neutral metal as with the more easily reduced oxides such as ZnO and CdO. While a complete discussion of the electronic theory of semiconductors is beyond the scope of this text some basic reviews are available for those needing more information.[4,5]

Basically, some oxides are semiconductors because they can have either a slight excess or deficiency of electrons. In the former case there is a net negative charge so the material is referred to as an n-type semiconductor. A net positive charge gives a p-type semiconductor. These two types are appreciably different in their adsorption and reaction characteristics

The most easily seen difference between n- and p-type oxides is the effect exerted by heating in air. n-Type oxides lose oxygen on heating while p-type oxides gain oxygen. This loss or gain of oxygen plays a significant role in their respective reaction characteristics. The loss of oxygen by an n-type oxide is made possible by the availability of a lower oxidation state of the metallic component. This can be a lower positive valence as with TiO_2, V_2O_5, CrO_3, CuO or Fe_2O_3 (Eqn. 10.1), or the zero valent metallic state as with CdO or ZnO (Eqn. 10.2). In these systems the excess electrons involved in electrical conduction are the unused valence electrons on the metal or lower valence ion.

$$2 \ Fe_2O_3 \longrightarrow 4 \ FeO \ + \ 3 \ O_2 \qquad (10.1)$$

$$2 \ ZnO \longrightarrow 2 \ Zn \ + \ O_2 \qquad (10.2)$$

The oxidation of carbon monoxide over an n-type semiconductor such as ZnO takes place by the initial reaction with a lattice oxygen to give a surface carbonate (Eqn. 10.3) and releasing two electrons that are used to reduce the zinc ion (Eqn. 10.4). Decomposition of the carbonate, (10.5) and re-oxidation of the metallic zinc with oxygen (10.6) completes the catalytic cycle and regenerates the ZnO surface.[4] This consecutive removal of an oxygen from the lattice by one reagent followed by its replenishment by another, usually oxygen, is referred to as an intrafacial process.[6] In many oxides, particularly the complex species discussed later, the oxygen vacancy in the lattice is filled by the migration of another lattice oxygen into the "hole" with the replacement made at a surface location removed from the 'active site.'

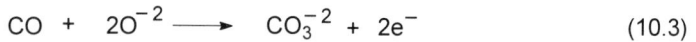

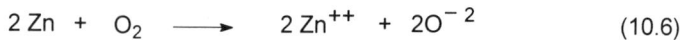

$$CO \ + \ 2O^{-2} \longrightarrow CO_3^{-2} \ + \ 2e^- \qquad (10.3)$$

$$Zn^{++} \ + \ 2e^- \longrightarrow Zn^0 \qquad (10.4)$$

$$2 \ CO_3^{-2} \longrightarrow 2 \ CO_2 \ + \ 2O^{-2} \qquad (10.5)$$

$$2 \ Zn \ + \ O_2 \longrightarrow 2 \ Zn^{++} \ + \ 2O^{-2} \qquad (10.6)$$

The adsorption of oxygen onto a p-type oxide is made possible by the availability of a higher oxidation state for the cationic component of the oxide. Some simple oxides having p-type character are Cu_2O, SnO, PbO, Cr_2O_3, CoO and NiO. Adsorption of oxygen on one of these oxides raises the oxidation state of some of the surface ions as shown in Eqn. 10.7 for NiO, where the extra oxygen is not part of the lattice. The Ni^{+3} ions in the lattice are referred to as positive holes and electrical conductivity takes place by electron migration from a neighboring Ni^{+2} ion converting it to Ni^{+3}. Successive electronic migrations moves the positive hole, Ni^{+3}, through the oxide lattice. The surface adsorbed oxygen is reactive and can combine with carbon monoxide, for instance, as shown in Eqn. 10.8.[4] This type of oxidation is referred to as a suprafacial process since it involves the activation of the oxygen on the surface of the oxide.[6]

Since the oxygen adsorbed on an oxide surface is more reactive than the lattice oxygen of an n-type oxide, p-type oxides are generally less selective in their reactions than are the n-type species. Oxidation of hydrocarbons over p-type oxides usually gives only carbon dioxide and water[7] while partial oxidation products can be obtained using n-type oxides, particularly the more complex materials.

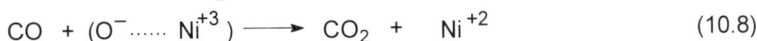

$$2Ni^{+2} + O_2 \longrightarrow 2\,(O^- \cdots\cdots Ni^{+3}) \qquad\qquad (10.7)$$

$$CO + (O^- \cdots\cdots Ni^{+3}) \longrightarrow CO_2 + Ni^{+2} \qquad\qquad (10.8)$$

Simple Oxides

A simple oxide catalyst can be used in either the bulk state or supported on an inert oxide support material. The bulk oxides are usually prepared using a precipitation-calcination sequence similar to those described in Chapter 9 for the preparation of support oxides.[8,9] In general, the simple semiconductor oxides are not very good catalysts for synthetic reactions. The insulator oxides, however, can be used as solid acids and bases for a number of reactions. Alumina has been used as an acid catalyst for the vapor phase rearrangement of cyclohexanone oxime to caprolactam (Eqn. 10.9). Modification of the γ-alumina surface by the addition of 10–20% of B_2O_3 increased its activity for this reaction, giving caprolactam in 80% selectivity even after several hours of continuous operation.[10,11]

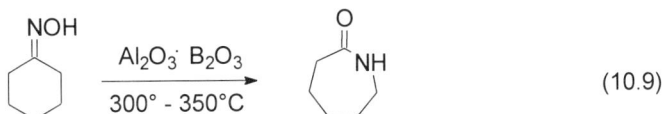

$$(10.9)$$

The acidic character of silica was used to promote the rearrangement of the epoxide **1** to the ketals, **2** and **3** (Eqn. 10.10).[12] Heating a mixture of **1** and silica gel to 150°–200°C gave **2** in quantitative yield. Similar results were also obtained using alumina or activated vermiculite as the catalyst. When a toluene solution of **1** was stirred with activated vermiculite at room temperature, **3** was produced in 90% yield along with 5% of **2**.[12] Adding about 2% Al_2O_3 to the silica greatly improved its activity for the vapor phase dehydration of alcohols.[13]

Even though TiO_2 is a semiconductor oxide, when treated with sulfuric acid it exhibits sufficient surface acidity to act as a solid acid catalyst in promoting the isomerization of α pinene to camphene (Eqn 10.11).[14] The reaction was run in the absence of solvent at 135°–160°C and gave camphene in high selectivity.

Basic oxides, particularly MgO and BaO, have been used to promote a number of base catalyzed reactions.[2,3] To be effective the oxide must first be heated to remove the water, carbon dioxide and oxygen that may have adsorbed

(10.10)

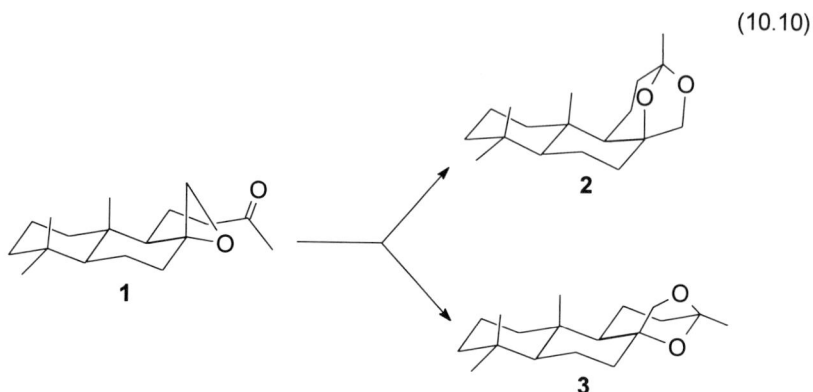

on the surface when the oxide was exposed to air. Without this activation, the oxides are relatively unreactive. The temperatures required for catalyst activation depend on the nature of the oxide and the reaction it is being used to catalyze. With MgO maximum activity for the vapor phase isomerization of 1-butene was observed with a catalyst preheated to 500°C[2] while the liquid phase aldol condensation between benzaldehyde and cyclohexanone (Eqn. 10.12) proceeded smoothly over a MgO catalyst that was heated to 390°C before being used in the reaction.[15]

(10.11)

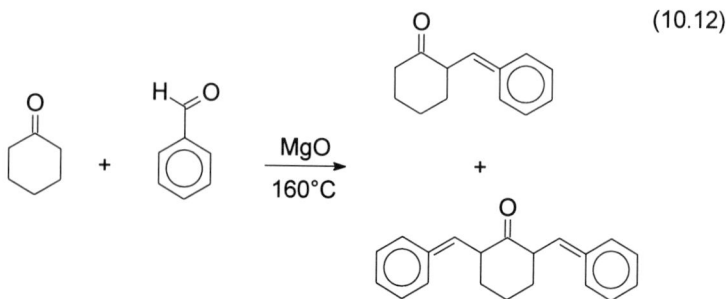

(10.12)

Supported Oxide Catalysts

Supported oxides have several advantages over unsupported materials.[16] Supports can be used to improve the mechanical strength, the thermal stability and the lifetime of the catalyst. They can also provide the means for increasing the surface area of the active species. In addition, data have shown that supported oxides frequently have structural features and chemical compositions different from those on the surface of the bulk oxide. This difference arises primarily from the interaction of the oxide with the support and the size and shape of the catalytic species. There are several general ways in which the active oxide can be found on the surface of the support material.[17] As shown in Fig. 10.1a. the active component and the support can be intimately mixed throughout the catalyst particle. This arrangement provides the maximum interaction between the support and the catalyst so it maximizes the catalyst support interaction. Unfortunately, it also leaves a large portion of the active material buried in the bulk of the particle. A close second, with respect to maximizing the catalyst support interaction, is the molecular dispersion depicted in Fig. 10.1b. Here individual 'molecules' of the catalytically active species are bound to the support surface as a localized binary oxide. The nature of the support can then have a significant influence on the reaction characteristics of the catalyst. Further deposition of the catalyst can give in a monomolecular layer of the catalyst on the support (Fig. 10.1c). Here, too, catalyst support interactions can be strong

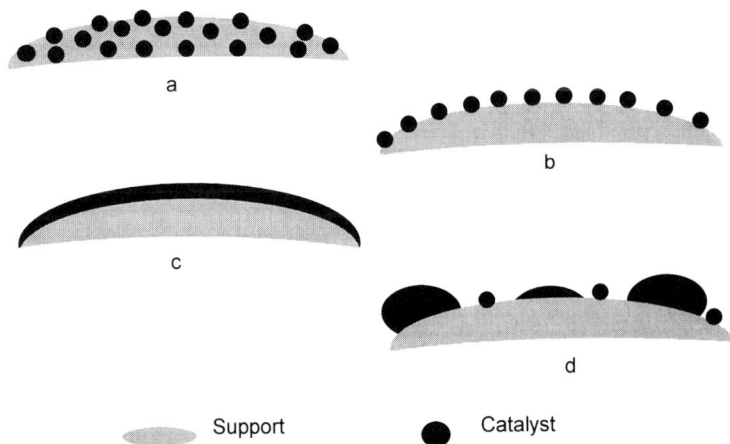

Fig. 10.1. Various arrangements of oxide catalysts on a support.
a) Coprecipitated material with the catalyst dispersed throughout the support; b) catalyst molecules on the surface of the support; c) monomolecular layer of catalyst on the support; d) mixture of molecules, monomolecular layers and larger particles on the support.

because of the large contact area between the support and the oxide relative to the surface area of the support. It is this monolayer coverage that is considered the best for a number of supported oxides[16] but, realistically, it is recognized that it is improbable that the entire surface of a high surface area porous support material will be uniformly covered by the catalytic species. Instead, there will be areas of monolayer coverage accompanied by some larger supported oxide particles and some uncovered support surface (Fig. 10.1d).

There are several methods commonly used to prepare supported oxide catalysts.[8,9,16] They differ primarily in the manner in which the active component or its precurser is placed in contact with the support. Once this is accomplished the supported species are then calcined at appropriate temperatures to give the active catalyst. Unfortunately, the methods that are used to disperse the active component on the support are not uniformly defined.

Coprecipitation of the support and the catalytic material is accomplished by treatment of a solution containing both the support precurser and the catalyst precurser with an appropriate precipitating agent. The resulting precipitate contains species either as a single phase or as multiple phases, from which the active component and the support material are eventually produced. This gives a material in which a large portion of the active species are buried inside the particle and, thus, unavailable for reaction. Coprecipitation has been used, for instance, to produce a highly dispersed MoO_3/SiO_2 catalyst that is effective for hydrodesulfurization reactions.[18] This catalyst was prepared by adding an ethylene glycol solution of ammonium paramolybdate, $(NH_4)_6Mo_7O_{24} \cdot 6H_2O$, to a slightly acidic aqueous ethanol solution of ethyl silicate at 80°C. The resulting gel was washed, dried and then calcined at 500°C to give the active catalyst.

The more common methods for preparing supported oxide catalysts involve the dispersion of the active species or its precurser onto the surface of a preformed support. This has been accomplished using procedures referred to as equilibrium adsorption, incipient wetness or dry impregnation.[16,19,20]

The equilibrium adsorption procedure involves the adsorption of the active species or its precurser from a solution in which the solid support is suspended. This process is also referred to as ion-exchange or as the grafting of the active species to the support surface.[16,19,20] After adsorption is complete, the modified support is removed from the suspension, washed, dried and calcined. As discussed in Chapter 9, the sign and extent of the charge on the support surface is determined by the isoelectric point (IEP) of the solid and the pH of the solution. For oxoanions to be adsorbed a positive support surface is needed. For instance, the molybdate solutions commonly used in catalyst preparations have a pH between 5 and 6. Since aluminas typically have IEPs between pH 7 and pH 9, the support will be positively charged at pH 5 and molybdate adsorption will take place giving well dispersed polymolybdate species anchored to the alumina by the formation of Al–O–Mo bonds. Silica, with an IEP of pH 1 to 2, will not adsorb molybdates to give a monolayer catalyst under these conditions. Instead, the molybdate, on drying, will merely crystallize in the pores of the support.[21] When the adsorptions were run at a pH corresponding to the IEP of the support the surface oxo-molybdenum species present was dependant on the surface pH at its IEP. Oxides with low IEPs such as SiO_2 had $Mo_7O_{24}^{6-}$ and/or $Mo_8O_{26}^{4-}$ anions on the surface. Amphoteric oxides such as TiO_2 and ZrO_2 had $Mo_7O_{24}^{6-}$ species at low molybdenum content and $Mo_8O_{28}^{4-}$ at higher molybdenum content. On Al_2O_3 both MoO_2^- and $Mo_7O_{24}^{6-}$ were found at low molybdenum content and $Mo_7O_{24}^{6-}$ and/or $Mo_8O_{26}^{4-}$ at high molybdenum content. Basic oxide supports such as MgO exhibited a preference for the isolated MoO_4^{2-} species regardless of the molybdenum content.[22]

Using a fresh solution of 1% molybdate at pH 1, though, provided a uniform distribution of molybdenum oxide on SiO_2 as well as on Al_2O_3, TiO_2, CeO_2, Cr_2O_3 and ZnO.[17] A MoO_3/SiO_2 catalyst was also prepared by impregnation of SiO_2 with an aqueous solution of ammonium paramolybdate followed by calcination at 500°C. This catalyst was effective for promoting the isomerization of dimethylethynylcarbinol to 3-methyl-2-butanal in 75% yield (Eqn. 10.13).[23]

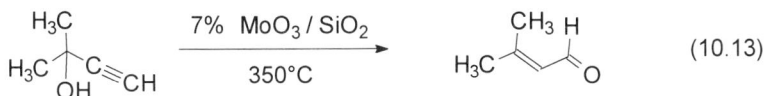

$$(10.13)$$

Titania supported vanadium oxide catalysts were prepared by the adsorption of ammonium metavanadate from an aqueous solution at pH 7.[19] This adsorption or 'grafting' has also been accomplished using more reactive vanadium reagents in non-aqueous media such as solutions of $VOCl_3$ in n-hexane[19] or benzene.[20] Hexane solutions of vanadyl trialkoxides were used to graft surface vanadyl groups to ZrO_2, TiO_2, TiO_2-SiO_2,[24] SiO_2[24,25] and Al_2O_3.[25] The surface reactions taking place using the alkoxide are depicted in Fig. 10.2.[26] Reaction of the surface hydroxy groups with a vanadyl alkoxide gave a near monolayer coverage since additional layers of vanadium were blocked from formation by the alkoxy groups on the modified surface. Heating removed the alkyl groups and generated surface hydroxy groups which could then react with more vanadium alkoxide in a separate reaction sequence to build up a second vanadium oxide layer. Further cycles could be used to add more oxide layers to the catalyst.[26] The activity of these monolayer catalysts are strongly influenced by the nature of the support. The interaction with titania is much stronger than with silica but neither catalyst is particularly active; on titania presumably because the interaction is too strong and on silica because it is too weak. By grafting V_2O_5 onto an 80:20 SiO_2–TiO_2 binary oxide support a catalyst with enhanced properties was obtained.[26]

The incipient wetness procedure involves the wetting of the support with just enough of a solution of the active species to fill the pores of the support material. This is usually accomplished by adding the solution slowly to the solid

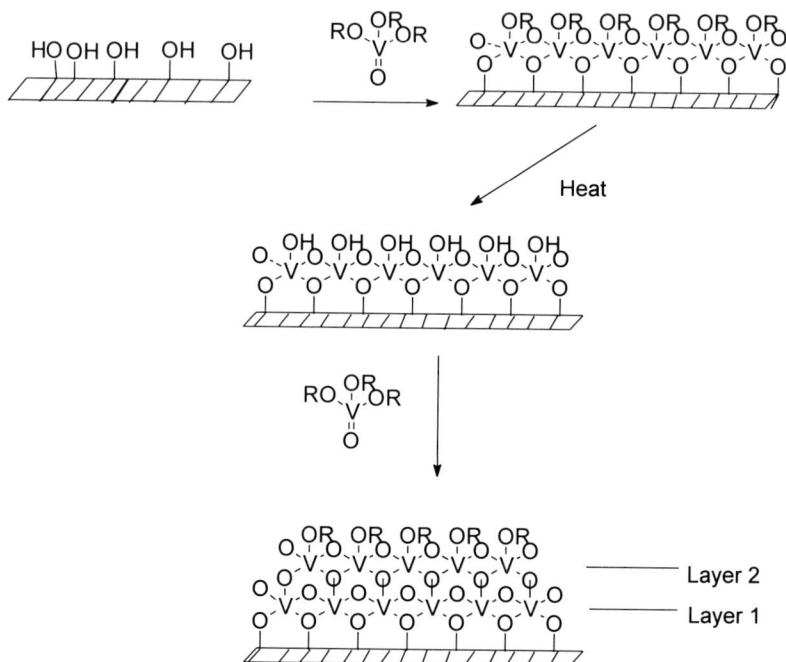

Fig. 10.2. Surface reactions taking place in the preparation of layered
supported vanadium oxide catalysts by the successive
reaction of vanadyl alkoxides with surface hydroxy groups.

with good stirring and continuing the addition until the mixture becomes slightly liquid. If one knows the volume of liquid needed to reach this point, the concentration of the solution can be adjusted to give the desired catalyst loading. The required volume can be determined by taking a weighed quantity of support, usually a gram, and adding the solvent to it until the incipient wetness point is reached. This gives the volume of liquid per gram of support, a factor that can be adjusted to any weight of support. Once the incipient wetness point has been reached, the solvent is then evaporated and the modified support calcined to give the catalyst.

This procedure is also referred to as a wet impregnation or sometimes simply as impregnation. This latter term, however, is ambiguous in that it has also been used to describe an equilibrium adsorption process and even one in which the support is added to an excess of solution and the solvent evaporated to give the supported catalyst precurser. In many cases catalysts are described as being prepared by impregnation without any further description being given. This is unfortunate since duplication of such preparations can become a serious problem.

Incipient wetness can be used with aqueous or organic solutions depending on the solubilities of the catalyst precurser. Evaporation of a high surface tension solvent such as water can result in pore disruption in the support which in turn causes a decrease in the surface area of the catalyst. Non-aqueous solvents do not cause pore disruption and should be used where possible.

The incipient wetness procedure has been used, for instance, to prepare catalysts from an aqueous solution of ammonium metavanadate on α alumina,[27] aqueous oxalic acid solutions of vanadyl oxalate onto α alumina[27] and titania,[20] and ethanol solutions of vanadyl triethoxide on titania[19] as well as a number of different catalytically active oxide precursors on alumina[28] and silica.[13] V_2O_5/Al_2O_3 catalysts were effective for the ammoxidation of 3-picoline, giving nicotinonitrile in 70-80% selectivity (Eqn. 10.14).[27] This reaction apparently proceeds by way of an initial oxidation of the aryl methyl group to give an adsorbed nicotinic acid which then reacts with ammonia to produce the nitrile. The most effective catalyst for this reaction was prepared using a non-aqueous solution of vanadyl acetyacetonate to impregnate α alumina followed by drying at 120°C and calcining at 500°C.[27]

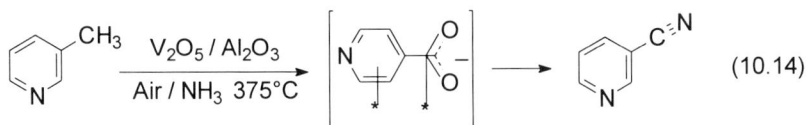

$$(10.14)$$

Heating a mechanical mixture of the support and the catalytic oxide can result in the dispersion of the active species over the support surface. This procedure, known as either dry impregnation[16,19] or solid-solid wetting[21] has been applied to the deposition of both V_2O_5 and MoO_3 on TiO_2 and Al_2O_3 and other supports. The presence of some water vapor during the heating facilitates the spread of the catalytic oxide over the support.[21]

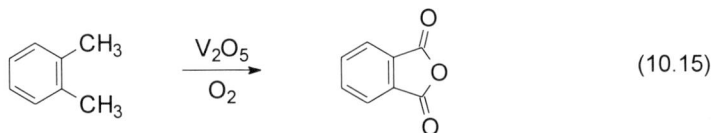

$$(10.15)$$

A series of V_2O_5/TiO_2 catalysts were prepared by coprecipitation, grafting, incipient wetness and dry impregnation. It was found that there were three types of vanadium species present on each of these catalysts with the relative amounts of each dependent on the method used for the preparation.[19,20] It was proposed that all three species are involved in the catalytic oxidation of o-xylene to phthalic anhydride (Eqn. 10.15), a reaction for which supported vanadium oxide catalysts are particularly useful. The catalyst prepared by coprecipitation was the most active for this reaction giving a 45% yield of anhydride at 100% conversion at temperatures at least 20°C lower than that needed by the other catalysts.[17] In a similar study of MoO_3/TiO_2 catalysts prepared in different ways it was found that the preparation method did not have a significant influence on the composition of the catalyst.[19]

Complex Oxides

Many oxide catalysts are materials having two or more cationic components. If the oxides are crystalline the crystal structure can determine the oxide composition. For instance perovskites have the general formula ABO_3, scheelites are ABO_4, spinels are AB_2O_4 and palmeirites are $A_3B_2O_8$.[29] In other cases where specific crystallinity is not observed the cationic ratio can vary over a wide range. The most common methods used to prepare complex oxides are coprecipitation followed by calcination and a solid state procedure involving the heating of precurser salts or oxides in appropriate ratios.

Coprecipitation occurs when a solution of two or more metal salts is treated with a precipitating agent, usually an alkali hydroxide, carbonate or bicarbonate. The resulting precipitate may contain not only the insoluble oxides, hydroxides and/or carbonates but also a mixed metal species if the solubility equilibria are favorable and the precipitation time is sufficient for the formation of such compounds. This material is then washed, dried and calcined at an elevated temperature to decompose any hydroxides or carbonates and give the desired mixed oxide. The critical factor with this procedure is the distribution of the components evenly throughout the solid. Unfortunately, this is usually not the case since the solubility constants of the individual components will not be identical so in the early stages of the precipitation one component may be present in excess while in the later phases the second or third component may predominate.[30,31] This difference in solubility can be magnified if the basic precipitant is added to the solution of metal salts. The initial quantity of base will be diluted by the metal solution so precipitation of the material with the lowest solubility will occur preferentially. This will also use up some of the base resulting in a further decrease in its concentration. As more base is added its concentration will increase and the more soluble species will then be precipitated.

A more uniform precipitate is obtained if the metal ion solution is added to the base but, even here some discrimination can be observed. As the metal salt solution is added the base concentration decreases. A further decrease is brought about by the precipitation of some of the basic anions. Homogeneous precipitation appears to take place when the metal salt solution and the base are added together at such a rate as to maintain the reacting mixture at a constant pH.[31]

Another way of effecting a nearly constant pH in the reaction mixture is to introduce the base by the aqueous hydrolysis of urea.[32] A TiO_2–SiO_2 catalyst was prepared by adding an acid solution of $TiCl_4$ and tetraethoxysilane to ammonium hydroxide. A second batch of catalyst was prepared by adding urea to the acid solution and heating the resulting mixture at 95°C for several hours. In both cases the resulting precipitates were washed, dried and calcined at 500°C. Even though the ammonium hydroxide precipitated material had a larger surface area than that prepared using urea hydrolysis as a source of the base, the later material was considerably more acidic, presumably because of the more homogeneous character of the precipitate.[32]

Another procedure that produces a more homogeneous mixed oxide involves the aqueous hydrolysis of metal alkoxides. Adding water to an alcoholic solution of tetraethoxysilane and titanium tetraisopropoxide gives a gel composed of the mixed oxides.[33] When an acid catalyst is used the hydrolysis of the alkoxides takes place very rapidly but the condensation of the resulting metal hydroxides is slow. This results in the formation of linear chains with little cross-linking and gives a material that on calcination collapses to a microporous solid of limited usefulness in catalysis. In base, the hydrolysis of the alkoxides takes place more slowly but the condensation of the hydroxides is rapid. Thus, as the hydrolyzed species is formed it becomes attached to a growing nucleus that is extensively cross-linked. On heating these clusters form mesoporous solids that are stable to elevated temperature treatment and, thus, useful for catalytic processes.[33] The addition of a small amount of ammonium hydroxide to an alcoholic solution of tetraethoxysilane and aluminum triisobutoxide gave an aluminosilicate gel that on washing, drying and calcination at 300°–900°C produced an amorphous aluminosilicate that was an effective solid acid catalyst.[34]

The advantage of using a coprecipitation procedure for the production of mixed oxides is that the ratio of the cations in the resulting mixed oxide can be varied over a wide range of compositions. This facilitates optimizing the oxide composition for a particular application. In the case of the TiO_2–SiO_2 mixed oxides, one containing 20% TiO_2 and 80% SiO_2 was shown to be particularly effective as a support for V_2O_5 oxidation catalysts. Changing the Ti:Si ratio from this value resulted in a decrease in the oxidation activity of the supported catalyst.[26,35]

With aluminosilicates, the acid strength of the mixed oxide is related to the Si:Al ratio. In an aluminosilicate both the silicon and the aluminum atoms are surrounded by four oxygens in a tetrahedral arrangement as depicted in Fig. 10.3.[36] Since an AlO_4^- species is unstable when bonded to another AlO_4^- it is necessary that they be separated by at least one SiO_4 unit. Thus, the Si:Al ratio cannot be lower than one. The negative charge on the aluminum ions is balanced by exchangeable extra lattice cations. The most common extra lattice cation is Na^+ with the sodium form of the aluminosilicate depicted in Fig. 10.3a. If a divalent cation is exchanged for the sodium ion, then it assumes a position somewhat midway between two AlO_4^- centers as shown in Fig. 10.3b. The hydrogen form is shown in Fig. 10.3c. Since treatment of an aluminosilicate with acid will result in its decomposition, H-aluminosilicates are produced by an initial exchange with ammonium ion followed by heating to remove ammonia. The

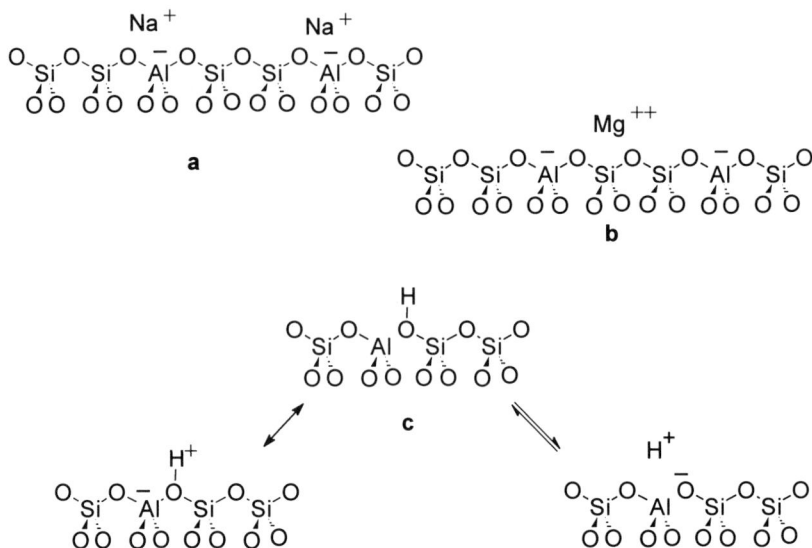

Fig. 10.3. Aluminosilicate structure. a) Na$^+$ form; b) Mg^{++} form;
c) H$^+$ form.

hydroxy groups in these systems are strong Bronsted acids. When there are a large number of aluminum ions present there are also a corresponding number of proton donor hydroxy groups. The strength of these acidic groups varies widely and appears to be related to the density of these hydroxy groups. The more isolated the silanol species, the stronger the acid so the acid strength of the aluminosilicate increases with decreasing aluminum content or, put another way, with increasing Si:Al ratios. These systems can also exert Lewis acid capability even in the metal cation exchanged form by using the electron accepting character of the aluminum. This is shown for the H$^+$ form by the resonance and equilibrium structures depicted in Fig. 10.3c.

Cation-exchanged aluminosilicates can also act as bases, particularly when the extra lattice cation is large. A low Si:Al ratio produces a more basic catalyst. A Cs$^+$ exchanged amorphous aluminosilicate can catalyze the liquid phase aldol condensation of benzaldehyde with cyclooctanone (Eqn. 10.16).[15] Both mono- and di-benzylidene products are formed over this amorphous basic catalyst. As discussed later, with the crystalline aluminosilicates, the zeolites, selectivity for mono-benzylidene product formation is increased.

As described above, the presence of the aluminum in the aluminosilicates imparts an increased acidity to the solid because of the replacement of a tetracoordinate silicon with a tricoordinate aluminum. The catalytic activity of many mixed oxide catalysts is also related to the presence of similar defect sites in

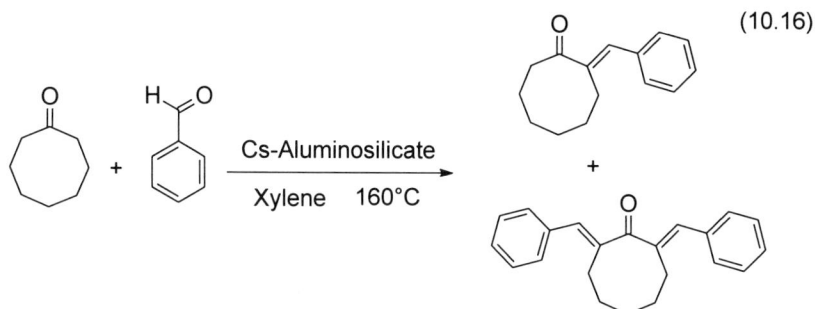

(10.16)

the solid.[29,37] They are produced by incorporating into the solid small amounts of cations having a higher or lower valence than those normally present. The means of producing these defect sites and the mode of catalytic action produced by them is probably best presented by a discussion of a specific material, the bismuth-molybdenum oxides. These oxides are used for selective oxidations of alkenes such as the conversion of propene to acrolein and 1-butene to butadiene as well as the ammoxidation of propene to acrylonitrile.[38]

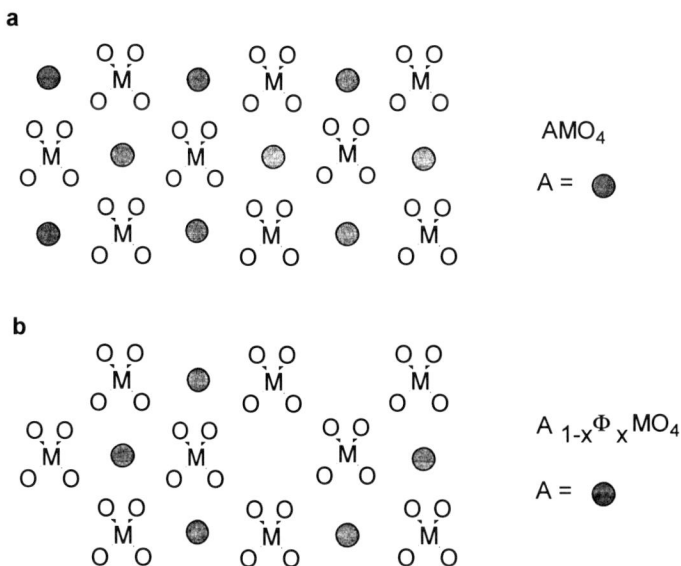

Fig. 10.4. a) Two dimensional representation of a sheelite crystal structure; b) sheelite structure with defect sites.

To understand the mechanism of these reactions some data on the crystal structure of the catalysts is needed. The catalysts used have the scheelite structure with the ideal formula, AMO_4. This structure is made up of A^{n+} cations and tetrahedral MO_4^{n-} anions. The A cation is coordinated to eight oxygens from eight different tetrahedra. All A cations are equivalent as are the M cations and the oxygens. A two dimensional schematic of this structure is given in Fig. 10.4a. Since the molybdate anion has a -2 charge, a typical scheelite would be $PbMoO_4$. When a cation, A, with a valence greater than +2 is used, fewer cations will be needed to neutralize the anionic charge and cation vacancies, such as those depicted in Fig. 10.4b, will be present. Such vacancies can be represented in a formula as ϕ with the general defect composition in scheelites given as $A_{1-x}\phi_x MO_4$.

Bismuth molybdate has the formula $Bi_2(MoO_4)_3$ but it can also be written as $Bi_{0.67}\phi_{0.33}MoO_4$. The value of 0.33 for the defect site is the maximum number permitted for this structural type. Values between 0 and 0.33 can be obtained by using mixtures of divalent and trivalent cations such as Pb^{+2} and Bi^{+3}. Such catalysts are prepared by calcining appropriate mixtures of oxides as illustrated in Eqn. 10.17.[38]

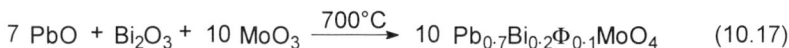

$$7\ PbO\ +\ Bi_2O_3\ +\ 10\ MoO_3\ \xrightarrow{700°C}\ 10\ Pb_{0.7}Bi_{0.2}\Phi_{0.1}MoO_4 \qquad (10.17)$$

The catalytic activity of bismuth-lead molybdate catalysts is related to the defect site densities. Olefin oxidations take place most rapidly when $\phi = 0.06$. Acrylonitrile is formed by the ammoxidation of propene in about 75% yield over these mixed oxides when ϕ is between 0.04 and 0.08.[38]

It is recognized that defect sites and bismuth ions are both needed for alkene oxidations over bismuth-lead molybdate catalysts. It is also generally accepted that the first step in these reactions is the dissociative adsorption of the olefin to form an allyl radical. It appears that the role of the defect is to promote this dissociative adsorption as shown in Fig. 10.5 for propene oxidation. The initial adsorption of the alkene occurs at a molybdate group near a defect site with the proton from the alkene attracted to one of the electron rich oxygens near the vacant site. Transfer of an electron to the oxide gives the adsorbed allyl radical. Transfer of one of the oxygens from the molybdate group and loss of another hydrogen follows along with the transfer of three more electrons to the catalyst. Desorption of acrolein and water gives an electron rich, oxygen deficient surface. An oxygen molecule is adsorbed on a surface bismuth ion where it takes up the four electrons from the catalyst and replenishes the surface oxide vacancies.

Thus, the defect site has adjacent electron rich species that are needed for the dissociative adsorption of the alkene. With 1-butene oxygen transfer from the molybdate is not favored. Instead, a hydrogen from C_4 is abstracted to give butadiene and a dihydroxy molybdenum from which water is eliminated to give an oxygen deficient species. Oxygen adsorption again takes place preferentially

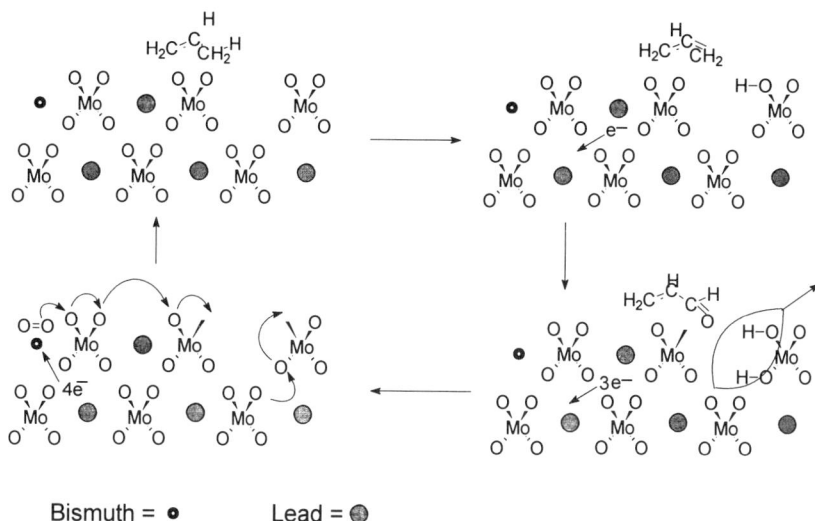

Bismuth = ●   Lead = ◉

Fig. 10.5. Mechanistic representation of propene oxidation over bismuth-
lead molybdate.

on a surface bismuth ion, which is a lone pair cation. This lone pair of electrons
attracts molecular oxygen that is seeking a center of electron density on the
reduced catalyst surface. The filling of the "oxygen holes" occurs by successive
transfer of adjacent oxygens until the vacancy appears in the vicinity of the
adsorbed oxide ions as depicted in Fig. 10.5.[38] This reaction, then, follows an
intrafacial mechanism.[6]

Heteropoly Acids

Heteropoly acids are polyprotic mixed oxides that are used both as solid acids and
selective oxidation catalysts.[39,40] They are generally composed of a central ion
bonded to an appropriate number of oxygens and surrounded by a near spherical
shell of octahedral oxometallic species joined together by shared oxygens as
illustrated in Fig. 10.6 for 12-tungstophosphoric acid.[40] The central atom or
'heteroatom" is commonly a cation having a +3 to +5 oxidation state such as P^{+5},
As^{+5}, Si^{+4} or Mn^{+4}. The metallic species associated with the octahedra are
usually molybdenum, tungsten or vanadium. The structure shown in Fig. 10.6
was first assigned by Keggin[41,42] and this type of oxide is referred to as a Keggin-
type heteropoly anion.

    These materials are thermally stable so they find use as solid acid catalysts
for a variety of vapor phase reactions. They can be used as the bulk oxyacid but
are more efficient when they are supported on materials such as alumina, silica or
carbon. The supported heteropoly acids are usually prepared by either incipient
wetness or equilibrium impregnation procedures.[43] Carbon supported heteropoly

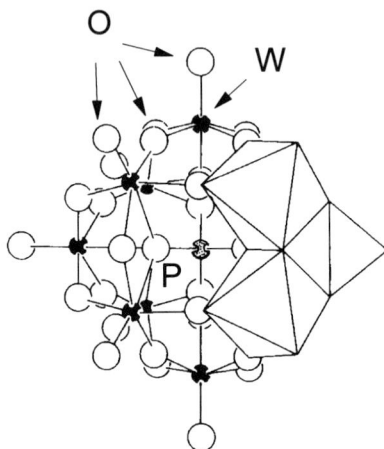

Fig. 10.6. Representation of the Keggin structure of the $PW_{12}O_{40}^{-3}$ anion. (Reproduced, with permission, from Ref. 40.)

acids can be prepared by suspending the support material in a methanol solution of the acid and evaporating the solvent under vacuum. The non-supported acid is removed by washing with water or methanol. The maximum loading possible using this procedure depends on the nature of both the carbon and the heteropoly acid but is usually in the 7%–14% range.[44]

$$(10.18)$$

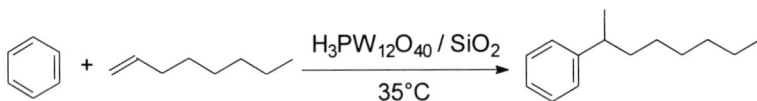

These oxyacids are soluble in water and, on crystallization from this solvent, retain several molecules of water of crystallization. The presence of water in the crystal facilitates their solution in polar solvents in which they act as homogeneous acid catalysts. Suspended in a non-polar solvent, however, they are heterogeneous catalysts which retain their Keggin structure[45] and acidic character. These heterogeneous acid catalysts have been used to promote a variety of reactions such as ether formation, esterification and Freidel-Crafts reactions. Supported acids are generally preferred for these reactions. The Freidel-Crafts alkylation of benzene with 1-octene (Eqn. 10.18) takes place over silica supported $H_3PW_{12}O_{40}$ at 35°C. The major product is 2-phenyloctane with some 3- and 4-phenyloctanes also formed. In an excess of benzene less than 5% of the dioctylbenzene is produced at 100% conversion of the alkene.[46] This catalyst must first be activated by heating to 150°C. Evidently this is necessary to remove the water of crystallization from the acid and, thereby, increase its acidity. Heating to 100°C failed to activate this acid while the use of temperatures above 150°C resulted in a decrease in catalytic surface area and a significant loss of activity.

Not all of the acid character of heteropoly acids is the result of the Bronsted acidity of the OH groups. Silica supported $H_3PMo_{12}O_{40}$ catalyzes the alkylation of benzene with benzyl chloride at 80°C. After three hours a 72% conversion was observed using a catalyst that had been activated by heating to 300°C.[46] Under the same conditions, $H_3PW_{12}O_{40}$ gave only a 27% conversion. Even though the acid strength of the tungstate is greater than the molybdate these results indicate that the molybdate has a greater Lewis acidity than does the tungstate.

When $H_3PMo_{12}O_{40}$ was used to catalyze acylation reactions some of the acid dissolved and the reactions were partially promoted by the acid in solution.[46] The $H_4SiW_{12}O_{40}$ and $H_3PW_{12}O_{40}$ catalysts, however, did not dissolve under the reaction conditions and were, therefore, effective promoters for liquid phase Freidel-Crafts acylation reactions.[46] Supported and bulk heteropoly acids have also been used to promote vapor phase Freidel-Crafts alkylations and esterifications.[47].

Heteropoly acids can also act as oxidation catalysts both in the vapor and liquid phase. In the vapor phase they are effective dehydrogenation catalysts for saturated carboxylic acids and aldehydes readily converting isobutyric acid to methacrylic acid (Eqn. 10.19).[48] Methacrylic acid is produced in 70% selectivity at 72% conversion over $(NH_4)_3PMo_{12}O_{40}$ at 260°C.[48] This reaction takes place only when there is a substituent α to the carbonyl group of the reactant.

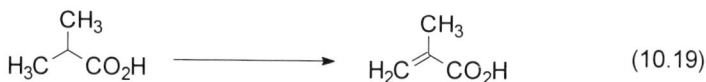

$$H_3C-\overset{\overset{\displaystyle CH_3}{|}}{C}H-CO_2H \longrightarrow H_2C=\overset{\overset{\displaystyle CH_3}{|}}{C}-CO_2H \qquad (10.19)$$

Zeolites

Zeolites are crystalline aluminosilicates that have a three dimensional polymeric arrangement of tetrahedral SiO_4 and AlO_4^- units joined through shared oxygens as depicted above in Fig. 10.3 for the amorphous aluminosilicates. A large number of zeolites are known, some of them are naturally occurring, but most have been prepared synthetically.[49,50] As with the amorphous aluminosilicates, the zeolites contain an extra framework cation, usually Na^+, to maintain electrical neutrality with the AlO_4^- species. These extra framework ions can be replaced by other cations using ion-exchange techniques. Since zeolites react with acids, the proton exchanged species are best prepared by way of an initial exchange to give the ammonium zeolite, which is then thermally dissociated to liberate ammonia and give the acid zeolite.

As described above for the amorphous aluminosilicates the acid strength of a zeolite increases with increasing Si:Al ratios. Species with particularly high ratios have been prepared by removing some of the aluminum by reaction with materials such as silicon tetrachloride. On the other hand, the basic character of these materials increases with increasing numbers of AlO_4^- species so aluminum-rich zeolites are better bases, particularly when the Na^+ is replaced with a large cation such as Cs^+.[51,52]

The most important aspect concerning the catalytic application of zeolites is not this range of potential acid-base properties since that is also available with the amorphous aluminosilicates. Instead it is the presence in these crystalline materials of molecular sized cavities and pores that make the zeolites effective as shape selective catalysts for a wide range of reactions.[51,53-59]

As these structures are commonly drawn they can be confusing to the organic chemist who is used to having the lines represent bonds between atoms located at the intersections of these lines. This is not the case with the zeolite structures. Here, the intersection of lines represents either a Si or an Al atom and the line represents the oxygen atom joining the two tetrahedra. The more detailed structure drawn in Fig. 10.7a is abbreviated as shown in Fig. 10.7b. This is commonly referred to as an 8-ring structure since there are eight oxygens in the ring system. It is the opening inside the oxygens that determines the size of the pores in a particular zeolite as depicted in Fig. 10.7c.

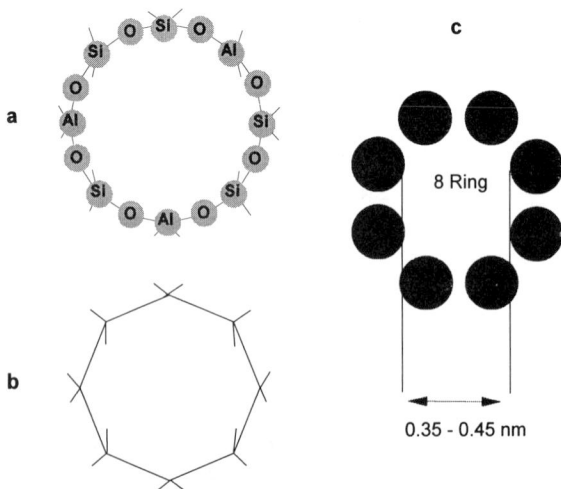

Fig. 10.7. a) Si-O-Al bonding in zeolites; b) representation of (a) with the intersections representing Si or Al atoms and the lines the O bonded to them; c) oxygen atoms in the ring.

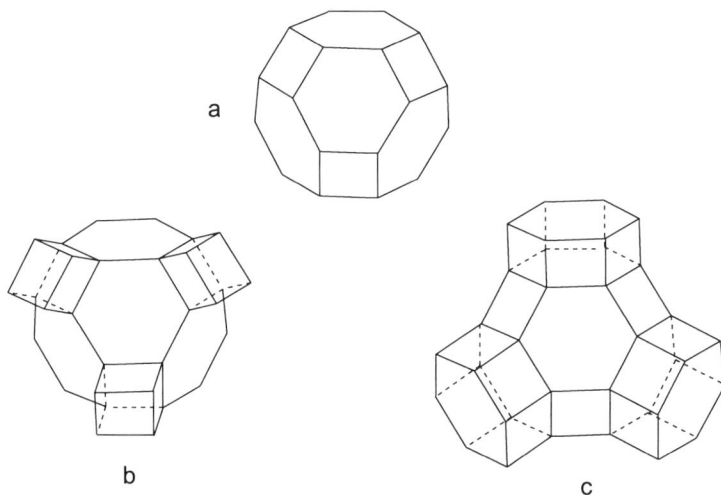

Fig. 10.8.   a) Sodalite cage with b) 4-ring extension and c) 6-ring
            extension.

The SiO_4 and AlO_4^- lattice units are combined by way of shared oxygens to form three dimensional arrangements which serve as the basic building blocks for the zeolites.   One of these basic units is the sodalite cage pictured in Fig. 10.8a.   This material is a truncated octahedron with four-membered rings made during the truncation and the six-membered rings as part of the side of the original octahedron.   Access to this cage, sometimes referred to as the β cage, is through the six-ring opening which has a diameter of about 0.25 nm.   The internal diameter of the cage itself is 0.66 nm.   Only very small molecules, such as H_2 and H_2O can enter this cage.[60]

The silica mineral, sodalite, is composed of these truncated octahedra joined through shared four-membered rings.   This gives a relatively close-packed arrangement with very small openings between the cages.   In the zeolites having the sodalite cage as a building unit these cages are joined together through extensions of either the 4- ring (Fig. 10.8b) or the 6-ring (Fig. 10.8c).   Zeolite A is composed of sodalite cages joined by extended 4-rings as shown in Fig. 10.9.   The center of this structure is a supercage that is surrounded by eight sodalite cages. The interior of this supercage has a diameter of 1.14 nm.   There are six mutually perpendicular 8-ring openings into the cage, each with a 0.42 nm diameter enabling organic molecules such as linear hydrocarbons to enter the supercage. The Si:Al ratio for A is between 1 and 1.2 making these materials among the least acidic and, potentially, the most basic of all zeolites after an appropriate ion exchange.[61]   However, the small size of the cage opening can hinder extensive

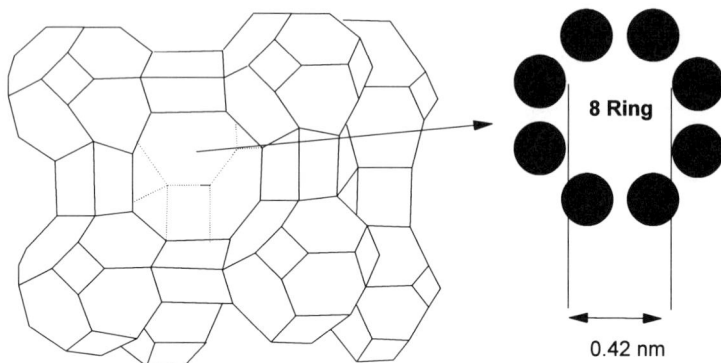

Fig. 10.9. Zeolite A with its 8-ring pore opening.

exchange with large cations such as Cs^+ so the full basicity of these materials is not easily realized.

Faujasites are naturally occurring zeolites in which the sodalite cages are joined through extensions of the 6-ring faces of the sodalite cage as depicted in Fig. 10.10. In this structure the internal supercage has a diameter of about 1.3 nm.[62] This has four tetrahedrally oriented 12-ring openings each with a diameter of 0.74 nm. This size permits larger molecules such as naphthalene and other relatively large aliphatic and aromatic compounds to enter the supercage. The synthetic zeolites X and Y have the same crystal structure as faujasite but differ from faujasite and each other in their Si:Al ratios. In faujasite this ratio is about 2.2, in X it is between 1 and 1.5 and in Y between 1.5 and 3.[62] The large cage openings and the low Si:Al ratio makes X an ideal candidate for exchange with large cations to produce a solid base catalyst. Of all of the common exchanged zeolites, CsX is the most basic.

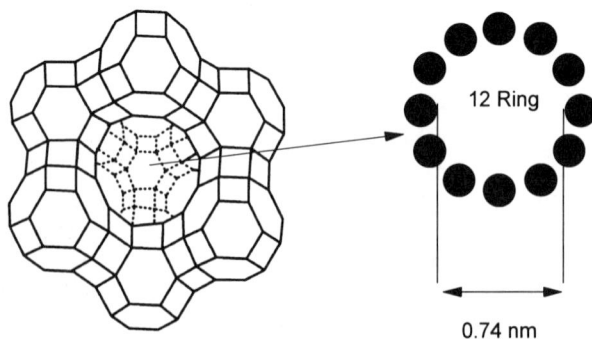

Fig. 10.10. Structure of faujasite, zeolite X and zeolite Y with the 12-ring pore opening.

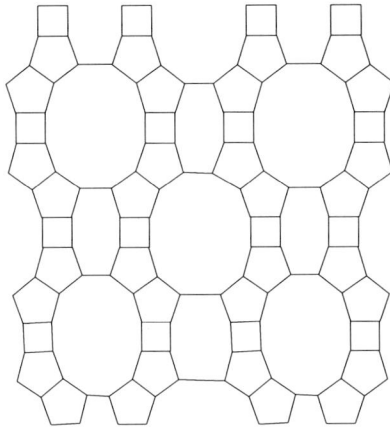

Fig. 10.11. Two dimensional view of the 12-ring and 8-ring pores of mordenite.

Mordenite is a naturally occurring zeolite that has a Si:Al ratio near ten.[63] Synthetic mordenites also have this same narrow range Si:Al ratio, a fact that suggests that the aluminum atoms are distributed in an orderly manner in this zeolite rather than in the presumed random orientation in the other zeolites. The central feature of mordenite is a slightly oval shaped 12-ring with a 0.67 x 0.7 nm

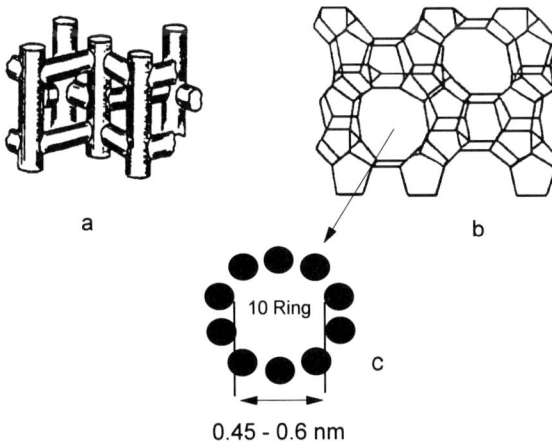

Fig. 10.12. Structural representation of ZSM-5. a)Schematic of the intersecting tunnel systems; b) framework structure; c) 10-ring tunnel size. (a and b reproduced, with permission, from Ref. 64.)

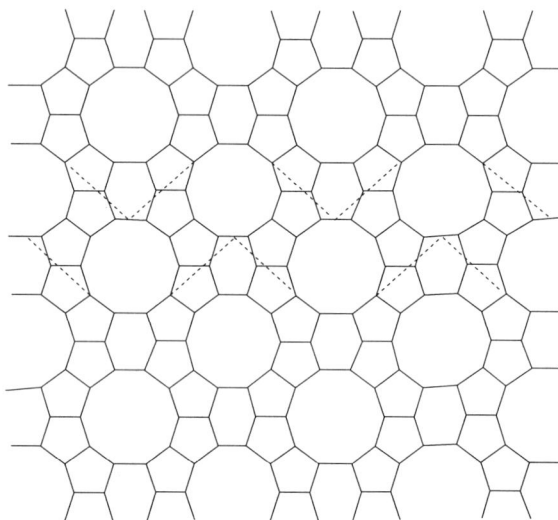

Fig. 10.13. Two dimensional view of ZSM-5 showing the top of the 10-
ring tunnels. The dashed lines represent the path of one of the
perpendicular sinusoidal tunnels.

size as pictured in Fig. 10.11. Near this is an 8-ring of 0.39 nm diameter. These
cavities extend through the zeolite framework. As seen from this structure every
framework atom is on the sides of either a 12-ring tunnel or an 8-ring tunnel, so
all of the aluminum atoms and their associated anion or acid sites are also on these
tunnel walls and accessible to reacting species migrating through them.

The most siliceous common zeolite is the synthetic material, ZSM-5,
which can have Si:Al ratios between 25 and 2000.[63] As shown in Fig. 10.12a,[58]
the ZSM-5 framework consists of two intersecting 10-ring tunnel systems with
0.55–0.6 nm diameters. One passes straight through the framework as do those in
mordenite while the other follows a sinusoidal path perpendicular to the first but
intersecting with them. The framework representation shown in Fig. 10.12b
shows the 10-ring openings of the straight tunnels. A top view of this framework
is shown in Fig. 10.13 in which the ends of the straight tunnels are surrounded by
5- and 6-rings. The path of one of the perpendicular sinusoidal tunnels is denoted
by the dashed lines in the figure. The structural unit of this zeolite is the pentasil
system pictured in Fig. 10.14a, which, as the name implies, is a composite of
fused 5-rings. A pentasil unit in the framework is illustrated by the partial
structure shown in Fig. 10.14b.

These are some of the more common examples of the more than 100
natural and synthetic zeolites described in the literature. They represent some of
the different pore sizes found in these materials and those zeolites having a more

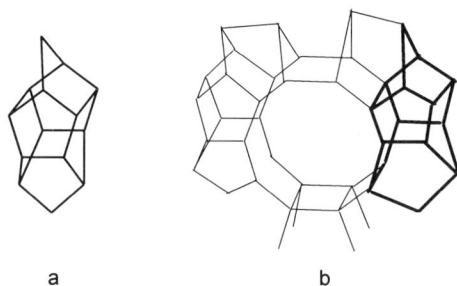

a   b

Fig. 10.14. a) Pentasil structural unit of ZSM-5; b) partial ZSM-5 structure
           showing a pentasil unit.

tortuous diffusion pathway such as the faujasites and those with the reactive
tunnels passing through the crystal as with A, mordenite and ZSM-5. Obviously,
the diffusion properties of these two types of systems can be different.

Further, the Si:Al ratios of these materials can be increased by removing
some of the aluminum from the zeolite. This results in an increase in the acidity
for the modified material. Acid leaching of mordenite gives a dealuminated
material in which X-ray diffraction patterns indicate that little, if any, structural
changes occurred.[64] It has been proposed that removal of the aluminum gives a
material in which the aluminum is replaced by a nest of four hydroxy groups (Fig.
10.15) a factor that changes the ionic character and catalytic behavior of the
dealuminated material. Dealumination with chelating agents such as
acetylacetone and ethylenediamine tetraacetic acid also results in the loss of
framework aluminum.[65-69] Spectral data have been obtained that indicates the
presence of hydroxy nests in Y that was treated with ethylenediamine tetraacetic
acid.[69]

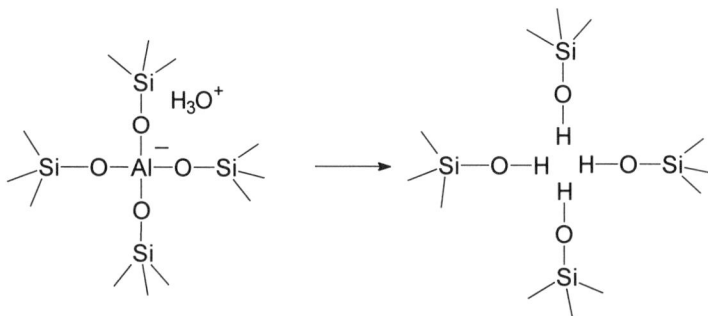

Fig. 10.15. "Hydroxy nest" produced on removal of aluminum from a
           zeolite.

Reaction of a zeolite with silicon tetrachloride removes some of the aluminum but instead of creating framework holes, the aluminum is replaced by silicon and the zeolite framework is retained.[70]

The wide range of Si:Al ratios permits an almost tailor-made acidity to be built into these systems particularly if dealumination is used. This and the molecular sized pores and cavities gives the zeolites a shape-selective capability not found in most other heterogeneous catalytic systems. This shape selectivity can be classified into three major categories.[65,71,72]

Reactant selectivity is operative when there are reagents of different size present in the reaction medium but access to the active site is restricted for one or more of them. In the narrow pore zeolite, CaA, n-butanol is dehydrated but the more bulky isobutanol is not (Eqn. 10.20). In the wider pore, CaX, isobutanol is dehydrated more readily than n-butanol as anticipated for normal, unrestricted acid catalyzed dehydration reactions.[72]

(10.20)

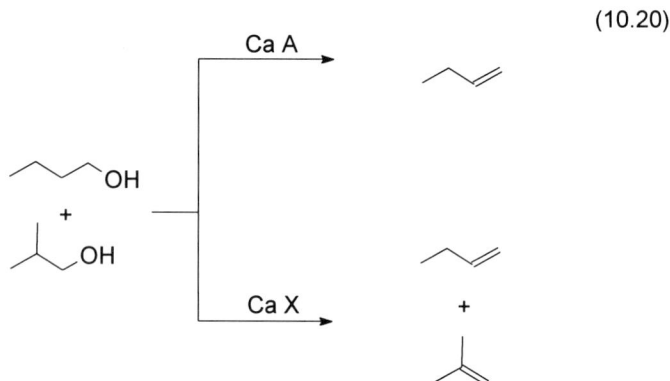

Product selectivity results from differences in the size of the products produced from a given reaction. In a homogeneous reactions methylation of toluene gives a mixture of ortho, meta and para xylenes but when H-ZSM-5 is used as the acid, p-xylene is the almost exclusive product (Eqn. 10.21) because the passage of this less bulky isomer through the pores of the zeolite is not restricted while the more bulky o- and m-isomers are too large to easily go through them.[72]

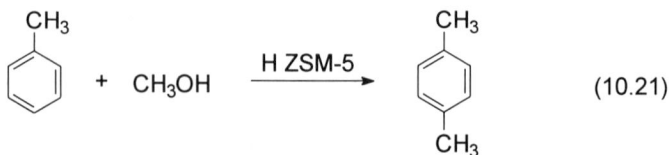

(10.21)

Restricted transition state selectivity is observed when the transition state leading to one product is more bulky than that leading to another. In this case the product from the less-restricted transition state will be favored. As an example, the disproportion of o-xylene to trimethyl benzene and toluene requires a bulky diaryl species as the transition state but isomerization to m- or p-xylene simply requires successive 1–2 methyl shifts, a process having a smaller transition state. As the size of the zeolite cavity decreases, the ratio of the rate of disproportionation to the rate of isomerization decreases because the larger transition state is not as easily accommodated in the smaller cavities.[72]

Since the shape selectivity is clearly related to reactant and/or product diffusion through the pores and cavities of the zeolite, the selectivity should increase as the size of the zeolite particles and, thus, the extent of diffusion, increases. This has been established using the H-ZSM-5 alkylation of toluene with methanol as the probe reaction. Three catalysts with particles ranging in size from 0.025 μm to 4.5 μm were used. The results listed in Table 10.1[73] for reactions run at various temperatures show that with the largest catalyst particles selectivity toward p-xylene formation was 100% at all temperatures. As the particle size decreased so did the reaction selectivity. Increasing the temperature increased the reaction selectivity with the smaller particle sized catalysts.[73]

Because of the extensive cavity and pore structure the zeolites have a high surface area but only about 5-10% of the surface area is not associated with the walls of the pores.[74]

Table 10.1

Xylene isomer distribution from toluene methylation over different sized zeolite particles[69]

Temperature °C	Particle Size (μm)	% para	% meta	% ortho
275°	0.025	63	18	19
	0.045	75	13	12
	4.5	100	0	0
300°	0.025	75	12	13
	0.045	85	8	7
	4.5	100	0	0
350°	0.025	84	10	6
	0.045	91	6	3
	4.5	100	0	0

Most of the catalytic activity of zeolites has been concerned with their ability to act as shape selective solid acids. The vapor phase Beckmann rearrangement of cyclohexanone oxime over HY zeolite at 300°C gave caprolactam in 80% selectivity at 82% conversion (Eqn. 10.22).[75]

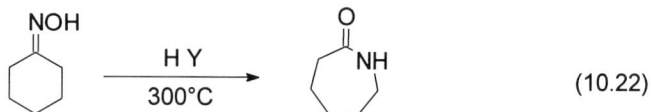

$$(10.22)$$

As described above, zeolites can also act as solid base catalysts when the Si:Al ratio is low and the extra framework cation is a large one such as Cs^+.[76-78] The most basic common exchanged zeolite is CsX. With this material the aldol condensation of cyclooctanone with benzaldehyde gave only the monobenzylidene compound (Eqn. 10.23)[15] while reactions using piperidine, magnesium oxide or amorphous cesium aluminosilicate gave both the mono- and di-benzylidene products(Eqn. 10.12). The smaller ketones, cyclohexanone and

$$(10.23)$$

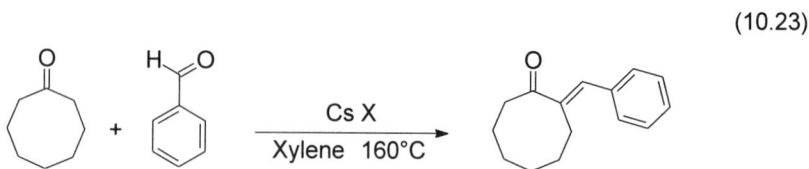

cyclopentanone gave primarily monosubstitution over CsX but some disubstitution was also observed because of the large pore size of this zeolite.[15] The more restrictive CsA gave almost exclusive monobenzylidene products from cyclohexanone and cyclopentanone, and exclusively the 6-benzylidene-3-methylcycohexanone from 3-methylcyclohexanone (Eqn. 10.24).

$$(10.24)$$

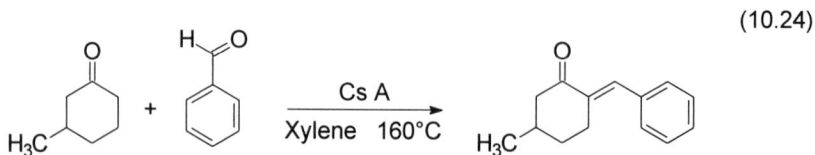

Interestingly, though, CsA also promoted the condensation with cyclooctanone, a ketone obviously too large to enter the cavity of this zeolite. These and other experimental data indicate that with the large pore zeolite, X, the condensation reaction takes place inside the cage and the reaction products are determined by their ability to migrate out through the pores. Reactions catalyzed by the small pore, CsA, however, appear to take place at the mouths of the pores and not inside the cavities.[15]   This still gives some shape selectivity to the reaction when compared to the product composition observed using the non-crystalline aluminosilicate as the catalyst.

The base strength of zeolites has been increased over that of the simple Cs^+ exchanged species by impregnating the zeolite with cesium acetate and then thermally decomposing the acetate inside the cages.[79-81]   These catalysts are capable of promoting the alkylation of toluene with methanol or formaldehyde at 437°C (Eqn. 10.25).[81]   This reaction is thought to proceed by way of an initial dehydrogenation of methanol to formaldehyde followed by reaction of the aldehyde with the benzyl anion produced by reaction of toluene with the strongly basic sites.[81]

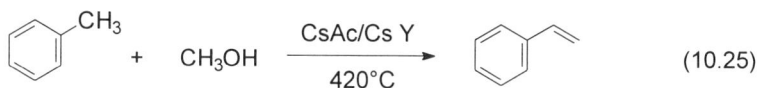

$$
\text{(structure: toluene)} + CH_3OH \xrightarrow[420°C]{CsAc/Cs\ Y} \text{(structure: styrene)} \qquad (10.25)
$$

Other Molecular Sieves

Replacing the SiO_4 or the AlO_4^- tetrahedra in the zeolite framework with other species can lead to the formation of molecular sieves with varying acidities and pore sizes. Aluminophosphates, called ALPOs, are molecular sieves consisting of AlO_4^- and PO_2^+ tetrahedra.[82-84]   The best known of these species is ALPO-5, a material having straight circular 12-ring channels similar to those found in mordenite with a pore diameter of 0.73 nm.[85]   A similar material, ALPO-31, has a more puckered 12-ring channel with an effective pore diameter of 0.53 nm.[86] ALPO-11 has elliptical 0.39 x 0.63 nm channels and ALPO-8 has larger elliptical channels of 0.79 x 0.87 nm dimensions.[84]   Considerable effort is being directed toward the synthesis of molecular sieves having larger pore sizes so shape selectivity in adsorption and reactions can be applied to larger, more complex molecules.[84]

Although ALPOs are neutral materials with a P:Al ratio near one, they do posses some Bronsted and Lewis acidity.[82]   Replacing some of the phosphorous and/or aluminum with other elements provides a means of increasing this acidity. Silicoaluminophosphates, SAPOs, are formed by the partial replacement of the lattice phosphorous by silicon.[87]   The resulting species have acidities higher than the corresponding ALPO but lower than the high silica zeolites such as H-ZSM-5.[82]   Incorporation of a small amount of manganese oxide,[86] zinc oxide[83] or titanium oxide[88] also increases catalyst acidity. But even with these materials, the acidity is still rather low making them desirable for reactions in which more strongly acidic catalysts could cause unwanted side reactions. $AlPO_4 \cdot Al_2O_3$ systems are the most acidic in this series.[89]

The titanium silicalites are composed of TiO_4 and SiO_4 tetrahedra joined through shared oxygens.   They are structurally isomorphous to the high silica zeolites but have little acid character.[90-92] Interestingly, though, the titanium silicalite, TS-2,[93] readily promotes the vapor phase Beckman rearrangement of cyclohexanone oxime to give caprolactam in about 90% selectivity at 95–100% conversion.[94] Acidic catalysts such as the Zeolite ZSM-5 were considerably less selective. These results agree with the proposal that the active sites for this rearrangement are the weak acid or neutral centers of the zeolite and that the strong acid sites promote unwanted side reactions.[95]

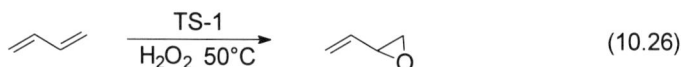

$$\text{(image of reaction)} \quad \xrightarrow[\text{H}_2\text{O}_2 \ 50°\text{C}]{\text{TS-1}} \quad \text{(product)} \qquad (10.26)$$

The primary use for the titanium silicalites is as shape selective catalysts for hydrogen peroxide oxidations.[96-99] Propylene is converted to propylene oxide at greater than 98% selectivity and 99% peroxide conversion at 50°C over TS-1.[92,97] Butadiene is oxidized to the monoepoxide (Eqn. 10.26), also in high selectivity, and primary alcohols are oxidized to the aldehydes in all cases with selectivites greater than 80%.[97]

Clays

Clays are naturally occurring minerals having a high aluminosilicate composition.[100] They have been used as solid acid catalysts particularly in petroleum refining from the 1930s to the mid 1960s when they were replaced by the zeolites. One of the most common clays, montmorillonite, has the general structure shown schematically in Fig. 10.16a. Each layer is composed of an octahedral sheet sandwiched between two tetrahedral sheets. Typically the octahedral sheet is composed of oxygens attached to Al^{+3} and some lower valence cations such as Mg^{+2}. The tetrahedral sheets are composed of Si^{+4} and some Al^{+3} oxides. The overall layer has a net negative charge with hydrated cationic

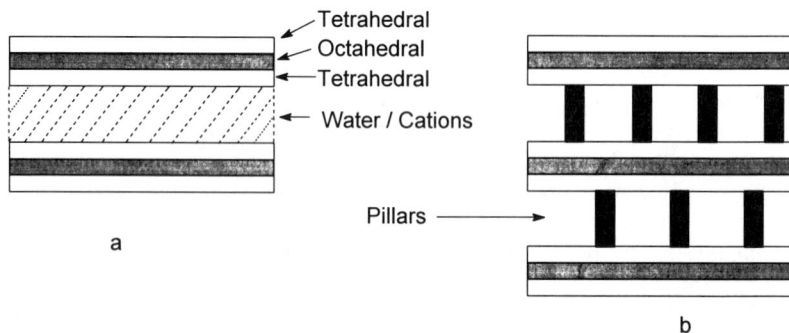

Fig. 10.16. Schematic representation of a) the layered structure of montmorillonite and b) pillared montmorillonite.

counterions occupying the spaces between the layers.[100-102] When immersed in water the mineral swells, exposing the intercalated cations to potential cation exchange. These materials exhibit substantial Bronsted acidity from the dissociation of surface Si-OH groups and from the hydrated interlayer cations. The more electronegative the cation, the stronger the acidity of the clay. An Al^{+3} exchanged montmorillonite is as active as concentrated sulfuric acid in promoting acid catalyzed reactions.[103] Other exchanged montmorillonites have been used to catalyze a number of useful reactions such as aldol condensations, Michael additions and other carbon-carbon bond formations[104] as well as alkene addition reactions, ether and ester formations and nitrile conversions to amides.[105] Aldehydes react over bentonite at room temperature to give good yields of trioxanes (Eqn 10.27).[106] Calcined Ti(IV) montmorillonite promotes the Diels Alder reaction between methyl acrylate and cyclopentadiene to give the endo (**4**) and exo (**5**) products in a 15:1 ratio with 95% conversion after two hours (Eqn. 10.28).[107] Unmodified kaolin also proved to be a good catalyst in promoting the Diels Alder reaction between methyl vinyl ketone and cyclopentadiene.[108]

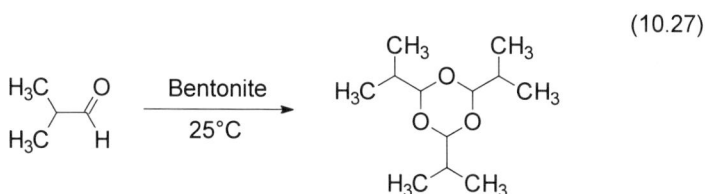

(10.27)

Montmorillonite supported zinc chloride, called Clayzic,[109] has become an important solid acid catalyst.[110] This material has both the Bronsted acidity associated with the clay as well as Lewis acidity from the zinc ions.[111] Thermal activation at 275°C gave a catalyst that promoted the benzylation of cumene at 40°C in near quantitative yield.[110]

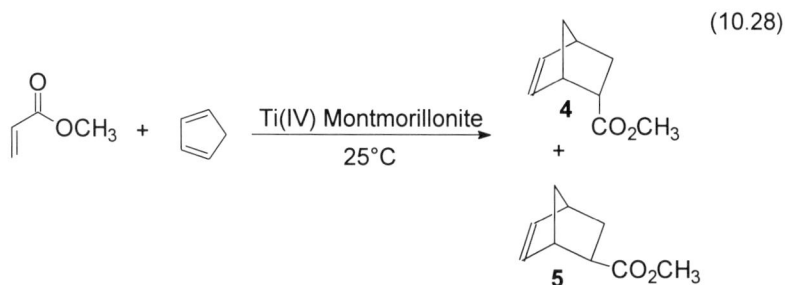

(10.28)

Generally speaking, however, heating exchanged clays results in a considerable loss of activity caused by the loss of water and subsequent collapse of the interlayer region so only the external surface of the clay is exposed to the reaction medium This collapse decreases the active surface area to less than 10% of that in the hydrated form. When the clay is treated with large inorganic cations, they react with the layer surfaces and remain as pillars or supports between the layers when the water is removed as depicted in Fig. 10.16b.[112-118] The most common pillaring agent[119] is the aluminum hydroxy acid, $Al_{13}O_4(OH)_{24}(H_2O)_{12}]^{+7}$. This is a Keggin-like oligomer having a central AlO_4^- tetrahedron surrounded by twelve aluminum oxyhydroxide octahedra[120] similar in structure to the Keggin ion pictured in Fig. 10.6. It is formed by the base hydrolysis of an aluminum chloride solution.[121,122] This pillaring provides a space of about 0.8 nm between the layers which remains after drying. It also increases the surface area of the dried, pillared material to a value about fifty times greater than the dried clay itself.[100] These pillared clays exhibit both the Bronsted acidity inherent to the silica layers and the Lewis acidity from the aluminum ions in the pillars.[123] They have an acidity somewhere between the H-ZSM-5 and the HY zeolites.[123] When the intercalation with the aluminum oxyhydroxide is run in the presence of other cations the acid strength of the resulting pillared clay can be modified.[115] For instance, Ce^{+3} exchange stabilizes Bronsted acidity and Zn^{+2} suppresses it.

The spacing between the layers can be regulated by the nature of the pillaring material. For instance, with titanium oxide oligomers as the pillaring agents, layer separations of 1.2 nm have been achieved.[124] An even larger, 1.5–1.9 nm, spacing was achieved using ferric oxyhydroxy pillars.[125]

References

1. J. M. Thomas, *Sci. Am.,* **266**, 112 (1992).
2. H. Hattori, *Mater. Chem. Phys.,* **18**, 533 (1988).
3. H. Hattori, *Stud. Surf. Sci. Catal.,* **78** (Heterog. Catal. Fine Chem., III), 35 (1993).
4. G. C. Bond, *Heterogeneous Catalysis, Principles and Applications*, Clarendon Press, Oxford, 1987, Chap. 4.
5. A. Holden, *The Nature of Solids*, Columbia Univ. Press, New York, 1965, Chap. XIV.
6. P. Mars and D. W. van Krevelen, *Chem. Eng. Sci.,* **41** (Suppl. 3), 41 (1954).
7. V. D. Sokolovskii, *Catal. Revs.,* **32**, 1 (1990).
8. N. Pernicone and F. Traina, *Appl. Ind. Catal.,* **3**, 1 (1984).
9. J. T. Richardson, *Principles of Catalyst Development*, Plenum Press, New York, 1989, Chap. 6.
10. T. Curtin, J. B. McMonagle and B. K. Hodnett, *Catal. Letters,* **17**, 145 (1993).
11. T. Curtin, J. B. McMonagle, M. Ruwet and B. K. Hodnett, *J. Catal.,* **142**, 172 (1993).
12. M. do Céu Costa, R. Tavares, W. Motherwell and M. J. M. Curto, *Stud. Surf. Sci. Catal.,* **78** (Heterog. Catal. Fine Chem., III), 639 (1993).
13. N. Cardona-Martinez and J. A. Dumesic, *J. Catal.,* **127**, 706 (1991).
14. A. Servino, J. Vital and L. S. Lobo, *Stud. Surf. Sci. Catal.,* **78** (Heterog. Catal. Fine Chem., III), 685 (1993).

15. L. Posner and R. L. Augustine, *Chem. Ind. (Dekker),* **62** (Catal. Org. React.), 531 (1995).
16. G. C. Bond, *Appl. Catal.,* **71**, 1 (1991).
17. T. Fransen, P. C. van Berge and P. Mars, *Stud. Surf. Sci. Catal.,* **1** (Prep. Catal.), 405 (1976).
18. J. Kobayashi, T. Shimuzu and K. Inamura, *Chem. Lett.,* 211 (1992).
19. T. Machej, J. Haber, A. M. Turek and I. E. Wachs, *Appl. Catal.,* **70**, 115 (1991).
20. F. Cavani, G. Centi, F. Parrinello and F. Trifiro, *Stud. Surf. Sci. Catal.,* **31,** (Prep. Catal., IV), 227 (1987).
21. R. Margraf, J. Leyrer, E. Taglauer and H. Knözinger, *React. Kinet. Catal. Lett.,* **35**, 261 (1987).
22. D. S. Kin, K. Segawa, T. Soeya and I. E. Wachs, *J. Catal.,* **136**, 539 (1992).
23. C. Mercier and P. Chabardes, *Stud. Surf. Sci. Catal.,* **78** (Heterg. Catal. Fine Chem., III), 677 (1993).
24. U. Scharf, M. Schrami-Marth, A. Wokaun and A. Baiker, *J. Chem. Soc., Faraday Trans.,* **87**, 3299 (1991).
25. M. Glinski and J. Kijenski, *Stud. Surf. Sci. Catal.,* **16** (Prep. Catal., III), 553 (1983).
26. A. Baiker and A. Wokaun, *Naturwissenschaften,* **76**, 168 (1989).
27. B. N. Reddy, B. M. Reddy and M. Subrahmanyam, *J. Chem. Soc., Faraday Trans.,* **87**, 1649 (1991).
28. A. M. Turek, I. E. Wachs and E. DeCanio, *J. Phys. Chem.,* **96**, 5000 (1992).
29. *Catalytic Chemistry of Solid-State Inorganics,* W. R. Moser and J. Happel, Eds., *Ann. N. Y. Acad. Sci.,* **272**, 1 - 73 (1976).
30. P. Courty and C. Marcilly, *Stud. Surf. Sci. Catal.,* **1** (Prep. Catal.),119 (1976).
31. P. Courty and C. Marcilly, *Stud. Surf. Sci. Catal.,* **16** (Prep. Catal., III), 485 (1983).
32. K. Tanabe, M. Itoh, K. Morishige and H. Hattori, *Stud. Surf. Sci. Catal.,* **1** (Prep. Catal.), 65 (1976).
33. M. Schraml-Marth, K. L. Walther, A. Wokaun, B. E. Handy and A. Baiker, *J. Non-Cryst. Solids,* **143**, 93 (1992).
34. T. López, M. Azomoza and R. Gómez, *J. Non-Cryst. Solids,* **147 & 148**, 769 (1992).
35. A. Baiker, P. Dollenmeier, M. Blinski and A. Reller, *Appl. Catal.,* **35**, 351 (1987).
36. B. C. Gates, *Catalytic Chemistry,* Wiley, New York, 1992, p 270.
37. P. L. Gai-Boyes, *Catal. Revs.,* **34**, 1 (1992).
38. A. W. Sleight and W. J. Linn, *Ann. N. Y. Acad. Sci.,* **272**, 22 (1976).
39. Y. Izumi, K. Urabe and M. Onaka, *Zeolite, Clay, and Heteropoly Acid in Organic Reactions,* VCH, New York, 1992, Chap. 3.
40. M. Misono, *Catal. Revs.,* **29**, 269 (1987).
41. J. F. Keggin, *Nature,* **131**, 908 (1933).
42. J. F. Keggin *Proc. Roy. Soc.,* **A144**, 75 (1934).
43. Y. Kera, Y. Ishihama, T. Kawashima, T. Kamada, T. Inoue and Y. Matsukaze, *Stud. Surf. Sci. Catal.,* **31** (Prep. Catal., IV) 259 (1987).
44. Ref. 39, p.130.
45. C. Rocchiccioli-Deltcheff, R. Thouvenot and R. Frank, *Spectrochim. Acta,* **32A**, 587 (1976).

46. Y. Izumi, N. Natsume, H. Takamine, I. Tamaoki and K. Urabe, *Bull. Chem. Soc., Japan*, **62**, 2159 (1989).
47. Y. Izumi, R. Hasebe and K. Urabe, *J. Catal.*, **84**, 402 (1983).
48. S. Albonetti, F. Cavani, M. Koutyrev and F. Trifirò, *Stud. Surf. Sci. Catal.*, **78** (Heterg. Catal. Fine Chem., III), 471 (1993).
49. S. L. Suib, *Chem. Rev.*, **93**, 803 (1993).
50   E. B. M. Doesburg *Stud. Surf. Sci. Catal.*, **79** (Catalysis), 309 (1993).
51. C. B. Dartt and M. E. Davis, *Catal. Today*, **19**, 151 (1994).
52. T. Yashima, K. Sato, T. Hayasaka and N. Hara, *J. Catal.*, **26**, 303 (1972).
53. Ref. 39, Chap. 1.
54. Ref. 36, Chap. 5.
55. F. R. Ribiero, *Catal. Letters*, **22**, 107 (1993).
56. J. Cornier, J.-M. Popa and M. Gubelmann, *Actual. Chim.*, 405 (1992).
57. F. G. Dwyer and A. B. Schwartz, in *Catalyst Supports and Supported Catalysts*, (A. B. Stiles, Ed.), Butterworths, Boston, 1987, Chap. 10.
58. W. Hölderich, M. Hesse and F. Näumann, *Angew. Chem. Int. Ed. Engl.*, **27**, 226 (1988).
59. C. S. John, D. M. Clark and I. E. Maxwell, *Prespectives in Catalysis* (J. M. Thomas and K. I. Zamaraev, Eds.), Blackwell, Oxford, UK, 1992, p 387.
60. Ref. 36, p.264.
61. D. W. Breck, W. B. Eversole, R. M. Milton, T. B. Reed and T. L. Thomas, *J. Am. Chem. Soc.*, **78**, 5963 (1956).
62. R. Rudham and A. Stockwell, *Catalysis (London)*, **1**, 87 (1977).
63. M. S. Spencer and T. V. Whittam, *Catalysis (London)*, **3**, 189 (1980).
64. G. T. Kokotailo, S. L. Lawton and D. H. Olson, *Nature*, **272**, 437 (1978).
65. J. W. Ward, *Appl. Ind. Catal.*, **3**, 271 (1984).
66. G. T. Kerr, *J. Phys. Chem.*, **72**, 2594 (1968).
67. G. T. Kerr, *J. Phys. Chem.*, **73**, 2780 (1969).
68. G. T. Kerr, A. W. Chester and D. H. Olson, *Acta Phys. Chem.*, **24**, 169 (1978).
69. J. Datka, W. Kolidziejski, J. Klinowski and B. Sulikowski, *Catal. Letters*, **19**, 159 (1993).
70. J. Klinowski, J. M. Thomas, M. W. Anderson, C. A. Fyfe and G. C. Gobbi, *Zeolites*, **3**, 5 (1983).
71. J. Dwyer, *Chem. and Ind., (London)*, 258 (1984).
72. P. B. Weisz, *Pure and Applied Chem.*, **52**, 2091 (1980).
73. K. Beschmann, L. Riekert and U. Müller, *J. Catal.*, **145**, 243 (1994).
74. A. Sayari, E. Crusson, S. Kaliaguine and J. R. Brown, *Langmuir*, **7**, 314 (1991).
75. S. Sato, K. Takematsu, T. Sodesawa and F. Nozaki, *Bull. Chem. Soc., Japan*, **65**, 1486 (1992).
76. A. Corma, *Mater. Res. Soc. Symp. Proc.*, **233** (Synthesis-Characterization and Novel Applications of Molecular Sieve Material), 17 (1991).
77. D. Barthomeuf and A. de Mallmann, *Stud. Surf. Sci. Catal.*, **37** (Innov. Zeol. Mater. Sci.), 365 (1988).
78. D. Barthomeuf, *Stud. Surf. Sci. Catal.*, **65** (Catal. Ads. Zeolites), 157 (1991).
79. P. E. Hathaway and M. E. Davis, *J. Catal.*, **116**, 262 (1989).

80. P. E. Hathaway and M. E. Davis, *J. Catal.*, **116**, 279 (1989).
81. P. E. Hathaway and M. E. Davis, *J. Catal.*, **119**, 497 (1989).
82. M. H. W. Burgers and H. van Bekkum, *Stud. Surf. Sci. Catal.*, **78** (Heterog. Catal. Fine Chem., III), 567 (1993).
83. M. R. Gelsthorpe and C. R. Theochris, *Catal. Today*, **2**, 613 (1988).
84. M. E. Davis, *Accounts Chem. Res.*, **26**, 111 (1993).
85. J. M. Bennett, J. P. Cohen, E. M. Flanigen, J. J. Pluth and J. V. Smith, *ACS Symp. Series*, **218**, 109 (1983).
86. H.-L. Zubowa, M. Richter, U. Roost, B. Barlits and R. Fricke, *Catal. Letters*, **19**, 67 (1993).
87. B. M. Lok, C. A. Messina, R. L. Patton, R. T. Gajek, T. R. Cannen and E. M. Flanigen, *J. Am. Chem. Soc.*, **106**, 6092 (1984).
88. F. M. Bautista, J. M. Campelo, A. Garcia, D. Luna, J. M. Marinas and M. S. Moreno, *Stud. Surf. Sci. Catal.*, **78** (Heterog. Catal. Fine Chem., III), 615 (1993).
89. J. M. Campelo, A. Garcia, D. Luna, J. M. Marinas and M. I. Martinez, *Mater. Chem. Phys.*, **21**, 409 (1989).
90. G. Deo, A. M. Turek, I. E. Wachs, D. R. C. Huybrechts and P. A. Jocobs, *Zeolites*, **13**, 365 (1993).
91. A. Tuel, Y. Ben Taârit and C. Naccache, *Zeolites*, **13**, 454 (1993).
92   B. Notari, *Catal. Today*, **18**, 163 (1993).
93. J. S. Reddy, R. Kumar and P. Ratnasamy, *Appl. Catal.*, **58**, L1 (1990).
94. J. S. Reddy, R. Ravishankar, S. Sivasanker and P. Ratnasamy, *Catal. Letters*, **17**, 139 (1993).
95. H. Sato, N. Ishii, K. Hirose and S. Nakamura, *Stud. Surf. Sci. Catal.*, **28** (New Devel. Zeo. Sci. Technol.), 755 (1986).
96. R. Millini, E. Previde Massara, G. Pergo and G. Bellussi, *J. Catal.*, **137**, 497 (1992).
97. B. Notari, *Stud. Surf. Sci. Catal.*, **37** (Innov. Zeol. Mater. Sci.), 413 (1988).
98. B. Notari, *Proc. Inter. Symp. Chem. Microporous Crystals*, 343 (1991).
99. B. Notari, *Catal. Today*, **18**, 163 (1993).
100. Ref. 39, Chap. 2.
101. R. E. Grim, *Clay Mineralogy*, McGraw-Hill, 1953.
102. P. Ravindranathan, P. B. Malla, S. Komarneni and R. Roy, *Catal. Letters*, **6**, 401 (1990).
103. Ref. 39, p. 53.
104. Ref. 39, p. 21.
105. J. H. Purnell, *Catal. Letters*, **5**, 203 (1990).
106. R. Camarena, A. C. Cano, R. Dalgado, N. Zúñiga and C. Alvares, *Tetrahedron Lett.*, **34**, 6853 (1993).
107. C. Cativiela, F. Figueras, J. M. Fraile, J. I. Garcia, M. Gil, J. A. Mayoral, L. C. de Ménorval and E. Pires, *Stud. Surf. Sci. Catal.*, **78** (Heterog. Catal. Fine Chem., III), 495 (1993).
108. C. Collet and P. Laszlo, *Tetrahedron Lett.*, **32**, 2905 (1991).
109. J. H. Clark, A. P. Kybett, D. J. Macquarrie, S. J. Barlow and P. Landon, *J. Chem. Soc., Chem. Commun.*, 1353 (1989).
110. S. J. Barlow, T. W. Bastock, J. H. Clark and S. R. Cullen, *J. Chem. Soc., Perkin Trans. 2*, 411, (1994).
111. S. J. Barlow, T. W. Bastock, J. H. Clark and S. R. Cullen *Tetrahedron Lett.*, **34**, 3339 (1993).
112. M. H. Stacey, *Catal. Today*, **2**, 621 (1988).

113.   J. P. Sterte and J.-E. Otterstedt, *Stud. Surf. Sci. Catal.*, **31** (Prep. Catal.,
       IV), 631 (1987).
114.   L. J. Michot and T. J. Pinnaviai, *Chem. Mater.*, **4**, 1433 (1992).
115.   D. Tichit, Z. Mountassir, F. Figueras and A. Auroux, *Stud. Surf. Sci.
       Catal.*, **63** (Prep. Catal., V), 345 (1991).
116.   F. Figueras, *Catal. Revs.*, 30, 457 (1988).
117.   T. J. Pinnavaia, M.-S Tzou and S. D. Landau, *J. Am. Chem. Soc.,* 107,
       4783 (1985).
118.   J. M. Thomas and C. R. Theocharis, *Prespectives in Catalysis* (J. M.
       Thomas and K. I. Zamaraev, Eds.), Blackwell, Oxford, UK, 1992, p
       465.
119.   J. Sterte, *Catal. Today*, **2**, 219 (1988).
120.   C. F. Baes, Jr. and R. E. Mesmer, *The Hydrolysis of Cations*, Wiley,
       New York, 1976, p 104.
121.   G. Johannson, *Acta Chem. Scand.*, **14**, 771 (1960).
122.   J. Y. Bottero, J. M. Cases, F. Fiessinger and J. E. Poirier, *J. Phys.
       Chem.,* **84**, 2933 (1980).
123.   A. Auer and J. Hofmann, *Appl. Catal., A,* **97**, 23 (1993).
124.   H. L. De. Castillo and P. Grange, *Appl. Catal., A,* **103**, 23 (1993).
125.   E. G. Rightor, M.-S. Tzou and T. J. Pinnavia, *J. Catal.*, **130**, 29 (1991).

Metal Catalysts

The metals shown in Fig. 11.1 are those most frequently used as heterogeneous catalysts. The more darkly-shaded elements are those generally used in synthetic reactions. The elements that are lightly shaded are less frequently utilized while the unshaded metals are either infrequently applied or promote only a specific reaction.

Nickel

The extensive use to which nickel has been put as a hydrogenation catalyst makes this metal one of the most common of all such catalysts. As a supported catalyst it is usually associated with processes such as the partial saturation of fats and oils and the hydrogenation of aromatic ring systems. More extensive synthetic use,

VIIa	VIII			Ib
		Co 58.9 fcc	Ni 58.7 fcc	Cu 63.5 fcc
	Ru 101.1 hcp	Rh 102.9 fcc	Pd 106.4 fcc	Ag 107.9 fcc
Re 186.2 hcp	Os 190.2 hcp	Ir 192.2 fcc	Pt 195.1 fcc	Au 197.0 fcc

Fig. 11.1. Catalytically active metals with their atomic weights and bulk metal crystal structures.

however, is made of skeletal nickel (e.g. Raney nickel), a form of catalyst that is produced by the action of base on a nickel–aluminum alloy.[1,2] With this more active form of nickel a large number of hydrogenations take place at low temperatures and pressures. The hydrogenation of practically all hydrogenateable functional groups can be affected over some form of a nickel catalyst. Raney nickel is also useful for the desulfurization of organic sulfur compounds (Eqn. 11.1)[3] even though this reaction requires a large amount of the catalyst.[4,5]

(11.1)

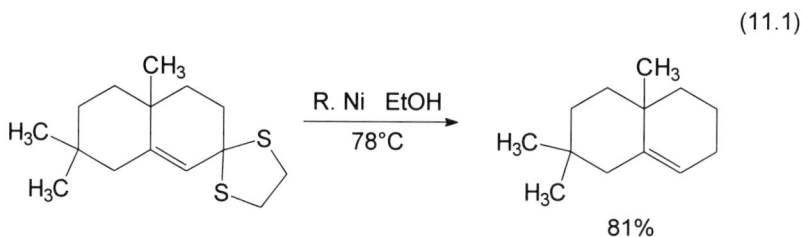

81%

Palladium

Palladium is one of the more important metals in the catalyst family. It is the best catalyst for the hydrogenation of olefins and acetylenes as well as for the hydrogenolysis of C–C, C–O, C–X and C–N bonds. It is also useful for the low pressure hydrogenation of many other functional groups and for the hydrogenation of phenols to the corresponding cyclohexanones[6] (Eqn. 11.2).[7] Palladium is commonly supported on a carrier, usually carbon. However, the finely divided palladium black obtained from the hydrogenation of palladium oxide is occasionally required for specific purposes such as the conversion of ketones to ethers by hydrogenation in an alcoholic solvent (Eqn. 11.3).[8]

(11.2)

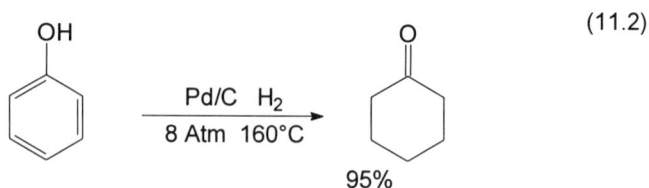

95%

(11.3)

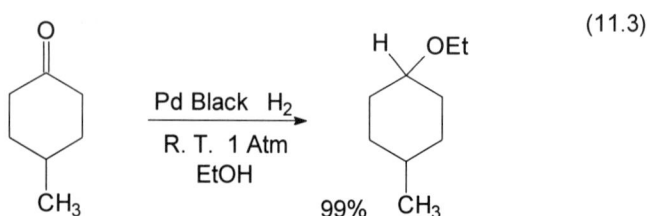

99%

Platinum

Platinum is capable of promoting the hydrogenation of most functional groups under relatively mild conditions, usually at temperatures below 70°C and hydrogen pressures of 3–4 atmospheres. These reactions are now usually run over supported platinum catalysts but much of the data reported in the earlier literature were obtained using reduced platinum oxide as the catalyst.[9] Esters, carboxylic acids, and amides are the only common functional groups not hydrogenated over this catalyst. Platinum is also a good catalyst for the oxidative dehydrogenation of alcohols[10] (Eqn 11.4).[11] Because of the general stability of platinum catalysts, these materials are those most frequently used for basic mechanistic studies.

(11.4)

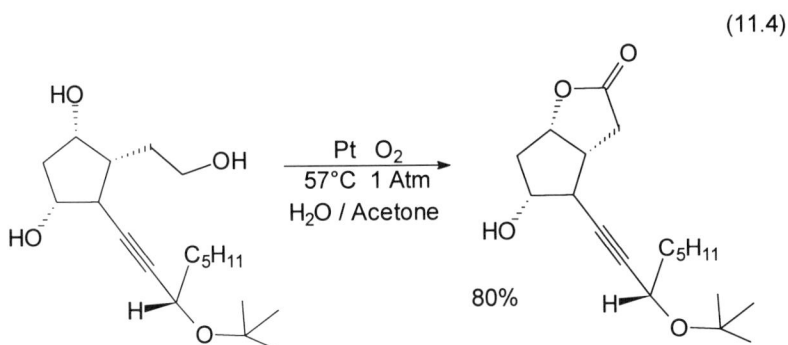

Rhodium

Like platinum, rhodium can be used for the hydrogenation of most functional groups. Most commonly rhodium is the catalyst for the hydrogenation of carbonyl groups, and carbocyclic aromatic and heteroaromatic systems at low temperatures and pressures.[12] This later reaction is accompanied by only a small amount of hydrogenolytic cleavage of oxygen and nitrogen substituents[13,14] (Eqn. 11.5).[15]

(11.5)

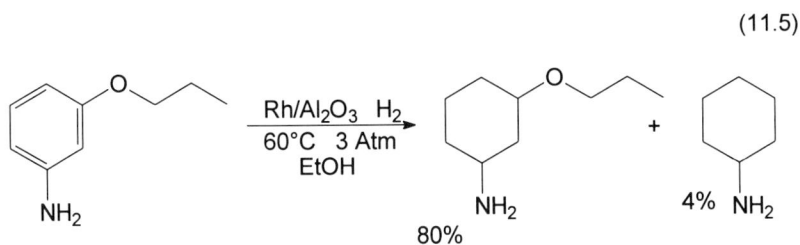

Ruthenium

The most important feature of ruthenium catalysts is their ability to promote the hydrogenation of aromatic rings without the hydrogenolysis of any amino and hydroxy groups present on the ring even though high temperatures and pressures are required for the reaction.[16,17] Ruthenium is also particularly effective for the low pressure hydrogenation of ketones and aldehydes especially in an aqueous environment.[18] Aldehydes and ketones are hydrogenated over ruthenium more readily in aqueous solutions than in non-aqueous media and alkenes are hydrogenated only when aqueous solutions are used even though they are insoluble in the medium.[19] The long induction period that is sometimes observed with ruthenium catalyzed hydrogenations is eliminated by presaturating the catalyst with hydrogen in an aqueous solvent before use. Ruthenium is also useful for the high pressure hydrogenation of carboxylic acids.[20] It is the only one of the platinum metal catalysts to catalyze this later reaction.

The data in Fig. 11.1 show that all of the shaded metals, except ruthenium, have an fcc crystal structure for the bulk metal. As implied above, ruthenium has some catalytic properties that are different from those of the other catalytically active metals. This may be the result of the hcp crystal structure of ruthenium which would give surface sites different from those shown in Schemes 4.1–4.4. It could also be a manifestation of the specific electronic configuration of ruthenium, or, it could arise from a combination of both factors.

Cobalt

Cobalt catalysts are reasonably active for hydrogenation reactions but they are particularly useful for the hydrogenation of nitriles and aldoximes to primary amines.[21,22] Other than for these reactions cobalt has not been generally used as a hydrogenation catalyst even though there is a similarity between the catalytic activity of cobalt and nickel. Raney cobalt, which is prepared from a commercially available aluminum–cobalt alloy,[21-23] was found to be somewhat less active than Raney nickel and more sensitive to variations in the reaction procedure and catalyst age than the nickel catalyst.[22,23]

Copper

Raney copper is another Raney type catalyst that is prepared from a copper–aluminum alloy. This catalyst has been used infrequently but does show some reaction selectivity not possible with other catalysts. Of particular interest is its use for the selective hydrogenation of substituted dinitrobenzenes (Eqn. 11.6).[24] This catalyst, as well as Raney cobalt, generally promotes fewer side reactions than does Raney nickel.[25]

The copper–chromium oxide ("copper chromite") catalyst, which was developed by Adkins,[1] has been considered to be a complement of Raney nickel for hydrogenation reactions.[26] This catalyst, which is, essentially, copper supported on chromium oxide or copper–chromium oxide,[27-29] is useful for the hydrogenation of esters and amides[30] but does not affect the saturation of aromatic ring compounds as readily as does Raney nickel (Eqn. 11.7).[31] The selectivity exhibited by this catalyst is quite interesting. In contrast to most other catalysts it will promote the hydrogenation of an unconjugated aldehyde group in preference to the more reactive vinyl group at lower temperatures (Eqn. 11.8).[32] In the hydrogenation of simpler α,β–unsaturated aldehydes, however, the primary product is the saturated aldehyde.[33]

(11.6)

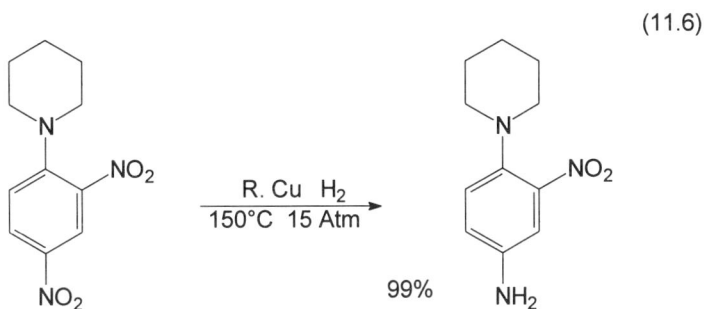

$$\xrightarrow[150°C \ \ 15 \ Atm]{R. \ Cu \ \ H_2}$$

99%

(11.7)

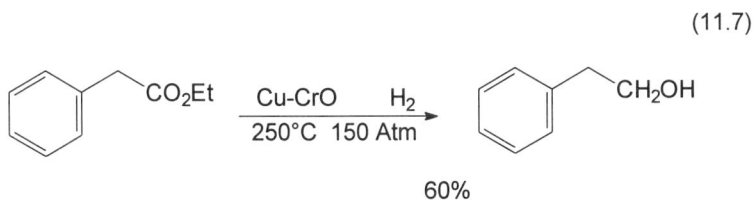

$$\xrightarrow[250°C \ \ 150 \ Atm]{Cu\text{-}CrO \ \ \ \ H_2}$$

60%

(11.8)

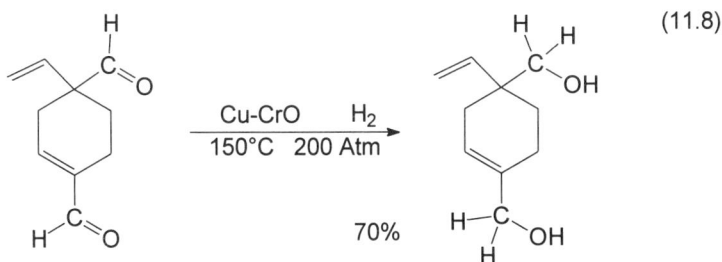

$$\xrightarrow[150°C \ \ \ 200 \ Atm]{Cu\text{-}CrO \ \ \ \ H_2}$$

70%

Rhenium

Rhenium, which has been used supported on controlled pore glass (CPG),[34] as the oxide,[35–38] the sulfide,[37] perrhennic acid[38] and rhenium black[39], is an effective catalyst for the hydrogenation of unsubstituted carboxylic acids and amides.[35,39] Highly branched acids, however, are not hydrogenated. The hydrogenation of maleic anhydride over perrhennic acid gives butyrolactone at lower temperatures and tetrahydrofuran at higher temperatures (Eqn. 11.9).[38] Rhenium is inferior to the other catalysts for the hydrogenation of other functional groups.

(11.9)

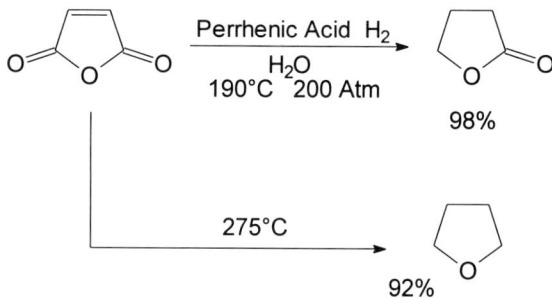

98%

275°C

92%

Osmium and Iridium

In general, osmium is a less active catalyst than the other platinum metals but it has some advantages in certain hydrogenations such as the conversion of α,β–unsaturated aldehydes to the corresponding unsaturated alcohols (Eqn. 11.10)[40] and the hydrogenation of halonitrobenzenes to the halo anilines with little dehalogenation.[40] Iridium is relatively inactive catalytically but it has been used for the selective hydrogenation of α,β–unsaturated aldehydes to the unsaturated alcohols.[41]

(11.10)

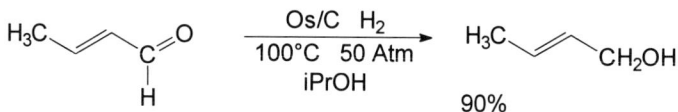

90%

Silver and Gold

About the only catalytic use for silver is in the conversion of ethylene to ethylene oxide by the action of oxygen.[42] This reaction is very specific. Other alkenes are oxidized to CO_2 and H_2O with no epoxide formation observed. Gold is generally inactive as a catalyst for most reactions but it has found some use in catalytic oxidations.[43]

Catalyst Additives

Substances that are deliberately or inadvertently added to a metal catalyst may have an influence on the activity of the catalyst and the selectivity of the reaction. These changes are brought about when the added material, a modifier, is adsorbed on the surface of the catalyst. Catalyst modification is brought about because the adsorbed species is either physically blocking the active sites on the metal surface or it is inducing a change in the electronic character or geometric structure of the catalyst surface.[44] When the chemisorption of the modifier is strong and all of the active sites are blocked or otherwise changed to prevent further reaction with the substrate, the catalyst is said to be poisoned. When catalyst activity is only diminished the modifier is described as an inhibitor. The concentration of the additive is frequently important in determining its effect on a specific reaction. Small amounts of a modifier may serve as an inhibitor while larger amounts may completely poison the catalyst.

Since catalyst modification is caused by the adsorption of the additive on the catalyst surface, the degree to which a specific modifier acts as a poison or inhibitor is influenced by the relative adsorption strength of the modifier relative to the reaction substrate and solvent.[44] As described in more detail later, sulfur compounds are among the strongest catalyst poisons but more sulfide is needed to inhibit the hydrogenation of nitrobenzene over a Pt/C catalyst than is needed for a similar degree of inhibition in the hydrogenation of acetophenone.[45] Since nitro compounds are strongly adsorbed on a catalyst they can compete more readily with the sulfide poison for adsorption on the active sites.

The nature of the metal and the conditions used for the reaction can also have an effect on this competitive adsorption. In addition, since these strongly adsorbed species seem to prefer to adsorb initially on the more coordinately unsaturated surface sites, the extent of inhibition can also depend on the morphology of the catalyst surface.[46,47]

As mentioned previously, a modifier induced decrease in catalytic activity is usually caused by the initial adsorption of the modifier on the most active surface sites.[48] With the most active sites blocked from further reaction, selectivity will frequently increase since there will be more of an opportunity for the remaining, less active, sites to differentiate between the various possible reaction paths available to the substrate. When this occurs, the additive is usually described as a promoter.[49] A promoter can also influence catalyst activity and/or selectivity by modifying the electronic character of the active sites.[50] This can result in either an increase or a decrease in activity. The addition of a small amount of an alkali metal ion to a catalyst develops a surface dipole which causes an increase in the rate of hydrogenation of the polar carbonyl group of aldehydes.[51] Promoters can also change the rate of a reaction by interacting with the substrate to alter its mode of adsorption. Small amounts of tertiary amines increase the rate of pyruvate hydrogenation presumably by interaction with the ketone carbonyl group to weaken its adsorption on the catalyst so hydrogenation can take place more easily.[52,53]

Catalyst poisons and inhibitors are usually added inadvertently to the reaction mixture by the use of impure solvents or substrates. Promoters, however, are generally added deliberately to enhance catalyst activity and/or reaction selectivity.

There are a number of different classes of catalyst modifiers: acids and bases, metal cations, nucleophilic species, and compounds with multiple bonds which are strongly chemisorbed.[44,54,55] A general description of each of these classes is given below. More detailed descriptions of the use of particular modifiers in specific reactions will be found in Section III.

Acids and Bases

Catalytic reactions are sometimes affected by the presence of small quantities of an acid or a base. Each may act as either an inhibitor or a promoter or it may simply counteract the effect of the other. If an inhibiting base is present in the reaction mixture the presence of an equivalent or more of an acid may be beneficial. The presence of acid or base can have a significant effect on the outcome of a catalytic reaction. The rates of all Raney nickel catalyzed hydrogenations, except those of nitro groups, are increased in basic media.[56-59] The rate of alkene hydrogenation over platinum and palladium catalysts and benzene hydrogenations over platinum catalysts increases with increasing acidity but base acts as a catalyst inhibitor in these reactions.[60,61] Small amounts of acid are frequently used as promoters for the hydrogenation of aldehydes and ketones over platinum catalysts.[62,63] It is thought that the acid protonates the oxygen of the carbonyl group and this changes its strength of adsorption. Pyridine,[64] triazines,[65] dry ammonia and other amines,[66] all act as inhibitors and, in some cases, poisons for platinum. The poisoning by amines can be minimized or even eliminated by the use of aqueous solutions or acidic media.[66] Palladium catalysts can be used effectively in either acidic or basic medium even though there is a slight deactivation by base.

(11.11)

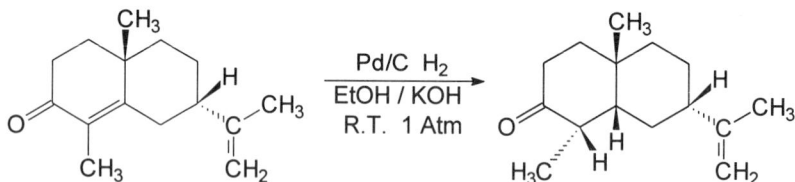

The rate of hydrogenation of the double bond of an α,β–unsaturated ketone is decreased in acidic and strongly basic media but is only slightly affected by moderate amounts of base. The selective hydrogenation of the conjugated double bond in *epi* cyperone in preference to the non-conjugated vinyl group (Eqn. 11.11)[67] occurs in a moderately basic solution. The base, apparently, reacts with the unsaturated ketone moiety to give a more readily adsorbed enolate.

Strong acid solutions cannot be used with nickel, cobalt or copper catalysts because of their reaction with the metal. It is possible to use weak acids or amine salt solutions in which the pH is kept greater than three with Raney nickel catalysts.[68] Rhodium catalysts are inhibited by the presence of strong bases that makes it necessary to use a large amount of this catalyst for the hydrogenation of pyridines and other heteroaromatic systems or to use acetic acid or other acidic solvent for any hydrogenation in which amines are present as products or reagents.[69] HBr and HI also poison rhodium catalysts, making rhodium undesirable for use in the hydrogenolysis of organic bromides and iodides. It is possible, however, to use rhodium for the cleavage of aromatic and aliphatic chlorides and aromatic fluorides. The rate of aromatic dehalogenation is approximately equal to that of ring hydrogenation with this catalyst and the product obtained is the corresponding dehalogenated cyclohexane.[70] Ruthenium is best used in neutral aqueous solutions since basic or non aqueous solvents result in an inhibition of the catalyst,[20,71] and the presence of acid almost completely deactivates it.[71]

Metal Cations

Metal cations of all types can cause changes in catalyst activity but there is no way of generalizing what effect specific ions will have on any particular metal catalyst. The usual approach is to use the cation of a heavy metal to deactivate a catalyst and, thus, hopefully increase the reaction selectivity. The addition of lead acetate to $Pd/CaCO_3$ gives Lindlar's catalyst[72,73] which is highly selective for the partial hydrogenation of acetylenes to the corresponding *cis* alkenes. It is thought that the lead ions block the most active sites, so the remaining sites are only able to react with the more strongly adsorbed triple bonds but are not sufficiently active for saturating a double bond.[74]

Chromium compounds increase the activity of platinum catalysts by increasing the electron densities of the active sites.[75] The addition of ferrous sulfate, which promotes the hydrogenation of carbonyl groups, and zinc acetate, which inhibits the hydrogenation of double bonds, to platinum gives a catalyst system capable of effecting the selective hydrogenation of an unsaturated aldehyde to an unsaturated alcohol[76,77] (Eqn. 11.12).[77]

(11.12)

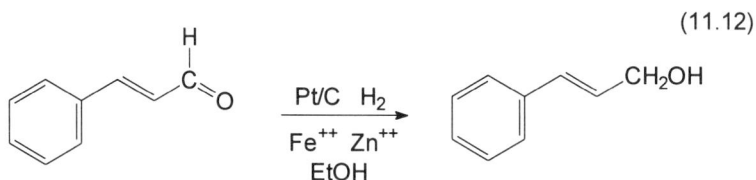

Metal cations can also affect the activity of Raney nickel catalysts. The presence of three to ten per cent of chromium or molybdenum in the Raney nickel alloy results in a general increase in the activity of the catalyst prepared from this alloy.[78] Activation of nickel catalysts has also been accomplished by treating them with aqueous solutions of chromium chloride.[79] Another effective promoter for Raney nickel is platinum which increases the activity of this catalyst for the hydrogenation of most functional groups. The best results are obtained when a small amount of platinum chloride is added to the Raney nickel immediately before starting the hydrogenation.[59,80,81] The further addition of a base such as triethyl amine, sodium hydroxide or, particularly, lithium hydroxide, gives a very effective catalyst system.[82,83]

The effect of added alkali metal ions, as mentioned above, is to develop a surface dipole which may be beneficial in reactions involving polarized groups.[51] These ions not only have an electronic effect on the catalyst but they can also be instrumental in determining the size, shape and morphology of the metal crystallites when they are present at the time the catalyze is prepared.[84]

Nucleophiles

Nucleophiles, such as halide ions and ions of the Groups VA and VIA elements have unshared electrons that interact with the catalyst surface leading to the poisoning or inhibition of a catalyzed reaction. Most catalysts are affected by the common catalyst poisons: sulfur, phosphorous, arsenic and bismuth compounds having unshared electron pairs.[84-86] Trisubstituted silanes and tetrasubstituted lead, tin and germanium compounds are catalyst poisons but tetrasubstituted silanes are not because the silicon atom in these compounds is sterically hindered from coordinating with the catalyst.[87] Platinum catalysts are particularly susceptible to poisoning by sulfur, phosphorous, or arsenic compounds having unshared electron pairs[84,88] as well as nucleophilic ions such as iodides, phenoxides, cyanates, or tosylates.[89] Rhodium is poisoned by sulfides, phosphenic acids and the other common catalyst poisons more readily than is palladium but to a lesser extent than platinum. Raney nickel is a successful desulfurizing agent only because large amounts of the catalyst are used so complete poisoning of the nickel is avoided.[4,5]

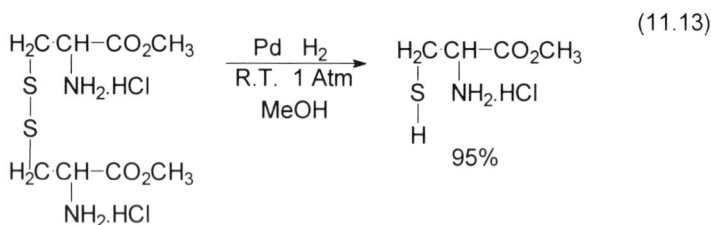

$$(11.13)$$

$$
\underset{\substack{\text{S} \\ \text{|} \\ \text{S} \\ \text{|} \\ H_2C{-}CH{-}CO_2CH_3 \\ \text{|} \\ NH_2 \cdot HCl}}{\overset{H_2C{-}CH{-}CO_2CH_3}{\underset{NH_2 \cdot HCl}{\text{|}}}}
\quad\xrightarrow[\substack{\text{R.T. 1 Atm} \\ \text{MeOH}}]{\text{Pd } H_2}\quad
\underset{\substack{\text{S} \\ \text{|} \\ \text{H} \\ \quad 95\%}}{\overset{H_2C{-}CH{-}CO_2CH_3}{\underset{NH_2 \cdot HCl}{\text{|}}}}
$$

Outside of some occasional catalyst deactivation, palladium catalysts are noted for their general resistance to catalyst poisons. They are also relatively unaffected by chloride and bromide ions and only moderately inhibited by iodide ions which makes palladium quite useful for the hydrogenolysis of carbon–halogen bonds. The most striking example of this inertness to catalyst poisons is illustrated by the hydrogenolysis of methyl cystine to methyl cysteine over palladium (Eqn. 11.13),[90] a reaction that occurs on one of the most powerful of all catalyst poisons, a sulfide. The hydrogenation of thiophene[91] can also be effected by palladium. Some inhibition by the sulfur is observed and a high catalyst to substrate ratio is required for these reactions. It does not appear that sulfur compounds act as poisons for a ruthenium catalyst, at least not when the hydrogenation is run at high temperatures.[92]

The catalytic activity of Raney nickel is also adversely affected by halide ions and organo–halides.[93,94] Chlorides and bromides are moderate inhibitors but iodides are highly poisonous. For this reason, it is not advantageous to use nickel for the hydrogenolysis of organo–halides unless a relatively large amount of catalyst is used. The inhibiting effect of halide ions can be diminished by the use of aqueous solvents. The hydrogenation of acetophenone over Pt/C is considerably less affected by iodide ions in 75% aqueous methanol than in 99% methanol.[45]

Halide ions can, sometimes, be promoters as in the use of chloride ion in the silver catalyzed epoxidation of ethylene. This oxidation takes place through a silver–oxygen surface complex that is charge deficient. On co-adsorption of the chloride ion, the presence of an oxychloride surface species optimizes the reaction between the electrophilic oxygen and the π electrons of the ethylene.[95–97]

The nature of the inhibition by these nucleophiles varies from metal to metal and modifier to modifier. The inhibition of platinum catalysts caused by amines is apparently the result of an electronic modification of the catalyst surface and not caused by site blocking.[98] An electronic effect also seems to be responsible for the sulfur inhibition of platinum.[99] The action of sulfur on nickel, though, depends on the coverage. At low coverage it appears that an electronic effect is operational while at high coverage site blocking is important.[44] The extent of site blocking is also regulated by the size of the groups on the inhibiting sulfide. The larger the group the less sulfide needed for complete site blockage because of the "umbrella effect" of the attached group.[45]

The electronic modification caused by a low coverage of the catalyst can sometimes be beneficial. An initial reaction of a catalyst with a small amount of a sulfur compound can prevent further reaction with additional quantities of sulfur compounds, within certain limits.[100] These sulfided catalysts have been used for reactions run in the presence of impurities such as sulfur and for the hydrogenations of sulfur containing compounds.[101] The relative inactivity of these catalysts for the hydrogenation of aromatic rings, ketones and nitriles can

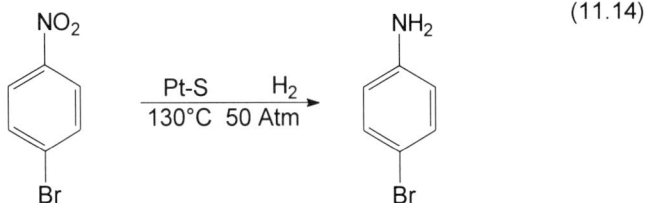

$$\text{(11.14)}$$

frequently be used to advantage for the selective hydrogenation of multifunctional compounds even though these catalysts require the use of higher temperatures and pressures than those commonly used with unsulfided catalysts. Sulfided platinum and rhodium catalysts have been used for the selective hydrogenation of halonitrobenzenes to the halo anilines (Eqn. 11.14)[101] and the conversion of dienes to monoolefins.[101]

CO and CO_2

Catalyst inhibition by carbon monoxide and carbon dioxide can be a problem in the hydrogenation of some compounds since decarbonylation or decarboxylation can take place during the hydrogenation of some functional groups. Palladium is a good decarbonylation catalyst for aldehydes[102–104] and decarboxylation can accompany the hydrogenation of some acids, particularly nicotinic acid.[105,106]

Detoxification

The detoxification of catalysts poisoned by Group V or VI compounds can be accomplished by reactions in which these inhibitors are converted to substances that do not have unshared electron pairs. For instance, bivalent sulfur compounds can be oxidized to sulfones or sulfonic acids by treatment with hypochlorite or hydroperoxides.[107,108] Thiophene, dimethyl sulfide and other sulfur and metal ion poisons[109] as well as phosphorous[110] and arsenic compounds[111] can be removed from platinum by washing the catalyst with acetic acid. This method for the reactivation of the catalyst is simpler than the oxidation techniques. Acidic or basic inhibitors are removed by the addition of an appropriate amount of base or acid, respectively. The effect of a small amount of inhibitor can frequently be overcome by the use of a larger amount of catalyst.

Unless the catalyst inhibitor is a reactant or product it is best to remove the offending material before it comes in contact with the catalyst. One widely used method for doing this is to stir the reaction mixture with some spent catalyst or a portion of Raney nickel so the inhibitor will be adsorbed at this stage, prior to contact with the reaction catalyst. Other treatment procedures that have been used include shaking or refluxing the reaction mixture, where appropriate, with sodium, zinc, sodium hydroxide, magnesium oxide, aluminum oxide or activated carbon; anything that may remove the offending material but not affect the reactants or the solvent.

References

1. H. Adkins, *Reactions of Hydrogen with Organic Compounds over Copper Chromium Oxide and Nickel,* The University of Wisconsin Press, Madison, WI, 1937, and later papers by this author.
2. R. Schrötter, *Newer Methods of Preparative Organic Chemistry,* Interscience, New York, 1948, p. 61.
3. F. Sondheimer and S. Wolfe, *Can. J. Chem.,* **37**, 1870 (1959).
4. G. R. Pettit and E. E. van Tamelen, *Org. Reactions,* **XII**, 356 (1962).
5. H. Hauptmann and W. F. Walter, *Chem. Rev.,* **62**, 347 (1962).
6. A. A. Wismeijer, A. P. G. Kieboom and H. van Bekkum, *Recl. Trav. Chim. Pays-Bas,* **105**, 129 (1986).

7. J. F. Van Peppen, W. B. Fisher and C. H. Chan, *Chem. Ind. (Dekker)*, **22** (Catal. Org. React.), 355 (1985).
8. S. Nishimura, S. Iwafune, T. Nagura, Y. Akimoto and M. Uda, *Chem. Lett.*, 1275 (1985).
9. R. Adams, V. Voorhees, and R. L. Shriner, *Org. Synthesis, Coll.* Vol. **1**, 452 (1941).
10. K. Heyns and H. Paulsen, *Newer Methods of Preparative Organic Chemistry*, Vol. II, Interscience, New York, 1963, p. 303.
11. J. Fried and J. C. Sih, *Tetrahedron Lett.*, 3899 (1973).
12. G. Gilman and G. Cohn, *Adv. Catal.*, **9**, 733 (1957).
13. H. A. Smith and R. G. Thompson, *Adv. Catal.*, **9**, 727 (1957).
14. R. Egli and C. H. Eugster, *Helv. Chim. Acta*, **58**, 2321 (1975).
15. M. Freifelder, Y. H. Ng and P. F. Helgren, *J. Org. Chem.*, **30**, 2485 (1965).
16. A. E. Barkdoll, D. C. England, H. W. Gray, W. Kirk, Jr. and G. M. Whitman, *J. Am. Chem. Soc.*, **75**, 1156 (1953).
17. R. E. Ireland and P. W. Schiess, *J. Org. Chem.*, **28**, 6 (1963).
18. P. N. Rylander, N. Rakoncza, D. Steele and M. Bolliger, *Engelhard Ind. Tech. Bull.*, **4**, 95 (1963).
19. L. M. Berkowitz and P. N. Rylander, *J. Org. Chem.*, **24**, 708 (1959).
20. J. E. Carnahan, T. A. Ford, W. F. Gresham, W. E. Grigsby and G. F. Hager, *J. Am. Chem. Soc.*, **77**, 3766 (1955).
21. W. Reeve and J. Christian, *J. Am. Chem. Soc.*, **78**, 860 (1956).
22. B. V. Aller, *J. Appl. Chem.*, **7**, 130 (1957); **8**, 163, 492 (1958).
23. A. J. Chadwell, Jr. and H. A. Smith, *J. Phys. Chem.*, **60**, 1339 (1956).
24. W. H. Jones, W. F. Benning, P. Davis, D. M. Mulvey, P. I. Pollak, J. C. Schaeffer, R. Tull and L. M. Weinstock, *Ann. N. Y. Acad. Sci.*, **158**, 471 (1969).
25. A. Sekera, *Chemie* (Prague), **5**, 5 (1949); *Chem. Abstr.*, **46**, 2717 (1952).
26. G. Grundmann, *Newer Methods of Preparative Organic Chemistry*, Interscience, New York, 1948, p. 103.
27. F. M. Capece, V. diCastro, C. Furlani, G. Mattogno, C. Fragale, M. Gargano and M. Rossi, *J. Electron Spect. and Rel. Phenomena*, **27**, 119 (1982).
28. A. Iimura, Y. Inoue and I. Yasumori, *Bull. Chem. Soc., Japan.*, **56**, 2203 (1983).
29. S. P. Tonner, M. S. Wainwright, D. L. Trimm and N. W. Cant, *Appl. Catal.*, **11**, 93 (1984).
30. H. Adkins, *Org. Reactions*, **VIII**, 1 (1954).
31. K. Folkers and H. Adkins, *J. Am. Chem. Soc.*, **54**, 1145 (1932).
32. R. Pummerer, F. Aldebert, F. Graser, and H. Sperber, *Liebigs Ann. Chem.*, **583**, 225 (1953).
33. R. Hubaut, M. Daage and J. P. Bonnelle, *Appl. Catal.*, **22**, 231 (1986).
34. T. H. Vanderspurt, *Catalysis in Organic Synthesis*, (W. H. Jones, Ed.) Academic Press, New York, 1980, p. 11.
35. H. S. Broadbent and T. G. Selin, *J. Org. Chem.*, **28**, 2343 (1963) and preceding papers.
36. W. E. Pascoe and J. F. Stenberg, *Catalysis in Organic Synthesis*, (W. H. Jones, Ed.) Academic Press, New York, 1980, p. 1.
37. H. S. Broadbent, *Ann. N. Y. Acad. Sci.*, **145**, 58 (1967).

38. H. S. Broadbent, V. L. Mylroie and W. R. Dixon, *Ann. N. Y. Acad. Sci.*, **172**, 194 (1970).
39. H. S. Broadbent G. C. Campbell, W. J. Bartley, and J. H. Johnson, *J. Org. Chem.*, **24**, 1847 (1959).
40. P. N. Rylander and D. R. Steele, *Engelhard Ind. Tech. Bull.*, **10**, 17 (1969).
41. P. N. Rylander and D. R. Steele, *Tetrahedron Lett.*, 1579 (1969).
42. R. W. Clayton and S. V. Norval, *Catalysis (London)*, **3**, 70 (1980).
43. J. Schwank, *Gold Bulletin*, **16**, 103 (1983).
44. C. H. Bartholomew, *Stud. Surf. Sci. Catal.*, **34** (Catal. Deact.), 81 (1987).
45. R. Baltzly, *J. Org. Chem.*, **41**, 928 (1976).
46. A. A. Sutyagine and G. D. Vovchenko, *Surface Technology*, **13**, 257 (1981).
47. G. A. del Angel, B. Coq, F. Figueras, S. Fuentes and R. Gomez, *Nouv. J. Chim.*, **7**, 173 (1983).
48. Z. Paal, G. Loose, G. Weinberg, M. Rebholz and R. Schlögl, *Catal. Letters*, **6**, 301 (1990).
49. G. C. Bond, M. R. Gelsthorpe, R. R. Rajaram and R. Yahya, *Stud. Surf. Sci. Catal.*, **48** (Struct. React. Surf.), 167 (1989).
50. R. A. Van Santen, *Surf. Sci.*, **251–252**, 6 (1991).
51. J. P. Hindermnn, A. Kinnemann, A. Chakor-Alami and R. Kieffer, *Proc. 8th Int. Catal. Conf., Berlin*, **2**, 163 (1984).
52. H. U. Blaser, H. P. Jalett, D. M. Monti, J. F. Reber and J. T. Wehrli, *Stud. Surf. Sci. Catal.*, **41** (Heterog. Catal. Fine Chem.), 153 (1988).
53. R. L. Augustine, S. K. Tanielyan and L. K. Doyle, *Tetrahedron: Asymmetry*, **4**, 1803 (1993).
54. R. Baltzly, *Ann. N. Y. Acad. Sci.*, **145**, 31 (1967).
55. P. N. Rylander, *Catalytic Hydrogenation over Platinum Metals*, Academic Press, New York, 1967, pp. 16–20.
56. H. Adkins and G. Krsek, *J. Am. Chem. Soc.*, **70**, 412 (1948).
57. H. E. Ungnade and D. V. Nightingale, *J. Am. Chem. Soc.*, **66**, 1218 (1944).
58. M. Paty, *Compt. rend.*, **220**, 827 (1945).
59. S. S. Scholnik, J. R. Reasenberg, E. Lieber and G. B. L. Smith, *J. Am. Chem. Soc.*, **63**, 1192 (1941).
60. B. Foresti, *Ann. chim. (Rome)*, **41**, 425 (1951).
61. B. Foresti, *Gazz. Chim. Ital.*, **66**, 455 (1936).
62. K. Kindler, H. G. Helling and E. Sussner, *Liebigs Ann. Chem.*, **605**, 200 (1957).
63. B. Foresti, *Gazz. Chim. Ital.*, **67**, 455 (1937).
64. T. S. Hamilton and R. Adams, *J. Am. Chem. Soc.*, **50**, 2260 (1928).
65. H. Brandenberger and R. Schwyzer, *Helv. Chim. Acta*, **38**, 1396 (1955).
66. E. B. Maxted and M. S. Biggs, *J. Chem. Soc.*, 3844 (1957).
67. R. Howe and F. J. McQuillan, *J. Chem Soc.*, 1194 (1958).
68. W. Wenner, *J. Org. Chem.*, **15**, 301 (1950).
69. M. Freifelder, *J. Org. Chem.*, **26**, 1835 (1961).
70. L. D. Freedman, G. O. Doak, and E. L. Petit, *J. Am. Chem. Soc.*, **77**, 4262 (1955).
71. E. Breitner, E. Roginski, and P. N. Rylander, *J. Org. Chem.*, **24**, 1855 (1959).
72. H. Lindlar, *Helv. Chim. Acta*, **35**, 446 (1952).

73. H. Lindlar and R. Dubois, *Org. Synth.*, **46**, 89 (1966).
74. R. Schloegl, K. Noack, H. Zbinden and A. Reller, *Helv. Chim. Acta*, **70**, 627 (1987).
75. L. T. Vlaev. M. M. Mohomed and D. P. Damyanov, *Appl. Catal.*, **63**, 293 (1990).
76. G. C. Bond, and D. E. Webster, *Proc. Chem. Soc.*, 398 (1964).
77. P. N. Rylander, N. Himelstein, and M. Kilroy, *Engelhard Ind. Tech. Bull.*, **4**, 49 (1963).
78. R. Paul, *Bull. soc. chim.*, *France*, 208 (1946).
79. J. Masson, P. Cividino, P. Fouilloux and J. Court, *C. R. Acad. Sci., Ser. II*, **317**, 33 (1993).
80. E. Lieber and G. B. L. Smith, J. *Am. Chem. Soc.*, **58**, 1417 (1936).
81. J. R. Reasenberg, E. Lieber, and G. B. L. Smith, J. *Am. Chem. Soc.*, **61**, 384 (1939).
82. D. R. Levering, F. L. Morritz, and E. Lieber, *J. Am. Chem. Soc.*, **72**, 1190 (1950).
83. R. B. Blance and D. T. Gibson, *J. Chem. Soc.*, 2487 (1954).
84. E. B. Maxted and R. W. D. Morrish, *J. Chem. Soc.*, 252 (1940).
85. A. G. Deem and J. E. Kaveckis, *Ind. Eng. Chem.*, **33**, 1373 (1941).
86. L. Horner, H. Reuter, and E. Herrmann, *Liebigs Ann. Chem.*, **660**, 1 (1962).
87. L. Spialter, G. R. Buell and C. W. Harris, *J. Org. Chem.*, **30**, 375 (1965).
88. E. B. Maxted and H. C. Evans, *J. Chem. Soc.*, 603 (1937).
89. R. Baltzly, *J. Am. Chem. Soc.*, **74**, 4586 (1952).
90. L. Zervas and D. M. Theodoropoulos, *J. Am. Chem. Soc.*, **78**, 1359 (1956).
91. R. Mozingo, S. A. Harris, D. E. Wolf, C. E. Hoffhine, Jr., N. R. Easton, and K. Folkers, J. *Am. Chem. Soc.*, **67**, 2092 (1945).
92. A. C. Cope and E. Farkas, *J. Org. Chem.*, **19**, 385 (1954).
93. R. Truffault, *Bull. soc. chim. France*, [5], **2**, 244 (1935).
94. J. N. Pattison and E. F. Degering, *J. Am. Chem. Soc.*, **73**, 611 (1951).
95. R. A. Van Santen, *Proc. 9th Int. Cong. Catal.*, **3**, 1152 (1988).
96. R. A. Van Santen and C. P. M. de Groot, *J. Catal.*, **98**, 503 (1986).
97. R. A. Van Santen and H. P. C. E. Kuipers, *Adv. Catal.*, **35**, 265 (1987).
98. J. P. Boitiaux, J. Cosyns and E. Robert, *Appl. Catal.*, **49**, 219 (1989).
99. E. Lamy-Pitara, Y. Tainon and J. Barbier, *Appl. Catal.*, **68**, 179 (1991).
100. P. C. L'Argentiere and N. S. Figon, *J. Chem. Technol. Biotechnol.*, **38**, 209 (1987).
101. H. Greenfield, *Ann. N. Y. Acad. Sci.*, **145**, 108 (1967).
102. J. O. Hawthorne and M. H. Wilt, *J. Org. Chem.*, **25**, 2215 (1960).
103. N. E. Hoffman, A. T. Kanakkannatt and R. F. Schneider, *J. Org. Chem.*, **27**, 2687 (1962).
104. N. E. Hoffman and T. Puthenpurackal, *J. Org. Chem.*, **30**, 420 (1965).
105. M. Freifelder and G. R. Stone, *J. Org. Chem.*, **26**, 3805 (1961).
106. M. Freifelder, *J. Org. Chem.*, **28**, 1135 (1963).
107. E. B. Maxted and R. W. D. Morrish, *J. Chem. Soc.*, 132 (1941).
108. E. B. Maxted, *J. Chem. Soc.*, 624 (1947).
109. E. B. Maxted and G. T. Ball, *J. Chem. Soc.*, 4284 (1952).
110. E. B. Maxted and G. T. Ball, *J. Chem. Soc.*, 1509 (1953).
111. E. B. Maxted and G. T. Ball, *J. Chem. Soc.*, 3153 (1953).

Unsupported Metals

Unsupported metals are found in a variety of forms; wires, ribbons, single crystals, colloids, powders, blacks and the so called skeletal species. A black is a metallic powder obtained by reduction of a metal salt or by condensation of a metal vapor.[1] Such materials are also sometimes referred to as powders. The blacks and powders are usually composed of relatively large particles having a low surface area. Skeletal metals are produced by leaching out one component of an alloy and leaving the active species behind in the form of a porous material having a high surface area.[2]

Massive Metals

Catalytic reactions can be run over large, massive metal particles as well as the much smaller, dispersed metal crystallites. The massive metal catalysts can be the single crystal catalysts such as those shown in Fig. 3.2 or polycrystalline forms of bulk metal such as wires, foils or ribbons. These latter materials were used somewhat routinely in the early catalytic research efforts that were involved with developing the mechanisms of vapor phase catalytic processes. These materials were considered to be analogs of the supported catalysts in which the effect of the support, if any, was eliminated.

At present, the most commonly used massive metal catalysts are the metal single crystals.[3-7] These materials have well-defined surfaces and can have varying amounts of corner, edge or face atoms on their surface depending on the angle of cleavage in producing the crystal. These catalysts can be particularly useful not only in determining the mechanism of vapor phase catalytic reactions but also in establishing the nature of the active site(s) on which a given reaction is taking place.[3,4]

While these massive metal catalysts have been useful for basic mechanistic and theoretical studies, they are not amenable for use in liquid phase reactions. Since most synthetically useful catalytic reactions are run in the liquid phase these massive metal catalysts find little use in the study of synthetic catalysis. Instead, synthetic reactions use the more dispersed forms of the metal catalysts.

Blacks and Powders

Noble metal blacks are usually prepared by the reduction of aqueous solutions of their salts with hydrogen, formaldehyde, formic acid or hydrazine. These blacks are large metal particles having a surface composed primarily of large areas of

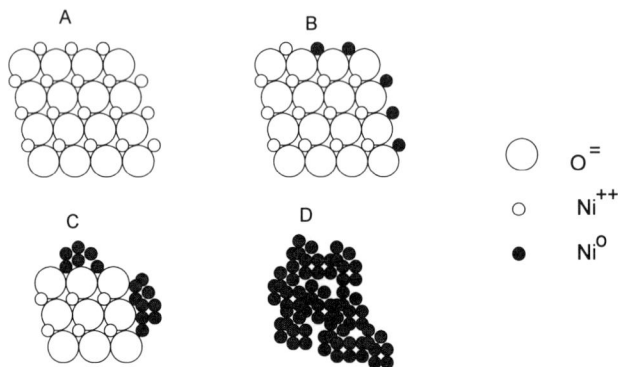

Fig. 12.1. The different stages in the hydrogenation of nickel oxide to
a nickel black catalyst.

face atoms so they can be useful for reactions that take place on ensembles of such
surface atoms. Such catalysts are also utilized for reactions, such as oxidations,
which are run under conditions which would more easily destroy the small
crystallites commonly present in most supported metal catalysts.[8]

A nickel powder catalyst is usually prepared by the high temperature
hydrogenation of bulk nickel oxide. The oxide is formed by calcining the nickel
hydroxide precipitate produced by the reaction of an aqueous solution of nickel
nitrate with base. As depicted in Fig. 12.1, the hydrogenation of the nickel oxide
(A) occurs by the initial production of some metallic nickel on the surface of the
agglomerated oxide crystallites (B). This metallic species then catalyzes the
reduction of the remaining oxide at the nickel–nickel oxide interface. The nickel
atoms then combine into small clusters (C) which grow and combine on further
reduction to give the larger crystallites which compose the nickel powder (D).[9]

Some metallic modifiers can be incorporated into the nickel oxide by
coprecipitation of a mixed hydroxide from a solution containing a salt of the
modifier along with the nickel nitrate.[9] When 1–3% of a more easily reduced
metal cation or oxide is present in the bulk nickel oxide, the temperature required
for reduction to nickel black is lowered. With additives such as platinum, which
readily dissociate hydrogen molecules, the reduction of the nickel oxide is
facilitated by the increased amount of reactive hydrogen atoms available for
reduction of the nickel oxide at the metal–oxide interface. Other metals, such as
silver and copper, which do not easily dissociate hydrogen molecules, apparently
facilitate the hydrogenation of nickel oxide by weakening the surface nickel–
oxygen bonds.[9] The presence of irreducible oxides, such as aluminum oxide,
increases the temperature required for the hydrogenation of the nickel oxide but
such materials may also inhibit the growth of the nickel crystallites thus leading to
higher metal dispersions and catalyst activities. Catalyst stability may also be
improved.[9]

Another form of active nickel catalyst has been prepared by the reaction of lithium iso-propoxide with nickel chloride to give nickel iso-propoxide. Heating this salt to 100°C in iso-propanol gives a black, finely divided nickel catalyst that is active for the hydrogenation of a variety of functional groups.[10]

Early synthetic reactions involving a platinum catalyst most commonly used platinum oxide, Adam's catalyst, which was hydrogenated in the reaction mixture to give a platinum black. This oxide is prepared by the fusion of chloroplatinic acid with sodium nitrite or nitrate.[11,12] Large quantities of the catalyst were generally used in these reactions and this increased the cost of the process. This factor was partially overcome by the relative ease with which the oxide could be regenerated. The spent catalyst was dissolved in aqua regia to give chloroplatinic acid, which after isolation and purification, was fused with sodium nitrite or nitrate to give the oxide.[12,13] It was thought that the fusion process introduced a sodium impurity into the oxide and that its presence deactivated the reduced platinum catalyst.[14] This presumption was supported by the finding that the activity of Adam's catalyst decreased as the amount of sodium present increased.[15] The reason for the common use of acetic acid as the solvent in platinum oxide catalysed reactions was, presumably, to dissolve this sodium impurity and keep it from deactivating the platinum. Hydrogenating the oxide in methanol gave a platinum black from which the offending sodium could be removed by further washings with methanol. The washed platinum black was a particularly active hydrogenation catalyst.[14]

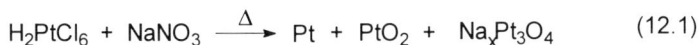

$$H_2PtCl_6 + NaNO_3 \xrightarrow{\Delta} Pt + PtO_2 + Na_xPt_3O_4 \qquad (12.1)$$

An X-ray diffraction analysis of several samples of commercially available platinum oxide has shown them to be a mixture of metallic platinum, α PtO$_2$ and a sodium platinum bronze, Na$_x$Pt$_3$O$_4$ (Eqn. 12.1).[16] The relative portions of each component varied from sample to sample. Each of these components, as well as some preformed platinum metal, were prepared individually and used for the hydrogenation of both cyclopentene and benzene. The α PtO$_2$ was easily reduced to a poorly crystalline platinum black that was an effective catalyst for both hydrogenations. The bronze was found to be stable under the conditions used for the hydrogenation, but while this material was not reduced by hydrogen to a platinum black, it was, itself, a good catalyst for these hydrogenations. The saturation of cyclopentene took place at about the same rate over the platinum metal, the black from the α PtO$_2$ and the bronze. The hydrogenation of benzene, though, was somewhat slower over the bronze than over the platinum blacks.[16]

These results make the previous conclusions concerning the nature of the sodium containing impurity a bit over-simplified. The presence of such an impurity cannot be denied. Since an excess of sodium nitrate or sodium nitrite is used in the fusion, some sodium oxide or similar material is probably formed along with the platinum oxide and some of this sodium containing species could be occluded in the platinum oxide particles.

When potassium nitrate and potassium chloroplatinate were used in the fusion, no bronze was formed.[16] The only species detected by X-ray diffraction were platinum metal and α PtO$_2$. A report that the use of ammonium chloroplatinate in the fusion gave a better catalyst than when chloroplatininic acid was used might also have been the result of an inhibition of bronze formation by the ammonium salt.[17]

The reduction of palladium oxide, rhodium oxide or ruthenium oxide gives the corresponding metal blacks generated by *in situ* hydrogenation in the reaction mixture. At present the use of these oxides, as well as Adam's catalyst, is not common because of the cost of the materials and the relatively large amounts which are required. These materials have been replaced by the more reactive and less expensive supported metal catalysts described in Chapter 13.

(12.2)

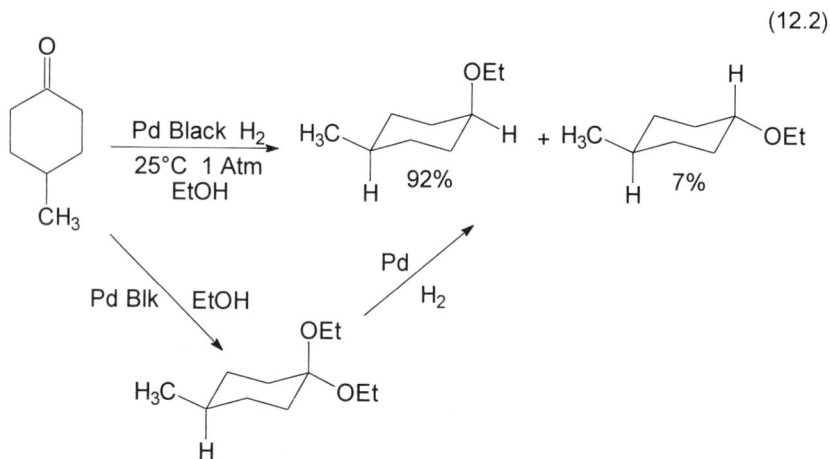

About the only specific use of palladium black is in the formation of ethers by the hydrogenation of ketones in the presence of an alcohol.[18,19] This reaction is unique in many respects. Palladium does not normally promote the hydrogenation of a carbonyl group unless it is conjugated with a benzene ring. In addition, only the pre-reduced palladium hydroxide or palladium oxide give the ether as the primary product in this reaction. With reduced platinum oxide or reduced rhodium oxide very little ether is produced and from reduced ruthenium oxide none is formed.[18,19] The mechanism put forth for this reaction involves the intermediate formation of a ketal which is then hydrogenolyzed to the ether (Eqn. 12.2).[19] Evaporated platinum and palladium blacks, which have no basic impurities, promoted facile acetal formation. Further hydrogenation over palladium gave the ether as the almost exclusive product at a rate four times faster than that observed when reduced palladium hydroxide was used as the catalyst. Over the evaporated platinum catalyst only moderate amounts of the ether were formed. The primary product was the alcohol accompanied by some alkane.

The surface areas of a number of commercial palladium blacks were measured using the BET procedure as well as hydrogen chemisorption, electron microscopy and X-ray diffraction analysis. These data showed that these blacks had particle sizes ranging from about 7 to 140 nm and surface areas between 70 and 4 m^2/g.[20] Ruthenium blacks prepared by the reduction of different samples of ruthenium oxide and ruthenium chloride were found to have surface areas ranging from 3–20 m^2/g.[21]

Borohydride Reduced Metal Salts

Catalytically active species have been prepared by the alkali borohydride reduction of a number of different metal salts. Some of these catalysts are metal borides while others appear to be the reduced metal.

Nickel Boride

The reaction of a nickel salt with sodium or potassium borohydride gives a nickel boride precipitate that is an active catalyst for the hydrogenation of a number of functional groups.[22-26] The composition of these borides as well as their catalytic activity depends on the method used for their preparation.[27] The initial reduction procedure involved the reaction between aqueous solutions of a nickel salt and the borohydride.[22] A granular precipitate was formed which, after washing with ethanol, gave what has been labeled the P-1 nickel boride that has an approximate Ni_2B stoichiometry.[22,25] The reduction of nickel acetate gave the most active catalyst. The borides formed by the reduction of nickel chloride or nickel sulfate were somewhat less active. Borohydride reduction of nickel nitrate gave a catalytically inactive material.[25] Almost all of the nickel boride catalysts reported in the literature have been prepared by the reduction of nickel acetate or nickel chloride.

Running the reduction in 95% ethanol gave the nearly colloidal P-2 nickel boride.[24,26] Both the P-1 and the P-2 catalysts are non-pyrophoric and non-ferromagnetic. The P-1 nickel boride is a good catalyst for alkene hydrogenations, particularly for those that also have an aromatic ring in the molecule.[25] Double bond isomerization is minimal over both the P-1 and P-2 catalysts.[25,26] These catalysts have also been used for the saturation of double bonds in molecules containing amine, amide and nitrile groups[28] as well as the hydrogenation of aldehydes and ketones.[29-32] The hydrogenation of nitro groups is inhibited apparently by some reaction intermediate.[30] While the presence of an aromatic ring facilitates double bond hydrogenation, these catalysts do not promote the hydrogenation of benzene rings nor the hydrogenolysis of benzyl oxygen bonds.[25,26]

The P-1 catalyst is more active than the P-2 but P-2 nickel boride is more selective in the hydrogenation of acetylenes and dienes (Eqns. 12.3 and 12.4).[26] When the reduction is run in aqueous ethanol, nickel borides with differing activities are produced with the composition of the catalyst and its activity depending on the water content of the reaction solvent.[30,33] The P-1.5 nickel boride that is precipitated from a 50% aqueous ethanol solution had an activity between those of the P-1 and P-2 catalysts.[30]

(12.3)

99%

(12.4)

98%

The borohydride reductions are commonly run in an inert atmosphere as contact with air during catalyst preparation leads to a significantly less active material. Brief exposure to air after formation of the catalyst, however, does not seem to significantly affect the activity of a P-2 catalyst.[26] More reproducible data were obtained when the reaction solvent was deaerated before the addition of the reducing agent.[34]

The more ethanol in the reduction solvent, the greater the amount of BO_2^- found in the resulting nickel boride precipitate.[33] This ion is formed by the hydrolysis of borohydride but, since it is not very soluble in ethanol, it is not kept in solution in alcoholic solvents. The BO_2^- is apparently responsible for the decreased activity and increased selectivity of the P-2 nickel boride.[33,35] When the P-2 catalyst was isolated and washed with water before use, the resulting P-2W catalyst was more active than both the P-2 and P-1 nickel borides but it became less selective in alkene hydrogenation than the P-2 catalyst.[36]

While various borohydride:nickel ratios have been used in these nickel boride preparations maximum P-2 catalytic activity was observed with a 2:1 $BH_4^-:Ni^{++}$ ratio.[37] However, when the reduction was run under a hydrogen atmosphere using a 4:1 $BH_4^-:Ni^{++}$ ratio, a hydrogenated nickel boride catalyst, the P-3 nickel boride, was obtained.[27] This catalyst was somewhat more active than the P-2 catalyst for alkene hydrogenation but it induced significantly more double bond isomerization during the reaction.

Using diborane as the reducing agent gave a nickel boride catalyst significantly different from those prepared using borohydride. This DBNi catalyst promotes the hydrogenation of ketones and nitro groups more effectively than do the P-1, P-2 or P-2W nickel borides. It is, though, only about as active as the P-1 catalyst for alkene saturation. The DBNi catalyst has a Ni:B ratio of 5.5 compared to the 2:1–3:1 ratios found for most of the borohydride reduced nickel borides. Another difference is the ferromagnetic character of DBNi, a property that can facilitate the separation of the catalyst from the reaction mixture but can be a problem if a magnetic stirrer is used for the agitation of the reaction medium.[38] All of the nickel boride catalysts show a resistance to poisoning by phosphines, phosphites and divalent sulfur compounds.[24,27,34]

Promoters

The borohydride reduction of a nickel salt solution containing small amounts of other metal salts can lead to co-reduced mixed metal borides that frequently have enhanced catalytic properties when compared to the unmodified nickel boride. The presence of about 2% chromium significantly increased the activity of a P-1 nickel boride toward aldehyde hydrogenation. Molybdenum, tungsten and vanadium modifiers were somewhat less effective than chromium while the presence of a small amount of cobalt had an inhibiting effect on the reaction.[22] As the data in Fig. 12.2 show, the amount of chromium responsible for optimum P-1 nickel boride activity depends on the substrate being hydrogenated.[39]

Substrate	Relative order of activity
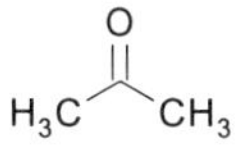 H_3C—CO—CH_3	25% Cr > 20% Cr > 15% Cr > 10% Cr > 0% Cr > 30% Cr
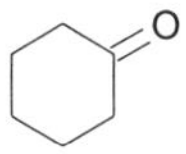 cyclohexanone	25% Cr > 20% Cr > 15% Cr > 10% Cr > 0% Cr > 30% Cr
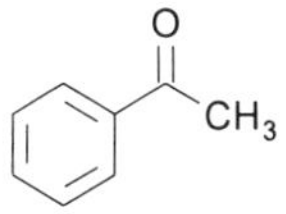 acetophenone (CH_3)	25% Cr > 20% Cr > 15% Cr > 10% Cr > 0% Cr > 30% Cr
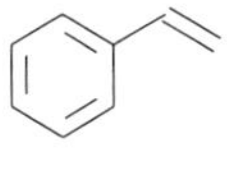 styrene	2% Cr > 5% Cr > 0% Cr > 10% Cr
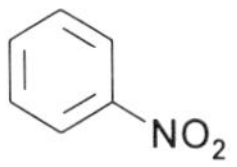 nitrobenzene (NO_2)	2% Cr > 5% Cr > 0% Cr > 10% Cr > 20% Cr

Fig. 12.2. The promoting effect of adding chromium to the P-1 nickel boride catalyst in the hydrogenation of various substrates.

This promoting effect is also somewhat dependent on the nickel boride preparation procedure. While chromium is the most effective of those additives used for the P-1 nickel boride catalyzed hydrogenations of carbonyl groups, a rhodium modified P-1 catalyst is best for alkene hydrogenation. Rhodium is also a significant promoter of alkene hydrogenation over P-2 nickel boride and it is almost as effective as chromium for ketone hydrogenation over this catalyst. With the P-2W catalysts, chromium and rhodium modifiers are equally effective for both alkene and ketone hydrogenations.[39] It was suggested that the rate increases observed using chromium promoted P-1 catalysts were due to an increase in catalyst surface area induced by the presence of the chromium.[40] Temperature programmed desorption (TPD) data obtained with these catalysts, however, indicated that the rate increase was the result of an increased strength of hydrogen chemisorption brought about by the promoting species.[39,41]

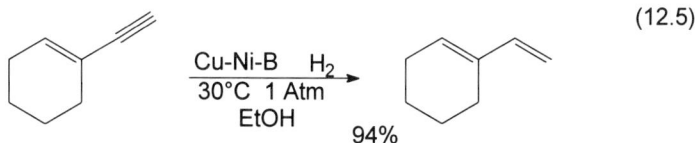

(12.5)

$$\text{Cu-Ni-B} \quad \text{H}_2 \atop \overline{30°C \ 1 \ Atm} \atop \text{EtOH} \qquad 94\%$$

Coprecipitated nickel–copper boride is somewhat less selective for the partial hydrogenation of phenylacetylene than is either the P-1 or P-2 catalyst. However, when a suspension of the P-1 nickel boride was stirred in a solution of cupric chloride in 99% ethanol, a copper modified nickel boride catalyst was produced. This material appears to be one of the most selective catalysts available for the semihydrogenation of alkynes.[42] In the hydrogenation of 1-ethynyl-cyclohexene over this copper modified nickel boride, hydrogen uptake ceased when all of the reactant was used up. The product composition at the completion of the reaction was 94% of the desired 1-vinylcyclohexene along with 6% of the various monoolefinic materials (Eqn. 12.5).[42] No fully saturated product was detected. For comparison, running this reaction over Lindlar's catalyst gave only 86% of the 1-vinylcyclohexene.[43] The hydrogenation of propargyl alcohol over the copper modified nickel boride catalyst gave only allyl alcohol with no hydrogenolysis of the hydroxy group (Eqn. 12.6).[42]

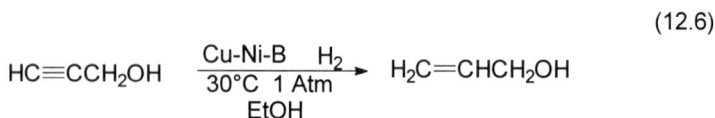

(12.6)

$$HC\equiv CCH_2OH \quad \xrightarrow[\substack{30°C \ 1 \ Atm \\ EtOH}]{\text{Cu-Ni-B} \quad H_2} \quad H_2C=CHCH_2OH$$

Reversed Micelle Stabilized Nickel Boride

The borohydride reduction of a nickel salt in the presence of reversed micelles provides a convenient method for the preparation of very small nickel boride

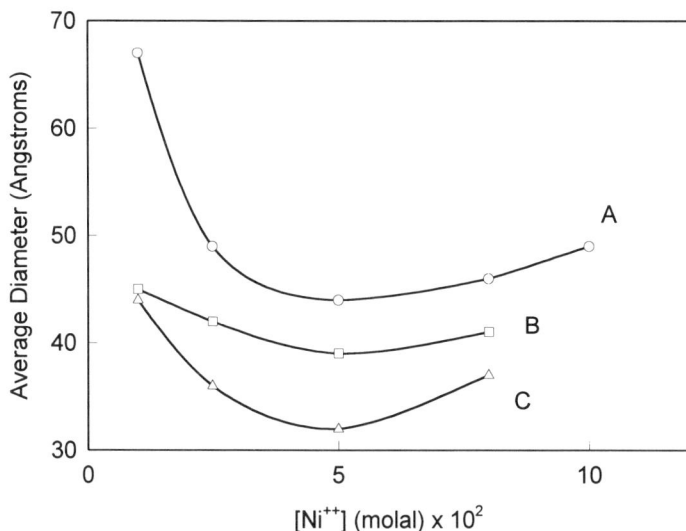

Fig. 12.3. Average diameter of nickel boride particles as a function of
           nickel ion concentration and micelle composition. The CTAB
           : hexanol : water ratios for the micelles were: A) 30:50:20;
           B) 24:60:16; C) 18:70:12. (Redrawn using data from Ref. 44.)

particles that are particularly active catalysts. The nickel borides prepared using
the cetyltrimethylammonium bromide (CTAB), 1-hexanol, water micelle system
have average particle sizes ranging from 2.5 nm to 7.0 nm. For comparison, P-1
nickel boride particles are 300 nm–400 nm in diameter while the P-2 catalyst
particles range from 250 nm to 300 nm.[44,45]

     In this system the average particle size depends on the $BH_4^-:Ni^{++}$ ratio,
the nickel salt concentration and the size of the inner water core of the reversed
micelle.[46] A $BH_4^-:Ni^{++}$ ratio of three gives the smallest size catalyst particles.
Lower ratios lead to the formation of larger particles while higher ratios have no
further effect on particle size. Micelles with smaller water cores produce smaller
catalyst particles.   The effect of nickel salt concentration on particle size is
complex; the smallest particles are formed with a Ni^{++} concentration of $5x10^{-2}M$.
Fig. 12.3 shows the relationship between the micelle composition, nickel salt
concentration and the nickel boride particle size.[44] For any given preparation the
catalyst particles are essentially uniformly sized with only a ±0.5 nm
distribution.[45]

     As shown by the data in Fig. 12.4 these micelle stabilized catalysts were
considerably more active for alkene hydrogenation than the P-2 nickel borides.[46]
The hydrogenation run using a preformed P-2 catalyst in a reversed micelle
medium proceeded at a faster rate than the reaction run over P-2 nickel boride in

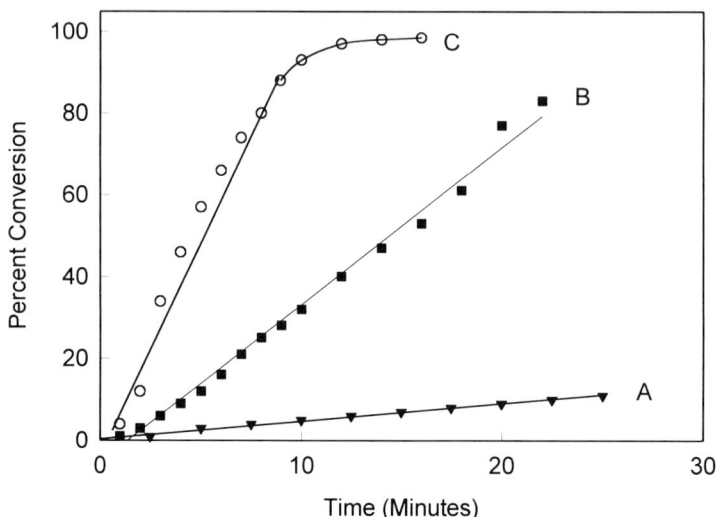

Fig. 12.4. Hydrogenation of 1-heptene over nickel boride catalysts.
A) P-2 nickel boride catalyst; B) preformed P-2 nickel boride
added to a micelle suspension; C) nickel boride formed in the
presence of a reversed phase micelle medium. (Redrawn
using data from Ref. 46.)

ethanol. This enhanced rate, though, was still only about half of that observed
using the micelle stabilized catalyst.

The borohydride reduction of nickel decanoate in cyclohexane also
produced a reversed micelle stabilized nickel boride catalyst.[37] The most active
catalyst was formed using a 3:1 BH_4^-:Ni^{++} ratio. This gave a catalyst that was
about as active as the P-2 catalyst for alkene hydrogenation. The use of other
BH_4^-:Ni^{++} ratios gave less active catalysts.

Polymer Stabilized Colloidal Nickel Boride

Running the borohydride reduction of nickel salts in the presence of a soluble
polymer gave a stabilized colloidal nickel boride.[47-50] Nickel chloride, acetate
and sulfate all give comparably active catalysts but nickel nitrate gave a colloidal
suspension with much lower activity. A BH_4^-:Ni^{++} ratio of two gave the most
effective catalyst system. Colloid stabilization was achieved using a number of
soluble polymers. Polyvinylpyrrolidone was the most commonly used polymeric
material but soluble nylon gave a more active catalyst. The polymer must be
present in at least 20–30% of the weight of the nickel salt to be effective. These
catalysts are generally prepared in a reaction flask for immediate use but they can
be stored without significant loss of activity in an oxygen free atmosphere that
contains at least several percent of hydrogen.[47,48] Not only are they effective for

alkene hydrogenations[49] but, the addition of base to pH 10–12 gives a catalyst system with which aldehydes and ketones are rapidly hydrogenated.[50] A comparably active colloidal nickel boride can be prepared in the absence of a polymer by using very dilute solutions of borohydride and the nickel salt.[51]

Other Reduced Nickel Salts

Reduction of nickel acetate with sodium hydride gave a nickel black catalyst that was active for the selective hydrogenation of acetylenes and dienes to the monoolefins. It was also capable of promoting the hydrogenation of carbonyl groups.[52]

Reaction of nickel chloride with aqueous NaH_2PO_2 gave a precipitate of a nickel phosphide labeled NiP-1. When a 50% aqueous ethanol solution was used, the NiP-2 catalyst was produced.[34] While these materials had some catalytic activity they were inferior to the nickel borides.

Cobalt Boride

Cobalt boride catalysts are prepared by the borohydride reduction of cobalt acetate using the procedures described above for the preparation of the various types of nickel boride catalysts.[53-55] The cobalt borides are, generally, less active catalysts than the corresponding nickel borides. They do have some advantages over the nickel borides in some reactions. The P-3 cobalt boride promotes significantly less double bond isomerization during the hydrogenation of an alkene than is observed with either P-3 nickel boride or Raney nickel.[55] Cobalt borides are also useful for the hydrogenation of nitro compounds[56] and nitriles to primary amines.[57]

The primary synthetic utility of these cobalt boride catalysts, however, lies in the selectivity they exhibit for a number of hydrogenations. While nickel boride catalysts are inactive for benzene hydrogenations, cobalt boride catalysts promote this reaction. The interesting thing about this reaction, though, is that the initial product is cyclohexene which is produced in greater than 95% selectivity at low conversion over a P-1 cobalt boride catalyst.[54] This catalyst also showed a reasonable degree of selectivity in the semihydrogenation of alkynes. Phenyl acetylene was hydrogenated to styrene with 79% selectivity and 2-heptyne gave 2-heptene in 97% selectivity.[53]

While a nickel boride catalyst preferentially saturates the carbon–carbon double bond of α,β-unsaturated aldehydes, the cobalt borides have a tendency to favor carbonyl group hydrogenation. Cinnamaldehyde was hydrogenated to cinnamoyl alcohol in 97% selectivity at 50% conversion and 86% selectivity at 74% conversion over a P-2 cobalt boride (Eqn. 12.7).[54] With a P-2W cobalt boride the unsaturated alcohol was produced in 97% selectivity at 73% conversion. The presence of the aromatic ring enhances selectivity in this reaction since the hydrogenation of crotonaldehyde to 2-buten-1-ol occurred with only about a 25% selectivity at under 20% conversion over either catalyst (Eqn. 12.8).[54]

(12.7)

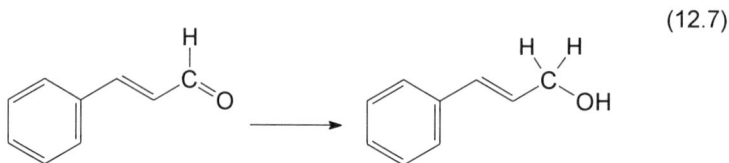

(12.8)

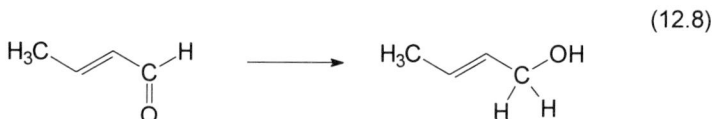

When the P-1 cobalt boride was modified by stirring it in a solution of chromium chloride before use, the selectivity for aldehyde hydrogenation was improved. Cinnamaldehyde was hydrogenated to the unsaturated alcohol in 90% selectivity at 65% conversion and crotonaldehyde was converted to 2-butenol in 44% selectivity at 52% conversion.[53] Reaction selectivity can also be improved by preparing the colloidal cobalt boride in reverse phase micelles. Reduction of cobalt chloride in a CTAB, 1-hexanol, water micelle medium gave cobalt boride catalysts in which the size of the catalyst particles were dependent on the composition of the micelle medium and the concentration of the cobalt salt.[44] The hydrogenation of crotonaldehyde using a micelle stabilized cobalt boride catalyst gave the unsaturated alcohol in 40% selectivity at 80% conversion.[45]

Borohydride Reduced Platinum Metals

The treatment of solutions of platinum metals with aqueous borohydride results in the formation of finely divided black precipitates that are active catalysts for alkene hydrogenations. The platinum black obtained in this way was twice as active as that obtained by the hydrogenation of platinum oxide. The borohydride reduced rhodium black is even more active.[58] While the borohydride reduction of base metals gives the corresponding metal borides, there is little, if any, boron incorporated into these platinum metal blacks. Analysis of the borohydride reduced palladium found that while the palladium:boron ratio in the bulk was 10:1, less than 1% of the surface was boron.[59] This small amount of boron, however, can impart a significant difference in catalytic activity to this catalyst as compared with other, more common, palladium catalysts. The most striking difference is the inability of the borohydride reduced palladium to promote the hydrogenolysis of activated C–O and C–N bonds,[60] a reaction that takes place readily over standard palladium catalysts.

Treatment of an aqueous solution of ruthenium chloride and stannous chloride with borohydride gave a precipitate of a ruthenium–tin–boride.[61] The ruthenium is present as the metal and the boron as the boride with electron donation to the ruthenium. The tin exists as a mixture of tin(II) and tin(IV) oxides which interact with the ruthenium through the oxygens. These catalysts are useful for the hydrogenation of carboxylic esters to the alcohols at high temperatures but moderate hydrogen pressures.[62,63] A ruthenium:tin ratio of one is optimum for this reaction.[63] A ruthenium–tin–boride catalyst having a ruthenium:tin ratio of two, however, was able to promote the selective hydrogenation of unsaturated esters to unsaturated alcohols. Methyl 9-octadecenoate gave 9-octaden-1-ol in 60% selectivity at 80% conversion over such a catalyst (Eqn. 12.9).[63]

$$\text{(12.9)}$$

$$H_3C(CH_2)_7HC{=}CH(CH_2)_7\underset{\underset{O}{\|}}{C}{-}OCH_3 \xrightarrow[270°C \quad 45 \text{ Atm}]{Ru\text{-}Sn\text{-}B} \Big|_{H_2} \downarrow$$

$$62\% \quad H_3C(CH_2)_7HC{=}CH(CH_2)_7CH_2OH$$

Skeletal Metal Catalysts

High surface area metal catalysts can be prepared by selectively removing one component of a bimetallic or polymetallic alloy or intermetallic compound. The remaining material has a microscopic spongy network of pores and is referred to as a skeletal metal catalyst.

Raney Nickel

The most common skeletal metal catalyst is that discovered by Murray Raney in 1927.[64] These Raney nickel catalysts are prepared by the action of sodium hydroxide on a powdered nickel aluminum alloy that leaves behind a high surface area nickel having a porous, spongelike microstructure.[2] The activity and composition of the Raney nickel catalysts depends primarily on the concentration of the base and the temperature of the reaction between the base and the alloy.[65]

Commercial Raney nickel alloy is about 50% nickel and 50% aluminum but alloys containing between 40% and 50% nickel have been used to prepare viable skeletal nickel catalysts. These alloys are composed of several phases, $NiAl_3$, Ni_2Al_3 and an $Al–NiAl_3$ eutectic.[65] The commercial alloy generally contains a small amount of the eutectic and somewhat more of the Ni_2Al_3 than the $NiAl_3$. The exact composition of the alloy depends not only on the Ni:Al ratio but also on the thermal conditions used in the preparation of the alloy.[66]

The eutectic and $NiAl_3$ are very reactive toward hydroxide and easily lose the aluminum to give the skeletal nickel. Ni_2Al_3 reacts more slowly with base but the aluminum can be removed at 50°C and the material completely decomposed in boiling alkali.[67,68] Catalysts prepared from the commercial alloy, pure $NiAl_3$ and pure Ni_2Al_3 all have about the same activity when the preparation conditions are such that most of the aluminum is removed from each material.[69]

The usual preparation procedure involves the addition of the alloy to a sodium hydroxide solution held at a specific temperature. This not only removes the aluminum but also generates an atmosphere of hydrogen that serves to activate the nickel catalyst. After the base reaction is complete the excess base and the sodium aluminate produced by the reaction are removed by an aqueous wash process. The water is then replaced by alcohol to give the active, highly pyrophoric, catalyst. An alternative procedure involves the addition of base to a suspension of the alloy.[70] Even though this procedure is said to give a somewhat more active catalyst it is not commonly used.

A number of different types of classic Raney nickel catalysts with varying activities have been prepared by the addition of the commercially available alloy to sodium hydroxide solutions.[2] These catalysts have been designated as W1, W2, W3, W4, W5, W6, W7 and W8. The procedures utilized to prepare these catalysts differ in the amount of sodium hydroxide used, the temperature at which the alloy is added to the basic solution, the temperature and duration of alloy digestion after addition to the base and the method used to wash the catalyst free from the sodium aluminate and the excess base. These differences are listed in Table 12.1.

W1 Raney nickel[71] is the least active of all of these catalysts with the possible exception of the W8 variety. This inactivity has been used to advantage, for instance, in the hydrogenation of 2,5-dihydroxy benzoic acid to cyclohexane carboxylic acid-3,5-dione (Eqn. 12.10).[72] The use of the more active W7 catalyst for this reaction, resulted in considerable over-reduction of the product.

(12.10)

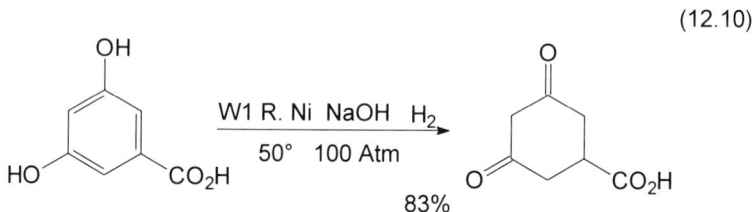

$$\text{OH} \quad \xrightarrow[\text{50° 100 Atm}]{\text{W1 R. Ni NaOH H}_2} \quad$$

83%

The W2[73] variety is the most commonly used of all of the Raney nickel catalysts. The commercially available Raney nickel is generally regarded as a W2 catalyst. The commercial catalyst is supplied as an aqueous suspension. Before the catalyst can be used, however, the water must be replaced by ethanol or other suitable solvent. This is done by decanting the water, washing the residual catalyst several times with 95% ethanol, and decanting the solvent after each wash. Several washes, again by decantation, with absolute ethanol removes the last of the water and gives an ethanol suspension of the active catalyst.

W2 Raney nickel is appreciably more active than W1 as shown by the fact that the hydrogenation of β-naphthol at a specific temperature and pressure required 240 minutes with the W1 catalyst but was complete in 15 minutes using the W2 Raney nickel. For this same reaction, W4 Raney nickel required only half as long as did the W2 (Eqn. 12.11).[74] This increased activity of the W4

Table 12.1

Preparation procedures for the different types of classic Raney nickel catalysts.

Type	Add'n. Temp[a]	NaOH:Alloy Ratio (W/W)	Digestion Temp. and Time	Washing Process	Relative Activity	Ref.
W1	0°C	1 : 1	115°-120°C 4 Hours	Filter, wash to neutral with H_2O. EtOH wash by decantation.	Least active $\cong$ W8	71
W2	25°C	4 : 3	Steam bath 8-12 Hours	H_2O wash to neutral by decantation. EtOH wash by decantation.	<W4 : >W1 Most common type.	73
W3	-20°C	4 : 3	50°C 50 Min	H_2O wash by decantation Several times. Continuous wash with a large volume of H_2O. EtOH wash without contact of catalyst with air.	Quite active >W2 : <W7	76 75 77
W4	50°C					
W5	50°C					
W6	50°C	4 : 3	50°C 50 Min	Continuous H_2O wash under a H_2 atmosphere. EtOH was without contact of catalyst with air.	Most active	77, 78
W7	50°C	4 : 3	50°C 50 Min	Three decantations with H_2O. EtOH wash without contact of catalyst with air.	Very active <W6 : >W4	77
W8	0°C	1 : 1	100°-105°C 4 Hours	Continuous H_2O wash to remove lighter Ni particles. Dioxane wash by decantation. Distill portion of the dioxane from the catalyst.	Least active $\cong$ W1 Useful for deuterations	84

[a] For addition of the alloy to the base.

catalyst[75], and the W3[76] and W5[77] as well, is due primarily to the continuous aqueous wash process used in their preparation and to the nonexposure of the catalyst to air during the ethanol wash sequence. The enhanced activity of these catalysts makes it possible to use them for the hydrogenation of esters to alcohols.[75] The introduction of the continuous wash process during the W1 or W2 preparation, results in an increase in the activity of these catalysts.[75]

(12.11)

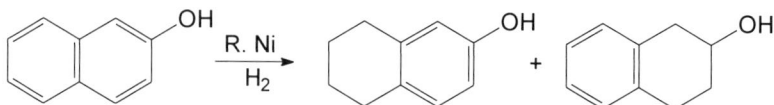

The most active of all Raney nickel catalysts is the W6 variety.[77,78] In fact, under mild conditions this catalyst is comparable to platinum and palladium for the hydrogenation of alkenes, alkynes, aldehydes, ketones, and aromatic nuclei. The distinguishing feature in the W6 preparation is a continuous aqueous wash under a slight pressure of hydrogen. Having the wash water under hydrogen pressure is not essential, however, since a catalyst of comparable activity can be more easily prepared by the use of the W4 wash procedure with hydrogen being bubbled through the water during the wash process.[70,80] It has been reported that in a number of hydrogenations in which the W6 catalyst was used at temperatures above 100°C, a sudden, rapid, and large increase in pressure was observed that necessitated immediate cooling and pressure release to avoid an explosion.[77] **It is recommended that this catalyst be used with caution and always at temperatures below 100°C.** This is not a serious limitation since the catalyst is so active that higher temperatures are not required.

The W7 catalyst[77] is prepared in the same way as the W4 but replacing the continuous wash by a brief decantation wash process. The resulting catalyst is quite alkaline and should be used only for those hydrogenations that are compatible with a strongly basic reaction medium. The basicity of the catalyst is probably responsible for its high activity.

The Raney nickel alloy will react with water at 70°–85°C to give the skeletal nickel in the presence of aluminum hydroxide. The resulting catalyst is active after washing with alcohol[81]. Since no base is present, the extensive water washings are not required. The reaction with water can be promoted by the addition of a small amount of base, and, further, some Bayerite, $Al(OH)_3$, to further facilitate the aluminum removal.[82]

Most Raney nickel applications involve the use of fine catalyst particles in a batch type stirred reactor. Raney nickel can, however, be adapted for use in a fixed bed reactor by filling the reactor with the commercially available large granules of the alloy. A 10% solution of sodium hydroxide is passed through the reactor for a sufficient time to remove about 10–12% of the aluminum from the granule surface. After washing the catalyst free of alkali, it is ready for use. This procedure gives a thin surface of Raney nickel on the outer layer of the granules with the unactivated alloy at the core of the granules acting as a support for the skeletal nickel. This surface Raney nickel exhibits all of the properties of a finely divided catalyst. After prolonged use the active surface may become deactivated, but activity can be regenerated by repeating the original base treatment to give a fresh surface of active catalyst.

Another process for preparing Raney nickel particles of sufficient size for use in a flow reactor involves mixing a Raney alloy powder with a polymer and plasticizer. This mixture is extruded to an appropriate shape, the plasticizer is removed and the extrudate calcined in air at high temperatures to form an α-alumina matrix to support the alloy. Reaction with base removes some of the surface aluminum to give an active surface of Raney nickel that has been shown to be active for vapor phase and trickle flow hydrogenations.[83]

One problem with the use of Raney nickel is measuring the amount of catalyst to be used. A general rule is that one teaspoonful of the catalyst suspended in ethanol contains about four grams of the catalyst. More precise weights can be determined by centrifuging a known volume of the suspended catalyst in a graduated centrifuge tube. The density of the catalyst can be calculated from the centrifuged volume and dried weight of the sample and this value can be used to determine the quantity of catalyst utilized in succeeding reactions.

Deuteration Catalysts

Because the Raney nickel catalysts are prepared in a hydrogen environment, before they can be used for the deuteration of organic substrates this hydrogen and all labile hydrogens present in the catalyst must be replaced by deuterium. This is easier to accomplish with the relatively inactive catalysts, such as the W8 Raney nickel, which was specifically developed for use in deuteration reactions.[84] The essential difference in its preparation lies in the use of dioxane instead of ethanol as the organic wash liquid and the distillation of a portion of the dioxane from the catalyst, a process that removes a large amount of the adsorbed hydrogen from the nickel. The exchange of deuterium for the rest of the adsorbed hydrogen is accomplished by suspending the catalyst in deuterium oxide and stirring the mixture under a deuterium atmosphere. Replacing the adsorbed hydrogen on W1, W2, and W3 catalysts by deuterium has been accomplished by agitating a deuterium oxide suspension of the catalyst in a deuterium atmosphere for a prolonged period of time with periodic replacement of the gas.[85]

A more direct approach is to use sodium deuteroxide in deuterium oxide to remove the aluminum from the Raney alloy.[86] The activity of the resulting catalyst depends on the temperature and duration of the leaching process. Washing the catalyst with deuterium oxide and/or dioxane gives an effective catalyst for deuteration reactions.

Composition

Raney nickel is a sponge-like material made up of 2.5–15 nm microcrystallites that are agglomerated into macroparticles several microns in diameter.[87] The surface area and composition of these particles depend on the base concentration and temperature used for the removal of the aluminum from the alloy. High temperature preparations that remove almost all of the aluminum generally have a surface area of 50–80 m[2]/g. Catalysts prepared at temperatures of 50°C or lower have more aluminum present and surface areas of 100–120 m[2]/g. These surface areas are related to the pore diameters and pore volume of the catalysts. The more extensive the base attack on the alloy the larger the average pore diameter and pore volume of these particles and the smaller their surface area.[88]

Chemically these catalysts contain metallic nickel along with 1–8% aluminum and up to 20% aluminum oxide or hydroxide.[88, 89] The surface is 50–100% metallic nickel which is present in an fcc crystalline lattice, the same crystal orientation found for the bulk nickel.[87] The use of low hydroxide ion concentrations and reaction temperatures in the reaction with the alloy gives catalysts containing more aluminum and aluminum oxide. The alumina in these preparations is occluded in the metallic skeleton and is difficult to remove even when later exposed to higher hydroxide concentrations.[66] High temperature digestion is needed to remove all of the alumina from these catalysts.

While catalyst activity is generally inversely related to the amount of aluminum and alumina present it is not desirable to remove all of these materials from the nickel. It has been proposed that some aluminum in the nickel crystal lattice creates the defect sites responsible for catalytic activity.[87] The alumina appears to prevent the sintering of the nickel particles. With the 20% alumina that is found in the commercial catalyst, heating to 500°C results in only a 20% reduction in surface area. When only 1% alumina is present there is a 50% reduction in surface area at this temperature but no change in surface area on heating to 250°C.

One reason for the observed activity of these Raney nickel catalysts is the large amount of hydrogen adsorbed on and in the metal particles during the preparation procedure. Amounts between 25 to 100 cc of hydrogen per gram of catalyst have been reported for the various classic preparations listed in Table 12.1. The commercially available W2 Raney nickel contains about 30 cc/g[90] while the very active W6 form is reported to have about 100 cc/g because of the hydrogen atmosphere in the wash procedure used in its preparation.[91] The deuteration catalyst, prepared using sodium deuteroxide in deuterium oxide, is reported to contain 45–50 cc of deuterium per gram of catalyst.[86]

Storage Limitations

Because of the difficulty in preparing reproducibly active duplicate samples of catalyst, it has been suggested that to have a uniform catalyst for a series of reactions it is best to prepare a larger amount of the Raney nickel and to use portions of it for the various individual reactions.[65] This, however, is not as straightforward as it might seem. The more active catalysts, W3–W7, lose their activity when stored over a period of time. It is not practical to store W6 for more than two weeks and the other catalysts should not be prepared more than a month before use. On the other hand, W1, W2, and W8 can be stored under solvent, in a full, closed container, for considerable lengths of time without an appreciable loss of activity. This stability, coupled with its activity, is undoubtedly the primary reason for the popularity of the W2 catalyst.

It should be noted, however, that some change does occur on aging these catalysts. Storage of Raney nickel in aqueous media promotes a slow surface oxidation.[92] This has been verified by the increasing amount of surface oxygen detected by X-ray photoelectron spectroscopic (XPS) analysis of aged catalysts.[93] An aged catalyst exhibits more selectivity for diene hydrogenation than does a freshly prepared catalyst.[94,95] An ultrasound treatment of an aged catalyst removes this surface oxide and restores its catalytic activity.[93] Raney nickel catalysts stored in deionized water under a hydrogen atmosphere have hydrogenation activity that increases for about thirty days and then decreases on further storage. Storage in deionized water gives a more active catalyst than storage in 95% ethanol or 5% aqueous sodium hydroxide. Catalysts stored under a hydrogen atmosphere at room temperature are more active than catalysts stored under air in the cold.[92]

Promoters

The activity of Raney nickel can sometimes be improved by adding a metallic promoter either to the starting alloy or to the activated catalyst. Stirring a Raney nickel catalyst in an aqueous solution of platinum chloride or chloroplatinic acid immediately before use greatly increases its activity in the hydrogenation of a number of different functional groups.[96-99] This effect appears to be specific to Raney nickel since a similar treatment of a supported nickel catalyst had no effect on its activity.[96]

A study of the effect of different promoters on the ability of Raney nickel to hydrogenate a variety of functional groups showed that the presence of 1–2% molybdenum in the alloy gave a catalyst that had enhanced activity for the hydrogenation of alkenes, ketones, nitriles and nitro groups. A catalyst containing 1–2% chromium was effective for the hydrogenation of ketones, nitriles and nitro groups, but the effect was not as pronounced as that observed with the molybdenum promoted catalysts. The hydrogenation of alkenes, though, was inhibited by the presence of chromium in the alloy used to prepare the catalyst. A Raney nickel catalyst containing 6–7% of iron is particularly useful for nitro group hydrogenations.[100]

An ESCA analysis of the molybdenum promoted Raney nickel showed that when a low molybdenum content alloy was used, the activated catalyst had greater amounts of nickel on the surface than the unpromoted active catalyst.[101] At a 2% molybdenum content the nickel surface area in the activated catalyst reached a maximum and then decreased with the presence of more molybdenum. These findings correlate with the observed maximum in catalytic activity observed with the 2% molybdenum Raney nickel catalyst. The initial increase in nickel surface area is brought about by a corresponding decrease in the amount of aluminum on the surface of the activated catalyst. The surface aluminum content reaches a minimum at about 2% molybdenum and then remains constant with higher molybdenum contents. The observed decrease in nickel surface area when higher amounts of molybdenum were present was caused by increasing amounts of molybdenum oxide being formed on the surface.[97,101]

In contrast to the molybdenum containing materials, alloys having chromium present give catalysts having higher amounts of aluminum. The more metallic aluminum present, the lower the turnover numbers for the hydrogenation of alkenes. Further, the chromium oxide which is produced during the reaction of the alloy with base, forms on the surface giving a lower nickel surface area.[102,103] It appears, however, that this surface oxide may have a promoting effect for carbonyl group hydrogenation.

Portions of activated Raney nickel were also stirred with aqueous solutions having different concentrations of chromium chloride. After removing the catalysts and calcining them, they were examined in the same way as those catalysts that were prepared from ternary alloys having differing amounts of chromium. It was found that increasing amounts of chromium oxide decreased the nickel surface area but had no effect on the turnover numbers for alkene hydrogenation. Since all of these catalysts had the same surface aluminum content, this finding reinforced the previous conclusion that the increased aluminum content was responsible for the decrease in the turnover number for alkene hydrogenation. The turnover number for carbonyl group hydrogenations, however, went through a maximum as the surface chromium oxide increased showing that this oxide was involved in the hydrogenation of aldehydes and ketones over this catalyst. The presence of chromium in these catalysts inhibited the hydrogenation of benzene rings and the hydrogenolysis of benzyl alcohols so the chromium modified Raney nickel could be used for the hydrogenation of acetophenone to phenethanol with the formation of a minimum amount of the by-products resulting from these other reactions (Eqn. 12.12).[104]

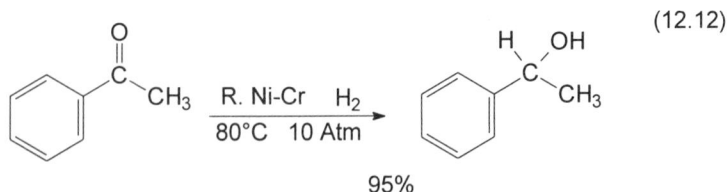

(12.12)

95%

Raney Cobalt

Raney cobalt, which is prepared from a commercially available cobalt aluminum alloy in the same way as is Raney nickel,[105-109] has been shown to have catalytic activity but it is generally less active than Raney nickel and more sensitive to variations in the reaction procedure and catalyst aging than the nickel catalyst.

Raney cobalt is useful for the hydrogenation of oximes[105] and nitriles[106] to primary amines.

Raney Copper

Raney copper is prepared from the commercially available copper aluminum alloy.[110] It does not have much to offer the synthetic chemist as only a few reactions are reported to be affected by this catalyst. Raney copper, as well as Raney cobalt, generally produces fewer side reactions than Raney nickel even though they usually require higher reaction temperatures for the same reaction. Raney copper is, however, quite useful for the selective hydrogenation of substituted dinitro benzenes (Eqn. 8.6)[111] with its activity apparently increasing with continued reuse. Raney copper can also be used for the catalytic hydrolysis of hindered nitriles to the amides (Eqn. 12.13).[112]

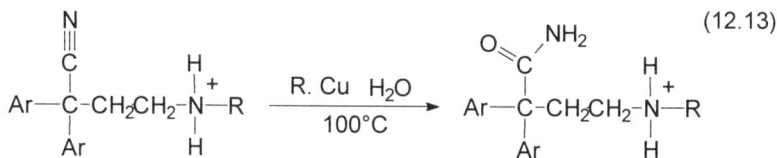

$$
\underset{\substack{\text{Ar} \\}}{\underset{\substack{| \\ \text{C} \\ |||\\ \text{N}}}{\text{Ar}-\overset{|}{\underset{|}{\text{C}}}-\text{CH}_2\text{CH}_2-\overset{\text{H}}{\underset{\text{H}}{\overset{|}{\text{N}}}}\text{--R}}} \quad\xrightarrow[\;100°\text{C}\;]{\text{R. Cu}\;\; \text{H}_2\text{O}}\quad \text{Ar}-\overset{|}{\underset{\text{Ar}}{\text{C}}}-\text{CH}_2\text{CH}_2-\overset{\text{H}}{\underset{\text{H}}{\overset{|}{\text{N}}}}\text{--R} \qquad (12.13)
$$

Urushibara Nickel

Another form of skeletal nickel catalyst is that known as Urushibara nickel. This material is somewhat analogous to Raney nickel but uses as the precurser a substance referred to as precipitated nickel that is prepared by the addition of metallic zinc to an aqueous solution of a nickel salt.[113-115] The precipitated nickel also contains zinc, zinc hydroxy halide and zinc oxide. The X-ray diffraction lines for nickel metal were not observed indicating that the nickel must be very finely divided. The precipitated nickel is analogous to the Raney nickel alloy in that it is not a catalyst itself but must be pretreated before use. These pretreatments consist of either acid or base washes. The base wash involves exposure to hydroxide for 25–30 minutes followed by a water washing to give a bulky hydrogenation catalyst containing 10–25% nickel. The extent of the water washing required depends on the desired pH of the resulting catalyst. If a neutral catalyst is needed, then extensive washing is required. These catalysts are termed U-Ni-B. The acid washed catalysts, called U-Ni-A are prepared by treating the precipitated nickel with acetic acid to give a less bulky catalyst containing 70–80% nickel which can more easily be washed to neutrality. When the precipitated nickel is activated by refluxing in an alcohol an active, neutral nickel catalyst labeled U-Ni-N is obtained.[116]

The precipitated nickel is composed of nickel on zinc particles with a covering of zinc oxide and other zinc compounds.[117] These zinc compounds are removed by the acid wash to expose the metallic nickel. Since the presence of the zinc prevents the oxidation of the nickel when exposed to mild oxidation conditions, the nickel is kept in the catalytically active metal state for a longer time when the zinc is present.[118] The precipitated nickel is stable for long periods of time as is the Raney nickel alloy. Urushibara nickel is reported to be insensitive to impurities and to be non-pyrophoric, which are some of the drawbacks to the use of Raney nickel. The time required for the preparation of the catalysts, particularly for the acid washed catalyst, from the precipitated nickel is rather short and the resulting catalyst can frequently be regenerated by an acetic acid wash after use.

Other Skeletal Metal Catalysts

A nickel catalyst with an activity comparable to that of W4 Raney nickel has been obtained by the reaction of dilute acetic acid on a nickel magnesium alloy.[119] After digestion, the residue is washed with dilute acetic acid, water and ethanol. The activity of this catalyst depends on the acetic acid wash and on the presence of 40–50% nickel in the alloy. While this catalyst would be useful for those hydrogenations needing slightly acidic media, it is not practical since the alloy required for its preparation is not generally available. A simpler expedient to use with those hydrogenations requiring a non-basic medium is the addition of a small amount of acetic acid to the Raney nickel catalyzed reaction mixture.

Other specialized alloys have also been used to prepare skeletal metal catalysts. Raney ruthenium has been prepared from the ruthenium aluminum alloy.[120] A colloidal platinum has been prepared by the action of acetic acid on a platinum lithium alloy.[121] Skeletal nickel catalysts have been made from a number of intermetallic compounds of nickel with the rare earth elements, lanthanum and samarium. The rare earth element is removed from the alloy by reaction with diiodoethane or dibromoethane which convert the rare earths to the soluble halide salts.[122] Several multicomponent catalysts have also been prepared from the corresponding aluminum alloys.[123-126]

Comparisons

The more common forms of Raney nickel have an activity comparable to that of supported nickel catalysts but usually promote similar reactions at temperatures 50°C lower. One reason for this apparent increased activity is the relatively large amounts of these catalysts that are generally used. It has been estimated that it takes about eight grams of W2 Raney nickel to equal the reactivity exhibited by 0.2g of a typical supported nickel catalyst. For preparative purposes the use of such a large amount of catalyst presents no particular problem since this material is readily available and relatively inexpensive.

Nickel boride catalysts have been considered viable alternatives to Raney nickel since they are non-pyrophoric, are easily prepared and the preparation procedure is reproducible. In addition these catalysts appear to be somewhat more active than W2 Raney nickel for the hydrogenation of alkenes and nitriles but both types of catalyst have about the same activity for carbonyl group hydrogenation.[30] While nickel boride and Raney nickel are both more resistant to

Table 12.2

Δq values for various nickel catalysts.[129]

Catalyst	Data Point in Fig. 9.5	A/Ni[a]	Δq
Ni-B (P-1)[b]	1	0.45	-0.11
Ni-B (P-2)[b]	2	0.21	-0.08
Ni-P-1[c]	3	0.52	+0.36
Ni-P-2[d]	4	0.31	+0.22
Raney Ni	5	0.4 ± 0.1	-0.07
U-Ni (A)[e]	6	0.5 ± 0.1	-0.06
D-Ni[f]	7	0	0

[a] Atomic ratio of the component element (B, P, Al or Zn) to nickel metal at the catalyst surface.
[b] Nickel boride prepared from nickel acetate.
[c] Nickel hydroxide reduced with NaH_2PO_2 in water.
[d] Nickel hydroxide reduced with NaH_2PO_2 in alcohol.
[e] Acid washed Urushibara nickel.
[f] Nickel formate decomposed at 300°C for 3 hr in a vacuum.

sulfur poisoning than unsupported or supported nickel catalysts,[127] the nickel borides are, themselves, more resistant to poisoning than is Raney nickel.[33]

X-ray photoelectron (XPS) studies of nickel boride, nickel phosphide, Raney nickel and Urushibara nickel showed that the electron density on the nickel was a function of the other metal present in these catalysts.[34,128,129] Boron, aluminum (Raney nickel) and zinc (Urushibara nickel) all increased the electron density on the nickel while phosphorous was an electron acceptor. Comparing the electron densities on the nickel in these catalysts with that on a nickel black prepared by the thermal decomposition of nickel formate (D-Ni) gave the series: Ni–B > Ni–Al > Ni–Zn > D–Ni > Ni–P.

A parameter, Δq, was defined as the relative change in electron density on nickel resulting from electron transfer between nickel and the second element as compared with that on nickel black. Table 12.2 shows the Δq values for these different nickel catalysts.[129] That for the P-1 nickel boride[25] is the most negative and that for the NiP-1 nickel phosphide[34] is the most positive. In Fig. 12.5 is shown the relationship between the areal turnover frequencies for the hydrogenation of styrene over these catalysts and their Δq values.[129] These data

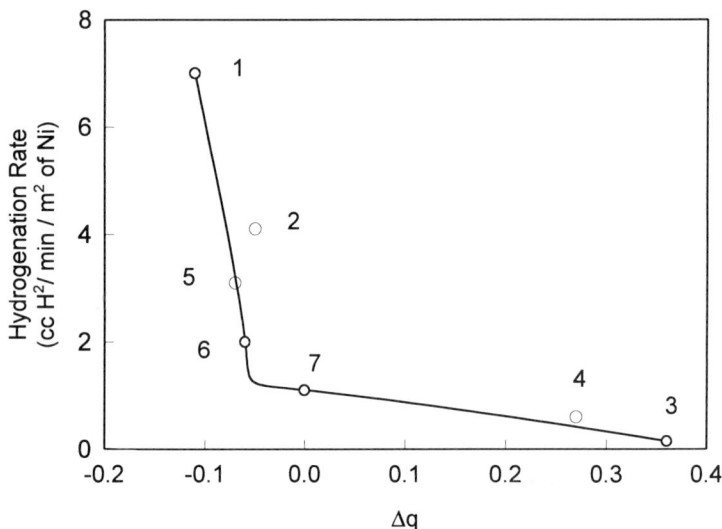

Fig. 12.5. Dependence of the rate of styrene hydrogenation on the Δq
of the nickel catalysts listed in Table 12.2. The numbers in
the figure correspond to the catalyst numbers in this table.
(Redrawn using data from Ref. 129.)

support the conclusion that as the electron density on the nickel increases there is
a corresponding decrease in the strength of alkene adsorption and/or an increase in
the strength of hydrogen chemisorption, either of which would result in an
increase in the rate of hydrogenation.[34] Similar trends have been observed for
carbon monoxide hydrogenations[130] so it would be expected that the
hydrogenation of aldehydes and ketones should also increase with an increase in
the negative value of Δq, but this has not been observed. Poisoning studies with
sulfides and phosphines showed that as Δq becomes more positive, the lower
electron density on the nickel increases the extent to which the electron rich
poison molecules are adsorbed so these catalysts are more easily poisoned.

Mixed Metals

The distinction between a catalyst promoted by a metal additive and a mixed
metal catalyst is vague. Generally, a promoted catalyst is one that contains a
small amount of a metal which is, itself, not catalytically active for the reaction
being investigated or it is a catalyst produced by adding the promoting metal to
the surface of an active catalyst. A mixed metal catalyst is usually an alloy or
intermetallic compound of two catalytically active metals or one in which both
components are present in relatively large amounts. There is a great deal of
overlap between these two definitions. Frequently the distinction between a

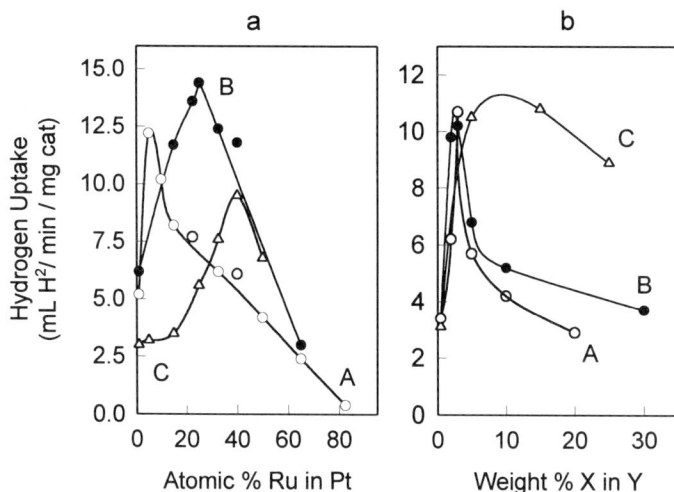

Fig. 12.6.   a) Effect of added ruthenium on the platinum catalyzed
             hydrogenation of: A) methylbutynol, B) maleic acid,
             C) cyclohexene. b) Rates of nitrobenzene hydrogenation
             over catalysts composed of X added to Y with A) X=Ni,
             Y=Pt, B) X=Co, Y=Pt, C) X=Ni, Y=Pd. (Redrawn using
             data from Ref. 134.)

promoted metal catalyst and a mixed metal catalyst depends on the nature of the
work in which they are involved.

Considerable interest has been shown in the use of mixed metal catalysts
to promote a variety of reactions.[131-141] This is based on the fact that a mixed
metal catalyst can frequently exhibit reactivities and/or selectivities considerably
greater than what is observed when either of the component metals is used
individually. An example of mixed metal synergism is given in Fig. 12.6.[134]
Fig. 12.6a shows the effect of adding ruthenium to platinum in the hydrogenation
of methylbutynol, maleic acid and cyclohexene. This synergism does not require
that both components be noble metals. Fig. 12.6b shows the effect of increasing
amounts of nickel and cobalt in the nickel–platinum, cobalt–platinum and nickel–
palladium catalysts used for the hydrogenation of nitrobenzene. In these, and all
other reported hydrogenations run over a mixed metal catalyst, there is a specific
metal composition at which an enhanced rate of reaction is observed. This
maximum composition varies with the nature of the reactant and the metal
components of the catalyst.[131-136].

The presence of a second metal can also have a significant impact on the
selectivity of a reaction. Recall the data shown in Fig. 3.1 illustrates the effect of
adding copper to a nickel catalyst used to promote both cyclohexane
dehydrogenation and ethane hydrogenolysis.[139] It was found that the addition of

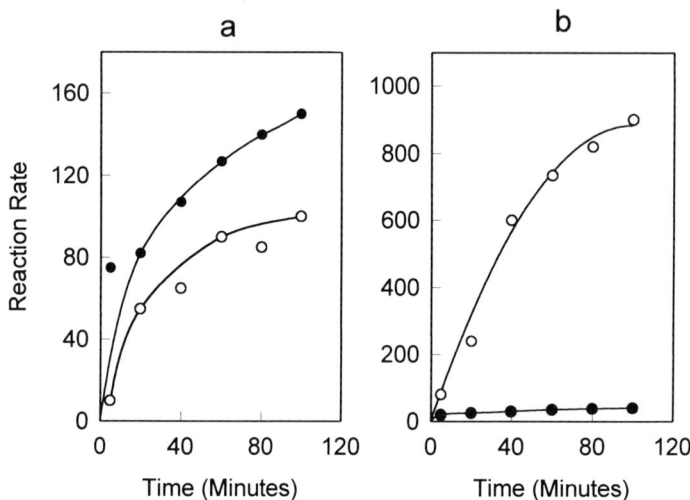

Fig. 12.7.   Reaction of cyclopentane with deuterium at 150°C over: a) a
            nickel catalyst containing no copper and b) a nickel catalyst
            containing 5% copper. Open circles represent the exchange
            reaction (C–H bond breaking); closed circles the hydrogenolysis
            (C–C bond breaking). (Redrawn using data from Ref. 142.)

a small amount of copper markedly suppressed the hydrogenolysis reaction but
had little effect on the dehydrogenation reaction.   Similar data was found in
studying the deuterium exchange and hydrogenolysis of cyclopentane over nickel
and 5% copper–nickel catalysts with the results depicted in Fig. 12.7.   The
breaking of the C–C bonds gave CD_4 as the primary product while C–H bond
breaking and exchange with deuterium gave C_5D_{10}. In the absence of copper in
the catalyst, the primary reaction is C–C bond breaking. The addition of 5%
copper to the nickel greatly enhances the C–H bond breaking process while
deactivating the catalyst for C–C bond hydrogenolysis.[142].

As described in Chapter 3, the reasons given to explain this dilution effect
of copper on nickel were that the hydrogenolysis reaction required a group or
ensemble of nickel atoms on the catalyst surface and the presence of copper
prevented the formation of the appropriately sized ensembles and the reaction was
inhibited. On the other hand, the reactions involving C–H bond breaking take
place on single atom sites so the surface dilution by copper has no effect until the
surface is almost completely covered by the copper. The rate increase observed in
the deuterium exchange on cyclopentane (Fig. 12.7b), however, is not easily
rationalized by this surface ensemble effect alone.

This so called ensemble or geometric effect, however, is not the only
reason for the distinct properties of mixed metal catalysts. The diluent metal can

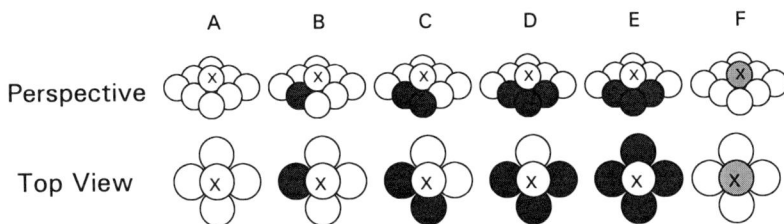

Fig. 12.8. Modification of a single atom active site (x) by the addition of successive amounts of an added metal (shaded circles). A) Active site, B) one added metal "ligand", C) two added metal ligands, D) three added metal ligands, E) four added metal ligands, and, F) the added metal as the active site with the first metal ligands.

also be seen as occupying near neighbor locations to the active sites so the nature of the "surface complex" composing this site is modified by the change in the electronic character of some of the "ligand" atoms. Fig. 12.8 shows how this ligand or electronic effect[143,144] can be pictured for an active site composed of an adatom on a 100 face, an octahedral corner (K in Figs. 4.4 and 4.5). In such a site the near neighbor atoms or "ligands" of the first metal can be replaced by one (B), two (C), three (D) or four (E) atoms of the diluent metal. It is apparent that such an incorporation could cause a change in the electronic character of the central active site atom and this could result in a modification of the adsorption and reaction characteristics of the site. Another possibility is that the diluent metal, itself, could become the central or active atom with its reaction characteristics determined by the presence of the first metal "ligands" (Fig. 12.8 F).

While there are proponents of both the geometric and electronic effect it is generally agreed that they are both important considerations in mixed metal catalysts. It would seem that the relative importance of each would be influenced primarily by the nature of the reaction being promoted. In those reactions that take place over groups or ensembles of surface atoms, the geometric effect should be more important but there is sufficient data available to indicate that electronic factors could also be significant. Unfortunately, detailed evaluations of single atom active sites on mixed metal catalysts has not been forthcoming, but it would be expected that in such cases the ligand or electronic effect should be the most important. One can imagine the synergism shown in Figs. 12.6a and b to be the result of the fine tuning of the strengths with which the reactants are adsorbed by the changing numbers of diluent ligands available. Such optimum adsorption characteristics would be expected to vary with changes in the types of reactants used as well as the conditions of the reaction. For those reactions in which the synergistic maximum is at a low concentration of added species, one might speculate that the active site may indeed be composed of a single atom of the added metal surrounded by the ligands of the first metal.

Unsupported mixed metal catalysts have been prepared in a number of ways, most of which are simply extensions of the procedures used to prepare single metal catalysts. Standard metallurgical methods have been used to prepare mixed metal foils, wires, granules, powders and blacks. Other procedures involve the reduction of an appropriate mixture of metal salts[145] or mixed oxides.[131-136] The best way to prepare a homogeneous mixture of metal salts for reduction is to utilize a solution of the appropriate salts. For hydrogen reduction techniques the solvent is rapidly evaporated to give a nearly homogeneous solid that is reduced at as low a temperature as possible. This reduction probably proceeds by way of the initial formation of some of the more easily reduced metal and this material, then, serves to catalyze the reduction of the rest of the mixed salt.[141]

Mixed oxides can also be reduced to the mixed metal catalysts by hydrogen reduction. Such oxides have been prepared by sodium nitrate fusion of mixed metal salts, primarily those of the noble metals, following the procedure used for the preparation of Adams' catalyst, platinum oxide.[131-136] Non-noble metal alloy powders are prepared by heating mixed nitrates, carbonates or hydroxides in air to convert them to the mixed oxides which are then reduced. The hydroxides can be prepared by coprecipitation from a mixed metal salt solution.[139]

A solution of appropriate salts can also be reduced in the liquid phase by the addition of an appropriate reducing agent. Sodium borohydride has been used but care must be taken to remove the boron from the catalyst, particularly for the mixed noble metals. This has been accomplished by adding a dilute borohydride solution to the mixed metal salt solution under rapid agitation followed by a thorough washing of the precipitated metal black with warm water.[146] The use of hydrazine, formaldehyde or formic acid is preferred to borohydride since the by-products of the reduction do not contaminate the catalyst.[147] Another procedure is to use a ternary alloy and to leach out one component as in the preparation of Raney nickel and similar catalysts.

All of the above procedures give crystalline metal alloy species. Metallic glasses or amorphous metal alloys, however, have been finding increasing use as catalysts.[148-150] The most common method for the preparation of such materials is by melt-quenching in which the melt of the components is so rapidly quenched that there is insufficient time for crystals to nucleate and grow. These amorphous materials are devoid of any long range ordering and have a predominance of low coordination and defect sites on their surfaces. As described previously, sites having significant coordinate unsaturation are those on which most synthetically useful catalytic reactions take place. These amorphous alloys, however, are limited in their use by their metastable structure which can usually revert to a crystalline material at sufficiently high temperatures. Another limitation is the general difficulty with which such species can be prepared, particularly on any relatively large scale.

Since the catalytic process takes place on the surface of the mixed metal catalysts, a knowledge of the surface composition of these materials is essential to

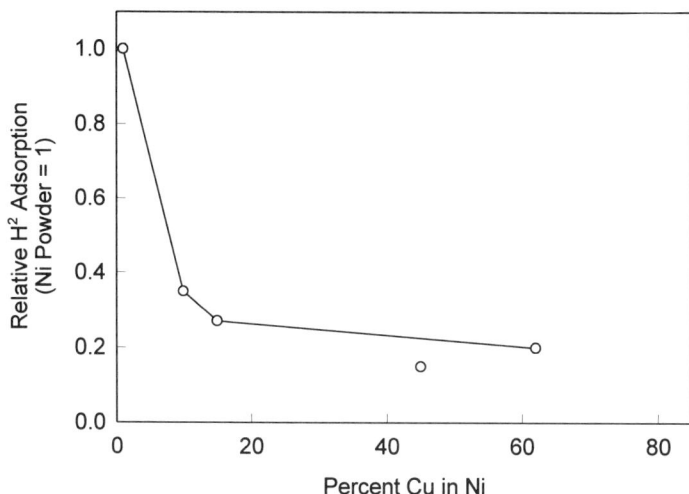

Fig. 12.9. Relative extent of hydrogen adsorption on copper-nickel
catalysts. The amount of hydrogen adsorbed on nickel powder
was arbitrarily set at 1. (Redrawn using data from Ref. 139.)

understanding their behavior.[151] It is generally agreed that the surface
composition is almost always different from the bulk composition of the alloy so
surface specific techniques have to be applied to determine the nature of the
species present on the surface of these catalysts. In general, two procedures are
used for this purpose but, as discussed later, they both present difficulties in
interpretation of the data. The more easily applied technique involves the
selective chemisorption of a gas that adsorbs on only one component of the
alloy.[138,140] These data, taken in combination with BET determined overall
surface area of the alloy gives a reasonable approximation of the surface
composition.

Fig. 12.9 shows the extent of hydrogen chemisorption on the copper–
nickel catalysts used for the hydrogenolysis of ethane and dehydrogenation of
cyclohexane.[139,152] Comparison with Fig. 3.1 shows that this curve corresponds
closely to the rate of ethane hydrogenolysis, a reaction governed by both
ensemble and ligand effects. The problem with this approach is that while the
ligand or electronic effect is thought to be operative by modifying the adsorption
characteristics of the surface sites, this rational does not seem to be applied to
selective chemisorption. It would seem that such modified adsorption
characteristics could relate to either enhanced or diminished adsorption on an
alloy surface, thus leading to a misrepresentation of the surface composition.
However, in the absence of specific data indicating the extent of such

Table 12.3

Comparison of surface compositions as measured by Auger electron spectroscopy (AES) and chemisorption.[153]

Alloy	Platinum atoms on the Surface (x 10^{17} m^{-2})		
	Theory	AES	Chemisorption
Pt$_3$Sn	49.1	68.3	44.1
PtSn	-	32.6	16.7

modification, if indeed it exists, this approach probably gives the best approximation of surface composition available at present. This approach works best when one component of the alloy does not adsorb the gas at all.

The second approach involves X-ray photoelectron spectroscopy (XPS) utilizing either ESCA or Auger electron spectroscopy (AES).[141] With these spectral techniques, though, the data represents a weighted average of the composition of several layers of the sample with the number of layers sampled depending on the escape depth of the measured electrons. Table 12.3 shows a comparison of the surface composition of two platinum–tin species as determined by both Auger electron spectroscopy and selective chemisorption.[153] With both catalysts the amount of platinum on the surface as determined by AES was higher than that measured by chemisorption because the AES data includes several layers in its analyses.

Several generalizations concerning the surface composition can be made.[138,140,141,154] In an alloy composed of a single phase the surface is enriched in that component which has the lower sublimation energy (lower surface energy or higher rate of diffusion). In the case of Group VIII–IB metal combinations, this is the IB metal. This does not mean that the outer layer is necessarily uniform. In a ruthenium–copper alloy, monolayer coverage would be expected to take place with a copper content of about 1.5 atom percent if all of the copper were present in a single surface layer. The available data, however, suggest that this catalyst consists of small clusters of copper on the surface of crystallites of essentially pure ruthenium.

If the alloy exists in two phases, that phase with the lower sublimation energy will form on the outer surface. If the nickel–copper system is equilibrated at temperatures near 200°C, it exists as two phases of constant composition in equilibrium.[155] The relative amounts of the phases depends on the overall

Type	Number of Phases		Composition
A	1		Homogeneous nickel-copper alloy. Surface enriched in copper.
B	2		Kernel of nickel or nickel-rich phase. Surface layer copper-rich phase.
C	2	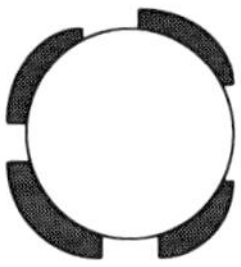	Kernel of nickel or nickel-rich phase with patches of copper-rich phase on the surface.
D	1	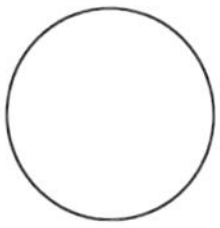	Homogeneous alloy rich in nickel. Surface concentration of copper is higher than in core.

Fig. 12.10. Location of the phases in nickel-copper alloys.

composition of the alloy. Fig. 12.10 shows the phase distribution of catalyst particles at different alloy compositions for the copper nickel system. When equilibrium is established after all of the nickel has dissolved, a copper rich homogeneous alloy is formed as represented by A. The surface concentration of copper will be greater than in the bulk because of the lower sublimation energy of copper.

When equilibrium is established after the copper has been consumed, the two phase system shown by Fig. 12.10B results. Here there is a kernel of almost pure nickel or, at least, a nickel rich phase surrounded by a surface layer of the copper rich phase. This has been referred to as the cherry pit model.[156]

When even more nickel is present, the surface of an inner nickel rich phase is covered with small patches of the copper rich phase which do not completely cover the inner core as depicted by Fig. 12.10C. With high quantities of nickel, another homogeneous alloy is formed that is uniformly rich in nickel

(Fig. 12.10D). Here, too, though the surface concentration of copper is higher than in the core.[155]

A third factor controlling surface composition is the atmosphere in which the catalyst is used. The surface will be enriched in that component of the alloy that has the highest heat of adsorption of the gas.[140] In an oxygen atmosphere the surface of a nickel–gold catalyst becomes enriched with nickel rather than gold.[157] In the presence of CO the surface of a palladium–silver alloy becomes enriched with palladium while, normally, silver would be the predominant surface component.[158] This enrichment of the surface by palladium should also be observed in a hydrogen atmosphere.

It can be seen, then, that while the surface composition of a mixed metal catalyst is of critical importance to the outcome of a given reaction, there is little that may be said concerning the optimum surface concentration for a particular reaction. Even if such a prediction could be made it would be difficult to design a catalyst having the prescribed surface composition under the reaction conditions used. Much more needs to be done to optimize the use of such mixed metal catalysts, particularly in synthetically useful reactions.

References

1. J. L. Cihonski, *Encyclopedia of Physical Science and Technology*, **9**, 15 (1987).
2. R. Schrötter, *Newer Methods of Preparative Organic Chemistry*, Interscience, New York, 1948, p. 61.
3. W. D. Gillespie, R. K. Herz, E. E. Petersen and G. A. Somorjai, *J. Catal.*, **70**, (1981) 147.
4. D. W. Goodman, *Chem. Ind. (Dekker)*, **22**, *(Catal. Org. React.)* (1985) 171.
5. G. A. Somorjai, *Catal. Revs.*, **7**, 87 (1972), **18**, 173 (1978).
6. G. A. Somorjai and S. M. Davis, *ChemTech.*, **13**, 508 (1983); *Plat. Met. Rev.*, **27**, 54 (1984).
7. D. W. Goodman, *Accounts Chem. Res.*, **17**, 194 (1984); *Ann. Rev. Phys. Chem.*, **37**, 425 (1986).
8. K. Heynes and H. Paulsen, *Newer Methods of Preparative Organic Chemistry*, **Vol. II**, Interscience, New York, 1963, p. 303.
9. J. T. Richarsdson, B. Turk, M. Lei, K. Forster and M. V. Twigg, *Appl. Catal. A*, **83**, 87 (1992).
10. G. P. Boldrini, D. Savoia, E. Tagliavini, C. Trombini and A. Umani-Ronchi, *J. Org. Chem.*, **50**, 3082 (1985).
11. R. Adams and V. Voorhees, *J. Am. Chem. Soc.*, **44**, 1683 (1922).
12. R. Adams, V. Voorhees and R. L. Shriner, *Org. Synth.*, **Coll. Vol. I** (2nd Ed.), 463 (1941).
13. V. L. Frampton, J. D. Edwards Jr. and H. R. Henze, *J. Am. Chem. Soc.*, **73**, 4432 (1951).
14. C. W. Keenan, B. W. Giesemann and H. A. Smith, *J. Am. Chem. Soc.*, **76**, 229 (1954).
15. P. N. Rylander, *Catalytic Hydrogenation over Platinum Metals*, Academic Press, New York, 1967, p. 312.
16. D. Cohen and J. A. Ibers, *J. Catal.*, **31**, 369 (1973).
17. W. F. Bruce, *J. Am. Chem. Soc.*, **58**, 687 (1936).

18. S. Nishimura, T. Itaya and M. Shiota, *J. Chem. Soc., Chem. Commun.*, 422 (1967).
19. S. Nishimura, S. Iwafune, T. Nagura, Y. Akimoto and M. Uda, *Chem. Lett.*, 1275 (1985).
20. P. A. Sermon, *J. Catal.*, **24**, 460, 467 (1972).
21. J. A. Don, A. P. Pijpers and J. J. F. Scholten, *J. Catal.*, **80**, 296 (1983).
22. R. Paul, P. Buisson and N. Joseph, *Ind. Eng. Chem.*, **44**, 1006 (1952).
23. H. C. Brown and C. A. Brown, *J. Am. Chem. Soc.*, **85**, 1003 (1963).
24. H. C. Brown and C. A. Brown, *J. Am. Chem. Soc.*, **85**, 1005 (1963).
25. C. A. Brown, *J. Org. Chem.*, **35**, 1900 (1970).
26. C. A. Brown and V. K. Ahuja, *J. Org. Chem.*, **38**, 2226 (1973).
27. D. G. Holah, I. M. Hoodless, A. N. Hughes and L. Sedor, *J. Catal.*, **60**, 148 (1979).
28. T. W. Russell, R. C. Hoy and J. C. Cornelius, *J. Org. Chem.*, **37**, 3552 (1972).
29. T. W. Russell and R. C. Hoy, *J. Org. Chem.*, **36**, 2018 (1971).
30. J. A. Schreifels, P. C. Maybury and W. E. Swartz, Jr., *J. Org. Chem.*, **46**, 1263 (1981).
31. S. Kishida, Y. Murakami, T. Imanaka and S. Teranishi, *J. Catal.*, **12**, 97 (1968).
32. T. W. Russell, D. M. Duncan and S. S. Hansen, *J. Org. Chem.*, **42**, 551 (1977).
33. J. A. Schreifels, P. C. Maybury and W. E. Swartz, *J. Catal.*, **65**, 195 (1980).
34. T. Imanaka, Y. Nitta and S. Teranishi, *Bull. Chem. Soc., Japan*, **46**, 1134 (1973).
35. Y. Okamoto, Y. Nitta, T. Inamaka and S. Teranishi, *J. Chem. Soc., Faraday I*, **75**, 2027 (1979).
36. L. L. Sheu, Y. Z. Chen and M. H. Rei, *J. Chin. Chem. Soc. (Taipei)*, **32**, 317 (1985).
37. G. Jannes, J.-P. Puttemans and P. Vanderwegen, *Catal. Today*, **5**, 265 (1989).
38. M. H. Rei, L. L. Sheu and Y. Z. Chen, *Appl. Catal.*, **23**, 281 (1986).
39. Y. Z. Chen, L. L. Sheu and M. H. Rei, *J. Chin. Chem. Soc. (Taipei)*, **36**, 67 (1989).
40. D. E. Mears and M. Boudart, *AIChE J.*, **12**, 313 (1966).
41. Y. Z. Chen and J. S. Wu, *J. Chin. Chem. Soc. (Taipei)*, **37**, 489 (1990).
42. Y. Nitta, T. Imanaka and S. Teranishi, *Bull. Chem. Soc., Japan*, **54**, 3579 (1981).
43. E. N. Marvell and J. Tashiro, *J. Org. Chem.*, **30**, 3991 (1965).
44. I. Ravet, J. B. Nagy and E. G. Derouane, *Stud. Surf. Sci. Catal.*, **31** (Prep. Catal., IV), 505 (1987).
45. J. B. Nagy, I. Bodart-Ravet and E. G. Derouane, *Faraday Discuss. Chem. Soc.*, **87**, 189 (1989).
46. J. B. Nagy, A. Gourgue and E. G. Derouane, *Stud. Surf. Sci. Catal.*, **16** (Prep. Catal., III), 193 (1983).
47. Y. Nakao and S. Fujishige, *Chem. Lett.*, 995 (1979).
48. Y. Nakao and S. Fujishige, *Bull. Chem. Soc., Japan*, **53**, 1267 (1980).
49. Y. Nakao and S. Fujishige, *J. Catal.*, **68**, 406 (1981).
50. Y. Nakao and S. Fujishige, *Chem. Lett.*, 673 (1980).
51. Y. Nakao and S. Fujishige, *Chem. Lett.*, 925 (1981).

52. J.-J. Brunet, P. Gallois and P. Caubere, *J. Org. Chem.*, **45**, 1937, 1946 (1980).
53. Y. Nitta, T. Imanaka and S. Teranishi, *Bull. Chem. Soc., Japan*, **53**, 3154 (1980).
54. Y. Z. Chen and K. J. Wu, *Appl. Catal.*, **78**, 185 (1991).
55. D. G. Holah, I. M. Hoodless, A. N. Hughes and L. Sedor, *J. Catal.*, **72**, 12 (1981).
56. T. Satoh, S. Suzuki, T. Kikuchi and J. Okada, *Chem. and Ind. (London)*, 1626 (1970).
57. C. Barnett, *Ind. Eng. Chem. Prod. Res. Dev.*, **8**, 145 (1969).
58. H. C. Brown and C. A. Brown, *J. Am. Chem. Soc.*, **84**, 1494 (1963).
59. C. Y. Che, H. Yamamoto and T. Kwan, *Chem. Pharm. Bull.*, **17**, 1287 (1969).
60. T. W. Russell and D. M. Duncan, *J. Org. Chem.*, **39**, 3050 (1974).
61. V. M. Deshpande, W. R. Patterson and C. S. Narasimhan, *J. Catal.*, **121**, 165 (1990).
62. C. S. Narasimhan, V. M. Deshpande and K. Ramnarayan, *Ind. Eng. Chem. Res.*, **28**, L110 (1989).
63. V. M. Deshpande, K. Ramnarayan and C. S. Narasimhan, *J. Catal.*, **121**, 174 (1990).
64. M. Raney, U. S. Patent 1,628,191 (1927).
65. P. Fouilloux, *Appl. Catal.*, **8**, 1 (1983).
66. J. Freel, W. J. M. Pieters and R. B. Anderson, *J. Catal.*, **16**, 281 (1970).
67. R. Sassoulas and Y. Trambouze, *Bull. Soc. Chim,, France*, **5**, 985 (1964).
68. M. L. Bakker, D. J. Young and M. S. Wainwright, *J. Mater. Sci.*, **23**, 3921 (1988).
69. S. Sane, J. M. Bonnier, J. P. Damon and J. Masson, *Appl. Catal.*, **9**, 69 (1984).
70. S. Nishimura and Y. Urushibara, *Bull. Chem. Soc., Japan,* **30,** 199 (1957).
71. L. W. Covert and H. Adkins, *J. Am. Chem. Soc.*, **54,** 4116 (1932).
72. E. E. Van Tamelen and G. T. Hildahl, *J. Am. Chem. Soc.*, **78,** 4405 (1956).
73. R. Mozingo, *Org. Synthesis,* **Coll. Vol. 3**, 181 (1955).
74. H. Adkins and G. Krsek, *J. Am. Chem Soc.*, **70,** 412 (1948).
75. A. A. Pavlic and H. Adkins, *J. Am Chem. Soc.*, **68,** 1471 (1946).
76. H. Adkins and A. A. Pavlic, *J. Am. Chem. Soc.*, **69,** 3039 (1947).
77. H. Adkins and H. R. Billica, *J. Am. Chem. Soc.*, **70,** 695 (1948).
78. H. R. Billica and H. Adkins, *Org. Synthesis,* **Coll. Vol. 3**,176 (1955).
79. H. A. Smith, A. J. Chadwell, Jr. and S. S. Kirslis, *J. Phys. Chem.*, **59,** 820 (1955).
80. R. J. Kokes and P. H. Emmett, *J. Am. Chem. Soc.,* **82,** 4497 (1960).
81. J. H. P. Tyman, *Chem. and Ind. (London)*, 404 (1964).
82. S. Nishimura, S. Ikeda, M. Kinoshita and M. Kawashima, *Bull. Chem. Soc., Japan*, **58**, 391 (1985).
83. W. C. Cheng, L. J. Czarnecki and C. J. Periera, *Ind. Eng. Chem. Res.*, **28**, 1764 (1989).
84. N. A. Khan, *J. Am. Chem. Soc.,* **74,** 3018 (1952).
85. N. A. Khan, *Science,* **117,** 130 (1953).
86. G. V. Smith, R. Song, J. M. Delich and M. Bartok, *Stud. Surf. Sci. Catal.*, **78** (Heterog. Catal. Fine Chem., III), 67 (1993).

87.   P. Fouilloux, G. A. Martin, A. J. Renouprez, B. Moraweck, B. Imelic and M. Prettre, *J. Catal.*, **25**, 212 (1972).

88.   J. Freel, W. J. M. Pieters and R. B. Anderson, *J. Catal.*, **14**, 247 (1969).

89.   S. D.Robertson and R. B. Anderson, *J. Catal.*, **23**, 286 (1971).

90.   I. Nicolau and R. B. Anderson, *J. Catal.*, **68**, 339 (1981).

91.   R. J. Kokes and P. H. Emmett, *J. Am. Chem. Soc.*, **81**, 5032 (1959).

92.   V. R. Choudary and M. G. Sane, *Indian Chem. Eng.*, **28**, 50 (1986).

93.   E. A. Cioffi, W. S. Willis and S. L. Suib, *Langmuir*, **4**, 697 (1988).

94.   G. Stork, *J. Am. Chem. Soc.*, **69**, 2936 (1947).

95.   F. J. Villani, M. S. King, and D. Papa, *J. Org. Chem.*, **18**, 1578 (1953).

96.   E. Lieber and G. B. L. Smith, *J. Am. Chem. Soc.*, **58**, 1417 (1936).

97.   J. R. Reasenberg, E. Lieber and G. B. L. Smith, *J. Am. Chem. Soc.*, **61**, 384 (1939).

98.   S. S. Scholnik, J. R. Reasenberg, E. Lieber and G. B. L. Smith, *J. Am. Chem. Soc.*, **63**, 1192 (1941).

99.   D. R. Levering, F. L. Morritz and E. Lieber, *J. Am. Chem. Soc.*, **72**, 1190 (1950).

100. S. Montgomery, *Chem. Ind. (Dekker)*, **5** (Catal. Org. React.), 383 (1981).

101. J. C. Klein and D. M. Hercules, *Anal. Chem.*, **56**, 685 (1984).

102. J. M. Bonnier, J. P. Damon, B. Delmon, B. Doumain and J. Masson, *J. Chim. Phys. Phys.-Chim. Biol.*, **84**, 889 (1987).

103. J. M. Bonnier, J. P. Damon and J. Masson, *Appl. Catal.*, **42**, 285 (1988).

104. T. Koscielski, J. M. Bonnier, J. P. Damon and J. Masson, *Appl. Catal.*, **49**, 91 (1989).

105. W. Reeve and J. Christian, *J. Am. Chem. Soc.*, **78**, 860 (1956).

106. W. Reeve and W. M. Earekson, *J. Am. Chem. Soc.*, **72**, 3299 (1950).

107. A. J.Chadwell, Jr. and H. A. Smith, *J. Phys. Chem.*, **60**, 1339 (1956).

109. B. V. Aller, *J. Appl. Chem.*, **7**, 130 (1957).

109. B. V. Aller, *J. Appl. Chem.*, **8**, 163, 492 (1958).

110. A. D. Tomsett, H. E. Curry-Hyde, M. S. Wainwright, D. J. Young and A. J. Bridgewater, *Appl. Catal.*, **33**, 119 (1987).

111. W. H. Jones, W. F. Benning, P. Davis, D. M. Mulvey, P. I. Pollak, J. C. Schaeffer, R. Tull and L. M. Weinstock, *Ann. N. Y. Acad. Sci.*, **158**, 471 (1969).

112. M. G. Scaros, J. P. Westrich, O. J. Goodmonson and M. L. Prunier, *Chem. Ind. (Dekker)*, **47** (Catal. Org. React.), 373 (1992).

113. Y. Urushibara, *Bull. Chem. Soc., Japan*, **26**, 280 (1952).

114. Y. Urushibara, *Ann. N. Y. Acad. Sci.*, **145**, Art. 1, 52 (1967).

115. K. Hata, *Urushibara Catalysts*, University of Tokyo Press, Tokyo, 1971.

116. M. Kajitani, J. Okada, T. Ueda, A. Sugimori and Y. Urushibara, *Chem. Lett.*, 777 (1973).

117. J. C. Klein and D. M. Hercules, *J. Catal.*, **82**, 424 (1983).

118. I. Jacob, M. Fisher, Z. Hadari, M. Herskowitz, J. Wisniak, N. Shamir and M. H. Mintz, *J. Catal.*, **101**, 28 (1986).

119. J. N. Pattison and E. F. Degering, *J. Am. Chem. Soc.*, **72**, 5756 (1950).

120. K. Aika, Y. Ogata, K. Takeishi, K. Urabe and T. Onishi, *J. Catal.*, **114**, 200 (1988).

121. C. P. Nash, F. M. Boyden and L. D. Whittig, *J. Am. Chem. Soc.*, **82**, 6203 (1960).

122. H. Imamura, Y. Kato, K. Yamada and S. Tsuchiya, *Appl. Catal.*, **27**, 243 (1986).
123. E. A. Vishnevetskii, S. D. Mikhailenko, N. A. Maksimova and A. B. Fasman, *React. Kinet. Catal. Lett.*, **31**, 445 (1986).
124. N. A. Maksimova, E. A. Vishnevetskii, V. S. Ivanov and A. B. Fasman, *Appl. Catal.*, **35**, 59 (1987).
125. A. B. Fasman, G. E. Bedelbaev, G. K. Alekseeva, V. N. Ermolaev and A. S. Kuanyshev, *React. Kinet. Catal. Lett.*, **39**, 81 (1989).
126. E. M. Moroz, S. D. Mikhailenko, A. K. Dzhunusov and A. B. Fasman, *React. Kinet. Catal. Lett.*, **43**, 63 (1991).
127. C. H. Bartholomew and A. H. Uken, *Appl. Catal.*, **4**, 19, (1982).
128. Y. Okamoto, Y. Nitta, T. Imanaka and S. Teranishi, *J. Chem. Soc., Faraday I*, **76**, 998 (1980).
129. Y. Okamoto, Y. Nitta, T. Imanaka and S. Teranishi, *J. Catal.*, **64**, 397 (1980).
130. Y. Okamoto, E. Matsunaga, T. Imanaka and S. Teranishi, *Chem. Lett.*, 565 (1983).
131. S. Nishimura, *Bull. Chem. Soc., Japan*, **33**, 566 (1960); **34**, 32, 1544 (1961).
132. G. C. Bond and D. E. Webster, *Plat. Met. Rev.* **9**, 12 (1965); **10**, 10 (1966); **15**, 57 (1969).
133. G. C. Bond and D. E. Webster, *Chem. and Ind., (London)*, 878 (1967).
134. G. C. Bond and D. E. Webster, *Ann. N. Y. Acad. Sci.*, **158**, Art. 2, 540 (1969).
135. P. N. Rylander, L. Hasbrouck, S. G. Hindin, I. Karpenko, G. Pond and S. Starrick, *Engelhard Ind. Tech. Bull*, **8**, 25 (1967).
136. P. N. Rylander, L. Hasbrouk, S. G. Hindin, R. Iverson, I. Karpenko and G. Pond, *Engelhard Ind. Tech. Bull.*, **8**, 93 (1967).
137. J. K. A. Clarke, *Chem. Revs.*, **75**, 291 (1975).
138. V. Ponec, *Catal. Revs.*, **11**, 41 (1975).
139. J. H. Sinfelt, J. L. Carter and D. J. C. Yates, *J. Catal.*, **24**, 283 (1972).
140. J. H. Sinfelt, *Accounts Chem. Res.*, **10**, 15 (1977).
141. J. H. Sinfelt and J. A. Cusumano, *Advanced Materials in Catalysis*, (J. J. Burton and R. L. Garten, Eds.) Academic Press, New York, 1977, p. 1.
142. A. Roberti, V. Ponec and W. M. H. Sachtler, *J. Catal.*, **28**, 381 (1973).
143. R. Burch, *J. Chem. Soc., Chem. Commun.*, 845 (1981).
144. R. Burch, *Accounts Chem. Res.*, **15**, 24 (1982).
145. D. W. McKee, *Trans. Faraday Soc.*, **61**, 2273 (1965).
146. J. L. Carter, J. A. Cusumano and J. H. Sinfelt, *J. Catal*, **20**, 263 (1971).
147. B. G. Allison and G. C. Bond, *Catal. Revs.*, **7**, 233 (1972).
148. A. Baiker, *Faraday Discuss. Chem. Soc.*, **87**, 239 (1989).
149. H. Yamashita, M. Yoshikawa, T. Funabiki and S. Yoshida, *J. Chem. Soc., Faraday Trans., I*, **83**, 2883 (1987).
150. M. Shibata and T. Masumoto, *Stud. Surf. Sci. Catal.*, **31** (Prep. Catal., IV), 353 (1987).
151. C. T. Campbell, *Ann. Rev. Phys. Chem.*, **41**, 775 (1990).
152. V. Ponec and W. M. H. Sachtler, *J. Catal.*, **24**, 250 (1970).
153. H. Verbeck and W. M. H. Sachtler, *J. Catal.*, **42**, 257 (1976).
154. W. M. H. Sachtler and R. A. Van Santen, *Adv. Catal.*, **26**, 69 (1977).
155. W. M. H. Sachtler and R. Jongepier, *J. Catal.*, **4**, 665 (1965).
156. W. M. H. Sachtler, *Vide*, **164**, 67 (1973).

157. F. L. Williams and M. Boudart, *J. Catal.*, **30**, 438 (1973).
158. R. Bouwman, G. J. M. Lippits and W. M. H. Sachtler, *J. Catal.*, **25**, 356 (1972).

13

Supported Metals

Since catalyzed reactions take place on the surface of the catalytically active material, the most efficient catalysts are those in which a high percentage of the active species is exposed to the reaction medium. Some metal catalysts, such as those discussed in Chapter 12, are composed of very finely divided metal particles that have the high ratio of surface to bulk atoms needed for good catalyst activity. These fine particles, however, sinter on heating. As discussed in Chapter 9, the most common way of minimizing metal catalyst sintering is to distribute the active component over a porous, thermostable, support.

The major objective in the preparation of supported catalysts is to have a high surface area of reduced metal deposited on a highly porous material. Supported metal catalysts are generally prepared by reducing a precurser metal salt which has been incorporated into or onto the support material. This supported salt is then dried and, possibly, calcined (heating in air) to modify its chemical composition. The final step is a reduction to give the supported metal. The primary difference in the various methods of preparation lies in the manner in which the support material and the precurser salt are brought together.[1]

Essentially, almost all of these preparation procedures can be divided into two general categories: coprecipitation and deposition.[2] Coprecipitation involves the addition of a precipitating agent to a solution containing both a support precurser and a catalyst precurser. The resulting precipitate contains species, either as a single phase or multiple phases, from which the active component and the support material are eventually produced. Deposition, on the other hand, describes the application of the catalytic component to a separately produced support. This can be accomplished in several ways. One of these is defined as a precipitation-deposition. In this procedure the active element is deposited on a suspension of the support by precipitation from the solution in which the support is suspended. Another method is a reduction-deposition which results when a solution of a metal salt, in which the support material is suspended, is treated with a reducing agent such as hydrogen, formaldehyde or borohydride. The adsorption of a preformed metal sol onto the surface of a support is another example of reduction-deposition, but in this instance the metal has been reduced to the colloidal state before coming into contact with the support material.

In most cases, however, the supported catalyst precursors are prepared by a procedure that can be described in general terms as an impregnation, contacting the support with a solution of the active element. The word, impregnation, though, has been used to describe a number of different preparation procedures and, while all of these may fall under the same general heading, they have significant differences so the generic description can be misleading. It is recommended that more specific terms be used for these different preparation procedures. Incipient wetness describes the impregnations of the support when the active elements are contained in a volume of solution corresponding to the pore volume of the support. This procedure has also been referred to as pore volume impregnation or dry impregnation.[2] When the volume of the solution exceeds the pore volume of the support, the process is called a wet impregnation or, simply, an adsorption but the descriptor, adsorption from solution, is more appropriate. If the surface of the support is modified to give a surface species that chemically reacts with the precurser salt, the process is called an ion-exchange.

There is an extensive literature concerned with the preparation of supported catalysts but, unfortunately, it is sometimes difficult to determine precisely what procedures were used because of the ambiguity of many of the experimental descriptions. While an IUPAC manual standardizing the terms used in catalysis has been published[2] the proposed usage is still not widely applied in the literature. The procedures described below are defined in these standard terms with some of the more commonly used descriptors also mentioned where appropriate.

Coprecipitation

The initial step in the preparation of a coprecipitated catalyst is the reaction between a solution of two or more metal salts and a base, generally a hydroxide, alkali carbonate or bicarbonate. The resulting precipitate may contain not only the insoluble hydroxides and/or carbonates but also a mixed metal compound if the solubility equilibria are favorable. Even if the formation of a mixed metal compound is not favorable, some of the support material is usually trapped in the active metal precipitate. This dilutes the precipitate and inhibits the formation of large crystals of the active metal compound. Smaller crystals are easier to reduce and give more finely divided metal particles.[3]

It is important that the ions of the active material and the support are intimately mixed prior to the precipitation step and that the precipitation conditions are such that the components are well mixed in the precipitate as well. Some of the problems associated with obtaining a precipitate having a uniform distribution of two components were discussed in Chapter 10. Since the most common catalyst prepared by coprecipitation is Ni/Al_2O_3, the factors associated with this procedure will be discussed in terms of this material. Adding base to a solution of nickel and aluminum salts gives a precipitate of, nominally, the mixed $Ni(OH)_2-Al(OH)_3$ (Eqn. 13.1). When the Ni:Al ratio is between two and three a single compound is produced on precipitation. Outside of this range the excess Ni^{+2} or Al^{+3} ions are present as separate phases of $Ni(OH)_2$ or $Al(OH)_3$.[4,5] Depending on the type of nickel and aluminum salts and the nature of the base used, the precipitate may also contain carbonate, nitrate, halide and/or sodium ions, all of which should be removed as thoroughly as possible from the final catalyst. Halide ions can be particularly troublesome so the use of halide salts is to be avoided if at all possible. While ammonium carbonate was found to be particularly effective for the precipitation-deposition of nickel hydroxide on silica,[6] sodium carbonate appears to be preferred for the preparation of Ni/Al_2O_3 by coprecipitation even though sodium ions are not easily washed out of the precipitate.

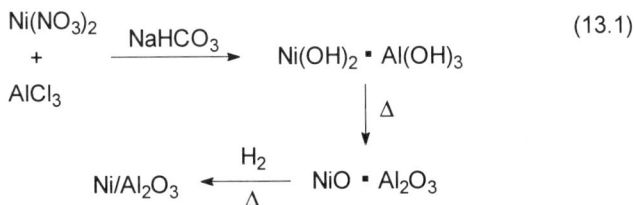

$$\begin{array}{c} Ni(NO_3)_2 \\ + \\ AlCl_3 \end{array} \xrightarrow{\ NaHCO_3\ } Ni(OH)_2 \cdot Al(OH)_3 \qquad (13.1)$$

$$Ni(OH)_2 \cdot Al(OH)_3 \xrightarrow{\ \Delta\ } NiO \cdot Al_2O_3$$

$$Ni/Al_2O_3 \xleftarrow[\Delta]{\ H_2\ } NiO \cdot Al_2O_3$$

The second phase in catalyst preparation by this process involves heating the precipitate. This takes place in two stages. In the first, the water of crystallization is lost and in the second, there is a loss of carbonate and other anions as well as the dehydration of the metal hydroxides. The first occurs on heating to about 250°C, while the second stage requires calcination usually at temperatures between 300° and 600°C[4] with 350°–400°C reported to be optimal.[7] At these temperatures metal carbonates and hydroxides are converted to an intimate mixture of the component oxides (Eqn. 13.1). The oxides of the active metals are generally more easily reduced than are other metal compounds.[3–5]

Reduction of the nickel oxide present in the mixed nickel–aluminum oxides is usually done in a flow of hydrogen at elevated temperatures (Eqn. 13.1). The extent of reduction depends on the conditions used. Fig. 13.1 shows that the most complete reduction and maximum nickel surface area were obtained when the reduction was run at 700°C.[4]

A small amount of copper in the mixed oxide facilitates the hydrogenation of the nickel oxide.[3] As discussed in Chapter 12, reduction of bulk nickel oxide was facilitated by the presence not only of copper but also of silver.[8] It would seem, then, that if silver were present in the coprecipitated nickel–aluminum oxide, it should also increase the rate of nickel oxide hydrogenation. High nickel areas were obtained when the partial pressure of water in the reducing gas was kept low.[3,9] The presence of a small amount of aluminum oxide in the reduced nickel particles distorts the crystallites and produces more of the catalytically active surface defect sites.[5]

One of the problems associated with the preparation of coprecipitated catalysts is found on drying and calcining these materials. The evaporation of the solvent creates a vapor–liquid interface inside the pores of the solid. With solvents having a high surface tension such as water, a partial collapse of the solid can occur giving a material with lower surface area.[10] One way of minimizing

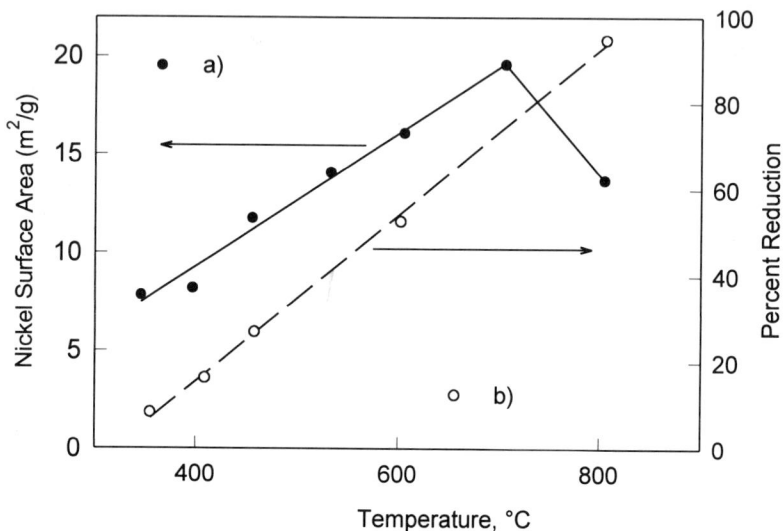

Fig. 13.1.   Dependence of a) nickel surface area and b) percent reduction
on the temperature used for the reduction of a coprecipitated
Ni/Al_2O_3 catalyst. (Redrawn using data from Ref. 4.)

this is to use a solvent having a lower surface tension than water. A particularly
useful procedure involves alcohol solutions of the active metal and support
precursors with the precipitation induced by the addition of stoichiometric
amounts of water. For instance a $Ni(OH)_2–Al(OH)_3$ coprecipitate was produced
by adding the stoichiometric amount of water to an alcoholic solution of
aluminum *sec*-butylate and nickel acetate according to Eqn. 13.2[10]

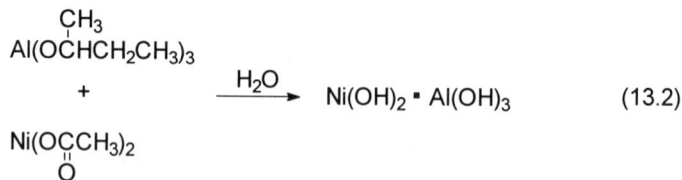

$$CH_3$$
$$Al(OCHCH_2CH_3)_3$$

$$+ \xrightarrow{\quad H_2O \quad} Ni(OH)_2 \cdot Al(OH)_3 \qquad (13.2)$$

$$Ni(OCCH_3)_2$$
$$O$$

This so-called 'sol-gel' procedure[11] has been used to prepare a number of
supported catalysts.[12]   Heating an aqueous alcoholic solution containing
tetraethoxysilane and $Pd(NH_3)_2Cl_2$ with ammonia gave a gelatinous palladium
ion–silica precipitate that was dried at 70°–100°C for an extended time and then

$$(13.3)$$

$$
\begin{array}{c}
\text{Si(OCH}_2\text{CH}_3)_4 \\
+ \\
\text{Pd(NH}_3)_4\text{Cl}_2
\end{array}
\xrightarrow{\text{NH}_4\text{OH}}
\text{Pd(NH}_3)_4\text{Cl}_2 \cdot \text{Si(OH)}_4
$$

$$\text{Pd/SiO}_2 \xleftarrow[\Delta]{\text{H}_2} \text{PdO} \cdot \text{SiO}_2$$

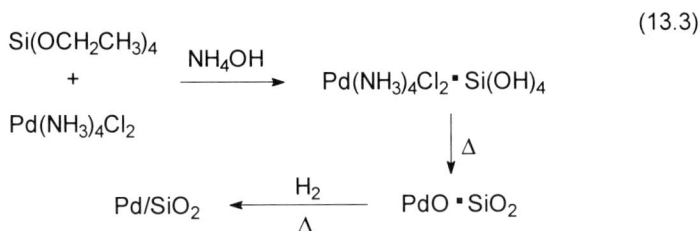

heated to 450°C for several hours.[13] Reduction gave a highly dispersed Pd/SiO$_2$ catalyst (Eqn. 13.3).[13] Pt/SiO$_2$[13-15] and Ru/SiO$_2$[16] catalysts were prepared in the same way. A highly dispersed Pt/Al$_2$O$_3$ catalyst was made by adding an alcoholic solution of aluminum tri-*sec*-butoxide containing the appropriate amount of chloroplatinic acid to aqueous ammonia. The resulting gel was dried, calcined and reduced to give the catalyst (Eqn. 13.4).[17] A Ru/Al$_2$O$_3$ catalyst was prepared in a similar way from the gel formed by the aqueous acid hydrolysis of an alcoholic solution of aluminum tri-*sec*-butoxide and RuCl$_3$.[16] Hydrolysis of titanium tetraethoxide solutions gave titania supported catalysts while magnesium ethoxide hydrolysis, produced MgO supported catalysts.[18] Rhodium supported on a mixed titanium–silicon oxide was prepared by adding an aqueous solution of rhodium nitrate to an alcoholic solution of tetraethoxysilane and tetrabutylorthotitanate.[19] Evaporating the organic solvent from the initially produced gel under hypercritical conditions increased the surface area of the support material.[10-12]

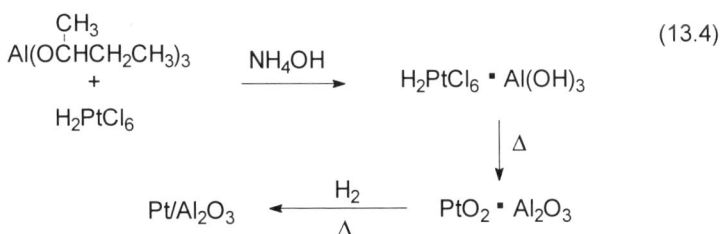

$$(13.4)$$

$$
\begin{array}{c}
\overset{\displaystyle \text{CH}_3}{\underset{\displaystyle}{\text{Al(OCHCH}_2\text{CH}_3)_3}} \\
+ \\
\text{H}_2\text{PtCl}_6
\end{array}
\xrightarrow{\text{NH}_4\text{OH}}
\text{H}_2\text{PtCl}_6 \cdot \text{Al(OH)}_3
$$

$$\text{Pt/Al}_2\text{O}_3 \xleftarrow[\Delta]{\text{H}_2} \text{PtO}_2 \cdot \text{Al}_2\text{O}_3$$

Another problem with the use of coprecipitation to prepare supported catalysts is that this procedure places a relatively large amount of the active metal inside the particles of the support material and, therefore, unavailable for reaction. A more efficient arrangement is to have the active material on or near the surface of the support particles. One way of accomplishing this is through a sequential precipitation procedure in which, for instance, a precipitate of Ni(OH)$_2$ is formed on a freshly prepared suspension of Al(OH)$_3$.[20] In coprecipitation the metal content of the precipitate is reasonably continuous throughout the precipitate. As illustrated in Fig. 13.2, though, in a sequential precipitation the first drops of

= Al(OH)$_3$ = Al-O-Ni-OH

= O-Ni-O-Ni-OH

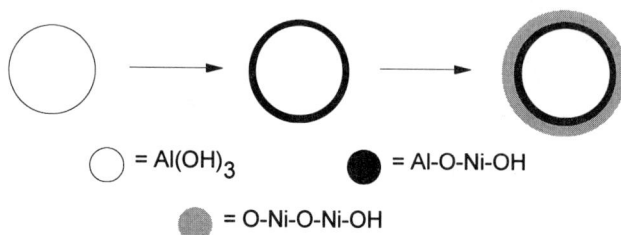

Fig. 13.2. Sequential precipitation of Ni(OH)$_2$ on a previously
precipitated Al(OH)$_3$ particle.

nickel produce Ni(OH)$_2$ adsorbed on the surface of the Al(OH)$_3$ while the later
ones give Ni(OH)$_2$ on a Ni(OH)$_2$ surface. Thus, there are two types of nickel
present; one having an aluminum environment and another in a more
homogeneous nickel environment. Calcination gives a mixture of nickel oxide
and a nickel oxide–aluminum oxide system that on reduction formed larger metal
crystallites than were observed on coprecipitated catalysts.[19]

Ternary systems have been prepared in this way as well. The sequential
precipitation of aluminum hydroxide, lanthanum hydroxide and, finally, nickel
hydroxide gave, after calcination and reduction, a lanthanum activated Ni/Al$_2$O$_3$
catalyst which had smaller metal crystallites and was somewhat more active than
a catalyst prepared by the simultaneous coprecipitation of the three
hydroxides.[21,22]

Another common catalyst prepared by coprecipitation is copper-chromium
oxide, also known as "copper chromite" or Adkins' catalyst.[23] This catalyst is
prepared by the addition of copper nitrate to a solution of ammonium dichromate
in ammonia giving a precipitate copper ammonium dichromate. This precipitate
is filtered, dried and then calcined at 650°–800°C, or more commonly, heated
with a flame to induce a thermal reaction (Eqn. 13.5). The resulting fine powder
is washed with acetic acid and dried to give the copper chromite catalyst.[23] A
more active catalyst is prepared by adding 10% barium nitrate by weight of
copper before precipitation.[24,25] Copper chromite catalysts containing calcium
and were found to be less effective than those having a barium promoter.[25]

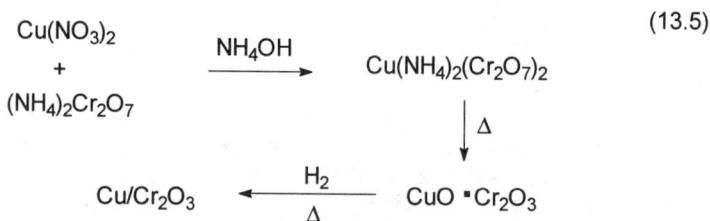

$$
\begin{array}{c}
Cu(NO_3)_2 \\
+ \\
(NH_4)_2Cr_2O_7
\end{array}
\quad\xrightarrow{\ NH_4OH\ }\quad
Cu(NH_4)_2(Cr_2O_7)_2
\tag{13.5}
$$

$$
Cu/Cr_2O_3 \quad\xleftarrow[\ \Delta\ ]{\ H_2\ }\quad CuO\cdot Cr_2O_3 \quad\xleftarrow{\ \Delta\ }
$$

The catalyst is activated by hydrogen reduction to give a highly dispersed copper metal supported on Cr_2O_3 and/or $Cu_2Cr_2O_4$. The temperature of the hydrogen reduction determines the extent to which the cupric ion is reduced. At temperatures below 200°C the primary conversion is Cu^{++} to Cu^+ with only a small amount of $Cu°$ found on the surface. As the temperature is increased to 500°C, more CuO is reduced to copper metal. [26,27] Cupric oxide and cuprous oxide are both hydrogenated more easily than is the copper chromite but at higher temperatures some reduction of Cr^{+6} to Cr^{+3} also takes place. The presence of Cr^{+3} deactivates the catalyst. However, barium ion inhibits the reduction to Cr^{+3} so loss of catalyst activity does not take place as easily with the Cu-Ba-CrO catalysts. [28,29]

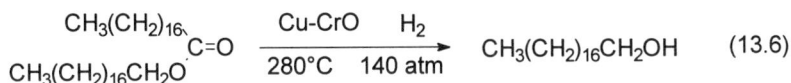

$$CH_3(CH_2)_{16} \!\!\diagdown \!\!\! \underset{CH_3(CH_2)_{16}CH_2O \diagup}{C\!=\!O} \quad \xrightarrow[280°C \quad 140\ atm]{Cu\text{-}CrO \quad H_2} \quad CH_3(CH_2)_{16}CH_2OH \qquad (13.6)$$

An alternate procedure involves the precipitation of mixed copper and chromium hydroxides from a solution of chromium nitrate and copper nitrate by the addition of an sodium bicarbonate and then heating the precipitate to 300°– 500°C. [30] In this procedure, the copper:chromium ratio can be varied over a wide range. A ratio between four and eight was optimum for use in the hydrogenation of esters to alcohols (Eqn. 13.6). [30] A related Zn–CrO catalyst prepared by the decomposition of precipitated zinc–copper hydroxides was effective in the hydrogenation of unsaturated esters to unsaturated alcohols (Eqn. 13.7). [30] The presence of a small amount of alumina increased catalyst activity and selectivity. Some of these catalysts, however, tend to become colloidal on use so they can present separation problems. [30]

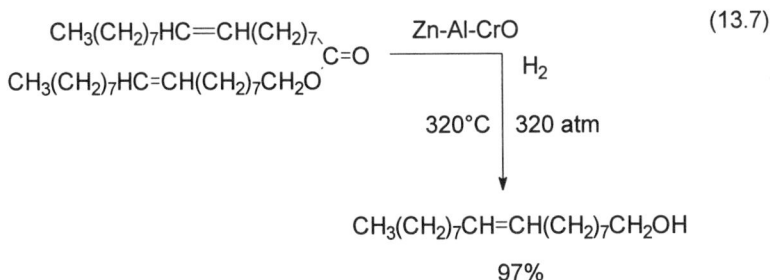

$$CH_3(CH_2)_7HC\!=\!CH(CH_2)_7 \!\!\diagdown \!\!\! \underset{CH_3(CH_2)_7HC\!=\!CH(CH_2)_7CH_2O \diagup}{C\!=\!O} \qquad \xrightarrow[\quad]{Zn\text{-}Al\text{-}CrO} \qquad (13.7)$$

$$\Big| \ H_2$$

$$320°C \ \Big| \ 320\ atm$$

$$\downarrow$$

$$CH_3(CH_2)_7CH\!=\!CH(CH_2)_7CH_2OH$$

$$97\%$$

Deposition

Coprecipitation gives a solid catalyst containing small metal particles imbedded throughout a porous oxide support. However, not only is this an inefficient use of the metal, it is also difficult to produce a support having a prescribed surface area and pore structure in this way. Because of this it is frequently more advantageous to add the active component to a separately produced carrier, especially since the structure of a pre-formed support is not usually affected by the thermal treatment required to produce a catalyst. This approach is also more general in that the precursor support material does not have to be formed in conjunction with its interaction with the active component.

Precipitation-Deposition

The addition of a precipitating agent to a suspension of the support in a solution of the active metal salt results in the precipitation of the active metal precurser onto the surface of the support provided there is a sufficiently strong interaction between the support and the precipitating compound.[31] The precipitate must be kept as homogeneous as possible. One way of accomplishing this is the slow injection of the precipitant into a well stirred suspension of the support in a solution of a salt of the active component. Slow addition of sodium hydroxide to a suspension of alumina in a nickel chloride solution was used to prepare Ni/Al_2O_3 catalysts.[32] The initial interaction was an adsorption of the nickel ions on the alumina giving, after calcining, a surface nickel aluminate. At high metal loadings a separate nickel oxide phase was also detected after calcining.[32]

Another way of obtaining a nearly homogeneous precipitate involves the aqueous hydrolysis of urea to liberate the base required for precipitation.[33,34] Here the support is suspended in an aqueous solution containing the nickel salt and urea. Heating promotes the hydrolysis of the urea which releases the hydroxide ion needed to precipitate the nickel hydroxide onto the support. On precipitation the nickel reacts with the silica to form a surface hydrosilicate that does not coalesce into larger particles and onto which additional nickel hydroxide is attracted. Ni/SiO_2 catalysts prepared in this way have, after reduction, high dispersions.[33] A further advantage to this procedure is the absence of strong base in the reaction mixture, something that can, at least partially, dissolve an alumina support.

After isolation the supported precipitate is washed, dried and usually calcined to produce a supported oxide which is then reduced, commonly in a hydrogen stream. Reduction of these supported oxides generally proceeds more readily than the mixed oxides produced by coprecipitation since there is only a monolayer in which there is a direct interaction of the active component with the support. This monolayer can be considered to be a silicate or aluminate which is more difficult to reduce than the oxide or hydroxide found in the outer metal-containing layers.[35] Precipitation-deposition gives catalysts having compositions similar to those produced by sequential precipitation as shown in Fig. 13.2.

Precipitation-deposition can be used to produce catalysts with a variety of supports, not only those that are formed from coprecipitated precursors. It has been employed to prepare nickel deposited on silica, alumina, magnesia, titania, thoria, ceria, zinc oxide and chromium oxide.[36] It has also been used to make supported precious metal catalysts. For example, palladium hydroxide was precipitated onto carbon by the addition of lithium hydroxide to a suspension of carbon in a palladium chloride solution to give $Pd(OH)_2/C$. This procedure was also used for the preparation of $Rh(OH)_3/C$ and $Ru(OH)_3/C$. These non-pyrophoric, "Pearlman's Catalysts", are dried and kept in the oxidized form until they are reduced, *in situ*, in the hydrogenation mixture.[37]

A simple procedure for the preparation of Ni/SiO_2 catalysts involves the precipitation of nickel dimethylglyoxime from a suspension of silica in an aqueous solution of a nickel salt.[38,39] Calcination is not required since the resulting supported precipitate is easily reduced in a stream of hydrogen at 200°–250°C, giving very highly dispersed supported nickel catalysts.

In many cases it is advantageous to keep the volume of the precurser salt solution to a minimum in order to have most of the active component absorbed in the pores of the support before precipitation. This is particularly true in the preparation of high metal loaded catalysts. The classic nickel on kieselguhr catalysts having 20%–50% nickel are prepared in this way.[23,40] A concentrated solution of nickel nitrate is added to an appropriate amount of kieselguhr ("Filter Cel", a form of silica) to form a thin paste which is then added slowly to a basic solution to precipitate the nickel hydroxide onto the support. Ammonium carbonate was originally preferred as the precipitant[23] but sodium bicarbonate and sodium carbonate are still commonly used even though sodium ions are not easily removed from the precipitate. After washing and drying, the supported precipitate is reduced, without calcining, at 450°C in a stream of hydrogen to give an effective hydrogenation catalyst.

Copper on kieselguhr catalysts containing about 50% copper have been prepared by the precipitation of copper hydroxide from a dilute solution of copper nitrate in which the keiselguhr was suspended. After filtering, washing and drying, the supported hydroxide was reduced in hydrogen at 225°C to give a catalyst effective at promoting the hydration of nitriles (Eqn. 13.8).[41]

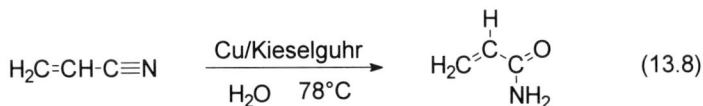

$$H_2C{=}CH{-}C{\equiv}N \quad \xrightarrow[{H_2O \quad 78°C}]{Cu/Kieselguhr} \quad H_2C{\underset{}{\overset{}{\diagdown}}}\overset{\overset{\textstyle H}{|}}{C}\diagdown\underset{\underset{\textstyle NH_2}{|}}{C}{\diagup}O \qquad (13.8)$$

Reduction-Deposition

Reduction-deposition describes a process by which a solution of a salt of the active material in which the support is suspended is treated with a reducing agent to give the supported reduced metal catalyst directly. The classic procedure for the preparation of Pd/C involves shaking a suspension of carbon in aqueous palladium chloride in a hydrogen atmosphere. Filtering, washing and drying gives an active Pd/C catalyst with the palladium load determined by the concentration of the palladium chloride and the amount of carbon used.[42] Pt/C catalysts can also be prepared using a similar procedure but frequently a small amount of palladium chloride is added to facilitate the hydrogen reduction of the chloroplatinic acid.[43]   Other reducing agents that can be used are formaldehyde, hydrazine and borohydride.[44]   This approach is, obviously, applicable to the preparation of a number of different supported metals and is limited only by the reducing agent and the conditions needed to reduce the salt to the metal. In most cases this is not strictly a reduction of the metal ion in solution but, rather, the reduction of the metal precurser already adsorbed on the support material.

A true reduction-deposition, however, occurs on reaction of palladium chloride with potassium graphite in 1,2-diethoxyethane.[45]   This gives a Pd/graphite catalyst that is active for the hydrogenation of nitro compounds and alkenes and is particularly useful for the semihydrogenation of alkynes (Eqn. 13.9).[45]   A Ni/graphite catalyst prepared by reaction of nickel bromide with potassium graphite is also effective for the stereoselective semihydrogenation of alkynes.[46,47]   Reaction of potassium graphite with bis-cyclopentadienyl nickel also gave a Ni/graphite catalyst.[48]

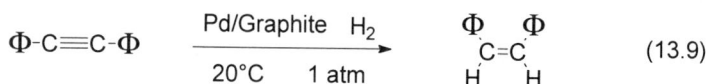

$$\Phi-C\equiv C-\Phi \quad \xrightarrow[\text{20°C} \quad \text{1 atm}]{\text{Pd/Graphite} \quad H_2} \quad \begin{matrix} \Phi & \Phi \\ C=C \\ H & H \end{matrix} \quad (13.9)$$

Mixing a preformed metal sol with a support material also provides a method for the preparation of supported catalysts with the colloidal metal particles attached to supports such as alumina[49,50], titania[51,52] and pumice.[53]   While this procedure gives catalysts having essentially a single size metal particle, the particles are not strongly bonded to the support which makes these materials primarily useful for vapor phase reactions. An added complication is that the citric acid commonly used to prepare the sols[49,51,52] or the micellar material in which they are stabilized,[50,53] can also be adsorbed on the support and, possibly, inhibit the activity of the resulting catalysts.

A final deposition procedure is the photodeposition of metals on semiconductor supports. Typically, the support is suspended in a metal salt solution and irradiated with a mercury lamp. This procedure was originally used to prepare an active Pt/TiO$_2$ catalyst.[54]   These catalysts have significantly higher dispersions than comparable catalysts prepared by the incipient wetness procedure described below. The small platinum crystallites prepared by photodeposition are uniformly distributed over the titania surface.[55]   The enhanced activity for CO/H$_2$ reactions suggests a strong metal support interaction is present on these catalysts.[55]   This procedure has also been explored as a method for producing a number of other metal/semiconductor catalysts with metals such as platinum, palladium, rhodium, silver and iridium, and supports such as ZnO, Nb$_2$O$_5$, ZrO$_2$, ThO$_2$,[56] and WO$_3$.[54]   A support consisting of titania grafted onto silica was also used for the photocatalytic deposition of rhodium giving a Rh/TiO$_2$ catalyst supported on silica.[57]   This method uses mild reaction conditions and is applicable to a number of different metal support combinations but, unfortunately, it is not readily adaptable to large scale preparations.

Impregnation

The term, impregnation, has taken on a number of different meanings in the catalyst preparation literature. As a result, it is sometimes difficult to determine exactly how a specific catalyst was made. Impregnation is properly defined as a means of catalyst preparation by the adsorption of a catalyst precurser salt from solution onto a support material. This process is sometimes referred to as wet impregnation[58] since the pores of the support are filled with the solvent before coming in contact with the precurser salt. It is also referred to as diffusional impregnation since the impregnation of the support is accomplished by the diffusion of the salt inside these solvent filled pores.[59] As discussed below for the ion exchange procedure, there is also a significant amount of interaction between the surface of the support and the salt, resulting in the adsorption of the salt onto the support. A more descriptive term for this process is adsorption from solution[60] since, generally, the procedure calls for stirring a suspension of the support in the salt solution for a prescribed length of time followed by the separation of the modified support by filtration or centrifugation. The supported salt is then dried and, frequently, calcined before the salt is reduced to the metal. Occasionally, the suspension of the support is evaporated to dryness giving, presumably, the dried supported salt. Care should be taken with this approach, however, since it could lead to the precipitation of unadsorbed salt along with the supported salt. Reduction, then, would give an unsupported metal black along with the supported metal, which is an undesirable mixture of active components.

Impregnation has been used to prepare a number of catalysts having different metal support combinations. Highly loaded nickel catalysts supported on alumina, titania, silica, niobia and vanadium pentoxide were prepared by adsorption of nickel nitrate from an ammoniacal solution onto the support material. The supported salts were dried at 120°C and calcined at 370°C before reduction to the supported metallic nickel. It was found that the ease of reduction depended on the crystallinity of the support. Amorphous or poorly crystalline supports made the reduction of the nickel oxide more difficult than on crystalline supports.[61] As examples of its generality, this procedure was also used to prepare Pd/polyethyleneimine,[62] Rh/TiO$_2$,[63] Rh/Al$_2$O$_3$,[64] Pt/C,[65-68] and Ru/MgO.[69]

An EXAFS study of the impregnation of γ alumina with chloroplatinic acid[70] and its subsequent treatment has provided some information concerning the nature of the platinum species present at the various stages of the impregnation process. After impregnation and separation of the impregnated alumina, the resulting material was composed of PtCl$_6^=$ complexes that were weakly held on the alumina. After calcining at 530°C the platinum atoms were surrounded, on average, by five oxygens and 2.5 chlorines. At this temperature the platinum oxide particles were relatively small but calcination at 700°C gave larger platinum oxide aggregates. Reduction of the 530°C calcined material in a hydrogen stream at 480°C gave platinum crystallites having an average coordination number of six as compared with twelve for a platinum atom in the bulk material. These data indicate that the reduction of the small platinum oxide particles leads to the formation of small platinum crystallites.[70]

A comparison of the various stages in the preparation of Pt/SiO$_2$ and Pt/Al$_2$O$_3$ catalysts formed by impregnation of the support with an aqueous solution of chloroplatinic acid also showed that the platinum was present in an octahedral Pt(IV) environment before reduction.[71] While the Pt/SiO$_2$ was essentially chloride free after reduction, a considerable amount of chloride was retained in the alumina supported catalyst.[71]

EXAFS studies were also undertaken on rhodium catalysts prepared by impregnation of silica and alumina.[72] The rhodium was highly dispersed with coordination numbers of 1.5–2 in the Rh/Al$_2$O$_3$ and about nine in Rh/SiO$_2$.[72] Similar data for Ni/Al$_2$O$_3$ catalysts showed that the coordination number of nickel increased with increasing metal loading and/or decreasing calcination temperature.[73]

In coprecipitation and deposition, the metal load is determined primarily by the concentration of the metal salt in solution before precipitation or deposition. A coprecipitated catalyst has the active component distributed throughout the resulting catalyst particles. With catalysts prepared by deposition, particularly precipitation-deposition, the active component can be found primarily on the surface of the supporting material. With impregnated catalysts, however, the situation is not as simple with the amount of salt adsorbed and its location on or in the support particles determined by the variables in the adsorption procedure. Thus, the concentration of the precurser salt, the type of salt, solvent, temperature, nature of the support, time of contact with the support and the presence of other materials can all influence both the metal load and the location of the metal in the support particle.

The distribution of the active component in the support particle can be described as one of the four major types shown in Fig. 13.3.[58,59] Each of these types can be particularly effective under different reaction conditions. If, for example, a catalytic reaction is mass-transfer limited, an "egg shell" or pellicular[59] type of catalyst having the active component on or near the surface of

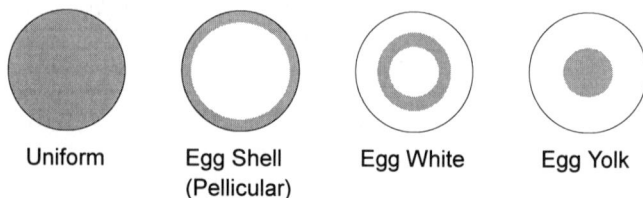

| Uniform | Egg Shell (Pellicular) | Egg White | Egg Yolk |

Fig. 13.3. Different modes of catalyst distribution within support particles.

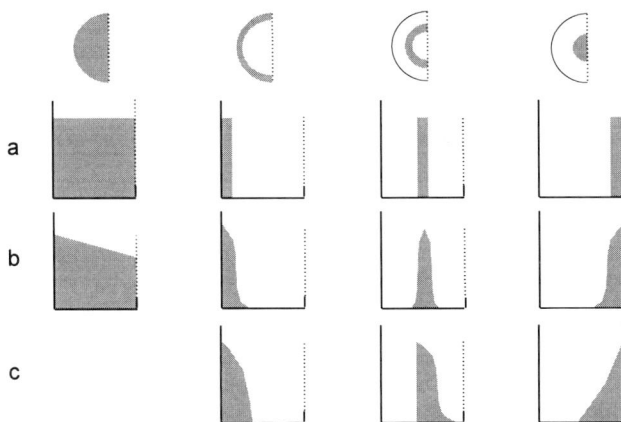

Fig. 13.4. Various distribution profiles of the catalyst in support
particles. a) Sharply defined; b) diffuse; c) degenerate.

the support would be the most efficient. A catalyst having a uniform distribution on the other hand, is preferred for reactions in which mass-transfer limitations have been removed. If the reaction medium contains a catalyst poison or inhibitor, then the use of an "egg white" or "egg yolk" catalyst is desirable so the exterior of the support can adsorb the poison before it can reach the active metal.[58,59] As depicted in Fig. 13.4[58], each of these distinct types can have varying concentration profiles. They can be sharply defined, diffuse or degenerate, with the exact nature depending on the conditions used for the catalyst preparation.[74]

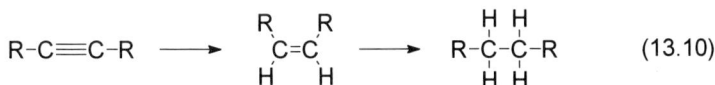

$$R-C\equiv C-R \longrightarrow \underset{H\;\;\;\;H}{\overset{R\;\;\;\;R}{C=C}} \longrightarrow \underset{H\;H}{\overset{H\;H}{R-C-C-R}} \qquad (13.10)$$

Reaction selectivity can also be enhanced by the proper choice of active component distribution. The selectivity observed in a sequential reaction (Chapter 3) such as the hydrogenation of an acetylene to an olefin and then to an alkane (Eqn. 13.10) is higher with eggshell catalysts than those with uniform palladium distributions. Alkynes are adsorbed more strongly than an alkene on palladium so the double bond is not hydrogenated while there are some molecules containing a triple bond accessible to the active sites. When the active metal is on or near the surface of the support particle, it is relatively easy for the alkyne to displace the alkene before it can be hydrogenated further to the alkane. With a uniform distribution of palladium, however, the alkyne is used up during the

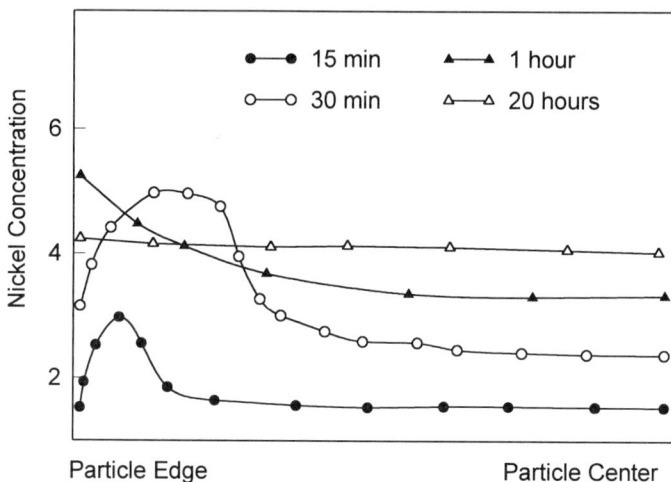

Fig. 13.5. Effect of impregnation time on the quantity of nickel salt
adsorbed and its distribution within the support particle on
impregnation of alumina with aqueous $Ni(NO_3)_2$. (Redrawn
using data from Ref. 77.)

diffusion to the center of the catalyst particle so the alkene is not desorbed and is hydrogenated instead.[75]

The strength of the interaction between the support and the catalyst precurser is a primary factor in determining the distribution of the catalytically active material on the support particle. Strong adsorption of the precurser on the support leads to the formation of egg shell catalysts. The concentration of the precurser salt in the impregnating solution simply determines the thickness of the shell. Weak adsorption leads to a more uniform distribution while intermediate strengths give degenerate egg shell profiles.[58,76] Thus, any factor that can influence the adsorption of the precurser salt on the support will have an effect on the metal distribution in the final catalyst. An obvious reaction parameter controlling this adsorption is the time of contact between the impregnating solution and the support. The data in Fig. 13.5[77] illustrate the effect of impregnation time on the nickel distribution in supported salts prepared by impregnating alumina with an aqueous solution of nickel nitrate. The material produced after a short time has a narrow egg shell distribution that becomes wider with longer contact time, eventually giving a uniform distribution.[77]

The solvent used to dissolve the precurser salt can also have a significant effect on the degree of impregnation and the metal distribution with specific supports. Chloroplatinic acid dissolved in water is adsorbed more strongly on alumina than on carbon, while in acetone the reverse is true.[65,66] As shown in

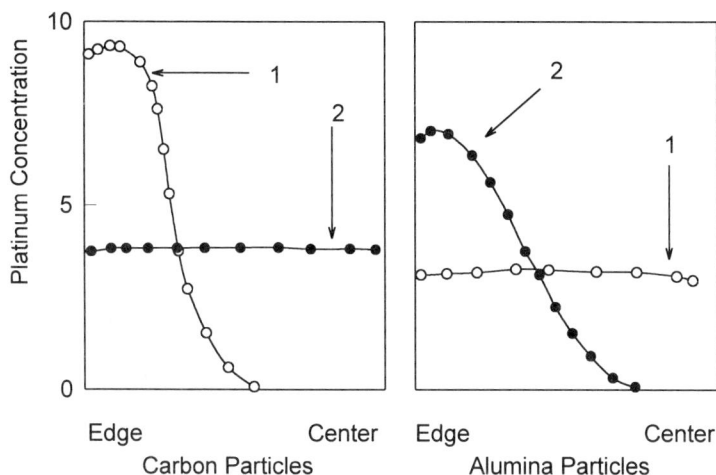

Fig. 13.6.   Concentration profiles of platinum distribution through support
             particles after impregnation of carbon and alumina with
             1) aqueous chloroplatinic acid solutions and 2) acetone
             chloroplatinic acid solutions. (Redrawn using data from Ref.
             65.)

Fig. 13.6,[65] strong adsorption leads to an egg shell distribution with aqueous solutions on alumina and acetone solutions on carbon. The use of aqueous solutions with carbon and acetone solutions with alumina promotes a weaker adsorption and a more uniform metal distribution. The data in Fig. 13.7 show that as the polarity of the solvent decreases the distribution of platinum on a carbon support becomes more diffused.[78]

The use of aqueous solutions with basic oxide supports such as magnesium oxide presents some difficulties because of the partial solubility of these oxides in water. To overcome this problem, highly dispersed Ru/MgO catalysts were prepared using ruthenium chloride dissolved in either anhydrous acetone or acetonitrile for the impregnation.[69]

The pH of the solution can be critical in the impregnation process. In the preparation of Rh/TiO_2 catalysts by impregnating titania with aqueous solutions of rhodium chloride it was found that at low pH the interaction between the salt and the support surface takes place in several steps. Initially the oxide particles become more positively charged and, thus, adsorb the anionic species such as $RhCl_4^-$. After adsorption, these anions are more firmly anchored to the support by the displacement of one or more chloro ligands by OH groups on the titania surface.[63]

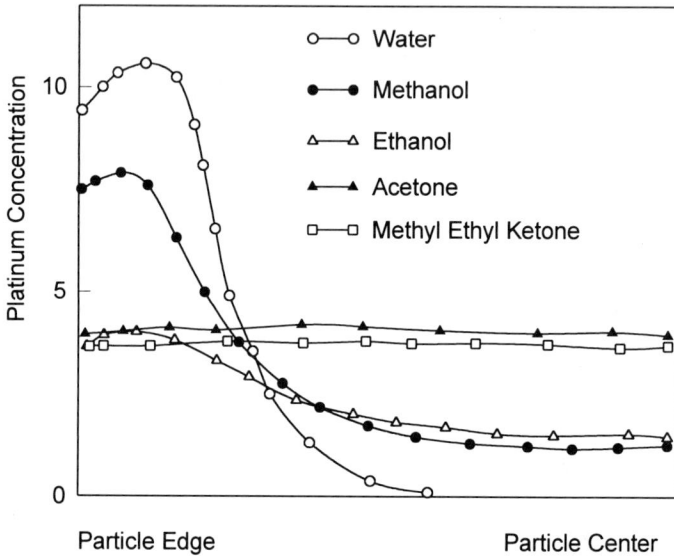

Fig. 13.7. Concentration profiles of platinum distributed through carbon
support particles after impregnation with chloroplatinic acid
dissolved in various solvents. (Redrawn using data from Ref. 78.)

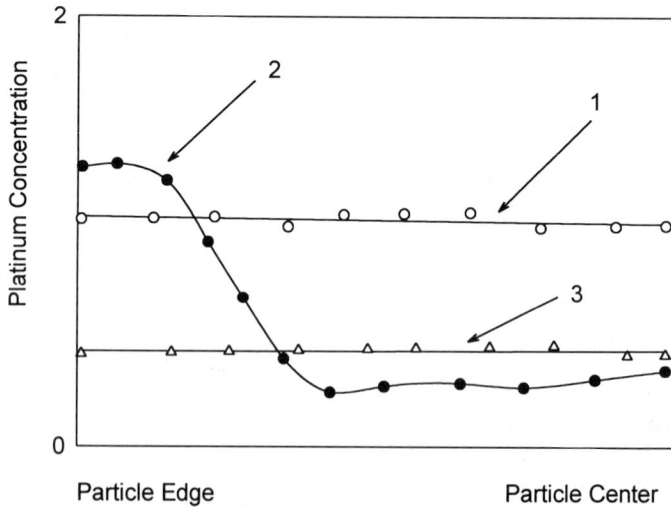

Fig. 13.8. Concentration profiles of platinum distributed through support
particles on impregnation. 1) $Pt(P\Phi_3)_4$ on carbon; 2) $Pt(P\Phi_3)_4$
on alumina, and 3) $Pt(NH_3)_4Cl_2$ on carbon. (Redrawn using
data from Ref. 65.)

The nature of the salt or complex used for the impregnation can influence the extent of adsorption and the properties of the resulting catalyst. Rh/Al_2O_3 catalysts prepared from rhodium nitrate are more highly dispersed than are those prepared using $Rh(NH_3)_5Cl_3$.[64] The data in Fig. 13.8[65] show that while both $Pt(\phi_3P)_4$ and $Pt(NH_3)_4Cl_2$ are adsorbed uniformly on both alumina and carbon particles, the extent of adsorption is significantly lower than that observed with chloroplatinic acid (Fig. 13.6).[65]   Adsorption of nickel ethylenediamine complexes onto silica gave, after reduction, very highly dispersed Ni/SiO_2 catalysts.[79]

The presence of other materials in the impregnating solution can have a marked effect on the location of the metal within the support particle. These additives have been conveniently divided into three classes.[80] Class 1 additives consist of simple inorganic electrolytes which influence the electrostatic interactions at the solution-support interface.[81]   Simple salts such as sodium nitrate, sodium chloride, or calcium chloride do not adsorb strongly enough on alumina to compete with platinum salts for adsorption. Fig. 13.9a[80] shows the concentration profile of platinum on an alumina particle when the impregnation of chloroplatinic acid was done in the absence of any additives. This a somewhat diffused egg shell profile. Fig. 13.9b[80] shows the adsorption profile for the catalyst prepared by impregnation in the presence of an amount of sodium nitrate equimolar to the chloroplatinic acid. Here the amount of platinum adsorbed decreases while the adsorption profile approaches a uniform distribution. It is

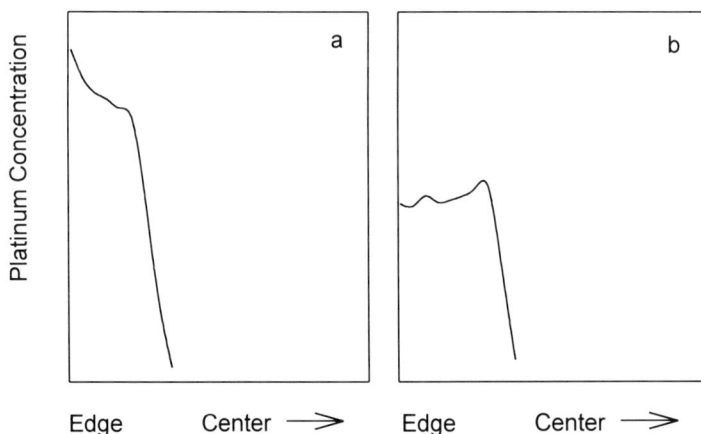

Fig. 13.9.   Concentration profiles of platinum adsorption on impregnation
            of γ alumina pellets with aqueous chloroplatinic acid.
            a) Standard adsorption, no additives and b) $NaNO_3$ added
            equimolar to the chloroplatinic acid. (Redrawn using data from
            Ref. 80.)

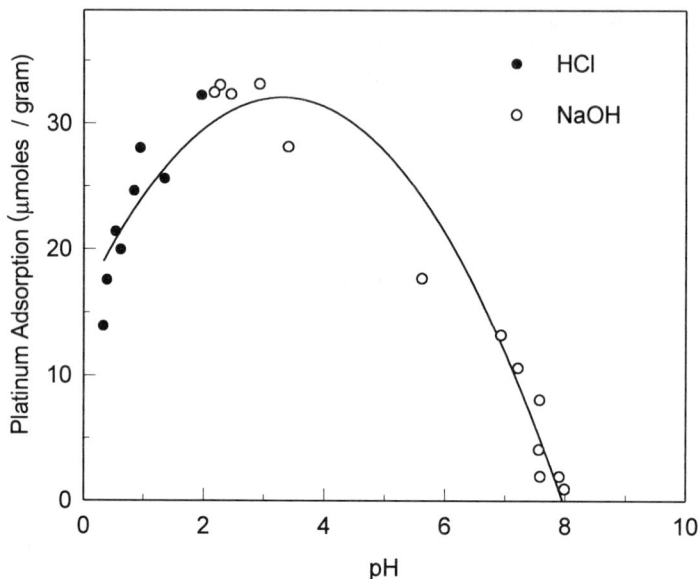

Fig. 13.10. Effect of pH on the extent of chloroplatinic acid adsorption on
alumina. (Redrawn using data from Ref. 80.)

thought that the electrolyte decreases the number of adsorption sites on the
alumina by screening them from the solution. Since this effect occurs uniformly
over the length of the pores in the support particle, the platinum coverage, while
lower, extends deeper into the pores giving a more uniform distribution.[80]

Class 2 modifiers are simple inorganic acids and bases that can alter the
chemistry of the support surface. The presence of acid significantly affects the
quantity of chloroplatinic acid adsorbed on alumina with the effect of solution pH
on the extent of adsorption (Fig. 13.10).[80] As discussed in Chapter 9 and
illustrated further in the discussion on ion exchange presented later, the isoelectric
point (IEP) for alumina occurs at a pH near 8 and no adsorption occurs at higher
pH values. The amount of nickel nitrate or nickel chloride adsorbed on γ alumina
is very low below pH 4, with the amount adsorbed increasing abruptly at a pH
near 5.2.[82,83]

Class 3 additives are materials such as phosphoric acid and citric acid that
can compete with the metal for adsorption sites. While Class 1 and Class 2
additives can control the depth and amount of metal adsorbed leading either to
uniform or egg shell catalysts, Class 3 species interfere with platinum adsorption
and can give entirely different adsorption profiles. This approach is used,
specifically, for the preparation of egg white and egg yolk type catalysts. Fig.
13.11[80] shows that the platinum distribution is displaced from the surface of the

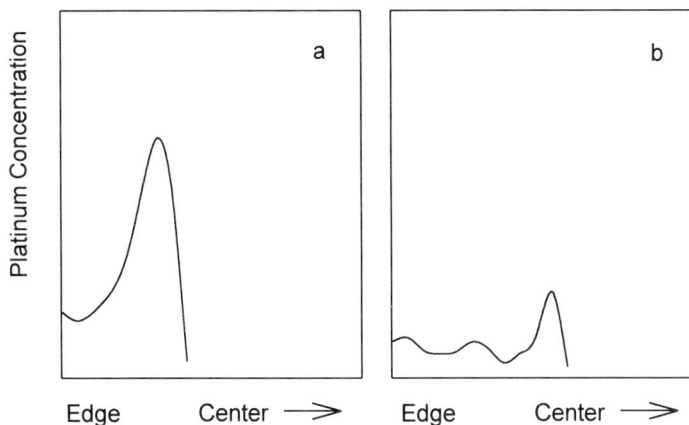

Fig. 13.11. Concentration profiles of platinum adsorbed by impregnation of
            γ alumina in the presence of phosphoric acid. a) Phosphoric
            acid concentration of 1.5 times the molar amount of
            chloroplatinic acid and b) phosphoric acid concentration 3 times
            the molar amount of chloro-platinic acid. (Redrawn using data
            from Ref. 80.)

support particle by the presence of phosphoric acid. This additive is more
strongly adsorbed than the platinum species so it is preferentially adsorbed on the
surface leaving the platinum no place to adsorb except through the pores toward
the center of the particle. A similar effect is observed using a citric acid
additive.[58] The more of the Class 3 additive present, the deeper its adsorption and
the closer the platinum is found to the center of the particle with the egg yolk
distribution obtained with high modifier concentrations.

On removal from the impregnating solution, the support contains the
precurser salt in two forms: some is adsorbed on the external and pore wall
surface and the rest is dissolved in the liquid filling the pores. The first type can
reasonably be expected to remain where it is on drying, particularly if the
interaction between the salt and the support is strong. The material in solution,
however, can be redistributed throughout the support particle during the drying
process.   Increasing the rate of drying minimizes the external surface
redistribution.[58] Under fast drying conditions the volume of solution in the pores
decreases with a concomitant increase in impregnant concentration so the salt may
begin to precipitate onto the pore wall when the concentration exceeds the
saturation point. The progressive deposition observed with rapid drying tends to
distribute the precipitated salt near the center of the support particle since that is
where the lower volumes in the pores will concentrate.[58] The more concentrated
the impregnating solution in the pores, however, the more uniform the salt

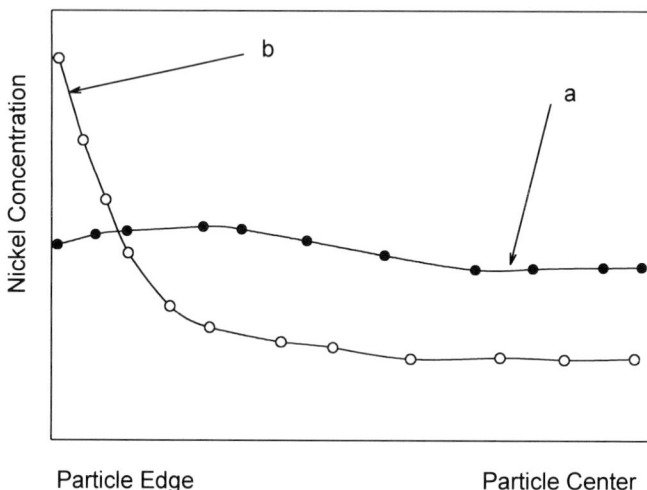

Fig. 13.12. Concentration profiles of nickel adsorbed on alumina. a) 1M
Ni(NO$_3$)$_2$ adsorption and b) after NaOH addition to (a) to give
Ni(OH)$_2$. (Redrawn using data from Ref. 77.)

deposition. Increasing the viscosity of the liquid retards the accumulation of the
salt at the periphery of the support particle and gives a more uniform distribution
of the metal.[84]

   After drying, the supported salt is generally calcined to convert the salt
into the oxide which is then reduced to give the supported metal catalyst. At
times the dried supported salt is reduced without the intermediate calcination step.
In general, a more dispersed catalyst is obtained by the reduction of a calcined
material than is found when the supported salt is reduced without prior
calcination.[85]

   It is possible to redistribute the metal by adding a precipitant to a support
that has already been impregnated. Fig. 13.12[77] shows the effect of adding
sodium hydroxide to an alumina support impregnated with one molar nickel
nitrate. While the initial impregnation showed a slight increase in nickel
concentration near the exterior of the support, the addition of base resulted in a
redistribution of the nickel with the formation of essentially a surface coverage of
nickel hydroxide.[77]

   In Ni/Al$_2$O$_3$ catalysts prepared by impregnation there is a weaker
interaction between the impregnating salt and the support than there is when the
catalyst is prepared by deposition-precipitation.[86,87] Calcination of the former
catalysts gave large nickel oxide crystallites and, on reduction, the supported
metal was rather easily sintered. Catalysts prepared by precipitation-deposition,
on the other hand, were resistant to sintering because of the strong interaction

between the metal and the support.[87] Ni/Al_2O_3 catalysts prepared by coprecipitation had higher metal surface areas than those prepared by impregnation even though the coprecipitation process placed a significant fraction of the metal in the bulk of the catalyst particle.[88] When the sol-gel method is used to prepare the coprecipitated Ni/Al_2O_3, the nickel crystallites have a more uniform particle size, a factor that is not present in catalysts prepared by impregnation. Coprecipitation produces a material that on calcination has small nickel clusters dispersed uniformly throughout the support material while large, variously sized nickel oxide particles are formed on calcining the impregnation produced supported salt.[89]

Incipient Wetness

Incipient wetness, also referred to as dry impregnation[58] or capillary impregnation[59], involves contacting a dry support with only enough solution of the impregnant to fill the pores of the support. The volume of liquid needed to reach this stage of 'incipient wetness' is usually determined by slowly adding small quantities of the solvent to a well stirred weighed amount of support until the mixture turns slightly liquid. This weight:volume ratio is then used to prepare a solution of the precurser salt having the appropriate concentration to give the desired metal loading. At the present time this is probably the most commonly used procedure for catalyst preparation. A partial list of catalysts prepared in this way includes Ni/Al_2O_3[88,90,91], Pt/SiO_2[92–94], Pt/Al_2O_3[94], Ni/SiO_2[95], Ni/C[96], Ru/SiO_2[97], Rh/SiO_2[98], Rh/Al_2O_3[99], Rh/V_2O_3[100], Pd/rare earth oxides[101] and Ru/C.[102] Since all of the impregnant solution is adsorbed into the pores of the support, this procedure can be used to prepare specific, predetermined metal loadings on the catalyst.

ESCA, XRD and TEM studies of Ni/Al_2O_3 catalysts prepared by incipient wetness has shown that after drying and calcining the catalyst is composed of nickel oxide particles within the alumina pores. As the metal load is increased the size of the nickel oxide particles increases. On reduction the relative size of the supported nickel crystallites is determined by the size of the nickel oxide particles.[91] An EXAFS study of some highly dispersed Rh/Al_2O_3 catalysts indicated that the rhodium atoms that are in contact with the support are surrounded by an average of two to three oxygens from the surface of the support. The Rh–O bond distance implies that the metal particle is held on the support by an interaction between metallic rhodium and a surface ionic oxygen resulting in an ion induced dipole bonding between the metal particle and the support rather than a formal Rh–O bond formation.[99]

The experimental conditions used to prepare the supported salt can have an impact on the dispersion and location of the metal in the resulting catalyst. The dispersions of Pt/SiO_2 catalysts prepared by incipient wetness using aqueous solutions of chloroplatinic acid are dependent more on the surface area of the support than on the average pore size distribution. There appears to be some pore blockage during the impregnation but this is somewhat reversed by calcination

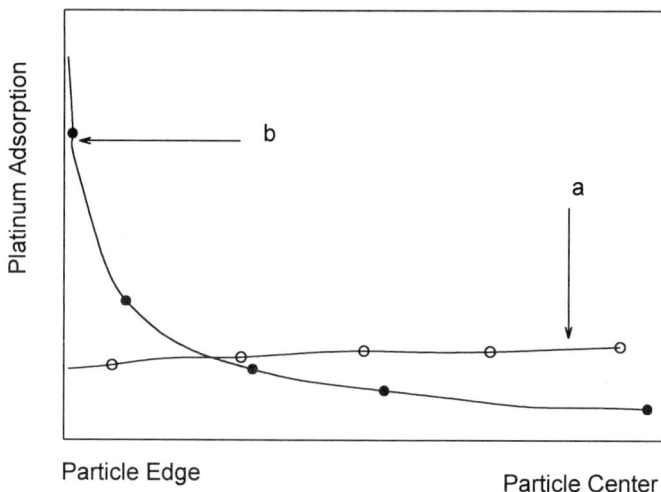

Fig. 13.13.   Concentration profiles of platinum adsorbed on alumina. a)
Adsorption of $Pt(NH_3)_4(NO_3)_2$ and b) adsorption of H_2PtCl_6.
(Redrawn using data from Ref. 103.)

and reduction.[93] It was also found in the preparation of Ni/Al_2O_3 catalysts that
the pH of the impregnant and the weight loading of the metal independently affect
the activity of the resulting catalysts. For example, a 0.9% Ni/Al_2O_3 catalyst
prepared at pH 5 had essentially the same dispersion as a 2.9% catalyst prepared
at pH 1 but the activity for CO hydrogenation was about 100% greater for the
lower loaded catalyst than that having the 2.9% nickel.[90] Ni/C catalysts prepared
from nickel nitrate are more highly dispersed than similar catalysts prepared from
nickel chloride.[96]

   As shown in Fig. 13.13[103] the use of an aqueous solution of chloroplatinic
acid as the impregnant on alumina gave an eggshell distribution of the platinum
while with an ammoniacal solution of $Pt(NH_3)_4(NO_3)_2$ a uniform distribution was
observed.[103] This is apparently due to the difference in the strength of adsorption
of the two platinum compounds on alumina. Other factors also play a role in
determining the metal distribution on the support. The conversion of a strongly
adsorbed impregnant that would normally give an egg shell distribution to a
species having a uniform distribution can frequently be accomplished by letting
the wet support stand for prolonged periods of time to permit the equilibration of
the adsorbed impregnant throughout the support pores.[104] Intermediate times will
lead to an expanded egg shell profile. This redistribution is only possible if an
adequate amount of solute is present. With low concentrations extended contact
times may not have any effect.[105] Sometimes even extended contact before
drying is not sufficient for the production of a uniform distribution as is frequently

observed with low concentrations of chloroplatinic acid on alumina. In this case, though, it was found that addition of hydrochloric acid to the impregnating solution resulted in a more uniform platinum distribution throughout the alumina.[106] Apparently the acid competes with the platinum salt for adsorption sites and causes the platinum to be distributed more evenly through the support particles.

The preparation of egg shell catalysts from more weakly adsorbed species that would normally give a uniform distribution has been accomplished by using volumes of the impregnating solution which are smaller than the pore volume of the support. A half pore volume, for instance, can give an egg shell distribution when a full pore volume gives a uniform profile.[107]

The problems associated with drying, calcining and reduction are the same as those described above for impregnated catalysts. The more significant of these are the possibility of pore disruption on evaporation of aqueous solutions and the migration of the metal salt within the pores as the solvent is removed. The former can be minimized by using organic solvents and a novel approach has been used to prevent the later. Metal salt migration can take place only as long as the salt is in solution so if it could be precipitated within the pores the metal would remain in place as the solvent was removed. This was accomplished by impregnation of alumina to incipient wetness with a solution of nickel nitrate and urea. Careful heating avoided solvent evaporation and hydrolyzed the urea to provide the base needed to precipitate nickel hydroxide inside the pores. After drying and reduction a Ni/Al_2O_3 catalyst was produced having very small nickel crystallites uniformly distributed throughout the alumina.[108]

Some difficulties have been observed in the preparation of supported ruthenium catalysts by incipient wetness using ruthenium chloride solutions with a silica support. The usual procedure of air calcination followed by reduction in hydrogen at 450–500°C gave a catalyst inferior to one in which the dried material was reduced directly in hydrogen at 700°C. This direct reduction did not result in any metal sintering, even though the temperature was high. The high temperature was needed, though, to remove the chlorine impurities from the ruthenium surface.[97] Application of this same procedure to a ruthenium chloride supported on carbon resulted in considerable hydrogenation of the support. This could be minimized by first heating the supported salt in a stream of nitrogen followed by high temperature hydrogen reduction with short contact times. This modified procedure produced a particularly active Ru/C catalyst.[102]

A Ni/SiO_2 catalyst prepared by incipient wetness was shown to consist, initially, of the silica support filled with the nickel salt. On reduction the nickel particles coalesced to form large crystallites because of their weak interaction with the support. Coprecipitated catalysts, however, were composed of layered silicate structures that were more difficult to reduce to metallic nickel. When reduction did take place, though, the reduced crystallites were fixed within the confines of the support so small metal particles with a narrow size range were produced.[95]

Table 13.1

Metal complexes commonly used to prepare supported catalysts by ion exchange.[109]

Metal	Anion	Cation
Cobalt		$Co(NH_3)_4^{++}$
Nickel		$Ni(NH_3)_4^{++}$
Copper		$Cu(NH_3)_4^{++}$
Ruthenium		$Ru(NH_3)_5Cl^{++}$
Rhodium	$RhCl_6^{3-}$	$Rh(NH_3)_5Cl^{++}$
Palladium	$PdCl_4^{2-}$	$Pd(NH_3)_4^{++}$
Silver		$Ag(NH_3)_2^{+}$
Iridium	$IrCl_6^{2-}$	$Ir(NH_3)_5Cl^{++}$
Platinum	$PtCl_6^{2-}$	$Pt(NH_3)_4^{++}$
Gold	$AuCl_4^{-}$	

Ion Exchange

As discussed in Chapter 9, the adsorption character of a support material is governed by the nature of its surface functionality. For oxides these are generally hydroxy groups and for carbon supports they are the acidic functions such as phenols and carboxylic acids. Depending on the acidity of these surface groups and the pH of the solution in which the support is suspended, the surface can be either positively or negatively charged as shown in Fig. 9.4. The pH at which the net charge on the support is zero is referred to as the isoelectric point (IEP) of the material. When the surface is negative, cationic species are attracted to it and become adsorbed. A positive surface interacts with negative species. The common anionic and cationic complexes used in catalyst preparation are listed in Table 13.1.[109]

For exchange on a support two conditions must be met: the pH of the impregnating solution must be either sufficiently high or low to provide the appropriate surface potential and the adsorbent must have the proper charge. The isoelectric points of a number of oxides are listed in Table 9.1[109] along with the

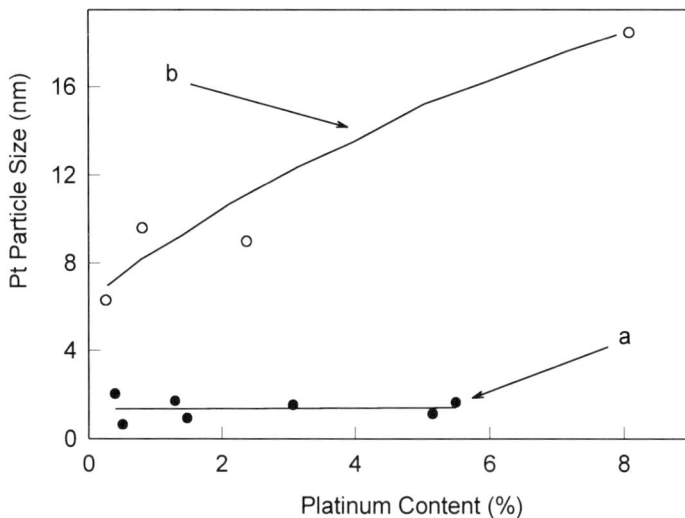

Fig. 13.14. Effect of platinum load on the mean metal crystallite size
of Pt/SiO_2 catalysts prepared by a) ion exchange and
b) incipient wetness. (Redrawn using data from Ref. 109.)

type of species that can be adsorbed on them. Basic oxides such as magnesium
oxide have a high IEP and can only adsorb anionic species such as $PtCl_6^=$.
Oxides such as alumina and titania have intermediate isoelectric points and, thus,
are amphoteric and can adsorb either anionic or cationic species depending on the
pH of the adsorbing solution. The zeta potential curve in Fig. 9.8 shows that even
though the IEP for silica is between 1 and 2, significant surface negativity is not
produced unless the pH of the solution is about six or higher. With this support
the metal precurser must be cationic.

The chloroplatinate ion is negative so this material does not adsorb on
silica. The only way of incorporating chloroplatinic acid onto silica is to use the
incipient wetness procedure.   This, however, gives much larger platinum
crystallites on reduction because the $PtCl_6^=$ is not held onto the silica and, thus,
the platinum particles are free to agglomerate. As the data in Fig. 13.14[109] show,
incipient wetness catalysts had progressively larger metal particles as the platinum
load increased.   However, catalysts prepared by ion exchange between the
cationic $Pt(NH_3)_4^{++}$ and the basic silica have a strong bond between the silica
surface and the platinum. On reduction these ion exchanged species gave highly
dispersed catalysts with small metal crystallites regardless of the platinum
load.[109]

Fig. 13.15[110] shows the effect of pH on the extent of $Pd(NH_3)_4^{++}$
adsorption on silica. There is a maximum exchange with this system at a pH of

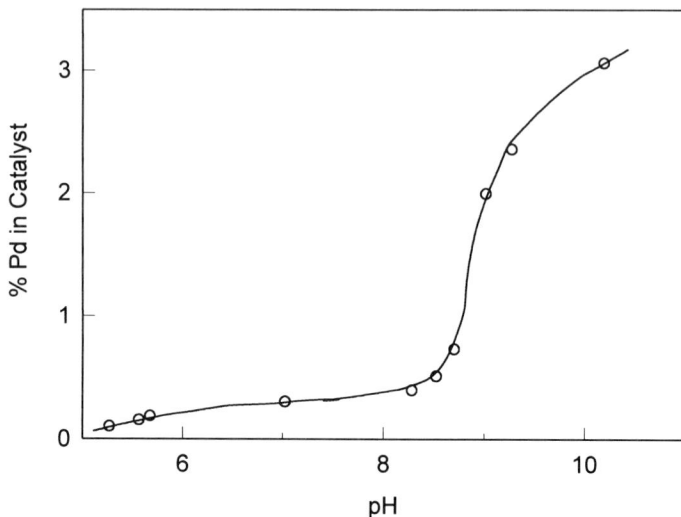

Fig. 13.15. Effect of impregnation pH on the extent of Pd(NH$_3$)$_4$$^{++}$
adsorption on silica. (Redrawn using data from Ref. 110.)

10.3.[111] Exchange of palladium chloride on silica in the presence of a small
amount of ammonia led to the formation of a more active and stable Pd/SiO$_2$
catalyst than those prepared using the more standard, larger amounts of
ammonia.[112]

The required conditions for adsorption on alumina are anionic precursors
at a pH lower than eight or cationic precursers at a pH higher than eight.
Adsorption of chloroplatinic acid falls into the first category and can take place
even in an acidic medium.[113] Interaction of alumina with Pt(NH$_3$)$_4$(OH)$_2$ and
Pt(NH$_3$)$_4$Cl$_2$ in ammonia fits in the second. A solution of sodium chloroplatinate
is, however, not sufficiently acidic so the pH is higher than eight and anionic
adsorption does not take place.[109]

The IEP of alumina, however, can be changed by the addition of
modifiers. It can be increased by the addition of Li$^+$ and Na$^+$. The presence of F$^-$
decreases the IEP. Adsorption of small amounts of Li$^+$ or Na$^+$ enlarges the pH
range in which negative species can be adsorbed on the surface. The addition of
fluoride ion as the modifier extends the pH range for adsorption of positive
species.[114] Fortunately only a relatively small amount of these modifiers are
needed since higher concentrations could lead to catalyst deactivation. Another
way of changing the IEP of alumina without the introduction of any modifier is to
increase the temperature of the alumina suspension. Increasing the temperature of
the alumina suspension to 50°C results in an increase in IEP to about nine so an

increase in the adsorption of negative ion results. On the other hand, cooling the suspension to 10°C extends the pH range for positive ion adsorption.[114]

Ion exchange can be used to prepare a variety of metal/support combinations providing the appropriate anion or cationic species is available for interaction with a specific support. $Cu(NH_3)_4^{++}$ readily interacts with silica in an ammoniacal solution.[115] The acidic functional groups on carbon supports are converted to reactive anionic species in base. These entities react with $Pt(NH_3)_4^{++}$ in basic medium While the number of such active groups is usually low in untreated carbon, reacting it with an oxidizing agent can greatly increase their surface concentration[116] as discussed in Chapter 9.

The preparation of Ru/SiO_2 by ion exchange, however, can present some problems using the standard procedure. Interaction of $Ru(NH_3)_6Cl_3$ with silica in a basic medium gave catalysts having very low metal dispersions.[117] The problem here is that the Ru(III) amines are unstable under the conditions used and are converted to hydroxy complexes at the pH needed to ionize the surface hydroxy groups on the silica. It is probable that this hydroxy complex then precipitates or agglomerates onto the surface in large particles which, after calcining and reduction, leads to the formation of large metal particles. An alternative is to use Ru(II) complexes which are more stable under the adsorption conditions but which must be prepared *in situ*. One way of accomplishing this is to add hydrazine to the $RuCl_3$ solution. When this was done in the presence of silica, highly dispersed Ru/SiO_2 catalysts of varying compositions were obtained.[118]

After separation the supported salt is then dried giving the stable supported complex.[119] This material is then either calcined and reduced or reduced directly to give the supported metal catalyst. Heating a silica supported palladium ammine complex in nitrogen resulted in the liberation of ammonia and the formation of a highly dispersed Pd/SiO_2 catalyst.[119,120] Heating in air, however, gave PdO and Pd^{+2} species supported on the silica.[119] A subsequent reduction was needed for catalyst preparation and this gave a moderately dispersed catalyst.[120] The optimum calcination temperature was found to be about 300°C.[121] Direct reduction with hydrogen gave a poorly dispersed Pd/SiO_2.[120] With other silica supported anionic complexes different results were observed.[122] Platinum ammines on direct reduction also gave poorly dispersed catalysts but with ruthenium and rhodium ammines high dispersion catalysts were obtained. The platinum ammine complex, on calcining, gave an amorphous platinum oxide similar to that found when the chloroplatinic acid impregnated onto alumina was calcined.[123,124] Heating the ruthenium and rhodium ammines gave poorly dispersed catalysts. Considerable metal loss was observed when the ruthenium containing species was heated to temperatures greater than 150°C.

Calcining a silica supported nickel ammine complex gave a nickel silicate like material with nickel ions present in an octahedral coordination surrounded by oxygens. Heating the titania supported species, however, resulted in the formation of Ni^{+2} and Ni^{+3} cations present in small patches of nickel oxide on the titania surface with some Ni–O interaction on the surface.[125] Pt/SiO_2 catalysts prepared by ion exchange are more highly dispersed than those prepared by incipient wetness.[92,126–128]

A comparison of Ni/TiO_2 catalysts prepared by ion exchange, impregnation and coprecipitation showed that on drying the first two had little evidence of interaction with the support but with the coprecipitated species there was some indication that titanate species had been formed.[129] On calcination at 300°C, nickel titanate was produced from the impregnated catalyst but not from that prepared using ion exchange.[130] The impregnated catalyst, after calcination, was much easier to reduce than the material produced by calcining the ion exchange produced species. Activation of this latter catalyst required heating in hydrogen at temperatures in excess of 350°C for prolonged times.[131] Rh/TiO_2 catalysts prepared by ion exchange have a higher dispersion than those prepared by impregnation.[132]

A procedure related to ion exchange involved the adsorption of neutral metal complexes onto a support surface, a process that results in the displacement of a ligand and the anchoring of the metal to the surface. Reaction of metal allyls with the hydroxy groups on the surface of silica results in the loss of propene and the formation of a surface diallyl as depicted in Fig. 13.16.[136] After hydrogenation the catalysts obtained from rhodium allyl[136–134] and nickel allyl[135] gave very highly dispersed catalysts but that prepared from palladium allyl was a poorly dispersed Pd/SiO_2.[137]

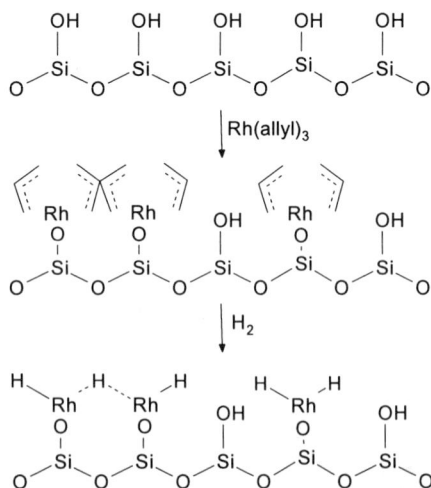

Fig. 13.16. Preparation of a Rh/SiO_2 catalyst by the reaction of rhodium allyl with surface Si-OH groups.

Calcined Pd/SiO_2 catalysts prepared from chloropalladic acid were easily reduced at low temperatures to give relatively highly dispersed catalysts but the dispersions decreased as the reduction temperature increased. The surface complex obtained by ion exchange, however, was not hydrogenated at temperatures lower than 100°C but at higher temperatures highly dispersed catalysts were obtained. It was proposed that at lower temperatures a palladium hydride species was formed and this material was more mobile on the silica surface so larger metal particles were produced. When higher temperatures are required for the reduction palladium hydride formation is minimized and smaller metal particles result.[137]

Another ion exchange procedure involves the interaction of a metal acetylacetonate (acac) with an oxide support. Virtually all acetylacetonate complexes, except those of rhodium and ruthenium, react with the coordinatively unsaturated surface sites of γ alumina to produce stable catalyst precursors. On thermal treatment and reduction these give alumina supported metal catalysts having relatively high dispersions.[138] Acetylacetonato complexes which are stable in the presence of acid or base such as $Pd(acac)_2$, $Pt(acac)_2$ and $Co(acac)_3$, react only with the Lewis acid, Al^{+3}, sites, on the alumina. Complexes which decompose in base but not in acid react not only with the Al^{+3} sites but also with the surface hydroxy groups. Complexes that are sensitive to acid but not to base react only slightly, if at all, with the hydroxy groups on the surface. It appears that this is the reason the rhodium and ruthenium complexes fail to adsorb on an alumina surface.[138]

This approach has even been used to prepare highly dispersed palladium catalysts supported on low surface area aluminas.[139] EXAFS analyses of the various stages in the interaction of $Pd(acac)_2$ with alumina has shown that the initial reaction involves an octahedral aluminum vacant site. On heating the resulting ionic species was converted to a supported crystalline oxide which was then reduced to the Pd/Al_2O_3 catalyst.[140]

The deposition of platinum, rhodium and ruthenium acetylacetonates on titania takes place by reaction with the surface hydroxy groups to give a supported complex. Thermal decomposition of these supported complexes in vacuum gave highly dispersed titania supported metal catalysts having metal particles about 2 nm in diameter.[141]

Reactions of acetylacetonates with silica are not as general as with alumina. $Cu(acac)_2$ is adsorbed on the silica in high concentrations while the platinum and palladium complexes do not interact with the silica at all and are easily removed by washing. No adequate explanation of these differences was proposed.[142] The palladium acetylacetonate complex was also used for the preparation of several palladium catalysts supported on rare earth oxides.[143]

Interaction of the dibenzylideneacetone complex of palladium with a high surface area carbon gave a supported complex which on heating produced a Pd/C catalyst having a uniform distribution of palladium metal particle sizes.[144] The rhodium carbonyls, $Rh_4(CO)_{12}$ and $Rh_6(CO)_{16}$, were adsorbed on silica to give supported carbonyl containing materials. Calcination and reduction gave Rh/SiO$_2$ catalysts that had much lower dispersions than comparable catalysts prepared by ion exchange using Rh(NH$_3$)$_6$Cl$_3$.[145]

Zeolite Supported Metals

Because of the special crystalline properties of zeolites which were described in Chapter 10, catalysts composed of a metal supported on a zeolite can have useful shape selective properties.[146,147] As shown in Figs. 13.17 and 13.18, the hydrogenation of a mixture of different sized olefins exhibits significant shape selectivity when a Rh/Y catalyst is used.[148] Zeolite supported metal catalysts are generally prepared by an ion exchange process,[149] but occasionally an incipient wetness procedure is also used as well as various forms of deposition.[149,150] Since zeolites are amenable to ion exchange, incorporation of the active metal precurser does not necessarily involve complex cations or anions. Nickel exchanged Y zeolites are prepared by exchanging Na/Y with an aqueous solution of nickel nitrate at a pH of 9.5. After drying, the Ni/Y species was reduced in hydrogen at 300°–700°C without a precalcination step. The resulting Ni/Y catalysts were effective catalysts for the hydrogenation of ketones (Eqn. 13.11).[151–153]

Exchanging NaY or NaX with Pt(NH$_3$)$_4$Cl$_2$ gave [Pt(NH$_3$)$_4$]$^{+2}$/NaY and [Pt(NH$_3$)$_4$]$^{+2}$/NaX having about a 4% platinum load. Heating to 300°C gave Pt^{+2}/NaY and Pt^{+2}/NaX. Treatment of Pt(NH$_3$)$_4$$^{+2}$/NaY with carbon monoxide at

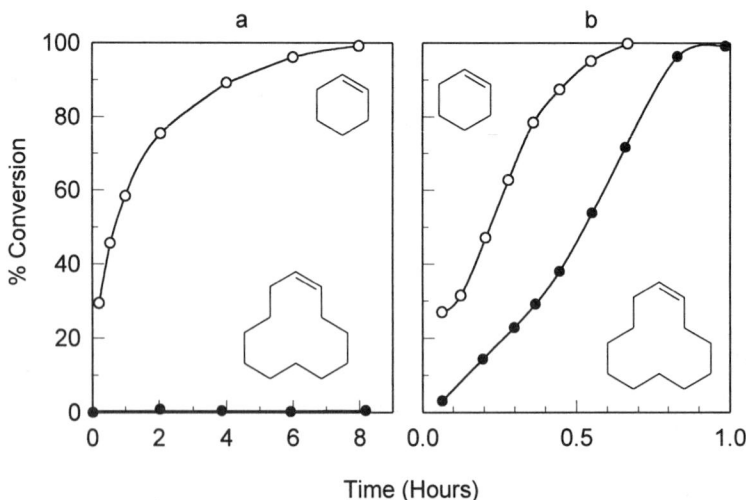

Fig. 13.17. Competitive hydrogenations between cyclohexene and cyclo-
decene over a)Rh/Y and b) Rh/C in hexane at 30°C and 1
atmosphere of hydrogen. (Redrawn using data from Ref. 148.)

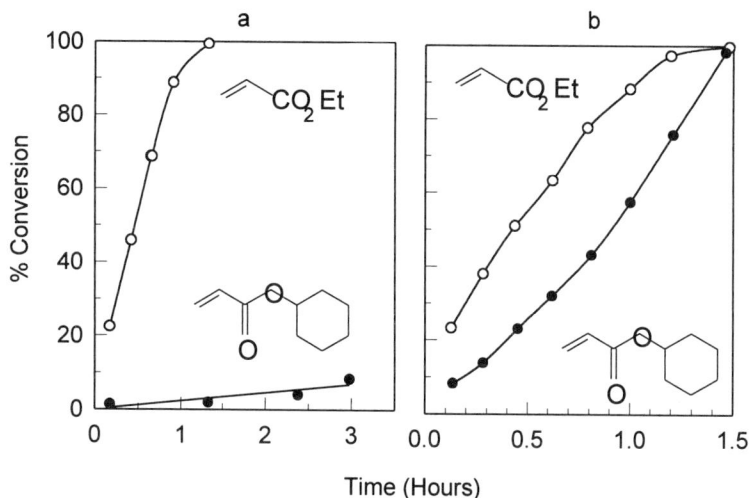

Fig. 13.18. Competitive hydrogenations between ethyl acrylate and cyclohexyl acrylate over a) Rh/Y and b) Rh/C in hexane at 30° and
one atmosphere hydrogen. (Redrawn using data from Ref. 148.)

100°C gave a $Pt_9(CO)_{18}^+$ cluster inside the Y cages. Carbonylation of the Pt^{+2}/NaY resulted in the formation of $Pt_{12}(CO)_{24}^+/NaY$ and a $Pt_{15}(CO)_{30}^+/NaX$ was prepared by carbonylation of the Pt^{+2}/NaX. These catalysts showed a more enhanced activity for the NO–CO reaction than was observed with conventional Pt/Al_2O_3 catalysts.[154] Exchange of $Rh(acac)_3$ with NaY produced an exchanged species which, on carbonylation, gave rhodium carbonyl clusters in the zeolite cages, a species particularly effective for CO hydrogenation reactions.[155] The migration of rhodium species from the external surface of the zeolite into the pores was observed when Rh/Y was treated with carbon monoxide. The reverse migration took place on exposure to water vapor.[156]

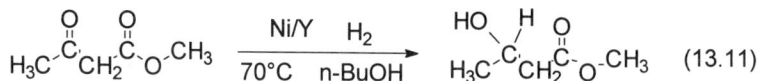

A Ru/NaY catalyst was prepared by the exchange of $Ru(NH_3)_6Cl_3$ with NaY followed by reduction in hydrogen at 350°C.[157] Exchange of $Pd(NH_3)_4(NO_3)_2$ with 5A gave the $Pd^{+2}/5A$ and with NaY gave the Pd^{+2}/NaY. Carbonylation of the palladium in the larger cage Y zeolite gave a $Pd_{13}(CO)_x$ cluster while with the smaller 5A a $Pd_6(CO)_x$ cluster was formed. The Pd/5A catalyst is more active for CO hydrogenation than the Pd/Y.[158] Heating of the

Pd/NaY in CO resulted in substantial migration and coalescence of the Pd clusters to give the $Pd_{13}(CO)_x$.[159] Calcining the exchanged $Pd(NH_3)_4^{+2}$/NaY resulted in the formation of Pd_{13} clusters in the zeolite cages.[160] Reaction of the Pd/NaY and Pd/HY with chlorine gave the mobile $PdCl_2$ which on reduction in hydrogen at 350°C produced highly dispersed palladium particles in the zeolite supercages.[155] The presence of protons in the zeolite results in the formation of smaller palladium particles. Pd/HY was more highly dispersed than Pd/NaY.[161]

A Pt/KL catalyst was prepared by exchange of $Pt(NH_3)_4(NO_3)_2$ with KL. After drying and calcining the resulting species were then reduced in hydrogen at 250°C. Similar catalysts were prepared by incorporating the precurser into the KL by incipient wetness followed by drying, calcining and reduction. The incipient wetness catalyst had smaller platinum particles located inside the zeolite channels while the ion exchange catalysts had larger particles, some of which migrated to the external surface of the zeolite.[162] When borohydride was used to reduce the platinum, the resulting catalyst was less stable than that produced by hydrogen reduction.[163]

The size of the zeolite pores determines the size of the complex that can be adsorbed. While octahedral complexes can be incorporated into the large cage X and Y faujasites, they are not exchanged into ZSM-5. Tetracoordinated complexes, though, can be exchanged into ZSM-5 giving, for instance, a $Pt(NH_3)_4^{+2}$ adsorbed complex that can be converted to metallic platinum by heating.[164] The size of the zeolite cage is also important with respect to the migratory ability of the metal intermediates. With a Y faujasite there is considerable migration of platinum, even at 350°C.[165] Small platinum particles can be obtained by slow heating in oxygen followed by hydrogen reduction.[166] The ZSM-5 supported platinum also shows some particle size dependence on calcining and reduction conditions.[167–169] The platinum particle size increases as the calcination temperature is increased. The coordination number increases to about two at 180°C and then goes from two to five with further increases in the calcination temperature. Direct reduction of the ammine precurser gives a more highly dispersed platinum than the catalyst prepared by reduction of a preoxidized material.[167,168] In contrast, though, with Pt/Y catalysts, calcination prior to reduction gives a more dispersed material than direct reduction.[166]

The reduction procedure can also influence the location of the metal particles in the zeolite. Reduction of $Pt(NH_3)_4^{+2}$/ZSM-5 in hydrogen at 300°C gave a catalyst which was as active as a Pt/Al_2O_3 catalyst for alkene hydrogenation. Neither catalyst, however, showed any selectivity in the hydrogenation of a mixture of 1-hexene and 4,4-dimethyl-1-hexene (Eqn. 13.12) indicating that the platinum particles in the Pt/ZSM-5 catalyst were primarily located on the exterior of the zeolite and not in its ten-ring tunnels. When the reduction was run using a hydrogen–alkene mixture at 400°C, the resulting catalyst showed marked selectivity with the hydrogenation of 1-hexene being favored over the saturation of 4,4-dimethyl-1-hexene by a 90:1 ratio (Eqn. 13.12). It was proposed that platinum migration and agglomeration during the reduction in hydrogen could be attributed to the formation of mobile platinum hydrides. The alkene present in the reduction mixture, though, can serve as a hydride trap and keep the platinum from moving out of the zeolite pores. Calcining the $Pt(NH_3)_4^{+2}$/ZSM-5 at 300°C followed by hydrogen reduction at 500°C gave a highly selective Pt/ZSM-5 catalyst.[171,172]

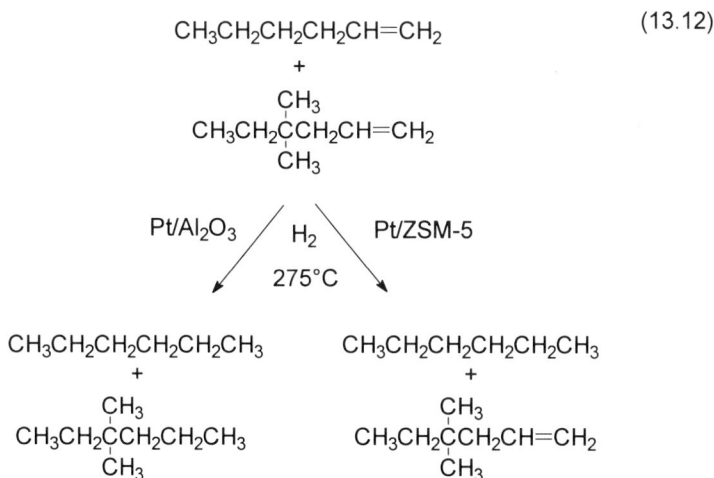

$$CH_3CH_2CH_2CH_2CH=CH_2 \qquad\qquad (13.12)$$

$$+$$

$$CH_3CH_2CCH_2CH=CH_2$$

Pt/Al$_2$O$_3$ / H$_2$ \ Pt/ZSM-5

275°C

$$CH_3CH_2CH_2CH_2CH_2CH_3 \qquad CH_3CH_2CH_2CH_2CH_2CH_3$$

$$+ \qquad\qquad\qquad\qquad +$$

$$CH_3CH_2CCH_2CH_2CH_3 \qquad CH_3CH_2CCH_2CH=CH_2$$

Reduction of the $Pt(NH_3)_4^{+2}$/Y with hydrazine gives a catalyst with a lower dispersion than hydrogen reduction,[150] probably because the hydrazine reduction, occurring in the liquid phase, takes place near the surface of the zeolite at the mouth of the pores so smaller particles are produced. These particles can block the access of hydrazine to the rest of the platinum complex inside the zeolite cages. Only about five percent of the platinum was accessible for hydrogen chemisorption. The resulting catalysts were considerably less effective for hexene hydrogenation than a standard Pt/Al$_2$O$_3$ catalyst.[150]

The small pores of zeolite A prevent exchange with the complex ions normally associated with the platinum metals. A Rh/A catalyst has been prepared, though, by adding aqueous rhodium chloride to the gel used to prepare the A. After crystallization a Rh/NaA was produced.[170] Reaction of rhodium triallyl with a zeolite resulted in an interaction with the hydroxy groups of the zeolite in the same way it reacted with silica. Hydrogenation gave the Rh/X in which the active species was probably in the cages since these catalysts exhibited a significant shape selectivity in olefin hydrogenation, molecules larger than cyclohexene were hydrogenated only very slowly. If the rhodium were on the external surface, this selectivity would not be observed.[173]

Other Supported Metals

Aluminum Phosphate Supported Metals

The neutral molecular sieve, aluminum phosphate (ALPO, Chapter 10) can also serve as a shape selective support for metal catalysts. These catalysts are normally prepared by the incipient wetness process followed by appropriate drying and reduction procedures. Fig. 13.19 illustrates the selectivity observed using a Ni/ALPO catalyst to hydrogenate a mixture of styrene and α methyl styrene (Eqn. 13.13).[174] The rates of hydrogenation of cyclic alkenes over a Rh/ALPO catalyst decreased in the order cyclopentene >> cyclohexene > cycloheptene > cyclooctene but no selectivity was observed in a competitive hydrogenation of a mixture of cyclohexene and cyclooctene.[175]

Clay Supported Metals

As described in Chapter 10, clays are laminar aluminosilicates. While these materials are most commonly used as solid acids their ion exchange capability provides a means of incorporating catalytically active metals into the laminar

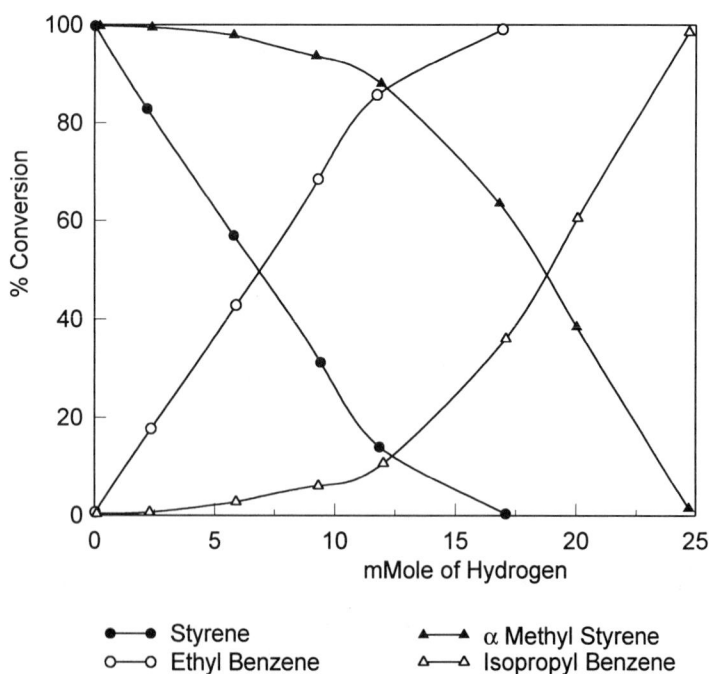

	Styrene	α Methyl Styrene
○—○	Ethyl Benzene	Isopropyl Benzene

Fig. 13.19. Course of the competitive hydrogenation of styrene and α methyl styrene over a Ni/AlPO$_4$ catalyst run in methanol at 15°C and 4 atmospheres of hydrogen. (Redrawn using data from Ref. 174.)

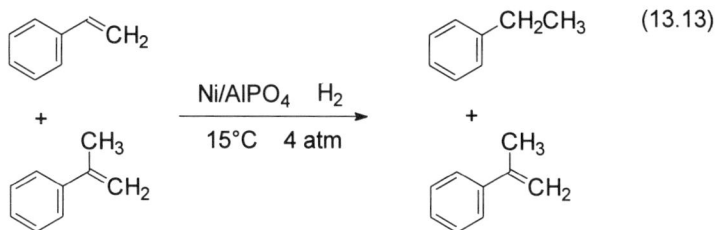

$$(13.13)$$

region to give potentially shape selective catalysts.   The most common preparation procedure involves the exchange of a cationic catalytic metal precurser for the Na^+ and Ca^{++} in the clay. The exchanged materials are then reduced. The preparation of clay supported platinum,[176] palladium,[177] nickel[180] and copper[179] catalysts have been reported. Most of the published data are concerned with catalyst characterization but it has been shown that a Pd/montmorillonite catalyst does possess some shape selectivity in alkene hydrogenation.[177]   With an average interlamellar spacing of 0.7 nm the hydrogenation of cyclopentene took place readily (Eqn. 13.14) but cycloheptene was not hydrogenated. Competitive hydrogenations were not reported.

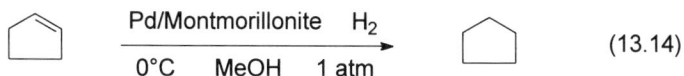

$$(13.14)$$

The use of pillared clays as metal supports has also been reported.[180,181] The more defined interlamelar spacing available with these supports should give a more predictable shape selectivity to the resulting supported metal catalysts. Further, since the pillars prevent the collapse of the layers on drying and further heating, the pillared clay supported metals salts can be calcined and reduced under conditions that can give the best metal dispersion without any concern for a change in the structure of the support.[180]

Supported Mixed Metal Catalysts

Supported mixed metal catalysts can be prepared by almost any of the procedures described above for the production of monometallic species. The discussion concerning the surface composition of these multimetallic species presented in Chapter 12 applies to supported catalysts as well but may be modified by the presence of very small crystallites in the supported catalysts.[182,183] The factors presented above concerning the location of the metal in the support particle are also applicable here as well and can be used to prepare bimetallic catalysts having specific distribution profiles.

Bimetallic catalysts prepared by coprecipitation or precipitation-deposition give an intimate mixture of the metallic species. Heating an aqueous suspension of silica in a solution containing nickel chloride, chloroplatinic acid and urea results, eventually, in the production of a highly dispersed Ni–Pt/SiO$_2$ catalyst.[184] In this case the addition of platinum results in a decrease in the methanation activity of the catalyst. A Pt–Sn/Al$_2$O$_3$ catalyst was prepared by using the sol gel method to produce a coprecipitated Sn–Al oxide on which chloroplatinic acid was adsorbed. This species had a very high platinum dispersion and exhibited enhanced activity toward carbon monoxide oxidation.[185]

Impregnation is also a general procedure for producing bimetallic catalysts. They can be prepared either by coimpregnation of both metal salts at the same time or by the sequential impregnation of first one salt followed by the second.[186–189] In coimpregnation the location and nature of the metals in the support particle depends on the type of precurser salt used and the tendency toward alloy formation of the two components. As shown in Fig. 13.20 coimpregnation of alumina using a solution containing chloroplatinic acid and palladium chloride resulted in an eggshell distribution with both metals present in the pellicular band. Using sequential impregnations under appropriate conditions it was possible to prepare catalysts having an eggshell platinum band and an egg white palladium band inside or an eggshell palladium band with an egg yolk platinum deposition inside.[187]

Rh–Pd/Al$_2$O$_3$ catalysts were prepared by the incipient wetness procedure using a solution containing both rhodium chloride and palladium chloride.[190] When a small amount of palladium was present, the palladium segregated toward

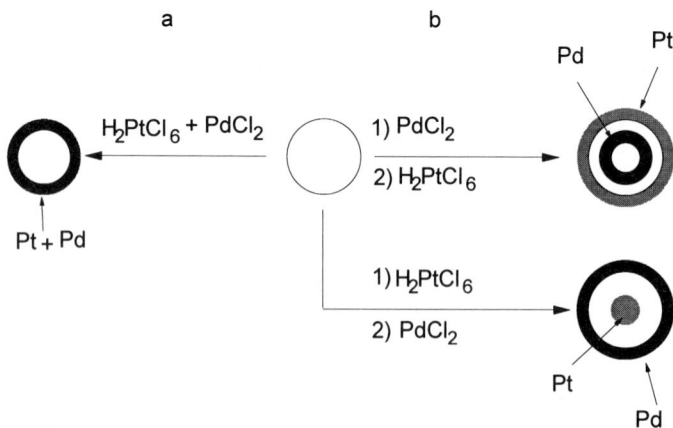

Fig. 13.20. Distribution of metals in bimetallic Pt–Pd/Al$_2$O$_3$ catalysts prepared by a) coimpregnation and b) sequential impregnations.

the surface of the small metal particles; with a low rhodium content no segregation was observed. These palladium–rhodium bimetallic catalysts are more resistant to sulfur poisoning than either monometallic catalyst.

Coimpregnation of silica with $Pt(NH_3)_4Cl_2$ and $Ru(NH_3)_6Cl_3$ using an ion exchange procedure gave a mixture of different types of bimetallic particles. The smaller particles were ruthenium rich and the larger ones rich in platinum.[191] Sequential impregnation of $Ru(NH_3)_6Cl_3$ on Pt/SiO_2 gave segregated platinum and ruthenium phases.[191] Coimpregnation of $Au(en)_2^{+3}$ and $Pd(NH_3)_4^{+2}$ followed by hydrogen reduction, however, gave alloy particles with a uniform composition overall and a surface composition similar to that of the bulk composition.[192]

A second metal can be reductively deposited onto a preformed supported metal catalyst to give a bimetallic species. Hydrogenation of a suspension of Ru/C in an aqueous solution of lead nitrate resulted in the deposition of lead onto the ruthenium surface to give a Ru–Pb/C catalyst in which the chemisorption data indicated a close interaction between the metals.[193] The hydrogenation of a Rh/Al_2O_3 suspended in a Cu^{++} solution gave a $Rh–Cu/Al_2O_3$ catalyst in which it appeared that the copper was deposited on the low coordination surface atoms of the rhodium.[194] Borohydride reduction of a $Ru^{+2}–Sn^{+4}$ suspension of alumina gave a supported Ru–Sn–boride catalyst.[195]

A common bimetallic catalyst is the Lindlar catalyst used for the selective hydrogenation of alkynes to alkenes (Eqn. 13.15)[196]. This catalyst is prepared by treating a calcium carbonate supported palladium catalyst with a solution of lead acetate to give the selective $Pd–Pb/CaCO_3$ catalyst.[197,198] The most selective catalysts are prepared by immersing an older, oxygen covered palladium catalyst in a solution of lead acetate, filtering and then thoroughly washing off the excess lead ions.[199] Another approach involved the treatment of a Pd/Al_2O_3 catalyst with tetrabutyl lead. The resulting surface complex, after hydrogen reduction, was a supported Pd–Pb catalyst in which the lead was found to be well dispersed and in close contact with the palladium.[200]

$$(13.15)$$

$$ClCH_2(CH_2)_5C\equiv CCH_2CH(OCH_3)_2 \quad \xrightarrow[\text{25°C \quad Pet. Ether \quad 2 atm}]{\text{Lindlar's Catalyst \quad H_2}}$$

$$ClCH_2(CH_2)_5 \quad CCH_2CH(OCH_3)_2$$
$$H \qquad H$$

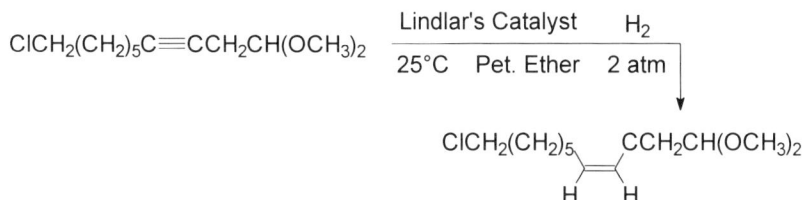

This organometallic approach to the preparation of mixed metals has also been used to prepare mixed metal species containing tin.[201–206] As shown in Fig. 13.21 the reaction of tetraalkyl tin with silica or alumina supported platinum,

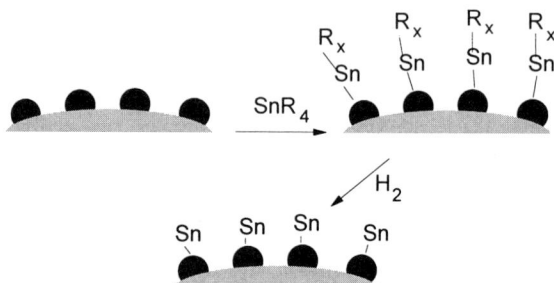

Fig. 13.21. Reaction of tetraalkyl tin with a supported metal to give
a tin containing supported mixed metal catalyst.

rhodium, palladium, ruthenium or nickel liberated one molecule of the alkane and
gave the corresponding tin containing bimetallic surface complexes still
containing alkyl radicals. Further reduction in hydrogen removes the hydrocarbon
entity as the alkane to give a bimetallic species. Rh–Sn/SiO$_2$[201] and Ni–
Sn/SiO$_2$[202] are both effective catalysts for the hydrogenation of esters to alcohols.
In the Rh-Sn catalyst there is a linear relationship between catalyst activity in this
reaction and the tin content up to a Sn:Rh ratio of about 0.3. This indicates that
the activity may be caused by a specific intermetallic phase on the catalyst
surface. With both the Rh–Sn and Ni–Sn catalysts, the presence of tin reduces
hydrogen and carbon monoxide chemisorption, isolates the rhodium or nickel
atoms from their neighbors and slightly increases the electron densities on the
rhodium or nickel.[201,202]

   Supported mixed metal catalysts are also prepared by other means such as
the deposition of bimetallic colloids onto a support[207] and the decomposition of
supported bimetallic cluster compounds.[208] The photocatalytic codeposition of
metals onto titania was also attempted with mixed results.[209] With a mixture of
chloroplatinic acid and rhodium chloride, very little rhodium was deposited on the
titania. With aqueous solutions of silver nitrate and rhodium chloride, more
rhodium was deposited but deposition was not complete. In aqueous ammonia,
though, deposition of both silver and rhodium was complete but the titania surface
was covered with small rhodium crystallites and larger silver particles containing
some rhodium. With a mixture of chloroplatinic acid and palladium nitrate both
metals were deposited but, while most of the resulting crystallites were bimetallic,
the composition varied from particle to particle.[209]

   Bimetallic species have also been placed into zeolites. A Pd–Ni/Y catalyst
was prepared by exchange with Pd(NH$_3$)$_4$$^{+2}$ and Ni^{++}. The presence of palladium
enhanced the reducibility of the nickel.[210,211] Impregnation at high pH placed the
nickel ions in the supercages where they were in close proximity to the reduced

palladium also found in the supercages. At low pH the nickel ions were found in the smaller cages and reduction was not facilitated. The ion exchange of a solution containing $Pt(NH_3)_4Cl_2$ and $Rh(NH_3)_5Cl_3$ gave bimetallic alloy particles in the zeolite supercage.[212]

References

1. J. A. Moulijn, P. W. N. M. van Leeuwen and R. A. van Santen, *Stud. Surf. Sci. Catal.*, **79** (Catalysis), 335 (1993).
2. J. Haber, *Pure and Appl. Chem.*, **63**, 1227 (1991).
3. S. P. S. Andrew, *Stud. Surf. Sci. Catal.*, **1** (Prep. Catal.), 429 (1976).
4. E. C. Kruissink, L. E. Alzamora, S. Orr, E. B. M. Doesburg, L. L. Van Reijen, J. H. H. Ross and G. Van Veen, *Stud. Surf. Sci. Catal.*, **3**, (Prep. Catal., II), 143 (1979).
5. D. C. Puxley, I. J. Kitchener, C. Komodromos and N. D. Parkyns, *Stud. Surf. Sci. Catal.*, **16**, (Prep. Catal., III), 237 (1983).
6. L. W. Covert, R. Connor and H. Adkins, *J. Am. Chem. Soc.*, **54**, 1651 (1932).
7. K. Ahmed and P. Mistry, *Coll. Czech. Chem. Commun.*, **57**, 2073 (1992).
8. J. T. Richarsdson, B. Turk, M. Lei, K. Forster and M. V. Twigg, *Appl. Catal. A*, **83**, 87 (1992).
9. J. Zielinski, *Catal. Lett.*, **13**, 389 (1992).
10. M. Astier, A. Bertrand, D. Bianchi, A. Chenard, G. E. E. Gardes, G. Pajonk, M. B. Taghavi, S. J. Teichner and B. L. Villemin, *Stud. Surf. Sci. Catal.*, **1** (Prep. Catal.), 315 (1976).
11. H. D. Gesser and P. C. Goswami, *Chem. Rev.*, **89**, 765 (1989).
12. M. A. Cauqui and J. M. Rodríguez-Izquierdo, *J. Non-Cryst. Solids*, **147–148**, 724 (1992).
13. T. López,, M. Morán, J. Navarrete, L. Herrera and R. Gómez, *J. Non-Cryst. Solids*, **147–148**, 753 (1992).
14. T. López,, A. Romero and R. Gómez, *J. Non-Cryst. Solids*, **127**, 105 (1991).
15. M. Azomoza, T. López,, R. Gómez and R. D. Gonzalez, *Catal. Today*, **15**, 547 (1992).
16. T. López, L. Herrera, J. Mendez-Vivar, P. Bosch, R. Gómez and R. D. Gonzalez, *J. Non-Cryst. Solids*, **147–148**, 773 (1992).
17. K. Balakrishnan and R. D. Gonzalez, *J. Catal.*, **144**, 395 (1993).
18. T. López, J. Mendez-Vivar and R. Juarez, *J. Non-Cryst. Solids*, **147–148**, 778 (1992).
19. H. G. J. Lansink Rotgerink, J. G. Van Ommen and J. R. H. Ross, *Stud. Surf. Sci. Catal.*, **31** (Prep. Catal., 4), 795 (1987).
20. M. A. Cauqui, J. J. Calvino, G. Cifredo, L. Esquivias and J. M. Rodriquez-Izquierdo, *J. Non-Cryst. Solids*, **147–148**, 758 (1992).
21. M. R. Gelsthorpe, B. C. Lippens, J. R. H. Ross and R. M. Sambrook, *Acta Simp. Iberoam.*, 1082 (1984).
22. B. C. Lippens, P. Fransen, J. G. van Ommen, R. Wijngaarden, H. Bosch and J. R. H. Ross, *Solid State Ionics*, **16**, 275 (1985).
23. H. Adkins and R. Connor, *J. Am. Chem. Soc.*, **53**, 1091 (1931).
24. R. Connor, K. Folkers and H. Adkins., *J. Am. Chem. Soc.*, **53**, 2012 (1931).

25. R. Connor, K. Folkers and H. Adkins, *J. Am. Chem. Soc.*, **54**, 1138 (1932).
26. A. Iimura, Y. Inoue and I. Yasumori, *Bull. Chem. Soc., Jpn.*, **56**, 2203 (1983).
27. S. P. Tonner, M. S. Wainwright, D. L. Trimm and N. W. Cant, *Appl. Catal.*, **11**, 93 (1984).
28. J. A. Schreifels, A. Rodero and W. E. Swartz. Jr., *Appl. Spectroscopy*, **33**, 380 (1979).
29. C. L. Bianchi, M. G. Cattania and V. Ragaini, *Surf. Interface Anal.*, **19**, 533 (1992).
30. H. Boerma, *Stud. Surf. Sci. Catal.*, **1** (Prep. Catal.), 105 (1976).
31. L. A. M. Hermans and J. W. Geus., *Stud. Surf. Sci. Catal.*, **3** (Prep. Catal., II), 113 (1979).
32. P. K. de Bokx, W. B. A. Wassenberg and J. W. Geus, *J. Catal.*, **104**, 86 (1987).
33. J. T. Richardson, R. J. Dubus, J. G. Crump, P. Desai, U. Osterwalder and T. S. Cale, *Stud. Surf. Sci. Catal.*, **3** (Prep. Catal., II), 131 (1979).
34. O. Clause, L. Bonneviot, M. Che and H. Dexport, *J. Catal.*, **130**, 21 (1991).
35. J. W. E. Coenen, *Stud. Surf. Sci. Catal.*, **3** (Prep. Catal., II), 89 (1979).
36. P. Turlier, H. Praliaud, P. Moral, G. A. Martin and J. A. Dalmon, *Appl. Catal.*, **19**, 286 (1985).
37. W. Pearlman, *Tetrahedron Lett.*, 1661 (1967).
38. E. Zagli and J. L. Falconer, *Appl. Catal.*, **4**, 135 (1982).
39. P. T. Cardew, R. J. Davey, P. Elliott, A. W. Nienow and J. M. Winterbottom, *Stud. Surf. Sci. Catal.*, **31** (Prep. Catal., IV), 15 (1987).
40. C. J. Song, T. J. Park and S. H. Moon, *Korean J. of Chem. Eng.*, **9**, 159 (1992).
41 E. Nino, A. Lapena, J. Martinez, J. M. Guteirrez, S. Mendioruz, J. L. G. Fierro and J. A. Pajarez, *Stud. Surf. Sci. Catal.*, **16** (Prep. Catal., III), 747 (1983).
42. R. Mozingo, *Org. Synthesis*, **Coll. Vol. 3**, 685 (1955).
43. R. Baltzly, *J. Am. Chem. Soc.*, **74**, 4586 (1952).
44. C. A. Brown and V. K. Ahuja, *J. Org. Chem.*, **38**, 2226 (1973).
45. D. Savoia, C. Tromboni, A. Umani-Ronchi and G. Verardo, *J. Chem. Soc., Chem. Commun.*, 540 (1981).
46. D. Savoia, E. Tagliavini, C. Trombini and A. Umani-Ronchi, *J. Org. Chem.*, **46**, 5340 (1981).
47. D. Savoia, E. Tagliavini, C. Trombini and A. Umani-Ronchi, *J. Org. Chem.*, **46**, 5344 (1981).
48. F. Béguin and A. Messaoudi, *J. Mater. Chem.*,,**2**, 957 (1982).
49. J. Turkevich and G. Kim, *Science*, **169**, 873 (1970).
50. M. Boutonnet, J. Kizling, R. Touroude, G. Maire and P. Stenius, *Catal. Lett.*, **9**, 347 (1991).
51. A. Mills, *J. Chem. Soc., Chem. Commun.*, 367 (1982).
52. W. Hoffmann, M. Graetzel and J. Kiwi, *J. Mol. Catal.*, **43**, 183 (1987).
53. M. Boutonnet, J. Kizling, V. Mintsa-Eya, A. Choplin, R. Touroude, G. Maire and P. Stenius, *J. Catal.*, **103**, 95, (1987).
54. B. Kraeutler and A. J. Bard, *J. Am. Chem. Soc.*, **100**, 4317 (1978).
55. S.-H. Chien, K.-N. Lu and C.-T. Chen, *Bull. Inst. Chem., Acad. Sin.*, **40**, 37 (1993).

56   J. M. Herrmann, J. Disdier, P. Pichat and C. Leclercq, *Stud. Surf. Sci. Catal.,* **31** (Prep. Catal., IV), 285 (1987).
57.   A. Fernández, A. R. González-Elepe, C. Real, A Caballero and G. Munuera, *Langmuir*, **9**, 121 (1993).
58.   S.-Y. Lee and R. Aris, *Catal. Rev.,* **27**, 207 (1985).
59.   M. Komiyama, *Catal. Rev.,* **27**, 341 (1985).
60.   R. L. Moss, in *Experimental Methods in Catalytic Research, Vol II* (R. B. Anderson and P. T. Dawson, Ed.), Academic Press, New York, 1976, p 43.
61.   O. Solcova, D.-C. Uecker, U. Steinike and K. Jiratova, *Appl. Catal.*, **94**, 153 (1993).
62.   G. P. Royer, W.-S. Chow and K. S. Hatton, *J. Mol. Catal.,* **31**, 1 (1985).
63.   R. J. Fenoglio, W. Alvarez, G. M. Nuñez and D. E. Resasco, *Stud. Surf. Sci. Catal.,* **63** (Prep. Catal., V), 77 (1991).
64.   E. A. Hyde, R. Rudham and C. H. Rochester, *J. Chem. Soc., Faraday Trans. I,* **79**, 2403 (1983).
65.   V. Machek, V. Ruzicka, M. Sourková, J. Kunz and L. Janácek, *Collect. Czech. Chem. Commun.,* **48**, 517 (1983).
66.   V. Machek, J. Hanika, K. Sporica,V. Ruzicka, J. Kunz and L. Janacek, *Stud. Surf. Sci. Catal.,* **16** (Prep. Catal., III), 69 (1983).
67.   V. Machek, V. Ruzicka, M. Sourkova, J. Kunz and L. Janacek, *React. Kinet. Catal. Lett.,* **21**, 13 (1982).
68.   J. |Ruzicka, J. Hanika, Z. Selacek and V. Ruzicka, *Coll. Czech. Chem. Commun.,* **52**, 2680 (1987).
69.   L. L. Murrell and D. J. C. Yates, *Stud. Surf. Sci. Catal.,* **3** (Prep. Catal., II), 307 (1979).
70.   P. Lagarde, T. Murata, G. Vlaic, E. Freund, H. Dexpert and J. P. Bournonville, *J. Catal.,* **84**, 333 (1983).
71.   S. D. Jackson, J. Willis, G. D. McLellan, G. Webb, M. B. T. Keegan, R. B. Moyes, S. Simpson, P. B. Wells and R. Whyman, *J. Catal.,* **139**, 191 (1993).
72.   G. H. Via, G. Meitzner, F. W. Lytle and J. H. Sinfelt, *J. Chem. Phys.,* **79**, 1527 (1983).
73.   R. B. Greegor, F. W. Lytle, R. L. Chin and D. M. Hercules, *J. Phys. Chem.,* **85**, 1232 (1981).
74.   Y. S. Shyr and W. R. Ernst, *J. Catal.*, **63**, 425 (1980).
75.   K. Tamaru, *Bull. Chem. Soc., Jpn.,* **24**, 177 (1951).
76.   R. W. Maatman and C. D. Prater, *Ind. Eng. Chem.,* **49**, 253 (1957).
77.   J. Cervello, E. Hermana, J. F. Jimeniz and F. Melo, *Stud. Surf. Sci. Catal.,* **1** (Prep. Catal.), 251 (1976).
78.   V. Machek, J. Hanika, K. Sporica,V. Ruzicka and J. Kunz, *Coll. Czech. Chem. Commun.,* **46**, 3270 (1981).
79.   Z. X. Cheng, C. Louis and M. Che, *Stud. Surf. Sci. Catal.*, **75** (New Frontiers in Catalysis, Pt. B), 1785 (1993).
80.   M. S. Heise and J. A. Schwarz, *Stud. Surf. Sci. Catal.,* **31** (Prep. Catal., IV), 1 (1987).
81.   M. S. Heise and J. A. Schwarz, *J. Colloid Interface Sci.,* **113**, 55 (1986).
82.   M. Komiyama, R. P. Merrill and H. F. Harsberger, *J. Catal.,* **63**, 35 (1980).

83. M. Komiyama and R. P. Merrill, *Bull. Chem. Soc., Jpn.,* **57**, 1169 (1984).
84. M. Kotter and L. Reikert, *Stud. Surf. Sci. Catal.,* **3** (Prep. Catal., II), 51 (1979).
85. J. Barbier, D. Bahloul and P. Marecot, *J. Catal.,* **137**, 377 (1992).
86. M. Montes, Ch. Penneman de Bosscheyde, B. K. Hodnett, F. Delannay, P. Grange and B. Delmon, *Appl. Catal.,* **12**, 309 (1984).
87. M. Montes, J. - B. Soupart, M. de Saedeleer, B. J. Hodnett and B. Delmon, *J. Chem. Soc., Faraday Trans. I,* **80**, 3209 (1984).
88. G. A. El-Shobaky, A. N. Al-Noimi and T. M. H. Saber, *Egypt. J. Chem.,* **31**, 341 (1988).
89. K. Tohji, Y. Udagawa, S. Tanabe and A. Ueno, *J. Am. Chem. Soc.,* **106**, 612 (1984).
90. Y.-J. Huang and J. A. Schwarz, *Appl. Catal.,* **32**, 45 (1987).
91. R. D. Srivastava, J. Onuferko, J. M. Schultz, G. A. Jones, K. N. Ral and R. Athappan, *Ind. Eng. Chem. Fundam.,* **21**, 457 (1982).
92. T. A. Dorling, B. W. J. Lynch and R. L. Moss, *J. Catal.,* **20**, 190 (1971).
93. M. A. Martin Luengo, P. A. Sermon and K. S. W. Sing, *Stud. Surf. Sci. Catal.,* **31** (Prep. Catal., IV), 29 (1987).
94. J. Sarkany and R. D. Gonzalez, *Ind. Eng. Chem. Prod. Res. Dev.,* **22**, 548 (1983).
95. D. G. Blackmond and E. I. Ko, *Appl. Catal.,* **13**, 49 (1984).
96. M. Domingo-García, L. Vincente-Gutiérrez and C. Moreno-Castilla, *React. Kinet. Catal. Lett.,* **43**, 93 (1991).
97. P. G. J. Koopman, A. P. G. Kieboom and H. van Bekkum, *Recl. Trav. Chim. Pays-Bas,* **102**, 429 (1983).
98. C. Wong and R. W. McCabe, *J. Catal.,* **107**, 535 (1987).
99. J. B. A. D. van Zon, D. C. Koningsberger, H. F. J. van't Blik and D. E. Sayers, *J. Chem. Phys.,* **82**, 5742 (1985).
100. Y.-J. Lin, R. J. Fenoglio, D. E. Resacso and G. L. Heller, *Stud. Surf. Sci. Catal.,* **31** (Prep. Catal., IV), 125 (1987).
101. M. D. Mitchell and M. A. Vannice, *Ind. Eng. Chem. Fundam.,* **23**, 88 (1984).
102. P. G. J. Koopman, A. P. G. Kieboom, H. van Bekkum and J. W. E. Coenen, *Carbon,* **17**, 399 (1979).
103. J. F. Roth and T. E. Reichard, *J. Res. Inst. Catalysis, Hokkaido Univ.,* **20**, 85 (1972).
104. P. Harriott, *J. Catal.,* **14**, 43 (1969).
105. R. W. Maatman, *Ind. Eng. Chem.,* **51**, 913 (1959).
106. R. W. Maatman and C. D. Prater, *Ind. Eng. Chem.,* **49**, 253 (1957).
107. P. B. Weisz and R. D. Goodwin, *J. Catal.,* **2**, 397 (1963).
108. L. M. Knijft, P. H. Bolt, R. van Yperen, A. J. van Dillen and J. W. Geus, *Stud. Surf. Sci. Catal.,* **63** (Prep. Catal., V), 165 (1991).
109. J. P. Brunelle, *Stud. Surf. Sci. Catal.,* **3** (Prep. Catal., II), 211 (1979).
110. C. Contescu, C. Sivaraj and J. A. Schwarz, *Appl. Catal.,* **74**, 95 (1991).
111. A. L. Bonivardi and M. A. Baltanas, *J. Catal.,* **125**, 243 (1990).
112. T. Fujitani and E. Echigoya, *Chem. Lett.,* 827 (1991).
113. M. A. Gutiérrez-Ortiz, J. A. González-Marcos, M. P. González-Marcos and J. R. González-Velasco, *Ind. Eng. Chem. Res.,* **32**, 1035 (1993).
114. L. Vordonis, A. Akratopulu, P. G. Koutsoukos and A. Lycorghiotis, *Stud. Surf. Sci. Catal.,* **31** (Prep. Catal., IV), 309 (1987).

115. J. C. Lee, D. L. Trimm, M. A. Kohler, M. S. Wainwright and N. W. Cant, *Catal. Today,* **2**, 643 (1988).
116. D. Richard and P. Gallezot, *Stud. Surf. Sci. Catal.,* **31** (Prep. Catal., IV), 71 (1987).
117. W. Zou and R. D. Gonzalez, *J. Catal.,* **133**, 202 (1992).
118. I. D. Gay, *J. Catal.,* **80**, 231 (1983).
119. A. L. Bonivardi and M. A. Baltanás, *Thermochim. Acta,* **191**, 63 (1991).
120. W. Zou and R. D. Gonzalez, *Catal. Lett.,* **12**, 73 (1992).
121. J. H. Sepúlveda and N. S. Figoli, *Appl. Surf. Sci.,* **68**, 257 (1993).
122. W. Zou and R. D. Gonzalez, *Catal. Today,* **15**, 443 (1992).
123. G. W. Graham, W. H. Weber and K. Otto, *Appl. Surf. Sci.,* **59**, 87 (1992).
124. R. W. Joyner, *J. Chem. Soc., Faraday I,* **76**, 357 (1980).
125. J. P. Espinos, A. R. González-Elipe, A. Caballero, J. Garcia and G. Munuera, *J. Catal.,* **136**, 415 (1992).
126. P. A. Sermon and J. Sivalingam, *Colloids Surf.,* **63**, 59 (1992).
127. T. A. Dorling, M. J. Eastlake and R. L. Moss, *J. Catal.,* **14**, 23 (1969).
128. H. A. Benesi, R. M. Curtis and H. P. Studer, *J. Catal.,* **10**, 328 (1968).
129. R. Burch and A. R. Flambard, *Stud. Surf. Sci. Catal.,* **16** (Prep. Catal., III), 311 (1983).
130. T. Arunarkavalli, G. U. Kulkarni, G. Sankar and C. N. R. Rao, *Catal. Lett.,* **17**, 29 (1993).
131. T. J. Gardner, C. H. F. Peden and A. K. Datye, *Catal. Lett.,* **15**, 111 (1992).
132. M. J. Holgado, A. C. Iñigo and V. Rives, *React. Kinet. Catal. Lett.,* **46**, 409 (1992).
133. M. D. Ward, T. V. Harris and J. Schwartz, *J. Chem. Soc., Chem Commun.,* 357 (1980).
134. M. D. Ward and J. Schwartz, *J. Am. Chem. Soc.,* **103**, 5253 (1981).
135. H. C. Foley, S. J. DeCanio, K. D. Tau, K. J. Chao, J. H. Onuferko, C. Dybowski and B. C. Gates, *J. Am. Chem. Soc.,* **105**, 3074 (1983).
136. G. Cocco, S. Enzo, L. Schiffini and G. Carturan, *J. Mol. Catal.,* **11**, 161 (1981).
137. G. Gubitosa, A. Berton, M. Camia and N. Pernicone, *Stud. Surf. Sci. Catal.,* **16** (Prep. Catal., III), 431 (1983).
138. J. A. R. van Veen, G. Jonkers and W. H. Hesselink, *J. Chem. Soc., Faraday Trans. I,* **85**, 389 (1989).
139. J. P. Boitaux, J. Cosyns and S. Vasudevan, *Stud. Surf. Sci. Catal.,* **16** (Prep. Catal., III), 123 (1983).
140. E. Lesage-Rosenberg, G. Vlaic, H. Dexpert, P. Lagarde and E. Freund, *Appl. Catal.,* **22**, 211 (1986).
141. J. A. Navio, M. Macias, F. J. Marchena and C. Real, *Stud. Surf. Sci. Catal.,* **72** (New Dev. Sel. Oxid. Heterog. Catal.), 423 (1992).
142. J. C. Kenvin, M. G. White and M. B. Mitchell, *Langmuir,* **7**, 1198 (1991).
143. C. Sudhakar and M. A. Vannice, *Appl. Catal,* **14**, 47 (1985).
144. A.S. Lisitsyn, S. V. Gurevich, A. L. Chuvilin, A. I. Boronin, V. I. Bukhtiyarov and V. A. Likholobov, *React. Kinet. Catal. Lett.,* **38**, 109 (1989).
145. Z. Karpinski, T.-K. Chuang, H. Katsuzawa, J. B. Butt, R. L. Burwell, Jr. and J. B. Cohen, *J. Catal.,* **99**, 184 (1986).

146. W. M. H. Sachtler, *Accounts Chem. Res.*, **26**, 383 (1993).
147. W. M. H. Sachtler and Z. Zhang, *Adv. Catal.*, **39**, 129 (1993).
148. I. Yamaguchi, T. Joh and S. Takahashi, *J. Chem. Soc., Chem. Commun.*, 1412 (1986).
149. W. M. H. Sachtler, *Catal. Today*, **15**, 419 (1992).
150. R. S. Miner, Jr., K. G. Ione, S. Namba and J. Turkevich, *J. Phys. Chem.*, **82**, 214 (1978).
151. M. A. Keane, *Zeolites*, **13**, 14 (1993).
152. M. A. Keane, *Zeolites*, **13,** 22 (1993).
153. B. Couglan and M. A. Keane, *Zeolites*, **11**, 2 (1991).
154. G.-J. Li, T. Fujimoto, A. Fukuoka and M. Ichikawa, *Catal. Lett.*, **12**, 171 (1992.
155. T. J. Lee and B. C. Gates, *J. Mol. Catal.*, **71**, 335 (1992).
156. N. Takahashi, K. Tanaka and I. Toyoshima, *J. Chem. Soc., Chem. Commun.*, 812 (1986).
157. I. R. Leith, *J. Chem. Soc., Chem. Commun.*, 93 (1983).
158. Z. Zhang, F. A. P. Cavalcanti and W. M. H. Sachtler, *Catal. Lett.*, **12**, 157 (1992).
159. Z. Zhang, H. Chen and W. M. H. Sachtler, *J. Chem. Soc., Faraday Trans.*, **87**, 1413 (1991).
160. W. Vogel, W. M. H. Sachtler and Z. Zhang, *Ber. Bunsen-Ges. Phys. Chem.,* **97**, 280 (1993).
161. L. Xu, Z. Zhang and W. M. H. Sachtler, *J. Chem. Soc., Faraday Trans.*, **88**, 2291 (1992).
162. G. Larsen and G. L. Haller, *Catal. Today*, 15, 431 (1992).
163. I. Manninger, Z. Zhan, X. L. Xu and Z. Paal, *J. Mol. Catal.*, **66**, 223 (1991).
164. W. J. Reagan, A. W. chester and G. T. Kerr, *J. Catal.*, **69**, 89 (1981).
165. P. Gallezot, *Catal. Revs.* **20**, 121 (1979).
166. R. A. DallaBetta and M. Boudart, *Proccedings of the 5th Intenational congress on Catalysis*, (J. W. Hightower, Ed.) **Vol. 2**, North Holland, Amsterdam, 1973, p. 1329.
167. M. B. T. Keegan, A. J. Dent, A. B. Blake, L. Conyers, R. B. Moyes and P. B. Wells., *Catal. Today*, **9**, 183 (1991).
168. E. S. Shpiro, R. W. Joyner, K. M. Minachev and P. D. A., Pudney, *J. Catal.*, **127**, 366 (1991).
169. G. N. Folefoc and J. Dwyer, *J. Catal.*, **136**, 43 (1992).
170. J. A. Rossin and M. E. Davis, *J. Chem. Soc., Chem. Commun.*, 234 (1986).
171. R. M. Dessau, *J. Catal.*, **77**, 304 (1982).
172. R. M. Dessau, *J. Catal.*, **89**, 520 (1984).
173. T.-N. Huang and J. Schwartz, *J. Am. Chem. Soc.*, **104**, 5244 (1982).
174. J. M. Campello, A. Garcia, D. Luna and J. M. Marinas, *Bull. Soc. Chim. Belg.*, **92**, 851 (1983).
175. J. M. Campello, A. Garcia, D. Luna and J. M. Marinas, *Appl. Catal.*, **10**, 1 (1984).
176. J. B. Harrison, V. E. Berkheiser and G. W. Erdos, *J. Catal.*, **112**, 126 (1988).
177. M. Crocker, R. H. M. Herold, J. G. Buglass and P. Companje, *J. Catal.*, **141**, 700 (1993).
178. M. Patel, *X-Ray Spectrom.*, **11**, 64 (1982).

179. P. Ravindranathan, P. B. Malla, S. Komarneni and R. Roy, *Catal. Letters*, **6**, 401 (1990).
180. V. Luca, R. Kukkadapu and L. Kevan, *J. Chem. Soc., Faraday Trans.*, **87**, 3083 (1991).
181. M. B. Logan, R. F. Howe and R. P. Cooney, *J. Mol. Catal.*, **74**, 285 (1992).
182. L. Guczi, G. Lu and Z. Zsoldos, *Catal. Today*, **17**, 459 (1993).
183. R. D. Gonzalez and H. Miura, *Catal. Revs.*, **36**, 145 (1994).
184. P. C. M. van Stiphout and J. W. Geus, *Appl. Catal.*, **25**, 19 (1986).
185. R. Gómez, V. Bertin, M. A. Ramirez, T. Zamudio, P. Bosch, I. Schifter and T. López, *J. Non-Cryst. Solids*, **147–148**, 748 (1992).
186. G. Blanchard, H. Charcosset, H. Dexpert, E. Freund, C. Leclercq and G. Martino, *J. Catal.*, **70**, 168 (1981).
187. J. C. Summers and L. L. Hegedus, *J. Catal.*, **51**, 185 (1978).
188. F. Melo, J. Cervello and E. Hermana, *Chem. Eng. Sci.*, **35**, 2175 (1980).
189. H.-C. Chen, G. C. Gillies and R. B. Anderson, *J. Catal.*, **62**, 367 (1980).
190. G. del Angel, B. Coq and F. Figueras, *J. Catal.*, **95**, 167 (1985).
191. S. Alerasool and R. D. Gonzalez, *J. Catal.*, **124**, 204 (1990).
192. Y. L. Lam and M. Boudart, *J. Catal.*, **50**, 530 (1977).
193. J. C. Ménézo, L. C. Hoang, C. Montassier and J. Barbier, *React. Kinet. Catal. Lett.*, **46**, 1 (1992).
194. J. M. Dumas, C. Geron, H. Hadrane, P. Marecot and J. Barbier, *J. Mol. Catal.*, **77**, 87 (1992).
195. V. M. Deshpande, W. R. Patterson and C. S. Narasimhan, *J. Catal.*, **121**, 165 (1990).
196. A. S. Bailey, V. G. Kendall, P. B. Lumb, J. S. Smith and Cl H. Walker, *J. Chem. Soc.*, 3027 (1957).
197. H. Lindlar, *Helv. Chim. Acta*, **35**, 446 (1952).
198. H. Lindlar and R. Dubuis, *Org. Synthesis*, **46**, 89 (1966)
199. T. Mallát, S. Szabó and J. Petró, *Appl. Catal.*, **29**, 117 (1987).
200. H. R. Adúriz, C. E. Gígola, A. M. Sica, M. A. Volpe and R. Touroude, *Catal. Today*, **15**, 459 (1992).
201. J. P. Candy, O. A. Ferretti, G. Mabilon, J. P. Bournonville, A. El Mansour, J. M. Basset and G. Martino, *J. Catal.*, **112**, 210 (1988).
202. M. Agnelli, J. P. Candy, J. M. Basset, J. P. Bournonville and O. A. Ferretti, *J. Catal.*, **121**, 236 (1990).
203. B. Coq, A. Bittar, R. Dutartre and F. Figueras, *J. Catal.*, **128**, 275 (1991).
204. O. A. Ferretti, L. C. Bettega de Pauli, J. P. Candy, G. Mabillon and J. P. Bournonville, *Stud. Surf. Sci. Catal.*, **31** (Prep. Catal., IV), 713 (1987).
205. J. Margitfalvi, E. Tálas, M. Hegedus and S. Gõbölös, *Proc. VIth Int. Symp. Heterogeneous Catalysis*, Sofia, Part 2, 346 (1987).
206. M. Agnelli, P. Louessard, A. El Mansour, J. P. Candy, J. P. Bournonville and J. M. Basset, *Catal. Today*, **6**, 63 (1989).
207. J. B. Michel and J. T. Schwartz, *Stud. Surf. Sci. Catal.*, **31** (Prep. Catal., IV), 669 (1987).
208. R. Lamber, N. I. Jaeger, A. Trunschke and H. Miessner, *Catal. Lett.*, **11**, 1 (1991).
209. J. M. Herrmann, J. Disdier, P. Pichat, A. Fernandez, A. Gonzalez-Elipe, G. Munuera and C. Leclercq, *J. Catal.*, **132**, 490 (1991).
210. J. S. Feeley and W. M. H. Sachtler, *Zeolites*, **10**, 738 (1990).

211. J. S. Feeley and W. M. H. Sachtler, *J. Catal.*, **131**, 573 (1991).
212. M. S. Tzou, K. Asakura, Y. Yamazaki and H. Kuroda, *Catal. Lett.*, **11**, 33 (1991).

SECTION THREE

Reactions

Hydrogenation I: Selectivity

The most common use of heterogeneous catalysts in organic synthesis is for the hydrogenation of functional groups. The number of books and reviews published in this area underscores the synthetic importance of these reactions.[1–11] In addition to these texts the proceedings of several conferences on catalytic hydrogenation and related pressure reactions[12–15] have also been published as have been those of a number of conferences on catalysis in organic synthesis,[16–18] catalysis in organic reactions[19–26] and heterogeneous catalysis and fine chemicals.[27–29] Most of the papers in these later series are concerned with catalytic hydrogenation.

While the hydrogenation of molecules containing a single functional group is a common synthetic transformation, selectivity in the hydrogenation of more complex substrates has become more important. These selective reactions can be classified as being chemoselective, regioselective or stereoselective. Selectivity is defined as the number of moles of the desired product divided by the number of moles of reactant consumed or the percent of desired product divided by the percent conversion (Eqn. 14.1). It is important that the degree of conversion be stated along with the selectivity since a 90% selectivity at 5% conversion will not necessarily convert to a 90% selectivity at 90% conversion.

$$\text{Selectivity} = \frac{\text{moles of desired product}}{\text{moles of reagent consumed}}$$

(14.1)

$$\% \text{ Selectivity} = \frac{\% \text{ desired product}}{\% \text{ conversion}}$$

Chemoselectivity

Chemoselectivity refers to the selective hydrogenation of one functional group in the presence of other groups that are potentially able to undergo hydrogenation. The degree of difficulty in obtaining such selectivity depends largely on the functional groups involved. Table 14.1 lists the common organic functional groups in an apparent order of decreasing ease of hydrogenation along with

315

Table 14.1

Hydrogenation of common organic functional groups.

Group	Product	Catalyst	Reaction Conditions
—C≡C—	—C=C— with H, H	Pd	Room temp, 1 atm, low catalyst ratio, deactivated catalyst
C=C—C=C	C—C=C—C with H, H	Pd	Room temp, 1 atm
C=C—C=C	C=C—C—C with H, H	Pd	Room temp, 1 atm Low catalyst ratio, deactivated catalyst
—NO$_2$	—NH$_2$	Pt, Pd, Rh	Room temp, 1 atm
C=C	CH-CH	Pd, Ni	Room temp, 1 atm
—C≡N	—CH$_2$NH$_2$	Raney Ni Raney Co	Room temp, 1–4 atm, NH$_3$ Room temp, 1–4 atm
C=N—	CH-NH—	Pt, Pd	Room temp, 1–4 atm
φ C=O	φ CH$_2$	Pd	Room temp, 1–4 atm
C=O	CH–OH	Pt, Rh Ru	Room. temp, 2–4 atm Room temp, 1–3 atm, H$_2$O
Heterocyclic and Carbocyclic Aromatics	X = C, N, O	Rh Raney Ni Ru	Room temp, 2–4 atm 100–120°C, 100 atm 150°C, 100 atm
—CO$_2$H (R)	—CH$_2$OH	Ru, CuCrO	High temperature and pressure
—CO$_2$NR$_2$	—CH$_2$NR$_2$	Ru, CuCrO	High temperature and pressure

typical laboratory reaction conditions. This order can frequently be modified by changes in the reaction conditions and/or the steric environment of the functional groups in the substrate molecule.

It is usually easy to affect the selective hydrogenation of a functional group listed near the top of Table 14.1 in the presence of one from the middle or near the bottom of the list. For example, the hydrogenation of a double bond in the presence of a ketone or ester is reasonably straightforward (Eqns. 14.2–3). Selectivity is more difficult to attain, however, when the two functional groups are close to each other in reactivity such as triple bonds and nitro groups or when the desired reaction involves the reduction of a group in preference to one that is listed above it in Table 14.1.

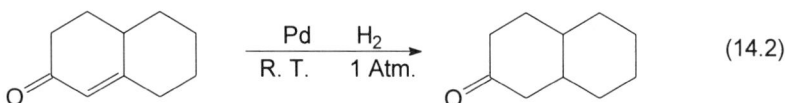

$$\text{(14.2)}$$

$$\text{(14.3)}$$

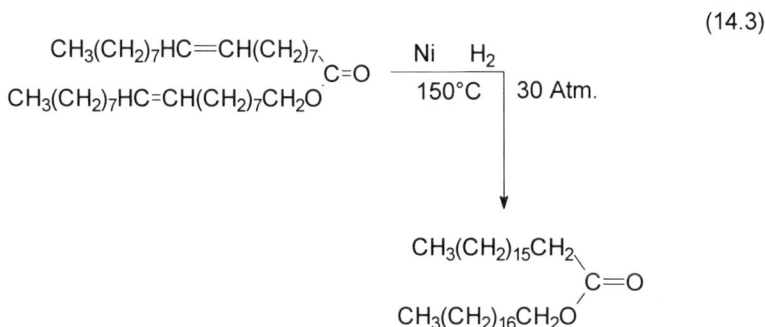

In this later instance, if the two groups have relatively similar activities, selectivity can sometimes be achieved by manipulating the reaction conditions. As mentioned in the discussion on reaction selectivity given in Chapter 5, the use of less active catalysts can frequently result in an increase in selectivity. Other modifications in reaction conditions are usually necessary to reverse the reactivities listed in Table 14.1. Normally, the hydrogenation of an α,β-unsaturated aldehyde such as **1** over a platinum catalyst results in the formation of either the saturated aldehyde, **2**, or, on further reduction, the saturated alcohol, **3**. Hydrogenation to the allyl alcohol, **4**, however, has been accomplished by modifying a platinum catalyst with Fe^{++} ions to facilitate carbonyl group hydrogenation and Zn^{++} ions to inhibit the saturation of the carbon-carbon double bond (Eqn. 14.4).[30]

Even though esters are relatively difficult to hydrogenate it is possible to selectively hydrogenate an ester group in the presence of a double bond using a

(14.4)

(14.5)

$$CH_3(CH_2)_7HC=CH(CH_2)_7$$
$$C=O \quad \xrightarrow[\text{Zn-Al-CrO}]{H_2}$$
$$CH_3(CH_2)_7HC=CH(CH_2)_7CH_2O$$

320°C | 320 atm

97% $CH_3(CH_2)_7CH=CH(CH_2)_7CH_2OH$

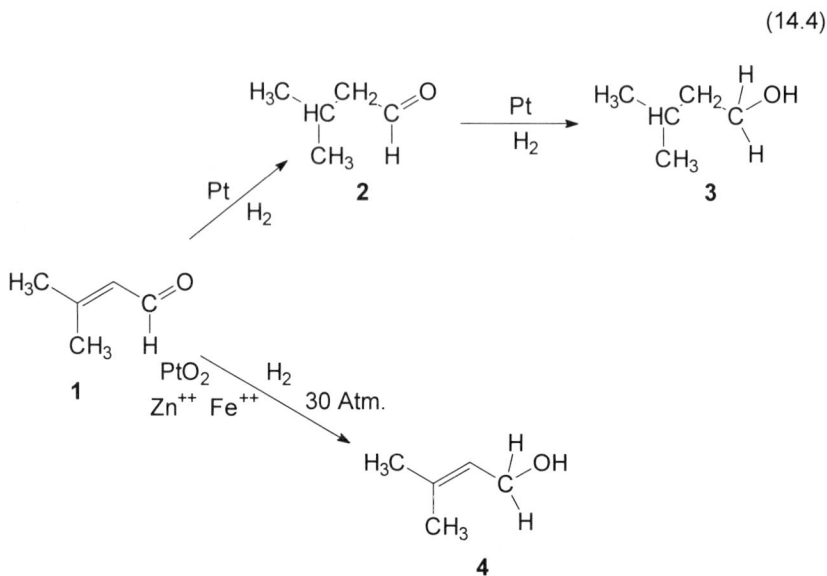

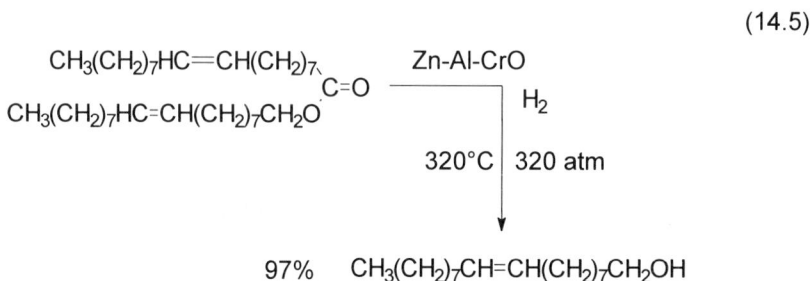

zinc chromate catalyst (Eqn. 14.5).[31] Here, advantage is taken of the fact that this catalyst is particularly ineffective for promoting the saturation of alkenes.

Regioselectivity

Regioselectivity is defined as the selective hydrogenation of one functional group in the presence of another, identical group in the same molecule as exemplified by the selective saturation of the monosubstituted double bond in 4-vinylcyclohexene (5) (Eqn. 14.6).[32] This is an example of the Type II selectivity defined in Chapter 5. Because both functional groups are the same, saturation of either one will have the same kinetic order so organic substrate diffusion is not an important factor in regioselective hydrogenations. It is important, however, to use a low hydrogen availability and a less active catalyst to minimize oversaturation. Another example of regioselectivity is the hydrogenation of the 4-nitro group in

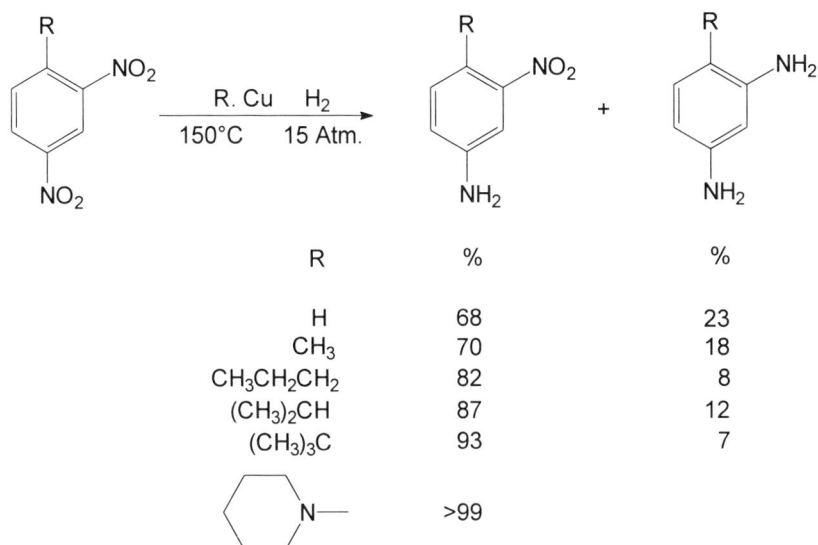

R	%	%
H	68	23
CH$_3$	70	18
CH$_3$CH$_2$CH$_2$	82	8
(CH$_3$)$_2$CH	87	12
(CH$_3$)$_3$C	93	7
piperidyl N—	>99	

Fig. 14.1. Steric effects in the selective hydrogenation of 1-substituted-2,4-dinitrobenzenes.

1-substituted-2,4-dinitro benzenes shown in Fig. 14.1.[33]   As the size of the substituent increases, reaction selectivity also increases reaching almost quantitative yields with the piperidyl dinitrobenzene. In this and the previous example the more sterically accessible group was preferentially hydrogenated but steric factors are not always the only ones involved. Aromatic ketones are readily hydrogenated and hydrogenolyzed to the methylene over palladium catalysts even though these catalysts are generally inert for the hydrogenation of aliphatic ketones. Thus, it is possible to remove an aryl carbonyl group selectively in the presence of an aliphatic ketone or aldehyde by hydrogenation over a palladium catalyst (Eqn. 14.7).[34]

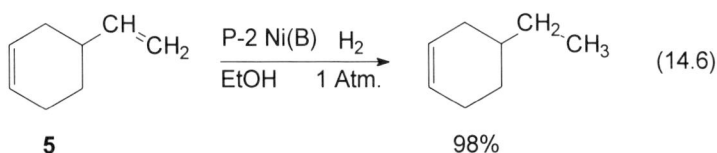

$$(14.6)$$

(14.7)

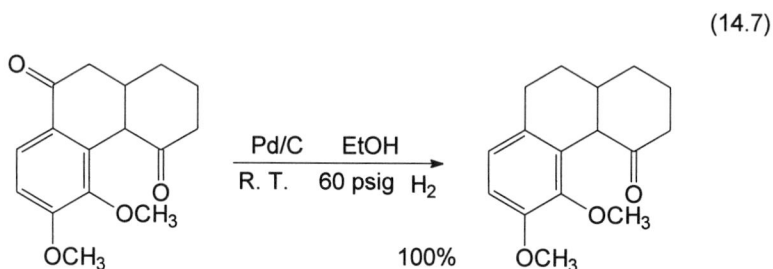

Pd/C   EtOH
R. T.   60 psig H$_2$

100%

Stereoselectivity

Stereoselectivity, the production of one stereoisomer of the product in preference to another, is one of the more important applications of a hydrogenation reaction.[11]   The predominant production of *cis* 1,4-disubstituted cyclohexanes (Eqn. 14.8)[35] on hydrogenation of 1,4-disubstituted cyclohexenes and the near exclusive formation of *cis* alkenes on hydrogenation of disubstituted acetylenes (Eqn. 14.9)[36] are examples of such selectivity. Since the hydrogen is added to the substrate from active sites on the surface of the catalyst, product stereochemistry is determined by the manner in which the substrate is adsorbed on the catalyst. The changes in product stereochemistry observed on hydrogenation of disubstituted cyclohexenes have been explained using the classic Horiuti-Polanyi mechanism[37] depicted in Fig. 14.2. As an example consider the hydrogenation of 4-tert-butyl methylenecyclohexane (**6**).[38] As shown in Fig. 14.3 adsorption of the π cloud of the double bond from direction A leads to the formation of the *trans* product, **7**, with the two hydrogens added from the active site of the catalyst. Adsorption and hydrogen transfer from side B gives the *cis* product, **8**.   The direction of adsorption depends on the relative steric hindrance on each side of the π cloud.

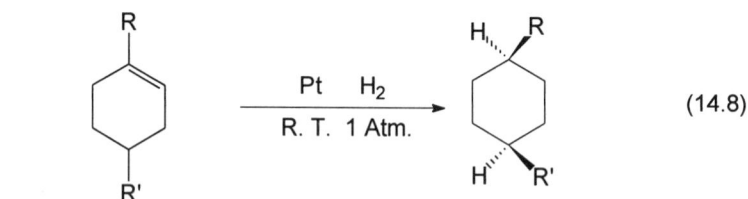

Pt   H$_2$
R. T. 1 Atm.

(14.8)

$R-C\equiv C-R'$

Lindlar's Cat.   H$_2$
R. T. 1 Atm.

(14.9)

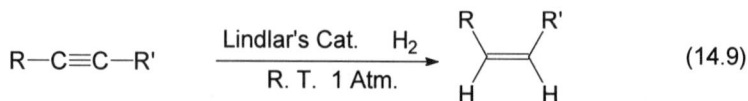

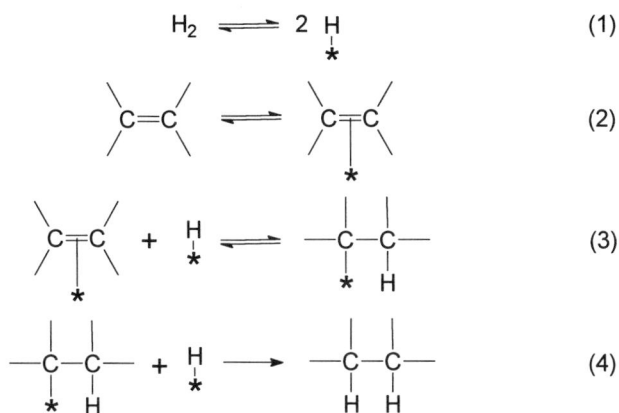

$$H_2 \rightleftharpoons 2 \overset{H}{\underset{*}{|}} \tag{1}$$

$$\overset{\diagdown}{\diagup}C=C\overset{\diagup}{\diagdown} \rightleftharpoons \overset{\diagdown}{\diagup}C=C\overset{\diagup}{\diagdown} \tag{2}$$

$$\overset{\diagdown}{\diagup}C=C\overset{\diagup}{\diagdown} + \overset{H}{\underset{*}{|}} \rightleftharpoons -\overset{|}{\underset{*}{C}}-\overset{|}{\underset{H}{C}}- \tag{3}$$

$$-\overset{|}{\underset{*}{C}}-\overset{|}{\underset{H}{C}}- + \overset{H}{\underset{*}{|}} \longrightarrow -\overset{|}{\underset{H}{C}}-\overset{|}{\underset{H}{C}}- \tag{4}$$

Fig. 14.2. Horiuti–Polanyi[37] mechanism for alkene hydrogenation.

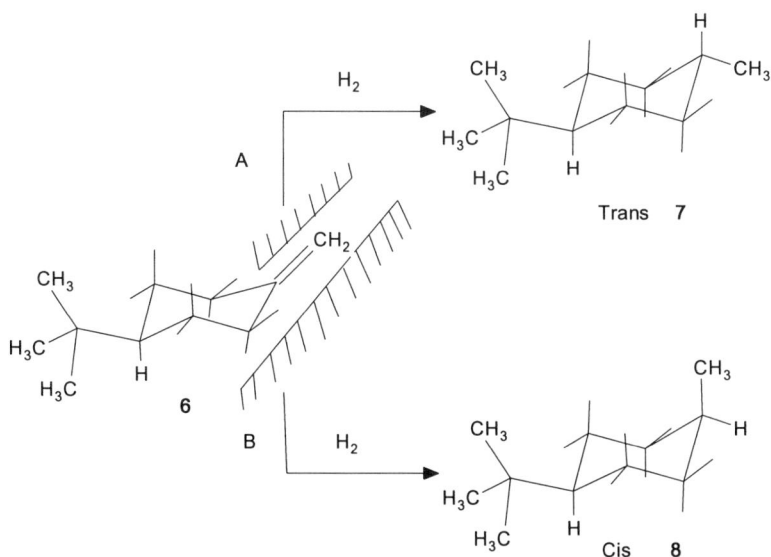

Fig. 14.3. Modes of 4-tert butyl-methylenecyclohexane (**6**) adsorption leading to *trans* (**7**) and *cis* (**8**) product formation.

It is sometimes possible to estimate the extent of hindrance by an examination of a molecular model of the substrate. While the sites that promote hydrogenation reactions are the corner atoms or adatoms on the catalyst surface, models of such surface species are not commonly available. The classic procedure for determining the relative steric hindrance to adsorption has been to place a model of the substrate on a flat surface with the π cloud of the alkene perpendicular to the surface as depicted in Fig. 14.3. In this case adsorption from side B is clearly favored. A similar conclusion can be drawn on examining the adsorption of the π cloud on a surface corner atom as depicted in Fig. 14.4. Here adsorption from the B direction is favored because of the interference of the 3,5-diaxial hydrogens to adsorption from side A, but the difference between the two modes of adsorption does not appear to be as great as that assumed from consideration of Fig. 14.3.

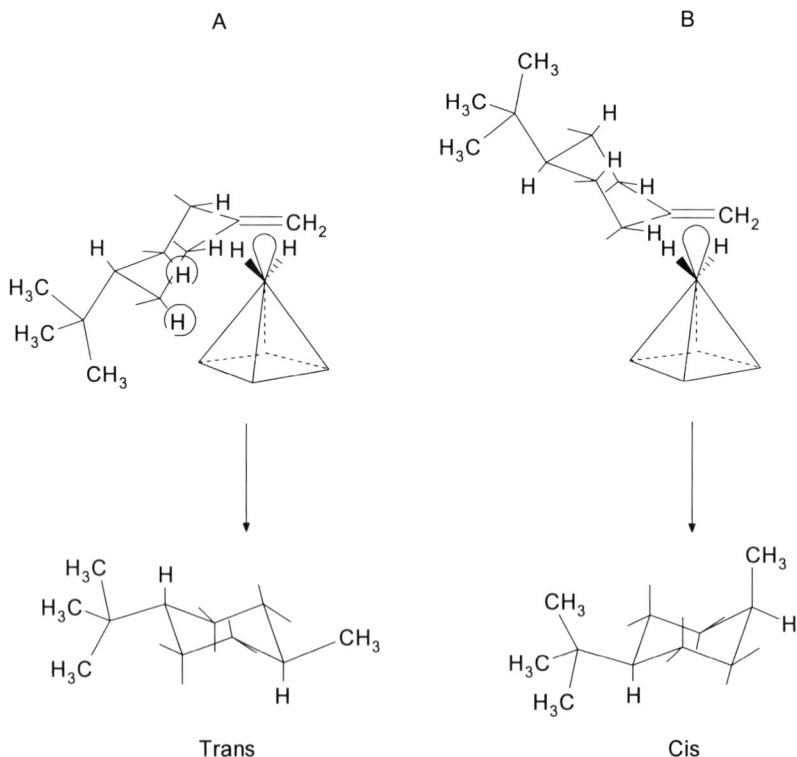

Fig.14.4.   Modes of adsorption of 4-tert butyl-methylenecyclohexane (6) on corner or adatom active sites.

According to the Horiuti-Polanyi mechanism[37] for alkene hydrogenation, in order for alkene adsorption (step 2) to be product determining, the addition of hydrogen (steps 3 and 4) should be essentially non-reversible so step 3 would become the rate determining step of the reaction. This condition is met when there is a high hydrogen availability to the catalyst as when the reaction is run under a high pressure of hydrogen and/or under sufficient agitation to ensure that hydrogen transport to the catalyst is not a determining factor. Under low hydrogen availability conditions, however, step 4 is rate determining and the product composition is regulated by the relative stability of the various half-hydrogenated species or surface metalalkyls that can be formed.

Consider, for instance, the effect of hydrogen pressure on the *cis/trans* product ratio observed on hydrogenation of 4-tert-butyl methylenecyclohexane (**6**) over a platinum catalyst which is depicted in Fig. 14.5.[38] As shown by Fig. 14.4, adsorption from side B should be favored to some extent so *cis* product formation should be somewhat preferred as is observed with the formation of about 65% of this material when the reaction was run under 30 atm of hydrogen. With lower hydrogen pressures, mass transport becomes important and the product ratio is determined primarily by the relative stabilities of the half-hydrogenated species, **9** and **10**, shown in Fig. 14.6. Since the metalalkyl, **10**, in which the adsorption occurs via an equatorial bond is the more stable,[35] increasing amounts of *cis* product are formed as the pressure decreases.

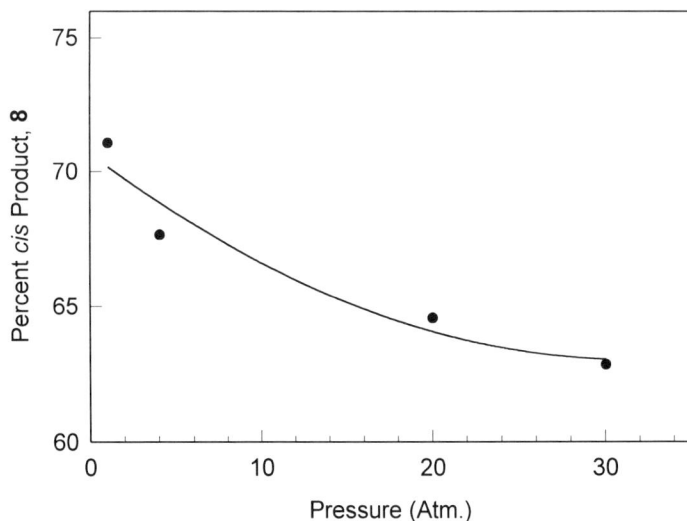

Fig. 14.5. Effect of pressure on product composition in the hydrogenation of 4-tert butyl-methylenecyclohexane (**6**). (Redrawn using data from Ref. 38.)

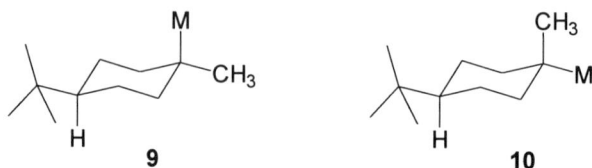

Fig. 14.6. Half-hydrogenated states involved in the hydrogenation of 4-
tert butyl-methylenecyclohexane (**6**).

The preferred direction of adsorption can also be influenced by the
presence of a non-reactive functional group on the molecule. The anchoring of
the substrate to the catalyst surface by the adsorption of such a non-reactive
functional group has been termed a haptophilic effect.[39–41] This factor can have a
significant effect in determining the stereoselectivity of the hydrogenation of
appropriately substituted substrates. Most reactions of this type involve the
hydrogenation of cyclic olefinic alcohols over nickel catalysts. In these reactions,
a strong interaction between a hydroxy group and the catalyst surface fixes the
alkene adsorption *cis* to the alcohol group giving the *trans* products as shown in
Eqns. 14.10–11.[42] This effect is most pronounced with axial or pseudoaxial
allylic and homoallylic secondary alcohols. The steric environment around a
tertiary alcohol, even one that is axial, can minimize its interaction with the
catalyst. Equatorial alcohols apparently do not easily adsorb on the catalyst
surface or if they do, the adsorption has little or no effect on product
stereochemistry. Raney nickel is more effective in these reactions than are
supported nickel catalysts.[42]

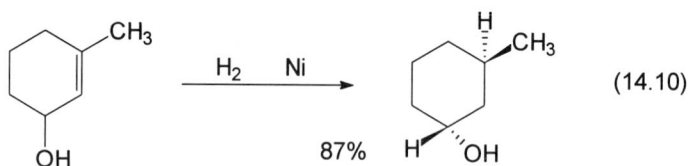

$$(14.10)$$

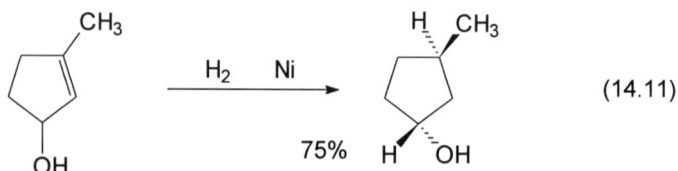

$$(14.11)$$

(14.12)

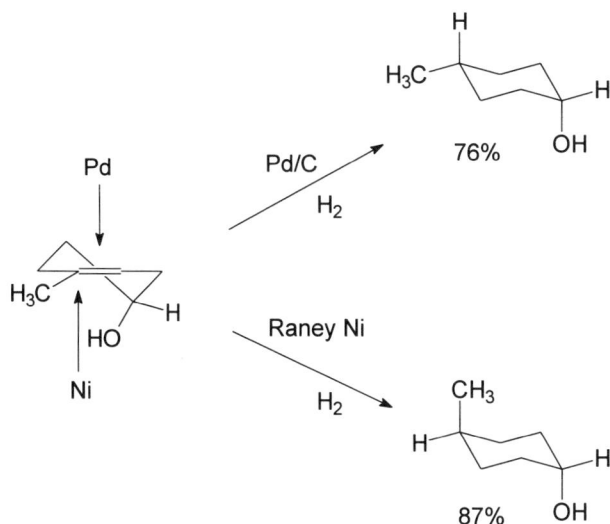

76%

Pd

Pd/C

H₂

H₃C

HO   H

Raney Ni

Ni

H₂

87%

(14.13)

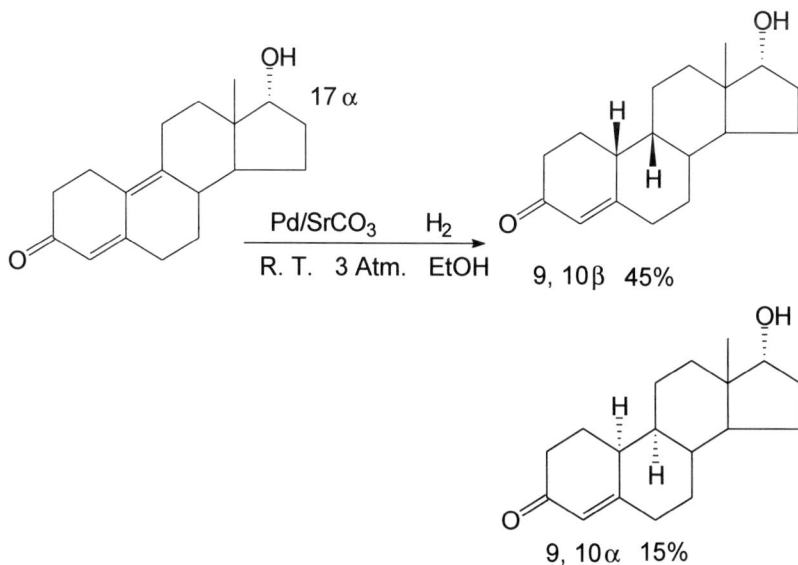

OH

17 α

Pd/SrCO₃   H₂

R. T.   3 Atm.   EtOH

9, 10β   45%

OH

9, 10α   15%

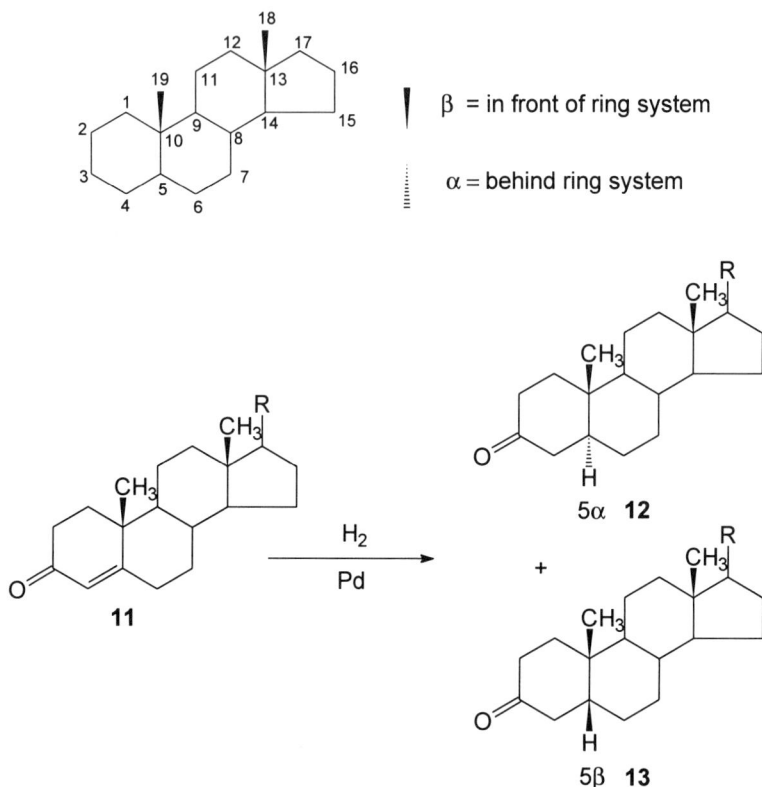

Fig. 14.7. Steroid numbering and stereochemical conventions.
Hydrogenation of Δ4-3-ketosteroids.

A hydroxy group haptophilic effect can also be operational with a
palladium catalyst but the reaction gives the opposite stereochemistry than is
observed with nickel catalysts.[42] This "hydroxy group effect"[2] has been observed
with many simple cyclic olefins as well as several steroidal systems.   The
difference between nickel and palladium catalysts is illustrated in Eqn. 14.12.[42]
This effect appears to be more pronounced in the hydrogenation of steroid double
bonds.   The hydrogenation of Δ4-3-ketosteroids, 11, in neutral media over
palladium gives a mixture of the 5α, 12, and 5β, 13, products as illustrated in Fig.
14.7.[2] When a 17α hydroxy group is present, however, hydrogenation over
palladium takes place from the top or β side of the ring system (Eqn. 14.13).[2]
Palladium catalyzed hydrogenation of a 17α-hydroxy-Δ4-3-ketosteroid, 14, gave
primarily the 5β product, 15.(Eqn. 14.14).[43] The presence of a 17β hydroxy

(14.14)

OH

17 α

CH3

15

5 β 70%

14

Pd/CaCO3   H2
R. T. 1 Atm.   THF

OH

CH3

5 α 30%

(14.15)

OH

17 β

Pd/SrCO3   H2
R. T. 1Atm.
Benzene

OH

H

H

9, 10 α
60 - 80%

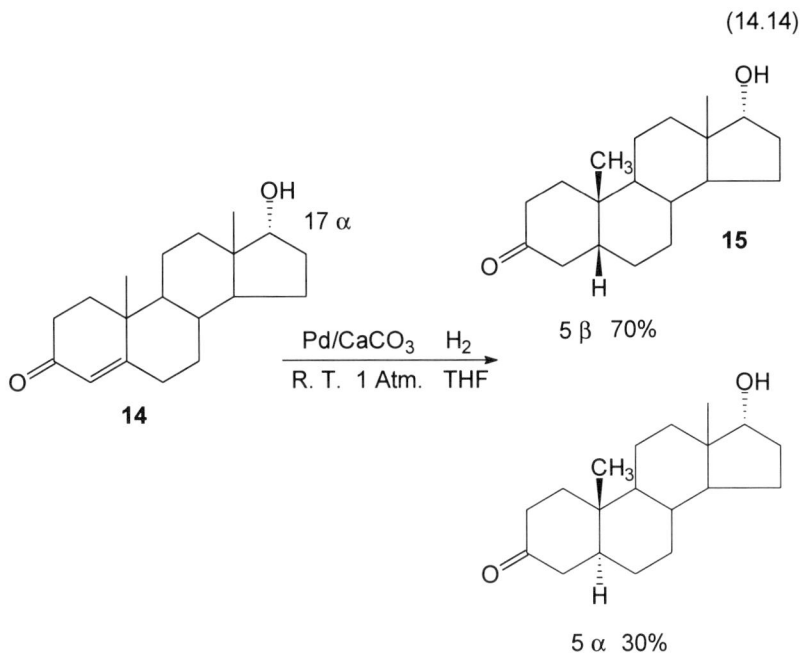

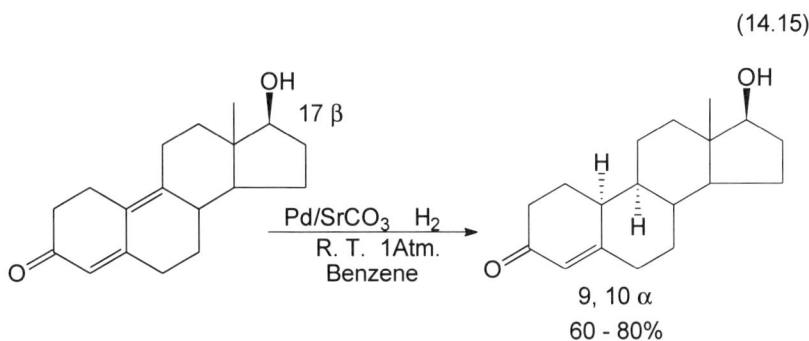

group promotes hydrogenation from the α side of the steroid (Eqns. 14.15–16).[2,43] Axial alcohols have a stronger directing influence than pseudoaxial alcohols with equatorial alcohols exerting only a minimal influence.

(14.16)

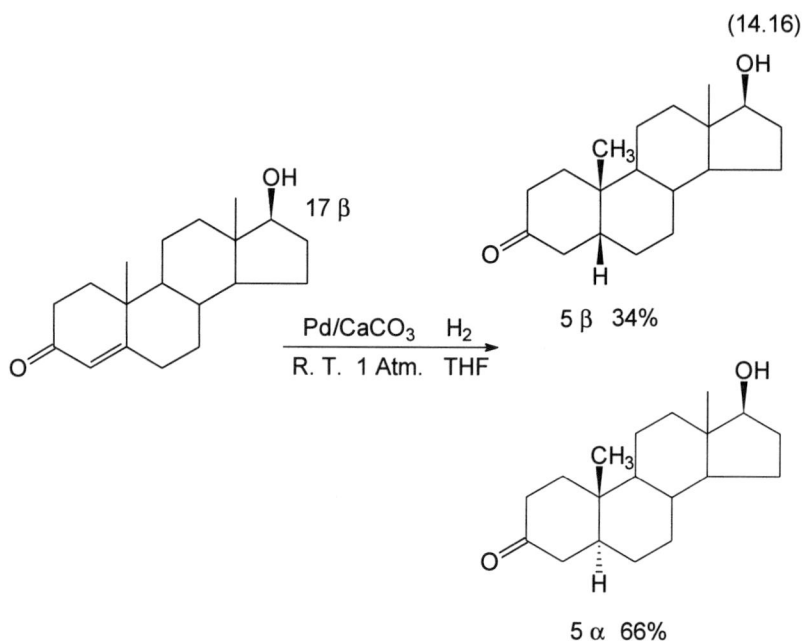

5 β 34%

5 α 66%

These results suggest that with a nickel catalyst the hydroxy group is adsorbed on the catalyst through an unshared electron pair on the oxygen atom. On palladium, however, it appears that the adsorption occurs at the carbinol carbon as depicted in Fig. 14.8.[2] The fact that the hydrogenolysis of substituted

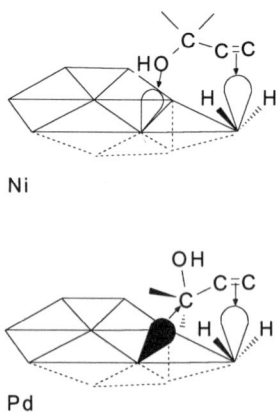

Fig. 14. 8. Proposed adsorption modes for unsaturated alcohols on nickel and palladium catalysts.

Table 14.2

Haptophilic effect in the hydrogenation of the substituted
tetrahydrofluorenones, **16**[40,41]

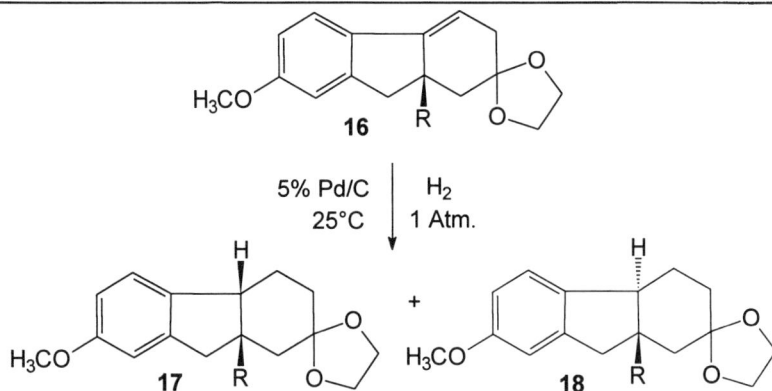

R	Product (%)	
	cis (**17**)	trans (**18**)
—CH$_2$OH	95	5
—C(=O)H	93	7
—C≡N	75	25
—C(=NOH)H	65	35
—C(=O)OH	18	82
—C(=O)OCH$_3$	15	85
—C(=O)CH$_3$	14	86
—C(=O)NH$_2$	10	90

benzyl alcohols proceeds with retention of configuration over nickel catalysts and inversion over palladium[44–47] supports these conclusions.

The haptophillic effect of other functional groups in directing product stereochemistry in the hydrogenation of the tetrahydrofluorenones, **16**, over a Pd/C catalyst is shown in Table 14.2.[40,41] Here the CH_2OH group is the most selective for the production of the *cis* product, **17**. However, because both the carbinol carbon and the oxygen are on the same side of the molecule the mode of haptophillic adsorption cannot be inferred. The aldehyde group also exerts a significant haptophilic effect but with other groups haptophillicity decreases. With the carboxylic acid and its derivatives, the stereochemistry of hydrogenation appears to be controlled not by the adsorption of these groups but by their steric bulk which leads to the primary formation of the *trans* isomer, **18**. It is interesting that the acid group does not appear to interact directly with the palladium catalyst. By analogy with the proposed alcohol adsorption on palladium through the carbinol carbon an acid, or its derivatives, may be expected to adsorb on palladium through an electron deficient α carbon atom so steric factors may not be the only explanation for *trans* product formation. Unfortunately, the unsubstituted material or one with a methyl substituent were not hydrogenated, so a true estimate of steric factors is not available.

Diastereoselectivity

Examination of the products obtained in Eqns. 14.10 and 11 shows that they are diastereomers. Since one of these diastereomers is the primary product, this

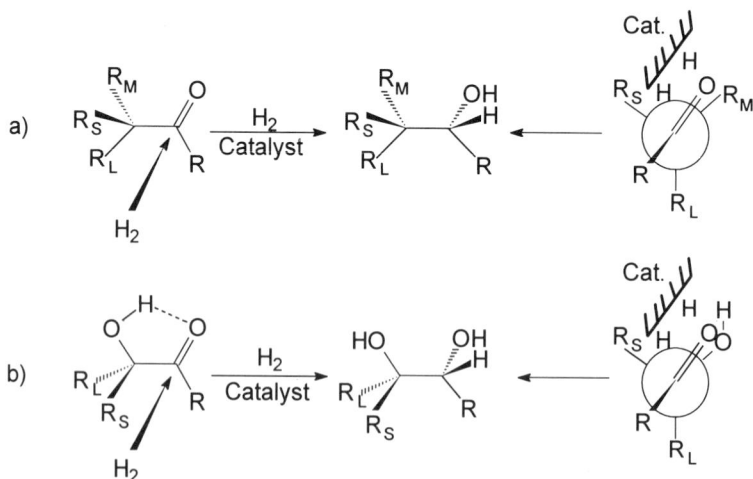

Fig. 14.9. a) Preferred conformation of ketones having a chiral center adjacent to the carbonyl group.[48] b) Preferred conformation of ketones having a chiral center containing a group capable of hydrogen bonding with the adjacent carbonyl group. [48]

reaction can also be referred to as being diastereoselective. The ease of determining the direction of hydrogenation in a diastereoselective reaction depends on the nature of the substrate. For cyclic systems the factors discussed above are usually sufficient. Acyclic molecules, however, are more conformationally flexible so determining the direction of attack is usually more difficult. The earliest systematic approach to the problem of asymmetric induction was concerned with the addition of Grignard reagents and hydride reducing agents to ketones having a chiral center in the molecule.[48–50] The most straightforward of these involves compounds in which the carbonyl group is adjacent to the chiral carbon. As depicted in Fig. 14.9a, the preferred conformation of the chiral carbon is that with the large group in the same plane and *anti* to the carbonyl group.[48] Attack on the carbonyl group takes place primarily from the less hindered side, that of the small group of the chiral carbon. The degree of selectivity depends on the relative sizes of the three groups on the chiral carbon. When one of these groups contains an OH or NH that can hydrogen-bond with the carbonyl group, the cyclic system shown in Fig. 14.9b is formed with primary attack again coming from the side of the smaller of the other two groups on the chiral carbon. Higher selectivities are generally observed with these *pseudo* ring systems.

(14.17)

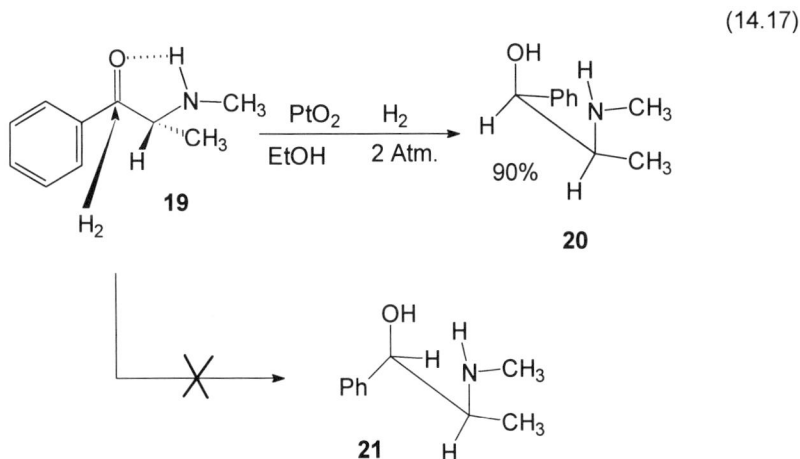

While this approach has been applied primarily to nucleophilic reactions of carbonyl compounds, it can also be used to determine the primary direction of hydrogenation since adsorption on a catalyst surface is preferred from the less hindered side of the carbonyl group. The synthesis of *dl* ephedrine (**20**) was accomplished by the hydrogenation of the amino ketone, **19**, one enantiomer of which is depicted in Eqn. 14.17.[51] The hydrogen bond to the carbonyl oxygen helps fix the conformation with hydrogenation taking place from the side of the hydrogen atom. Excellent selectivity was observed in this reaction with the *erythro dl* ephedrine obtained essentially pure after one recrystallization. None of the diastereomeric, *threo*, pseudoephedrine (**21**) was found.

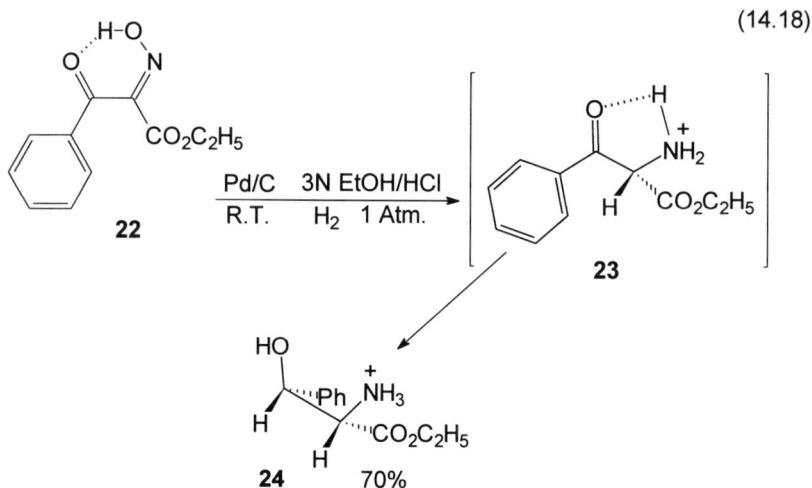

(14.18)

The hydrogenation of the oximino ketone, **22** (Eqn. 14.18), to give only the *erythro* amino alcohol follows this same pathway.[52] Hydrogen bonding between the oxime OH and the carbonyl oxygen fixes the initial conformation as the planar species. Hydrogenation of the oxime gives the amino ketone, **23**, having a chiral carbon. The acidic solvent protonates the amine to prevent condensation with the ketone group on another molecule. Hydrogen bonding again fixes the conformation so hydrogen is added to the carbonyl group from the side of the hydrogen on the chiral carbon giving the *erythro* product, **24**.

This approach should be useful in determining the direction of hydrogenation for molecules in which the carbinol group is replaced by carbon–carbon or carbon–nitrogen double bonds. With an alkene, though, the simple conformational model would have to be used and the hydrogenation should be run under conditions that do not promote double bond isomerization, that is, not with palladium or nickel catalysts. With carbonyl compounds the preferred conditions for selective reaction involve platinum, rhodium or ruthenium catalysts under non-diffusion control conditions. The use of nickel catalysts, especially Raney nickel, with its basic components, can cause an equilibration of the alcohol product.

Conformational preferences have been proposed for acyclic ketones having the chiral center further removed from the carbonyl group[50] but the extent of asymmetric induction decreases as the chiral center is removed further from the reacting functional group. The best diastereoselectivity is obtained when these two entities are adjacent to each other.

Enantioselectivity

A common method for introducing chirality into a molecule is to combine a prochiral species with a chiral auxiliary and using this chiral entity to direct the reaction on the prochiral center. This reaction is diastereoselective in that one of the two possible diastereomers is preferentially produced. If, however, the chiral auxiliary can be removed from the diasteriomeric products the resulting material is a mixture of enantiomers with one predominating. This is an example of an indirect enantioselective synthesis. Since this process involves optically active materials, the degree of enantioselectivity can be determined by measuring the optical yield of the product which is related to the optical yield as described by Eqn. 14.19. With the advent of chiral shift reagents for nmr analysis and chiral chromatographic techniques, however, the amounts of each enantiomer present in the reaction mixture can frequently be measured directly. The degree of enantioselectivity is more readily stated as the enantiomeric excess (ee) of the reaction as calculated using Eqn. 14.20.

$$\text{Optical Yield} = \frac{\text{Optical rotation of product}}{\text{Optical rotation of pure material}} \qquad (14.19)$$

$$\text{Enantiomeric excess} = ee = \frac{(R-S)}{(R+S)} \times 100 \qquad (14.20)$$

Common chiral auxiliaries are optically active alcohols that are used to esterify acids containing a prochiral group. The conformational preference for a chiral alcohol ester of an α keto acid is shown in Fig. 14.10.[49,53] Here the carbonyl groups of the ketone and ester are oriented *trans* to each other to minimize the electrostatic repulsion between the two carbonyl oxygens. The preferred arrangement of the groups on the chiral alcohol is one in which the large group eclipses the keto carbonyl group and the angle between the small and medium groups is bisected by the ester carbonyl. Preferred attack at the keto carbonyl takes place from the side of the small group on the alcohol portion of the ester.[49,53–56]

The catalytic hydrogenation of a number of α ketoesters of various optically active alcohols have been reported.[57] The degree of enantioselectivity of the resulting α hydroxyacids depends not only on the nature of the chiral alcohol used and the group attached to the carbonyl carbon but also on the reaction conditions. Examples of the hydrogenation of l-menthyl esters of α ketoacids are listed in Table 14.3.[58–61] According to the drawing the preferred direction for adsorption on the catalyst is from the side of the carbonyl group that also has the hydrogen on the menthyl carbinol carbon. After ester hydrolysis, this

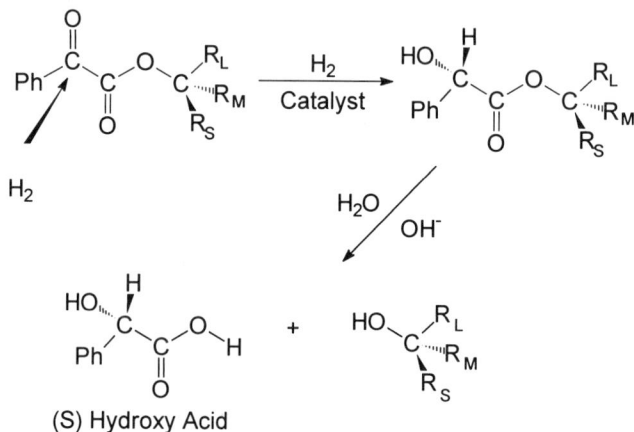

Fig. 14.10. Preferred conformation of α keto esters of chiral alcohols.[49,53]

leads to (R) hydroxy acid formation. When the R group is methyl or tert-butyl the (R) hydroxy acid is produced with all catalysts used. However, when the R group is aromatic, platinum and palladium give the (R) hydroxy acid but nickel catalysts give the (S) enantiomer, possibly the result of a nickel induced alcohol isomerization. In none of these hydrogenations, though, was the ee high enough for this approach to be a practical method for preparing chiral α hydroxy acids.

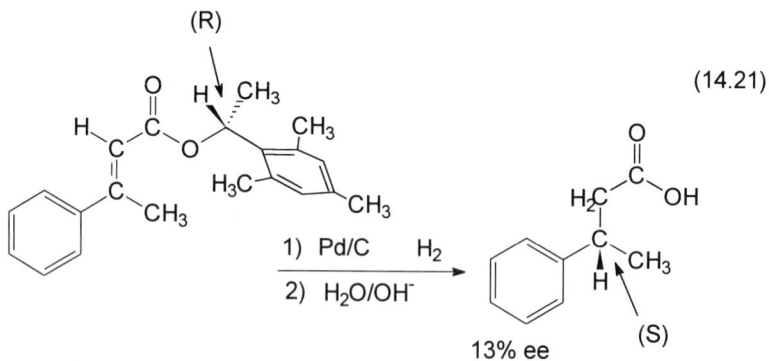

(14.21)

The ketone carbonyl can be replaced by a substituted carbon-carbon double bond as in the β methyl cinnamic esters shown in Eqn. 14.21,[61] or by an α imino (Eqn. 14.22) or oximino group (Eqn. 14.23).[62] Hydrogenation of these compounds gave the corresponding chiral materials with ee's ranging from 5 – 35%.

Table 14.3

Asymmetric hydrogenation of some chiral α keto esters.[57]

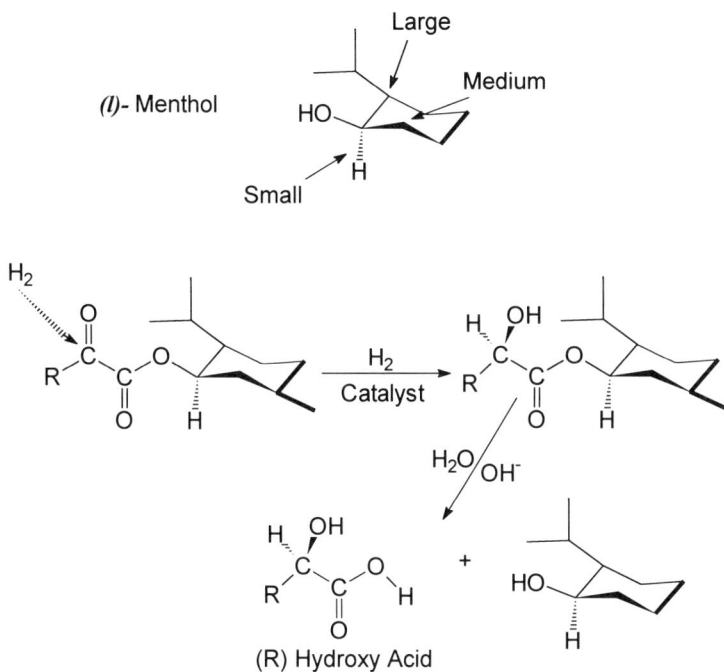

R	Catalyst	Product Configuration	ee (%)	Reference
Me	Pt, Pd, Ni	R	21–33	58
t-Bu	Pt, Pd, Ni	R	17–41	58
Phenyl	Ni	S	14	59
α-Naphthyl	Pt, Pt	R	14–55	60
α-Naphthyl	Ni	S	22	60

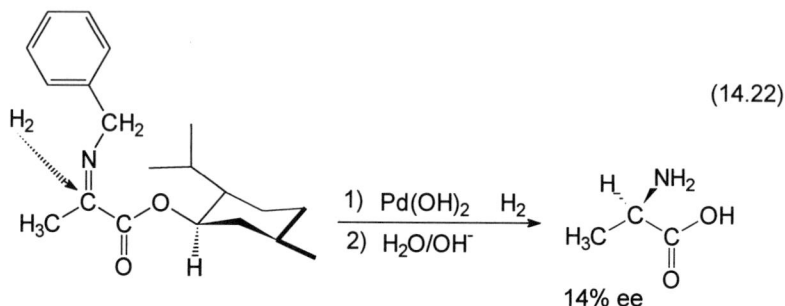

(14.22)

14% ee

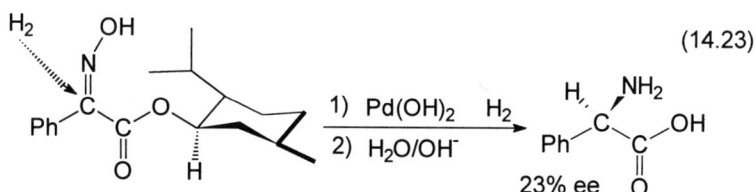

(14.23)

23% ee

Another approach to the enantioselective synthesis of amino acids involves the reductive amination of α keto esters using a chiral amine.[63–69] The chiral amines commonly used are (R) or (S) methylbenzylamine, **25** and **26**,[63–67] and (R) and (S) phenylglycine, **27** and **28**.[68,69] The benzyl group on the nitrogen can be removed by hydrogenolysis after the reductive amination. The optical yields in these reactions are generally better than those obtained from the procedures described above using a chiral alcohol as a chiral auxiliary. (R) Alanine (**29**) was obtained in 70% ee using (R) methylbenzyl amine (**25**) (Eqn. 14.24) and (S) alanine in 81% ee using (S) methylbenzyl amine.[63]

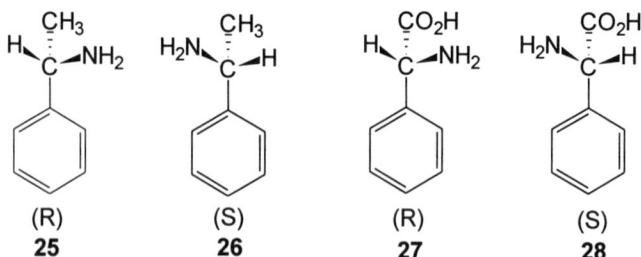

| (R) | (S) | (R) | (S) |
| **25** | **26** | **27** | **28** |

A more direct approach to the enantioselective hydrogenation of a prochiral group involves the use of a chiral catalyst for the reaction.[70–73] At present most of these reactions are run using chiral homogeneous catalysts but the

(14.24)

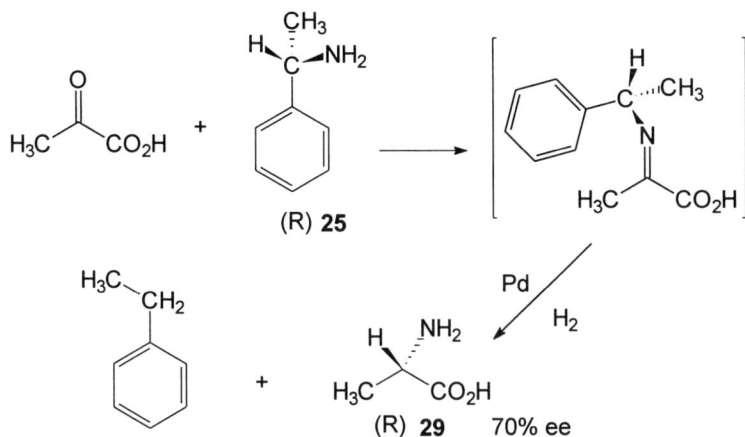

(R) 25

(R) 29   70% ee

earliest examples of enantioselective hydrogenations involved a metal catalyst on a chiral support such as quartz[74] or silk fibroin.[75] Another approach was to adsorb a chiral modifier onto the catalyst surface. The former concept has not been very successful but the use of chiral modifiers has proved to be quite effective in some catalytic hydrogenations. Two of these systems give products with consistently high ee's but they are very substrate specific.

Platinum catalysts modified with members of the cinchona alkaloids are effective enantioselective catalysts for the hydrogenation of α keto esters giving the chiral α hydroxy esters with ee's as high as 95%.[71,72] Of the cinchona alkaloids, cinchonidine (30) and cinchonine (31) appear to be more effective than quinine (32) and quinidine (33). With cinchonidine the (R) lactate, 34, is

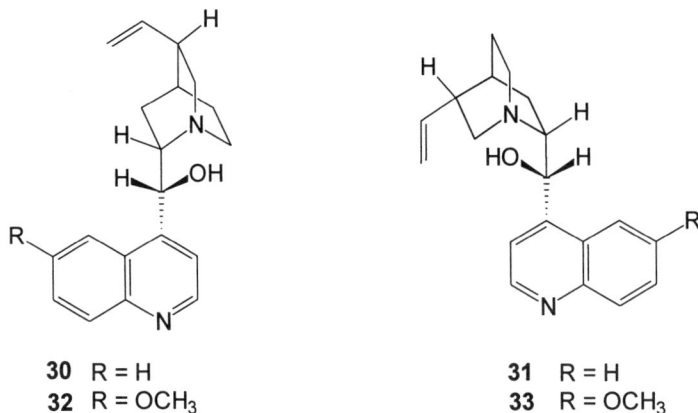

30 R = H
32 R = OCH$_3$

31 R = H
33 R = OCH$_3$

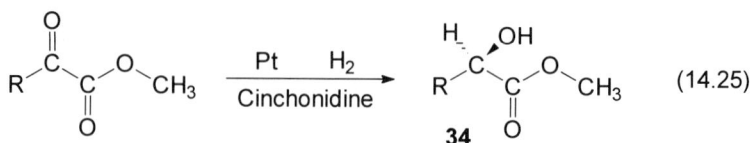

$$(14.25)$$

preferentially produced (Eqn. 14.25) while with cinchonine as the modifier the (S) lactate is formed with a somewhat lower ee. A number of proposals have been made concerning the nature of the chiral environment on the platinum surface[72,76-79] but at the present time nothing definitive has been forthcoming. Some recent work has shown that the quinoline amino alcohol, **35**, which incorporates the chiral environment of the cinchona alkaloids is also an effective chiral modifier for this reaction.[80] This reaction is generally specific for the hydrogenation of α keto esters with optimum optical yields obtained using cinchonidine as the modifier in an acetic acid solvent under at least 10 – 14 atmospheres of hydrogen.

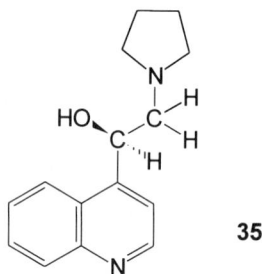

35

    Cinchonidine modified catalysts, though, have been effectively used for several other enantioselective hydrogenations. Platinum-cinchonidine catalysts have been used for the hydrogenation of the α ketolactone, **36**, to D-pantolactone (**37**) in 35% ee at complete conversion (Eqn. 14.26)[81] while palladium-cinchonidine catalysts have been used for the enantioselective dehydrohalogenation of α,α-dichloro-2-benazapinone (**38**) (Eqn. 14.27)[82] and the hydrogenation of (E) α phenylcinnamic acid (**39**) to (S) 2, 3-diphenylpropionic acid (**40**) in a 44% ee (Eqn. 14.28).[83]

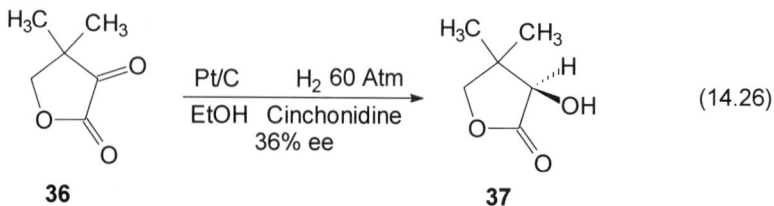

$$(14.26)$$

36   **37**

(14.27)

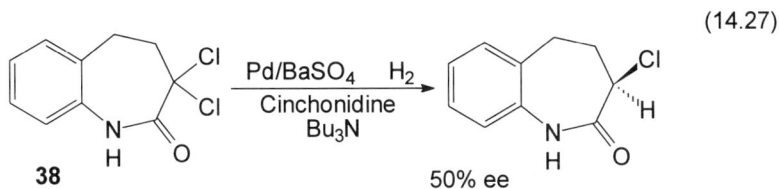

38   50% ee

(14.28)

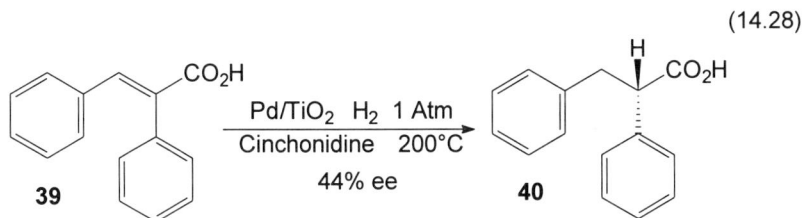

39   44% ee   40

The second, effective heterogeneous enantioselective catalytic system is nickel modified with tartaric acid and sodium bromide. This system is most effective for the hydrogenation of β keto esters giving chiral β hydroxy esters, **41**, with ee's as high as 95% (Eqn. 14.29).[70,72,84,85] It has also been used for the enantioselective hydrogenation of β diketones (Eqn. 14.30)[86] and methyl ketones.[84]

(14.29)

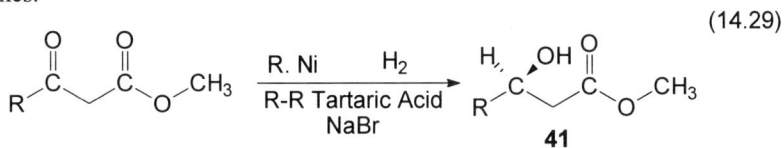

41

(14.30)

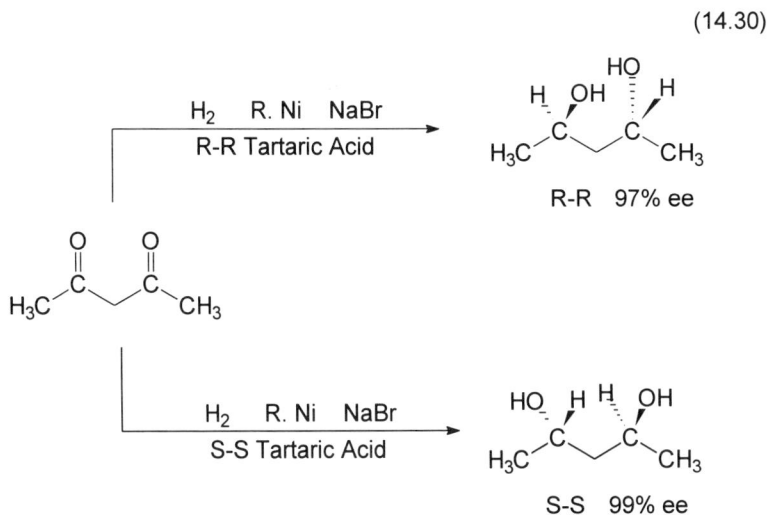

R-R   97% ee

S-S   99% ee

In the hydrogenation of β keto esters (R, R) tartaric acid gives the (R) alcohols and the (S, S) tartaric acid gives the (S) enantiomers. Raney nickel is more effective than supported nickel catalysts. It appears that the active catalyst is nickel tartrate which is adsorbed on the catalyst surface and that the sodium bromide is adsorbed on the non-chiral active sites, thus, keeping them from promoting the non-selective hydrogenations.[72,84] This procedure has been used to prepare the chiral intermediate in the synthesis of the Pine Sawfly sex pheromone.[87]

References

1. R. L. Augustine, *Catalytic Hydrogenation, Techniques and Applications in Organic Synthesis,* Dekker, New York, 1965.
2. R. L. Augustine, "Steroid Hydrogenation" in *Organic Reactions in Steroid Chemistry*, Vol. 1 (J. Fried and J. A. Edwards, Eds.) Van Nostrand Reinhold, New York, 1972, p 111.
3. R. L. Augustine, *Catal. Revs.*, **13**, 285 (1976).
4. P. N. Rylander, *Catalytic Hydrogenation over Platinum Metals*, Academic Press, New York, 1967.
5. P. N. Rylander, *Catalytic Hydrogenation in Organic Synthesis*, Academic Press, New York, 1979.
6. P. N. Rylander, *Hydrogenation Methods*, Academic Press, New York, 1985.
7. M. Freifelder, *Practical Catalytic Hydrogenation*, Wiley-Interscience, New York, 1971.
8. A. P. G. Kieboom and F. van Rantwijk, *Hydrogenation and Hydrogenolysis in Synthetic Organic Chemistry*, Delft University Press, Delft, 1977.
9. R. J. Peterson, *Hydrogenation Catalysts* (Chemical Technology Review, Vol. 94) Noyes Data Corp., Park Ridge, NJ, 1977.
10. *Catalytic Hydrogenation* (L. Cerveny, Ed.) *Stud. Surf. Sci. Catal.*, Vol. 27, Elsevier, Amsterdam, 1986.
11. M. Bartok, J. Czombos, K. Felfoldi, L. Gera, Gy. Gondos, A. Molinari, F. Notheisz, I. Palinko, Gy. Wittmann and A. G. Zsigmond, *Stereochemistry of Heterogeneous Metal Catalysis*, Wiley, New York, 1985.
12. *Catalytic Hydrogenation and Analogous Pressure Reactions*, (J. M. O'Connor, Ed.), *Ann. N. Y. Acad. Sci.*, **145**, 1–206 (1967).
13. *Second Conference on Catalytic Hydrogenation and Analogous Pressure Reactions*, (J. M. O'Connor, Ed.),*Ann. N. Y. Acad. Sci.*, **158**, 439–588 (1969).
14. *Third Conference on Catalytic Hydrogenation and Analogous Pressure Reactions*, (M. A. Rebensdorf, Ed.), *Ann. N. Y. Acad. Sci.*, **172**, 151–276 (1970).
15. *Fourth Conference on Catalytic Hydrogenation and Analogous Pressure Reactions*, (P. N. Rylander, Ed.), *Ann. N. Y. Acad. Sci.*, **214**, 1–275 (1973).
16. *Catalysis in Organic Synthesis, 1976*, (P. N.Rylander and J. Greenfield, Eds.), Academic Press, New York, 1976.
17. *Catalysis in Organic Synthesis, 1977*, (G. V. Smith, Ed.), Academic Press, New York, 1977.

18. *Catalysis in Organic Synthesis*, (W. H. Jones, Ed.), Academic Press, New York, 1980.
19. *Catalysis of Organic Reactions [Chem. Ind. (Dekker)*, **5**], (W. R. Moser, Ed.), Dekker, New York, 1981.
20. *Catalysis of Organic Reactions [Chem. Ind. (Dekker)*, **18**],(J. R. Kosak, Ed.), Dekker, New York, 1984.
21. *Catalysis of Organic Reactions [Chem. Ind. (Dekker)*, **22**], (R. L. Augustine, Ed.), Dekker, New York, 1985.
22. *Catalysis of Organic Reactions [Chem. Ind. (Dekker)*,**33**], (P. N. Rylander, H. Greenfield and R. L. Augustine, Eds.) Dekker, New York, 1988.
23. *Catalysis of Organic Reactions [Chem. Ind. (Dekker)*, **40**], (D. W. Blackburn, Ed.), Dekker, New York, 1990.
24. *Catalysis of Organic Reactions [Chem. Ind. (Dekker)*, **47**], (W. E. Pascoe, Ed.), Dekker, New York, 1992.
25. *Catalysis of Organic Reactions [Chem. Ind. (Dekker)*, **53**], (J. R. Kosak and T. A. Johnson, Eds.), Dekker, New York, 1994.
26. *Catalysis of Organic Reactions [Chem. Ind. (Dekker)*, **62**], (M. G. Scaros and M. L. Prunier, Eds.), Dekker, New York, 1995.
27. *Heterogeneous Catalysis and Fine Chemicals (Stud. Surf. Sci. Catal.*, **41**), (M. Guisnet, J. Barrault, C. Bouchoule, D. Duprez, C. Montassier and G. Perot, Eds.), Elsevier, Amsterdam, 1988.
28. *Heterogeneous Catalysis and Fine Chemicals, II (Stud. Surf. Sci. Catal.*, **59**), (M. Guisnet, J. Barrault, C. Bouchoule, D. Duprez, G. Perot, R. Maurel and C. Montassier, Eds.), Elsevier, Amsterdam, 1991.
29. *Heterogeneous Catalysis and Fine Chemicals, III (Stud. Surf. Sci. Catal.*, **78**), (M. Guisnet, J. Barbier, J. Barrault, C. Bouchoule, D. Duprez, G. Perot and C. Montassier, Eds.), Elsevier, Amsterdam, 1993.
30. W. F. Tuley and R. Adams, *J. Am. Chem. Soc.*, **47**, 3061 (1925).
31. H. Boerma, *Stud. Surf. Sci. Catal.*, **1** (Prep. Catal.), 105 (1976).
32. C. A. Brown, *J. Org. Chem.*, **35**, 1900 (1970).
33. W. H. Jones, W. F. Benning, P. Davis, D. M. Mulvey, P. I. Pollak, J. C. Schaeffer, R. Tull and L. M. Weinstock, *Ann. N. Y. Acad. Sci.*, **158**, 471 (1969).
34. D. Ginsburg and R. Pappo, *J. Chem. Soc.*, 938 (1951).
35. S. Siegel, *Adv. Catal.*, **16**, 123 (1966).
36. E. N. Marvell and T. Li, Synthesis, 457 (1973).
37. I. Horiuti and M. Polanyi, *Trans. Faraday Soc.*, **30**, 1164 (1934).
38. R. L. Augustine, F. Yaghmaie and J. F. van Peppen, *J. Org. Chem.*, **49**, 1865 (1984).
39. H. W. Thompson, *J. Org. Chem.*, **36**, 2577 (1971).
40. H. W. Thompson, *Ann. N. Y. Acad. Sci.*, **214**, 195 (1973).
41. H. W. Thompson and R. E. Naipawer, *J. Am. Chem. Soc.*, **95**, 6379 (1973).
42. S. Mitsui, M. Ito, A. Nanbu and Y. Senda, *J. Catal.*, **36**, 119 (1975).
43. M. G. Combe, H. B. Henbest and W. R. Jackson, *J. Chem. Soc., C*, 2467 (1967).
44. W. A. Bonner, J. A. Zderic and G. A. Casaletto, *J. Am. Chem. Soc.*, **74**, 5086 (1952).
45. S. Mitsui, Y. Senda and K. Konno, *Chem. and Ind., (London)*, 1345 (1963).
46. D. J. Cram and J. Allinger, *J. Am. Chem. Soc.*, **76**, 4516 (1954).

47. D. Y. Curtin and S. Schumnkler, *J. Am. Chem. Soc.*, **77**, 1105 (1955).
48. D. J. Cram and F. A. Abd Elhafez, *J. Am. Chem. Soc.*, **74**, 5828, 5851 (1952).
49. V. Prelog, *Helv. Chim. Acta*, **36**, 308 (1953).
50. J. D. Morrison and H. S. Mosher, *Asymmetric Organic Reactions*, American Chemical Society, Washington, D.C., 1976.
51. J. F. Hyde, E. Browning and R. Adams, *J. Am. Chem. Soc.*, **50**, 2287 (1928).
52. Y. T. Chang and W. H. Hartung, *J. Am. Chem. Soc.*, **75**, 89 (1953).
53. V. Prelog, *Bull. soc. chim., France*, 987 (1956).
54. V. Prelog, O. Ceder and M. Wilhelm, *Helv. Chim. Acta*, **38**, 303 (1955).
55. V. Prelog, E. Philbin, E. Watanabe and M. Wilhelm, *Helv. Chim. Acta*, **39**, 1086 (1956).
56. A. McKenzie and P. D. Ritchie, *Biochem. Z.*, **237**, 1 (1931).
57. Ref. 50, pp70–71.
58. S. Mitsui and A. Kanai, *Nippon Kagaku Zasshi*, **87**, 179 (1966); *Chem. Abstr.*, **65**, 17006 (1966).
59. M. Kawana and S. Emoto, *Bull. Chem. Soc., Japan*, **41**, 259 (1968).
60. S. Mitsui and Y. Imai, *Nippon Kagaku Zasshi*, **88**, 86 (1967); *Chem. Abstr.*, **67**, 43934 (1967).
61. V. Prelog and H. Scherrer, *Helv. Chim. Acta*, **42**, 2227 (1959).
62. K. Matsumoto and K. Harada, *J. Org. Chem.*, **31**, 1956 (1966).
63. R. G. Hiskey and R. C. Northrop, *J. Am. Chem. Soc.*, **83**, 4798 (1961).
64. R. G. Hiskey and R. C. Northrop, *J. Am. Chem. Soc.*, **87**, 1753 (1965).
65. K. Harada and K. Matsumoto, *J. Org. Chem.*, **32**, 1794 (1967).
66. K. Harada and K. Matsumoto, *J. Org. Chem.*, **33**, 4467 (1968).
67. K. Matsumoto and K. Harada, *J. Org. Chem.*, **33**, 4526 (1968).
69. K. Harada and Y. Kataoka, *Chem. Lett.*, 791 (1978).
70. Y. Izumi, *Angew. Chem., Int. Ed.*, **10**, 871 (1971).
71. H. U. Blaser and M. Muller, *Stud. Surf. Sci. Catal.*, **59** (Heterog. Catal. Fine Chem., II) 73 (1991).
72. G. Webb and P. B. Wells, *Catal. Today*, **12**, 319 (1992).
73. R. Noyori, *Asymmetric Catalysis in Organic Synthesis*, Wiley, New York, 1994, p 346.
74. A. P. Terent'ev, E. I. Klabunovskii and V. V. Patrikeev, *Doklady Acad. Nauk, S.S.S.R.*, **74**, 947 (1950); *Chem Abstr.*, **45**, 3798 (1951).
75. S. Akabori, Y. Izumi, Y. Fuji and S. Sakurai, *Nature*, **178**, 323 (1956).
76. I. M. Sutherland, A. Ibbotson, R. B. Moyes and P. B. Wells, *J. Catal.*, **125**, 77 (1990).
77. G. Bond, P. A. Meheux, A. Ibbotson and P. B. Wells, *Catal. Today*, **10**, 371 (1991).
78. M. Garland and H. U. Blaser, *J. Am. Chem. Soc.*, **112**, 7048 (1990).
79. R. L. Augustine, S. K. Tanielyan and L. K. Doyle, *Tetrahedron: Asymmetry*, **4**, 1803 (1993).
80. G. Wang, T. Heinz, A. Pfaltz, B. Minder, T. Mallat and A. Baiker, *J. Chem. Soc., Chem. Commun.*, 2047 (1994).
81. S. Niwa, Y. Imamura and K. Otsuka, *Jpn. Kokai Tokkyo Koho*, JP 62,158,268; *Chem. Abstr.*, **109**, 128815 (1988).
82. H. U. Blaser, S. K. Boyer and U. Pettelkow, *Tetrahedron: Asymmetry*, **2**, 721 (1991).
83. Y. Nitta, Y. Ueda and T. Imanaka, *Chem. Lett.*, 1095 (1994).
84. Y. Izumi, *Adv. Catal.*, **32**, 215 (1983).

85. W. M. H. Sachtler, *Chem. Ind. (Dekker)*, **22** (Catal. Org. React.) 189 (1995).

86. J. Bakos, I. Tóth and L. Markó, *J. Org. Chem.*, **46**, 5427 (1981).

87. A. Tai, N. Morimoto, M. Yoshikawa, K. Uehara, T. Sugimura and T. Kikukawa, *Agric. Biol. Chem.,* **54** 1753 (1990).

Hydrogenation II:
Alkenes and Dienes

Monoalkenes

A carbon–carbon double bond is one of the more easily hydrogenated functional groups. An isolated double bond is more readily hydrogenated than any other functional group with the exception of acetylenes, polyenes and, under certain conditions, aromatic nitro groups. This reaction is normally run at room temperature under an atmospheric pressure of hydrogen. With simple alkenes, ease of hydrogenation depends primarily on the ability of the olefinic π electrons to adsorb on the catalyst surface. This is influenced by the size and number of substituents on the double bond. In general, the ease of hydrogenation decreases in the order: monosubstituted > 1,1-disubstituted ≈ 1,2-*cis* disubstituted > 1,2-*trans* disubstituted > trisubstituted > tetrasubstituted. Strained double bonds are hydrogenated more readily than unstrained alkenes and exocyclic double bonds are generally more easily hydrogenated than are endocyclic olefins. Tetrasubstituted alkenes such as those that are common to two rings as in $\Delta^{1,9}$-octalin (**1**) as well as double bonds at certain positions in the steroid or similar ring systems, are hydrogenated with difficulty and then only under conditions that promote isomerization of the double bond to a more accessible, less substituted position in the substrate molecule[1] (Eqn. 15.1).[2] This sometimes requires the use of higher temperatures (50°–75°C) and pressures (2–4 atmospheres) as well as the addition of a small amount of hydrochloric or perchloric acid to the reaction mixture. The factors associated with the stereochemistry of alkene hydrogenation have been discussed in Chapter 14.

(15.1)

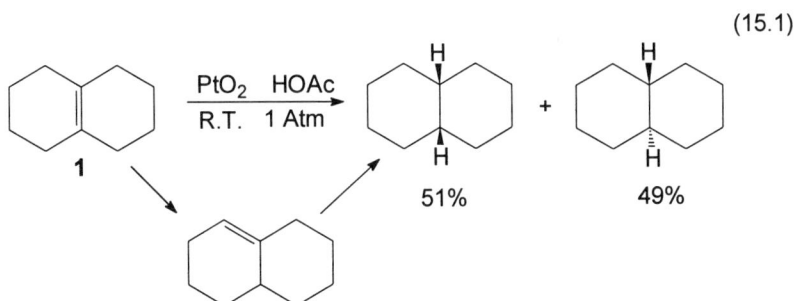

51%   49%

Virtually all of the common hydrogenation catalysts promote alkene hydrogenation. For a given olefin the ease of hydrogenation over metal catalysts decreases in the order: Pd ≥ Rh > Pt ≥ Ni >> Ru. While palladium is the most common catalyst for most synthetic applications, platinum has been used primarily for mechanistic studies so it is the standard against which other catalysts are compared. The mechanism generally accepted for platinum-catalyzed alkene hydrogenations is the Horiuti–Polanyi[3] reaction sequence depicted in Fig. 15.1. Available data indicate that rhodium catalyzes alkene hydrogenations in essentially the same way as does platinum but rhodium appears to be more susceptible to steric factors associated with the olefin.[4] Ruthenium is not very effective for alkene hydrogenations.

The initial steps in this reaction sequence involve the activation of both the hydrogen (Step 1) and the alkene (Step 2). Hydrogen transfer to the double bond occurs in two stages. One adsorbed hydrogen is first transferred to the adsorbed double bond to give the metalalkyl or "half-hydrogenated" species, **2**, (Step 3) which then reacts with a second adsorbed hydrogen to give the alkane (Step 4). The last step of this mechanism is essentially irreversible under common preparative hydrogenation conditions but the other steps are all potentially reversible. This reversibility accounts for the double bond isomerization which frequently occurs on hydrogenation of alkenes. This isomerization results from the reversal of Step 3 using an allylic carbon–hydrogen bond, that is, one that was adjacent to the double bond in the starting olefin as depicted in Step 3b in Fig. 15.1. The *cis–trans* equilibration of the double bond can also take place as shown in Step 3c. The extent of isomerization depends both on the reaction conditions and the catalyst used.

Under conditions of low hydrogen availability such as low pressure, slow agitation and/or a large quantity of catalyst, insufficient hydrogen is present at the catalyst surface so rapid transfer of both adsorbed hydrogens is not possible. Under these mass transport limiting conditions Steps 2 and 3 can reverse to produce an isomerized olefin. Under high hydrogen availability such as high pressure, rapid agitation and/or a small amount of catalyst, Step 4 proceeds with sufficient speed that the lifetime of the "half-hydrogenated" species is not long enough for reversal of Step 3 to be significant. Under these conditions, double bond isomerization is minimal. Palladium and nickel catalysts, however, promote double bond hydrogenations through a π-allyl intermediate, **3** (Fig. 15.2), and not by the classic Horiuti–Polanyi mechanism.[4]

The extent of double bond isomerization also varies with the nature of the catalyst. The degree of isomerization over metal catalysts usually decreases in the order: Pd > Ni > Rh > Ru > Os ≈Ir ≈Pt.[5,6] The extensive double bond isomerization observed with palladium and, to some extent, with nickel catalysts can be attributed to the formation of the adsorbed π-allyl species with these catalysts. While double bond isomerization may not be important in a routine alkene hydrogenation, it may influence a selective hydrogenation because the isomerized olefin can have different adsorption characteristics from those of the

$$H_2 \;\rightleftharpoons\; 2\,\overset{|}{\underset{*}{H}} \tag{1}$$

$$\tag{2}$$

$$\tag{3}$$

$$\tag{3b}$$

$$\tag{3c}$$

$$\tag{4}$$

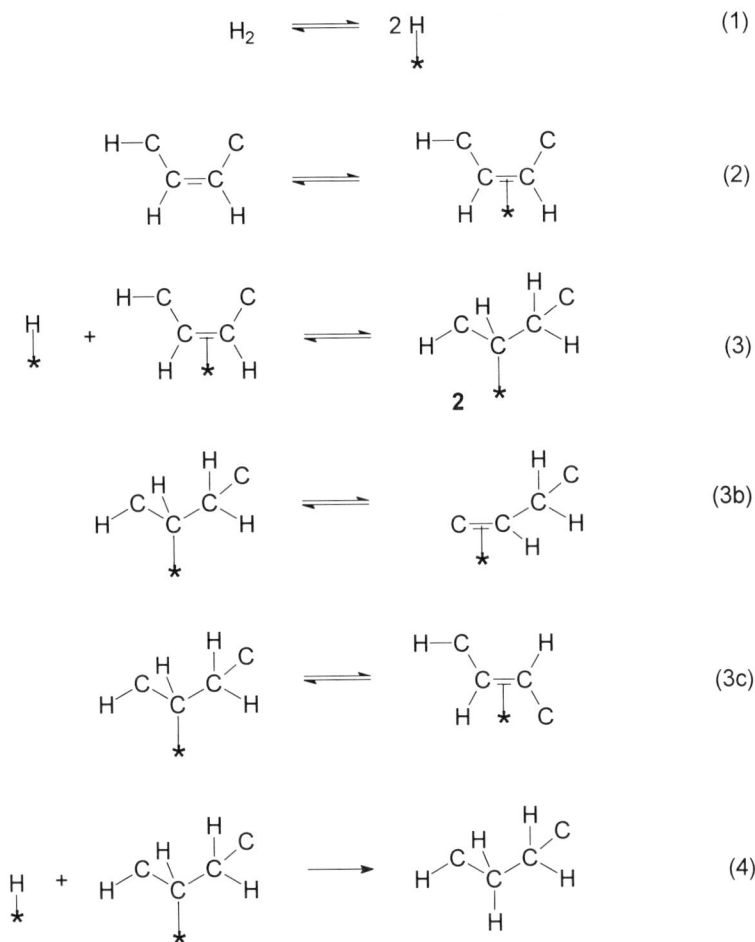

Fig. 15.1. Horiuti-Polanyi[3] alkene hydrogenation mechanism showing double bond isomerization steps.

starting alkene. Catalysts with low isomerization capabilities such as platinum, ruthenium and P-2 Ni(B) are most effective for selective alkene hydrogenations. Rhodium supported on $AlPO_4$–SiO_2 has been reported to be effective for the hydrogenation of a variety of unsaturated compounds at 20°C and five atmospheres of hydrogen with no double bond isomerization taking place.[7]

In many cases double bond isomerization is only observed when the reaction is interrupted before complete saturation takes place, but sometimes the isomerized product is sufficiently difficult to saturate so it simply accumulates in

$$H_2 \rightleftharpoons 2 \underset{*}{H}$$

$$
\begin{array}{ccc}
C=C-C & & C-C=C \\
\quad \backslash H & & H \nearrow \\
\end{array}
$$

$$\underset{*}{C}{-}C{-}\underset{H}{C} \rightleftharpoons C{-}\underset{*}{C}{-}C + \underset{*}{H}\; \underset{*}{H} \rightleftharpoons \underset{H}{C}{-}C{-}\underset{*}{C}$$

3

$$\underset{*}{H} \qquad \underset{*}{H}$$

$$\underset{H}{C}{-}\underset{*}{C}{-}\underset{H}{C}$$

$$\underset{*}{H}$$

$$\underset{H}{C}{-}\underset{H}{C}{-}\underset{H}{C}$$

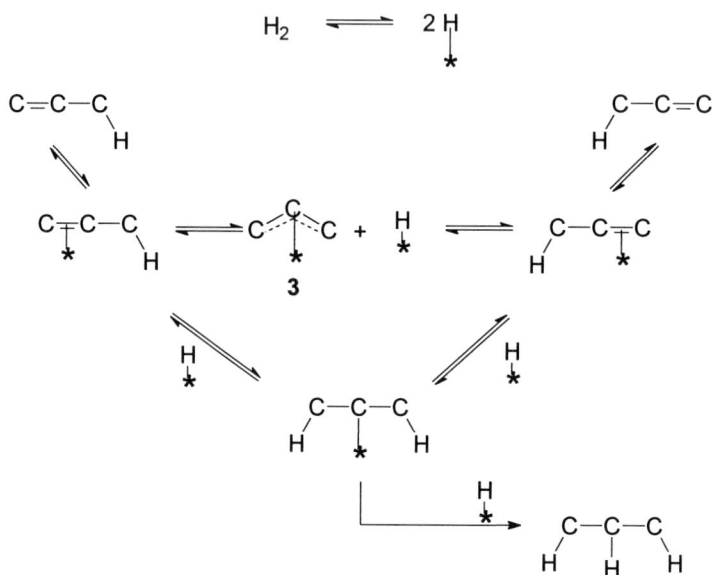

Fig. 15.2. Alkene hydrogenation mechanism proceeding through a π-allyl intermediate.

the reaction mixture. One such isomerization is the migration of the Δ^7 or $\Delta^{8,9}$ steroidal double bonds to the $\Delta^{8,14}$ position (Fig. 15.3). This isomerization takes place over palladium catalysts[8] or platinum in acetic acid[9] but not over Raney nickel[10] or platinum in neutral solvents.[8] The addition of piperidine or pyridine to the palladium catalyzed reaction mixture also inhibits double bond isomerization.[11] The conformational drawings in Fig. 15.3 show the accessibility to the catalyst of the allylic hydrogen which is removed during the isomerization in the 9α or $\Delta^{8,9}$ steroids.[12] With a 9β steroid, the allylic hydrogen at C–14 is not available to the catalyst so isomerization to the inert $\Delta^{8,14}$ position does not take place and the Δ^7 double bond is saturated instead (Fig. 15.4).[13]

Electronic effects can also influence the ease of double bond hydrogenation. Compounds such as Δ^3 cyclohexenecarboxaldehyde (4) and other 3-carbonyl substituted cyclohexenes, such as 5 and 6, are more difficult to hydrogenate than the non-carbonyl containing materials.[8] This decrease in activity has been attributed to an interaction between the carbonyl carbon and the π cloud of the double bond, shown by 7, which has been termed a supra-annular effect.[9-11]

α = below the plane ················
β = above the plane ▬▬▬

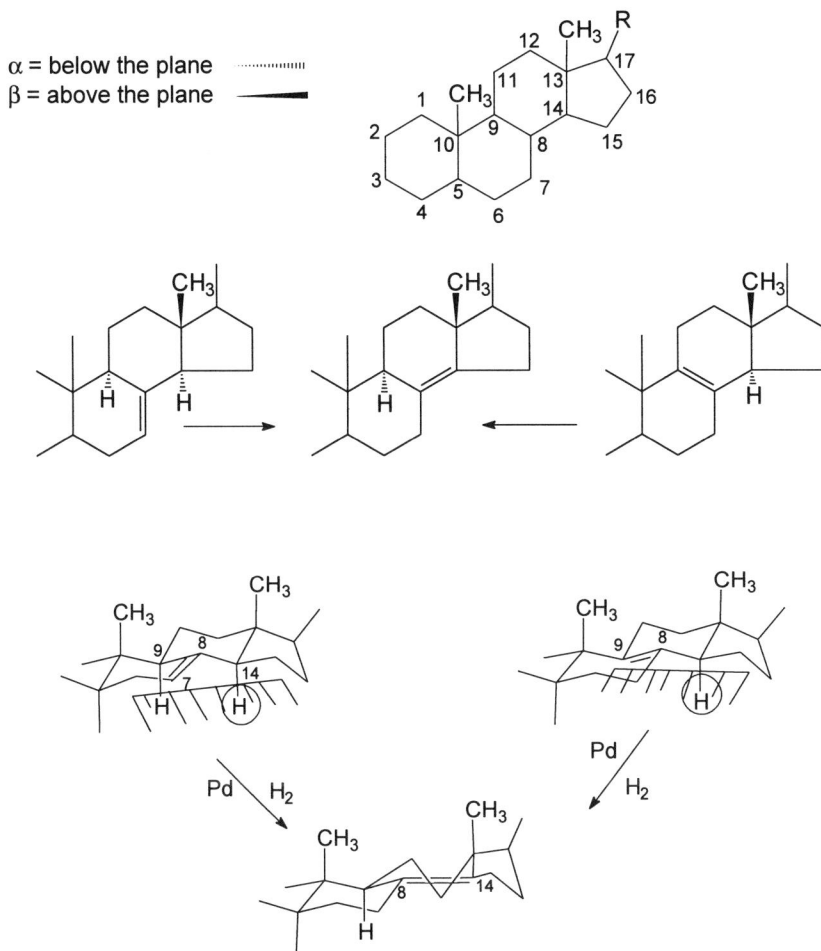

Fig. 15.3. Double bond isomerization in a 9α steroid showing accessibility of the 14α hydrogen to the catalyst as needed for double bond migration.

For those hydrogenations in which the hydrogenolysis of another functional group in the molecule can occur along with the saturation of a double bond, palladium should not be used if bond cleavage is to be avoided. The palladium catalyzed hydrogenation of unsaturated compounds which also contain a cyclopropane ring results in the opening of the cyclopropane along with the alkene saturation (Eqns. 15.2–3).[18,19] With platinum,[18] Raney-nickel[19] or

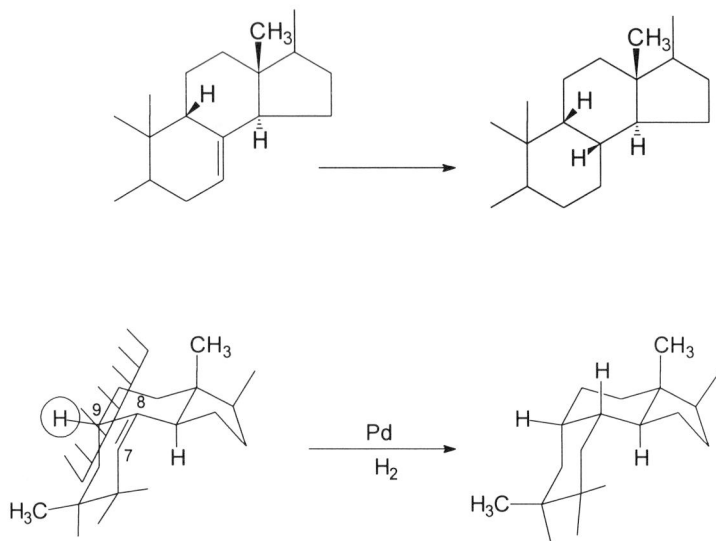

Fig. 15.4. Hydrogenation of Δ^7-9β steroid showing the orientation of
the 14α hydrogen away from the catalyst so no double bond
isomerization can take place.

copper–barium-chromium oxide,[20] the cyclopropane ring remains intact.   The
hydrogenolysis of alkyl, aryl and vinyl halides, which is also readily catalyzed by
palladium, can be avoided by using platinum (Eqns. 15.4–5)[21,22] or rhodium
catalysts.[23]

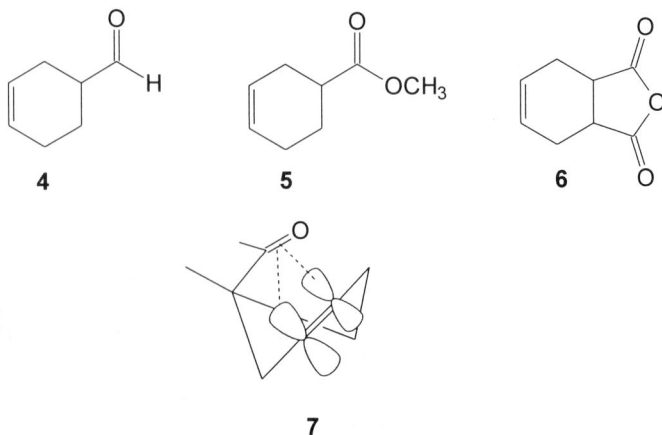

4   5   6

7

(15.2)

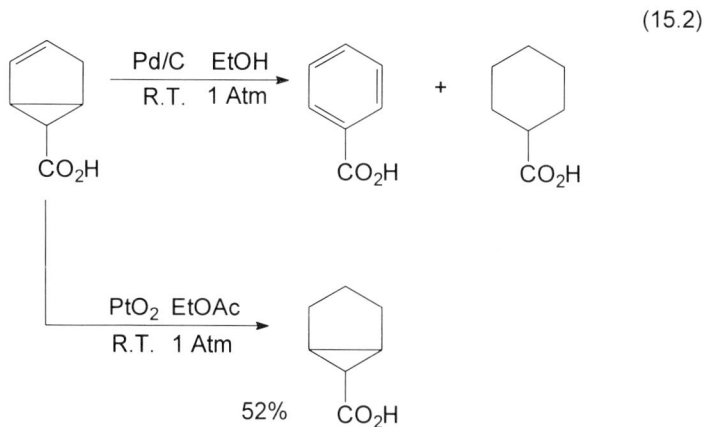

(15.3)

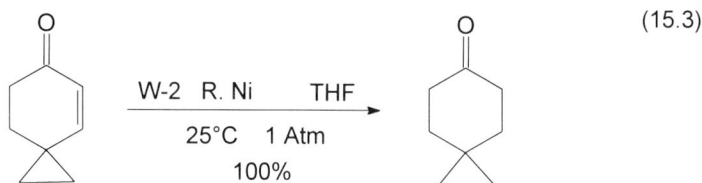

(15.4)

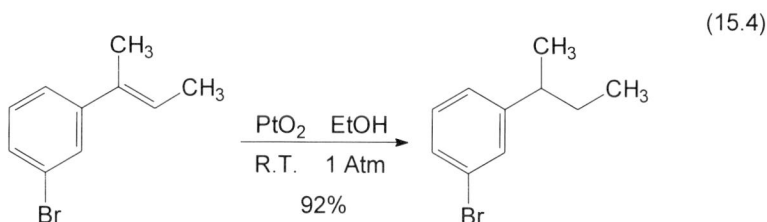

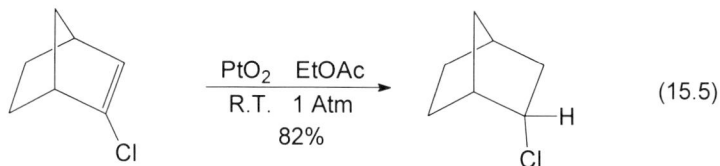

(15.5)

The hydrogenolysis of allylic carbon–oxygen bonds during the saturation of the double bond can be minimized by the use of rhodium (Eqn. 15.6), ruthenium[24] or P-1 Ni(B).[25] Borohydride reduced palladium catalysts, the Pd(B) species, have also been successfully used for the selective hydrogenation of allyloxy compounds with no hydrogenolysis of the oxygen containing groups detected.[26]

(15.6)

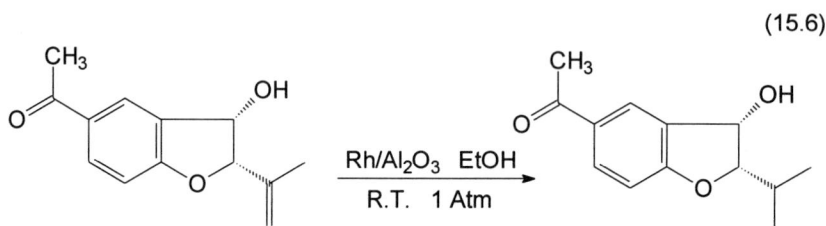

Enol Ethers, Esters, Enamines and Enamides

The hydrogenation of enol ethers, esters, enamines and enamides is commonly run using a palladium catalyst at room temperature and atmospheric pressure. Isolated double bonds [27-29] and double bonds conjugated with the enol[30] are, however, preferentially hydrogenated (Eqn. 15.7). The hydrogenation of *trans* l-methoxy-1,3-butadiene (**8**) over Lindlar's catalyst was 70% selective at 50% conversion with predominantly 1,2-addition at the unsubstituted double bond rather than the less active enol ether moiety (Eqn. 15.8). P-2 Ni(B) also gave this product but the selectivity was considerably lower. With PtO_2 the reaction was essentially nonselective.[31] Hydrogenation of the isomeric enol ethers, **9** and **10**, over palladium gave predominantly the *trans* product, **11**, while over platinum and rhodium, the *cis* ether, **12**, predominated.[32] With a ruthenium catalyst, **12** is formed almost exclusively.[33] The major by-product from palladium-, platinum-, or rhodium-catalyzed hydrogenations run in ethanol was the ketal, **13**.[32] Little ketal was formed during the ruthenium catalyzed hydrogenation.

(15.7)

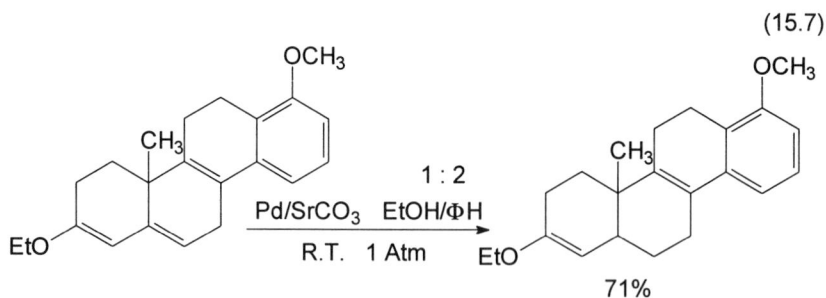

(15.8)

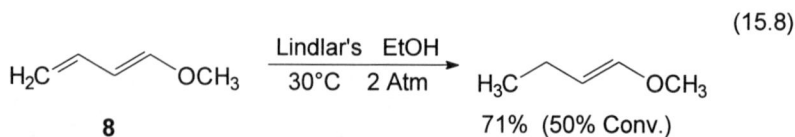

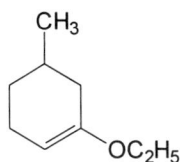

9

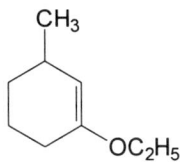

10

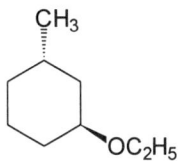

11

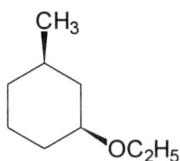

12

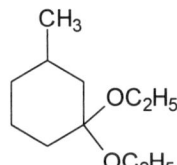

13

(15.9)

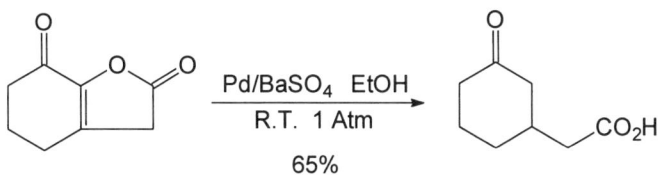

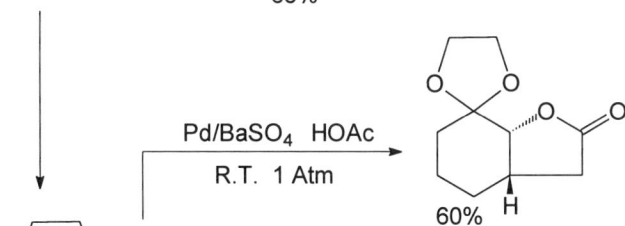

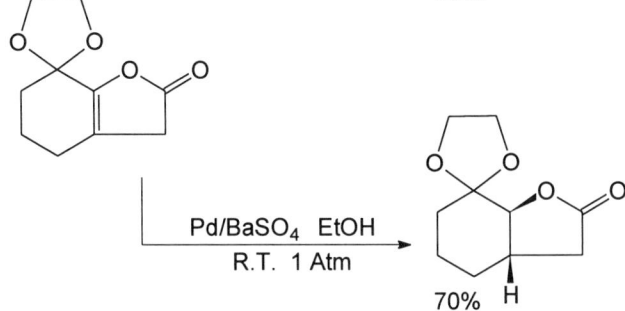

Enol lactones obtained from δ or γ keto acids are hydrogenated to the saturated lactone unless the enol C–O bond is activated for hydrogenolysis by an α keto or aryl group. When this is the case, the product is the saturated desoxy acid. If a ketone is the activating group, converting it to the ketal will prevent hydrogenolysis (Eqn. 15.9).[34] The addition of a small amount of ferric chloride to the reaction mixture also inhibits hydrogenolysis.[35]

The hydrogenation of unactivated enol acetates of cyclopentanones takes place without hydrogenolysis over either platinum or palladium catalysts. Cyclohexenyl acetates, however, are hydrogenolyzed over platinum, but not over palladium (Eqn. 15.10).[36,37] It is interesting to note that this is one of the very few reactions for which palladium is inferior to another catalyst for a hydrogenolysis reaction (see Chapter 20). Little, if any, hydrogenolysis is observed during the hydrogenation of enol ethers. The enamine, **14**, was hydrogenated over Pd/C at room temperature and atmospheric pressure to give the *cis* saturated product, **15**, in 90% isolated yield (Eqn. 15.11).[38] The enantioselective preparation of amino acids by the hydrogenation of the corresponding enamides has been discussed in Chapter 14.

(15.10)

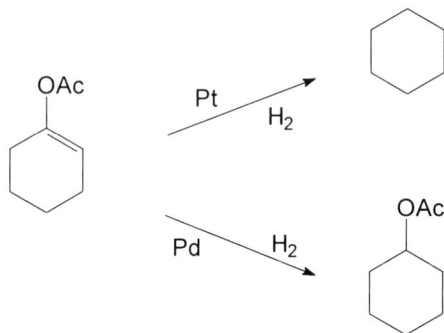

Steroidal Double Bonds

Because of the shape and rigidity of the steroid ring system, (See Fig. 15.3) double bonds at the various positions are hydrogenated with different degrees of difficulty.[39] Δ^{14}, Δ^{16}, $\Delta^{17,20}$ and Δ^{20} double bonds are hydrogenated over palladium catalysts more readily than a non-conjugated Δ^5 alkene.[40,41] The Δ^5 olefin in cholesterol (**16**) is hydrogenated over palladium at room temperature and 2–3 atmospheres of hydrogen giving the 5α product, **17**, in >90% yield (Eqn. 15.12).[42] The 5α product is commonly obtained on hydrogenation of Δ^5 steroids[43] unless a 3α substituent is present, then the 5β product, **18**, predominates (Eqn. 15.13).[44,45]

(15.11)

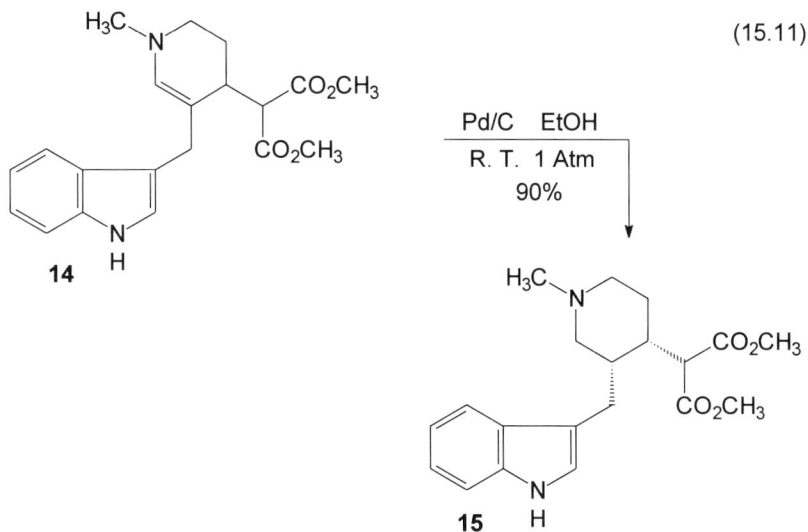

(15.12)

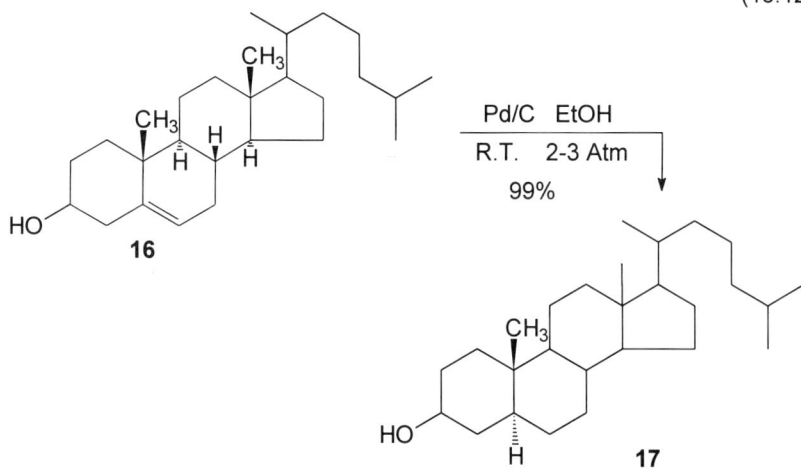

The 17-keto group in **19** is hydrogenated over W-4 Raney nickel in preference to a Δ^5 double bond (Eqn. 15.14)[46] but not the $\Delta^{17,20}$ enol acetate, **20**, over palladium (Eqn. 15.15).[27] The saturation of a 3β hydroxy Δ^5 steroid is accomplished more readily after conversion to the 3β acetate.[47]

Δ^{14}, Δ^8, $\Delta^{9,11}$, Δ^{27}, Δ^7 and $\Delta^{8,14}$ double bonds are progressively more difficult to hydrogenate than the Δ^5 with the $\Delta^{8,14}$ olefin inert to hydrogenation except under drastic conditions. The $\Delta^{9,11}$-12-keto steroid, **21**, is hydrogenated to

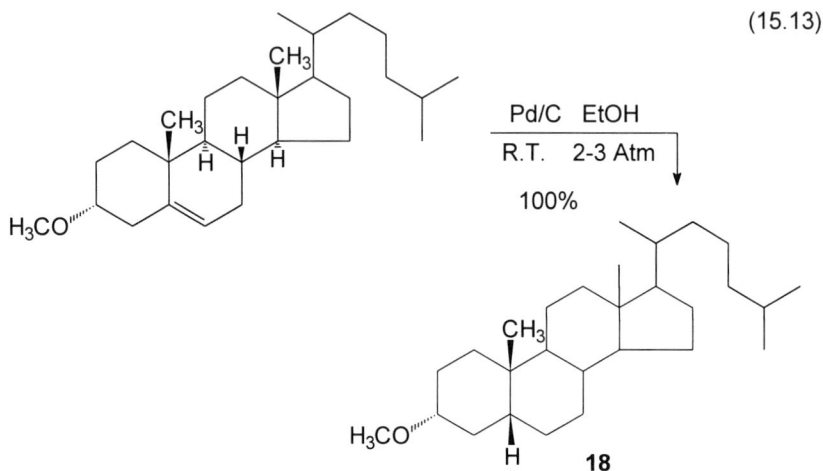

(15.13)

Pd/C   EtOH

R.T.   2-3 Atm

100%

18

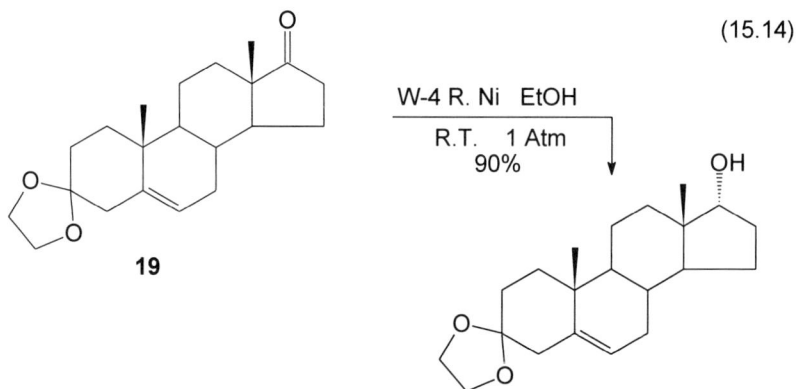

(15.14)

19

W-4 R. Ni   EtOH

R.T.   1 Atm

90%

the $\Delta^{9,11}$-12 hydroxy compound over platinum oxide in acetic acid[48] again illustrating the inertness to hydrogenation exhibited by some steroidal double bonds (Eqn. 15.16).

Unsaturated Carbonyl Compounds

The selective hydrogenation of the double bond of an α,β-unsaturated carbonyl compound is rather easily accomplished over most metal catalysts under moderate conditions. Because double bond isomerization does not take place in these systems, palladium catalysts are often used in the liquid phase at ambient temperature and atmospheric pressure. An added advantage here is that palladium is essentially inert for aliphatic aldehyde and ketone hydrogenations under these conditions. Vapor phase hydrogenations should be run at temperatures as low as possible to minimize carbonyl group hydrogenations. Catalysts such as Ni(B) are

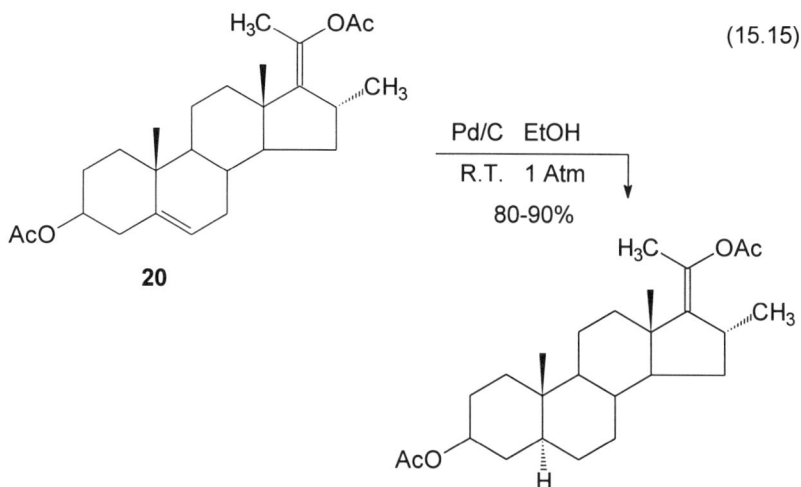

(15.15)

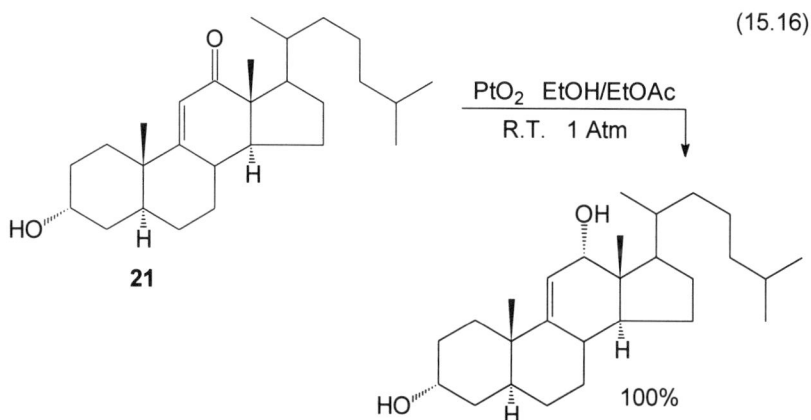

(15.16)

also effective for the selective hydrogenation of the double bond of α,β-unsaturated ketones.[49,50] Verbinone (22) was completely converted to verbanone (23) at 100% conversion over Ni(B) at 100°C and 70 atmospheres. When the reaction was run at 125°C for an extended period of time, verbanol (24) was produced in 95% yield (Eqn. 15.17).[49] Carbonyl conjugated double bonds are selectively hydrogenated in preference to the saturation of an isolated double bond over Cu/SiO$_2$ catalysts (Eqn. 15.18)[51] or over palladium catalysts in basic medium (Eqn. 11.11).[52]

In liquid phase hydrogenations of the double bond of α,β-unsaturated carbonyl compounds, mechanistic considerations are complicated because the

(15.17)

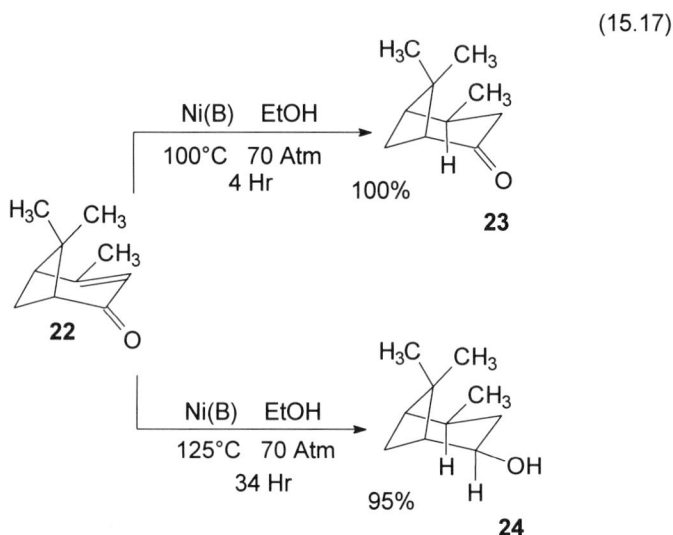

(15.18)

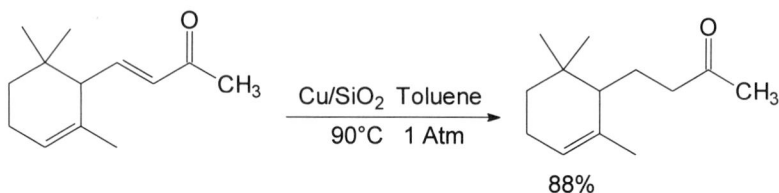

product composition is affected by the hydrogen availability to the catalyst as well as by the nature of the solvent used in the reaction.[53,54] In a neutral solvent, the reaction can proceed by either a 1,2-addition (**25**) or a 1,4-addition (**26**) process as depicted in Scheme 15.1.

In aprotic solvents of low dielectric constant, hydrogenation takes place by 1,2-addition to the double bond with product stereochemistry related to the mode of the 1,2-adsorption and the hydrogen availability to the catalyst. In aprotic solvents of high dielectric constant, 1,4-addition predominates giving the enol, **27**, which ketonizes to give the product. Hydrogen availability is also a factor in determining product stereochemistry. However, because the steric factors of 1,4-adsorption, **26**, are different from those associated with 1,2-adsorption, **25**, product stereochemistry is influenced by the dielectric constant of the solvent as well. In hydroxylic solvents the reverse effect is noted; solvents of low dielectric constant promote 1,4-addition and those with high dielectric constants give products arising from the 1,2-addition process. Some stereochemical results

Scheme 15.1

C—C—C—C=O
   | |
   H H
27

2 H

C—C—C—C=O
  * * **25**

C—C—C=C—OH
   | ↑ 2 H
   H

C—C—C=C—O
    *   **26** *

C—C=C—C=O

H⁺

OH⁻

C—C—C—C—O⁻
  29 *   *   *

C—C—C=C—OH
 28 * *

H⁻ from Cat.

H⁻ from Cat.

C—C—C=C—O⁻
   H *   *   *
          30

H⁺
from Sol'n

C—C—C—C=O
   | |
   H H

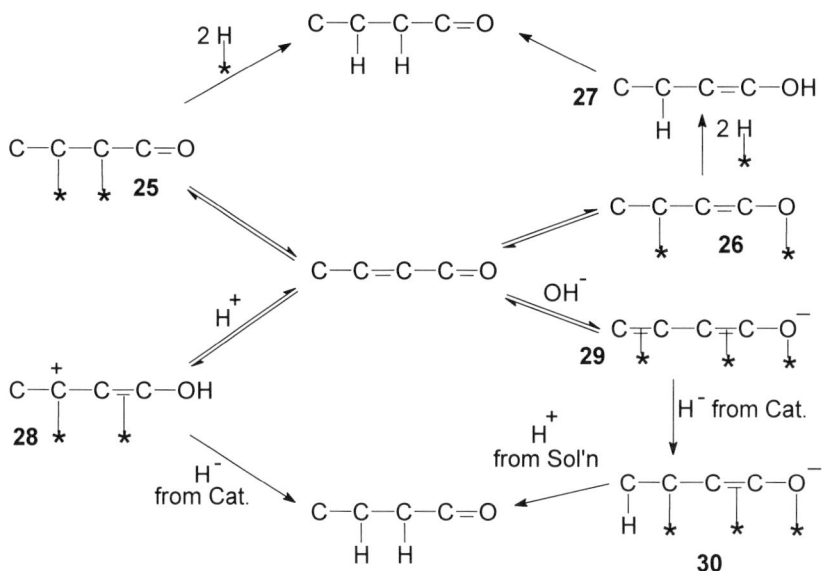

obtained on hydrogenation of $\Delta^{1,9}$-2-octalone (**31**) in different solvents are listed in Table 15.1.[53-55] Generally, if the solvent influences the stereoselectivity of a reaction, extremes in dielectric constant should show the greatest selectivity differences.

In acidic solvents, the unsaturated ketone is first protonated to give the carbocation, **28**, shown in Scheme 15.1. Hydride ion transfer from the catalyst completes the reaction. Product stereochemistry is determined by the steric factors associated with the adsorbed cation. In basic media, the enolate ion, **29**, is formed and adsorbed on the catalyst from the least hindered side. Hydride ion transfer from the catalyst followed by protonation from the solvent gives the product. Product stereochemistry is determined by the direction of protonation of the adsorbed intermediate, **30**.[54]

In the hydrogenation of bicyclic unsaturated ketones such as **31**, a *cis* ring fused product, **32**, is most readily formed using an acidic solvent such as 10% ethanolic HCl with a palladium catalyst at room temperature and atmospheric pressure.[53] The *cis* A/B ring fused steroids, **34** (Eqn. 15.19) and analogous tri- and tetra-cyclic compounds are most easily obtained in high selectivity by hydrogenating the Δ^4-3-keto starting material, **33**, over palladium at room temperature and atmospheric pressure in an alcoholic solution that is about 0.3N

Table 15.1

Effect of solvent and hydrogen pressure on the stereochemistry of the hydrogenation of $\Delta^{1,9}$-octalone-2 (**31**).[53-55]

31   **32**

Solvent	Percent *cis* Product (**32**)	
	Low H$_2$ Pressure	High H$_2$ Pressure
n-Hexane	77	48
Dimethylformamide	86	75
tert-Butyl Alcohol	91	48
Ethanol	78	48
0.3N EtOH/HCl	91	80
0.3N EtOH/NaOH	50	50

in NaOH or KOH.[56-58] The corresponding *trans* ring fused products are not produced with any significant selectivity by any catalytic hydrogenation process. Such compounds are best produced in high selectivity using chemical reducing agents such as lithium in liquid ammonia.[59]

    The stereochemistry at the α-carbon on hydrogenation of α,β-unsaturated carbonyl compounds in neutral solvents depends primarily on the nature of the

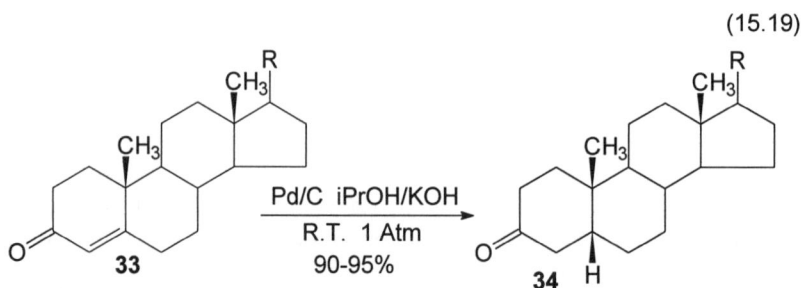

(15.19)

(15.20)

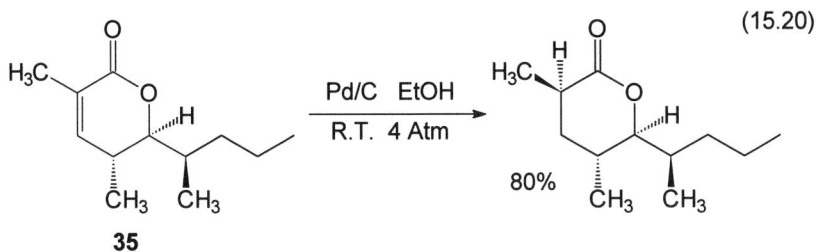

35

species adsorbed on the catalyst. The hydrogenation of both **35** (Eqn. 15.20)[60] and **36** (Eqn. 15.21)[61] takes place from the least hindered side of the substrate. In acidic or basic media, product stereochemistry is determined by the keto-enol equilibrium of the product.

(15.21)

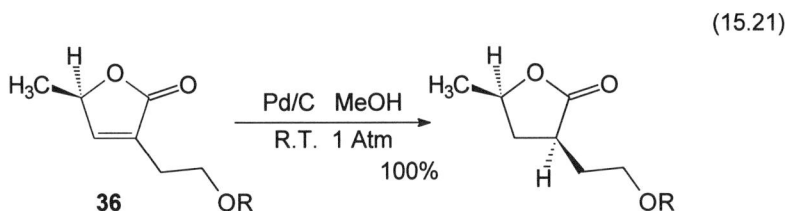

36

Vinylogous Esters and Amides

Vinylogous esters, **37**, and amides, **38**, are derivatives of β-dicarbonyl compounds that are, in effect α,β-unsaturated carbonyl compounds having an oxygen or nitrogen atom on the double bond. Because the resonance between the O or N and the carbonyl group involves the double bond such species are more difficult to hydrogenate than simple enones.[62,63] Hydrogenation of the vinylogous ester **39**[64] using standard enone hydrogenation conditions gave only the vinylogous transesterification product, **40** (Eqn.15.22). Double bond saturation was accomplished using a Pd/C catalyst in tetrahydrofuran at room temperature under four atmospheres of hydrogen (Eqns. 15.23-5).[64]

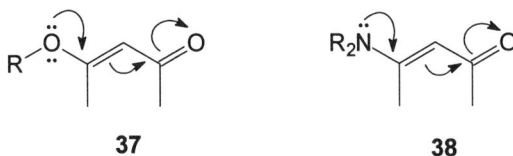

37   **38**

The vinylogous urethane, **41**, was even more difficult to hydrogenate with saturation reported only by the use of a large amount of a Rh/Al$_2$O$_3$ catalyst at 85°C and thirty atmospheres pressure (Eqn. 15.26).[65] These conditions, however,

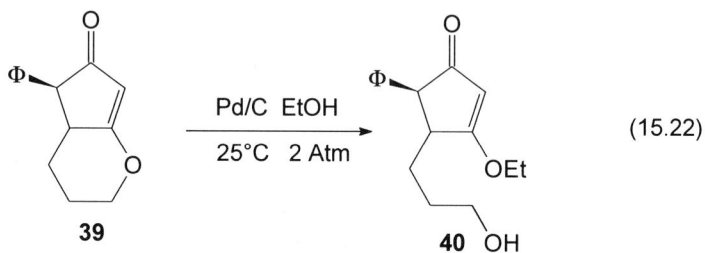

(15.22)

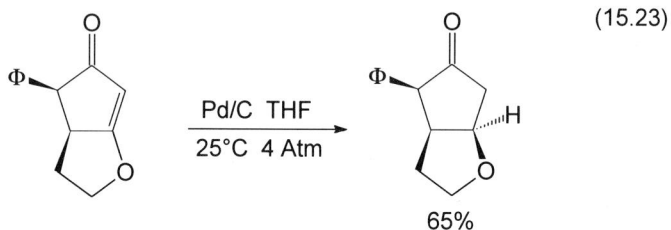

(15.23)

65%

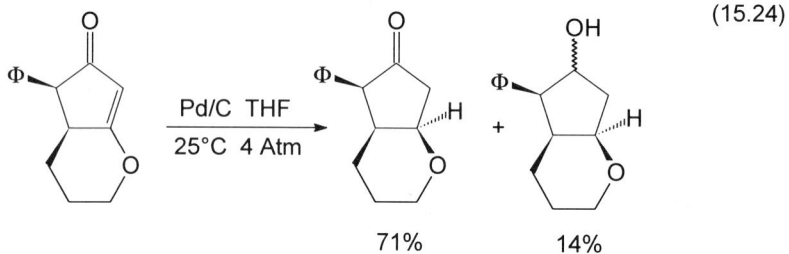

(15.24)

71%   14%

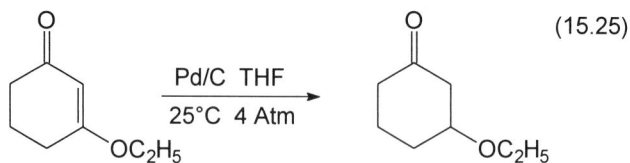

(15.25)

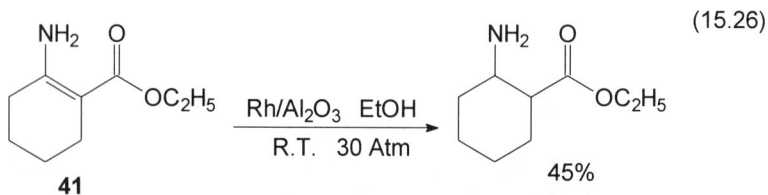

(15.26)

45%

were not effective for the saturation of the diester, **42**, in which a supra-annular effect was also present. More general conditions for the hydrogenation of vinylogous urethanes such as **41, 42,** or **43,** were the use of a Pd/C catalyst at 85°C and 100 atmospheres of hydrogen (Eqns. 15.27).[66]

(15.27)

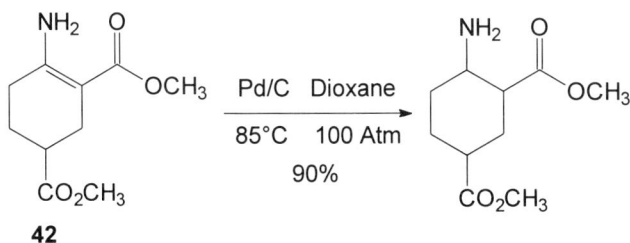

42

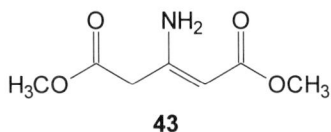

43

Monoalkene Mixtures

The selective hydrogenation of one alkene in the presence of another is an example of *Type I* selectivity (Chapter 5). The olefin that is hydrogenated faster will be favored for saturation. However, with this type of selectivity the faster reaction is more influenced by diffusion limitations, so it is best to use a relatively unreactive catalyst. It is particularly important that the migration of the organic substrates to the catalyst is beyond mass transport limitations so the catalyst will have equal access to both alkenes. As the reaction proceeds, the amount of the more readily adsorbed species decreases so reaction selectivity may also decrease.

Because the rate of alkene hydrogenation generally decreases as the number and size of the substituents on the double bond increase, the least hindered double bond should be hydrogenated preferentially in the competitive hydrogenation of olefin mixtures. This selectivity, however, is not always observed either because of diffusion constraints or the presence of concurrent double-bond isomerization. Isomerization modifies the olefin structure, which changes the alkene reactivity and makes reaction selectivity difficult to attain.

Because ruthenium catalysts are relatively unreactive for alkene hydrogenation and they are poor for double-bond isomerization, these catalysts are particularly effective for the selective hydrogenation of monosubstituted alkenes in the presence of di- and tri-substituted olefins at ambient temperature under 2–3 atmospheres of hydrogen (Eqn. 15.28). Water in the reaction medium

(15.28)

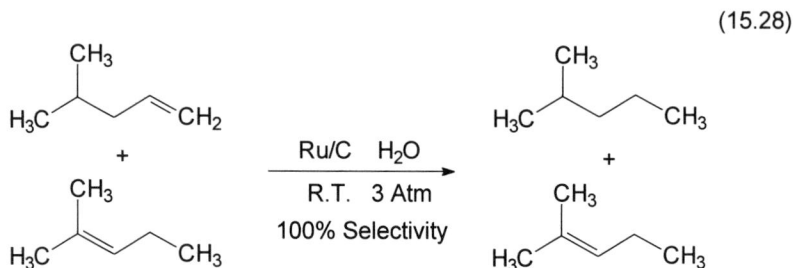

facilitates alkene hydrogenation over ruthenium as does the presaturation of the catalyst with hydrogen.[67] This system, however, is nonselective in the hydrogenation of di- and tri-substituted alkenes.

The P-2 Ni(B) is an effective catalyst for the saturation of double bonds.[68,69] With this catalyst, terminal alkenes are hydrogenated with greater than 95% selectivity in the presence of di- and tri-substituted olefins in ethanol solution at room temperature and atmospheric pressure.[69]

Solvent Effect

As discussed in Chapter 13, the hydrogenation of mixtures of simple alkenes and olefinic alcohols in non-polar solvents results in the selective saturation of the hydroxy olefin because of the enhanced adsorption of this material caused by the presence of the hydroxy group.[70] In alcoholic solvents, however, this haptophilic assistance is negated by the adsorption of the solvent molecules on the catalyst, and hydrogenation selectivity is controlled by the steric environments of the competing olefins.[71,72]

Shape-selective Hydrogenations

Several reports have been published on the use of zeolite-supported metals for the shape-selective hydrogenation of mixtures of monoolefins. An early report[73] discussed the preparation of A zeolites in the presence of a cationic platinum complex. After reduction with hydrogen, these materials showed significant differences in the rates of hydrogenation for variously substituted alkenes. On passing a mixture of hydrogen, propene, and isobutene over this catalyst at a flow rate of 30 cm^3/min at atmospheric pressure and 343°C, only the propene was hydrogenated; no isobutane was detected. This catalyst, however, was rapidly deactivated under these reaction conditions.

Pt/ZSM-5   catalysts   were   used   for   effective   shape-selective hydrogenations.[74,75] With these materials, prepared by the ion exchange of ZSM-5 with dilute aqueous Pt(NH$_3$)$_4$Cl$_2$, the selectivity depended on the reduction procedure used to produce the Pt(0) catalyst. Reduction in a stream of hydrogen at 300°C for one hour gave a supported catalyst that had an activity and selectivity similar to that of Pt/Al$_2$O$_3$. Neither catalyst was selective in the hydrogenation of a mixture of 1-hexene and 4,4-dimethyl-1-hexene and both exhibited thermal deactivation, presumably caused by platinum sintering. Because of the similarity

Table 15.2

Shape-selective competitive hydrogenation over Pt/ZSM-5.[75]

Linear Alkene	Branched Alkene	Temp. (°C)	Percent Hydrogenation	
			Linear	Branched
1-Pentene	4,4-dimethylpentene-1	300	97	2
1-Heptene	4,4-dimethylpentene-1	300	91	1.1
1-Hexene	6-Methylheptene-1	300	25	2
Styrene	2-Methylstyrene	425	50	2

between these catalysts, the Pt/ZSM-5 was considered to be composed of active platinum sites located mainly on the exterior of the zeolite crystals and not in the cavities,.

Reduction of Pt^{2+}/ZSM-5 in a hydrogen/olefin mixture at 400°C, however, produced a selective catalyst that showed increased activity with increasing temperature. Table 15.2 illustrates this selectivity, and Table 15.3 compares these hydrogenations with those run using Pt/Al$_2$O$_3$.[75] Why these catalyst pretreatment procedures gave such different results is not understood. It was proposed that platinum migration and agglomeration during hydrogen reduction could be attributed to the formation of unstable, mobile platinum hydrides.[75] The olefin may serve as a hydride trap to decompose the hydride

Table 15.3

Shape-selective hydrogenation over Pt/ZSM-5.[75]

Catalyst	Temp. (°C)	Percent Hydrogenation	
		1-Hexene	4,4-Dimethylhexene-1
0.5% Pt/Al$_2$O$_3$	275	27	35
Pt/ZSM-5	275	90	<1
		Styrene	2-Methylstyrene
0.5% Pt/Al$_2$O$_3$	400	57	58
Pt/ZSM-5	400	50	<2

before it can migrate out of the cavity. This premise was supported by decomposing the zeolite-containing the platinum tetraamine complex in flowing oxygen at 300°C, followed by hydrogen reduction at 500°C, to produce a highly selective Pt/ZSM-5 catalyst.[75]

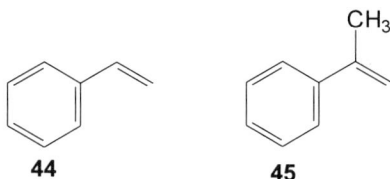

44   **45**

The data in Tables 15.2–3 indicate that the catalytically active metal in the zeolite catalysts is present primarily in the zeolite cavities. This conclusion is further substantiated by selective poisoning experiments. With an unmodified catalyst, the hydrogenation of a mixture of styrene (**44**) and 2-methylstyrene (**45**) gave about a 90:10 mixture of the two respective saturated products, ethylbenzene and cumene. The saturation of 2-methylstyrene indicated that some of the platinum was on the exterior of the zeolite. Pretreatment of the catalyst with tritolylphosphine decreased the overall activity of the catalyst but raised the selectivity to 93:1. The bulky phosphine was too large to enter the zeolite cavities, but was able to deactivate the exterior platinum sites, which were thought to be primarily responsible for the hydrogenation of the 2-methylstyrene. Similar selectivities were observed using either large or small particles of ZSM-5 in the catalyst preparation, but platinum exchange with the cesium form of the zeolite gave a more active catalyst than did exchange with the ammonium ion-containing species.[75]

Shape-selective hydrogenations over metal/zeolite catalysts were also reported for liquid-phase reactions. Triallyl rhodium interacts with the hydroxyl groups on zeolites primarily within the cavities of partially proton-exchanged X-type zeolites[76,77] to give supported diallyl rhodium complexes containing two equivalents of rhodium per unit cell of the zeolite. Reduction with hydrogen gave zeolite-bound rhodium-hydrogen species, which were active for the hydrogenation of double bonds. The hydrogenation rates were influenced by the size of the substituents on the olefin.

The addition of trimethylphosphine to these rhodium/zeolite catalysts destroyed all catalytic activity because the phosphine was small enough to fit into the zeolite cavity and could deactivate all of the rhodium in the catalyst.[77] The bulky tributylphosphine, however, could not enter the cavity and, thereby, only blocked the external rhodium from further reaction. This specific blocking enhanced the selectivity in the hydrogenation of a mixture of cyclopentene and 4-methylcyclohexene over a Rh/ZSM-11 catalyst. After treatment of the catalyst with tributylphosphine to block the external catalytically active sites, only the cyclopentene was saturated.[78]   The hydrogenation selectivity with these zeolite catalysts depends also on the amount of water in the crystals.[78]   With Rh/X zeolites, hydration states of 2%, 5%, and 32% gave selectivity ratios for the cyclopentene:methylcyclohexene hydrogenation of 1, 6, and 30, respectively. With Rh/ZSM-11, a maximum selectivity ratio of 47 was obtained at about a 25% hydration level. This effect can be rationalized as follows. In the absence of water, the size selectivity is based on the zeolite framework alone. With X-zeolites, this size is such that little selectivity is observed between cyclopentene and methylcyclohexene. Water coats the internal pores and cavities of the zeolite, reduces the internal pore diameter, and increases the reaction selectivity. Hydration also significantly decreases the overall hydrogenation activity.

Reaction of a Pt/A or Rh/A catalyst with diphenyldiethoxysilane significantly increased reaction selectivity.   Hydrogenation of a mixture of 1-nonene and *trans* 4-nonene over the non-silanized catalysts showed only a modest selectivity toward 1-nonene hydrogenation. With the silanized catalyst selectivity increased about ten fold.[79]

Unconjugated Dienes

The selective hydrogenation of one double bond of a nonconjugated diene belongs in the *Type II* selectivity category (Chapter 5) as exemplified by the selective hydrogenation of 4-vinylcyclohexene (**46**) to 4-ethylcyclohexene (**47**) (Eqn. 15.29). Because the saturation of either double bond is of the same kinetic order, organic substrate diffusion factors are not important in attaining reaction selectivity. But they are important, along with low hydrogen availability and the use of less active catalysts, in minimizing oversaturation.   Isomerization-promoting catalysts should be avoided.

(15.29)

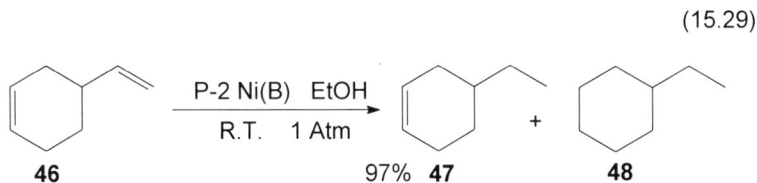

Since this reaction is a unimolecular example of the selective hydrogenation of an alkene mixture, the successful saturation of the less-substituted double bond should take place most readily over those catalysts that are most effective for the preferential saturation of one olefin in a mixture. Ruthenium has not been used extensively for such hydrogenations, but P-2 Ni(B) has been effective in promoting the selective hydrogenation of one of the double bonds in **46**, methylene norbornene (**49**) (Eqn. 15.30), dicyclopentadiene (**50**) (Eqn. 15.31), and 2-methyl-1,5-hexadiene (**51**) (Eqn. 15.32).[68,69,80,81]

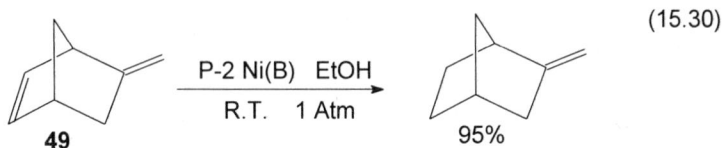

(15.30)

P-2 Ni(B)   EtOH

R.T.   1 Atm

49   95%

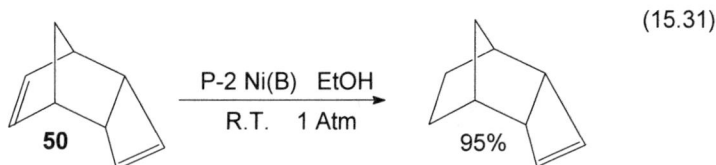

(15.31)

P-2 Ni(B)   EtOH

R.T.   1 Atm

50   95%

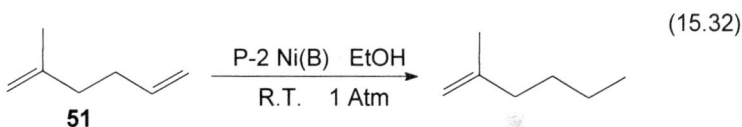

(15.32)

P-2 Ni(B)   EtOH

R.T.   1 Atm

51

The hydrogenation of 4-vinylcyclohexene (**46**) to 4-ethylcyclohexene (**47**) was also reported to take place over a supported nickel arsenide, Ni-As(B), which was prepared by the sodium borohydride reduction of nickel arsenate supported on either silica[82,83] or alumina.[84] These catalysts, however, function best in the presence of additives. When the reaction was run in pentane at 125°C and 25 atmospheres of hydrogen in the presence of a small amount of acetone, the product mixture at 96% conversion was 96% 4-ethylcyclohexene (**47**) and 4% ethylcyclohexane (**48**). No isomeric olefins were detected.[84]

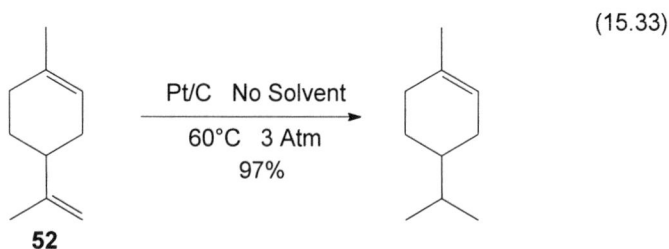

(15.33)

Pt/C   No Solvent

60°C   3 Atm

97%

52

The selective hydrogenation of the disubstituted double bond of limonene (**52**) took place over a platinum catalyst at 60°C and 3 atmospheres pressure (Eqn. 15.33).[85] 1,5-Undecadiene was hydrogenated to 5-undecene with 78% selectivity at 97% conversion over a Pt/A zeolite catalyst that was treated with diphenyldiethoxysilane.[79]

As will be described in the next section, the selective hydrogenation of conjugated dienes is accomplished more readily than the partial hydrogenation of unconjugated dienes.   Consequently, double-bond isomerization can be advantageous for the selective hydrogenation of unconjugated dienes such as 1,5-cyclooctadiene (53) in which the symmetry dictates that only one conjugated diene and mono-unsaturated product is possible.   In this reaction the isomer, 1,3-cyclooctadiene (54), is more easily hydrogenated to cyclooctene than the unconjugated cyclooctadienes. The use of Ni-As/Al$_2$O$_3$ for the hydrogenation of 1,5-cyclooctadiene in cyclohexane at 70 atmospheres of hydrogen, 200°C gave, at 95% conversion, a 66% selectivity for the formation of cyclooctene. The addition of carbon monoxide in a 3–7% mole ratio with the 1,5-cyclooctadiene increased the selectivity to 97% (Eqn. 15.34). Similar results were obtained using acetone, acetaldehyde, or acetic acid as a modifier.[83]

(15.34)

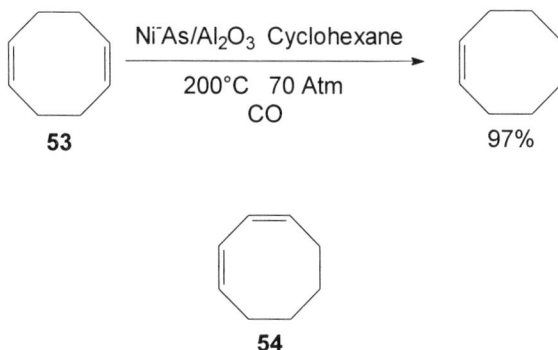

53   Ni̅As/Al$_2$O$_3$ Cyclohexane
      ⎯⎯⎯⎯⎯⎯⎯⎯⎯⎯⎯⎯⎯⎯→
      200°C   70 Atm
      CO   97%

54

Effect of Modifying Compounds and Solvents

The effect of added phenylacetaldehyde on the selectivity of the palladium-catalyzed hydrogenation of methyl linoleate (55) is shown in Fig. 15.5.[86] A similar effect was also found in the hydrogenation of 1,5-cyclooctadiene.[86] A 99% selectivity was reported for methyl oleate (56) formation when phenylacetaldehyde was added to the hydrogenation mixture (Eqn. 15.35). A 97% selectivity for cyclooctene formation was observed on hydrogenation of 1,5-cyclooctadiene in the presence of phenylacetaldehyde.   From competitive hydrogenations, the relative strengths of adsorption were estimated to be 1,3-cyclooctadiene >> 1,5-cyclo-octadiene > methyl linoleate > (phenylacetaldehyde) >> cyclooctene > methyl oleate.[86] Phenylacetaldehyde was also observed to increase the rate of isomerization of the unconjugated dienes, so the increase in selectivity found in the presence of this additive was attributed to two factors: (1) the increase in isomerization activity facilitated monoene formation; and. (2) the more pronounced adsorption capability of phenylacetaldehyde with respect to the monoenes resulted in their displacement from the catalyst before they could be further saturated. In agreement with this competitive adsorption explanation, the reaction selectivity was found to be influenced by the solvent. In solvents that do

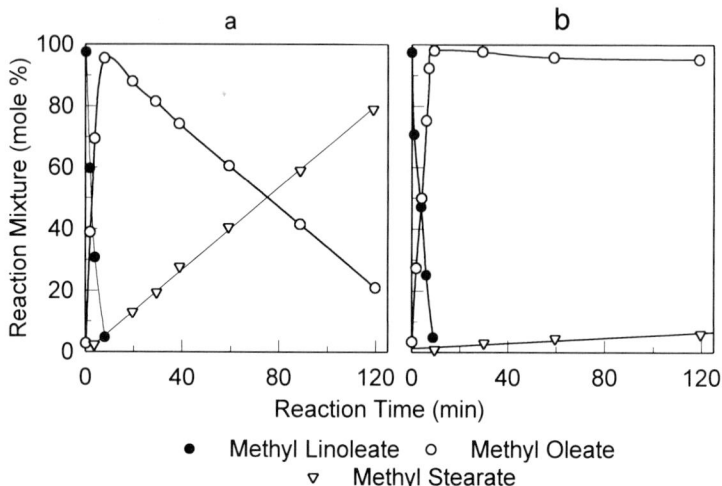

Fig. 15.5.   Hydrogenation of methyl linoleate with Pd/CaCO$_3$.
a) Without an additive and b) with phenylacetaldehyde
added. (Redrawn using data from Ref. 86)

not interact appreciably with the catalyst [e.g., hydrocarbons or tetrahydrofuran
(THF)], phenylacetaldehyde is capable of interacting more readily with the
catalyst, and monoene saturation is very low even after prolonged reaction at
room temperature and atmospheric pressure. In *tert* butyl alcohol, which can
adsorb on the catalyst surface,[87] the effect of the phenylacetaldehyde is not as
pronounced and monoene saturation occurs.[86]

(15.35)

$$H_3C(CH_2)_4HC=HCCH_2HC=CH(CH_2)_7CO_2CH_3 \quad \xrightarrow[25°C \quad 1 \; Atm]{Pd/CaCO_3 \; Cyclohexane}$$

55

99% Selectivity at 99% Conversion $H_3C(CH_2)_7HC=CH(CH_2)_7CO_2CH_3$

56

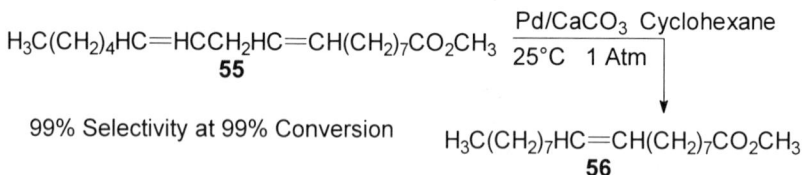

The selective hydrogenation of non-isomerizable dienes takes place over a
variety of catalysts. The saturation of m-vinyl isopropenylbenzene (**57**) in the
presence of m-diisopropenylbenzene (**58**) takes place over 5% Rh/C at room
temperature and 4 atmospheres with about 75% selectivity.[88] The hydrogenation
of the more accessible double bond in **59** (Eqn. 15.36) takes place with 40:1
selectivity in 90% yield at 0°C and one atmosphere over 5% Pd/BaSO$_4$ in a 1:1
mixture of pyridine and ethanol.[89]

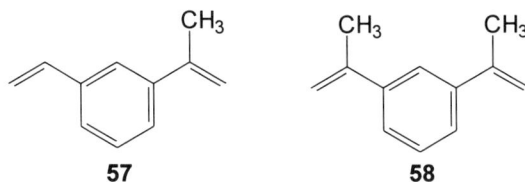

57

58

(15.36)

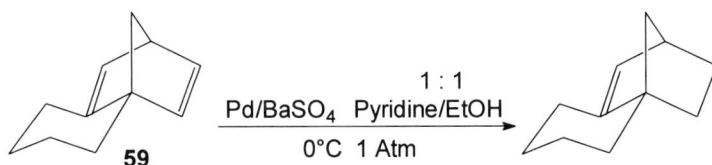

59

$$\xrightarrow[\text{0°C 1 Atm}]{\substack{1:1 \\ \text{Pd/BaSO}_4 \text{ Pyridine/EtOH}}}$$

Selective Reduction of the More Highly Substituted Olefinic Site

In all the examples mentioned above, whenever selectivity was observed in the hydrogenation of a diene, it was the less substituted double bond that was saturated. A more difficult procedure is obtaining the reverse selectivity in these reactions, that is, saturating the more substituted double bond and leaving the less substituted one intact. This task was accomplished by using the dicarbonyl-cyclopentadienyl iron (Fp') entity as a blocking group (Eqns. 15.37-39).[90] The Fp'-olefin-BF_4^- complexes are prepared by exchange of the diene with Fp(i-butylene)$^+BF_4^-$. The reaction takes place at the more reactive, less substituted double bond. These complexes are sufficiently stable so they can survive the hydrogenation of the remaining double bond using Pd/C at room temperature and atmospheric pressure. Decomposition of the saturated complex by reaction with NaI in acetone regenerates the monoalkene with the less substituted double bond intact.

(15.37)

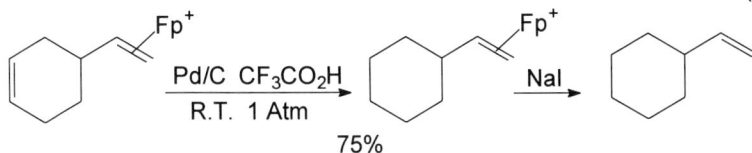

$$\xrightarrow[\text{R.T. 1 Atm}]{\text{Pd/C CF}_3\text{CO}_2\text{H}}$$

75%

$$\xrightarrow{\text{NaI}}$$

(15.38)

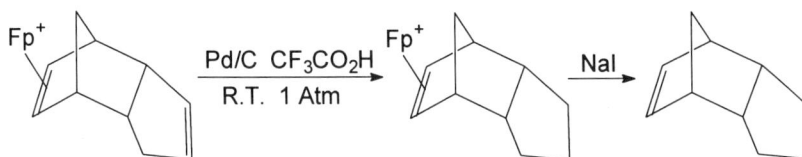

$$\xrightarrow[\text{R.T. 1 Atm}]{\text{Pd/C CF}_3\text{CO}_2\text{H}}$$

$$\xrightarrow{\text{NaI}}$$

(15.39)

$$\text{Fp}^+ \quad -C\equiv C- \quad \xrightarrow[\text{R.T. 1 Atm}]{\text{Pd/C CF}_3\text{CO}_2\text{H}} \quad \text{Fp}^+ \quad -\text{CH}_2\text{CH}_2-$$

$$\downarrow \text{NaI}$$

$$-\text{CH}_2\text{CH}_2-$$

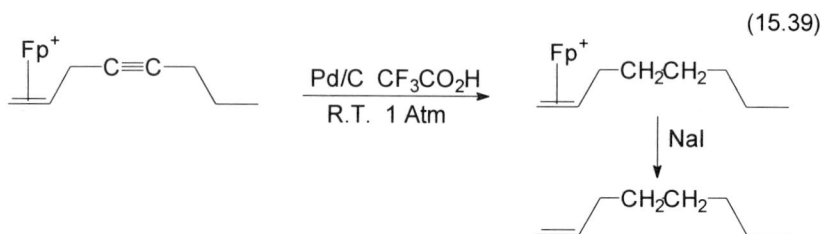

Conjugated Dienes

Cyclic Conjugated Dienes

The partial hydrogenation of conjugated dienes generally takes place more easily than the saturation of a monoalkene, which improves the prospect for selective hydrogenation of the conjugated species. This reaction is an example of *Type III* selectivity (Eqn. 15.40) (Chapter 5). Active catalysts can enhance organic substrate diffusion limitations and decrease the selectivity of monoalkene formation. Low hydrogen availability will minimize oversaturation.

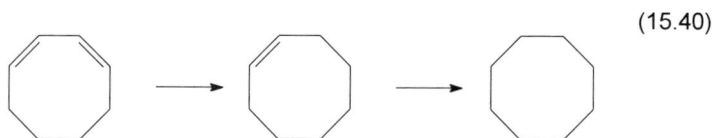

(15.40)

Hydrogenation studies of conjugated dienes show that the diene is more strongly adsorbed than the monoene produced on partial hydrogenation. This means that the monoolefin is displaced from the catalyst surface by the diene. Also, the double bond of the monoene does not isomerize in the presence of the diene.[91-94]   Therefore, maximum selectivity is obtained when the reaction conditions are such that sufficient diene is available to the catalyst to displace the monoene product. Low hydrogen availability can enhance selectivity at higher conversions.

In general, palladium is the preferred catalyst for this reaction. For example, the partial hydrogenation of 1,3-cyclooctadiene to cyclooctene took place with 99% selectivity at 99% conversion over both Pd/C and Pd/Al$_2$O$_3$ at 20°C and one atmosphere in isooctane solution.[95]   The hydrogenation of 1,5-cyclooctadiene under these conditions gave cyclooctene in 92% selectivity at 96% conversion with Pd/C and 96% selectivity at 98% conversion over Pd/Al$_2$O$_3$. Because palladium is a good catalyst for double-bond isomerization, the selectivity observed on hydrogenation of 1,5-cyclooctadiene is attributed, at least in part, to isomerization to the 1,3-isomer. Decreased activity was reported when silver was added to these palladium catalysts. Although the hydrogenation of

1,5-cyclooctadiene over the Pd-Ag catalyst decreased in selectivity, that of 1,3-cyclooctadiene did not change.[95] This is presumably due to the inhibition of isomerization by the Pd-Ag catalysts. If the 1,5-cyclooctadiene isomerization were lessened, selectivity should decrease. The 1,3-cyclooctadiene was conjugated, so the isomerization capability of the catalyst was not important and the reaction selectivity remained essentially unchanged.

A similar comparison in the selectivity observed on hydrogenation of 1,3-cyclooctadiene and 1,5-cyclooctadiene was reported for hydrogenations run over colloidal palladium supported on poly(N-vinyl-2-pyrrolidone) (Pd/PVP) at 30°C and atmospheric pressure in alcoholic solvents.[96] Under these conditions the hydrogenation of 1,3-cyclooctadiene was rapid up to the absorption of one equivalent of hydrogen, then the rate decreased by a factor of almost 600 with cyclooctene produced in 99.9% yield. With 1,5-cyclooctadiene, the change in rate at equimolar hydrogen uptake was smaller by a factor of about 20 and more gradual, but even then cyclooctene was produced in a 98% yield. The hydrogen uptake curves for these reactions are shown in Fig. 15.6.[96] For the hydrogenation of 1,5-cyclooctadiene, Pd/PVP has about the same activity as Pd/C, but over the latter catalyst the selectivity is somewhat lower (a 94% yield). The Pd/PVP catalyst was also reported to be selective for the partial hydrogenation of cyclopentadiene.[96]

One reason for the high selectivities observed in cyclooctadiene hydrogenations is the unusually low rate of hydrogenation of cyclooctene caused by the transannular interactions present in cyclooctane.[97] A better test of reaction selectivity would be the hydrogenation of cyclopentadiene where the monoolefin product, cyclopentene, is also easily hydrogenated. As mentioned above, the yields of cyclooctene obtained by the partial hydrogenation of 1,3-cyclooctadiene and 1,5-cyclooctadiene over Pd/PVP and Pd/C were 98% and 94%, respectively, at 100% conversion.[98] When these same catalysts were used for the hydrogenation of cyclopentadiene, the yields of cyclopentene at 100% conversion were 99% and 83%, respectively.[96] In this case the superiority of the Pd/PVP catalyst for the selective hydrogenation of conjugated dienes is more clearly seen. A similar effect was also noted in the use of colloidal palladium supported on a chelate resin composed of iminodiacetic acid groups attached to a styrene-divinylbenzene copolymer. This catalyst gave a cyclopentene yield of 97% at 100% conversion at 30°C and atmospheric pressure in methanol solution. Under the same conditions, the selectivity observed with Pd/C was 88%.[99] However, with a 0.5% Pd/Al$_2$O$_3$ catalyst, cyclopentene was obtained in 100% selectivity on hydrogenation of cyclopentadiene in toluene solution at 20°C and atmospheric pressure. In methanol the selectivity was lower and some cyclopentane was also formed.[100] Selectivity in the hydrogenation of cyclopentadiene increases with decreasing particle size for the catalysts Pd black < 5% Pd/C < colloidal Pd/PVP.[96]

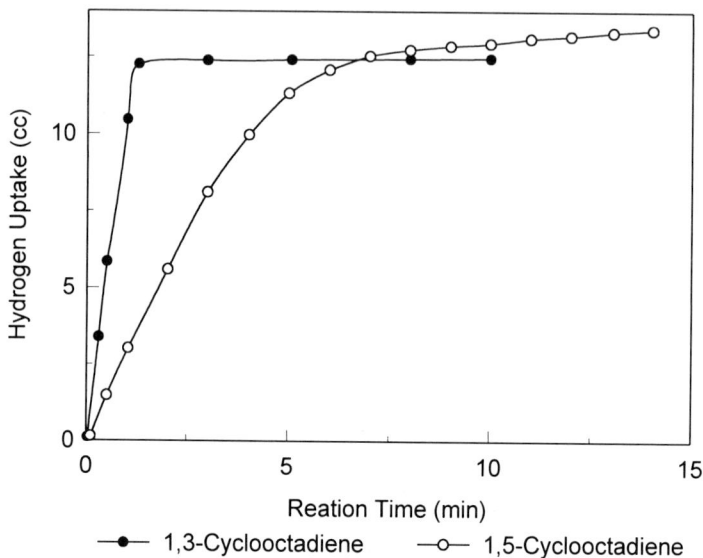

Fig. 15.6. Hydrogenation of the isomeric cyclooctadienes over Pd/PVP.
(Redrawn using data from Ref. 96)

Acyclic Conjugated Dienes

With the cyclic dienes discussed above, molecular symmetry dictates that regardless of the manner in which the hydrogen is added, only a single monoalkene is produced so these materials provide no evidence concerning the extent to which 1,2- and 1,4-addition of hydrogen are taking place. To obtain this information, it is necessary to examine the hydrogenation of straight-chain conjugated dienes, the simplest of which is 1,3-butadiene. The 1,2-addition of hydrogen to this molecule gives 1-butene (Eqn.15.41). The 1,4-addition products are the *cis* and *trans* 2-butenes (Eqn. 15.42) with the *cis/trans* ratio dependent on the diene conformation that is adsorbed on the catalyst.

(15.41)

(15.42)

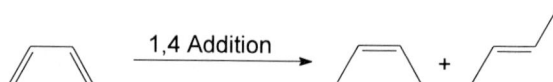

It has been proposed that 1,2-addition occurs through a bis π-adsorbed species, **60,** whereas 1,4-addition occurs through an adsorbed π-allyl intermediate, **61.** Catalysts that promote primarily 1,2-addition were referred to as Type A catalysts and those that react via a 1,4-addition mechanism as Type B catalysts.[92] The 2-butenes formed over Type B catalysts have low *cis/trans* ratios with the *trans* 2-butene usually the predominant alkene present. Some 2-butenes are produced with Type A catalysts, but these have *cis/trans* ratios near unity.

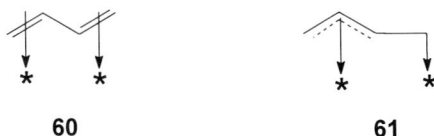

60 **61**

The vapor-phase hydrogenation of 1,3-butadiene at was run at 0–100°C over noble metal catalysts present as either metal wires or films[91] and on noble metal catalysts dispersed on oxide supports.[101,102] In all cases, palladium was the most selective catalyst giving essentially 100% conversion to the isomeric butenes. Reaction selectivity was somewhat lower over rhodium and platinum catalysts and even lower with ruthenium, but even these selectivities were between 80–95%. Iridium and osmium catalysts were essentially nonselective.[101] All of these catalysts, except palladium, fall into the Type A category with primary 1-butene formation and *cis/trans* 2-butene ratios near unity. Palladium is a Type B catalyst giving *cis/trans* 2-butene ratios less than 0.4 for hydrogenations run over films and wires[91] and less than 0.07 for reactions run over Pd/Al$_2$O$_3$[101] The supported catalysts gave 1-butene as the primary product, while the polycrystalline wires and films gave *trans* 2-butene as the primary product so the use of 1-butene formation data to classify the catalyst type can be ambiguous.

Vapor-phase 1,3-butadiene hydrogenations were also run over a number of base metal catalysts. Of these, the copper catalysts were, by far, the most selective. Copper oxide reduced at 180°C gave, at 95% conversion, 92% 1-butene, 3% *trans* 2-butene, and 5% *cis* 2-butene with no butane formation observed for hydrogenations run at 65°C. Similar results were observed with a Cu/Al$_2$O$_3$catalyst.[92] Alumina-supported nickel, cobalt, and iron catalysts also proved to be reasonably selective, but the 1-butene yields obtained using these catalysts were not nearly as high as those found using copper. Iron exhibited Type B behavior with the primary product being the *trans* 2-butene.[92]

Both cobalt and nickel catalysts could be modified to give either Type A or Type B behavior depending on the temperature used to reduce the catalyst.[92] Cobalt catalysts reduced at lower temperatures (250–300°C) gave primarily 1-butene as the product, with the *cis* and *trans* 2-butenes formed in a ratio of about 0.5. Reduction at temperatures greater than 400°C resulted in the formation of a catalyst that gave *trans* 2-butene as the primary product with a *cis/trans* 2-butene ratio of about 0.12. The same catalyst could be changed from one type to the other depending on its treatment before use in the hydrogenation. A powdered nickel catalyst showed Type B behavior after reduction at 400°C. After air oxidation at room temperature for one week followed by re-reduction at 195°C, this same catalyst showed Type A behavior. Further treatment with hydrogen at 400°C restored the Type B characteristics. The overall selectivity toward partial hydrogenation was not changed by these treatments.[92]

Electron microscopic examination of these catalysts showed that the powders that exhibited Type A behavior had many particles less than 10 nm in diameter, whereas the powders showing Type B characteristics had relatively few of these small particles.[92] A similar size effect was reported for Ni(P) catalysts prepared by the reduction of nickel orthophosphate.[93] When the catalyst was reduced at temperatures near 400°C, Type A behavior was observed, whereas Type B behavior occurred with catalysts prepared at higher reduction temperatures. Here, too, a particle size effect was proposed to explain these results. Lower temperature reduction gives fine clusters of Ni interspersed with the Ni(P), and it was considered that these small particles were more effective for 1,2-addition. High-temperature reduction produced primarily the Ni_2P species, which was catalytically less active and interacted by a 1,4-hydrogen addition mechanism to give the *trans*-2-butene. In contrast to reports on the use of Ni/Al_2O_3 and Co/Al_2O_3 catalysts, the selectivity for partial hydrogenation was higher with the larger particle, Type B, catalysts than those with smaller particles.[92]

It is interesting, though, that amorphous Ni(P) and Ni(B) alloy ribbons prepared by rapid quenching were active catalysts and quite selective in 1,3-butadiene hydrogenations, which occurred by way of a Type A process. Heating these ribbons to about 500°C resulted in crystallization. The crystallized catalysts were less active for hydrogenation, but their selectivities and mode of hydrogenation remained the same.[103]

As the dispersion of the supported palladium catalysts increased above 20%, the rate of 1,3-butadiene hydrogenation decreased.[104] The selectivity for 1-butene formation at 90% and 99% conversion decreased with increasing dispersion. The 1-butene/2-butene ratio and the *cis/trans* 2-butene ratio, however, remained constant regardless of the catalyst dispersion, support, or metal load.[104] The decrease in rate was ascribed to an electronic effect: the smaller the particle size, the higher the ratio of surface atoms having high-coordinate unsaturation and, thus, the stronger the interaction with electron-donating species particularly those having a high electron density such as dienes and alkynes. This strong coordination would decrease the reaction rate. It was considered that interaction of the catalyst with an electron-donating species should modify this strong coordination and increase the rate of the reaction. This was observed for the rates of hydrogenation of 1,3-butadiene and 1-butyne when piperidine was added to the reaction mixture.[105] However, data on the product composition from the 1,3-butadiene hydrogenation were not reported for the piperidine-modified catalyst.

An electronic effect was also used to explain the difference in 1,3-butadiene hydrogenation selectivity observed over various types of nickel catalysts such as Ni(B), Raney nickel, nickel powder from the decomposition of nickel formate, Ni(P), and Ni(S).[106] As discussed in Chapter 12, chemical shifts in XPS binding energies (Δq) for the various nickel species were compared with that of the decomposed nickel catalyst to determine the extent of 1-butene formation as related to the electron density on the metal. The higher the electron density, the more 1-butene formation was favored.[106]

These seemingly contradictory observations can be reconciled by a third type of catalyst behavior. As stated previously, Type A catalysts direct primary 1,2-addition and Type B catalysts react by 1,4-addition with large amounts of *trans* 2-butene being formed. It would be better to split this latter category into two subdivisions, Type B and Type B', both of which promote primarily 1,4-addition as shown by the preponderant formation of the 2-butenes but which differ in *cis/trans* 2-butene ratios. All of the Type B catalysts discussed above except palladium gave *cis/trans* ratios of approximately 0.4 or more. With palladium, these ratios are much lower. Because double-bond hydrogenation over palladium catalysts takes place via a π-allyl adsorbed species whereas only π- and σ-adsorbed species are involved with the other noble metal catalysts,[4] it is reasonable to place palladium into a separate category regarding diene hydrogenation as well. Thus, Type A catalysts can promote 1,2-addition as shown by the primary formation of 1-butene, Type B catalysts can give 1,4-addition with the primary formation of the 2-butenes having a high *cis/trans* ratio, and Type B' catalysts can react by 1,4-addition and produce 2-butenes with a low *cis/trans* ratio. That palladium falls into the Type B' category is probably the result of extensive π-allyl adsorption in the hydrogenation mechanism.

Reaction Products and the Single-Atom Site Model

These catalytic properties can be explained considering the active sites in alkene hydrogenation as the $^3MH_2$ and ^3MH and the isomerization sites as the ^2MH (Chapters 3 and 4). The enhanced adsorption capability of dienes may be the result of the increased electron density in these species. The 1,2-addition mechanism is the only one that can reasonably be expected to affect $^3MH_2$ sites. Thus, on this type of site the diene can adsorb either as a monoadsorbate (**62**) with the enhanced adsorption caused by the electronic nature of the diene or as a diadsorbate (**63**) with one double bond adsorbed on an unreactive face or edge site to facilitate the overall adsorption. Presently there is no way of determining between these possibilities. The observation that 1,2-addition increases with decreasing particle size fits with the idea that such addition takes place over the $^3MH_2$ sites because the smaller the metal crystallite, the higher the percent of corner or adatom sites present among the surface atoms. Such small particles, though, will also have fewer face atoms so the increased 1,2-addition is more likely due to enhanced monoadsorption, as depicted by **62**, than by the assisting diadsorbed model, **63**.

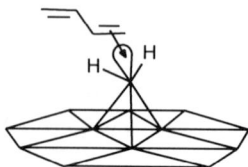
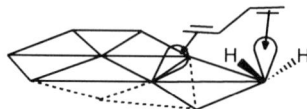

62   **63**

On the other hand, 1,4-addition must involve a diadsorbed species such as **64**. The surface atoms can be ^2MH, ^3MH, or $^3MH_2$, but only one can be the $^3MH_2$ type. The transfer of one hydrogen atom from each surface site gives the adsorbed olefin with its stereochemistry determined by the mode of adsorption of the diene. The s-*cis* form of butadiene (**65**) is about 2.3 kcal/mol less stable than the s-*trans* form (**66**).[107] This corresponds to about a 30:70 ratio, which agrees quite well with the observed *cis/trans*-2-butene ratios formed over Type B catalysts. The likelihood of 1,4-addition is increased when larger catalyst particles are present because more neighboring edge and corner atoms are also present.

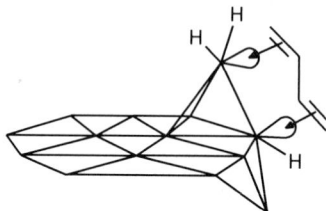

64   **65**   **66**

With a palladium catalyst in the hydrogenation of butadiene, the reaction mechanism probably takes place by way of a π-allyl intermediate such as **67** or **68**. The steric constraints of the *cis* form, **68**, are such that the *trans* species, **67**, is considerably more stable. The low *cis/trans* 2-butene ratios observed with palladium are a consequence of the difference in energy between these two adsorbed species.

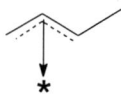

67   **68**

It is not clear why selectivity in butadiene hydrogenation decreases with increasing palladium catalyst dispersion when selectivity appears to increase with increasing palladium dispersion in the hydrogenation of cyclopentadiene. The cyclic diene may be more strongly adsorbed and, thus, more capable of displacing the monoene as it is formed. This difference would become more important as the electronic characteristics change with decreasing particle size.

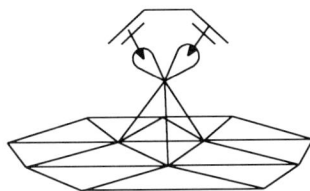

69

The chelate diadsorbed entity, **69**, is not a very likely surface moiety in these reactions. The 1,4-addition to such species would lead to exclusive *cis* 2-butene formation. The *cis* 2-butene is never the predominant product of any metal catalyzed hydrogenation since even the most coordinatively unsaturated metal surface atoms do not have the unsaturation required to adsorb the two double bonds of the diene and a hydrogen molecule, the orientation needed for the single-atom 1,4-hydrogenation mechanism. However, hydrogenation of 1,3-butadiene over a molybdenum subcarbonyl species in a Y zeolite cage gave *cis* 2-butene in greater than 96% selectivity.[108]

Substituted Acyclic Conjugated Dienes

The hydrogenation of substituted butadienes introduces another element into the problem of selectivity, that is the influence of the substituents on the degree of hydrogenation and the direction of hydrogen addition. The hydrogenation of 1,3-butadiene-2,3-dicarbonitrile (**70**) over Pd/C in THF at atmospheric pressure and room temperature gave almost exclusively the *cis*- and *trans*-2-butene-2,3-dicarbonitriles (**71** and **72**), the products of 1,4-addition (Eqn. 15.43).[109]

(15.43)

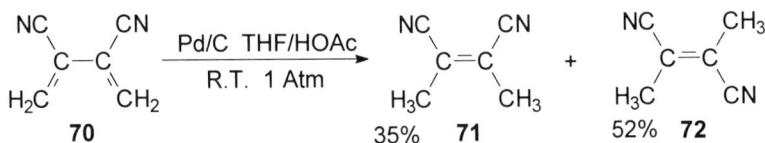

The hydrogenation of 1,3-pentadiene either neat or in methanol solution over noble metal catalysts gave results[5] comparable to those observed in the vapor-phase hydrogenation of 1,3-butadiene over these same catalysts.[101] Palladium provided a selectivity for monohydrogenation of 98-99% at 100% conversion with the *trans* 2-pentene the predominant product regardless of the stereochemistry of the starting material. These results agree with the assignment of palladium to the Type B' catalyst category. The other noble metals showed significantly lower selectivities, but in all cases in which 1,2-addition was operative it was the 3,4-double bond that was saturated as shown in Eqn. 15.44.

(15.44)

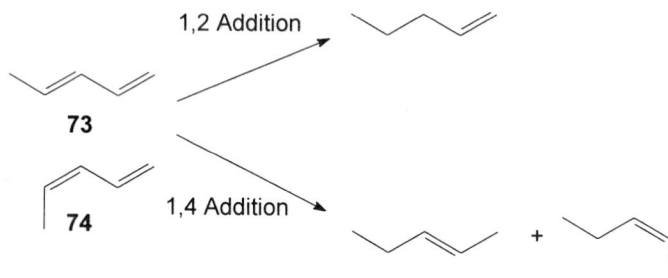

The results of the vapor-phase hydrogenation of 1,3-pentadiene[110] also paralleled those of the selective hydrogenation of 1,3-butadiene.[92,101] Palladium exhibited Type B' behavior with greater than 99% selectivity and a *cis/trans* 2-pentene ratio of 0.17 when the starting material was the *trans* isomer, **73**, and a ratio of 0.4 when the starting material was the *cis* isomer, **74**. A Cu/Al$_2$O$_3$ catalyst gave a 70% yield of 1-pentene at 100% selectivity regardless of starting material stereochemistry. A Co/Al$_2$O$_3$ catalyst that was reduced at a higher temperature gave almost exclusive 1,4-addition with a *cis/trans*-2-pentene ratio between 0.07 and 0.12.[110] Hydrogenation over copper chromite took place by way of a Type A process to give mainly 1-pentene.[111] With a ZnO catalyst an 87% yield of 1-pentene was produced, the highest reported for this reaction.[112] The selective hydrogenation of the 3,4-double bond in 1,3-pentadiene appears to

(15.45)

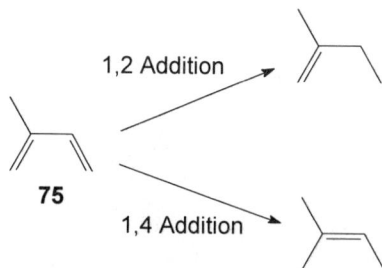

be contrary to the observed saturation of the least substituted double bond in unconjugated dienes discussed previously. Apparently the enhanced adsorption capability of a 3,4-disubstituted alkene is more important than steric factors with this conjugated molecule.

The hydrogenation of isoprene, **75**, however shows the reverse trend with the primary product from 1,2-addition being 2-methyl-l-butene (Eqn. 15.45), which is formed in about 75% yield over copper chromite catalysts.[111] The hydrogenation of isoprene over a sulfided Raney nickel catalyst[113] took place primarily by way of a 1,4-addition process to give 2-methyl-2-butene (Eqn. 15.45).

Apparently with isoprene, steric factors are more important than electronic considerations although the reason is not obvious because the steric differences between 1,3-pentadiene and isoprene are not great. It would seem from these results that some factors controlling the hydrogenation of these substituted conjugated dienes can be rather subtle.

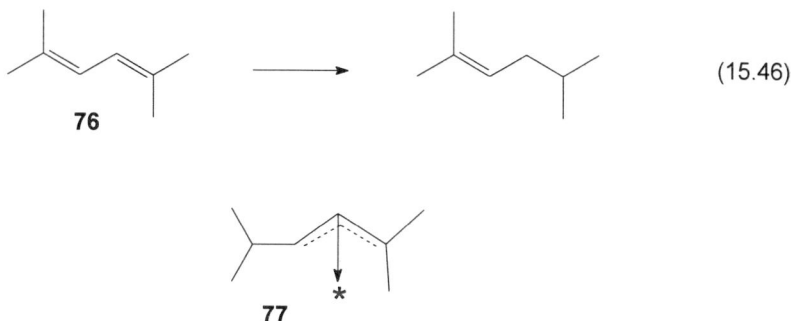

(15.46)

76

77

Further increases in the steric environment can give even more peculiar results. The hydrogenation of 2,5-dimethyl-2,4-hexadiene (**76**) proceeded by way of a 1,2-addition mode over palladium catalysts (Eqn. 15.46).[114] In this case the π-allyl intermediate (**77**) required for 1,4-addition is evidently too sterically crowded because of the additional alkyl groups present, so this species is not a viable surface moiety and 1,2-addition is favored by default.

Cumulative Double Bonds

The hydrogenation of cumulative double bonds (allenes) takes place at a rate comparable to that of the corresponding acetylene but no isomerization to the acetylene is observed during the reaction.[115] Selective hydrogenation to the alkene occurs since the allene readily displaces the product olefin from the catalyst before the hydrogenation can take place.[115] Selectivities for monoalkene formation of greater than 95% were commonly obtained in the hydrogenation of various substituted allenes.[116] This reaction selectivity is probably facilitated by the fact that the π electron clouds of the two double bonds are perpendicular to each other so after the saturation of one double bond the electrons of the other are not in a position to easily interact with the catalyst surface. Palladium is the catalyst of choice for allene hydrogenation. The initial hydrogenation takes place primarily on the less substituted double bond of polysubstituted allenes to give the cis-alkenes (Eqns. 15.47–49).[116]

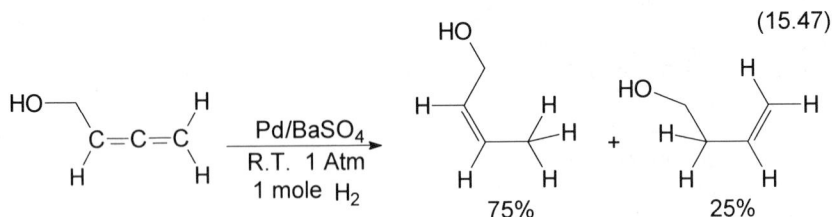

(15.47)

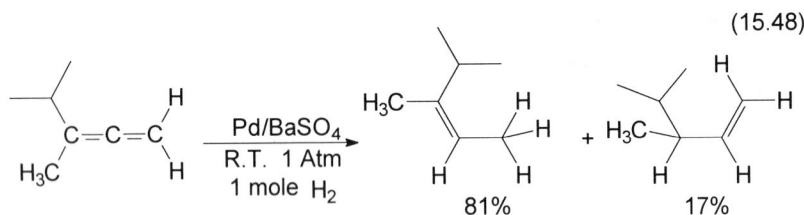

(15.48)

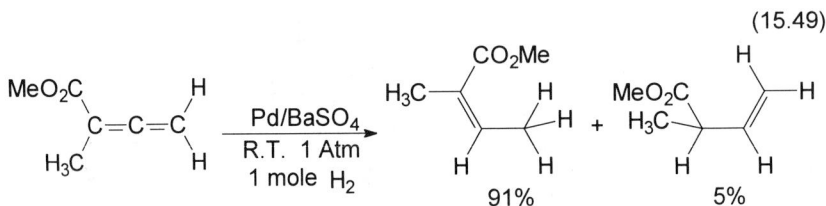

(15.49)

References

1. G. V. Smith and R. L. Burwell, Jr., *J. Am. Chem. Soc.,* **84**, 925 (1962).
2. E. Kovats, A. Fürst and H. H. Gunlliered, *Helv. Chim. Acta,* **37**, 534 (1954).
3. L. Horiuti and M. Polanyi, *Trans. Faraday Soc.,* **30**, 1164 (1934).
4. R. L. Augustine, F. Yaghmaie and J. F. Van Peppen, *J. Org. Chem.,* **49**, 4853 (1984).
5. G. C. Bond and J. S. Rank, *Proc. 3rd. Int. Cong. Cat.,* **2**, 1225 (1965).
6. G. C. Bond and P. B. Wells, *Adv. Catal.,* **15**, 91 (1964).
7. J. M. Campelo, A. Garcia, D. Luna and J. M. Marinas, *React. Kinet. Catal. Lett.,* **21**, 209 (1982).
8. G. D. Lauback and K. J. Brunings, *J. Am. Chem. Soc.,* **74**, 705 (1952).
9. D. H. R. Barton and J. D. Cox, *J. Chem. Soc.,* 1354 (1948).

10. P. Bladon, J. M. Fabian, H. B. Henbest, H. P. Kich and G. W. Wood, *J. Chem. Soc.,* 2402 (1951).
11. C. Djerassi, J. Romo and G. Rosenkranz, *J. Org. Chem.,* **16**, 754 (1951).
12. J. B. Bream, D. C. Eaton and H. B. Henbest, *J. Chem. Soc.,* 1974 (1957).
13. J. Elks, R. M. Evans, C. H. Robinson, G. H. Thomas and L. J. Wyman, *J. Chem. Soc.,* 2933 (1953).
14. Y. C. Zunt and F. Sondheimer, *J. Chem. Soc.,* 2957 (1950).
15. G. P. Kugatova-Shemyakina and Yu. A. Ovinnikov, *Tetrahedron,* **18**, 697 (1962).
16. G. P. Kugatova-Shemyakina, G. M. Nikolaev and V. M. Andreev, *Tetrahedron,* **23**, 2721 (1967).
17. G. P. Kugatova-Shemyakina and G. M. Nikolaev, *Tetrahedron,* **23**, 2987 (1967).
18. J. Meinwald, S. S. Labana and M. S. Chadha *J. Am. Chem. Soc.,* **85**, 582 (1963).
19. M. T. Wuesthoff and B. Rickborn, *J. Org. Chem.,* **33**, 1311 (1968).
20. V. A. Slabey, P. H. Wise and L. C. Gibbons, *J. Am. Chem. Soc.,* **71**, 1518 (1949).
21. C. S. Marvel, R. E. Allen and C. G. Overberger, *J. Am. Chem. Soc.,* **68**, 1088 (1946).
22. H. C. Brown and C. G. Rao, *J. Org. Chem.,* **44**, 1348 (1979).
23. G. E. Ham and W. P. Cocker, *J. Org. Chem.,* **29**, 194 (1964).
24. W. A. Bonner, N. I. Burke, W. E. Fleck, R. K. Hill, J. A. Joule, B. Sjoberg and J. H. Zalkow, *Tetrahedron,* **20**, 1419 (1964).
25. T. W. Russell and R. C. Hoy, *J. Org. Chem.,* **36**, 2018 (1971).
26. T. W. Russell and D. M. Duncan, *J. Org. Chem.,* **39**, 3050 (1974).
27. K. Heusler, J. Kebrle, C. Meystre, H. Ueberwasser, P. Wieland, G. Anner and A. Wettstein, *Helv. Chim. Acta,* **42**, 2043 (1959).
28. F. Kogl and O. A. deBruin, *Recl. Trav. Chim. Pays-Bas,* **69**, 729 (1950).
29. L. J. Sargent and L. F. Small, *J. Org. Chem.,* **16**, 1031 (1951).
30. W. S. Johnson, *J. Am. Chem. Soc.,* **78**, 6278 (1956).
31. J. M. Bell, R. Garrett, V. A. Jones and D. G. Kubler, *J. Org. Chem.,* **32**, 1307 (1967).
32. S. Nishimura, K. Kagawa and N. Sato, *Bull. Chem. Soc., Japan.,* **51**, 3330 (1978).
33. S. Nishimura and Y. Kano, *Chem. Lett.,* 565 (1972).
34. K. W. Rosenmund, G. Kositzke and H. Bach, *Chem. Ber.,* **92**, 494 (1959).
35. K. W. Rosenmund, H. Herzberg and H. Schütt, *Chem. Ber.,* **87**, 1258 (1954).
36. H. H. Inhoffen, G. Stoeck, G. Kolling and U. Stoeck, *Liebigs Ann. Chem.,* **568**, 52 (1950).
37. J. Fajkos, *Coll. Czech. Chem. Commun.,* **25**, 1978 (1960).
38. D. S. Grierson, J.-L. Bettiol, I. Buck, H.-P. Husson, M. Rubiralta and A. Diez, *J. Org. Chem.,* **57**, 6414 (1992).
39. R. L. Augustine, "Steroid Hydrogenation", in *Organic Reactions in Steroid Chemistry*, Vol. 1, (J. Fried and J. A. Edwards, Eds.), Van Nostrand, Reinhold, New York, 1972, p. 111.

40. H. Heusser, V. Frick and P. A. Plattner, *Helv. Chim. Acta*, **33**, 1260 (1950).
41. E. B. Hershberg, E. P. Oliveto, C. Gerold and L. Johnson, *J. Am. Chem. Soc.*, **73**, 3073 (1951).
42. R. L. Augustine and E. J. Reardon, Jr., *Org. Prep. and Proc.*, **1**, 107 (1969).
43. C. W. Shoppee, B. D. Agashe and G. H. R. Summers, *J. Chem. Soc.*, 3107 (1957).
44. J. R. Lewis and C. W. Shoppee, *Chem. and Ind., (London)*, 897 (1953).
45. J. R. Lewis and C. W. Shoppee, *J. Chem. Soc.*, 1365 (1955).
46. H. J. Dauben, Jr., B. Loken and J. H. Ringold. *J. Am. Chem. Soc.*, **76**, 1359 (1954).
47. H. R. Nace, *J. Am. Chem. Soc.*, **73**, 2379 (1951).
48. B. V. McKenzie, V. R. Mattox and E. C. Kendall, *J. Biol. Chem.*, **175**, 249 (1948).
49. B. Torok, A. Molnar, K. Borszeky, E. Toth-Kadar and I. Bakonyi, *Stud. Surf. Sci. Catal.*, **78** (Heterog. Catal. Fine Chem., III), 179 (1993).
50. C. M. Belisle, Y. M. Young and B. Singaram, *Tetrahedron Lett.*, **35**, 5595 (1994).
51. N. Ravasio, M. Antenori and M. Gargano, *Stud. Surf. Sci. Catal.*, **78** (Heterog. Catal. Fine Chem., III), 75 (1993).
52. E. B. Maxted and M. S. Biggs, *J. Chem. Soc.*, 3844 (1957).
53. R. L. Augustine, D. C. Migliorini, R. E. Foscante, C. S. Sodano and M. J. Sisbarro, *J. Org. Chem.*, **34**, 1075 (1969).
54. R. L. Augustine, *Adv. Catal.*, **25**, 56 (1976).
55. R. L. Augustine, *Catal. Revs.*, **13**, 285 (1976).
56. G. Slomp, Jr., Y. F. Shealy, J. L. Johnson, R. A. Donia, B. A. Johnson, R. P Holysz, R. L. Pederson, A. O. Jensen and A. C. Ott, *J. Am. Chem. Soc.*, **77**, 1216, (1955).
57. F. J. McQuillin, W. O. Ord and P. L. Simpson, *J. Chem. Soc.*, 5996 (1963).
58. S. Nishimura, M. Shimahara and M. Shiota, *J. Org. Chem.*, **31**, 2394 (1966).
59. M. Smith, "Dissolving Metal Reductions", in *Reductions, Techniques and Applications in Organic Synthesis* (R. L. Augustine, Ed.) Dekker, New York, 1968, p 95.
60. R. A. Pilli and M. M. Murta, *J. Org. Chem.*, **58**, 338 (1993).
61. D. Cortes, S. M. Myint, J. C. Harmange, S. Sahpaz and B. Figadere, *Tetrahedron Lett.*, **33**, 5225 (1992).
62. A. C. Day, J. Nabney and A. I. Scott, *J. Chem. Soc.*, 4067 (1961).
63. B. R. Baker, R. E. Schaub, M. V. Querry and J. H. Williams, *J. Org. Chem.*, **17**, 97 (1952).
64. J. W. Herndon and J. J. Matasi, *Tetrahedron Lett.*, **33**, 5725 (1992).
65. K. J. Liska, *J. Pharm. Sci.*, **53**, 1427 (1964).
66. R. L. Augustine, R. F. Bellina and A. J. Gustavsen, *J. Org. Chem.*, **33**, 1287 (1968).
67. L. M. Berkowits and P. N. Rylander, *J. Org. Chem.*, **24**, 708 (1959).
68. C. A. Brown and H. C. Brown, *J. Am. Chem. Soc.*, **85**, 1003 (1963).
69. C. A. Brown and V. K. Ahuja, *J. Org. Chem.*, **38**, 2226 (1973).
70. L. Cerveny, R. Junova and V. Ruzicka, *Coll. Czech. Chem. Commun.*, **44**, 2378 (1979).
71. L. Cerveny and V. Ruzicka, *React. Kinet. Catal. Lett.*, **17**, 161 (1981).

72. L. Cerveny and V. Ruzicka, *Catal. Revs.*, **24**, 503 (1982).
73. P. B. Weisz, V. J. Frilette, R. W. Maatman and E. B. Mower, *J. Catal.*, **1**, 307 (1962).
74. R. M. Dessau, *J. Catal.*, **77**, 304 (1982).
75. R. M. Dessau, *J. Catal.*, **89**, 520 (1984).
76. M. D. Ward, T. V. Harris and J. Schwartz, *J. Chem. Soc., Chem. Commun.*, 357 (1980).
77. T.-N. Huang and J. Schwartz, *J. Am. Chem. Soc.*, **104**, 5244 (1982).
78. D. R. Corbin, W. C. Seidel, L. Abrams, N. Herron, G. D. Stucky and C. A. Tolman, *Inorg. Chem.*, **24**, 1800 (1985).
79. H. Kuno, M. Shibagaki, K. Takahashi, I. Honda and H. Matsushita, *Bull. Chem. Soc., Japan*, **65**, 1240 (1992).
80. H. C. Brown and C. A. Brown, *Tetrahedron, Suppl. 8, Part 1*, 149 (1966).
81. C. A. Brown, *J. Chem. Soc., Chem. Commun.*, 952 (1969).
82. M. M. Johnson and G. P. Nowack, U. S. Patent, 4,716,256 (1987).
83. M. M. Johnson, G. P. Nowack and D. C. Tabler, *J. Catal.*, **27**, 397 (1972).
84. T. H.Cymbaluk, U. S. Patent, 4,659,687 (1987).
85. W. F. Newhall, *J. Org. Chem.*, **23**, 1274 (1958).
86. S. Nishmura, M. Ishibashi, H. Takamiya, N. Koike and T. Matsunaga, *Chem. Lett.*, 167 (1987).
87. R. L.Augustine, R. M. Warner and M. J. Melnick, *J. Org. Chem.*, **49**, 4853 (1984).
88. H. A. Colvin and J. Muse, Jr., U. S. Patent, 4,390,741 (1982).
89. J. E. Nystrom, T. D. McCanna, P. Helquist and R. S. Iyer, *Tetrahedron Lett.*, **26**, 5393 (1985).
90. K. M. Nicholas, *J. Am. Chem. Soc.*, **97**, 3254 (1975).
91. P. B. Wells and A. J. Bates, *J. Chem. Soc.*, 3064 (1968).
92. J. J. Phillipson, P. B. Wells and G. R. Wilson, *J. Chem. Soc. A,* 1351 (1969).
93. E. Nishimura, Y. Inoue and I. Yasumori, *Bull. Chem. Soc., Japan.*, **48**, 803 (1975).
94. F. Wozak and R. Adachi, *J. Catal.*, **40**, 166 (1975).
95. C. Fragale, M.Gargano, N. Raviso, M. Rossi and I. Santo, *J. Mol. Catal.*, **24**, 211 (1984).
96. H. Hirai, H. Chawanya and N. Toshima, *React. Polym., Ion Exch., Sorbents*, **3**, 127 (1985).
97. I. Jardine and F. J. McQuillin, *J. Chem. Soc., C*, 458 (1966).
98. H. Hirai, H. Chawanya and N. Toshima, *Bull. Chem. Soc., Japan.*, **58**, 682 (1985).
99. H. Hirai, S. Komatsuzaki and N. Toshima, *Bull. Chem. Soc., Japan.*, **57**, 488 (1984).
100. L. Cerveny, J. Bopatova and V. Ruxicka, *React. Kinet. Catal. Lett.*, **19**, 223 (1982).
101. G. C. Bond, G. Webb, P. B. Wells and J. M. Winterbottom, *J. Chem. Soc.*, 3218 (1965).
102. A. J. Bates, Z. K. Leszczynski, J. J. Phillipson, P. B. Wells and G. R. Wilson, *J. Chem. Soc. A,* 2435 (1970).
103. S. Yoshida, H. Yamashita, T. Funabiki and T. Yonezawa, *J. Chem. Soc., Chem. Commun.*, 964 (1982).
104. J. P. Boitiaux, J. Cosyns and S. Vasudevan, *Appl. Catal.*, **6**, 41 (1983).

105. J. P. Boitiaux, J. Cosyns and S. Vasudevan, *Appl. Catal.*, **15**, 317 (1985).

106. Y. Okamoto, K. Fukino, T. Imanaka and S. Teranishi, *J. Catal.*, **74**, 173 (1982).

107. R. G. Parr and R. A. Mulliken, *J. Chem. Phys.*, **18**, 1338 (1950).

108. Y. Okamoto, A. Maezawa, H. Kane and T. Imanaka, *J. Chem. Soc., Chem. Commun.*, 380 (1988).

109. R. L. Cobb, B. C. Vives and J. E. Mahan, *J. Org. Chem.*, **43**, 926 (1978).

110. P. B. Wells and G. R. Wilson, *J. Chem. Soc., A,* 2442 (1970).

111. M. Daage and J. P. Bonnelle, *Appl. Catal.*, **16**, 355 (1985).

112. T. Okuhara and K. Tanaka, *J. Chem. Soc., Chem. Commun.*, 53 (1978).

113. J. P. Boitiaux, J. Cosyns and G. Martino, *Stud. Surf. Sci. Catal.*, **11** (Met. Support-Met. Addit. Eff. Catal.), 355 (1982).

114. L. Cerveny, I. Paseka, K. Surma, T. T. Nquyen and V. Ruzicka, *Coll. Czech. Chem. Commun.*, **50**, 61 (1985).

115. G. C. Bond and J. Sheridan, *Trans. Faraday Soc.*, **48**, 651 (1952).

116. L. Crombie, P. A. Jenkins and D. A. Mitchard, *J. Chem. Soc., Perkin I.,* 1081 (1975).

Hydrogenation III: Alkynes

The catalytic hydrogenation of acetylenes is the most facile of all hydrogenation reactions, generally taking place in the presence of all other functional groups including conjugated dienes and aromatic nitro groups. The ease of saturation is enhanced if the triple bond is terminal. The complete hydrogenation of alkynes to alkanes occurs over all common hydrogenation catalysts at room temperature and atmospheric pressure. This reaction and the related semihydrogenation to form alkenes is important in synthetic procedures because the acetylenic group can participate in a wide range of substitution reactions and, thus, can join two pieces of a carbon chain to produce a variety of aliphatic species. The selective modification of the triple bond to form either a double or single bond further enhances the synthetic utility of acetylenic materials.[1]

The selective hydrogenation of a triple bond to give an alkene without concomitant positional or geometric isomerization is particularly important in synthetic procedures and many industrial processes. In the absence of any isomerization, selective partial hydrogenation of a disubstituted alkyne produces the *cis* alkene. Small amounts of the *trans* alkene are sometimes formed in these reactions, but catalytic processes do not lead to the production of the *trans* olefin as the primary product. The *trans* alkenes can be produced as a primary product by metal-ammonia reduction of disubstituted alkynes.[2]

(16.1)

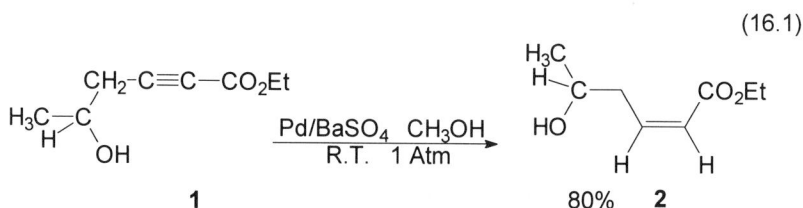

While the hydrogenation of the propargylic ester, **1**, gave the *cis* alkene, **2**, (Eqn. 16.1) partial hydrogenation of the hydroxy propargylic acid, **3**, gave the lactone, **4**, (Eqn. 16.2).[3] The semihydrogenation of the hydroxy acetal, **5**, in the presence of a trace of acid gave the dihydropyran, **6**, (Eqn. 16.3).[4]

The hydrogenation of an alkyne is an example of a *Type III* selectivity (Chapter 5) (Eqn. 16.4). As a result, when diffusion, particularly that of the

(16.2)

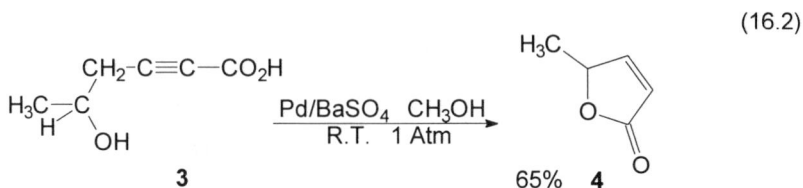

3   Pd/BaSO₄ CH₃OH
                                          R.T.   1 Atm

65%   4

(16.3)

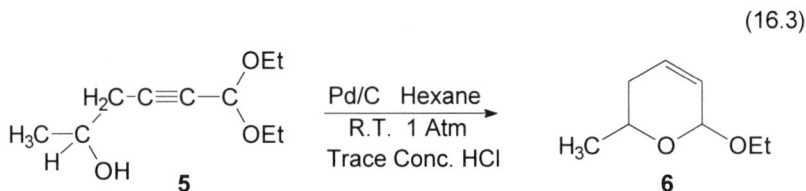

5   Pd/C   Hexane
                        R.T. 1 Atm
                     Trace Conc. HCl   6

organic substrate, is the limiting factor, reactions can exhibit decreased selectively toward alkene formation. It is generally accepted that at least part of the reaction selectivity occurs because of the strong chemisorption of alkynes, so the alkyne displaces the alkene, which is produced by partial hydrogenation, from the catalyst surface. Thus, for maximum selectivity, the catalyst system must show a considerable difference between the strengths of chemisorption of alkynes and alkenes.   Further, reaction conditions must promote a facile exposure of the catalyst surface to alkyne molecules so the displacement of the alkene will occur faster than the further saturation or isomerization of the double bond, reactions that can lead to a significant lowering of selectivity at high conversions. This problem may be overcome by terminating the reaction prior to the uptake of one equivalent of hydrogen or by using low hydrogen availability conditions.

(16.4)

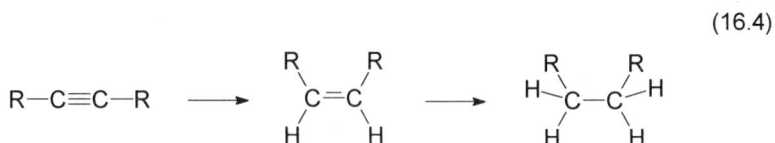

It was shown that with a Pd/C catalyst in the liquid phase terminal triple bonds were saturated faster than internal ones, and both hydrogenated faster than terminal or internal double bonds in competitive processes (Eqn. 16.5).[5] Further, alkene isomerization generally does not take place over palladium catalysts when alkynes are present.   This selective hydrogenation depends on the stronger adsorption of an alkyne compared to an alkene. It is also possible that steric factors can influence the selectivity in the competitive semihydrogenation of an acetylene and an olefinic group in the same molecule. When the double bond and the triple bond are *cis* to each other as in **7**, selective adsorption of the acetylene

(16.5)

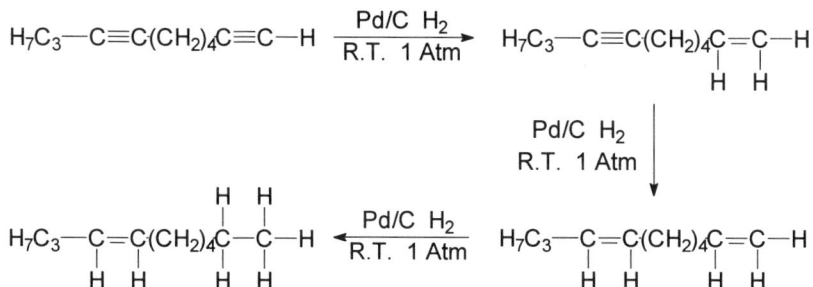

without having the olefinic group also in close proximity to the catalyst is not possible, so double bond saturation occurs along with the semihydrogenation of the triple bond (Eqn. 16.6). When the two groups are *trans* to each other as in **8**, selective adsorption and semihydrogenation of the acetylene can more easily take place and the original olefinic group is not saturated (Eqn. 16.7).[6]

(16.6)

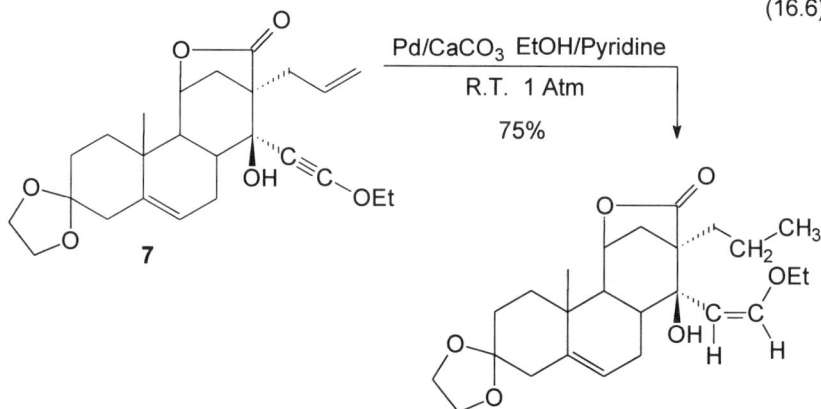

Catalysts

All of the common hydrogenation catalysts can effect the complete saturation of the alkyne to alkane, but all catalysts are not equally effective in the selective hydrogenation to produce alkenes. Selectivities for *cis* 2-pentene formation from 2-pentyne decreased in the order Pd > Rh > Pt > Ru > Ir.[7] Palladium is the most selective of the noble metal catalysts for alkyne semihydrogenation with respect

(16.7)

Pd/CaCO$_3$ EtOH/Pyridine
R.T. 1 Atm
90%

8

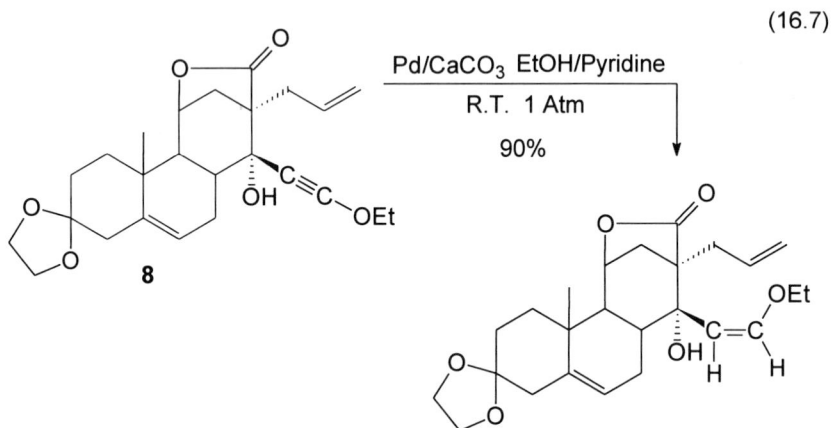

to both overall alkene formation and the production of the *cis* olefin. Nickel,[8-14] copper[8,10] and cobalt[15] catalysts are also effective catalysts for alkyne semihydrogenation. Copper catalysts are the most stereoselective giving *cis* alkenes as the sole product.[8]

Palladium Catalysts

Although palladium catalysts are preferred for the selective partial hydrogenation of alkynes, they are usually used with a variety of modifiers to improve reaction selectivity. Metal salts, amines, amine oxides, sulfur compounds, hydroxides, and carbon monoxide are among the modifiers that have been used. The variety of these modifiers suggests that they may function, at least in part, by decreasing the overall catalyst activity and, thereby, improving reactant diffusion to the remaining sites. The metal ions used to increase semihydrogenation selectivity include zinc, zirconium, tin, ruthenium, mercury, lead, gold, silver, and copper.

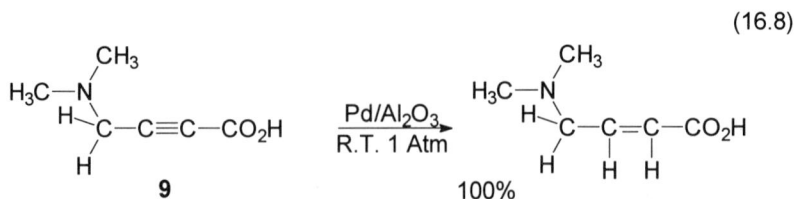

(16.8)

Pd/Al$_2$O$_3$
R.T. 1 Atm
9   100%

While unmodified Pd/Al$_2$O$_3$ was an effective catalyst for the semihydrogenation of the amino alkyne, **9**, (Eqn. 16.8)[16], the hydrogenation of the amino diyne, **10**, over palladium in absolute ethanol gave the products resulting from the cyclization of the partially saturated triple bonds on the catalyst surface (Eqn. 16.9).[17]

(16.9)

HC≡C—C—NH—C—C≡CH

10

Pd/C EtOH
─────────→
0°C 1 Atm

48% (H₃C / CH₃ pyrrole structure)

+

15% (H₂C / CH₃ pyrrole structure)

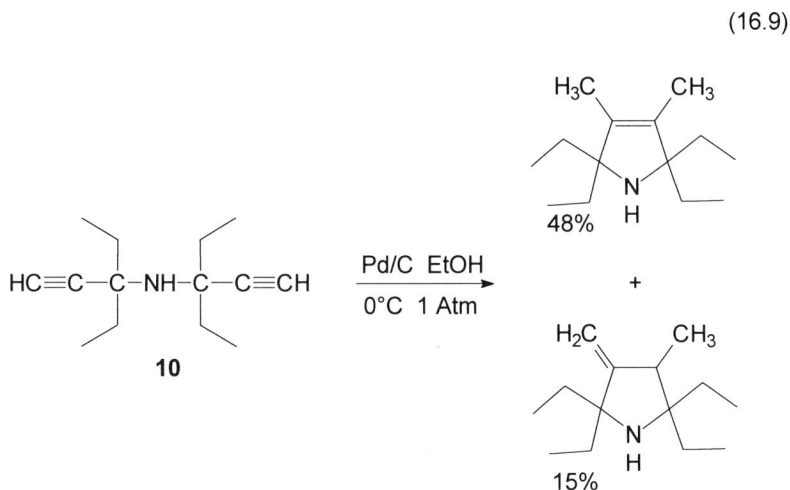

A commonly used catalyst for the semihydrogenation of alkynes is Lindlar's catalyst, Pd/CaCO₃, modified by the addition of lead acetate.[18,19] Recent work indicated that the lead ions in the Lindlar catalyst block the more active hydrogenation sites on the palladium, thus inhibiting alkene hydrogenation.[20] The hydrogenation of halo-alkynes takes place over Lindlar's catalyst without the hydrogenolysis of the carbon-halogen bond (Eqn. 16.10).[21] Semihydrogenation of perfluoroacetylenic esters over Lindlar's catalyst gave the perfluorosubstituted *cis* acrylates in 75%–85% yield (Eqn. 16.11).[22]

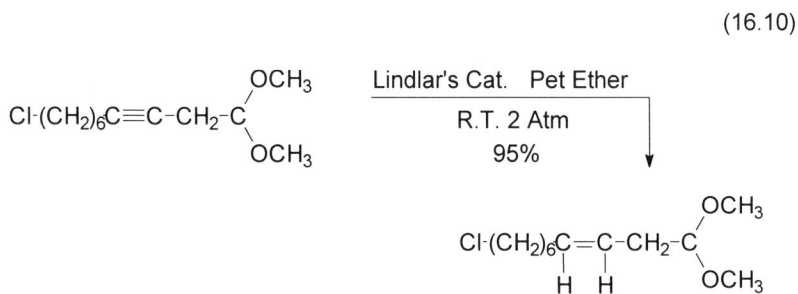

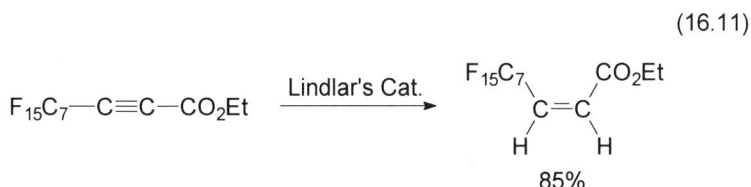

(16.10)

$Cl\text{-}(CH_2)_6C{\equiv}C\text{-}CH_2\text{-}C$ (OCH₃)(OCH₃)

Lindlar's Cat.   Pet Ether
─────────────────
R.T. 2 Atm
95%

$Cl\text{-}(CH_2)_6C{=}C\text{-}CH_2\text{-}C$ (H)(H)(OCH₃)(OCH₃)

(16.11)

$F_{15}C_7\text{—}C{\equiv}C\text{—}CO_2Et$

Lindlar's Cat.
────────────→

$F_{15}C_7$ \ C=C / CO_2Et (H)(H)

85%

(16.12)

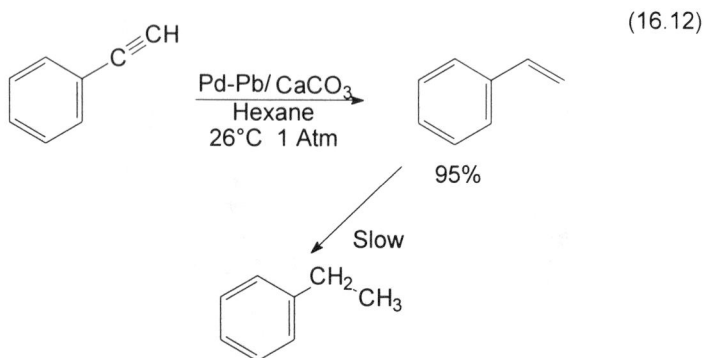

A CaCO$_3$ supported Pd-Pb alloy catalyst was found to be more selective in alkyne hydrogenation than the Lindlar catalyst.[23] Styrene was produced in over 95% selectivity by the hydrogenation of phenyl acetylene over this catalyst (Eqn. 16.12). Further hydrogenation to ethyl benzene was significantly less than that observed using Lindlar's catalyst. The Z (*cis*) alkene was formed in >99% selectivity at 100% conversion in the hydrogenation of 11-hexadecynyl acetate Eqn. 16.13).[23]

(16.13)

$$H_3C(CH_2)_3—C≡C—(CH_2)_9CH_2OAc \xrightarrow[\text{R.T. 1 Atm}]{Pd\text{-}Pb/\,CaCO_3 \quad MeOH}$$

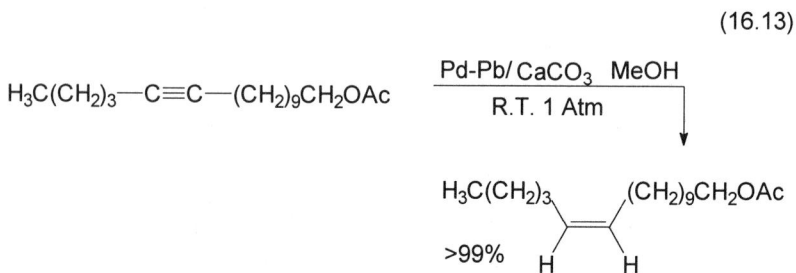

Secondary modifiers are commonly used with Lindlar's catalyst. The usual secondary modifier is quinoline, which is reported to enhance the reaction selectivity for the semihydrogenation of a variety of substituted acetylenes.[1] Very good yields (80%) of the *cis* previtamin D, **12,** were obtained on hydrogenation of the disubstituted alkyne, **11,** using quinoline modified Lindlar's catalyst followed by removal of the blocking groups (Eqn. 16.14).[24] Aromatic amine oxides were also used to increase reaction selectivity in the hydrogenation of 1-bromo-11-hexadecyne **(13)** over a commercially available Lindlar's catalyst (Eqn. 16.15).[25,26] Quinoline has also been used to increase the selectivity of other palladium catalysts in the semihydrogenation of alkynes. Hydrogenation of the alkyne diester, **15,** over Pd/BaSO$_4$ modified with quinoline gave the *cis* alkene, **16,** in excellent yield (Eqn. 16.16).[27] Hydrogenation of the unconjugated diyne,

(16.14)

(16.15)

H$_3$C(CH$_2$)$_3$—C≡C—(CH$_2$)$_9$CH$_2$Br

13

Lindlar's Cat. Hexane
30°C 7 Atm

H$_3$C(CH$_2$)$_3$ $\diagdown$ $\diagup$ (CH$_2$)$_9$CH$_2$Br
              C=C
97%   H   H

14

(16.16)

MeO$_2$C(CH$_2$)$_3$C≡C(CH$_2$)$_3$CO$_2$Me

15

Pd/BaSO$_4$ MeOH
Quinoline
R.T. 1 Atm

MeO$_2$C(CH$_2$)$_3$ $\diagdown$ $\diagup$ (CH$_2$)$_3$CO$_2$Me
              C=C
97%   H   H

16

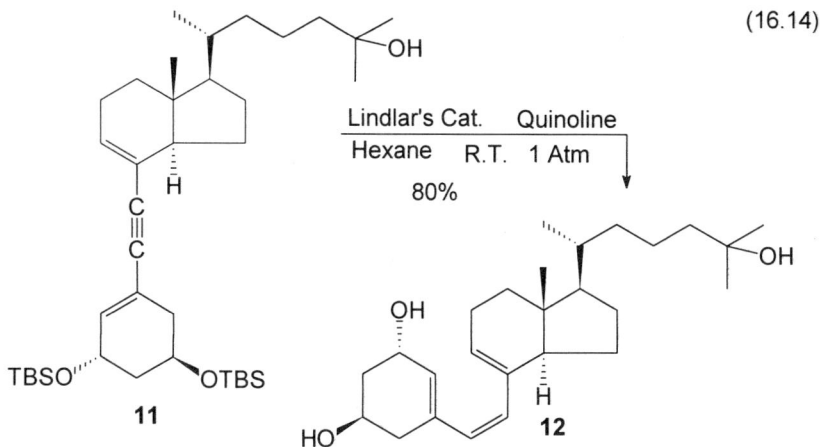

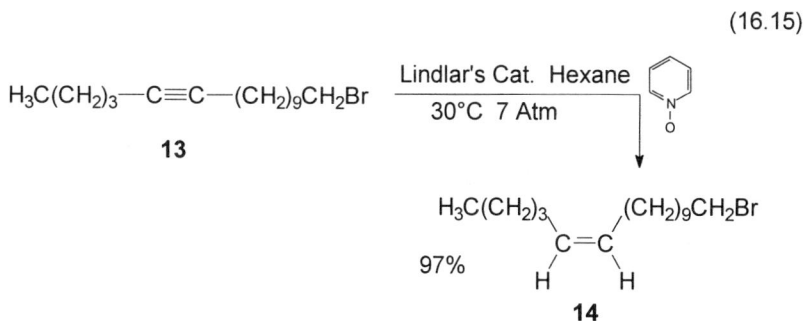

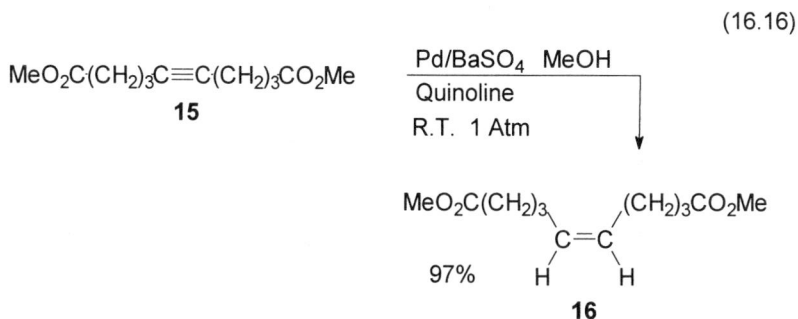

17, over quinoline modified Pd/C gave the Z-Z unconjugated diene, **18** (Eqn. 16.17).[28]

Catalysts prepared by the reduction of palladium acetate with sodium hydride in alcoholic media, referred to as Pd-c, were more selective for the partial

(16.17)

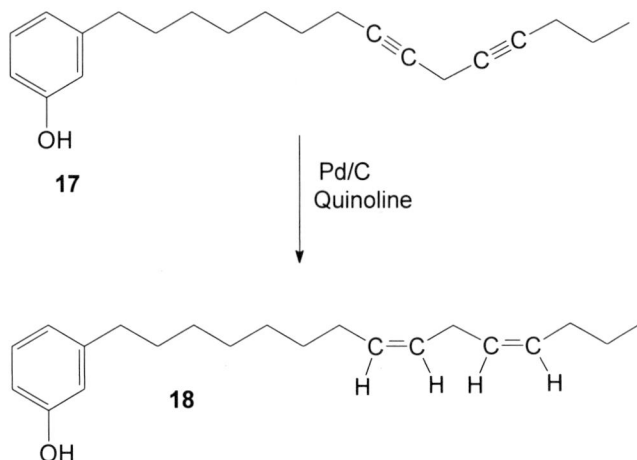

17

Pd/C
Quinoline

18

hydrogenation of variously substituted alkynes in the presence of quinoline than the Lindlar catalyst/quinoline combination.[29]   In the semihydrogenation of diphenylacetylene, Lindlar catalyst with quinoline gave 95% of the alkene with a *cis/trans* ratio of 93:2 at 100% conversion. The Pd-c catalyst with quinoline gave better than 99% conversion to stilbene with a *cis/trans* ratio of 97:2 (Eqn. 16.18).

(16.18)

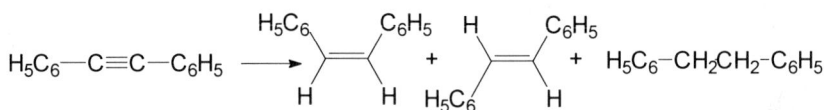

Lindlar's Cat. Quinoline MeOH 20°C 1 Atm	93%	2%	5%
Pd-c Quinoline MeOH 20°C 1 Atm	97%	<2%	<1%

A Pd/Zeolite A catalyst that was treated with diphenyldiethoxysilane was effective for the semihydrogenation of alkynes, particularly disubstituted acetylenes.[30] Hydrogenation of 3-nonyne gave 3-nonene in 97% yield over this catalyst (Eqn. 16.19). Using the non-silanized catalyst the alkene was formed only in 40% yield.[30]

(16.19)

$$H_3C(CH_2)_4C\equiv CCH_2CH_3 \xrightarrow[25°C\ 1\ Atm]{Pd/A\ Silane}$$

H₃C(CH₂)₄ and CH₂CH₃ groups on a cis alkene, $C=C$ with H and H, 97%

Nickel Catalysts

The P2 nickel boride [Ni(B)] catalyst is also effective for the semihydrogenation of alkynes. With this catalyst reaction selectivity is significantly improved by the addition of ethylenediamine in 2–3 molar amounts of the catalyst.[9] Selectivity toward *cis* alkene formation is also influenced by the substrate to catalyst ratio. Hydrogenation of 3-hexyne using an 8:1 substrate:catalyst ratio gave 3-hexene in a 95% yield with a 100:1 *cis:trans* ratio. Using a 20:1 substrate:catalyst ratio doubled the stereoselectivity with a 200:1 ratio of *cis* and *trans* 3-hexene observed. There was more reactant diffusion control in reactions having larger amounts of catalyst, so isomerization of the initial product, the *cis* alkene, was possible before displacement by the alkyne.

Incorporation of copper chloride in about 10% of the molar quantity of nickel before the borohydride reduction step further improved the selectivity of the Ni(B) catalysts.[10] Copper improves the selectivity of Raney nickel catalysts in phenylacetylene semihydrogenation but not to the extent obtained using the copper-modified Ni(B) catalysts. The Cu-Ni(B) catalysts were more selective than the Cu(B) catalysts prepared by the borohydride reduction of copper chloride.[10] Adding zinc or iron salts to a Ni(B) catalyst had only a slight effect on phenylacetylene semihydrogenation selectivity.[10]

A Ni-As(B) catalyst prepared by the borohydride reduction of alumina-supported nickel arsenate gave, on hydrogenation of 1-bromo-11-hexadecyne (13) in the presence of a small amount of acetone, a 97% yield of the alkene (14) having a 92:5 *cis/trans* ratio. No hydrogenolysis of the carbon-bromine bond occurred.[11,12] Borohydride reduction of cobalt acetate gave a Co(B) catalyst that was somewhat less active than Ni(B) but that was quite selective in alkyne semihydrogenations (Eqn. 16.20).[15]

(16.20)

$$H_3C(CH_2)_3C\equiv CCH_3 \xrightarrow[30°C\ 1\ Atm.]{Co(B)\ EtOH}$$

H₃C(CH₂)₃ and CH₃ groups on a cis alkene, $C=C$ with H and H, 97%

As with the palladium salts, nickel acetate could be reduced by sodium hydride in alcoholic media to give the catalyst termed Ni-c, which is comparable to P-2 Ni(B) in semihydrogenation selectivity.[13]   Reduction of bis-(dimethoxyethane) dibromonickel, NiBr$_2$-2DME, with potassium graphite gave a nickel catalyst, which on hydrogenation of diphenylacetylene gave at 86% conversion a 72% yield of stilbene having a 94:6 *cis/trans* ratio. The use of ethylenediamine as a modifier in this reaction raised the stilbene yield to 92% at 96% conversion, but the *cis/trans* alkene ratio remained the same.[14]

Active Site

The nature of the surface site where acetylene hydrogenation occurs has been discussed extensively. It was proposed that two types of active sites are present on a Pd/Al$_2$O$_3$ catalyst.[31] One type promotes the hydrogenation of both triple and double bonds, whereas the other catalyzes only double-bond hydrogenation. This latter site can be poisoned by the addition of carbon monoxide to the reaction mixture as evidenced by the marked increase in reaction selectivity observed in the hydrogenation of acetylene in the presence of carbon monoxide.[32,33] Other modifiers can presumably act in the same way.

The rate of the vapor phase hydrogenation of 1-butyne decreased significantly with a series of Pd/Al$_2$O$_3$ catalysts having dispersions >20%[34] but the effect of dispersion on the reaction selectivity is not as clear.[35] At 90% conversion only 2%–3% of the saturated product, butane, is formed regardless of the catalyst dispersion. At 99% conversion, however, the decrease in selectivity with increasing dispersion is significant with over 25% butane formation over the most highly dispersed catalysts. In contrast, in the semihydrogenation of phenylacetylene, selectivity increased with increasing palladium dispersion.[36] A catalyst having a 100% dispersion showed 100% selectivity.

It was suggested that these differences were the result of different preparation conditions used for the catalysts with the ratios of the two different types of reactive sites being affected by the preparation techniques.[37] Further, the substrates and reaction conditions were different and it was possible that some degree of reactant diffusion control could have been present in some of the reactions, particularly those run over the more highly dispersed catalysts. Some support for this idea comes from the different selectivities reported for the butyne hydrogenations. At 90% conversion, the alkyne present is sufficient to displace the alkene product even on the more highly dispersed catalysts, which, presumably, have a higher percent of the more active corner or adatoms on the surface of the metal particles. At 99% conversion, unreacted alkyne is minimal so selectivity is decreased. This is because the diffusion of the reactant to the catalyst becomes the limiting factor, particularly with the more dispersed catalysts that have a high number of active sites. The degree of conversion at which phenylacetylene semihydrogenation selectivity was determined was not reported.[36] It is reasonable to conclude, therefore, that as long as sufficient alkyne is present, reaction selectivity either increases with increasing catalyst dispersion or remains essentially unchanged.

Solvent

The nature of the solvent in liquid-phase alkyne hydrogenations and the extent to which it can influence the competitive adsorption factors needed to attain selectivity should also be considered. The semihydrogenation of 1-octyne over a series of Pd/Nylon-66 catalysts of varying metal load gave 1-octene with a selectivity of 100% over a wide range of metal loads when the reaction was run in heptane.[38] In n-propanol, however, reaction selectivity increased with decreasing metal load. Apparently the alcohol interacted with the catalyst to modify the active sites[39] and influenced the relative adsorption characteristics of the acetylenic and olefinic species. This can affect reaction selectivity particularly if reactant diffusion assumes some importance in the reaction.

Acetylenic Carbinols and Glycols

Acetylenic carbinols and glycols appear frequently in synthetic processes because they are readily prepared by condensation of an aldehyde or ketone with an acetylide anion. Hydrogenation of such species, however, is generally less selective than the hydrogenation of the hydrocarbon analog. The selective semihydrogenation of acetylenic carbinols or glycols is complicated by the potential for hydrogenolysis of the initially formed allyl alcohol if it is not displaced rapidly from the catalyst. This displacement is complicated by the presence of the hydroxy group on the alkene. As with other alkynes, palladium is the catalyst of choice for these semihydrogenations, although palladium is also a superior catalyst for allylic alcohol hydrogenolysis.

(16.21)

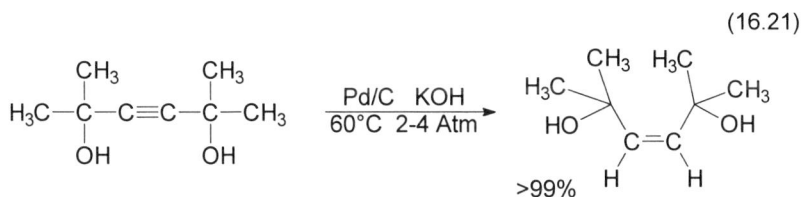

Small amounts of sodium or potassium hydroxide added to the reaction mixture eliminate the hydrogenolysis side reaction and produce near quantitative yields of the olefinic alcohols or diols (Eqn. 16.21).[40] With potassium hydroxide and a Pd/BaCO$_3$ catalyst, most acetylenic carbinol hydrogenations stop selectively after the uptake of one equivalent of hydrogen to give quantitative yields of the alkene.[41] The role of the hydroxide is not completely understood. It may simply neutralize any acidic species that might be present in the system and, thus, inhibit acid-promoted hydrogenolysis. Although this may be a partial explanation it was shown that potassium hydroxide and acetylenic carbinols react to form 1:1 adducts.[42] It is possible that this complex is more easily adsorbed on the catalyst surface than the acetylenic carbinol and much more readily than the product allyl alcohol. If this were the case, the complex would be more capable

(16.22)

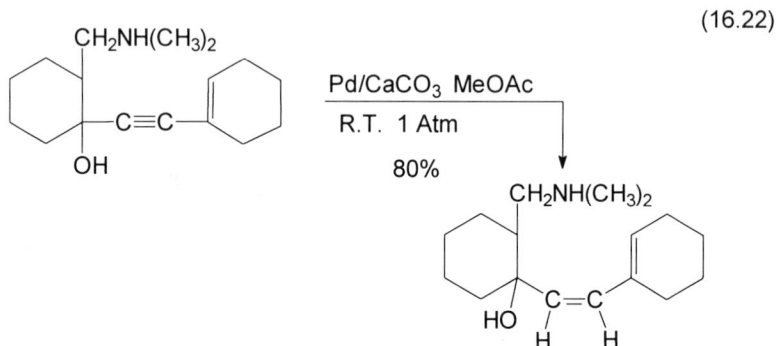

of desorbing the allyl alcohol and would do so rapidly enough that hydrogenolysis could not take place, thus increasing reaction selectivity.

Basic supports for the palladium catalyst can also be used to inhibit the hydrogenolysis of allyl alcohols (Eqn. 16.22)[43] and acetates (Eqn. 16.23).[44] The addition of quinoline to the reaction mixture has the same effect (Eqn. 16.24).[45]

(16.23)

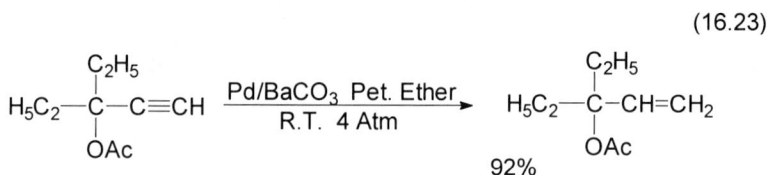

(16.24)

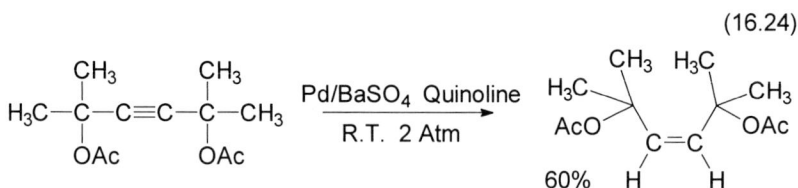

While potassium hydroxide is beneficial in catalysts containing palladium loads near 5%; with low palladium loads it appears that potassium hydroxide is not always necessary. The semihydrogenation of dehydrolinalool (**19**) gives 100% of linalool (**20**) over 0.5% Pd/Al_2O_3 in alcoholic solvents as shown in Eqn. 16.25.[46] Perhaps with lower palladium-loaded catalysts reactant diffusion is not a factor in the reaction, so the presumed enhanced adsorption of the potassium hydroxide complex is not as important.

Even though palladium catalysts are generally used for these reactions, Raney nickel has been somewhat effective (Eqn. 16.26)[47] as have $Ni/AlPO_4$ catalysts (Eqn. 16.27).[48]

(16.25)

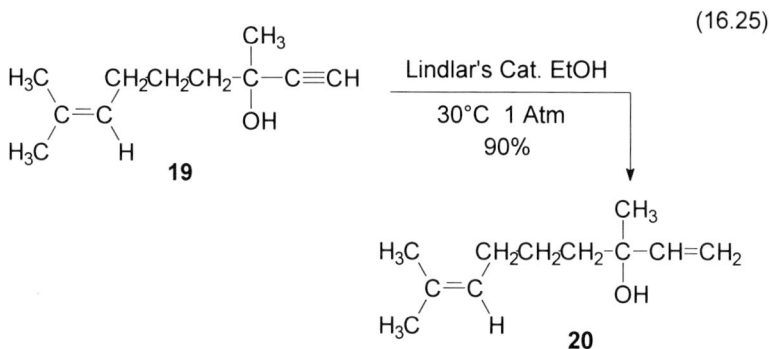

(16.26)

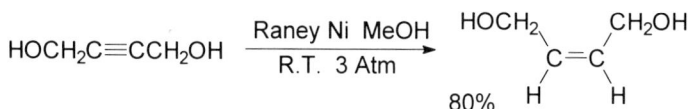

(16.27)

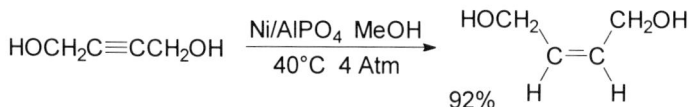

Conjugated Acetylenes

Selecting the catalyst and reaction conditions for partial reduction of a triple bond situated in a conjugated system is a challenge. Where the hydrogenation of an isolated alkyne can proceed with nearly complete selectivity, the partial saturation of an enyne takes place selectively with much more difficulty. Selectivities of 85–90% in these latter reactions are common and are considered to be reasonable for synthetic applications.

(16.28)

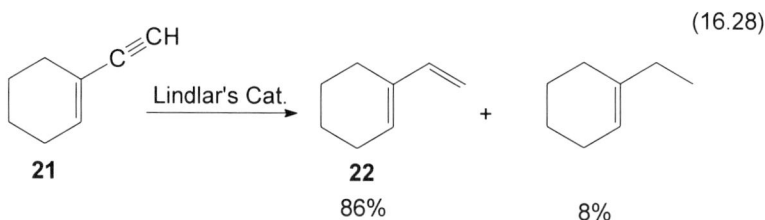

(16.29)

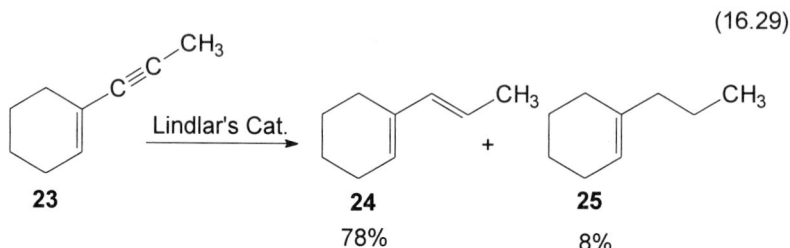

	23		24		25
			78%		8%

The extent of semihydrogenation selectivity of conjugated enynes depends on the nature of the triple bond in the system. Hydrogenation of ethynyl cyclohexene (**21**) over Lindlar's catalyst gave, at 94% conversion, the vinylcyclohexene, **22**, in 86% yield along with 8% of ethyl cyclohexene (Eqn. 16.28).[49] Placing a substituent on the triple bond decreased reaction selectivity. Semihydrogenation of **23** over Lindlar's catalyst gave, at 86% conversion, 78% of the diene, **24**, along with 8% of propyl cyclohexene, **23** (Eqn. 16.29).[49]

References

1. E. N. Marvell and T. Li, *Synthesis*, 457 (1973).
2. M. Smith, "Dissolving Metal Reductions", in *Reductions, Techniques and Applications in Organic Synthesis* (R. L. Augustine, Ed.) Dekker, New York, 1968, p 95.
3. L. J. Haynes and E. R. H. Jones, *J. Chem. Soc.*, 954 (1946).
4. H. Newman, *Chem. and Ind., (London)*, 372 (1963).
5. N. A. Dobson, G. Egliinton, M. Krishnamurti, R. A. Raphael and R. G. Willis, *Tetrahedron*, **16**, 16 (1961).
6. A. Lardon, O. Schindler and T. Reichstein, *Helv. Chim. Acta*, **40**, 666 (1957).
7. G. C. Bond and J. S. Rank, *Proc. 3rd Int. Cong. Cat.*, **2**, 1225 (1965).
8. J. J. Phillipson, P. B. Wells and G. R. Wilson, *J. Chem. Soc., A*, 1351 (1969).
9. C. A. Brown and V. K. Ahuja, *J. Chem. Soc., Chem. Commun.*, 553 (1973).
10. Y. Nitta, T. Imanaka and S. Teranishi, *Bull. Chem. Soc., Japan*, **54**, 3579 (1981).
11. T. H. Cymbaluk, U. S. Patent 4,659,687 (1987).
12. T. H. Cymbaluk, U. S. Patent 4,748,290 (1988).
13. P. Ballois, J.-J. Brunet and P. Baubere, *J. Org. Chem.*, **45**, 1946 (1980).
14. D. Savoia, E. Tagliavini, C. Trombini and A. Umani-Ronchi, *J. Org. Chem.*, **46**, 5340 (1981).
15. Y. Nitta, T. Imanaka and S. Teranishi, *Bull. Chem. Soc., Japan*, **53**, 3154 (1980).
16. I. Marszak and M. Oloniucki, *Bull. soc. chim., France*, 182 (1959).
17. I. E. Kopka, Z. A. Fataftah and M. W. Rathke, *J. Org. Chem.*, **45**, 4616 (1980).
18. H. Lindlar, *Helv. Chim. Acta*, **35**, 446 (1952).
19. H. Lindlar and R. Dubuis, *Org. Synthesis*, **46**, 89 (1966)

20. R. Schlogo, K. Noack, H. Zbinden and A. Reller, *Helv. Chim. Acta*, **70**, 627 (1987).
21. A. S. Bailey, V. G. Kendall, P. B. Lumb, J. C. Smith and C. H. Walker, *J. Chem. Soc.*, 3027 (1957).
22. M. Lanier, M. Haddach, R. Pastor and J. G. Riess, *Tetrahedron Lett.*, 2469 (1993).
23. J. Sobczak, T. Boleslawska, M. Pawlowska and W. Palczewska, *Stud. Surf. Sci. Catal.*, **41** (Heterog. Catal. Fine Chem.), 197 (1988).
24. L. A. Sarandeses, J. L. Mascareñas, L. Castedo and A. Mouriño, *Tetrahedron Lett.*, **33**, 5445 (1992).
25. C. A. Drake, U. S. Patent 4,596,783 (1986).
26. C. A. Drake, U. S. Patent 4,605,797 (1986).
27. D. J. Cram and N. L. Allinger, *J. Am. Chem. Soc.*, **78**, 2518 (1956).
28. J. Caplin and J. H. P. Tyman, *J. Chem. Research*, 34 (1982).
29. J.-J. Brunet and P. Caubere, *J. Org. Chem.*, **49**, 4058 (1984).
30. H. Kuno, M. Shibagaki, K. Takahashi, I. Honda and H. Matsushita, *Chem. Lett.*, 1725 (1992).
31. A. S. Al-Ammar, S. J. Thomson and G. Webb, *J. Chem. Soc., Chem. Commun.*, 323 (1977).
32. A. H. Weiss, S. LeViness, V. Nair, L. Guczi, A. Sarkany and Z. Schay, *Proc. 8th Int. Cong. Catal.*, **5**, v-591 (1984).
33. S. Leveness, V. Nair and A. H. Weiss, *J. Mol. Catal.*, **85**, 405 (1984).
34. J. P. Boitiauz, J. Cosyns and S. Vasudevan, *Appl. Catal.*, **15**, 317 (1985).
35. J. P. Boitiauz, J. Cosyns and S. Vasudevan, *Appl. Catal.*, **6**, 41 (1983).
36. G. Carturan, G. Faccin, G. Cocco, S. Enzo and G. Navazio, *J. Catal.*, **76**, 405 (1982).
37. A. Molnar, G. V. Smith and M. Bartok, *J. Catal.*, **101**, 67 (1986).
38. C. Bowden, R. J. Mclellan, R. P. MacDonald and J. M Winterbottom, *J. Chem. Tech. Biotechnol.*, **32**, 567 (1982).
39. R. L. Augustine, R. W. Warner and M. J. Melnick, *J. Org. Chem.*, **49**, 4853 (1984).
40. R. Tedeschi, *J. Org. Chem.*, **27**, 2398 (1962).
41. R. Tedeschi, H. C. McMahon and M. S. Pawlak, *Ann. N. Y. Acad. Sci.*, **145**, 91 (1967).
42. R. Tedeschi, M. F. Wilson, J. Scanlon, M. Pawlak and V. Cunicella, *J. Org. Chem.*, **28**, 2480 (1963).
43. R. A. Raphael and F. Sondheimer, *J. Chem. Soc.*, 2693 (1951).
44. G. F. Hennion, W. A. Schroeder, R. P. Lu and W. B. Scanlon, *J. Org. Chem.*, **21**, 1142 (1956).
45. D. R. Paulson, L. S. Gilliam, V. O. Tery, S. M. Farr, E. J. Parker, F. Y. N. Tang, R. Ullman and G. Ribar, *J. Org. Chem.*, **43**, 1783 (1978).
46. A. M. Pak, O. I. Kartonozhkina, A. P. Pogorelskii, E. M. Tsai and D. V. Sokolskii, *React. Kinet. Catal. Lett.*, **16**, 265 (1981).
47. C. S. Marvel and C. H. Young, *J. Am. Chem. Soc.*, **73**, 1066 (1951).
48. F. M. Bautist, J. M. Campelo, A. Garcia, R. Guardeno, D. Luna and J. M. Marinas, *Stud. Surf. Sci. Catal.*, **59** (Heterog. Catal. Fine Chem., II), 269 (1991).
49. E. N. Marvell and J. Tashiro, *J. Org. Chem.*, **30**, 3991 (1965).

Hydrogenation IV:
Aromatic Compounds

Carbocyclic Aromatics

Benzene and Substituted Benzenes

The catalytic hydrogenation of substituted benzenes is a classic method for the preparation of substituted cyclohexanes. The ease of ring hydrogenation depends on the substituents present with the order of catalyst reactivity different for each substituent.[1] For example, catalyst activity for the hydrogenation of benzene and alkylbenzenes decreased in the order Rh > Ru > Pt > Ni > Pd > Co.[2] In general, the hydrogenation of substituted benzenes is best accomplished using a rhodium catalyst at 20°–60°C and 3–4 atmospheres of hydrogen. The hydrogenation of strained ring systems such as **1**, however, takes place under more mild conditions, with a palladium catalyst at room temperature and atmospheric pressure (Eqn. 17.1).[3]

(17.1)

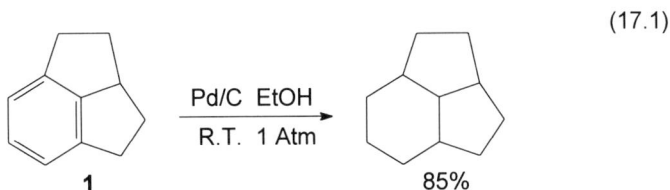

$$\text{Pd/C EtOH}$$
$$\text{R.T. 1 Atm}$$

1 85%

The presence of hydroxy or amine groups generally has little effect on the reaction. Hydrogenolysis of the aryl C–O or C–N bonds is not usually observed under the standard conditions mentioned above unless a carboxyl group is present in the para position. If C–O or C–N bond hydrogenolysis is a problem with the rhodium catalyst, it can be minimized by running the hydrogenation over ruthenium in aqueous ethanol at 100°–150°C and 100 atmospheres. The carbon-halogen bond of aryl halides, however, is always hydrogenolyzed along with the hydrogenation of the ring regardless of the reaction conditions.

Hydrogenolysis of the C–O bond is not observed when the benzene rings of benzyl alcohols and ethers are hydrogenated under these conditions, particularly when a small amount of acetic acid is added to the reaction mixture.[4,5] Under these conditions the hydrogenation of the benzene ring of benzyl amines also takes place without cleavage of the benzyl-nitrogen bond. There is no loss of optical activity when the aromatic rings of optically active α-alkylbenzyl amines are hydrogenated using the acetic acid modified reaction conditions.

The selective hydrogenation of other functional groups in the presence of a benzene ring is normally preferred unless the functional group is a carboxylic acid or acid derivative. About the only time a benzene ring is saturated in preference to one of the more readily hydrogenatable groups is when the benzene is part of a strained-ring system. Another interesting example of preferential hydrogenation of a benzene ring is that of o-methoxyphenyl propanone (**2**) to the methoxycyclohexyl propanone, **3** (Eqn. 17.2).[6] This reaction takes place only over rhodium catalysts in an acetic acid solvent. In methanol, some saturation of the carbonyl group also occurs. The reason for this selectivity is not apparent. The solvent effect seems to rule out any steric factors and a protonated, stabilized intermediate does not seem probable. If either oxygen were protonated, hydrogen bonding with the remaining oxygen would involve a seven-membered ring, something that is not usually involved in the stabilization of reaction intermediates.

(17.2)

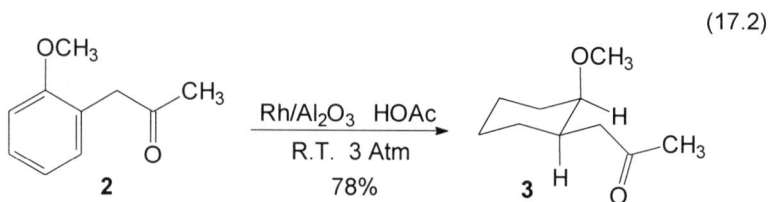

Product Stereochemistry

If the hydrogenation of the benzene ring were to take place during a single period of residence on the catalyst, all six of the hydrogens would be added to the chemisorbed face of the ring and only the *cis* substituted cyclohexane would result. Although the *cis* products are generally predominant in these reactions, *trans* substituted materials are formed in varying amounts probably because of the intermediate formation of substituted cyclohexadienes and cyclohexenes, that are desorbed and subsequently readsorbed on the catalyst to be saturated further.[7-10] Although cyclohexadiene intermediates have not been isolated from these reactions, cyclohexenes can be produced.[7,11,12]

Product stereochemistry depends on the catalyst used and the nature of the substituents on the ring. Of the common catalysts, nickel is generally the least stereoselective, usually giving product mixtures in which the thermodynamically favored product (the one having the most equatorial substituents on the cyclohexane ring) predominates. Ruthenium generally gives the highest yields of *cis* substituted cyclohexanes particularly in the hydrogenation of ortho and meta disubstituted benzenes. In the hydrogenation of the para isomers, rhodium ordinarily gives more of the *cis* product.[13]

Selective Hydrogenation of Benzene to Cyclohexene

Obtaining *trans* substituted cyclohexanes suggested that desorbed cyclohexenes were intermediates in the hydrogenation of benzenes. The isolation of cyclohexene and substituted cyclohexenes from the hydrogenation of benzene and substituted benzenes was first reported for hydrogenations run over a Ru/C catalyst, but the maximum olefin concentrations observed in this early work were only 0.2–3.4%.[7]

$$(17.3)$$

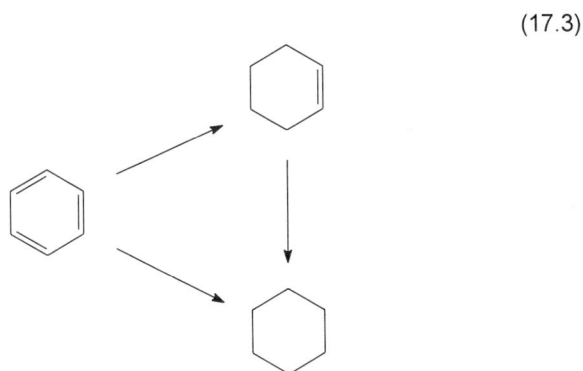

Selective conversion of benzene to cyclohexene can be considered either a *Type II* or *Type III* selectivity (Chapter 5) (Eqn. 17.3). Kinetic data suggest that there are two types of sites on a ruthenium catalyst that are used for benzene hydrogenation.[8,11] One type presumably promotes the direct hydrogenation of benzene to cyclohexane and the other gives the cyclohexene intermediate. Thus, the catalyst is considered to act as a *Type II* system. However, because cyclohexene can be further hydrogenated to cyclohexane, *Type III* selectivity is also involved. The influence of reactant diffusion on the selectivity in such a system is, therefore, difficult to predict. However, cyclohexene formation is enhanced by a decrease in the amount of hydrogen present on the catalyst. Presumably, a low hydrogen availability permits the cyclohexene to be displaced by the benzene before it can be saturated.[12] Increasing the reaction temperature or the quantity of catalyst used for the hydrogenation, two factors that decrease the hydrogen availability to the catalyst, resulted in an increase of the reaction selectivity for cyclohexene formation.[11]

Most partial benzene hydrogenations use a ruthenium catalyst since it is a good species for promoting the hydrogenation of aromatic systems but not very effective for double bond saturation. The reaction is commonly run at 150°–200°C under 10–70 atmospheres of hydrogen. The maximum cyclohexene formation of about 40–70% selectivity is usually observed at low to moderate conversions (i.e., 10–50%). This suggests that competitive adsorption between benzene and cyclohexene, while taking place, does not occur as readily as, for instance, the displacement of an alkene by an alkyne. Thus, for a benzene hydrogenation to give a significant conversion to cyclohexene, the process should involve a partial hydrogenation and the recycling of the unreacted benzene.

It is essential that water be present in these reactions.[14] No cyclohexene formation is observed when water is absent, even in vapor phase hydrogenations run over ruthenium catalysts. When the benzene vapor was saturated with water vapor the selectivity for cyclohexene formation increased to 14% while the reaction rate decreased. Pre-adsorption of water on the catalyst gave a 41% selectivity at 20% conversion.[15,16] It was suggested that the effect exhibited by water vapor was not due to a selective site blocking or an electronic effect but, rather, (1) to an adsorption mode of water that facilitated the displacement of cyclohexene from the catalyst as it was formed[15] or (2) to a decrease in hydrogen chemisorption caused by the lower solubility of hydrogen in water than in benzene.[11] Neither suggestion is consistent with all of the facts. Water has no effect on the platinum-catalyzed reaction selectivity, or the lack of it, as would be expected from these concepts.[15] But it is possible that water adsorption on ruthenium is more facile than it is on platinum. In fact, water vapor has little, if any, effect on platinum-catalyzed olefin hydrogenations.[17] Further, water increases the activity of ruthenium toward olefin hydrogenation.[18] This increased activity is common in most ruthenium-catalyzed reactions, a finding that is inconsistent with previous suggestions concerning the effect of water on benzene hydrogenation selectivity. However, a special relationship may exist between water and ruthenium enabling the adsorption of water on a ruthenium surface to change the electronic character of the surface atoms or, possibly, to restructure the surface. Either possibility could result in a modification of the properties of the catalyst.

The presence of other modifiers also contribute to an increase in cyclohexene formation. Sodium hydroxide,[11] ferrous sulfate,[14] cadmium sulfate[19] and other salts[20] have been used to increase the selectivity for partial benzene hydrogenation over ruthenium catalysts. Since it was found that optimal conversions to cyclohexene were produced when these modifiers were present in a specific amount it was considered reasonable that these materials block the more active sites on the catalyst and, thus, decrease the extent to which either the cyclohexene intermediate is saturated further or the cyclohexane is produced directly from benzene. Interestingly, running the hydrogenation in ethylene glycol resulted in a fairly good selectivity for cyclohexene formation without the presence of any of these modifiers.[19,21] Hydrogenations of aqueous benzene were even more selective when run over catalysts that were prepared in ethylene glycol.[22,23] The use of La_2O_3 and other rare earth oxides as supports for the ruthenium provided reasonably selective catalysts.[24]

Zinc salts have also been employed as modifiers.[24-26] A very selective catalyst was prepared by treating a mixture of $RuCl_3$ and $ZnCl_2$ with NaOH and then drying and reducing the residue with hydrogen. This catalyst, in the presence of $ZnSO_4$, promoted the selective hydrogenation of benzene at 150°C and 50 atmospheres. A 48% yield of cyclohexene was obtained at 60% conversion (79% selectivity).[25] The use of zinc with Ru/La_2O_3 and other rare earth oxide supports increased reaction selectivity somewhat.[24] Using an Fe-Ru catalyst in aqueous sulfuric acid in the presence of $ZnSO_4$ gave a selectivity of 75% corresponding to a 41% yield of cyclohexene at 55% conversion.[25] Cobalt[27,28] and copper[22,29] were also used as potential modifiers, but cobalt was not very effective. Copper, however, does enhance reaction selectivity. The most effective copper containing catalysts were mixed $Ru-Cu/SiO_2$ species that were prepared by coprecipitation from a solution of the metal salts and tetraethylsiloxane in ethylene glycol.[29]

Modifiers have also been used to influence the selectivity of vapor-phase partial hydrogenations of benzene.[16] The presence of ethylene glycol increased reaction selectivity with a ruthenium black catalyst from 7% to 41% while the turnover frequency (TOF) decreased from 31 to 3. Pyridine also increased selectivity in the short term, but prolonged use poisoned the catalyst. Passivating a ruthenium black catalyst with caprolactam not only stabilized the catalyst toward deactivation but also increased reaction selectivity from 7% to 20%.

Although the hydrogenation of benzene by standard supported platinum or rhodium catalysts does not lead to cyclohexene formation, platinum/polyamide

Table 17.1

Partial hydrogenation of alkyl substituted benzenes.[24]

Substituted Benzene	Rxn Time (min)	Conversion (%)	Olefin Selectivity (%)	Product Distribution	
toluene (CH₃)	50	32	75	methylenecyclohexene (CH₃) 72%	methylcyclohexene (CH₃) 28%
p-xylene (H₃C–, –CH₃)	80	28	72	(H₃C, CH₃) 35%	(H₃C, CH₃) 65%
ethylbenzene (C₂H₅)	75	28	79	(C₂H₅) 76%	(C₂H₅) 24%

catalysts do show a selectivity for the formation of cyclohexene[30-32] but only in the early stages of the reaction. A cobalt-boride catalyst has been used to hydrogenate benzene to cyclohexene in the vapor phase. Better than 95% selectivity was obtained but only at 10% conversion.[33] A 95%–100% selectivity for the vapor phase partial hydrogenation of benzene was reported using a 15% Yb/SiO$_2$ catalyst at 25°C but the extent of conversion was not stated.[34]

The selective hydrogenation of substituted benzenes to produce substituted cyclohexenes is also possible using ruthenium catalysts under the conditions required for the selective hydrogenation of benzene.[24,26] The difficulty in most of these reactions is that a mixture of cyclohexenes is formed. Table 17.1 shows the data obtained in the partial hydrogenations of substituted benzenes run using a reduced RuCl$_3$ catalyst in the presence of ZnO, water, and sodium hydroxide at 150°C and 50 atmospheres.[24] Hydrogenation of 1,2-di-tert-butylbenzene (**4**) in ethanol over Rh/C at 25°C and atmospheric pressure gave 2,3-di-tert-butylcyclohexene, **5**, in 30% yield at 100% conversion. With acetic acid as the solvent the cyclohexene yield increased to 45% (Eqn. 17.4).[35] Little selectivity for cyclohexene formation was observed using platinum catalysts in this reaction.[35]

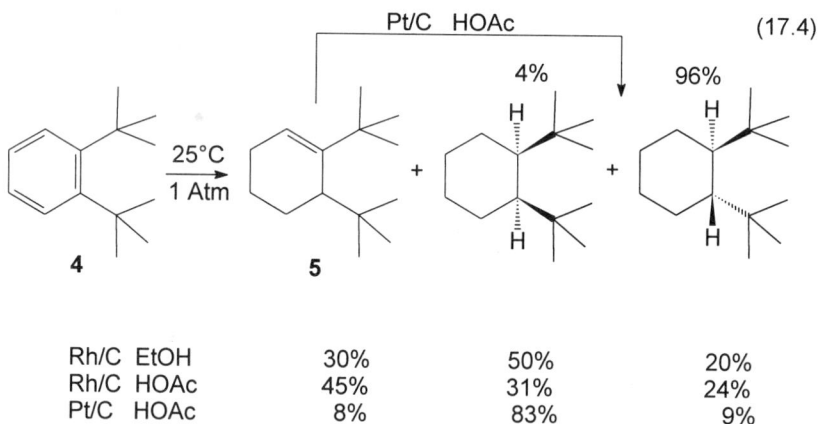

(17.4)

Rh/C EtOH	30%	50%	20%
Rh/C HOAc	45%	31%	24%
Pt/C HOAc	8%	83%	9%

The partial hydrogenation of 1,3,5-trisubstituted benzenes gives only a single cyclohexene product. When the substituents are large, as with 1,3,5-tri-tert-butylbenzene (**6**), selective formation of the cyclohexene, **7**, occurs even over unmodified Pt/C and Rh/C catalysts when the reaction is run in heptane at room temperature and atmospheric pressure. With Pt/C, 45–50% selectivity toward the cyclohexene was found at 50% conversion while with Rh/C 60–65% selectivity was obtained at 60% conversion (Eqn. 17.5).[36]

(17.5)

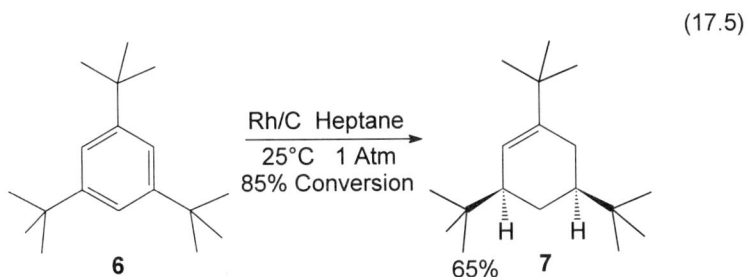

6   65%   **7**

Phenols and Phenyl Ethers

The hydrogenation of a phenol or phenyl ether is best accomplished using a ruthenium catalyst in aqueous ethanol at 100°–125°C under 70–100 atmospheres of hydrogen (Eqn. 17.6).[37] The choice of ruthenium is based on the fact that little, if any, hydrogenolysis of the C–O bond occurs over ruthenium even though moderately high temperatures and pressures are required for this reaction. Hydrogenation of phenols over Pd/SrCO$_3$ at 150°–200°C and 125–150 atmospheres of hydrogen also gives the cyclohexanols in very good yields.[38,39] Varying amounts of hydrogenolysis have been observed using Raney nickel, particularly at higher temperatures.[40] Extensive hydrogenolysis generally occurs with platinum[41,42] and somewhat less with rhodium,[43] particularly when there is an activating para carboxy group present.

(17.6)

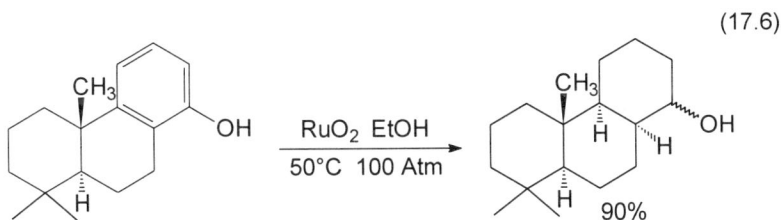

The hydrogenation of phenols can give a variety of products depending on the catalyst and reaction conditions used (Eqn. 17.7).[44] Complete saturation followed by the oxidation of the resulting cyclohexanol is a reasonable method for the preparation of substituted cyclohexanones. It is possible, however, to convert phenol to cyclohexanone directly by hydrogenation in a basic medium using mild conditions (Eqn. 17.8). This hydrogenation can take place in either the vapor-phase or liquid-phase generally using a palladium catalyst. Apparently palladium is selective in this reaction because it is ineffective for the hydrogenation of aliphatic ketones. Hydrogenation of phenol with Pd/Al$_2$O$_3$ alone gave only a 20% selectivity to cyclohexanone but the addition of Na$_2$CO$_3$ increased the selectivity of cyclohexanone formation to 85%.[45]

(17.7)

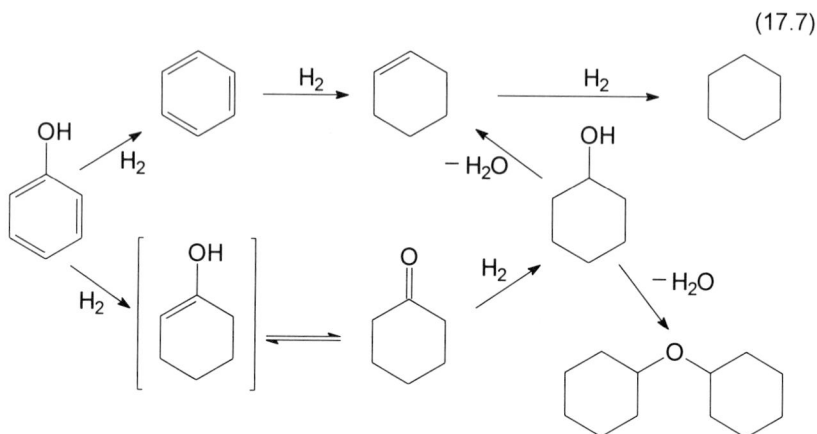

The amount of palladium in the catalysts ranged between 0.5–5% with 2–60% of a basic alkali or alkaline earth metal ion modifier (e.g., sodium or calcium hydroxide or carbonate). Vapor-phase reactions are usually run at 80°–200°C with an excess of hydrogen.[44,46,47] Liquid-phase hydrogenations are run in a solvent at 30°–60°C and 1–3 atmospheres of hydrogen[48-50] or without a solvent at 150°–180°C and 5–20 atmospheres of hydrogen.[51,52] High activity and selectivity are required of the catalyst as it is difficult to separate cyclohexanone from large amounts of phenol because of the formation of phenol-cyclohexanone and phenol-cyclohexanol azeotropes.[46] Typical product mixtures obtained from these reactions are composed of 90–98% cyclohexanone, 1–4% cyclohexanol, 0.5–4% phenol and some cyclohexane.

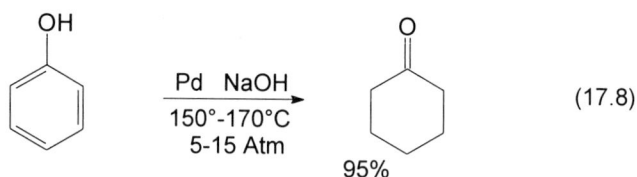

(17.8)

Covering a supported metal catalyst with a thin liquid film that differs from the bulk solvent affects both the reaction rate and selectivity.[48] Treating a Pd/C catalyst with 2M KOH before using it in a phenol hydrogenation with a heptane solvent gave cyclohexanone in 97% yield. Apparently, the distribution of phenol and phenolate ions between the thin aqueous film around the catalyst and the bulk hydrocarbon solvent enhanced the adsorption of the phenolate ion on the catalyst and the facile transfer of the neutral cyclohexanone to the heptane after the selective hydrogenation was completed (Figure 17.1).

The use of a palladium membrane catalyst was not very effective for this reaction, giving only a 60% yield of cyclohexanone at about 80% conversion.[53]

Heptane

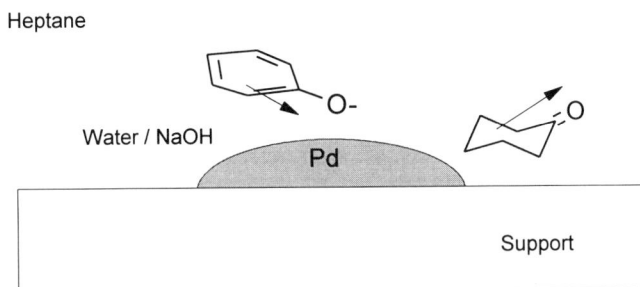

Water / NaOH

Pd

Support

Fig. 17.1   Depiction of a thin film of alkaline water on the metal
            catalyst and its effect on selectivity in phenol
            hydrogenation.[48]

Vapor phase hydrogenations using Pd/Zeolite catalysts gave cyclohexanone at 94–99% selectivity at 40–45% conversion. With Pt/Zeolite catalysts the primary product formed was cyclohexanol, only a small amount of cyclohexanone was produced,[54] but liquid phase hydrogenations over Pt/Al$_2$O$_3$ gave cyclohexanone in about 50% selectivity.[50] Near 100% selectivity for cyclohexanone formation at 55% conversion was reported for the vapor phase hydrogenation of phenol over a Pt-Cr/C catalyst.[47]

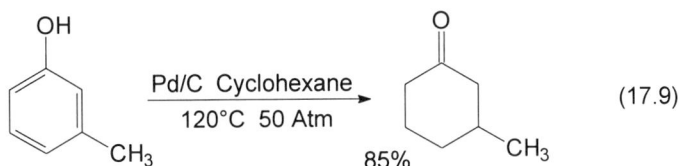

$$\text{(17.9)}$$

85%

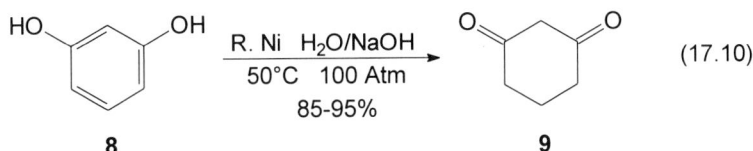

$$\text{(17.10)}$$

8   9

The partial hydrogenation of substituted phenols gives the substituted cyclohexanones. Hydrogenation of the isomeric cresols over Pd/C gave the corresponding methylcyclohexanones in 75–85% yields (Eqn. 17.9).[49] Resorcinol, **8**, was hydrogenated in aqueous base to 1,3-cyclohexanedione, **9**, over Raney nickel at 50°C and 100 atmospheres of hydrogen (Eqn. 17.10).[55,56] Similar conditions were also used to convert 5-carboxyresorcinol to 1,3-cyclohexanedione-5-carboxylic acid in 83% yield.[57] The use of higher temperatures, a more active catalyst, or no base resulted in either continued hydrogenation or condensation of the product. Palladium on charcoal in basic media was used for the hydrogenation of β-naphthol, **10**, to β-tetralone, **11**, at 175°C and 160 atmospheres.[58] With a rhodium catalyst in acetic acid the hydrogenation of either α- or β-naphthol at room temperature and four atmospheres of hydrogen gave fairly good yields of the corresponding decalones along with some decalol. If ethanol or methanol was substituted for the acetic acid as the solvent, the *cis* decalols, **12**, were the primary products (Eqn. 17.11).[59]

(17.11)

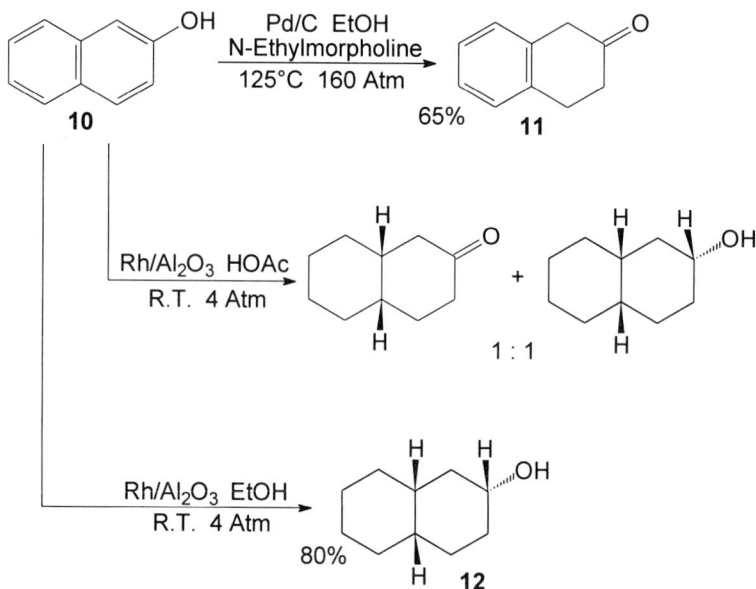

Anilines

Anilines are hydrogenated under the same conditions used to hydrogenate phenols but they are generally less susceptible to hydrogenolysis than are phenols (Eqn. 17.12).[60-62] An alternative procedure for the hydrogenation of anilines uses a rhodium catalyst in ethanol at 50°–100°C and 50–70 atmospheres.[63] Raney nickel generally gives lower yields of the cyclohexyl amines than are obtained with rhodium or ruthenium.[64] The hydrogenation of aniline and substituted anilines is used primarily to prepare cyclohexylamines. The major difficulty encountered in this reaction, besides the potential hydrogenolysis of the C–N bond, is the formation of varying amounts of a dicyclohexylamine by-product as depicted in Eqn. 17.13.[65] The amount of coupled product obtained depends on a

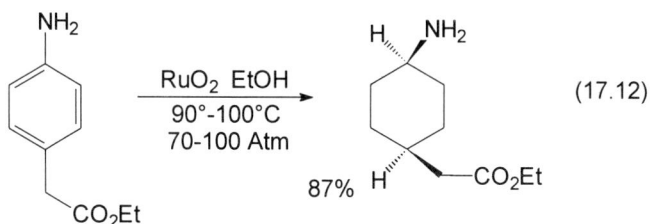

(17.12)

number of reaction variables. Increasing the temperature increases coupling.[2] With various catalysts coupling increases in the order Ru < Rh < Pd - Pt[66] so ruthenium and rhodium are preferred for this hydrogenation, not only because of the low hydrogenolysis observed using these catalysts but also for the minimal coupling produced.

(17.13)

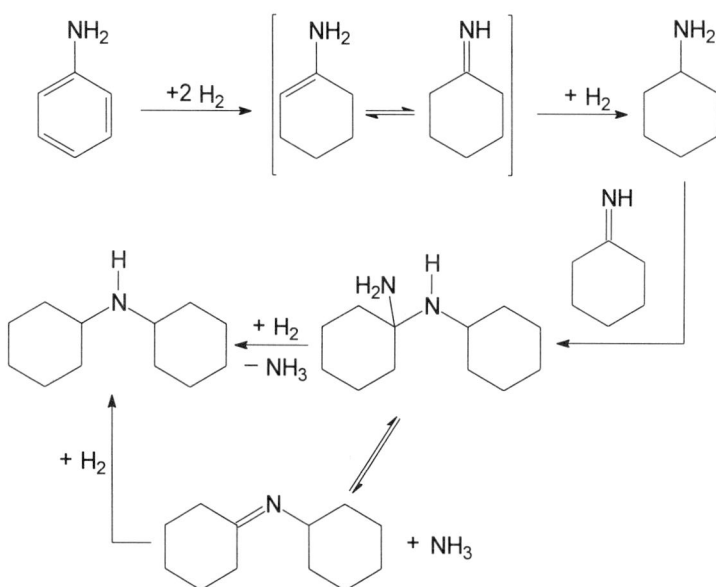

With various supported rhodium catalysts, coupling was found to be influenced by the support used in the order C > BaCO$_3$ > Al$_2$O$_3$ > BaSO$_4$ > CaCO$_3$. When carbon was added to a Rh/Al$_2$O$_3$ catalyzed hydrogenation, the amount of coupling increased, suggesting that carbon could promote the formation of intermediate condensation products. It is possible that the promoting effect of carbon is due to the presence of carboxylic acid groups in the material because the use of carboxylic acid solvents also increases the amount of coupling.[67] This would also explain why basic additives decrease coupling[65,68] as well as increase the rate of hydrogenation of anilines and substituted anilines.[69] The amount of dicyclohexylamine formed on hydrogenation of aniline over ruthenium oxide decreased with changes in solvent in the order methanol > ethanol > iso-propanol > tert-butanol. The reverse order was found for the reaction rate.[65]

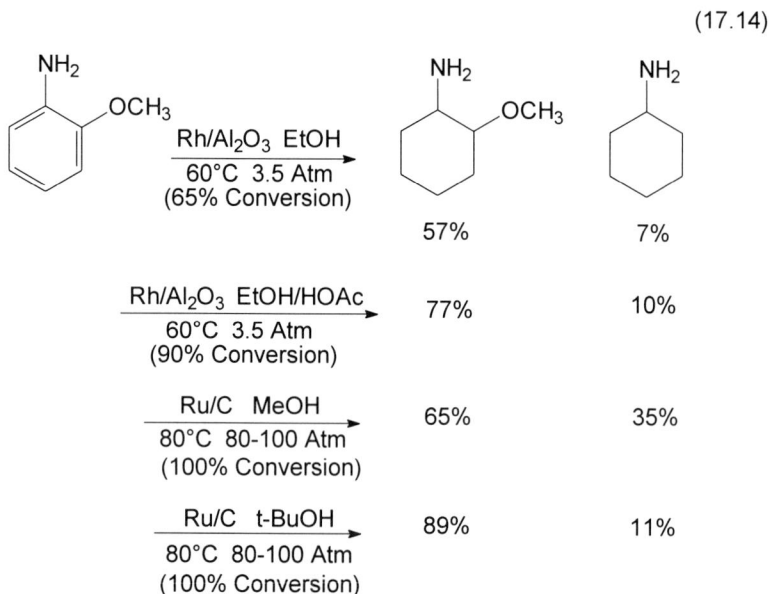

(17.14)

Conditions	OCH_3 product	NH_2 product
Rh/Al$_2$O$_3$ EtOH 60°C 3.5 Atm (65% Conversion)	57%	7%
Rh/Al$_2$O$_3$ EtOH/HOAc 60°C 3.5 Atm (90% Conversion)	77%	10%
Ru/C MeOH 80°C 80-100 Atm (100% Conversion)	65%	35%
Ru/C t-BuOH 80°C 80-100 Atm (100% Conversion)	89%	11%

Hydrogenation of alkoxy anilines takes place over a rhodium catalyst at 60°C and 3.5 atmospheres (Eqn. 17.14).[70,71] Hydrogenation over a ruthenium catalyst occurs at 80°–90°C and 75–100 atmospheres of hydrogen. The amount of alkoxy group hydrogenolysis observed under these conditions decreased from about 35% in methanol to about 10% in tert-butyl alcohol.[72]

Hydrogenation of p-aminobenzoic acid (**13**) over a rhodium catalyst at 60°C and 4 atmospheres pressure gave a mixture of the *cis,* **14**, and *trans,* **15**, aminocyclohexane carboxylic acids (Eqn. 17.15).[73] Heating this mixture of amino acids, or esters, at 260°C gave the isoquinuclidone, **16**.[73,74] However, hydrogenation of either the 3- or 4-aminobenzoic acid over ruthenium at 150°C and 100 atmospheres of hydrogen gave the bicyclic lactams directly (Eqns. 17.15–16).[75] Hydrogenation of bicyclic aminobenzoic acids gave, under these conditions, the tricyclic lactams shown in Eqns. 17.17–18.[76,77] Hydrogenation of the aminobenzoic acid esters over ruthenium at 125°–150°C and 150 atmospheres also gave the bicyclic lactam (Eqn. 17.19).[78,79]

(17.15)

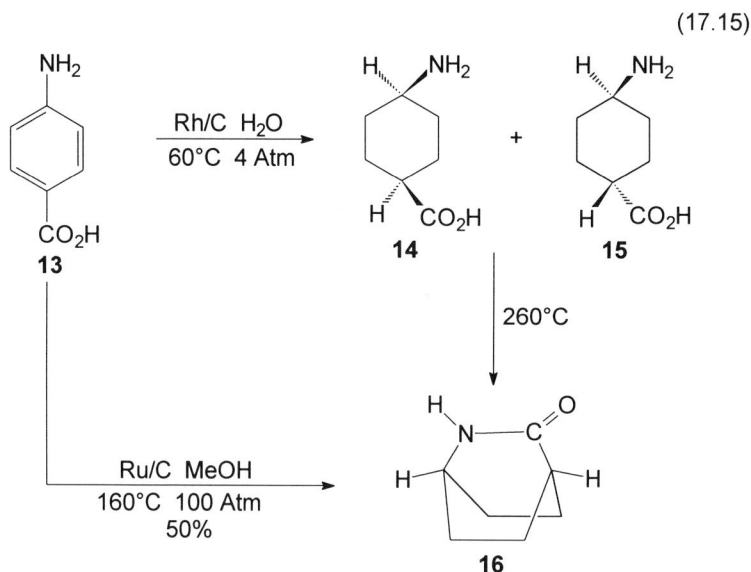

13

14   **15**

260°C

Ru/C MeOH
160°C 100 Atm
50%

16

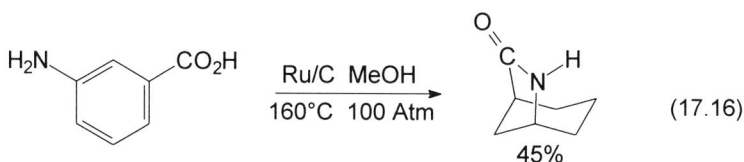

Ru/C MeOH
160°C 100 Atm

45%

(17.16)

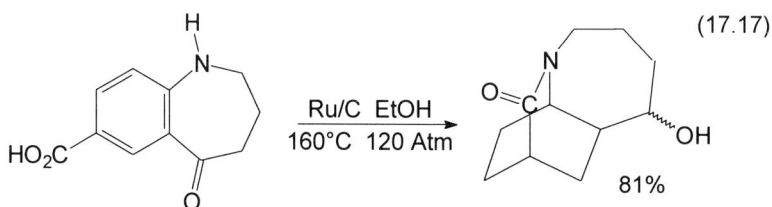

Ru/C EtOH
160°C 120 Atm

81%

(17.17)

Benzoic Acids and Esters

The hydrogenation of benzoic acids and esters proceeds with about the same facility as does the saturation of alkylbenzenes. The isomeric phthalic acids and pyromellitic acid (**17**) were each hydrogenated over Rh/C at 60°–70°C and 3 atmospheres pressure to give the corresponding cyclohexane polycarboxylic acids in 90–95% yields (Eqn. 17.20). The *cis* product predominated in each case.[80] Pyromellitic acid has also been hydrogenated over Raney nickel at 150°C and 200 atmospheres pressure.[81]

HO₂C ... (structure)

Ru/C EtOH
160°C 120 Atm

(17.18)

75%

(17.19)

MeO₂C ... C—CH₃
NH₂
Et

RuO₂ iPrOH
150° 120 Atm

HO
H₃C—CH
55%

(17.20)

HO₂C ... CO₂H
HO₂C ... CO₂H
17

Rh/C H₂O
60°C 3 Atm

HO₂C.... ....CO₂H
HO₂C.... ....CO₂H
75-80%

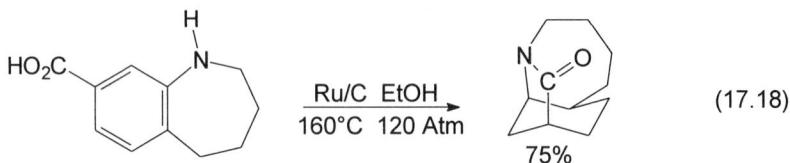

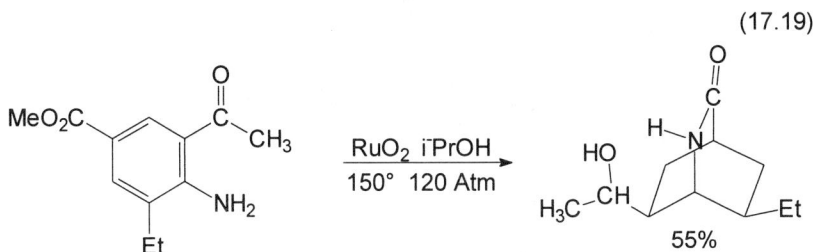

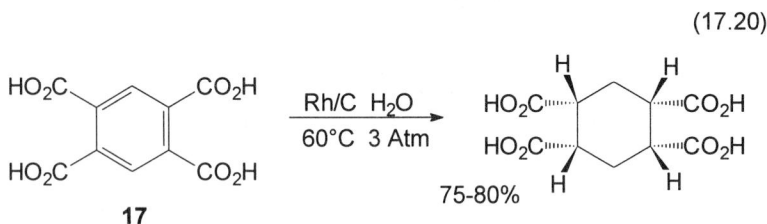

Tropones and Tropolones

In contrast to benzene rings, tropones, **18**, and tropolones, **19**, are easily hydrogenated over palladium catalysts at low temperatures and pressures in neutral solvents (Eqns. 17.21–22).[82,83] In acetic acid the ketone group of the saturated tropolone is also reduced.[83]

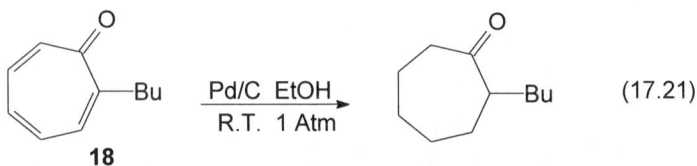

(structure) —Bu

Pd/C EtOH
R.T. 1 Atm

(structure) —Bu   (17.21)

18

Polycyclic Aromatics

The initial rate of hydrogenation of fused polycyclic aromatics increases with the number of rings present; phenanthrene > naphthalene > benzene. Only one ring is generally saturated at a time. This partial hydrogenation is accomplished because

(17.22)

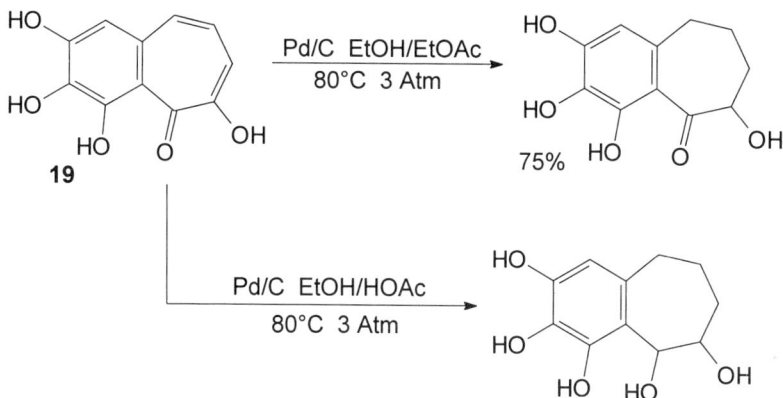

the preferred adsorption of the parent polycyclic aromatic precludes the saturation of the partially saturated species until almost all of the parent material has reacted. Palladium, platinum, rhodium and Raney nickel are all effective catalysts for the saturation of polycyclic aromatics. Partial hydrogenation of naphthalene takes place using a platinum catalyst at room temperature and four atmospheres of hydrogen. Continued hydrogenation of tetralin takes place under the same conditions used for the hydrogenation of alkyl benzenes. *cis* Decalin is the primary product obtained from the complete saturation of naphthalene (Eqn. 17.23).

(17.23)

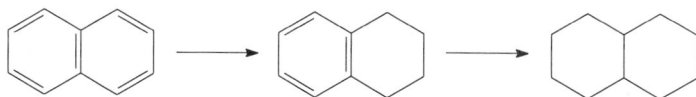

The direction of the partial hydrogenation of substituted naphthalenes is dependent on the nature and position of the substituent and the conditions used to effect the reaction. In the hydrogenation of 2-alkylnaphthalenes, there is a slight preference for the saturation of the unsubstituted ring.[84]   With 1-alkyl-naphthalenes the regioselectivity depends on the size of the alkyl group. The more bulky the substituent the greater the saturation of the substituted ring as shown in Table 17.2.[84] The regioselectivity in the hydrogenation of the bulky 1-alkylnaphthalenes is probably caused by the peri interaction between the 1-alkyl group and the 8-hydrogen, a significant interaction that is relieved by the saturation of the substituted ring.

Table 17.2

Regioselectivity in the partial hydrogenation of 1-alkylnaphthalenes.[84]

R		
Methyl	66%	34%
Ethyl	55%	45%
iso-Propyl	32%	68%
tert-Butyl	3%	97%

Hydrogenation of β-naphthol (**20**) over Raney nickel in neutral media results in the saturation of the non-substituted ring, while in base the phenolic ring is saturated (Eqn. 17.24).[85,86] The hydrogenation of β-naphthyl methyl ether, **21**, over Raney nickel gave the phenyl ether, **22**, under slightly acidic conditions.[85] Tetralin, **23**, was the primary product when base was added to the reaction mixture[85] and the alkyl ether, **24**, resulted when the hydrogenation was run in a neutral solvent (Eqn. 17.25).[86,87] A number of other examples of the effect of catalyst and reaction conditions on the regioselectivity of the hydrogenation of substituted naphthalenes have been reported.[88-94]

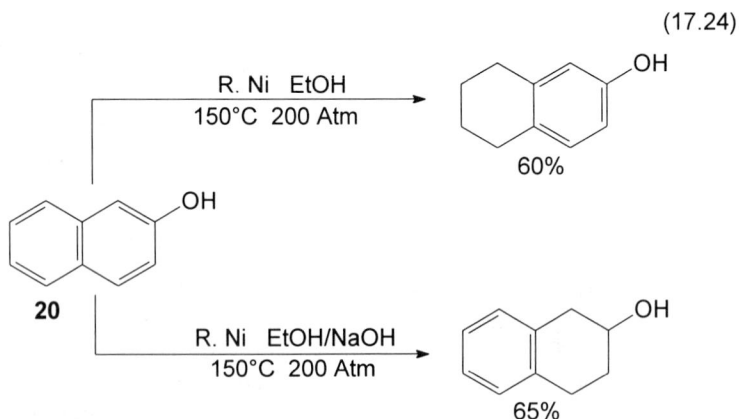

(17.24)

The hydrogenation of biphenyls over Raney nickel proceeds in a manner similar to that observed for the hydrogenation of naphthalenes.[87] The hydrogenation of p-phenylphenol (**25**) takes place over a Pd/C catalyst at 125°C

(17.25)

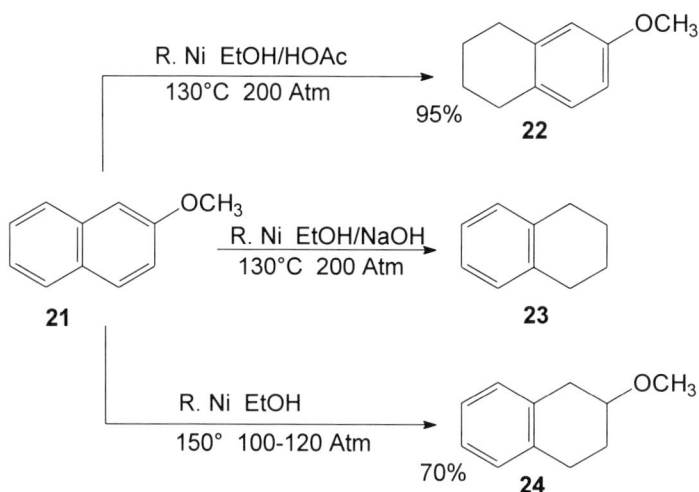

and 70 atmospheres of hydrogen. In an aqueous acetic acid solution p-cyclohexylphenol (**26**) is produced in greater than 90% yield while in a cyclohexane solvent, the primary product is 4-phenylcyclohexanol (**27**) (Eqn. 17.26).[95]

(17.26)

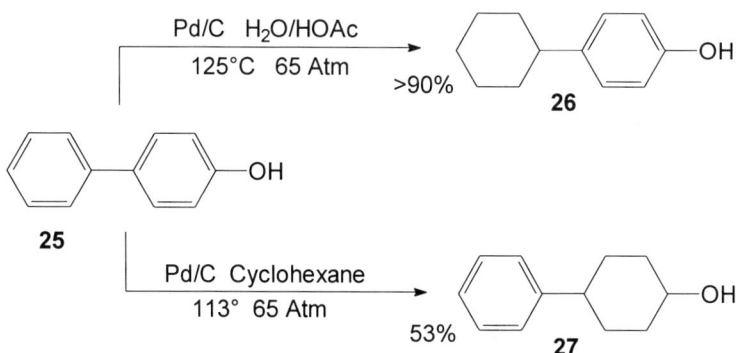

The regioselectivity of the partial hydrogenation of polycyclic aromatics depends on both the substrate structure and the catalyst used.[96-102] The hydrogenation of phenanthrene, (**28**) over Raney nickel[96] or palladium[98] takes place by way of the initial formation of the 9,10 dihydro species, **29** (Eqn. 17.27).

(17.27)

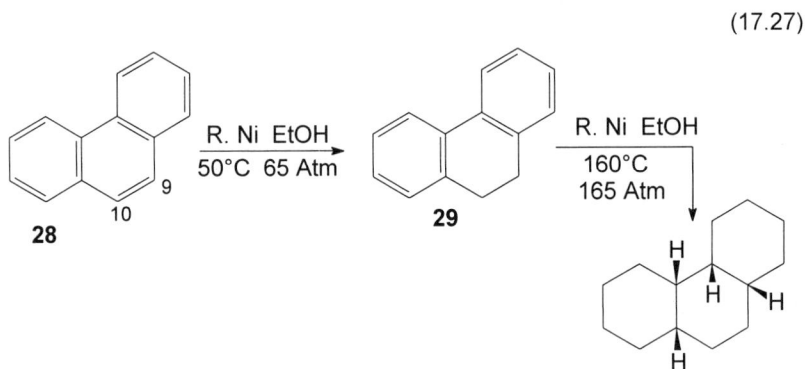

Hydrogenation of anthracene (**30**) over Raney nickel,[96] platinum or palladium[100] also gives 9,10 dihydroanthracene (**31**) as the initial product but a rhodium catalyst gives the octahydro product, **32**, in which both terminal rings are saturated (Eqn. 17.28).[100] Further hydrogenation takes place under the conditions used to saturate alkylbenzenes with the predominant product the *cis* ring junction isomer.

(17.28)

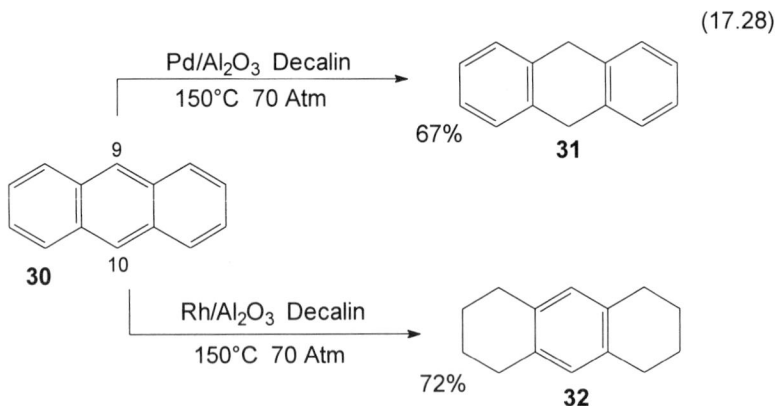

Partial hydrogenation of either the ortho or meta terphenyl (**33**) over platinum or copper chromite results in the saturation of the internal ring (Eqn. 17.29) but hydrogenation of para terphenyl (**34**) takes place primarily on a terminal ring (Eqn. 17.30).[101]

In the partial hydrogenation of benz[a]anthracene (**35**) palladium promotes the regioselective production of the internally saturated dihydroarene, **36**, while a platinum catalyst saturates a terminal ring to give the tetrahydroarene, **37**, as

(17.29)

33 $\xrightarrow[\text{25°C 5 Atm}]{\text{PtO}_2 \text{ HOAc}}$

(17.30)

34 $\xrightarrow[\text{25°C 5 Atm}]{\text{PtO}_2 \text{ HOAc}}$

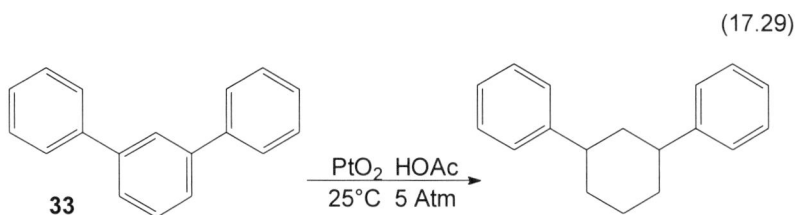

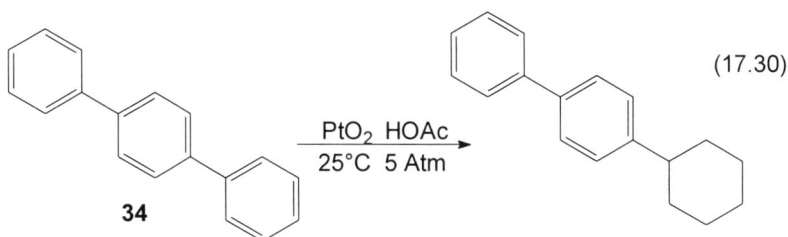

illustrated in Eqn. 17.31.[98] This generality has been shown to hold true for a number of polycyclic aromatic compounds.[97,98]

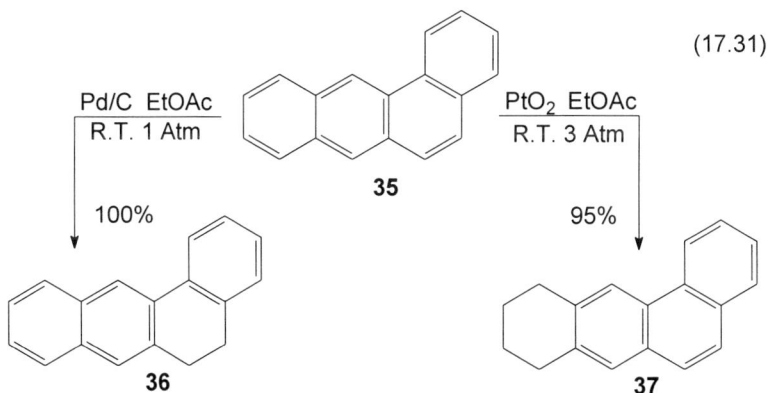

(17.31)

$\xleftarrow[\text{R.T. 1 Atm}]{\text{Pd/C EtOAc}}$ **35** $\xrightarrow[\text{R.T. 3 Atm}]{\text{PtO}_2 \text{ EtOAc}}$

100%   95%

36   **37**

Heterocyclic Aromatics

Pyridines

Piperidines are formed by the hydrogenation of pyridines over platinum catalysts in acid solution, generally at room temperature under 3–4 atmospheres of hydrogen (Eqn. 17.32). Acid is used in this reaction since the pyridinium salt is more easily hydrogenated than the free base.[103] A quaternary pyridinium salt gives the N-alkyl piperidine in ethanol, without added acid, under these same

(17.32)

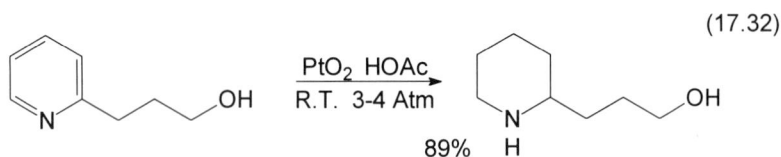

conditions (Eqn. 17.33).[104] The selective conversion of pyridine-N-oxide to pyridine also takes place under these standard conditions.[105] Other catalysts can also be used[106] but they are generally less effective than platinum. Raney nickel requires higher temperatures (125°–175°C) and pressures (70–150 atmospheres) and N-alkylation of the piperidine can occur under these reaction conditions, particularly at temperatures exceeding 150°C, if an alcohol is used as the solvent. Ruthenium in aqueous solvents also requires at least 100°C and 70 atmospheres of hydrogen but N-alkylation does not take place over this catalyst.[107] The use of rhodium is limited to acidic media since this catalyst is poisoned by the piperidine formed in the reaction.[108,109] The conditions required for rhodium catalysis are the same as those used for platinum. The use of palladium requires higher temperatures (70°–80°C) than the other precious metals.[110]

(17.33)

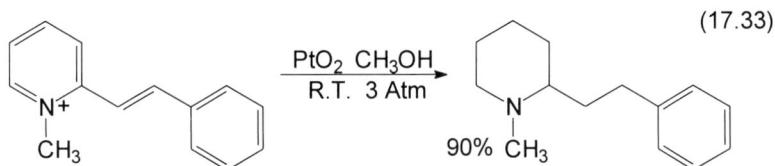

        The saturation of the pyridine ring over platinum in acetic acid occurs less readily than the hydrogenation of ketones,[111] oximes,[112] nitriles, olefins, acetylenes and nitro groups as well as the cleavage of aryl halides, particularly bromides and iodides.[106] If the pyridine is present as the hydrochloride, hydrobromide or quaternary salt, the ease of hydrogenation over platinum is increased and this reaction takes place in preference to the saturation of carbonyl,[113,114] and cyano groups.[115] Double bonds and nitro groups are still preferentially reduced. Under the standard reaction conditions the hydrogenation of a pyridine ring can be effected in preference to the saturation of indole[116] or benzene rings. Hydrogenations run in acetic acid over Pd/C at 70°C–80°C and 3–4 atmospheres of hydrogen saturate a pyridine ring in the presence of an aldehyde or cyclopropyl group as well as a benzene ring (Eqn. 17.34). Primary and secondary benzyl alcohols, however, are hydrogenolyzed along with the saturation of the pyridine ring under these reaction conditions.[110] Over rhodium, however, there is very little selectivity between pyridine and benzene ring hydrogenation.[108]

(17.34)

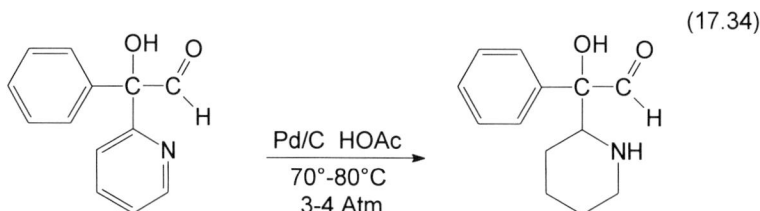

3-Aminopyridine was hydrogenated to 3-aminopiperidine in ethanolic hydrochloric acid over platinum under the usual conditions.[117] The 2-amino compound under these conditions was initially hydrogenated only to the tetrahydro stage. Further hydrogenation resulted in the loss of ammonia to give piperidine.[118] The 2-mono-or dialkylaminopyridines react in the same manner as the 2-aminopyridines.[119]   But the 1-substituted-2-aminopyridines gave the 1-alkyl-2-aminopiperidines when hydrogenated over Rh/C in ethanolic hydrochloric acid at room temperature and 3–4 atmospheres of hydrogen (Eqn. 17.35).[120] When the free base was hydrogenated the loss of ammonia occurred. Pyridines with tertiary amine groups in the 4-position were hydrogenated without excessive deamination over Ru/C in water at 100°C and 70–100 atmospheres.[108]

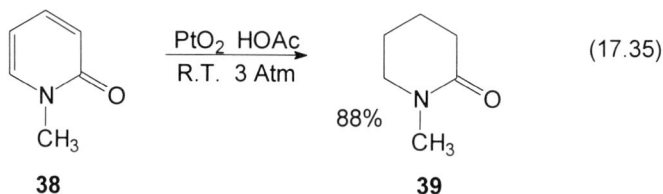

(17.35)

The hydrogenation of 3- and 4-hydroxy- or alkoxypyridines to the hydroxy- or alkoxypiperidines was accomplished using platinum catalysts in acetic acid at room temperature and 3–4 atmospheres of hydrogen.[121,122] The lactam, N-methyl-2-piperidone, 39, was obtained by the hydrogenation of N-methyl-2-pyridone, 38, under any of the standard conditions (Eqn. 17.35).[123] The 2-alkoxypyridines are hydrogenolyzed to give the desoxypiperidines on hydrogenation over platinum.[124]

The hydrogenation of 2- and 4-pyridinecarboxylic acids proceeded best in water over a platinum catalyst at room temperature and 3–4 atmospheres of hydrogen.   Nicotinic acid (3-pyridinecarboxylic acid) was, however, decarboxylated under these conditions.[125]   This compound, as well as the pyridine alkanoic acids (except 2-pyridine acetic acid which is easily decarboxylated), was hydrogenated at room temperature and 3–4 atmospheres of hydrogen over a rhodium catalyst in water to which had been added a slight excess of ammonium hydroxide. Product isolation merely involved the removal

of the catalyst and evaporation of the solvent.[126,127] The esters of these acids
were hydrogenated in acetic acid over platinum under the usual conditions.

(17.36)

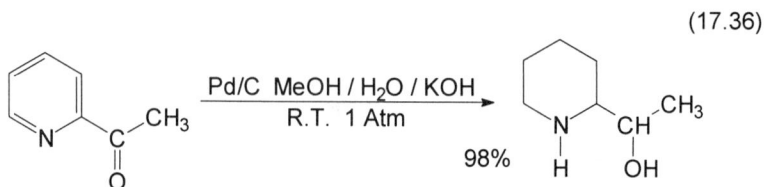

The 2-and 4-acylpyridines can be hydrogenated to the hydroxypiperidines
over Rh/Al_2O_3 at low temperatures and pressures or Pd/C in basic aqueous
methanol at room temperature and atmospheric pressure (Eqn. 17.36).[128]
Hydrogenation of 3-acylpyridines, 3-pyridine carboxylic esters or 3-cyano-
pyridines or their quaternary salts, using the standard reaction conditions, gave the
tetrahydropyridines almost exclusively (Eqn. 17.37)[129,130] because of the
inherent stability and difficulty in hydrogenating this vinylogous amide (Chapter
15).   Hydrogenation of 3-acylpyridine over Pd/C in basic aqueous methanol,
however, gave the piperidyl alcohol (Eqn. 17.38).[128]

(17.37)

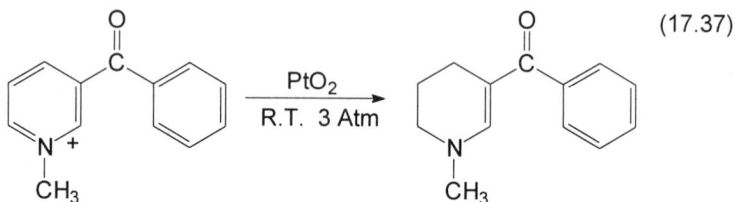

(17.38)

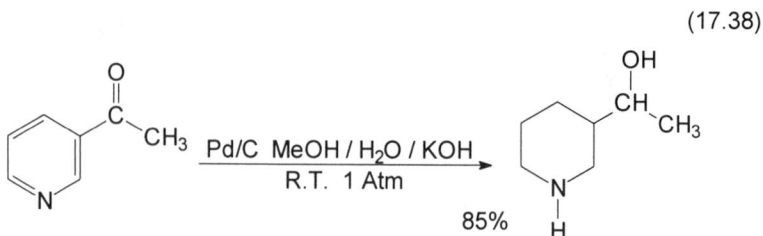

Quinolines and Isoquinolines

The hydrogenation of quinolines and isoquinolines takes place in two distinct
steps. The formation of the *py*-tetrahydro compound, **40** or **41**, occurs under
conditions used to saturate a nonfused pyridine ring (Eqn. 17.39).[131-136] Further
hydrogenation of the tetrahydroquinoline takes place using the standard
conditions   for   the   hydrogenation   of   aniline   and   substituted   anilines.

(17.39)

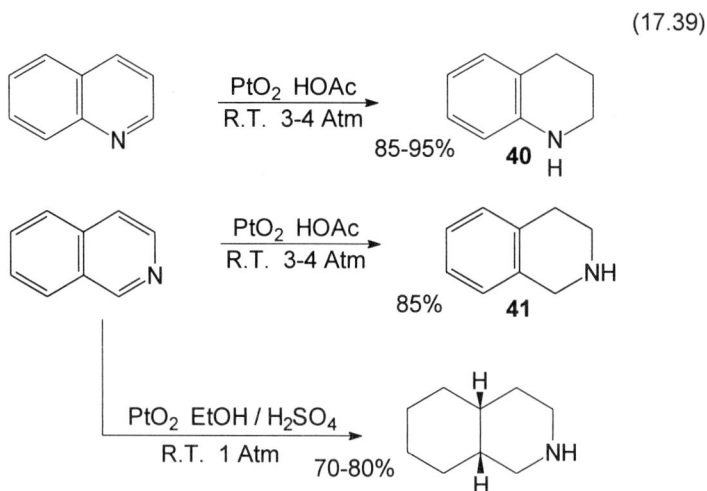

Hydrogenation of quinoline or isoquinoline over platinum in methanolic hydrogen chloride,[137] or other strong acids such as trifluoroacetic or sulfuric acids,[138-140] gives the *bz*-tetrahydro product (Eqn. 17.40)

(17.40)

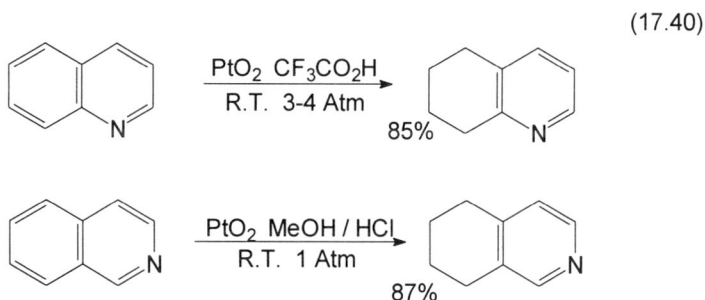

py-Tetrahydroisoquinoline is difficult to hydrogenate as the free base.[106] Acylation of the amine, though, makes it possible to continue the hydrogenation under the same conditions used to prepare the decahydroquinolines.[141] Decahydroisoquinoline is also obtained by the hydrogenation of isoquinoline using an equal weight of platinum oxide in a sulfuric acid medium at room temperature and atmospheric pressure (Eqn. 17.39).[144] Deviation from these conditions, however, results in the formation of only *py*-tetrahydroisoquinoline. These conditions have also been used to fully saturate some quinoline esters that were difficult to hydrogenate under normal conditions.[143]

The stereochemistry of the ring juncture of the decahydro compounds is dependent on the catalyst and conditions used for the reactions. Hydrogenations over nickel catalysts give primarily the *trans* products[131] and those hydrogenations run in acid medium lead to a predominance of the *cis* isomer.[142,144] The use of ruthenium should also lead to preferential formation of the *cis* ring fused products.

Substituted quinolines and isoquinolines are hydrogenated in the same way as are the substituted pyridines to give the *py*-tetrahydro compounds. Further hydrogenation under conditions used for the saturation of substituted benzenes, normally gives the decahydro compounds provided that any tetrahydro-isoquinoline which may be present has been acetylated prior to the reaction.

Pyrroles

Pyrrolidines are obtained by the hydrogenation of pyrroles in glacial acetic acid over platinum, rhodium or palladium catalysts at 25°–50°C and 3–4 atmospheres of hydrogen.[145,146] The *cis* isomer is formed on hydrogenation of disubstituted pyrroles under these conditions (Eqn. 17.41). The N-substituted pyrroles are more easily hydrogenated than the N-unsubstituted compounds.

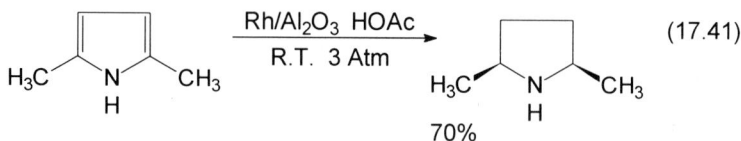

$$\text{(17.41)}$$

$$\text{(17.42)}$$

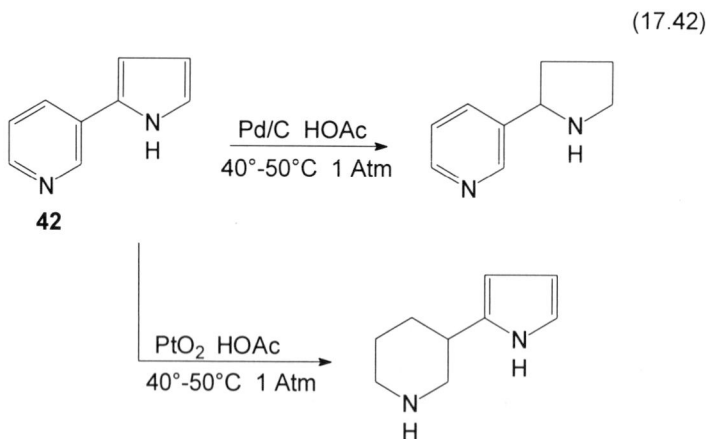

The selectivity observed in the hydrogenation of pyrroles is strongly dependent on the catalyst and conditions used for the reactions. The pyrrole ring in nicotyrine, **42**, was hydrogenated preferentially over palladium in acetic acid,[147] but over platinum the pyridine ring was saturated first (Eqn. 17.42).[148,149] Keto and ester groups on an N-unsubstituted pyrrole were hydrogenated more readily than the ring over copper chromite or Raney nickel. The ring of N-substituted pyrroles, however, was hydrogenated more easily than the ester group over Raney nickel.[150]

Indoles

The pyrrole ring of an indole is saturated with more difficulty than an unfused pyrrole. The 2,3-dihydro compound is usually prepared by the hydrogenation of the indole over Raney nickel at 90°–100°C and 70–100 atmospheres of hydrogen (Eqn. 17.43)[152] or over platinum in aqueous alcoholic HBF_4 at room temperature and atmospheric pressure.[151] Raising the temperature to 150°–160°C resulted in the formation of good to excellent yields of the N-ethyl octahydroindole when ethanol was used as the solvent.[152] N-Alkylation could be avoided by the use of dioxane as the solvent. The hydrogenation of some 2,3-disubstituted indoles over platinum in acetic acid at room temperature and atmospheric pressure resulted in the formation of the 4,5,6,7-tetrahydroindole (Eqn. 17.44).[153] Similar results were observed in the hydrogenation of the nitroindole, **43**, over an excess of palladium in acetic acid (Eqn. 17.45).[154] Steric factors apparently hindered attack at the nitro group and the pyrrole ring, with the benzene ring being preferentially hydrogenated.

(17.43)

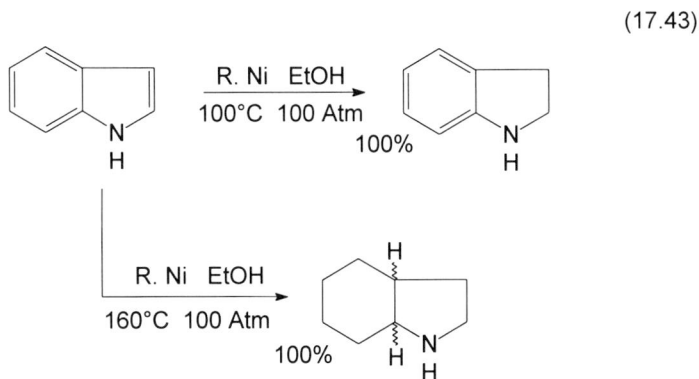

Acridines, Carbazoles and Other Polycyclic Nitrogen Heterocycles

Hydrogenation of acridine (**44**) in decalin over platinum or palladium catalysts at 150°C and 70 atmospheres of hydrogen gave primarily the 9,10 dihydro product, **45**. Over rhodium the fully saturated compound, **46**, was obtained (Eqn. 17.46).[100] Hydrogenation of acridine over platinum in strongly acidic media gave the benzene ring saturated material, **47**, in quantitative yield.[139,155] Hydrogenation of benzo[h]quinoline (**48**) under these strongly acidic conditions also produced the benzene ring saturated material as the primary product (Eqn. 17.47).[139]

(17.44)

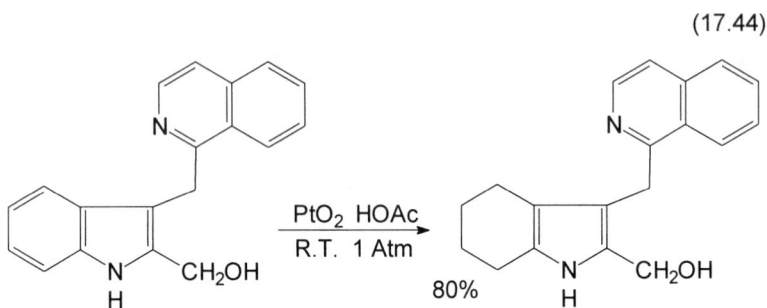

(17.45)

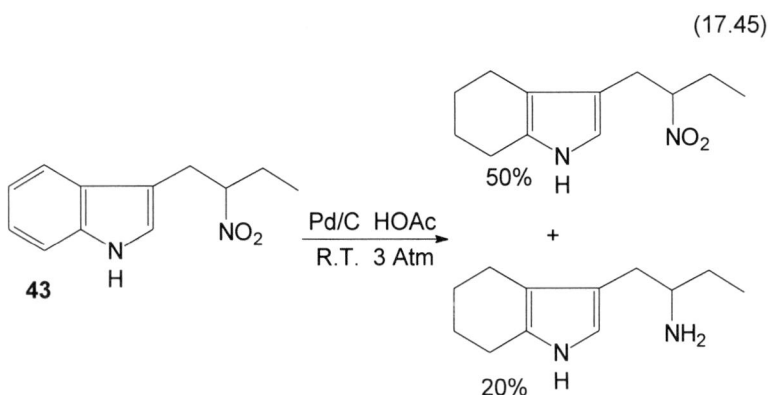

(17.46)

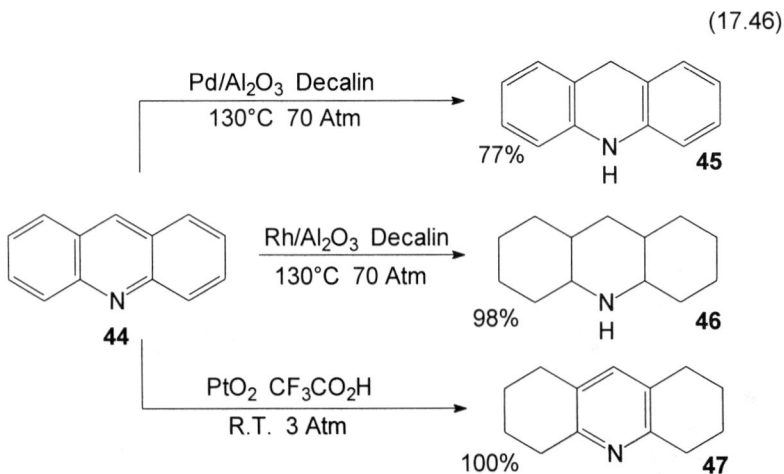

(17.47)

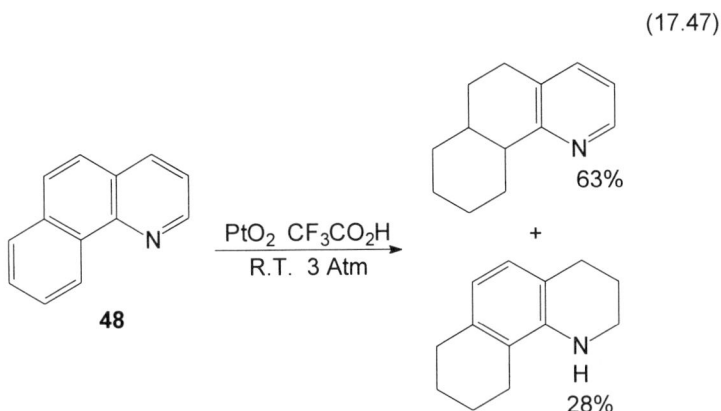

Hydrogenation of carbazole (**49**) over palladium resulted in the saturation of one of the benzene rings while over rhodium the fully saturated product, **50**, predominated (Eqn. 17.48).[100]

(17.48)

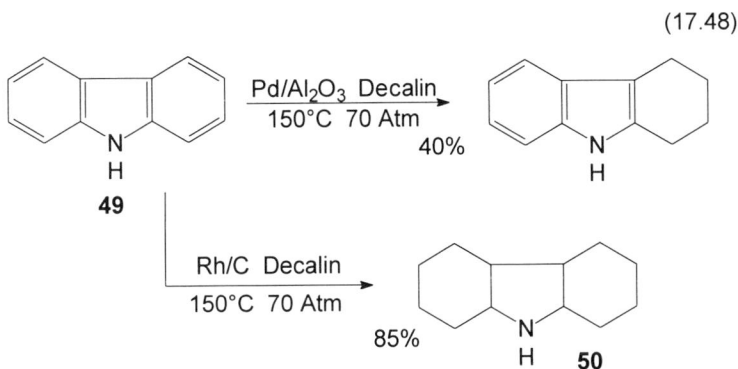

Furans

The saturation of a furan ring takes place over Raney nickel at 25°–50°C and 3–4 atmospheres of hydrogen (Eqn. 17.49).[156,157] The more highly substituted furans are more difficult to hydrogenate and may require temperatures of 100°–125°C and pressures of 70–100 atmospheres.[158,159] Ring opening, however, can occur at these higher temperatures. Alternative procedures in which hydrogenolysis of the ring does not occur involves the use of Ru/C in dioxane at 120°–160°C and 150–200 atmospheres of hydrogen[160] or Rh/C at room temperature and 3 atmospheres pressure.[161]

The hydrogenation of 2-methylfuran (**51**) over palladium in acidic acetone can be stopped at the dihydro stage to give very good yields of the hydroxy

(17.49)

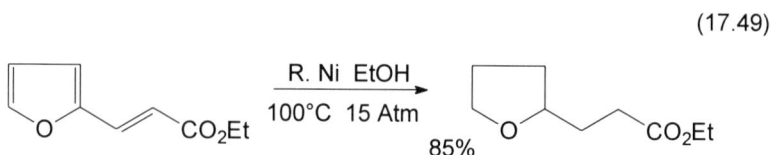

85%

ketone, **52** (Eqn. 17.50).[162]   Hydrogenation of furfural (**53**) gave the tetrahydrofurfuryl alcohol on hydrogenation over Ni/SiO$_2$ at 130° and 40 atmospheres of hydrogen (Eqn. 17.51).[163]   Similar results were obtained for hydrogenations run over palladium, ruthenium or rhodium catalysts.[164,165]

(17.50)

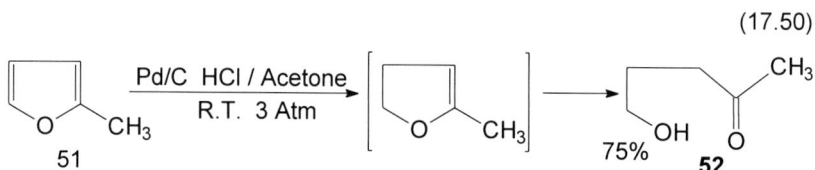

(17.51)

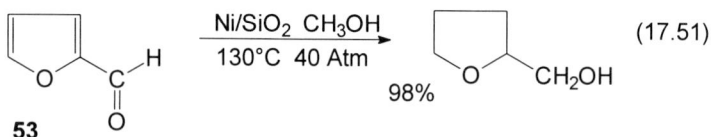

Pyrans, Pyrones, Chromones and Coumarins

Pyrans, pyrones and dihydropyrans are hydrogenated best over Raney nickel under the same conditions used to saturate the furan ring. The more substituted compounds require higher temperatures and, as with furans, ring opening can occur at temperatures above 125°–150°C. The hydrogenation of 4-pyrones can be stopped at either the tetrahydro (keto) or hexahydro (alcohol) stage (Eqn. 17.52)[166-169] but 2-pyrones give only the lactone[170] which is resistant to further hydrogenation under these conditions (Eqn. 17.53).   Platinum or palladium catalysts can also be used for these reactions at 25°–50°C and 3–4 atmospheres of hydrogen.

Any of these conditions can also be used for the preparation of dihydro-coumarins, **54**, (Eqn. 17.54) and dihydrochromones, **55** (Eqn. 17.55).[171-173] With a palladium catalyst at 150°C and 100 atmospheres of hydrogen in alcoholic solution, the ring-opened hydroxy ester was produced. This material was reclosed to the dihydrocoumarin.[174] The use of copper chromite at high temperatures and pressures resulted in lactone hydrogenation as well.[175] Hydrogenation of the chroman, **56**, over Ru/C at 100°–110°C and 100–150 atmospheres of hydrogen gave the hexahydrochromane, **57** (Eqn. 17.56).[176]

(17.52)

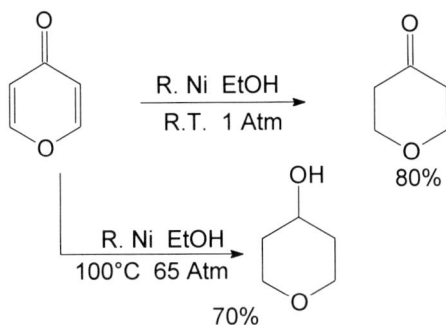

(17.53)

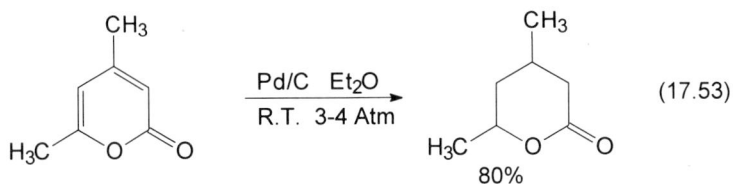

(17.54)

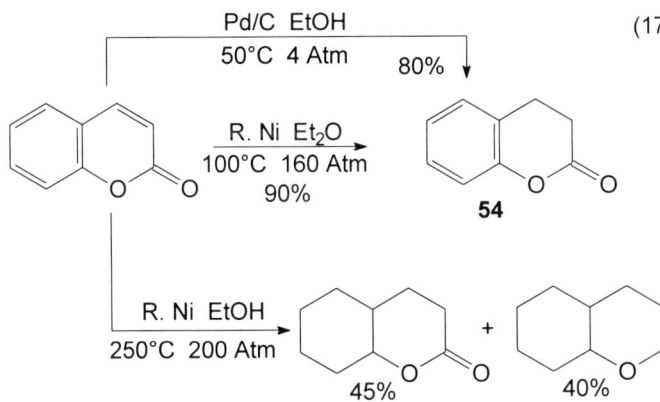

(17.55)

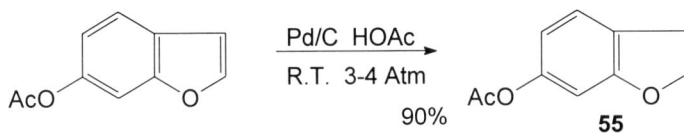

(17.56)

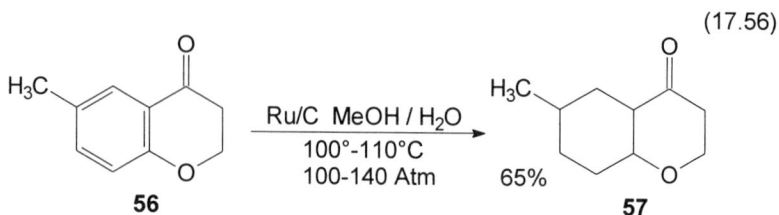

56 → **57**

Ru/C MeOH / H₂O
100°-110°C
100-140 Atm   65%

Thiophenes

Since sulfur is the most effective of all catalyst poisons, the hydrogenation of sulfur containing heterocycles is not easily accomplished unless there are no unshared electron pairs on the sulfur atom or the catalyst used is not affected by the poison. The hydrogenation of the cyclic sulfone, **58**, takes palace over an excess of palladium in acetic acid at room temperature and atmospheric pressure (Eqn. 17.57).[177] Thiophene, itself, can be hydrogenated to tetrahydrothiophene over rhenium heptasulfide at 250°C and 300 atmospheres of hydrogen[178] or over a large excess of palladium in methanolic sulfuric acid at room temperature and 3–4 atmospheres.[179] No hydrogenolysis of the carbon-sulfur bond was observed in these reactions.

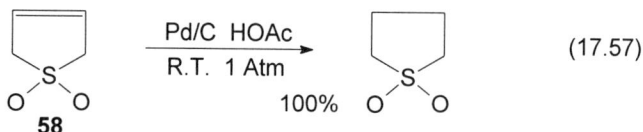

Pd/C HOAc
R.T. 1 Atm   100%   (17.57)

58

References

1. P. N. Rylander and D. R. Steele, *Engelhard Ind. Tech. Bull.*, **3**, 19 (1962).
2. H. Greenfield, *Ann. N. Y. Acad. Sci.*, **214**, 233 (1973).
3. H. Rapoport and J. Z. Pasky, *J. Am. Chem. Soc.*, **78**, 3788 (1956).
4. J. H. Stocker, *J. Org. Chem.*, **29**, 3593 (1964).
5. C. H. Hurd and H. Jenkins, *J. Org. Chem.*, **31**, 2045 (1966).
6. S. E. Cantor and D. S. Tarbell, *J. Am. Chem. Soc.*, **86**, 2902 (1964).
7. F. Hartog and P Zwietering, *J. Catal.*, **2**, 79 (1963).
8. M.M. Johnson and G. P. Nowack, *J. Catal.*, **38**, 518 (1975).
9. S. Siegel, V. Ku and W. Halpern, *J. Catal.*, **2**, 348 (1963).
10. J. F. Outlaw, Jr., J. R. Cozort, N. Garti and S. Siegel, *J. Org. Chem.*, **48**, 4186 (1983).
11. C. U. I. Odenbrand and S. T. Lundin, *J. Chem. Technol. Biotechnol.*, **30**, 677 (1980).
12. C. U. I. Odenbrand and S. L. T. Andersson, *J. Chem. Technol. Biotechnol.*, **32**, 691 (1982).
13. P. N. Rylander and D. Steele, *Engelhard Ind. Tech. Bull.*, **3**, 91, 125 (1962); **4**, 20 (1963).

14. C. U. I. Odenbrand and S. T. Lundin, *J. Chem. Technol. Biotechnol.*, **31**, 660 (1981).
15. J. A. Don and J. J. F. Scholten, *Faraday Discuss. Chem. Soc.*, **72**, 146 (1981).
16. P. J. Van der Steen and J. J. F. Scholten, *Int. Cong. Catal., Proc., 8th*, **2**, II 659 (1984).
17. R. L. Augustine, R. M. Warner and M. J. Melnick, *J. Org. Chem.*, **49**, 4853 (1984).
18. L. M. Berkowitz and P. N. Rylander, *J. Org. Chem.*, **24**, 708 (1959).
19. S. Niwa, F. Mizukami, J. Imamura, K. Shimizu and T. Tsuchiya, *Jpn. Kokai Tokkyo Koho* JP 60 199,837 (1985); *Chem. Abstr.*, **104**, 109172 (1986)
20. J. Struijk, R. Moene, T. van der Kamp and J. J. F. Scholten, *Appl. Catal. A*, 89, 77 (1992).
21. H. Ishihashi, K. Obata, Y. Mori and H. Yoshioka, *Jpn. Kokai Tokkyo Koho* JP 61 85,334 (1984); *Chem. Abstr.*, **106**, 4566 (1987).
22. S. Niwa, F. Mizukami, M. Kuno, K. Takeshita, H. Nakamura, T. Tsuchiya, K. Shimizu and J. Imamura, *J. Mol. Catal.*, 34, 247 (1986).
23. S. Niwa, J. Mizukami, S. Isoyama, T. Tsuchiya, K. Shimizu, S. Imai and J. Imamura, *J. Chem. Technol. Biotechnol.*, 36, 236 (1986).
24. O. Mitsui and Y. Fukuoka, U S Patent 4,678,861 (1987).
25. H. Nagahara and M. Konishi, *Jpn. Kokai Tokkyo Koho* JP 62 45,544 (1987); *Chem. Abstr.*, **107**, 77344 (1987).
26. M. Soede, E. J. A. X. van de Sandt, M. Makkee and J. J. F. Scholten, *Stud. Surf. Sci. Catal.*, **78** (Heterog. Catal. Fine Chem., III), 345 (1993).
27. F. Mizukami, *Senryo to Yakuhin*, **31**, 297 (1986); *Chem. Abstr.*, **106**, 195922 (1987).
28. H. Ichihashi and H. Yoshioka, EP 214,530 (1985); *Chem. Abstr.*, **107**, 9339 (1987).
29. F. Mizukami, S. Niwa, S. Ohkawa and A. Katayama, *Stud. Surf. Sci. Catal.*, **78** (Heterog. Catal. Fine Chem., III), 337 (1993).
30. T.-N. Huang and J. Schwartz, *J. Am. Chem. Soc.*, **104**, 5244 (1982).
31. S. J. Tiechner, C. Hoang-Van and M. Astier, *Stud. Surf. Sci. Catal.*, **11** (Metal-Supp. and Metal-Add. Effects in Catal.), 121 (1982).
32. P. Dini, D. Dones, S. Montelatici and N. Giordano, *J. Catal.*, **30**, 1 (1973).
33. Y. Z. Chen and K. J. Wu, *Appl. Catal.*, **78**, 185 (1991).
34. H. Imamura, T. Konishi, Y. Sakata and S. Tsuchiya, *J. Chem. Soc., Chem. Commun.*, 1852 (1993).
35. B. van de Graaf, H. van Bekkum and B. M. Wepster, *Recl. Trav. Chim. Pays-Bas*, **87**, 777 (1968).
36. H. van Bekkum, H. M. A. Buurmans, G. van Minnen-Pathuis and B. M. Wepster, *Recl. Trav. Chim. Pays-Bas*, **88**, 779 (1969).
37. R. E. Ireland and P. W. Schiess, *J. Org. Chem.*, **28**, 6 (1963).
38. R. H. Martin and R. Robinson, *J. Chem. Soc.*, 491 (1943).
39. J. W. Cornforth and R. Robinson, *J. Chem. Soc.*, 1855 (1949).
40. W. H. Clingman, Jr. and F. T. Wadsworth, *J. Org. Chem.*, **23**, 276 (1958).
41. R. Lukes, J. Trojanek and K. Blaha, *Coll. Czech. Chem. Commun.*, **20**, 1136 (1955).
42. R. H. Levin and J. H. Pendergrass, *J. Am. Chem. Soc.*, **69**, 2436 (1947).
43. H. A. Smith and R. G. Thompson, *Adv. Catal.*, **9**, 727 (1957).

44. G. Neri, A. M. Visco, A. Donato, C. Milone, M. Malentacchi and G. Gubitosa, *Appl. Catal. A*, **110**, 49 (1994).
45. T. P. Murtha, U. S. Patent 4,409,401 (1983).
46. J. R. Gonzalez-Velasco, J. I. Gutierrez-Ortiz, M. A. Guiierrez-Ortiz, M. A. Martin, S. Mendioroz, J. A. Pajares and M. A. Folgado, *Stud. Surf. Sci. Catal.*, **31** (Prep. Catal., IV), 619 (1987).
47. S. T. Srinivas and P. K. Rao, *J. Chem. Soc., Chem. Commun.*, 33 (1993).
48. A. A. Wismeijer, A. P. G. Kieboom and H. van Bekkum, *Recl. Trav. Chim. Pays-Bas*, **105**, 129 (1986).
49. M. Higashijima and S. Nishimura, *Bull. Chem. Soc., Japan.*, **65**, 2955 (1992).
50. M. A. Gutierrez-Ortiz, A. Castano, M. P. Gonzaled-Marcos, J. I. Gutierrez-Ortiz and J. R. Gonzalez-Velasco, *Ind. Eng. Chem. Res.*, **33**, 2571 (1994).
51. J. F. Van Peppen, W. B. Fisher and C. H. Chan, *Chem. Ind. (Dekker)*, **22** (Catal. Org. React.), 355 (1985).
52. D. S. Thakur, T. J. Sullivan, J. F. Van Peppen and W. B. Fisher, *Chemical Catalyst News* (Engelhard Corp.), Feb. 1992.
53. N. Itoh and W.-C. Xu, *Appl. Catal. A*, **107**, 83 (1993).
54. A. K. Talukdar and K. G. Bhattacharyya, *Appl. Catal. A*, **96**, 229 (1993).
55. R. B. Thompson, *Org. Synth.*, **27**, 21 (1947).
56. F. E. King and D. G. I. Felton, *J. Chem. Soc.*, 1371 (1948).
57. E. E. Van Tamelen and G. T. Hildahl, *J. Am. Chem. Soc.*, **78**, 4405 (1956).
58. G. Stork and E. L. Foreman, *J. Am. Chem. Soc.*, **68**, 2172 (1946).
59. A. I. Meyers, W. Bevering and G. Garcia-Munoz, *J. Org. Chem.*, **29**, 3427 (1964).
60. M. Friefelder and G. R. Stone, *J. Am. Chem. Soc.*, **80**, 5270 (1958).
61. M. Friefelder and G. R. Stone, *J. Org. Chem.*, **27**, 3568 (1962).
62. L. C. Behr, J. E. Kirby, R. N. MacDonald and C. W. Todd, *J. Am. Chem. Soc.*, **68**, 1296 (1946).
63. R. E. Malz, Jr. and H. Greenfield, in *Catalysis in Organic Synthesis,1976,* (P. N. Rylander, Ed.) Academic Press, New York, 1976, p. 343.
64. A. E. Barkdoll, D. C. England, H. W. Gray, W. Kirk, Jr. and G. M. Whitman, *J. Am. Chem. Soc.*, **75**, 1156 (1953).
65. S. Nishimura, T. Shu, T. Hara and Y. Takagi, *Bull. Chem. Soc., Japan.*, **39**, 329 (1966).
66. P. N. Rylander, L. Hasbrouk and I. Karpenko, *Ann. N. Y. Acad. Sci.*, **214**, 100 (1973).
67. S. Nishimura and H. Taguchi, *Bull. Chem. Soc., Japan.*, **36**, 873 (1963).
68. D. V. Sokol'skii, A. Ualikhanova and A. E. Temirbulatova, *React. Kinet. Catal. Lett.*, **20**, 35 (1982).
69. S. Nishimura, Y. Kono, Y. Otsuki and Y. Fukaya, *Bull. Chem. Soc., Japan.*, **44**, 240 (1971).
70. M. Friefelder, Y. H. Ng and P. F. Helgren, *J. Org. Chem.*, **30**, 2485 (1985).
71. R. Egli and C. H. Ergster, *Helv. Chim. Acta*, **58**, 2321 (1975).
72. S. Nishimura and H. Yoshino, *Bull. Chem. Soc., Japan.*, **42**, 499 (1969).

73. W. M. Pearlman, *Org. Synth.*, **Coll. Vol. 5**, 670 (1969).
74. M. G. Scaros, H. L. Dryden, Jr, J. P. Wastrich and O. J. Goodnomson, *Chem. Ind. (Dekker)*, **18** (Catal. Org. React.), 279 (1984).
75. R. L. Augustine and L. A. Vag, *J. Org. Chem.*, **40**, 1074 (1975).
76. R. L. Augustine and W. G. Pierson, *J. Org. Chem.*, **34**, 1070 (1969).
77. R. L. Augustine and W. G. Pierson, *J. Org. Chem.*, **34**, 2235 (1969).
78. V. A. Snieckus, T. Onouchi and V. Boekelheide, *J. Org. Chem.*, **37**, 2845 (1972).
79. J. Witte and V. Boekelheide, *J. Org. Chem.*, **37**, 2849 (1972).
80. M. Freifelder, D. A. Dunnigan and E. J. Baker, *J. Org. Chem.*, **31**, 3438 (1966).
81. D. T. Longone, *J. Org. Chem.*, **28**, 1770 (1963).
82. W. v. E. Doering and C. F. Hiskey, *J. Am. Chem. Soc.*, **74**, 5688 (1952).
83. G. N. Walker, *J. Am. Chem. Soc.*, **77**, 6699 (1955).
84. T. J. Nieuwstad, P. Klapwijk and H. van Bekkum, *J. Catal.*, **29**, 404 (1973).
85. G. Stork, *J. Am. Chem. Soc.*, **69**, 576 (1947).
86. H. A. Arbit, *J. Am. Chem. Soc.*, **68**, 1662 (1946).
87. D. M. Musser and H. Adkins, *J. Am. Chem. Soc.*, **60**, 664 (1938).
88. H. Adkins and H. Krsek, *J. Am. Chem. Soc.*, **70**, 412 (1948).
89. H. J. Dauben, B. C. McKusick and G. B. Mueller, *J. Am. Chem. Soc.*, **70**, 4179 (1948).
90. W. E. Bachmann and J. Contoulis, *J. Am. Chem. Soc.*, **73**, 2636 (1951).
91. L. F. Fieser and R. N. Jones, *J. Am. Chem. Soc.*, **60**, 1940 (1938).
92. W. G. Dauben, C. F. Hiskey and A. H. Markhart, Jr, *J. Am. Chem. Soc.*, **73**, 1393 (1951).
93. H. Adkins and E. E. Burgoyne, *J. Am. Chem. Soc.*, **71**, 3528 (1949).
94. S. Nishimura, S. Ohbuchi, K. Ikeno and Y. Okada, *Bull. Chem. Soc., Japan.*, **57**, 2557 (1984).
95. P. N. Rylander and D. R. Steele, *Engelhard Ind. Tech. Bull.*, **5**, 113 (1964).
96. J. R. Durland and H. Adkins, *J. Am. Chem. Soc.*, **60**, 1501 (1938).
97. P. P. Fu and R. G. Harvey, *Tetrahedron Lett.*, 415 (1977).
98. P. P. Fu, H. M. Lee and R. G. Harvey, *J. Org. Chem.*, **45**, 2797 (1980).
99. S. Friedman, S. Metlin, A. Svedi and I. Wender, *J. Org. Chem.*, **24**, 1287 (1959).
100. K. Sakanishi, M. Okira, I. Mochida, H. Okazaki and M. Soeda, *Bull. Chem. Soc., Japan.*, **62**, 3994 (1989).
101. D. A. Scola, J. S. Adams, C. I. Tewksbury and R. J. Wineman, *J. Org. Chem.*, **30**, 384 (1965).
102. M. Minabe, S. Urushibara, F. Mishina, T. Kimura and M. Tsubota, *Bull. Chem. Soc., Japan*, **66**, 670 (1993).
103. R. R. Burtner and J. M. Brown, *J. Am. Chem. Soc.*, **69**, 630 (1947).
104. A. P. Philips, *J. Am. Chem. Soc.*, **72**, 1850 (1950).
105. A. R. Katritzky and A. M. Monro, *J. Chem. Soc.*, 1263 (1958).
106. M. Freifelder, *Adv. Catal.*, **14**, 203 (1963).
107. M. Friefelder and G. R. Stone, *J. Org. Chem.*, **26**, 3805 (1961).
108. M. Freifelder, R. M. Robinson and G. R. Stone, *J. Org. Chem.*, **26**, 284 (1962).
109. D. F. Barringer, G. Berkelhammer, S. D. Caarater, L. Goldman and A. E. Lanzilotti, *J. Org. Chem.*, **38**, 1933 (1973).
110. G. N. Walker, *J. Org. Chem.*, **27**, 2966 (1962).

111.   M. Freifelder, *J. Org. Chem.*, **29**, 2895 (1964).
112.   A. Dornow and K. Bruncken, *Chem. Ber.*, **83**, 189 (1950).
113.   D. R. Howton and D. R. V. Golding, *J. Org. Chem.*, **15**, 1 (1950).
114.   F. Krohnke and K. Fasold, *Ber*, **67**, 656 (1934).
115.   V. Boekelheide, W. J. Lonn, P. O'Grady and M. Lamborg, *J. Am. Chem. Soc.*, **75**, 3243 (1953).
116.   A. P. Gray and H. Kraus, *J. Org. Chem.*, **26**, 3368 (1961).
117.   H. Nienburg, *Chem. Ber.*, **70**, 635 (1937).
118.   A. E. Chichibabin and M. P. Gertchuk, *Chem. Ber.*, **63**, 1153 (1930).
119.   M. Freifelder, R. W. Mattoon and Y. H. Ng, *J. Org. Chem.*, **29**, 3730 (1964).
120.   S. L. Shapiro, H. Soloway and L. Freedman, *J. Org. Chem.*, **26**, 818 (1961).
121.   H. K. Hall, *J. Am. Chem. Soc.*, **80**, 6412 (1958).
122.   K. Stach, M. Thiel and F. Bickelhaupt, *Monatsh.*, **93**, 1090 (1962).
123.   N. J. Leonard and E. Barthel, Jr., *J. Am. Chem. Soc.*, **71**, 3098 (1949).
124.   T. B. Grave, *J. Am. Chem. Soc.*, **46**, 1460 (1924).
125.   M. Freifelder, *J. Org. Chem.*, **27**, 4046 (1962).
126.   M. Freifelder, *J. Org. Chem.*, **28**, 602 (1963).
127.   M. Freifelder, *J. Org. Chem.*, **28**, 1135 (1963).
128.   R. Fornasier, F. Marcuzzi and D. Zorzi, *J. Mol. Catal.*, **43**, 21 (1987).
129.   U. Wlsner, *J. Chem. Soc., Chem. Commun.*, 1348 (1969).
130.   R. E. Lyle and S. E. Mallett, *Ann. N. Y. Acad. Sci.*, **145**, 83 (1967).
131.   B. Witkop, *J. Am. Chem. Soc.*, **71**, 2559 (1949).
132.   N. J. Leonard and G. W. Leubner, *J. Am. Chem. Soc.*, **71**, 3408 (1949).
133.   T. Momose, S. Uchida, N. Yamaashi and T. Imanishi, *Chem. Pharm. Bull*, **25**, 1436 (1977).
134.   J. E. Shaw and P. R. Stapp, *J. Heterocycl. Chem.*, **24**, 1477 (1987).
135.   H. Okazaki, M. Soeda, Y. Ikefuji and R. Tamura, *Appl. Catal.*, **43**, 71 (1988).
136.   H. Okazaki, M. Soeda, Y. Ikefuji and R. Tamura, *Bull. Chem. Soc., Japan*, **62**, 3622 (1989).
137.   J. Z. Ginos, *J. Org. Chem.*, **40**, 1191 (1975).
138.   F. W. Vierhapper and E. L. Eliel, *J. Am. Chem. Soc.*, **96**, 2256 (1974).
139.   F. W. Vierhapper and E. L. Eliel, *J. Org. Chem.*, **40**, 2739 (1975).
140.   M. Hoenel and F. W. Vierhapper, *J. Chem. Soc., Perkin Trans. I*, 2607 (1982).
141.   R. B. Woodward and W. v. E. Doering, *J. Am. Chem. Soc.*, **67**, 860 (1945).
142.   B. Witkop, *J. Am. Chem. Soc.*, **70**, 2617 (1948).
143.   J. A. Hirsch and G. Schwartzkopf, *J. Org. Chem.*, **39**, 2044 (1974).
144.   W. Hückel and F. Stepf, *Liebigs Ann. Chem.*, **453**, 163 (1927).
145.   C. G. Overberger, L. C. Palmer, B. S. Marks and N. R. Byrd, *J. Am. Chem. Soc.*, **77**, 4100 (1955).
146.   H.-P. Kaiser and J. M. Muchowski, *J. Org. Chem.*, **49**, 4203 (1984).
147.   E. Späth and F. Kuffner, *Chem. Ber.*, **68**, 484 (1935).
148.   E. Ochiai, K. Tsuda and S. Ikuma, *Chem. Ber.*, **69**, 2238 (1936).
149.   J. Overhoff and J. P. Wibaut, *Recl. Trav. Chim. Pays-Bas*, **50**, 957 (1931).
150.   F. K. Signaigo and H. Adkins, *J. Am. Chem. Soc.*, **58**, 709 (1936).
151.   A. Smith and J. H. P. Utley, *J. Chem. Soc., Chem. Commun.*, 427 (1965).

152. F. E. King, J. A. Barltrop and R. J. Walley, *J. Chem. Soc.,* 277 (1945).
153. V. Boekelheide and C.-T. Liu, *J. Am. Chem. Soc.,* **74**, 4920 (1952).
154. D. V. Young and H. R. Snyder, *J. Am. Chem. Soc.,* **83**, 3160 (1961).
155. K. Sakanishi, I. Mochida, H. Okazaki and M. Soeda, *Chem. Lett.*, 319 (1990).
156. D. S. Tarbell and C. Weaver, *J. Am. Chem. Soc.,* **63**, 2939 (1941).
157. H. Hinz, G. Meyer and G. Schucking, *Chem. Ber.*, **76**, 676 (1943).
158. R. Paul and G. Hilly, *Compt. rend.*, **208**, 359 (1939).
159. W. N. Haworth, W. G. M. Jones and L. F. Wiggins, *J. Chem. Soc.,* 1 (1945).
160. I. D. Webb and G. T. Borcherdt, *J. Am. Chem. Soc.,* **73**, 752 (1951).
161. J. A. Moore and J. B. Kelly, *Org. Prep. Proced. Int.,* **4**, 289 (1972).
162. T. E. Londergan, N. L. Hause and W. R. Schmitz, *J. Am. Chem. Soc.,* **75**, 4456 (1953).
163. N. Merat, C. Godawa and A. Gaset, *J. Mol. Catal.,* **57**, 397 (1990).
164. N. Merat, C. Godawa and A. Gaset, *J. Chem. Technol. Biotechnol.,* **48**, 145 (1990).
165. T. B. L. w. Marinelli, V. Ponec, C. G. Raab and J. A. Lercher, *Stud. Surf. Sci. Catal.*, **78** (Heterog. Catal. Fine Chem., III), 195 (1993).
166. R. Cornubert, R. Delmas, S. Montiel and J. Viriot, *Bull. soc. chim., France,* 36 (1950).
167. R. I. Longley, Jr., W. S. Emerson and T. C. Shafer, *J. Am. Chem. Soc.,* **74**, 2012 (1952).
168. R. Mozingo and H. Adkins, *J. Am. Chem. Soc.,* **60**, 669 (1938).
169. J. Attenburrow, J. Elks, D. F. Elliot, B. A. Hems, J. O. Harris and C. I. Brodrick, *J. Chem. Soc.,* 571 (1945).
170. R. H. Wiley and A. J. Hart, *J. Am. Chem. Soc.,* **77**, 2340 (1955).
171. P. L. deBenneville and R. Connor, *J. Am. Chem. Soc.,* **62**, 283 (1940).
172. W. Gruber and K. Horvath, *Monatsch.*, **81**, 828 (1950).
173. S. Isaac, *Compt. rend.*, **240**, 2534 (1955).
174. F. D. Mills, *J. Heterocycl. Chem.*, **17**, 1597 (1980).
175. G. Chatelus, *Compt. rend.*, **224**, 201 (1946).
176. J. A. Hirsch and G. Schwartzkopf, *J. Org. Chem.*, **38**, 3534 (1973).
177. C. C. Bolt and H. J. Backer, *Recl. Trav. Chim. Pays-Bas*, **55**, 898 (1936).
178. H. S. Broadbent and C. W. Whittle, *J. Am. Chem. Soc.,* **81**, 3587 (1959).
179. R. Mozingo, S. A. Harris, D. E. Wolf, C. E. Hoffhine, Jr., N. R. Easton and K. Folkers, *J. Am. Chem. Soc.,* **67**, 2092 (1945).

Hydrogenation V: Carbonyl Compounds

Aldehydes and Ketones

Aldehydes and ketones are usually hydrogenated to the alcohol over platinum, rhodium or ruthenium catalysts at 25°–60°C and 1–5 atmospheres pressure. Platinum catalyzed hydrogenations are generally best run in acidic media while with rhodium or ruthenium neutral or basic solvents are preferred.[1] Hydrogenations run over ruthenium catalysts are facilitated by the presence of water which makes ruthenium a particularly effective catalyst for the hydrogenation of sugars and other water soluble aldehydes and ketones.[2]

(18.1)

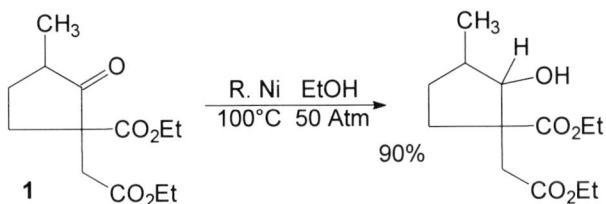

Nickel catalysts have also been used effectively for the hydrogenation of aldehydes and ketones. Commercial or W2 Raney nickel generally requires temperatures near 100°C and about 50 atmospheres of hydrogen.[3,4] Under these conditions, heptaldehyde was converted to heptyl alcohol in quantitative yield[4] and the keto-diester, **1**, to the alcohol in about 90% yield (Eqn. 18.1).[3] The more active W6 and W7 Raney nickel catalysts can promote aldehyde and ketone hydrogenations at room temperature and atmospheric pressure.[5] Care should be taken in the use of these basic catalysts as well as other basic reaction conditions[6] since they can promote the condensation of the starting carbonyl compound, particularly if the hydrogenation is not very fast. An increase in the hydrogen pressure can increase the rate of an otherwise sluggish hydrogenation and this will usually minimize any condensation problems. W2 Raney nickel is particularly useful for the hydrogenation of α,α-disubstituted-β-amino aldehydes. No hydrogenolysis of the amine group was observed when this catalyst was used to hydrogenate the amino aldehyde hydrochloride, **2**, in aqueous solution at 40°–

50°C and 25–40 atmospheres of hydrogen (Eqn. 18.2). The use of the free base or other catalysts resulted in the production of varying amounts of the amine cleavage product.[7] The presence of molybdenum[8,9] or chromium[9-11] increases the activity of a Raney nickel catalyst toward aldehyde and ketone hydrogenation.

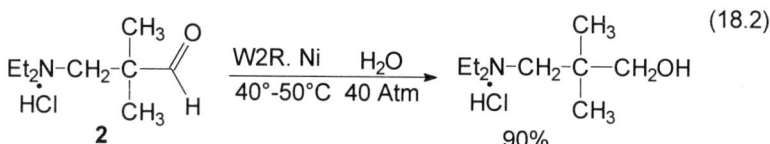

$$(18.2)$$

Nickel boride, Ni(B), catalysts are also effective for promoting the hydrogenation of aldehydes and ketones at room temperature and 2–3 atmospheres of hydrogen.[12,13] When these reactions were run in basic media at a pH between 10 and 12 catalyst activity increased significantly and the hydrogenation proceeded smoothly at atmospheric pressure.[6] As with Raney nickel, chromium containing Ni(B) catalysts are more active than the simple nickel boride for carbonyl group hydrogenations. In general, Ni(B) catalysts containing about 25% chromium were found to be the most effective.[14] A Ni/Y zeolite catalyst promoted the hydrogenation of methyl acetoacetate in n-butanol solution but the zeolite also catalyzed the transesterification of the product with the solvent alcohol giving varying amounts of n-butyl 3-hydroxybutryate.[15] Raney copper has been used to catalyze the hydrogenation of fructose in aqueous solvents at a pH between 6 and 8 under 50–60 atmospheres of hydrogen and 60°–70°C.[16]

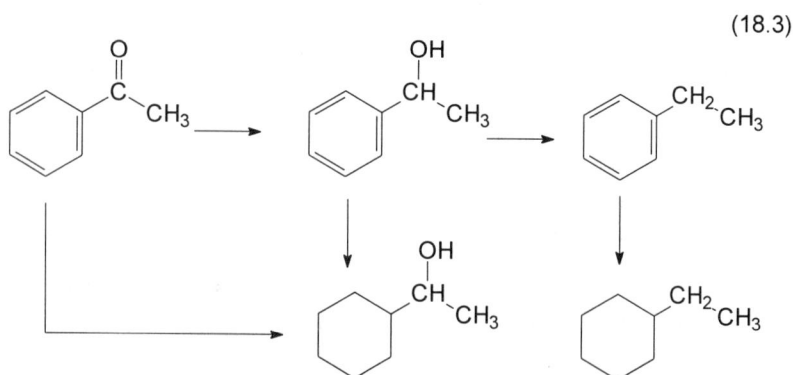

$$(18.3)$$

The hydrogenation of aryl aldehydes and ketones is complicated by the potential for the hydrogenolysis of the resulting benzyl alcohol as well as benzene ring hydrogenation (Eqn. 18.3). With the proper selection of reaction conditions

(18.4)

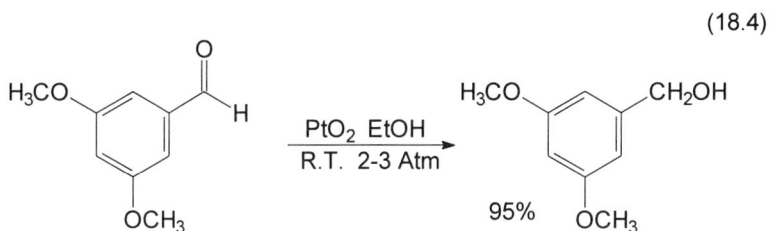

these competing reactions can be minimized, if not completely eliminated. The use of platinum in neutral media at room temperature and 2–3 atmospheres pressure gave the benzyl alcohol in 95% yield (Eqn. 18.4).[17] The addition of a small amount of ferrous sulfate[18] or ferric chloride[19] to the reaction mixture facilitates carbonyl group hydrogenation (Eqn. 18.5). The presence of acid increases the extent of alcohol hydrogenolysis. Hydrogenation of aqueous ethanolic solutions of substituted benzaldehydes over platinum supported on a weakly acidic titania gave the corresponding alcohols in near 95% yields at 150°C and 10 atmospheres of hydrogen. The use of more acidic supports decreased the yield of alcohol and increased the amount of substituted toluenes obtained.[20] The vapor-phase hydrogenation of acetophenone over a Pt/TiO_2 catalyst that was reduced under hydrogen at high temperatures (SMSI state) showed a high selectivity for the formation of the hydrogenolyzed product, ethyl benzene.[21]

(18.5)

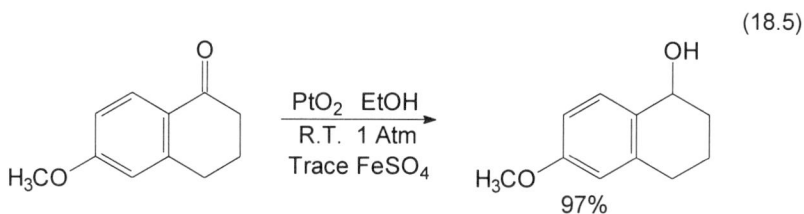

Hydrogenations using a ruthenium catalyst are generally run in aqueous alcohol at 25°–50°C and 2–3 atmospheres of hydrogen. Good yields of the alcohol are usually obtained under these conditions. Hydrogenations run over rhodium catalysts, however, can be complicated by the competitive hydrogenation of the aromatic ring.[1,22] Hydrogenation of acetophenone over unpromoted Raney nickel at 80°C and 9 atmospheres gave 1-phenylethanol in 97% yield using aqueous iso-propanol as the solvent.[23]

Hydrogenolysis of the benzyl alcohol occurs readily during the hydrogenation of o- and p- hydroxyphenyl aldehydes and ketones, even over platinum and nickel catalysts. This can be minimized by using an activated Cu-CrO catalyst at 80°–90°C and 165–200 atmospheres of hydrogen. The catalyst was activated by refluxing it in cyclohexanol under nitrogen for four

(18.6)

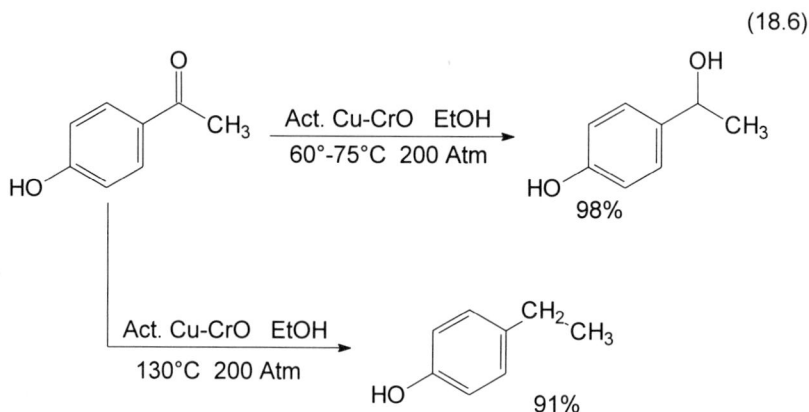

hours.[24]   Using this activated catalyst at 130°C[24] or an unactivated Cu-CrO catalyst at temperatures above 100°C[25] resulted in extensive hydrogenolysis (Eqn. 18.6).   Hydrogenation of piperonal (3) over Cu-CrO at 160°–175°C and 300 atmospheres pressure gave the alcohol in 98% yield but at 280°C and 375 atmospheres the substituted toluene was the primary product (Eqn. 18.7).[26]

(18.7)

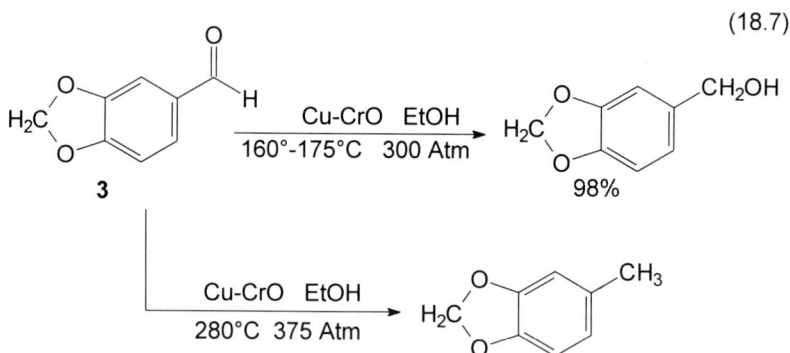

    While the hydrogenation of unactivated aliphatic aldehydes and ketones does not generally take place over palladium, this catalyst readily promotes the hydrogenolysis of aryl aldehydes and ketones to the methylene at room temperature and 1–4 atmospheres pressure.   The use of an acidic solvent facilitates this reaction[27-29] but is not essential for obtaining good to excellent yields of the desoxy product (Eqn. 18.8).[30] With a basic substrate such as a 2- or 4-acyl-pyridine, however, the alcohol product was obtained (Eqn. 18.9).[31]

(18.8)

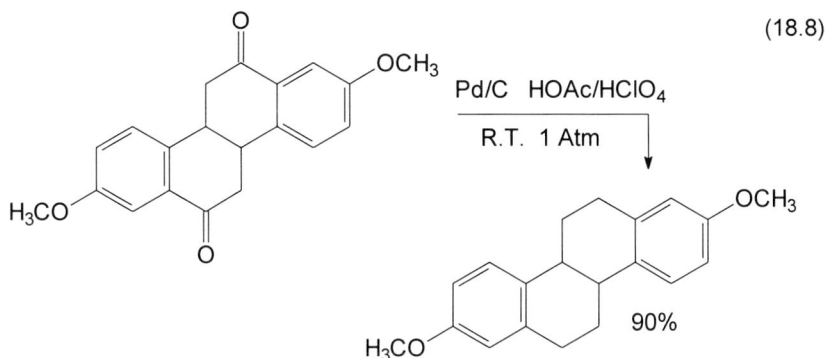

90%

(18.9)

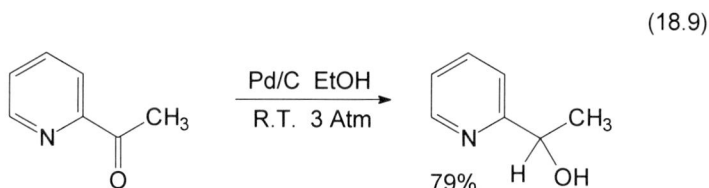

79%

Since aliphatic aldehydes and ketones are not hydrogenated over palladium, this reaction provides a means of selectively removing an aromatic carbonyl group in the presence of an aliphatic aldehyde or ketone (Eqn. 18.10).[32] The palladium catalyzed hydrogenolysis of aryl aldehydes and ketones is preferable to any of the chemical reduction procedures such as the Wolff-Kishner or Clemmenson reactions for the removal of an aryl carbonyl group.

(18.10)

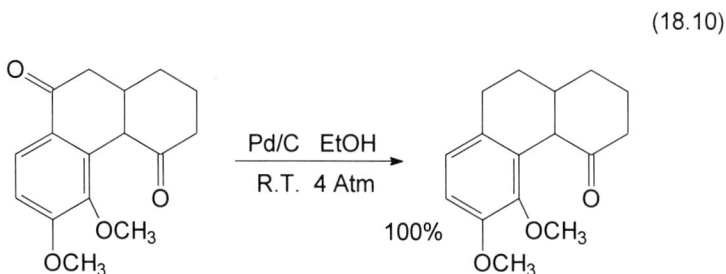

100%

Hydrogenolysis of some azacyclic ketones such as **4** to the methylene compound, **5**, has been accomplished using an equal weight of PtO_2 in 6N HCl at room temperature and 3 atmospheres of hydrogen (Eqn. 18.11).[33] Similar results

(18.11)

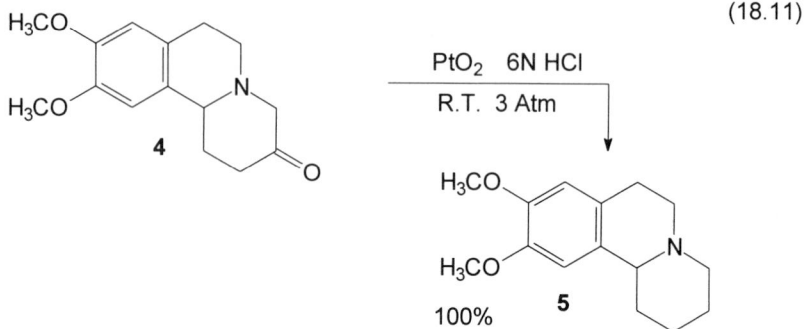

were observed on hydrogenation of γ-amino ketones in strongly acidic media.[34] Hydrogenation to the alcohol predominated when less catalyst or weaker acid was used.

As described in Chapter 12, hydrogenation of cyclohexanone or related cyclic ketones over pre-reduced palladium hydroxide or palladium oxide in alcoholic solvents gave the ether as the primary product.[35] The mechanism put forth for this reaction involved the intermediate formation of a ketal which was then hydrogenolyzed to the ether (Eqn. 18.12)[36]. Evaporated platinum and palladium blacks, which have no basic impurities, promoted facile acetal formation. Further hydrogenation over palladium gave the ether as the almost exclusive product at a rate four times faster than that observed when reduced palladium hydroxide was used as the catalyst. Over the evaporated platinum catalyst only moderate amounts of the ether were formed. The primary product was the alcohol accompanied by some alkane.[36] Ether formation was also observed on hydrogenation of cyclic ketones over PtO_2 in ethanolic-HCl at room temperature and atmospheric pressure.[37] Acetal formation occurred on

(18.12)

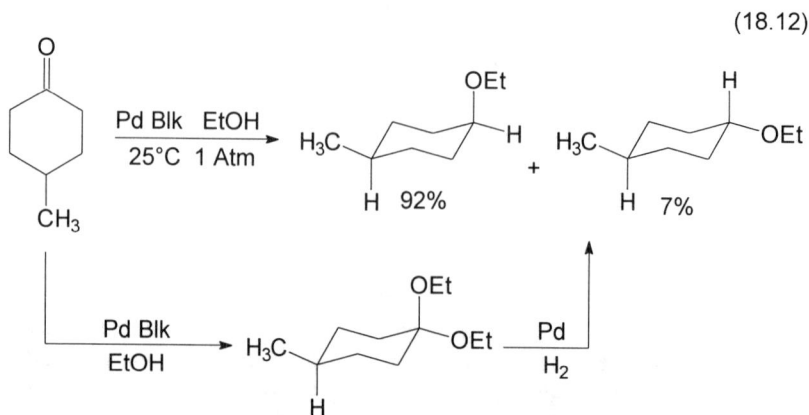

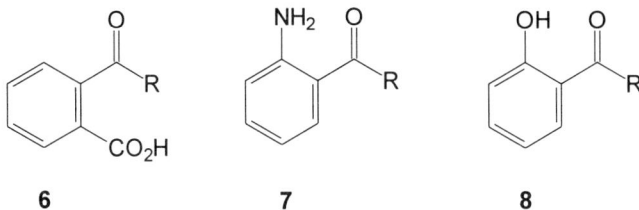

hydrogenation of unsaturated cyclic ketones over palladium in alcoholic solvents.[38]

There are some types of aryl ketones that are hydrogenated to the alcohol over palladium catalysts without further hydrogenolysis. The o-acylbenzoic acids, **6**, are hydrogenated to the lactone. Since a lactone is hydrogenolyzed less easily than an alcohol, it can be isolated at this stage. If the reaction were run in neutral or slightly alkaline media, lactone formation does not take place and the alkylbenzoic acid is formed.[39,40] The hydrogenation of o-aminoaryl, **7**, and o-hydroxyaryl ketones, **8**, gave the amino and hydroxy benzyl alcohols which were resistant to hydrogenolysis over palladium in neutral media.[41,42] These alcohols, however, could be hydrogenolyzed if the hydrogenation were run in acid medium or if Raney nickel were used at 40°–50°C and 100–135 atmospheres of hydrogen.[43]

The ease with which the carbonyl group is hydrogenated depends on its environment. Steric factors are more important in the hydrogenation of aldehydes than with ketones. Bulky substituents hinder the adsorption of the carbonyl group on the catalyst surface, thus decreasing the rate of hydrogenation. In general aldehydes are hydrogenated more easily than are ketones. Aryl aldehydes and ketones are more readily hydrogenated than the corresponding aliphatic carbonyl compounds. Other compounds in which the aldehyde or ketone is α to another carbonyl function or a trifluoromethyl group have enhanced adsorption. They are, thus, hydrogenated more readily than an unsubstituted aldehyde or ketone. Cyclic ketones, particularly cyclohexanones, are more easily hydrogenated than are acyclic ketones.

The competitive hydrogenations of cyclohexanone with 2-, 3-, and 4-substituted cyclohexanones run in cyclohexane at 30°C and atmospheric pressure showed that the observed relative reactivities depended on both the position of the substituent and the catalyst used.[44] The 2-substituted cyclohexanones were less easily hydrogenated in the presence of cyclohexanone than were the 3- and 4-substituted cyclohexanones. As the size of the 2-substituent increased, selectivity toward cyclohexanone hydrogenation increased. The best selectivities were obtained using Ru/Al$_2$O$_3$ and the worst with Pt/Al$_2$O$_3$.[45] This decreased selectivity with platinum catalysts is probably related to the enhanced activity of such catalysts for carbonyl group hydrogenation. Steroidal ketones are hydrogenated in the following order: C-3 >

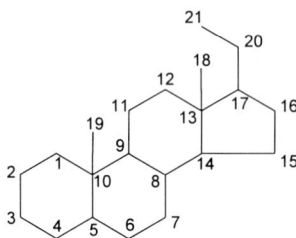

Fig. 18.1. Steroid numbering system.

C-16, C-17, C-20 > C-7, C-6 > C-12 > C-11 (Fig. 18.1).[46-49] Hindered ketones usually require higher temperatures and pressures.

Double and triple bonds as well as aromatic nitro groups are hydrogenated in neutral media over platinum or palladium in preference to the saturation of an aryl or alkyl carbonyl group. Acid facilitates carbonyl group hydrogenation to the extent that this reaction can take place in competition with the saturation of these generally more easily saturated functional groups. Acid, however, also enhances hydrogenolysis, especially of aryl carbonyl groups, and complete deoxygenation can result from the hydrogenation of such groups in acidic media, particularly if a palladium catalyst is used.

Even though aromatic fluorides and chlorides are not hydrogenolyzed to any great extent under the conditions used to completely reduce an aryl ketone over palladium, the addition of acid to the reaction mixture further inhibits halogen hydrogenolysis while promoting carbonyl hydrogenation (Eqn. 18.13).[50] Aryl bromides and iodides, however, are at least partially, if not completely, hydrogenolyzed under these conditions even in the presence of acid.

(18.13)

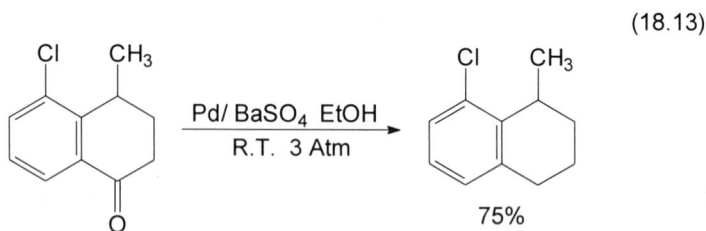

As discussed in Chapter 13 the competitive hydrogenation of ketones and alkenes over ruthenium is influenced by the solvent. There is virtually no selectivity observed on hydrogenation of a mixture of cyclohexene and cyclohexanone over Ru/SiO_2 in water. However, when the solvent was cyclohexane, containing a small amount of water, selective hydrogenation of the ketone was observed. In this latter instance the water was attracted to the

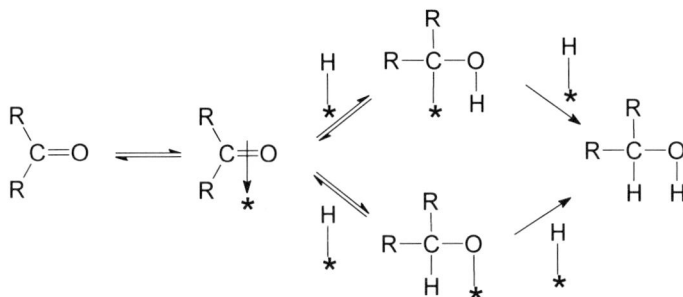

Fig. 18.2. Modified Horiuti-Polanyi mechanism for the hydrogenation
of carbonyl groups.[52]

hydrophilic surface of the catalyst support, providing a higher concentration of polar cyclohexanone near the catalyst. At the same time, the cyclohexene was concentrated in the cyclohexane of the multiphase system and, thus, kept from the catalyst by the water barrier. When the catalyst surface was silanized to make it less hydrophilic or when the hydrophobic Ru/C was used, the reverse selectivity was noted.[51] This is similar to the multiphase approach used to enhance the selectivity in the hydrogenation of phenol to cyclohexanone discussed in Chapter 17. This type of system may also be applicable to other selective hydrogenations.

Stereochemistry

The hydrogenation of carbonyl groups in neutral media is thought to take place by way of the modified Horiuti-Polanyi mechanism shown in Fig. 18.2.[52] As described in Chapter 14 for the hydrogenation of alkenes, the stereochemistry of the alcohol obtained on hydrogenation of ketones in neutral solvents depends on the relative ease with which the two faces of the carbonyl moiety can be adsorbed on the catalyst surface. Hydrogen availability to the catalyst has an effect on the stereochemistry of carbonyl group hydrogenations in the same way it has on alkene hydrogenations but the effect is somewhat less for ketones.

The stereochemistry of the alcohols resulting from the hydrogenation of cyclic ketones is dependent on the steric features of the ketone, the nature of the catalyst and the reaction medium. In neutral media, the addition of hydrogen usually takes place from the least hindered side of the molecule. Hydrogenation of ketones such as the methylcyclopentanones,[53] the methylcyclohexanones,[54] and the substituted cycloheptanones,[55] in which there is little difference in hindrance to adsorption from either side of the molecule, results in the formation of nearly equal amounts of the stereoisomeric alcohols. Product stereochemistry is also influenced by the nature of the catalyst used. Hydrogenation of 2-methylcyclohexanone over platinum oxide at room temperature and atmospheric pressure gave nearly equal amounts of the compounds containing the

(18.14)

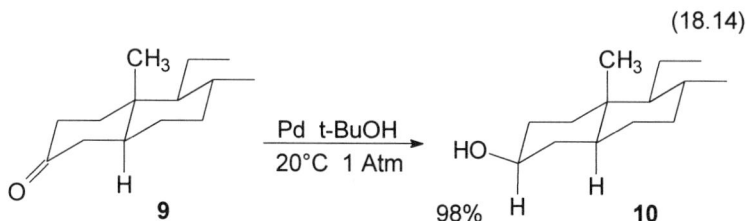

axial and equatorial hydroxy group.[56]   With Raney nickel, 69% of the axial alcohol was obtained. Rhodium gave more of the axial alcohol products than any of the other catalysts.[57] Hydrogenation of the reactive 5α-3-ketosteroids, **9**, over palladium at room temperature and atmospheric pressure gave the equatorial, 3β, alcohol, **10**, as the predominant product (Eqn. 18.14) while the hydrogenation of 5β-3-ketosteroids, **11**, gave the axial, 3β, alcohol, **12**, (Eqn. 18.15).[58,59]

(18.15)

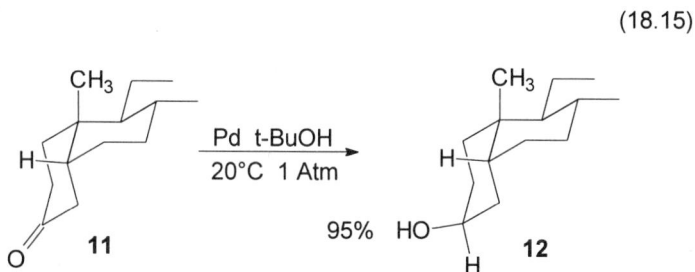

As described in Chapter 14, the presence of other functionality in the molecule can influence the direction of adsorption of the ketone on the catalyst surface. Hydrogenation of the aminoketone, **13**, over Raney nickel at 60°C and 120 atmospheres of hydrogen gave high yields of the alcohol resulting from adsorption on the catalyst from the side of the amino group (Eqn. 18.16).[60] A similar effect was found on hydrogenation of the amino ketone, **14**, over a platinum catalyst (Eqn. 18.17),[61] but when the amine group of **14** was quaternized, preventing assisted adsorption through the electron pair on the nitrogen, hydrogenation took place from the other side of the molecule.[61]

Since the carbonyl group is a polar species, the stereochemistry of the alcohols obtained on hydrogenation is dependent on the nature of the reaction medium used in much the same way as discussed for the hydrogenation of α,β-unsaturated ketones in Chapter 15. The hydrogenation of cyclohexanones in acidic media gives almost exclusively the axial alcohol product (Eqn. 18.18).[62] Generally, these reactions are run using a platinum catalyst at 25°–50°C and 2–3 atmospheres pressure. Acetic acid containing a small amount of concentrated

(18.16)

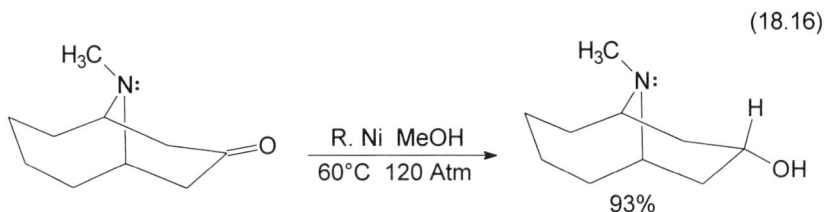

93%

hydrochloric or perchloric acid is the usual solvent. The use of acetic acid alone as the solvent affords the equatorial alcohol as the main product from both 5α-3-[63] and 5α-7-keto[64] steroids. With the strong acid present, the axial alcohol, **16**, is the primary, if not the exclusive, product (Eqn. 18.19).[65] The use of the strong acid in alcoholic solvents can lead to ether formation.[66] The hydrogenation of cholestanone (**15**) in methanolic HBr gave β-cholesteryl methyl ether (**17**) in good yield.[67] The configuration of the methoxy group is opposite from that of the alcohol formed under acidic hydrogenation conditions. The axial alcohols are formed by an initial protonation of the carbonyl group to give the cation, **19** (Scheme 18.1), which is then adsorbed on the catalyst from the least hindered, equatorial side as in **20**. Hydride ion transfer from the catalyst gives the axial alcohol, **21**.[68]

(18.17)

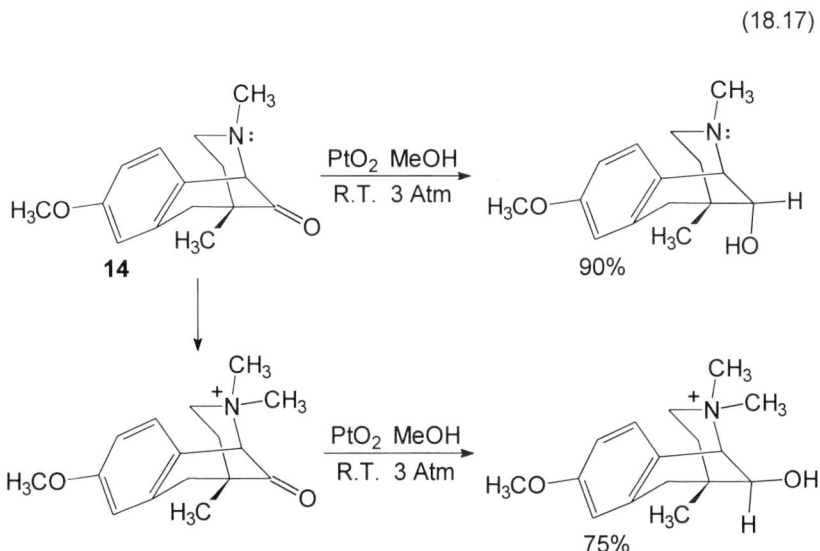

(18.18)

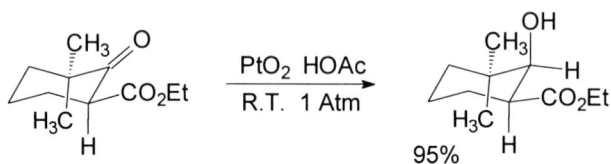

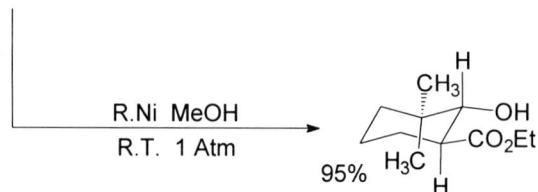

(18.19)

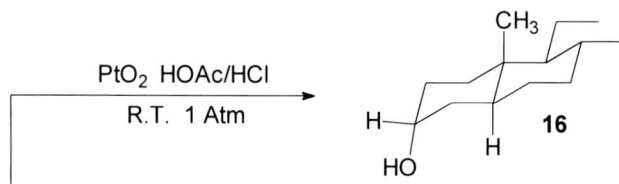

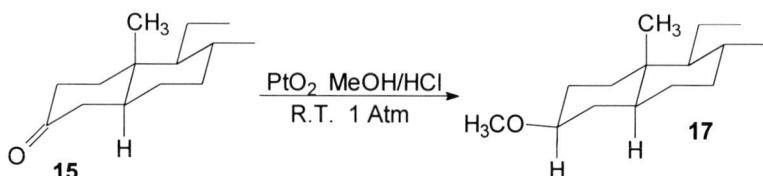

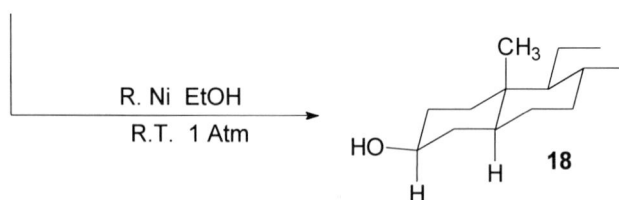

Hydrogenations of cyclohexanones in basic media leads to the formation of the more stable, equatorial alcohol, **18**, as the primary product (Eqns. 18.18–19). With basic Raney nickel catalysts, the predominance of equatorial alcohol has been shown to arise from the isomerization of the initially produced axial alcohol under hydrogenation conditions.[69] The use of platinum oxide with its alkaline impurity also leads to extensive equatorial alcohol formation unless acetic acid is used as the solvent. This basic impurity was shown to be

Scheme 18.1

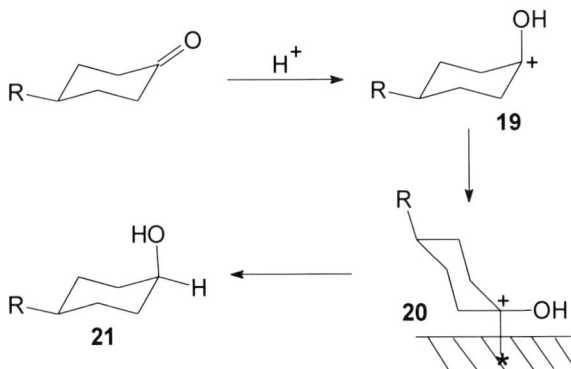

responsible for the formation of equatorial alcohols in the hydrogenation of substituted cyclohexanones mentioned above.[57] When the platinum oxide was pre-reduced and washed thoroughly with water and ethanol, the resulting platinum black catalyst gave almost 80% of the axial alcohol on hydrogenation of 2-methylcyclohexanone.[57] When sodium hydroxide was added to the reaction medium, the platinum black catalyzed hydrogenation of substituted cyclohexanones gave predominantly the equatorial alcohol.[70] Hydrogenations in base to give the equatorial alcohol presumably take place through the initial

Scheme 18.2

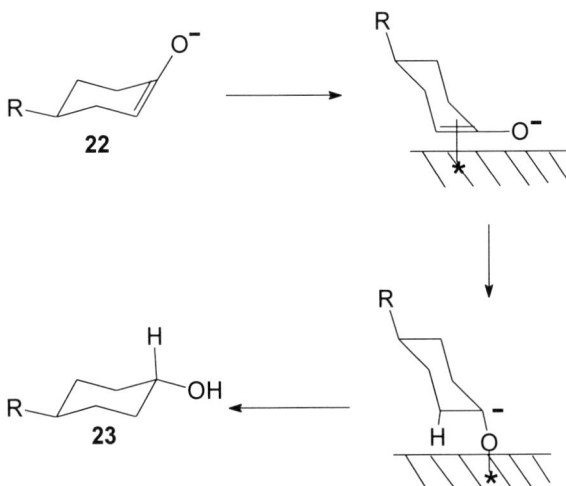

formation of the enolate ion, **22** (Scheme 18.2), which is adsorbed on the catalyst from the least hindered side. Hydride ion transfer from the catalyst followed by protonation from the solution gives the equatorial alcohol, **23**.[68]

(18.20)

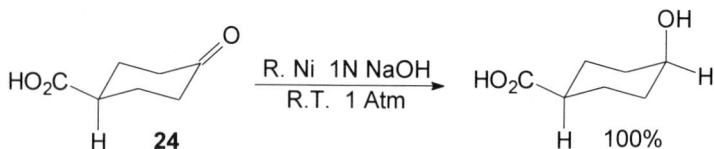

One exception to this generalization is the formation of the *cis* hydroxyacid by the hydrogenation of 4-carboxycyclohexanone (**24**) over Raney nickel in base (Eqn. 18.20).[57] The production of this stereoisomer is probably the result of transannular participation by the carboxy group.

Dicarbonyl Compounds

The selective hydrogenation of one carbonyl group in preference to another group in the same molecule is an example of *Type II* selectivity (Chapter 5). However, the experimental techniques used to evaluate catalytic conditions to effect such selectivity generally involve mixtures of monocarbonyl compounds, an example of *Type* I selectivity. Although it is probably safe to extrapolate the general findings from these mixed-compound studies to the monomolecular example, a direct comparison may not be possible because the intramolecular adsorption mode of the dicarbonyl compounds probably differs from the competitive adsorption exhibited by two separate monocarbonyl compounds. Whether the intramolecular adsorption would lead to more or less selectivity than that exhibited by a mixture is not easily predictable because little information is available concerning the nature of the adsorption of polyfunctional groups other than dienes and related species.

Since aldehydes are hydrogenated more readily than ketones, α-hydroxy-methylene ketones (β ketoaldehydes) are converted to the keto alcohols on hydrogenation over platinum at room temperature and atmospheric pressure. Further hydrogenation under these conditions leads to the formation of the 1,3 glycol. Some hydrogenolysis to the monohydric alcohol can also occur but this is usually a minor side reaction. If the hydroxymethylene group is esterified, hydrogenation under these conditions results in the hydrogenolysis of the acyl group to give the methyl ketone (Eqn. 18.21).[71]

The selective hydrogenation of dicarbonyl compounds involves the saturation of one of the keto groups to produce a ketol. The α-di- and polyketonic[72-74] compounds as well as acyloins[75,76] and polyhydroxyaldehydes, such as glyceraldehyde and the sugars,[2,77,78] can be hydrogenated either selectively or completely to the polyhydroxy materials under mild reaction conditions. The

(18.21)

H3C—N (ring) ... =CHOH $\xrightarrow[\text{R.T.   1 Atm}]{\text{PtO}_2 \quad \text{EtOH}}$ H3C—N (ring) ... CH2OH, O

↓

H3C—N (ring) ... HCO—O (R) $\xrightarrow[\text{R.T.   1 Atm}]{\text{PtO}_2 \quad \text{EtOH}}$ H3C—N (ring) ... CH3, O

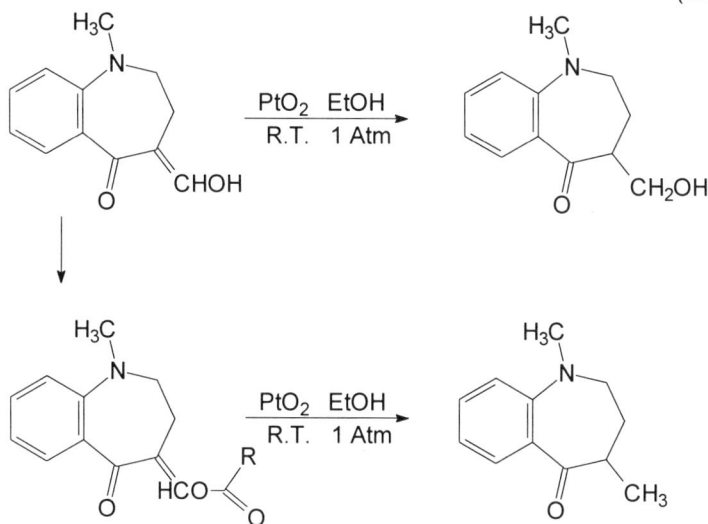

hydrogenation of 2,3-butanedione to 3-hydroxy-2-butanone occurred at room temperature and atmospheric pressure over platinum, palladium, rhodium, ruthenium, nickel and cobalt catalysts. Further hydrogenation to 2,3-butanediol took place over platinum, ruthenium, nickel and cobalt but not over palladium or rhodium.[79]

The conversion of β-diketones to the glycols is also accomplished using these mild conditions but hydrogenation over palladium in acetic acid containing some sulfuric acid at 50°–60°C and 3–4 atmospheres pressure resulted in the hydrogenolysis of one of the carbonyl groups giving a monocarbonyl product. 2-Acetylcyclohexanone (25) gave 2-ethylcyclohexanone (26) under these conditions but hydrogenation of 2-acetylcyclopentanone (27) produced an equal mixture of 2-ethylcyclopentanone (28) and acetylcyclopentane (29) (Eqn. 18.22).[80]   This hydrogenolysis is thought to occur through the initial hydrogenation of one of the carbonyl groups to give the ketol, followed by acid-catalyzed dehydration and hydrogenation of the resulting enone. No satisfactory explanation, however, was provided for the selectivity observed in these reactions.[80]

The selective partial hydrogenation of symmetric cyclic diketones was accomplished by interrupting the reaction after the consumption of one equivalent of hydrogen, which indicates a stronger mode of adsorption for the diketone than for the ketol product.[81,82] The hydrogenation of 1,4-cyclohexanedione (30) to the ketol (Eqn. 18.23) was catalyzed by nickel, copper, palladium, platinum, iridium, and ruthenium. Iridium was the most active and selective of these catalysts. Hydrogenations run over this catalyst in iso-propanol at 20°C and 6 atmospheres

(18.22)

25 → 26

27 → 28 + 29

30 → 78% (18.23)

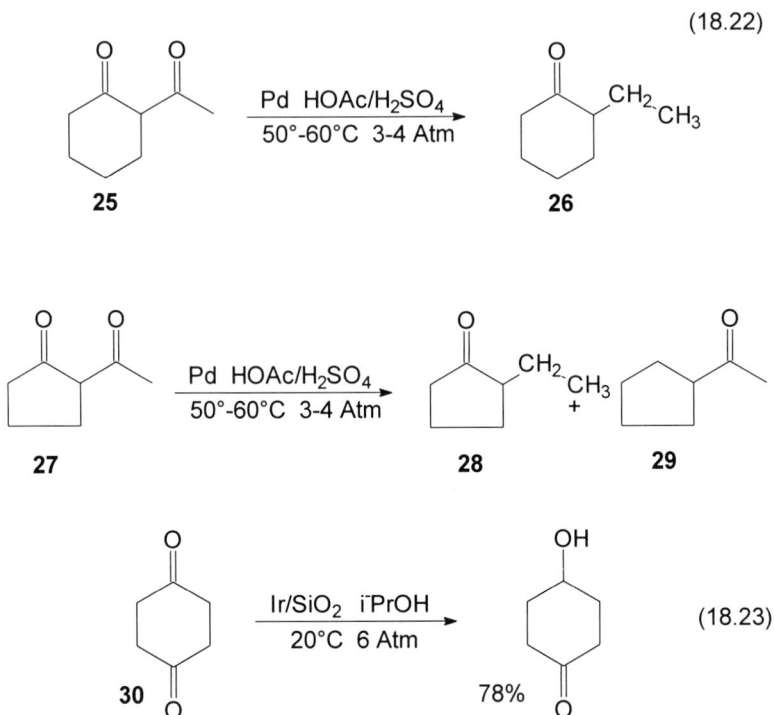

of hydrogen occurred at a rate ten times faster than that obtained using platinum under identical conditions. The reaction with iridium showed a selectivity for ketol formation of 78% on adsorption of one equivalent of hydrogen (100% conversion). With platinum, a maximum ketol formation of 48% occurred at 87% conversion. Ruthenium gave, at a somewhat higher pressure, a maximum yield of 73% at 94% conversion. With Ru/SiO_2 it was found that selectivity decreased significantly with increasing temperature, but it was only slightly affected by changes in the amount of substrate, amount of catalyst, or hydrogen pressure. The use of iso-propanol as the solvent gave higher selectivity than was found with either methanol or water.[82] The less hindered carbonyl group in methylcyclohexan-1,4-dione was selectively hydrogenated over a platinum catalyst.[83]

Unsaturated Carbonyl Compounds

A more common hydrogenation selectivity involving carbonyl groups is the hydrogenation of an aldehyde or a ketone in the presence of a carbon-carbon double bond. In general, the olefinic functionality, whether conjugated or not, is more easily hydrogenated in preference to the reduction of the carbonyl group, particularly if the double bond is not overly hindered. The selective hydrogenation of the double bond in unsaturated aldehydes and ketones is most readily accomplished using palladium under mild conditions. With Raney nickel at 100°C and 15 atmospheres pressure, hydrogenation proceeds to the saturated alcohol.[84]

α, β-Unsaturated Aldehydes

The selective hydrogenation of an α,β-unsaturated carbonyl compound can be considered an example of *Type II* selectivity (Chapter 5), but the reaction is more complex as shown in Scheme 18.3. Not only is there potential selectivity between double bond and carbonyl group hydrogenation, but both partially hydrogenated products can be reduced further to give the saturated alcohol, **34**. In addition, the hydroxy group in the unsaturated alcohol, **31**, is allylic and, thus, capable of being hydrogenolyzed to the alkene, which can be further saturated to the alkane. The double bond in **31** can also be isomerized to give the enol, **32** of the saturated aldehyde, **33**, providing an alternative route to **33** involving the initial selective hydrogenation of the carbonyl group. As if this were not enough, two modes of hydrogen addition are possible. Both functional groups can be saturated by a 1,2-addition to give either the allyl alcohol, **31**, or the saturated aldehyde, **33**. 1,4-Addition of hydrogen to this conjugated system is also possible and this gives the enol, **32**, and provides still another route to the saturated aldehyde, **33**.

With these different ways available for the production of the saturated aldehyde, it is not surprising that the selective hydrogenation of the carbonyl group to give the allyl alcohol is not a straightforward matter. Not only is it necessary to find the proper catalyst and reaction conditions for allyl alcohol formation, but the elimination of side reactions is required because the yield of the allyl alcohol is diminished by subsequent hydrogenation, hydrogenolysis or isomerization.   Further, many factors are involved in attaining the desired selectivity. Although the catalyst is an important factor, the structure of the substrate is also critical. Steric hindrance around the double bond facilitates the selective hydrogenation of the carbonyl group, whereas a lack of substituents on

Scheme 18.3

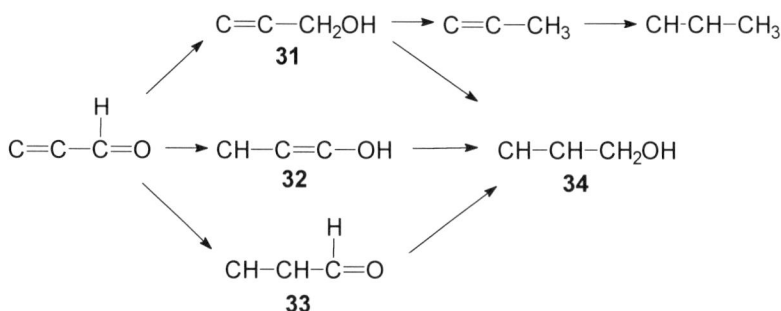

the double bond makes this selective hydrogenation more difficult. Acrolein (**35**) is the most difficult of the α,β-unsaturated aldehydes to hydrogenate selectively. Placing another substituent on the carbonyl carbon (conversion of the aldehyde to a ketone) increases the steric hindrance to carbonyl group adsorption and essentially precludes any selective hydrogenation of the keto group. A number of catalysts are available for the hydrogenation of α,β-unsaturated aldehydes to allyl alcohols, but none have been reported for the competitive hydrogenation of α,β-unsaturated ketones.

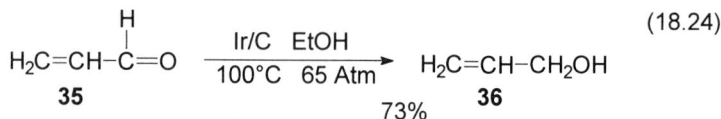

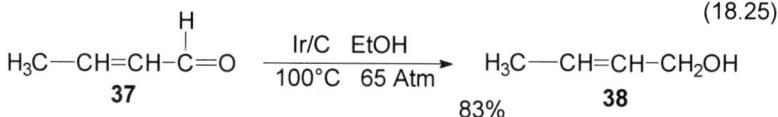

$$
\underset{\textbf{35}}{H_2C=CH-\overset{\overset{\textstyle H}{|}}{C}=O} \quad \xrightarrow[\text{100°C \quad 65 Atm}]{\text{Ir/C \quad EtOH}} \quad \underset{\underset{\textbf{36}}{73\%}}{H_2C=CH-CH_2OH} \tag{18.24}
$$

$$
\underset{\textbf{37}}{H_3C-CH=CH-\overset{\overset{\textstyle H}{|}}{C}=O} \quad \xrightarrow[\text{100°C \quad 65 Atm}]{\text{Ir/C \quad EtOH}} \quad \underset{\underset{\textbf{38}}{83\%}}{H_3C-CH=CH-CH_2OH} \tag{18.25}
$$

The common hydrogenation catalysts promote the saturation of the double bond to give the saturated aldehyde or complete saturation to give the saturated alcohol. The more uncommon catalysts, osmium and iridium, however, do show some selectivity toward the hydrogenation of the carbonyl group of acrolein (**35**) at 100°C and 65 atmospheres to give a 73% yield of allyl alcohol (**36**) (Eqn.18.24). As an illustration of the substituent effect mentioned above, the use of these same conditions for the hydrogenation of crotonaldehyde (**37**) (Eqn. 18.25) and cinnamaldehyde (**39**) gave, respectively, 83% of crotyl alcohol (**36**) and 96% of cinnamoyl alcohol (**40**).[85] The use of Ir/C for the hydrogenation of acrolein and cinnamaldehyde in ethanol at room temperature and atmospheric pressure gave a 60% yield of allyl alcohol[86] and a 100% yield of cinnamoyl alcohol (Eqn. 18.26).[87]

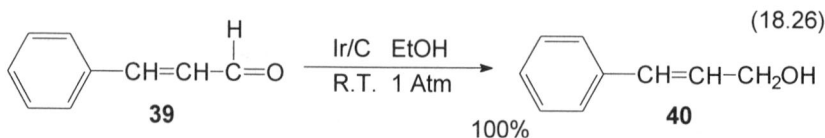

$$
\underset{\textbf{39}}{\text{Ph}-CH=CH-\overset{\overset{\textstyle H}{|}}{C}=O} \quad \xrightarrow[\text{R.T. \quad 1 Atm}]{\text{Ir/C \quad EtOH}} \quad \underset{\underset{\textbf{40}}{100\%}}{\text{Ph}-CH=CH-CH_2OH} \tag{18.26}
$$

In general, Pt/C catalysts show little selectivity in the hydrogenation of the carbonyl group of α,β-unsaturated aldehydes but a Pt/graphite catalyst promoted the hydrogenation of cinnamaldehyde at 60°C and 40 atmospheres in an aqueous iso-propyl alcohol solution containing some sodium acetate. Cinnamyl alcohol was obtained in 83% selectivity at 50% conversion.[88,89] Comparable selectivities were also observed using Ru/ZrO$_2$[90,91] Rh/graphite[90] and Ru/TiO$_2$ which had been reduced at high temperatures (SMSI state).[90] Hydrogenation of cinnamaldehyde over a Pt/Y catalyst in iso-propanol at 60°C and 40 atmospheres pressure gave cinnamyl alcohol in 97% selectivity at 25% and 50% conversions.[89] This selectivity has been attributed to the size of the zeolite pores which permitted the adsorption of only the carbonyl group at the end of the unsaturated aldehyde into these pores and placing it on the active metal surface. In this system the C–C double bond cannot easily reach the metal particles inside the zeolite cavities so it is not saturated.[89] The selectivity observed in the vapor phase hydrogenation of 3-methylcrotonaldehyde over Ru/Y catalysts has been explained in the same way.[92]

Selectivity in the hydrogenation of cinnamaldehyde to cinnamyl alcohol over platinum, rhodium[89,93] or ruthenium[94,95] catalysts increases as the metal particle size increases but the selective hydrogenation of citral (41) to geraniol (43) over ruthenium was not influenced by the size of the catalytically active species.[95] The selectivity in the hydrogenation of cinnamaldehyde was attributed to the need for a sufficient number of face atoms for the benzene ring to adsorb adjacent to the active site on which the carbonyl group is hydrogenated. This is not required in the hydrogenation of citral.

While some vapor phase hydrogenations of unsaturated aldehydes have shown good initial selectivities,[96-99] in most cases the reaction selectivity decreases significantly with prolonged reaction time. The vapor phase hydrogenation of crotonaldehyde over a Pt/TiO$_2$ catalyst in the SMSI state, however, showed reasonable selectivities that remained relatively constant over the entire time of the reaction.[100]

In most cases a modifier is used with the catalyst to enhance hydrogenation selectivity. One of the earliest successful catalysts for this reaction was the Pt-Zn-Fe moiety.[101,102] It was proposed that with this catalyst system, the iron complexed with the carbonyl group and enhanced the hydrogenation of the carbonyl group, whereas the zinc deactivated the double bond toward saturation. More recent work showed that the presence of zinc was not essential. A supported platinum catalyst treated with FeCl$_3$ gave an 80% selectivity of cinnamyl alcohol at 35% conversion of cinnamaldehyde.[103,104] The addition of Fe$_2$O$_3$ to a platinum black catalyst in aqueous medium markedly changed the direction of hydrogenation of crotonaldehyde. With platinum black alone there was a 91% selectivity toward double bond saturation. However, in the presence of Fe$_2$O$_3$ only small amounts of butyraldehyde were formed and the reaction gave an 80% selectivity for carbonyl group saturation. The presence of Fe$_2$O$_3$ also increased selectivity for carbonyl group hydrogenation over osmium and iridium black catalysts.[105] Ferric chloride doped Ru/C was used for the hydrogenation of (Z) and (E) citral (41 and 42) (Eqn. 18.27) at 100°C and 50 atmospheres of hydrogen in a methanol solution containing a small amount of trimethylamine. The reaction gave a 97% yield of geraniol (43) and nerol (44) along with a small amount of citronellol (45).[62]

(18.27)

CH₃ H ... C=O ... H ... H₃C CH₃ **41** + CH₃ H C=O H H₃C CH₃ **42**   Ru/C/FeCl₃ MeOH/Et₃N   100°C 50 Atm

CH₃ CH₂OH H H₃C CH₃ **43** + CH₃ H CH₂OH H₃C CH₃ **44** + CH₃ CH₂OH H₃C CH₃ **45**

97%

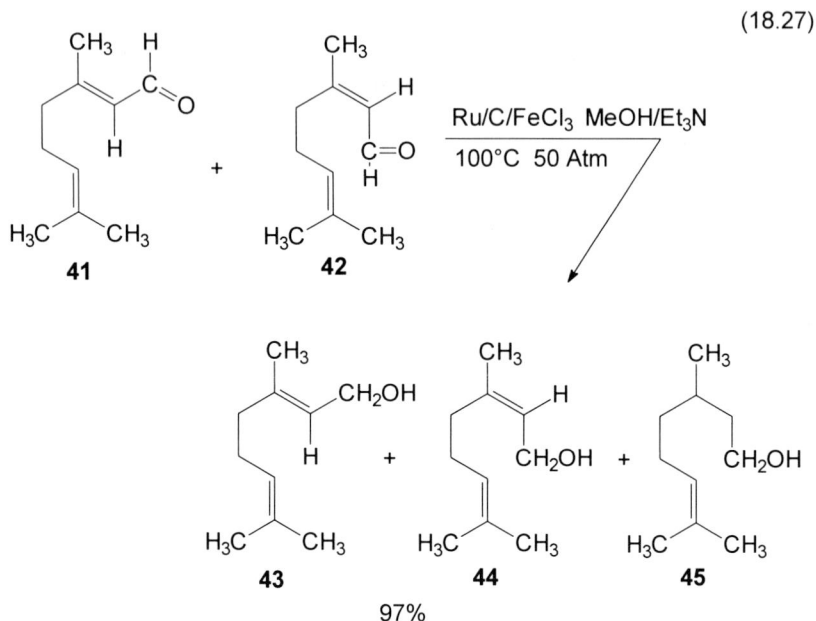

   If the only purpose of the iron as a modifier was to complex with the carbonyl group to make it easier to hydrogenate, one might expect that its presence would enhance carbonyl group hydrogenation selectivity at least to a reasonable extent with all metals, but this is not the case. Iron oxide modified ruthenium and rhodium show only a slight selectivity in this reaction.[105] Another likely role of the modifier could be to induce an electronic change in the surface atoms of the catalyst, thereby modifying their catalytic properties. For this to take place, an intimate contact between the metal and modifier ions must take place. The enhanced selectivity reported for FeCl₃ doped Ru/C[62] compared to the lack of selectivity observed using a Ru black/ Fe₂O₃ mixture[105] supports this premise.

   If this electronic effect were important, one should see comparable results using other, electronically similar, metal ion modifiers. The addition of tin to a Pt/C[106] or a Pt/nylon[107] catalyst significantly increased reaction selectivity. Maximum selectivity was attained at 20–25% tin levels with higher amounts of tin having little effect. The rate of reaction went through a maximum at 20–25% tin and then decreased significantly with the addition of more tin. Similar effects were observed with Ru-Sn catalysts[108] as well as when germanium was added to the Pt/nylon catalysts.[109] Rh-Sn catalysts[110-113] were shown to be effective for the selective hydrogenation of citral (**41**) with geraniol (**43**) obtained in 96% selectivity at 100% conversion. Ru-Sn catalysts, however, gave only an 80–90% selectivity in citral hydrogenations.[95,114]

It appears that an electronic effect may be modifying the nature of the active sites on the catalyst. The increased selectivity observed on adding alkali ions to ruthenium catalysts supports this conclusion.[115-118]

Besides the platinum metals, other catalytically active species show some success for selective carbonyl group hydrogenation of α,β-unsaturated aldehydes. Cobalt boride catalysts gave 40% butenol (**38**) at 80% conversion in the hydrogenation of crotonaldehyde (**37**)[119] and allyl alcohol (**36**) was formed in 30% selectivity at 60% conversion on hydrogenation of acrolein (**35**) over Co/Al$_2$O$_3$ at 30° and 10 atmospheres of hydrogen.[120]   An 82% yield of crotyl alcohol was obtained on hydrogenation of crotonaldehyde over a Co-Zr/Al$_2$O$_3$ catalyst at 50°C and 10 atmospheres pressure.[121]   Raney cobalt modified with CoCl$_2$ promoted the hydrogenation of citral (**41** and **42**) to geraniol (**43**) and nerol (**44**) in 92% selectivity at 30% conversion when the reaction was run in a propanol solution at 30°–65°C and atmospheric pressure.[122]   Silver-iron[123] and Ag-Zn[124] catalysts have been used for the vapor-phase hydrogenation of acrolein to allyl alcohol. The Ag-Fe catalyst gave 52% selectivity at 48% conversion for reactions run at 230°C. The addition of a small amount of Fe^{+3} to the Ag-Zn catalyst gave allyl alcohol in 95% selectivity at 70% conversion when the reaction was run at 120°C.[124]

Rhenium catalysts have been somewhat effective for the vapor-phase hydrogenation of acrolein. The use of controlled pore glass (CPG) as the support for the rhenium stabilized the catalyst as compared to Re/C. The Re/CPG catalyst prepared by the partial decomposition of Re$_2$(CO)$_{10}$ on the support showed 60% selectivity for allyl alcohol formation at 13% conversion.[125,126] The use of pre-reduced Re$_2$O$_7$ in the presence of pyridine in the liquid-phase hydrogenation of crotonaldehyde (**37**) at 100°C and 170 atmospheres of hydrogen gave crotyl alcohol (**38**) in 86% yield at 96% conversion.   The hydrogenation of cinnamaldehyde (**39**) under these conditions gave cinnamyl alcohol (**40**) in 99% yield, whereas the hydrogenation of citral (**41**) gave 66% geraniol (**42**) and 33% citronellol (**45**).[127]

The use of copper chromite at 40°C and atmospheric pressure was not very effective for selective carbonyl group hydrogenation. Unsaturated alcohols were produced from unsaturated aldehydes in low yields at low conversions and not at all from methyl vinyl ketone.[128]   With unconjugated, unsaturated aldehydes, copper chromite is effective as a selective hydrogenation catalyst. Hydrogenation of **46** at 140°–160°C and 200 atmospheres gave better than 70% of the diene diol, **47**. Increasing the temperature to 240°C resulted in the complete saturation of **46** (Eqn. 18.28).[129]

Unconjugated Unsaturated Carbonyl Compounds

The selective hydrogenation of the carbonyl group of an unconjugated unsaturated aldehyde or ketone is simpler than the process involving conjugated systems. 1,4-Addition of hydrogen is not a viable mechanistic consideration and the intermediate unsaturated alcohol is not an allyl alcohol so hydrogenolysis of the hydroxy group generally does not take place. Although the double bond can

(18.28)

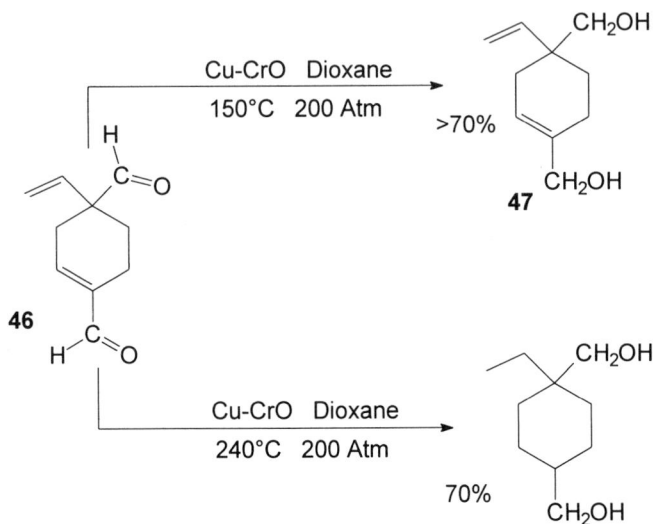

isomerize, the enol of the saturated aldehyde or ketone is not a direct isomerization product.   Because none of these yield-decreasing steps are operative, the reaction becomes a straightforward *Type II* selectivity (Chapter 5) with saturation of either the double bond or the carbonyl group. The only further reaction to consider is the continued hydrogenation to give the saturated alcohol (Eqn. 18.29). Double-bond isomerization prior to saturation will decrease reaction selectivity.

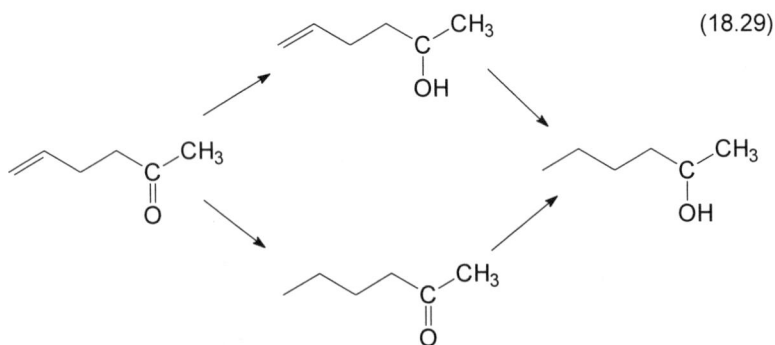

(18.29)

The selective hydrogenation of the aldehyde group of citronallal (**48**) to give citronallol (**45**) was accomplished in good yield over Raney nickel at 100°C

(18.30)

48   **45**

R.Ni EtOH
100°C   7 Atm

70%

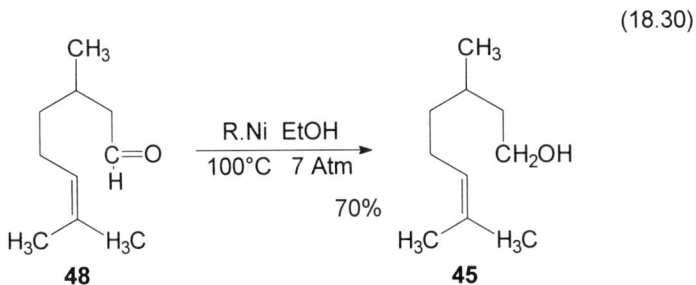

and seven atmospheres pressure (Eqn. 18.30).[130]   A Ru/TiO$_2$ catalyst, when subjected to a high temperature reduction, also hydrogenated citronellal to citronellol at 30°C and 100 atmospheres of hydrogen.[131] One of the important factors in obtaining high selectivity in this later reaction was the temperature used for the hydrogen activation of the catalyst.   When the hydrogen activation of Ru/TiO$_2$ was run at a temperatures above 400°–450°C, typical conditions for preparing catalysts in the SMSI state, the resulting catalyst was very selective for this reaction. However, when the SMSI state was reversed by oxidation, carbonyl group hydrogenation activity decreased but reaction selectivity remained about the same. Because of this and other data, it was proposed that the primary factor for the increase in selectivity with thermal treatment was caused by the loss of chloride ions in the catalyst.[131]

Although selective hydrogenation of the carbonyl group of α,β-unsaturated ketones does not appear possible at this time, the keto group can be selectively hydrogenated with unconjugated, unsaturated ketones provided the

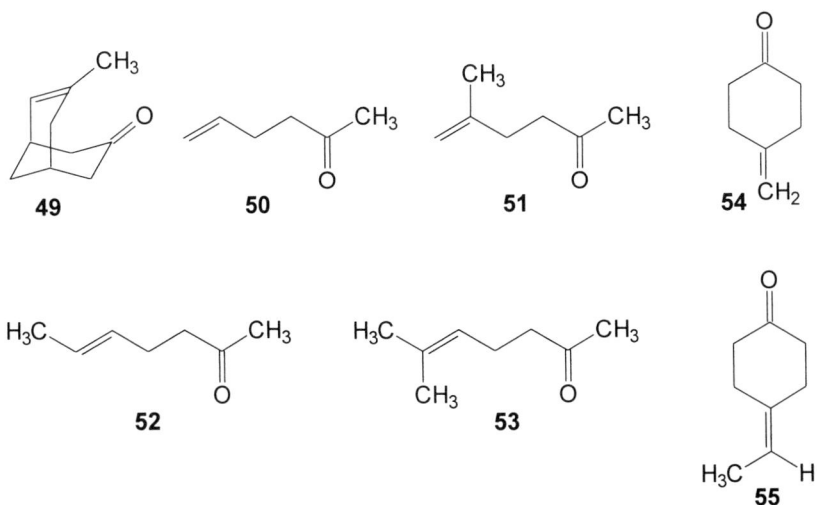

49   **50**   **51**   **54**

52   **53**   **55**

steric hindrance around the double bond is reasonably great. The hydrogenations of **49**,[132] the series of substituted 5-hexene-2-ones, **50–53**, and the 4-alkylidene-cyclohexanones, **54** and **55**,[133] were run over a cobalt catalyst at room temperature and atmospheric pressure. The unsaturated alcohols were obtained in 100% selectivity from the hindered species **49**, **53**, and **55**. Moderate selectivity was obtained on hydrogenation of **54**, but essentially no unsaturated alcohol was obtained from the other materials, again showing the importance of steric effects in the selective hydrogenation of unsaturated carbonyl compounds.

Carboxylic Acids

Carboxylic acids are one of the most difficult of all functional groups to hydrogenate. They can be hydrogenated to the alcohol, however, over ruthenium[134] or rhenium heptoxide[135,136] at 150°–200°C and 300–400 atmospheres pressure. Activated Cu-Ba-CrO is also effective at 250°–300°C and 300–400 atmospheres (Eqn. 18.31).[137] A Re-Os bimetallic catalyst has been used to hydrogenate carboxylic acids at 100°–120°C and 20–100 atmospheres of hydrogen. High conversions to the alcohol were reported.[138]

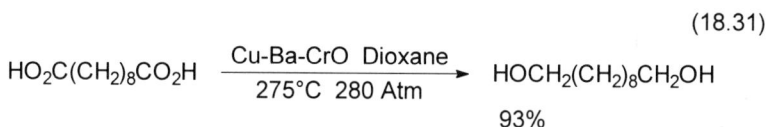

(18.31)

$$HO_2C(CH_2)_8CO_2H \xrightarrow[\text{275°C 280 Atm}]{\text{Cu-Ba-CrO Dioxane}} HOCH_2(CH_2)_8CH_2OH$$

93%

Since carboxylic acids are so difficult to hydrogenate it should be expected that virtually all other groups should hydrogenate in preference to an acid. Normally, this is the case but there are exceptions. A supported Re-Sn catalyst was reported to promote the hydrogenation of oleic acid (**56**) to 9-octadecen-1-ol (**57**) at 200°–275°C and 10–40 atmospheres pressure but the best yield was about 30% at 80% conversion.[139] Better results were obtained with a Ru-Sn/ Al_2O_3 catalyst at 250°C and 55 atmospheres of hydrogen. Under these conditions oleic acid was converted to **57** in about 82% selectivity at 97% conversion (Eqn. 18.32).[140] When ruthenium was precipitated onto the support first followed by the tin, the resulting catalyst was less active but more selective than one prepared by precipitating the metal salts in the reverse order. The most active and selective catalyst was one that was prepared by coprecipitation of ruthenium, tin and aluminum hydroxides followed by appropriate calcination.[140]

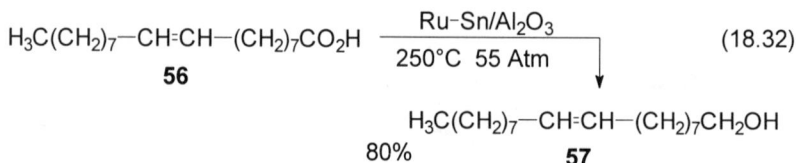

$$H_3C(CH_2)_7-CH{=}CH-(CH_2)_7CO_2H \xrightarrow[\text{250°C 55 Atm}]{\text{Ru-Sn/Al}_2O_3} \quad (18.32)$$

56

$$H_3C(CH_2)_7-CH{=}CH-(CH_2)_7CH_2OH$$

80%   **57**

(18.33)

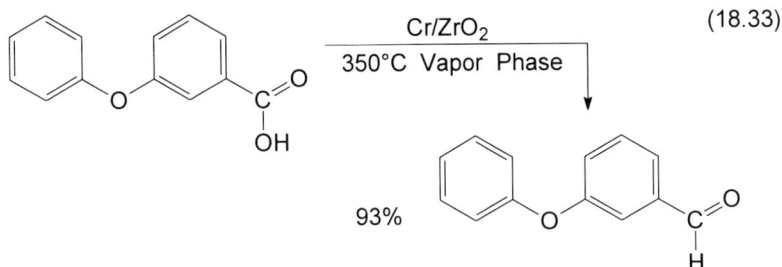

93%

The partial hydrogenation of aromatic acids to the aldehydes has been accomplished by passing a mixture of hydrogen and the acid vapor over a ZrO_2 catalyst at 350°C. Benzoic acid was converted to benzaldehyde in 97% selectivity at 50% conversion in this reaction.[141] Adding chromium to the ZrO_2 increased catalyst activity and selectivity with benzaldehyde being formed with this system in 96% selectivity at 98% conversion. Partial hydrogenation of substituted benzoic acids gave the corresponding aldehydes in similar yields (Eqn. 18.33).[141] Some aliphatic aldehydes have been prepared in the same way (Eqn. 18.34).[141,142]

(18.34)

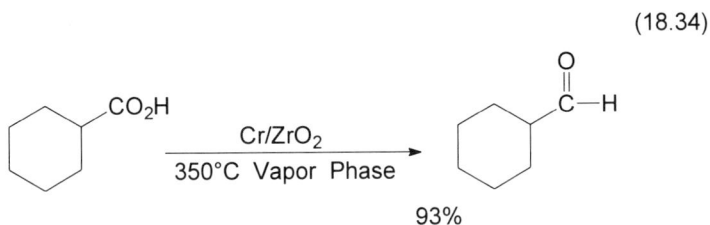

93%

Esters and Lactones

Esters are hydrogenated to the alcohol over Cu-Ba-CrO at 175°–200°C and 300–400 atmospheres of hydrogen.[143-145] A large catalyst:substrate ratio is commonly used in order to run the hydrogenation successfully at 75°–100°C lower than would be the case if the usual 5–10% of the weight of the substrate were used.[146] This lowering of the reaction temperature makes possible the hydrogenation of β-hydroxy, β-keto and β-carboalkoxy esters to the glycols and aromatic esters to the benzyl alcohols in good yields. It is sometimes necessary to acylate an amino nitrogen, if present, in order to prevent N-alkylation by the product alcohol. These reaction conditions are also used to hydrogenate lactones to glycols.[144] Carboxylic acids, benzene rings and amides are the only functional groups not hydrogenated under the conditions used to hydrogenate an ester (Eqn. 18.35).[147] The more active forms of Raney nickel are also effective catalysts for ester hydrogenations at 50°–100°C and 50–80 atmospheres.[146,148] Again, a large

(18.35)

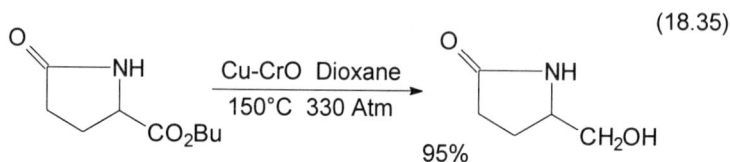

quantity of catalyst is used to hydrogenate β-substituted esters with a minimum of hydrogenolysis. Saturation of an aromatic ring takes place in preference to the hydrogenation of an ester with these catalysts but amides are not affected. Lower yields of the alcohol are generally obtained using Raney nickel than with a Cu-Ba-CrO catalyst.

Sugar lactones are hydrogenated to the aldoses over platinum at room temperature and atmospheric pressure.[149,150] Other δ-lactones have been hydrogenated to the cyclic ethers in very good yields at room temperature and atmospheric pressure over platinum oxide in acetic acid containing a small amount of perchloric acid (Eqn. 18.36).[151,152]

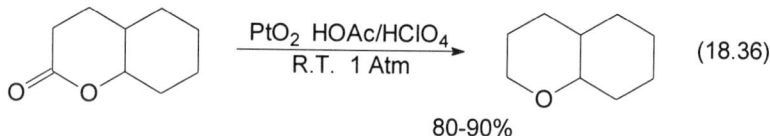

Ru-Sn (B) catalysts have been used for the hydrogenation of fatty acid esters to give good yields of the long chain alcohols at 250°–300°C and 40–45 atmospheres.[153,154] With Zn-CrO catalysts unsaturated esters are hydrogenated to the unsaturated alcohols in good yields.[145] Methyl benzoate was converted to benzaldehyde by a vapor-phase hydrogenation over ZrO_2 at 350°C and atmospheric pressure. The aldehyde was formed in 63% selectivity at 75% conversion.[155]

Amides and Imides

Amides are hydrogenated to amines over Cu-CrO[156] or ruthenium[157] catalysts under the same conditions used for the hydrogenation of carboxylic acids (Eqn. 18.37).[158] Amides are hydrogenated more slowly than are acids and all other functional groups are generally hydrogenated in preference to an amide.

Some lactams, such as α-nor lupenone (**58**) and N-methylpyridone (**59**) are, however, more easily hydrogenated. Compounds such as these can be reduced using platinum oxide in 3N HCl at room temperature and atmospheric pressure (Eqns. 18.38-39).[159] This hydrogenation does not take place without the strong acid and a large amount of catalyst is needed.

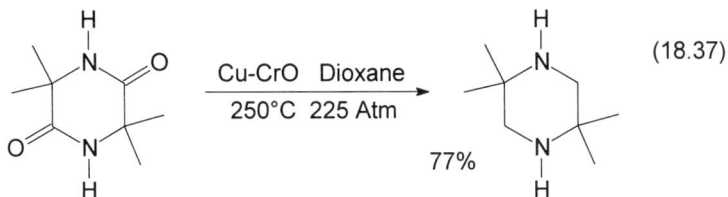

(18.37)

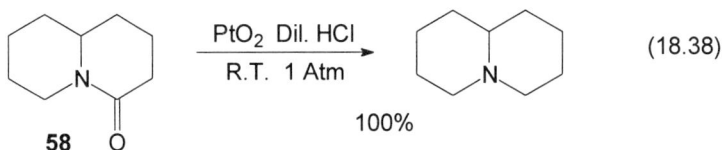

(18.38)

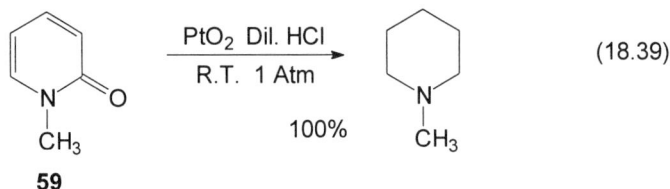

(18.39)

Phthalimide (**60**) was hydrogenated to the lactam, **61**, over Cu-CrO at 250°C and 200 atmospheres of hydrogen (Eqn. 18.40).[160] This reaction is also catalyzed by Pd/C at room temperature and atmospheric pressure in the presence of trifluoroacetic acid.[161]

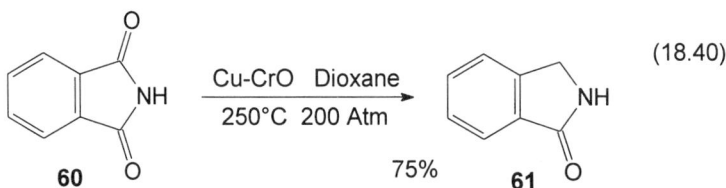

(18.40)

Anhydrides

The complete hydrogenation of an anhydride over Cu-CrO at high temperature and pressure results in glycol formation but partial hydrogenations are also possible. Phthalic anhydride (**62**) is hydrogenated to phthalide (**63**) in excellent yields over Raney nickel at 150°C and 165–200 atmospheres (Eqn. 18.41).[162] Hydrogenation over a large amount of palladium catalyst at room temperature and atmospheric pressure gives o-toluic acid (**64**).[161,163]

(18.41)

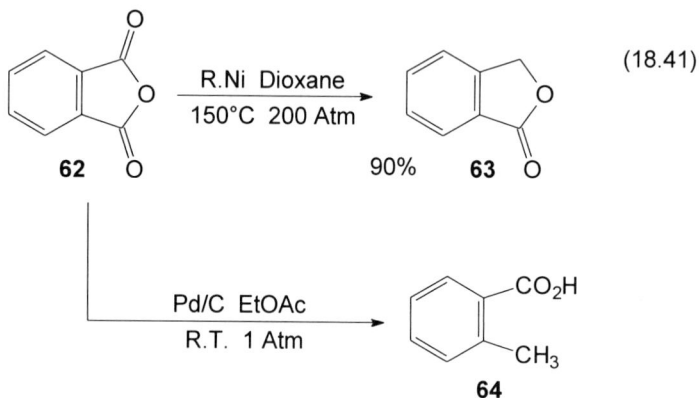

The hydrogenation of substituted succinic anhydrides such as **65**, over platinum oxide at room temperature and 3–4 atmospheres leads to the initial formation of the hydroxy lactone, **66**. Further hydrogenation in acetic acid at the same temperature and pressure converts **65** into a 2:1 ratio of the lactone, **67** and the methyl acid, **68**. Extended hydrogenation of the anhydride in ethyl acetate gave almost equal amounts of **67** and **68** (Eqn. 18.42).[164] Since anhydrides such as **65** are available from Diels-Alder reactions with maleic anhydride, this procedure has synthetic utility since the hydrogenations take place exclusively on the least hindered carbonyl group of the anhydride.

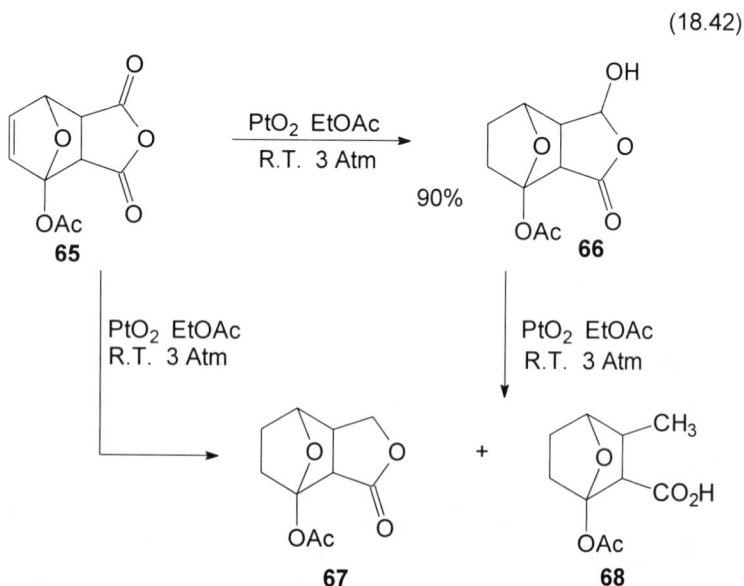

(18.42)

(18.43)

Hydrogenation of maleic anhydride (**69**) over Cu-CrO catalysts modified with Zn^{+2} and Mg^{+2} gave a 75% yield of γ-butryolactone (**70**) when the reaction was run in the vapor phase at 250°C (Eqn. 18.43).[165,166]

References

1. E. Breitner, E. Roginski and P. N. Rylander, *J. Org. Chem.*, **24**, 1855 (1959).
2. G. G. Boyers, U. S. Pat. 2,868,847; *Chem. Abstr.*, 53, 9084 (1959).
3. L. N. Owen, *J. Chem. Soc.,* 1582 (1949).
4. P. Vijayalakshmi and R. Subbarao, *J. Am. Oil Chem. Soc.*, **70**, 435 (1993).
5. H. Adkins and H. R. Billica, *J. Am. Chem. Soc.*, **70**, 695 (1948).
6. Y. Nakao and S. Fujishige, *Chem. Lett.*, 673 (1980).
7. W. Wenner, *J. Org. Chem.*, **15**, 301 (1950).
8. Y.-J. Lin, *Chem. Ind. (Dekker)*, **47** (Catal. Org. React.) 19 (1992).
9. J. Court, J. P. Damon, J. Masson and P. Wierzchowski, *Stud. Surf. Sci. Catal.*, **41** (Heterog. Catal. Fine Chem.), 189 (1988).
10. J. M. Bonnier, J. P. Damon and J. Masson, *Appl. Catal.*, **42**, 285 (1988).
11. T. Koscielski, J. M. Bonnier, J. P. Damon and J. Mason, *Appl. Catal.*, **49**, 91 (1989).
12. T. W. Russell, D. M. Duncan and S. C. Hansen, *J. Org. Chem.*, **42**, 551 (1977).
13. S. Kishida, Y. Murakami, T. Imanaka and S. Teranishi, *J. Catal.*, **12**, 97 (1968).
14. Y.-Z. Chen, L.-L, Sheu and M-H. Rei, *J. Chin. Chem. Soc. (Taipei)*, **36**, 67 (1989).
15. M. A. Keane, *Zeolites*, **13**, 330 (1993).
16. M. Hegedus, S. Gobolos and J. L. Margitfalvi, *Stud. Surf. Sci. Catal.*, **78** (Heterog. Catal. Fine Chem., III), 187 (1993).
17. R. Adams, S. McKenzie, Jr. and S. Loewe, *J. Am. Chem. Soc.*, **70**, 664 (1948).
18. L. Long, Jr. and A. Burger, *J. Org. Chem.*, **6**, 852 (1941).
19. N. Campbell, W. Anderson and J. Gilmore, *J. Chem. Soc.,* 819 (1940).
20. M. Bankmann, R. Brand, A. Freund and T. Tacke, *Stud. Surf. Sci. Catal.*, **78** (Heterog. Catal. Fine Chem., III), 91 (1993).
21. S. D. Lin, D. K. Sanders and M. A. Vannice, *Appl. Catal. A*, **113**, 59 (1994).
22. P. N. Rylander and L. Hasbrouck, *Engelhard Ind. Tech. Bull.*, **8**, 148 (1967).
23. J. Masson, P. Cividino, J. M. Bonnier and P. Fouilloux, *Stud. Surf. Sci. Catal.*, **59** (Heterog. Catal. Fine Chem., II), 245 (1991).
24. W. R. Nes, *J. Org. Chem.*, **23**, 899 (1958).
25. J. L. R. Williams, *J. Org. Chem.*, **19**, 1205 (1954).

26. W. Reeve and J. D. Sterling, Jr., *J. Am. Chem. Soc.*, **71**, 3657 (1949).
27. H. Oelschlager, *Chem. Ber.*, **89**, 2025 (1956).
28. R. H. Baker and W. W. Jenkins, *J. Am. Chem. Soc.*, **68**, 2102 (1946).
29. J. Novak, F. Sorm and J. Sicher, *Coll. Czech. Chem. Commun.*, **19**, 1265 (1954).
30. A. L. Wilds and R. E. Sutton, *J. Org. Chem.*, **16**, 1371 (1951).
31. M. Freifelder, *J. Org. Chem.*, **29**, 2895 (1964).
32. D. Ginsburg and R. Pappo, *J. Chem. Soc.*, 938 (1951).
33. L. P. Reiff and H. S. Aaron, *Tetrahedron Lett.*, 2329 (1967).
34. W. Wysoka, *Bull. Acad. Pol. Aci., Ser. Sci. Chim.*, **28**, 263 (1980).
35. S. Nishimura and T. Itaya, *J. Chem. Soc., Chem. Commun.*, 422 (1967).
36. S. Nishimura, S. Iwafune, T. Nagura, Y. Akimoto and M. Uda, *Chem. Lett.*, 1275 (1985).
37. M. Verzele, M. Acke and M. Anteunis, *J. Chem. Soc.*, 5598 (1963).
38. P. Hudson and P. J. Parsons, *Synlett*, 867 (1992).
39. J. Kollonitsch, H. E. Mertel and V. F. Verdi, *J. Org. Chem.*, **27**, 3362 (1962).
40. R. Huisgen and E. Ravenbusch, *Liebigs Ann. Chem.*, **641**, 51 (1961).
41. H. Bretschneider, *Monatsch.*, **77**, 385 (1947).
42. Y. T. Chang and W. H. Hartung, *J. Am. Chem. Soc.*, **75**, 89 (1953).
43. H. E. Ungnade and A. D. McLaren, *J. Am. Chem. Soc.*, **66**, 118 (1944).
44. K. Tanaka, Y. Takagi, O. Nomura and I. Kobayashi, *J. Catal.*, **35**, 24 (1974).
45. T. Chihara and K. Tanaka, *Bull. Chem. Soc., Japan.*, **52**, 512 (1979).
46. C. Djerassi, E. Batres, M. Velasco and G. Rosenkranz, *J. Am. Chem. Soc.*, **74**, 1712 (1953).
47. S. Kaufman and G. Rosenkarnz, *J. Am. Chem. Soc.*, **70**, 3502 (1948).
48. L. Lieberman, D. K. Fukushima and K. Dobriner, *J. Biol Chem.*, **182**, 299 (1950).
49. H. J. E. Loewenthal. *Tetrahedron*, **6**, 269 (1959).
50. K. Kindler, H. Oelschlater and P. Henrich, *Chem. Ber.*, **86**, 501 (1953).
51. P. G. J. Koopman, A. P. G. Kieboom and H. van Bekkum, *Recl. Trav. Chim. Pays-Bas*, **100**, 156 (1981).
52. K. Tanaka, *Stud. Surf. Sci. Catal.*, **27** (Catalytic Hydrogenation), 79 (1986).
53. J. P. Wibaut and H. P. L. Gitsels, *Recl. Trav. Chim. Pays-Bas*, **60**, 577 (1941).
54. P. N. Rylander and D. R. Steele, *Engelhard Ind. Tech. Bull.*, **3**, 125 (1963).
55. M. Muhlstadt, R. Borsdorf and F. J. Struber, *Tetrahedron Lett.*, 1879 (1966).
56. S. Mitsui, H. Saito, Y. Yamashita, M. Kaminaga and Y. Senda, *Tetrahedron*, **29**, 1531 (1973).
57. E. Hardegger, P. A. Plattner and F. Blank, *Helv. Chim. Acta*, **27**, 793 (1944).
58. S. Nishimura, M. Ishige and M. Shiota, *Chem. Lett.*, 535 (1977).
59. S. Nishimura, M. Murai and M. Shiota, *Chem. Lett.*, 1239 (1980).
60. K. Alder, H. Wirtz and H. Koppleberg, *Liebigs Ann. Chem.*, **601**, 138 (1956).
61. H. Kugita and E. L. May *J. Org. Chem.*, **26**, 1954 (1961).
62. R. Helg and H. Schinz, *Helv. Chim. Acta*, **35**, 2406 (1952).

63. G. Vavon and B. Jakubowicz, *Bull. soc. chim., France,* **53**, 1111 (1933).
64. O. Wintersteiner and M. Moore, *J. Am. Chem. Soc.*, **65**, 1503 (1943).
65. J. T. Edward and J. M. Ferland, *Can. J. Chem.*, **44**, 1311 (1966).
66. M. Verzele, M. Acke and M. Antennis, *J. Chem. Soc.*, 5598 (1963).
67. J. C. Babcock and L.F. Fieser, *J. Am. Chem. Soc.*, **74**, 5472 (1952).
68. R. L. Augustine, *Adv. Catal.*, **25**, 56 (1976).
69. R. J. Wicker, *J. Chem. Soc.*, 2165 (1956).
70. M. Sugahara, S. Tsuchida and I. Anazawa, *Chem. Lett.*, 1389 (1974).
71. B. D. Astill and V. Boekelheide, *J. Am. Chem. Soc.*, **77**, 4079 (1955).
72. P. Ruggle and A. H. Lutz, *Helv. Chim. Acta*, **30**, 1070 (1947).
73 L. A. Bigelow, H. G. Rule and W. A. P. Black, *J. Chem. Soc.*, 83 (1935).
74. M. Orchin and L. W. Butz, *J. Am. Chem. Soc.*, **65**, 2296 (1943).
75. G. F. Hennion and E. J. Watson, *J. Org. Chem.*, **23**, 656 (1958).
76. E. E. Turner and L. Turner, *J. Chem. Soc.*, 1761 (1952).
77. F. Weygand, *Chem. Ber.*, **73**, 1259 (1940).
78. M. L. Wolfrom, B. W. Lew, R. A. Hales and R. M. Goepp, Jr., *J. Am. Chem. Soc.*, **68**, 2342 (1946).
79. J. Ishiyama, T. Juhita and S. Imaizumi, *Nippon Kagaku Kaishi*, 771 (1991); *Chem. Abstr.*, **115**, 91633 (1991).
80. R. A. Cormier and M. D. McCauley, *Synth. Commun.*, **18**, 675 (1988).
81. H. van Bekkum, A. P. G. Kieboom and K. J. G. van de Putte, *Recl. Trav. Chim. Pays-Bas*, **88**, 52 (1969).
82. M. Bonnet, P. Beneste and M. Rodriguez, *J. Org. Chem.*, **45**, 40 (1980).
83. T. Chihara, M. Shinzawa, Y. Yokoyama, K. Taya and H. Ogawa, *React. Kinet. Catal. Lett.*, **39**, 181 (1989).
84. P. Hazra, A. P. Kahol, K. K. Aggarwal and S. Mandal, *Res. Ind.*, **35**, 118 (1990).
85. P. N. Rylander and D. R. Steele, *Tetrahedron Lett.*, 1579 (1969).
86. R. N. Bakhanova, A. S. Astakhova, K. A. Brikenshtein, V. G. Dorokhov, V. I. Savchenko and M. L. Khidekel, *Isv. Akad. Nauk SSSR, Ser. Khim.*, 1993 (1972); *Chem. Abstr.*, **78**, 15967 (1973).
87. M. L. Khedekel, R. N. Bakhonova, A. S. Astakhova, K. A. Brikenshtein, V. I. Savchenko, I. S. Monakhova and V. G. Dorokhov, *Isv. Akad. Nauk SSSR, Ser. Khim.*,499 (1970); *Chem. Abstr.*, **73**, 3130 (1970).
88. A. Giroir-Fendler, D. Richard and P. Gallezot, *Stud. Surf. Sci. Catal.*, **41** (Heterog. Catal. Fine Chem.), 171 (1988).
89. P. Gallezot, A. Giroir-Fendler and D. Richard, *Chem. Ind. (Dekker)*, **47** (Catal. Org. React.), 1 (1992).
90. B. Coq, P. S. Kumbhar, C. Moreau, P. Moreau an M. G. Warawdekar, *J. Mol. Catal.*, **85**, 215 (1993).
91. B. Coq, P. S. Kumbhar, C. Moreau, P. Moreau and F. Figueras, *J. Phys. Chem.*, **98**, 10180 (1994).
92. D. G. Blackmond, A. Waghray, R. Oukaci, B. Blanc and P. Gallezot, *Stud. Surf. Sci. Catal.*, **59** (Heterg. Catal. Fine Chem., II), 145 (1991).
93. A. Giroir-Fendler, D. Richard and P. Gallezot, *Catal. Letters*, **5**, 175 (1990).
94. S. Galvagno, C. Milone, A. Donato, G. Neri and R. Pietropaolo, *Catal. Letters*, **18**, 349 (1993).

95.   S. Galvagno, C. Milone, G. Neri, A. Donato and R. Pietropaolo, *Stud. Surf. Sci. Catal.*, **78** (Heterog. Catal. Fine Chem., III), 163 (1993).
96.   C. A. Raab and J. A. Lercher, *Catal. Letters*, **18**, 99 (1993).
97.   C. G. Raab, M. Englisch, T., B. L. W. Marinelli and J. A. Lercher, *Stud. Surf. Sci. Catal.*, **78** (Heterog. Catal. Fine Chem., III), 211 (1993).
98.   R. Makouangou-Mandilou, R. Touroude and A. Dauscher, *Stud. Surf. Sci. Catal.*, **78** (Heterog. Catal. Fine Chem., III), 219 (1993).
99.   B. Coq, F. Figueras, P. Geneste, C. Moreau, P. Moreau and M. Warawdekar, *J. Mol. Catal.*, **78**, 211 (1993).
100. M. A. Vannice and B. Sen., *J. Catal.*, **115**, 65 (1989).
101. W. F. Tuley and R. Adams, *J. Am. Chem. Soc.*, **47**, 3061 (1925).
102. P. N. Rylander, N. Himelstein and M. Kilroy, *Engelhard Ind. Tech. Bull.*, **4**, 49 (1963).
103. J. Sabadie and G. Descotes, *Bull. soc. chim., France,* 515 (1977).
104. D. Richard, J. Ockelford, A. Giroir-Fendler and P. Gallezot, *Catal. Letters*, **3**, 53 (1989).
105. D. V. Sokolskii, N. V. Anisimova, A. K. Zharmagambetova, S. G. Mukhamedzhanova and L. N. Edygenova, *React. Kinet. Catal. Lett.*, **33**, 399 (1987).
106. G. Neri, C. Milone, A. Donato, L. Mercadante and A. M. Visco, *J. Chem. Technol. Biotechnol.*, 60, 83 (1994).
107. Z. Poltarzewski, S. Galvagno, R. Pietropaolo and P. Staiti, *J. Catal.*, **102**, 190 (1986).
108. S. Galvagno, A. Donato, G. Neri, R. Pietropaolo and G. Capannelli, *J. Mol. Catal.*, **78**, 227 (1993).
109. S. Galvagno, Z. Poltarzewski, A. Donato, G. Neri and R. Pietropaolo, *J. Chem. Soc., Chem. Commun.*, 1729 (1986).
110. B. Didillon, A. El Mansour, J. P. Candy, J. P. Bournonville and M. M. Basset, *Stud. Surf. Sci. Catal.*, **59** (Heterg. Catal. Fine Chem., II), 137 (1991).
111. B. Didillon, A. El Mansour, J. P. Candy, J. M. Basset, F. Le Peletier and J. P. Boitiaux, *Stud. Surf. Sci. Catal.*, **75** (New Frontiers in Catalysis, Pt. C), 2371 (1993).
112. B. Didillon, J. P. Candy, F. Le Peletier, O. A. Ferretti and J. M. Basset, *Stud. Surf. Sci. Catal.*, **78** (Heterog. Catal. Fine Chem., III), 147 (1993).
113. B. Didillon, J. P. Candy, A. El Mansour, C. Houtmann and J. M. Basset, *J. Mol. Catal.*, **74**, 43 (1992).
114. S. Galvagno, C. Milone, A. Donato, G. Neri and R. Poetropaolo, *Catal. Letters*, **17**, 55, 1993).
115. D. G. Blackmond, R. Oukaci, B. Blanc and P. Gallezot, *J. Catal.*, **131**, 401 (1991).
116. A. Waghray, R. Oukaci and D. G. Blackmond, *Stud. Surf. Sci. Catal.*, **75** (New Frontiers in Catalysis, Pt. C), 2479 (1993).
117. D. G. Blackmond and A. Waghray, *Chem. Ind. (Dekker)*, **62** (Catal. Org. React.), 295 (1995).
118. A. Waghray, R. Oukaci and D. G. Blackmond, *Chem. Ind. (Dekker)*, **53** (Catal. Org. React.), 569 (1994).
119. J. B. Nagy, I. Bodart-Ravet and E. G. Derouane, *Faraday Discuss. Chem. Soc.*, **87**, 189 (1989).
120. Y. Nitta, T. Kato and T. Imanaka, *Stud. Surf. Sci. Catal.*, **78** (Heterog. Catal. Fine Chem., III), 83 (1993).

121. H. Miura, M. Saito, J. Watanage, K. Ichioka and T. Matsuda, *Nippon Kagaku Kaishi*, 487 (1994); *Chem. Abstr.*, **121**, 107947 (1994).
122. K. Hotta, S. Watanabe and T. Kubomatsu, *Nippon Kagaku Kaishi*, 352 (1982); *Chem. Abstr.,* **96**, 199900 (1982).
123. Y. Nagase and K. Wada, *Ibaraki Daigaku Kogakubu Kenkyu Shuho*, **33**, 223 (1985); *Chem. Abstr.*, **106**, 178436 (1987).
124. Y. Nagase, H. Hattori and K. Tanabe, *Chem. Lett.*, 1615 (1983).
125. T. H. Vanderspurt, *Ann. N. Y. Acad. Sci.*, **333**, 155 (1980).
126. T. H. Vanderspurt, *Catalysis in Organic Synthesis* (W. H. Jones, Ed.), Academic Press, New York, 1980, p. 11.
127. W. E. Pascoe and J. F. Stenberg, *Catalysis in Organic Synthesis* (W. H. Jones, Ed.), Academic Press, New York, 1980, p. 1.
128. R. Hubaut, M. Daage and J. P. Bonnelle, *Appl. Catal.*, **22**, 231 (1986).
129. R. Pummerer, F. Aldebert, F. Graser and H.Sperber, *Liebigs Ann. Chem.*, **583**, 225 (1953).
130. P. Hazra, A. P. Kahol and S. Mondal, *Res. Ind.*, **37**, 106 (1992).
131. A. A. Wismeijer, A. P. G. Kieboom and H. van Bekkum, *Appl. Catal.*, **25**, 181 (1986).
132. J. Ishiyama, Y. Senda and S. Imaizumi, *Chem. Lett.*, 1243 (1983).
133. J. Ishiyama, S. Maeda, K. Takahashi, Y. Senda and S. Imaizumi, *Bull. Chem. Soc., Japan.*, **60**, 1721 (1987).
134. J. E. Carnahan, T. A. Ford, W. F. Gresham, W. E. Grigsby and G. F. Hager, *J. Am. Chem. Soc.*, **77**, 3766 (1955).
135. H. S. Broadbent, *J. Org. Chem.*, **24**, 1847 (1959); **28**, 2343 (1963).
136. H. S. Broadbent, V. L. Mylroie and W. R. Dixon, *Ann. N. Y. Acad. Sci.*, **172**, 194 (1970).
137. A. Guyer, A.Bieler and M. Sommaruga, *Helv. Chim. Acta*, **38**, 976 (1955).
138. K. Yoshino, Y. Kajiwara, N. Takaishi and Y. Inamoto, *J. Am. Oil Chem. Soc.*, **67**, 21 (1990).
139. T.-S. Tang, K.-Y. Cheah, F. Mizukami, S. Niwa, M. Toba and Y.-M. Choo, *J. Am. Oil Chem. Soc.*, **70**, 601 (1993).
140. T.-S. Tang, K.-Y. Cheah, F. Mizukami, S. Niwa andToba, *J. Am. Oil Chem. Soc.*, **71**, 51 (1994).
141. T. Yokoyama, T. Setoyama, N. Fujita, M. Kakajima and T. Maki, *Appl. Catal. A*, **88**, 149 (1992).
142. N. Ding, J. Kondo, K. Maruya, K. Domen, T. Yoloyama, N. Fujita and T. Maki, *Catal. Letters*, **17**, 309 (1993).
143. H. Adkins, *Org. Reactions*, **8**, 1 (1954).
144. K. Folkers and H. Adkins, *J. Am. Chem. Soc.*, **54**, 1145 (1932).
145. H. Boerma, *Stud. Surf. Sci. Catal.*, **1** (Prep. Catal.)105 (1976).
146. H. Adkins and H. R. Billica, *J. Am. Chem. Soc.*, **70**, 3118, 3121 (1948).
147. E. Segal, *J. Am. Chem. Soc.*, **74**, 851 (1952).
148. H. Adkins and A. A. Pavlic, *J. Am. Chem. Soc.*, **69**, 3039 (1948).
149. J. W. E. Glattfeld and G. W. Schmipff, *J. Am. Chem. Soc.*, **57**, 2204 (1935).
150. O. T. Schmidt and H. Muller, *Chem. Ber.*, **76**, 344 (1943).
151. J. T. Edward and J. M. Ferland, *Chem. and Ind., (London)*, 975 (1964).
152. J. T. Edward and J. M. Ferland, *Can. J. Chem.*, **44**, 1249 (1966).
153. C. S. Narasimhan, V. M. Deshpande and K. Ramnarayan, *Ind. Eng. Chem. Res.*, **28**, 1112 (1989).

154. V. M. Deshpande, K. Ramnarayan and C. S. Narasimhan, *J. Catal.*, **121**, 174 (1990).
155. A. Aboulay, A. Chambellan, M. Marzin, J. Saussey, R. Mauge, J. C. Lavalley, C. Mercier and R. Cacquot, *Stud. Surf. Sci. Catal.*, **78** (Heterog. Catal. Fine Chem., III), 131 (1993).
156. H. Adkins, et al., *J. Am. Chem. Soc.*, **56**, 2419 (1934); **58**, 2487 (1936); **60**, 402 (1938); **61**, 1675 (1939).
157. A. Guyer, A. Bieler and G. Gerliczy, *Helv. Chim. Acta*, **38**, 1649 (1955).
158. S. M. McElvain ande. H. Pryde, *J. Am. Chem. Soc.*, **71**, 326 (1949).
159. F. Galinovsky and E. Stern, *Chem. Ber.*, **76**, 1034 (1943).
160. A. Dunet, R. Rollet and A. Willemart, *Bull. soc. chim., France*, 877 (1950).
161. A. J. McAtees, R. McCrindle and D. W. Sneddon, *J. Chem. Soc., Perkin I*, 2030 (1977).
162. W. Theilacker and H. Kalenda, *Liebigs Ann. Chem.*, **584**, 87 (1953).
163. R. Kuhn and J. Butula, *Liebigs Ann. Chem.*, **718**, 50 (1968).
164. R. McCrindle, K. H. Overton and R. A. Raphael, *J. Chem. Soc.*, 4798 (1962).
165. G. L. Castiglioni, M. Gazzano, G. Stefani and A. Vaccari, *Stud. Surf. Sci. Catal.*, **78** (Heterog. Catal. Fine Chem., III), 275 (1993).
166. G. L. Castiglioni, C. Fumagalli, R. Lancia, M. Messori and A. Vaccari, *Chem. and Ind., (London)*, 510 (1993

Hydrogenation VI:
Nitrogen Containing Functional Groups

Aromatic Nitro Groups

An aromatic nitro group is generally the most easily hydrogenated of all functional groups. Reduction to the aniline takes place under mild conditions over virtually all hydrogenation catalysts. This hydrogenation usually proceeds best in neutral solution. In acidic or basic media it takes place more slowly. Selectivity in the hydrogenation of aromatic nitro groups can be manifested in two ways: (1) the nitro group can be selectively hydrogenated to the amine in the presence of almost all other hydrogenateable functional groups and (2) the conversion to the aniline is a multistep process. Under the proper conditions the reaction can be stopped at a partially hydrogenated stage to give one of the intermediates as the product.

Selective Hydrogenation

The nitro group of 4-nitropyridine-N-oxide (**1**) was hydrogenated preferentially over palladium in neutral medium[1,2] but in acid the N-oxide was also reduced (Eqn. 19.1).[2] Hydrogenation of **1** or 4-nitroquinoline-N-oxide over Raney nickel at elevated temperatures and pressures resulted in the reduction of both the nitro group and the N-oxide.[3,4] Hydrogenation of 2-nitropyridine-N-oxide gave only 2-aminopyridine.[5]

(19.1)

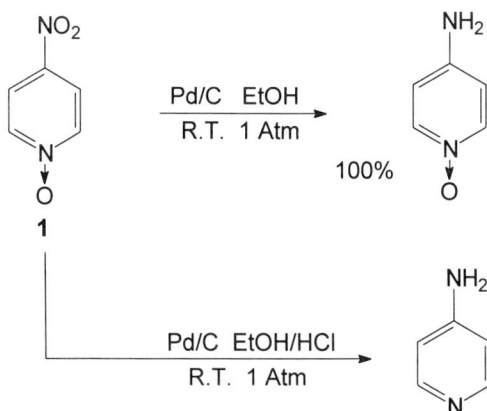

(19.2)

80%   +   2%

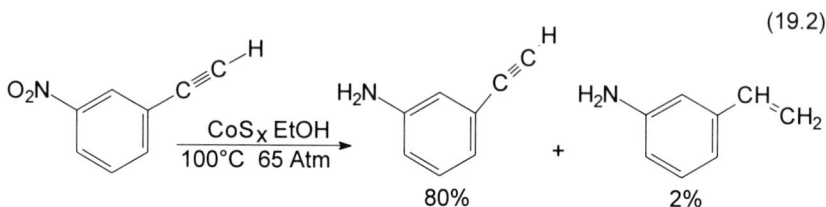

Nitrophenyl Acetylenes

The selective hydrogenation of an aryl nitro group has been accomplished even in the presence of an easily hydrogenated acetylene group. Cobalt polysulfide and ruthenium sulfide have been reported to be effective for the selective hydrogenation of the nitro group in m-nitrophenyl acetylenes giving the amino-acetylenes in 75–85% yields (Eqn. 19.2).[6] Ruthenium catalysts were also effective for this reaction but since the order of reactivity over these catalysts was found to be:

$$RC \equiv CH \rangle ArC \equiv CH \approx ArNO_2 \rangle ArC \equiv CR$$

the selectively for aminophenyl-acetylene formation improved when the acetylene was substituted.[7,8] To take advantage of this, terminal acetylenes were condensed with acetone to give the tertiary carbinol prior to hydrogenation. This bulky substituent improved reduction selectivity and could be removed after hydrogenation by treatment with base to give the amino acetylene in very good overall yield (Eqn. 19.3).[7,8]

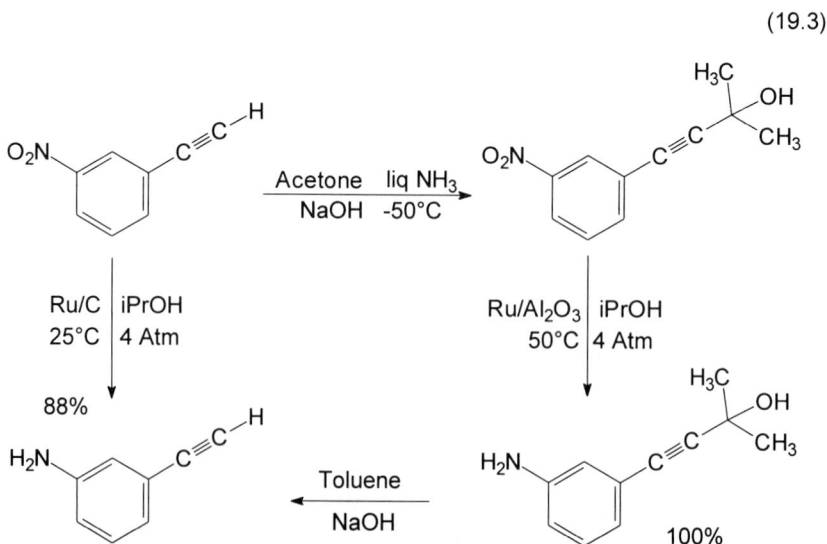

(19.3)

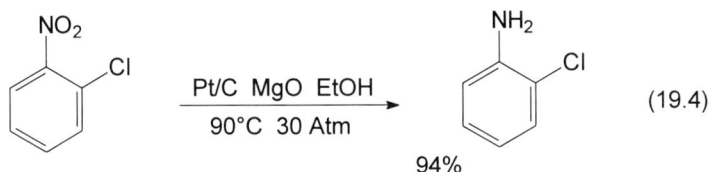

$$(19.4)$$

94%

Halonitrobenzenes

In the hydrogenation of halonitrobenzenes to haloanilines the extent of dehalogenation depends on the halogen (I > Br > Cl > F) and its position (ortho > para > meta).   Platinum is one of the best catalysts to use for minimum dehalogenation combined with a relatively rapid rate of nitro group hydrogenation especially when the reaction is run in the presence of a trace of magnesium oxide or similar basic material which keeps the reaction mixture slightly basic (Eqn. 19.4).[9]   Ruthenium catalysts gave excellent yields of p-chloroaniline on hydrogenation of p-chloronitrobenzene at 85°C and 100 atmospheres pressure.[10] With a Pt/TiO$_2$ catalyst that was activated by high temperature treatment with hydrogen, p-chloronitrobenzene was hydrogenated to p-chloroaniline in methanol at 30°C and atmospheric pressure in 99% selectivity at 98% conversion. The use of a Pt/Al$_2$O$_3$ catalyst gave p-chloroaniline in 85% selectivity while with Pt/MgO and Pt/graphite p-chloroaniline was obtained in 92% and 98% selectivities, respectively. In all three cases the conversion was near 100%.[11] Loss of halogen is favored by higher temperatures and retarded by increasing the hydrogen pressure.   Although rhenium,[12] the noble metal sulfides[13] and sulfided metal oxide catalysts[14] are also effective, their use requires more vigorous reaction conditions.   Hydrogenation of p-chloronitrobenzene in cyclohexane over P-1 Ni(B) gave the chloroaniline exclusively at 120°C and 6 atmospheres of hydrogen.[15]

The best catalyst for the selective hydrogenation of the three isomeric bromonitrobenzenes was 5% Rh/C, which promoted the hydrogenation of the nitro group with less than 2% debromination in all cases. Ethanol was the solvent of choice for the hydrogenation of the meta isomer while the use of THF as the solvent gave better results in the hydrogenation of the ortho and para isomers (Eqn. 19.5).[16] Palladium catalysts generally promote extensive dehalogenation.

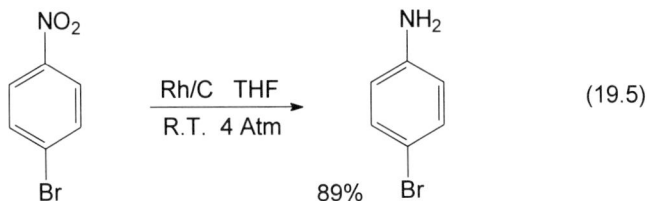

$$(19.5)$$

89%

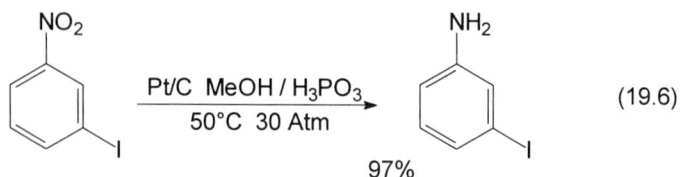

$$(19.6)$$

97%

    The presence of various modifiers has been reported to increase reaction selectivity. With platinum catalysts, the most common dehalogenation inhibitors are alkali and alkaline earth hydroxides or oxides and organic bases.[9] The effectiveness of organic bases is structure dependent. Morpholine is particularly effective, permitting only 0.5% dehalogenation of p-chloronitrobenzene using Pt/C at 95°C and 25–30 atmospheres. Piperizine was also effective, but the open-chain amines were catalyst poisons. Triphenylphosphite and phosphorous acid were found to be the most effective dehalogenation inhibitors for the platinum catalyzed reactions. Only 2.5% dehalogenation was observed on hydrogenation of m-iodonitrobenzene over Pt/C in the presence of phosphorous acid (Eqn. 19.6).[17] Thiocyanates,[18] cyanamide,[19] amidines[20] and thiourea[21] were effective dehalogenation inhibitors for the selective Raney nickel catalyzed hydrogenations of chloronitrobenzenes.

Dinitrobenzenes

The selective hydrogenation of one nitro group of a dinitrobenzene represents yet another type of reaction selectivity observed with aromatic nitro groups. This reaction, which follows a combination of *Type II and Type III* selectivity (Chapter 5) (Eqn. 19.7) is influenced by a number of factors. It was found that in the hydrogenation of 2,6-dinitrotoluene to 2-amino-6-nitrotoluene (Eqn. 19.8), platinum and rhodium were more selective than palladium. With Pt/Al$_2$O$_3$

$$(19.7)$$

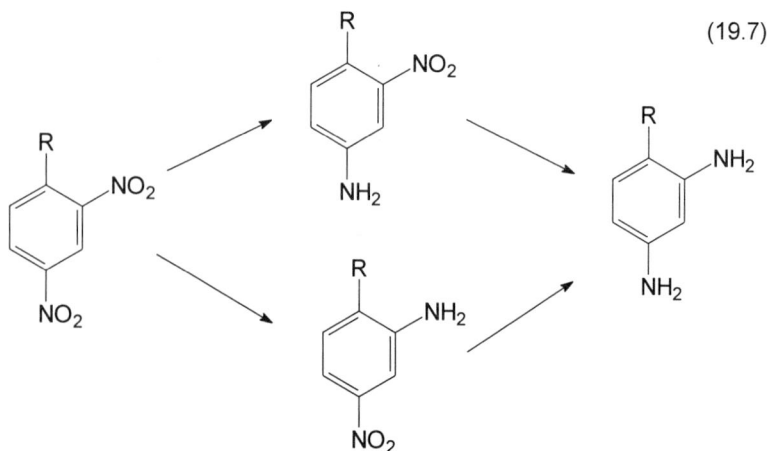

(19.8)

$$O_2N \underset{\displaystyle CH_3}{\diagdown} NO_2 \quad \xrightarrow[\text{40°C 100 Atm}]{\text{Pt/Al}_2\text{O}_3 \quad \text{EtOH}} \quad O_2N \underset{\displaystyle CH_3}{\diagdown} NH_2$$

95%

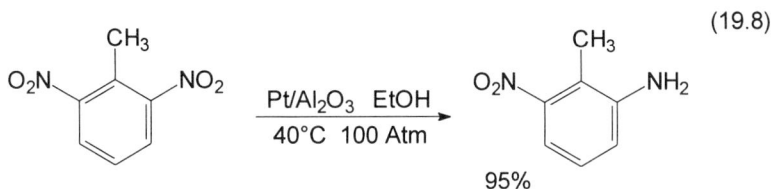

catalysts, selectivity increased with decreasing metal load going from 46% with a 5% Pt/Al$_2$O$_3$ catalyst to 83% using 0.1% Pt/Al$_2$O$_3$. At 40°C a maximum yield of about 95% of the monoamine was obtained when heat and mass transfer effects were eliminated.[22] With a copper chromite catalyst the monoamine was obtained in 91% yield when the hydrogenation was run in ethanol at 150°C and 25 atmospheres.[23] 2,6-Dinitro-anilines having a 4-methyl, 4-carboxy or 4-carboethoxy group were selectively hydrogenated to the mononitro compound in good yields using Pd/C in a mixture of 1,2-dimethoxyethane and chloroform at room temperature and three atmospheres pressure. In ethanol/chloroform complete reduction to the triamine was observed (Eqn. 19.9).[24]

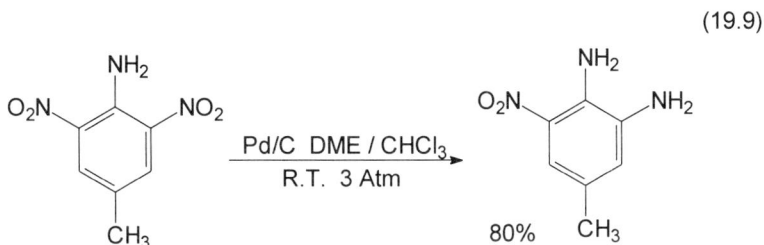

(19.9)

$$O_2N \underset{\displaystyle CH_3}{\overset{\displaystyle NH_2}{\diagdown}} NO_2 \quad \xrightarrow[\text{R.T. 3 Atm}]{\text{Pd/C DME / CHCl}_3} \quad O_2N \underset{\displaystyle CH_3}{\overset{\displaystyle NH_2}{\diagdown}} NH_2$$

80%

In the selective hydrogenation of 2,4-dinitroalkylbenzenes (Eqn. 19.7), regioselectivity is also a factor. Raney copper was found to be the most selective catalyst for the hydrogenation of the 4-nitro group.[25] Even though large amounts of catalyst were needed, its activity increased on reuse up to six times. This improved performance was believed to be caused by the presence of some of the diamino compound that apparently activates the catalyst. An initially active catalyst could be produced by using it to reduce a small amount of the dinitro compound to the diamine. As might be expected, selectivity for the hydrogenation of the 4-nitro group increased with increasing size of the alkyl group, going from 68% of the mono-nitro compound with the unsubstituted material (R=H) to 70% of the 2-nitro-4-amino species when R was a methyl group to 93% selectivity for the 4-nitro group hydrogenation when R was a tert-butyl group.[25]

Although the effect of alkyl substituents is rather straightforward, that of substituents having nonbonded electron pairs is more complex. The hydrogenation of the dinitroanisole, **2**, over Raney copper was completely non-

(19.10)

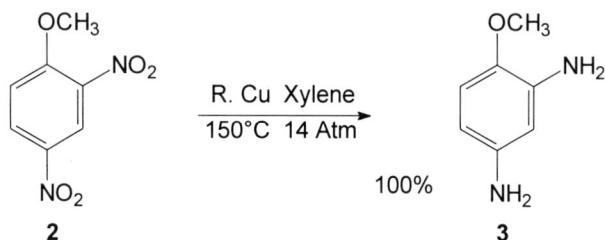

2   100%   **3**

selective giving, as the only product, **3**, in which both nitro groups were reduced (Eqn. 19.10). However, the dinitro piperidyl compound, **4**, was selectively hydrogenated at the 4-nitro group over this catalyst in near quantitative yield (Eqn. 19.11).[25] N-Substituted 2,4-dinitroanilines were selectively hydrogenated to the 2-nitro-1,4-benzenediamines over Pt/C in aqueous H_2SO_4/HOAc at 85°C and two atmospheres of hydrogen. The yields for the unsubstituted species, **5**, (Eqn. 19.12), and the N-phenyl and N-methyl materials, however, were only about 45–50%.[26]

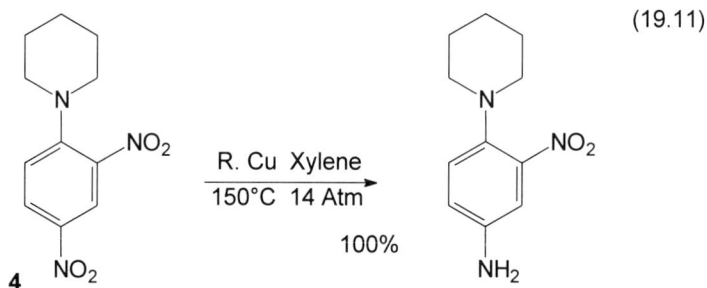

4

The unsubstituted dinitroaniline (**5**) was selectively converted to 2-nitro-1,4-benzene diamine (**6**) in 70% yield over Pt/C in acidic ethanol.[27] Hydrogenation of **5** to 4-nitro-1,2-benzenediamine (**7**), however, was more difficult to accomplish. Platinum and palladium catalysts, in either ethanol or DMF containing a small amount of ammonium hydroxide, were non-selective giving a nearly 50:50 mixture of the 2 and 4-hydrogenated materials on interruption of the reaction after the uptake of the amount of hydrogen required to saturate one nitro group. The hydrogen uptake was rapid and showed no decrease in rate after one nitro group had been reduced. With Rh/Al_2O_3 or Rh/C, though, selectivity toward 2-nitro group hydrogenation was observed when the hydrogenation was run in the presence of a small amount of concentrated ammonium hydroxide.[28] In ethanol a 6:94 ratio of **6** to **7** was obtained when the reaction was stopped after the reduction of one nitro group. In DMF a 15:85 ratio

(19.12)

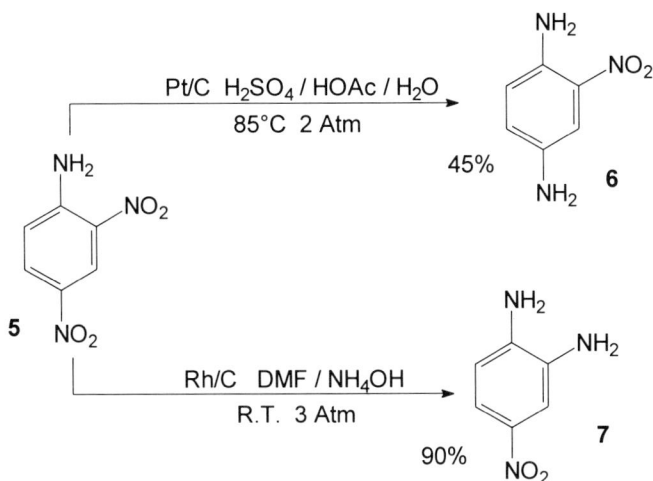

was obtained but the reaction ceased after the reduction of one nitro group. The use of Raney nickel under strongly alkaline conditions also provided good selectivity for the reduction of the 2-nitro group but, again, the reaction had to be interrupted after one nitro group had been hydrogenated.[28]

The selective hydrogenation of the nitrophenyl-nitroimidazole, **8**, was accomplished in 85-90% yield by careful attention to reaction conditions.[29] The maximum yield of **9** (Eqn. 19.13) was obtained using a Pd/C catalyst with concentrated aqueous ammonia as the solvent at 20-35°C and 2.7 atmospheres. Variation from these conditions resulted in a significant decrease in selectivity. The need for these exacting conditions arose because of two critical requirements: (1) the imidazole ring should exist as its anion to inhibit the hydrogenation of the imidazole nitro group and (2) the concentration of reactant in solution should be

(19.13)

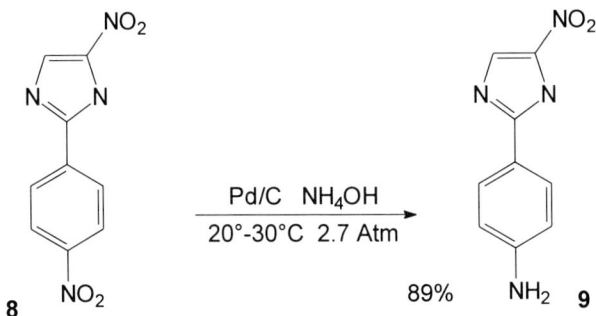

Scheme 19.1

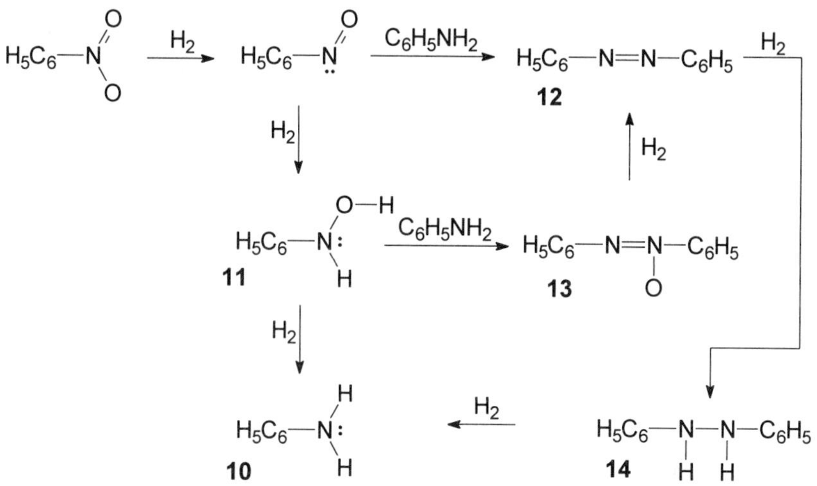

limited to prevent the formation of dimeric products. Concentrated aqueous ammonia was an effective solvent, but alcoholic ammonia was not because the concentration of the substrate was too high. Increasing substrate solubility by heating or agitation before the addition of hydrogen also decreased the yield of 9.[29] The meta isomer of **8** was also selectively hydrogenated to the aniline in about 70% yield using these reaction conditions.[29]

Partial Hydrogenation

The reaction sequence involved in the hydrogenation of aromatic nitro groups is shown in Scheme 19.1. This can be classed as a complex *Type III* selectivity. The end product from all paths is the aniline (**10**), but intermediates such as hydroxylamines (**11**), azo (**12**), azoxy (**13**), and hydrazo (**14**) compounds are present and can sometimes be isolated under the proper reaction conditions. In general, the dimeric products usually form in alkaline media, the partially reduced monomeric species form in neutral solutions, and anilines are produced in acid. The best yields of partially reduced products are obtained when the reaction is interrupted before it stops spontaneously and when it is carried out in the presence of various modifiers.[30]

Hydroxylamines

Aromatic hydroxylamines are frequently produced during the hydrogenation of aryl nitro groups. Generally, these are undesired intermediates because when they are present in excessive amounts, a potentially explosive situation caused by the exothermic disproportionation of the hydroxylamine can result.[31,32] This is usually not a problem, but **care should be exercised to prevent the accumulation of large amounts of the hydroxylamine**, particularly when the

(19.14)

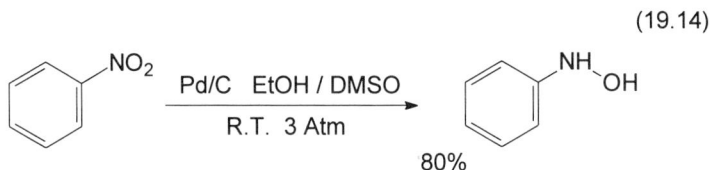

80%

reaction temperature nears 250°C which is close to the autodecomposition temperature of these materials.

Under mild conditions, high yields of aryl hydroxylamines can be obtained by the hydrogenation of aromatic nitro compounds using Pt/C catalysts in aqueous alcohol containing a small amount of dimethyl sulfoxide (DMSO). The DMSO acts as an inhibitor for further hydrogenolysis of the hydroxylamine entity (Eqn. 19.14).[30,33]   A maximum yield of over 80% phenylhydroxylamine was reported when the hydrogenation of nitrobenzene over platinum in the presence of DMSO was interrupted after the absorption of two equivalents of hydrogen.[30] Further hydrogen uptake resulted in the hydrogenolysis of the phenyl hydroxylamine to give aniline. As with most *Type III* selective reactions, running the hydrogenation of nitrobenzene under mass transport control conditions decreases the yield of the hydroxylamine. Phenylhydroxylamine formation is favored by low metal loading, small amounts of catalyst, rapid agitation and higher pressures; all parameters supporting extensive reactant diffusion through the reaction media. This rapid diffusion also favors the displacement of the phenylhydroxylamine by nitrobenzene before hydrogenolysis can take place.[30]

Hydrogenation of 2,5-dichloronitrobenzene (**15**) over Pt/C in pyridine gave very good yields of the dichlorophenylhydroxylamine (**16**) (Eqn. 19.15).[34] Evidently the pyridine inhibits both dehydrochlorination and the further hydrogenolysis of the hydroxylamine.

(19.15)

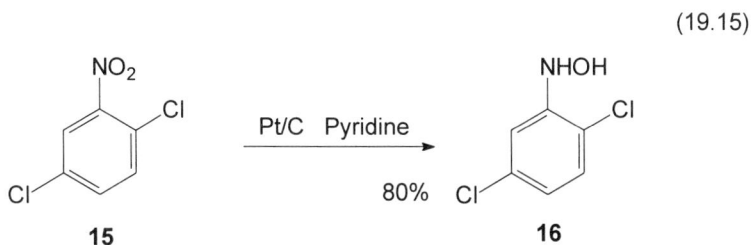

Because of their reactivity, aryl hydroxylamines can sometimes be trapped by further reaction with another material present in the hydrogenation system. Aryl nitrones such as **17** are produced by the *in situ* condensation of the aryl hydroxylamine with an aldehyde which is present in the reaction medium (Eqn. 19.16).[35]   These reactions take place in good yield over a Urushibara nickel catalyst (Chapter 12) in aqueous alcohol at room temperature and four

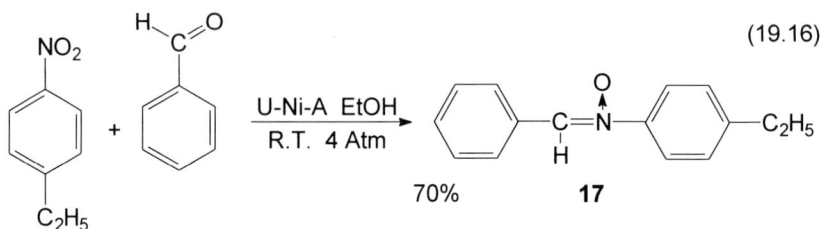

(19.16)

70%   **17**

atmospheres pressure.[36] The hydrogenation of nitrobenzenes over platinum oxide in anhydrous hydrogen fluoride gave p-fluoroaniline which was formed by the rearrangement of the intermediate hydroxylamine (Eqn. 19.17).[37]   If other reactive species, such as phenols, were also present, substituted diphenylamines were produced (Eqn. 19.18).[38]

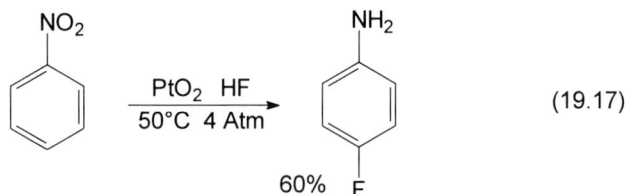

(19.17)

60%

(19.18)

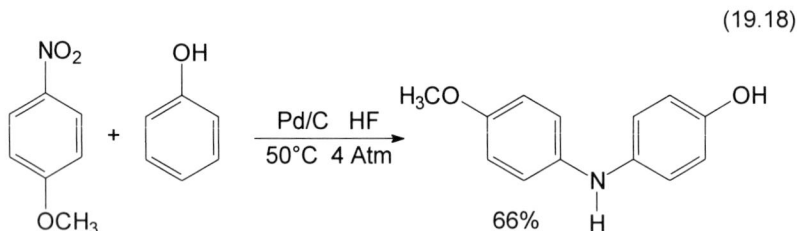

66%

The synthesis of p-aminophenol was accomplished by the platinum catalyzed hydrogenation of nitrobenzene in aqueous sulfuric acid with concomitant   acid   catalyzed   rearrangement   of   the   intermediate phenylhydroxylamine (Scheme 19.2).[39] The addition of quaternary ammonium salts[40] or DMSO[30,33] decreased the rate of phenylhydroxylamine hydrogenolysis to give more time for the rearrangement to take place.

Hydrazobenzenes

The undesired production of o,o'-dimethoxyhydrazobenzene (**18**) (Eqn. 19.19) during the hydrogenation of o-nitroanisole was promoted by the presence of chlorine containing impurities in the feed stock, higher temperatures and lower

Scheme 19.2

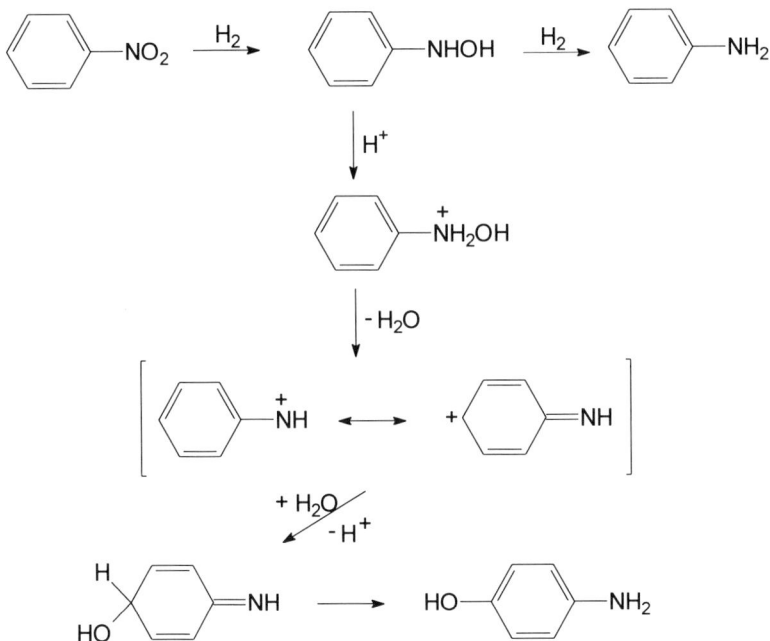

hydrogen pressures. Platinum is essentially insensitive to the presence of the chlorine containing impurities and gives high yields of o-anisidine, **19**, with minimal hydrazo formation.[41] Palladium, however, was more sensitive to these impurities and gave o-anisidine only on hydrogenation of purified o-nitroanisole. Otherwise, significant yields of the hydrazo compound were obtained.[41] For this reason a palladium catalyst was chosen for the selective production of o-hydrazotoluene (**20**) by the partial hydrogenation of o-nitrotoluene (Eqn. 19.20).[42] This reaction was carried out at 60°C and 4 atmospheres in the presence of sodium hydroxide and a quinone modifier. The optimum reaction conditions were the use of a THF/aqueous ethanol solvent to keep both products and reactants in solution, a 1:1 quinone to catalyst ratio and about a 6% NaOH concentration.

Since the condensation of the hydroxylamine intermediate with its precursor nitrosobenzene is a key step in generating coupled intermediates the presence of DMSO in the reaction mixture increased hydrazo compound formation but did so at the expense of a significant decrease in reaction rate.[42,43]

(19.19)

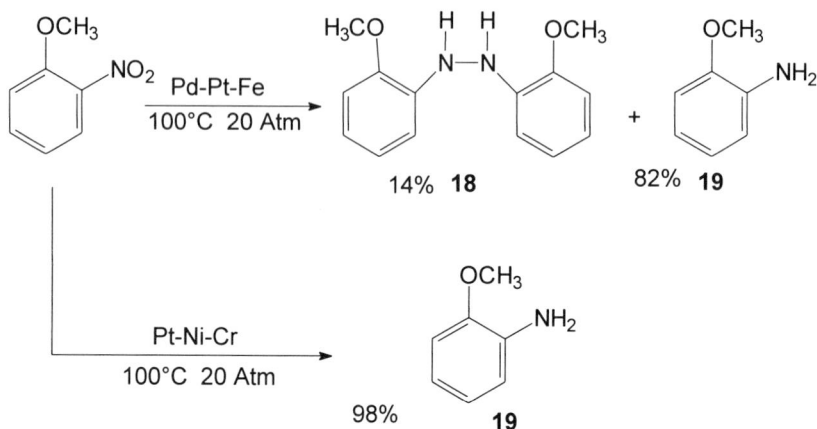

14% **18**   82% **19**

Pt-Ni-Cr

100°C 20 Atm

98%   **19**

(19.20)

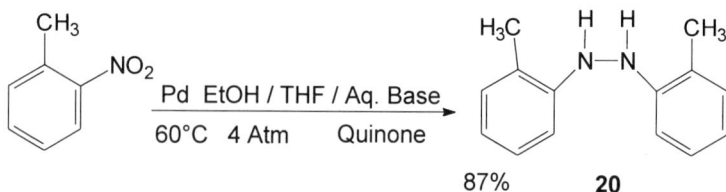

87%   **20**

The use of platinum catalysts in the presence of DMSO and aqueous alkali resulted in increased hydrazobenzene selectivity. Electron releasing substituents on the aromatic ring decreased selectivity toward hydrazobenzene formation.[43]

Aliphatic Nitro Groups

Aliphatic nitro compounds are hydrogenated under the same conditions used for the reduction of aromatic nitro groups. They follow the same reaction path as illustrated in Scheme 19.1 for aromatic nitro compounds,[44] but they are hydrogenated somewhat less readily than the aromatic species (Eqn. 19.21).[45] Double bonds are hydrogenated in preference to the aliphatic nitro group but not carbonyl or oximino groups.[46,47]

   Use is made of the facile hydrogenation of nitro groups to prepare α-hydroxy primary amines from nitromethane-ketone addition products (Eqn. 19.22).[48,49] The addition of a little acetic acid to the reaction medium can sometimes improve the yield. The β-hydroxyamines are more easily prepared in this way than by the hydrogenation of the corresponding cyanohydrin. α-Chloro nitro compounds are selectively hydrogenated to the oxime over palladium under

(19.21)

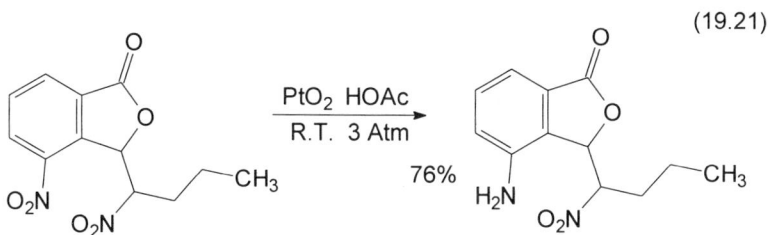

(19.22)

(19.23)

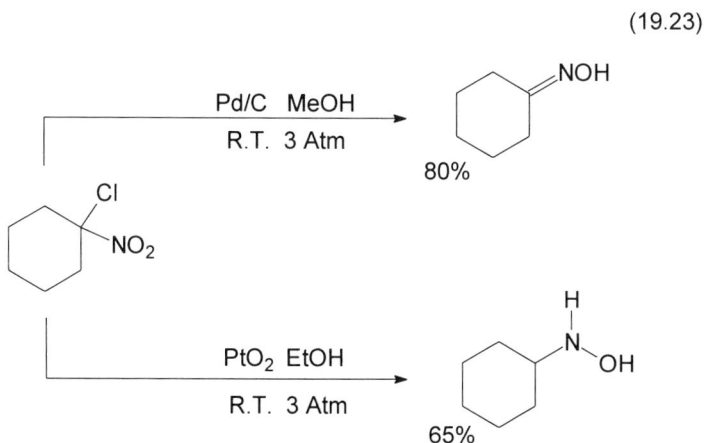

mild conditions if the reaction is stopped after two equivalents of hydrogen have been absorbed (Eqn. 19.23.).[50]   Hydrogenation over PtO$_2$ gave the hydroxylamine.[50]

The hydrogenation of aliphatic nitro groups takes place with retention of configuration when the reaction is run under the normal conditions. The basic nature of Raney nickel does not promote epimerization of the nitro group before hydrogenation at low temperatures but the alkaline impurity found in platinum oxide does. The use of platinum oxide in ethanol resulted in the formation of a mixture of product stereoisomers but, in acetic acid, retention of configuration was observed. Higher temperatures also promoted inversion (Eqn. 19.24).[51]

(19.24)

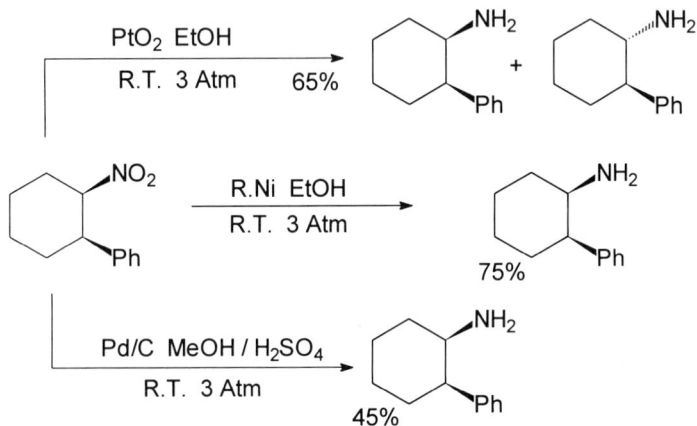

As with aryl nitro groups, aliphatic nitro compounds can be partially hydrogenated to nitrosamines which can rearrange to oximes and further hydrogenated to produce hydroxylamines. The hydrogenation of nitrocyclohexane over platinum gave cyclohexanone oxime in 70% yield while the use of a palladium catalyst led to cyclohexylhydroxylamine formation.[52] Hydrogenation of the nitroazaadamantane, **21**, over Raney nickel gave the amine, **22**, in 88% yield but hydrogenation over palladium gave the hydroxylamine, **23** (Eqn. 19.25).[53]

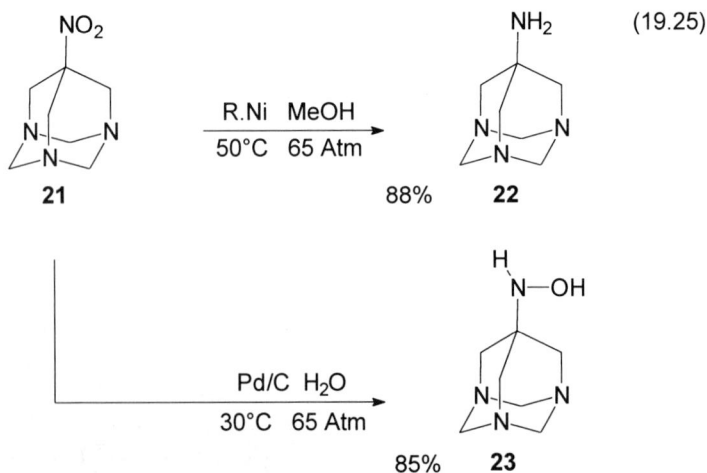

(19.26)

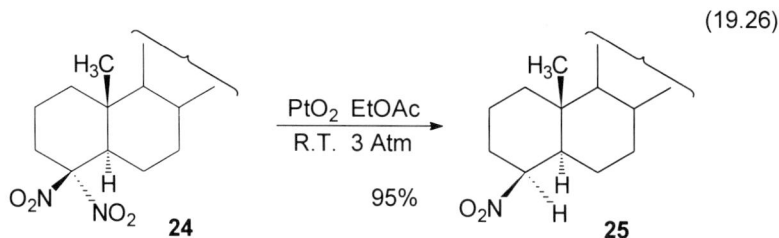

The partial hydrogenation of gem-dinitro compounds presumably proceeds through the formation of a non-isolated α-amino nitro moiety which is then hydrogenolyzed to give the mononitro species.[54] Hydrogenation of 4,4-dinitro-chlolestane (**24**) over platinum at room temperature and three atmospheres gave 4β-nitrocholestane (**25**) in 90% yield (Eqn. 19.26).[54]   The product stereochemistry depended on the position of the dinitro groups on the steroid ring.[54]

Excellent yields of the oximes are formed by the hydrogenation of vinyl or styryl nitro compounds under the usual mild conditions over a palladium catalyst deactivated with an equal weight of pure pyridine (Eqn. 19.27)[55] or in methanolic HCl.[56]   The reaction must be stopped after the absorption of two molar equivalents of hydrogen since further hydrogenation can take place under these conditions to give the saturated amine. Running the reaction in aqueous acetic acid, though, resulted in hydrolysis of the oxime to give the ketone.[57] Hydrogenation of steroidal vinyl nitro compounds over platinum, however, resulted in preferential double bond saturation giving the saturated nitro compound.[54]

(19.27)

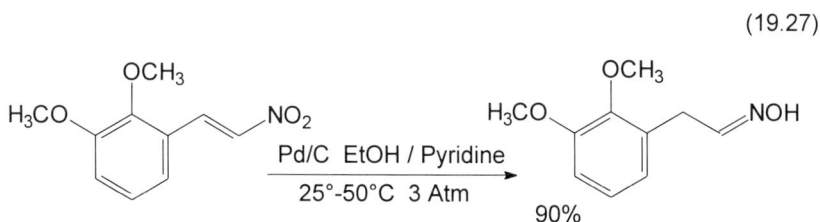

Hydrogenative Cyclization

If other active groups are γ or δ to the nitro group, hydrogenation can result in the formation of cyclic products. Esters condense with the product amine to give lactams. Ketones or aldehydes produce imines that can be hydrogenated further to an amine. The hydrogenation of γ- or δ-nitro carboxylic acids under mild conditions, however, gives the amino acids in excellent yields.[58,59]   The 2,2'-nitrobiphenyl esters[60] and related compounds[61,62] give the cyclic lactams on hydrogenation under mild conditions (Eqn. 19.28).[61] The γ-nitronitrile, **26**, was

(19.28)

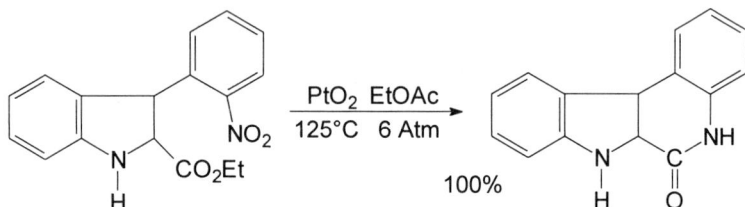

hydrogenated over Raney nickel or palladium under mild conditions to give the hydroxy-amidine, **27**. Further hydrogenation of **27** under more severe conditions led to the formation of the pyrrolidine, **28**.[63] Hydrogenation of **26** over Raney nickel at high temperature and pressure gave the pyrrolidone, **29** (Eqn. 19.29).[64]

(19.29)

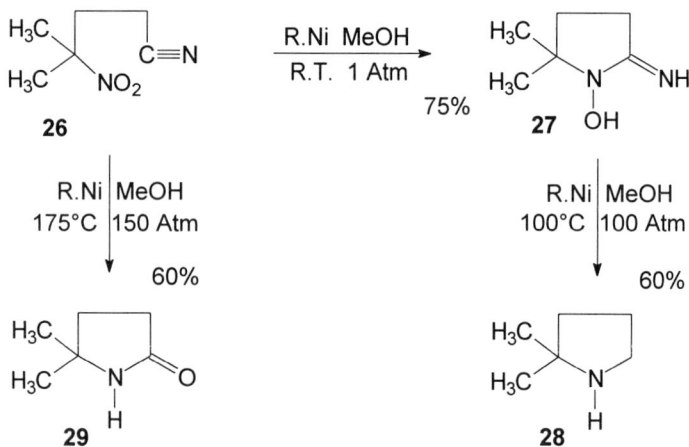

Nitroso, Azoxy and Azo Compounds

Nitroso, azoxy and azo groups are also reduced under the conditions used to effect the hydrogenation of nitro compounds. Nitroso compounds form the amine with about the same ease as is observed with the corresponding nitro compound.[65] α-Nitroso-chlorides are hydrogenated to oximes over platinum if the reaction is stopped when the color disappears (Eqn. 19.30).[66] No other catalyst is effective for this reaction, which is also superior to all chemical reduction methods. The use of other catalysts, as well as further hydrogenation over platinum, results in amine formation.[66]

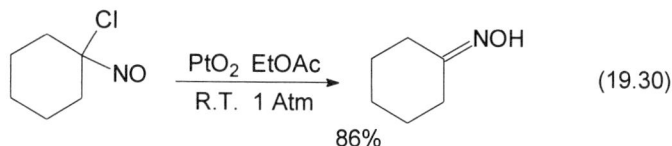

$$(19.30)$$

86%

The hydrogenation of azoxy benzenes occurs in a stepwise fashion to give the azo and hydrazo compounds with final hydrogenolysis to the amine. The reaction can be stopped at any of these stages over W2 Raney nickel at room temperature and atmospheric pressure.[67,68] With platinum oxide the hydrazine is obtained in very good yield.[69] Hydrogenolysis to the amine occurs over all common metal catalysts but takes place most readily with palladium.

Olefins, acetylenes, nitro, and nitroso groups are saturated during this reaction but ketones are not affected. Aromatic chlorides are also cleaved[70] but this can be inhibited by running the hydrogenation in acid over a platinum catalyst. Even under these conditions, however, aryl bromides and iodides are lost, at least to some extent.

Azides and Diazo Compounds

Primary amines result from the hydrogenation of azides at room temperature and 3–4 atmospheres over platinum oxide or palladium catalysts.[71,72] The addition of acid is sometimes beneficial.[71] Raney nickel is a less efficient catalyst for this reaction.

Azides are hydrogenated with retention of configuration (Eqn. 19.31).[73] This, in conjunction with the nucleophilic character of the azide ion, provides an efficient method for the stereospecific introduction of a primary amine group into a molecule.

$$(19.31)$$

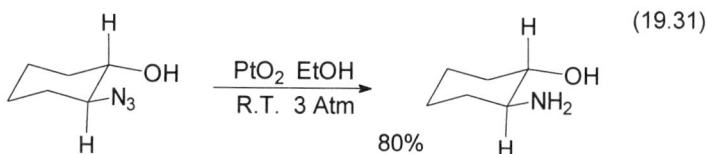

80%

Aryl diazomethanes and ethyl diazoacetate lose nitrogen completely when hydrogenated over palladium in neutral medium at room temperature and atmospheric pressure. Aryl diazoketones, however, under the same conditions, give the α-amino ketones which condense directly to the dihydropyrazines. The addition of acetic acid to the solvent for this later reaction permits the isolation of the amino ketone (Eqn. 19.32). With dilute hydrochloric acid in the reaction mixture, complete loss of the nitrogen was observed.[74] The hydrogenation of alkyl diazoketones in neutral or acetic acid solution results in hydrazone formation (Eqn. 19.33).[74]

(19.32)

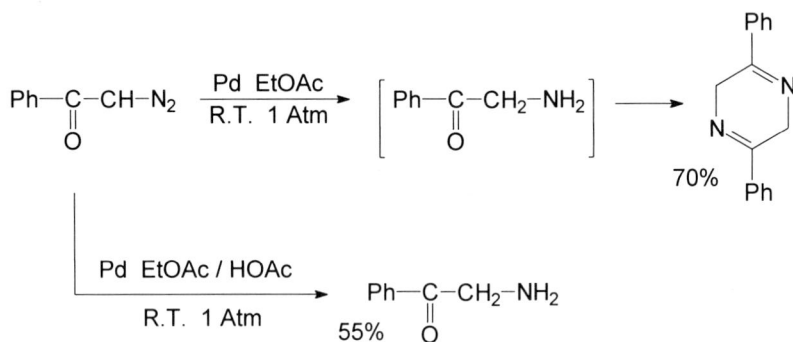

(19.33)

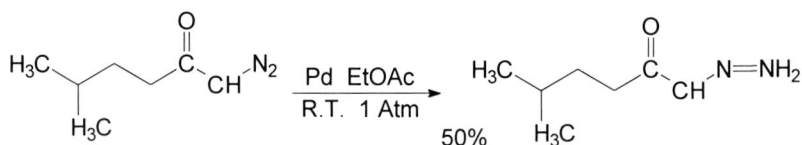

Nitriles

Acetylenes, alkenes, and nitro groups are usually more easily hydrogenated than nitriles, but aldehydes, ketones, and benzene rings are generally more difficult to reduce under normal reaction conditions. The hydrogenation of the double bond in acrylonitrile and other unsaturated nitriles is best accomplished using palladium

(19.34)

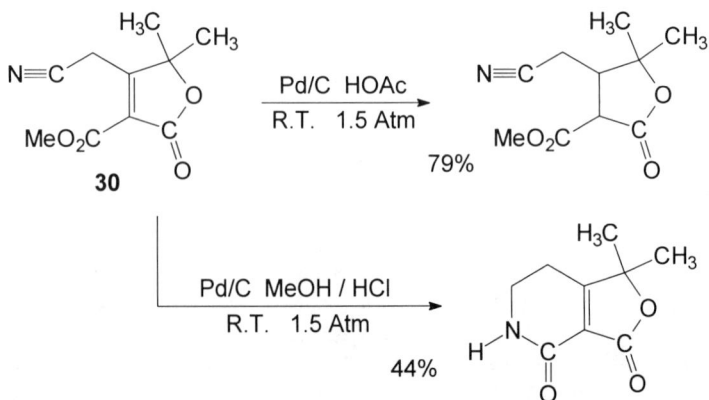

or nickel catalysts. Vapor-phase hydrogenation of acrylonitrile over Ni/SiO$_2$-Al$_2$O$_3$ at 100°C gave only propionitrile as the product. Nitrile hydrogenation to the amine was observed at higher temperatures.[75] The selective hydrogenation of the tetrasubstituted double bond in **30** occurred over Pd/C in acetic acid but when the hydrogenation was run in methanolic HCl, the nitrile was preferentially saturated (Eqn. 19.34).[76] Hydrogenation of oleic acid nitrile over Raney nickel in aqueous alcohol containing some sodium carbonate gave the unsaturated amine in very good yield (Eqn. 19.35).[77]

(19.35)

$$H_3C(CH_2)_7HC{=}CH(CH_2)_7C{\equiv}N \quad \xrightarrow[\text{60°C 1 Atm}]{\text{R.Ni EtOH / H}_2\text{O / Na}_2\text{CO}_3}$$

90% $H_3C(CH_2)_7HC{=}CH(CH_2)_7CH_2NH_2$

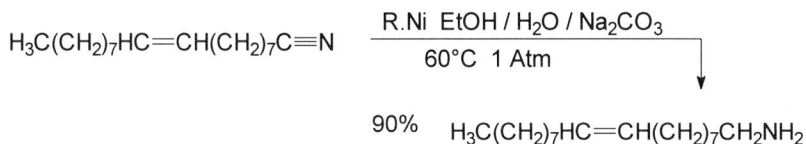

No selectivity was observed in the hydrogenation of β-iminonitriles over Raney nickel at 120°–150°C and 200–300 atmospheres with the diamines obtained in fair to good yields.[78] β-Ketonitriles were hydrogenated to the amino ketones,[79] but γ-ketonitriles gave the pyrrolidines,[80] and δ-ketonitriles gave the piperidines[81-84] as a result of the hydrogenative cyclization of the initially formed keto amines. The hydrogenation of oximino nitriles over platinum in acetic acid at room temperature and atmospheric pressure led to the preferential saturation of the cyano group (Eqn. 19.36).[85] In acetic anhydride, however, the oxime acetate was selectively hydrogenated.[86]

(19.36)

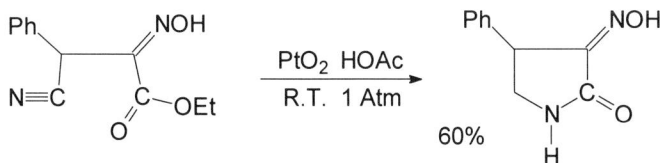

Mononitriles

Although the hydrogenation of nitriles is generally expected to produce the primary amine, selectivity in the formation of such products is frequently lowered by the production of secondary and tertiary amines. As depicted in Scheme 19.3, these products are formed by the condensation of an imine intermediate (**31**) with the product amines to give 1,1-diamino species (**32** and **33**), which are hydrogenolyzed to give the secondary (**34**) or tertiary amine, **35**.

Scheme 19.3

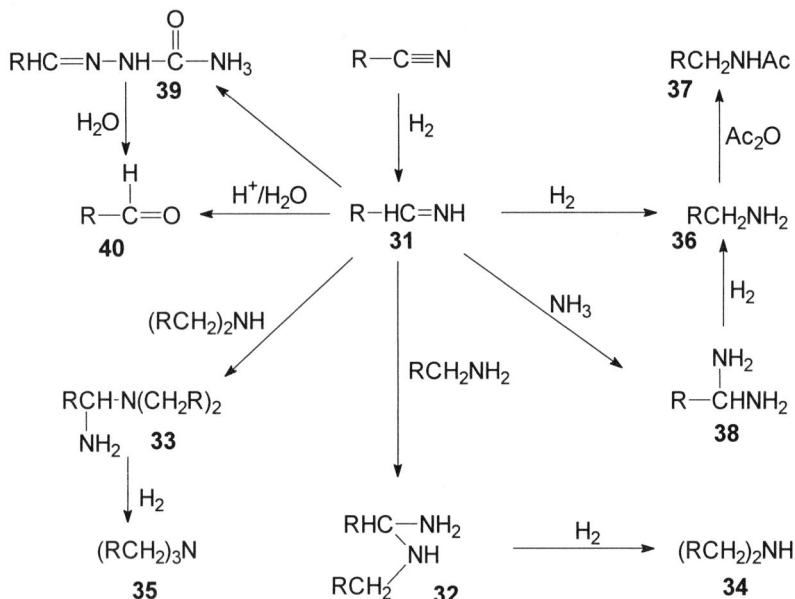

Primary Amine Formation

The direct formation of the primary amine, **36**, is best accomplished using Raney cobalt at 25°–50°C and 3–5 atmospheres.[87] Apparently the imine is not desorbed from Raney cobalt, so the imine reduces rapidly and is not available for condensation with other species. Nickel boride and cobalt boride catalysts behave in a similar way for the direct production of primary amines by the hydrogenation of nitriles.[88,89] Platinized Raney nickel is particularly effective for the hydrogenation of hindered nitriles.[90] High hydrogen pressure or other factors leading to high hydrogen availability to the catalyst also increase the possibility of imine saturation before condensation and, thus, lead to the formation of more primary amine.[91,92]

Other approaches to primary amine formation have been developed. Running the hydrogenation in strong acid solutions gives a protonated primary amine that cannot react further.[93,94] The use of $NaOAc/Ac_2O$ in the reaction medium to trap the primary amine, **36**, as the amide, **37**, accomplishes the same thing.[95] In hydrogenations run over Raney nickel in alcoholic ammonia solutions, the imine reacts with the ammonia to form the diamine, **38**, which is hydrogenolyzed to give the primary amine.[96] Aqueous ammonium hydroxide can also be used.[97] The ammonia competes favorably with the primary amine product for reaction with the imine so secondary amine formation is suppressed.[98-103]

The use of ammoniacal solutions in rhodium catalyzed hydrogenations of nitriles also leads to primary amine formation.[104] In the absence of ammonia, secondary amine formation predominates in rhodium catalyzed reactions. The hydrogenation of valeronitrile over Raney nickel at 80°C and 30 atmospheres gave a mixture of pentyl amine, dipentyl amine and tripentyl amine. Primary amine formation was favored by running the reaction at lower pressures and temperatures in the presence of large amounts of ammonia.[105]   Doping the Raney nickel with chromium also improved the yield of the primary amine.[106] In the vapor-phase hydrogenation of acetonitrile over Ni/Al_2O_3 selectivity toward primary amine formation was improved by adding an alkali modifier such as Na_2CO_3 to the catalyst.[107]

Cyanohydrins and α-Amino Nitriles

Cyanohydrins can be converted to the β-hydroxyamines by hydrogenation over platinum oxide in acetic acid at 30°–50°C and 2–3 atmospheres of hydrogen (Eqn. 19.37).[108]   The addition of a small amount of a strong acid is sometimes beneficial.[109] β-Hydroxy primary amines, however, are more easily obtained by the hydrogenation of nitromethane-ketone adducts as described above. 1,2-Diamino compounds have been prepared in good yields by the hydrogenation of α-amino nitriles under these same conditions.[110] With a Raney nickel catalyst the poison, HCN, was formed by hydrogenolysis. This difficulty was overcome by first acetylating the amino group and then hydrogenating the acetamido nitrile. The imidazoles that were formed were hydrolyzed to the diamine in good overall yields (Eqn. 19.38).[111] The presence of secondary or tertiary amine groups next to the nitrile results in an increase in amine hydrogenolysis under these reaction conditions.[110]

(19.37)

H₃C–N OH [structure] –CN → PtO₂ HOAc, 40°C 2 Atm → H₃C–N OH [structure] –CH₂NH₂

72%

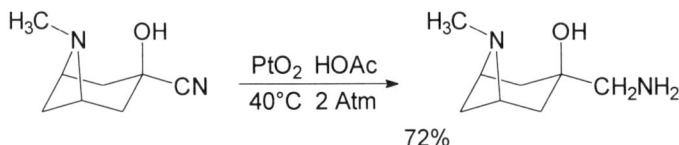

Secondary and Tertiary Amine Formation

Nitrile hydrogenation can also be used for the selective formation of secondary and tertiary amines when the proper reaction conditions are used.[92] With short chain aliphatic nitriles a rhodium[91,112] or rhodium boride[88] catalyst gave the secondary amines as the predominant products, while palladium, palladium boride, platinum or platinum boride produced the tertiary amines.[88,113,114] In the hydrogenation of long-chain nitriles, steric factors apparently become important in inhibiting amine-imine reactions. Lower yields of secondary and tertiary amines are obtained in these reactions.[91] The hydrogenation of lauronitrile over a cobalt catalyst at 160°C and 40 atmospheres of hydrogen gave the primary amine in high

(19.38)

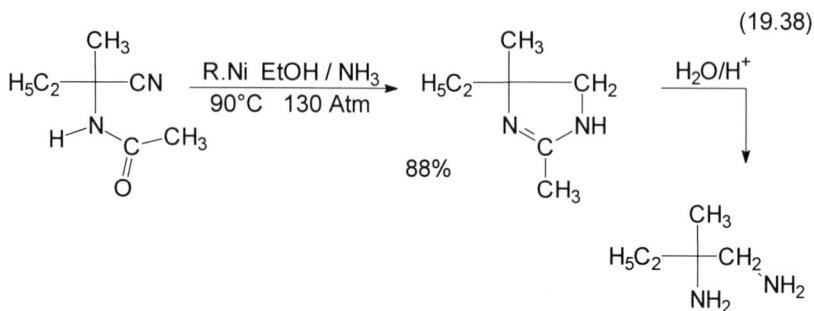

selectivity while over nickel the product was composed of a mixture of the primary, secondary and tertiary amines.[115]

Hydrogenation of aromatic nitriles over platinum or rhodium catalysts produced the secondary amines with good selectivity. With palladium the primary amine was the main product when octane or ethanol was used as the solvent, but in benzene, palladium gave the secondary amines.[91]

(19.39)

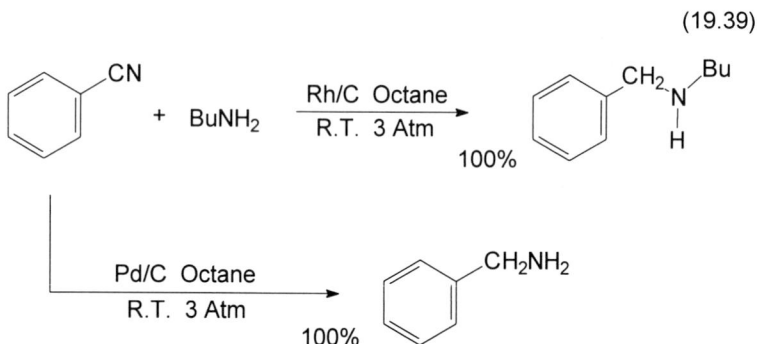

In the presence of an added amine, mixed secondary amines are the predominant or even exclusive product. Hydrogenation of benzonitrile in the presence of butylamine in octane gave butylbenzylamine over Pt/C or Rh/C but only benzylamine over Pd/C (Eqn. 19.39).[91] Solvents such as ethanol or benzene can interact with the catalyst and thus change its adsorption characteristics as well as influence the reactivity of the imine intermediate. The product composition is not straightforward in these solvents.[91] The hydrogenation of the aliphatic nitrile, valeronitrile, in the presence of butylamine gave mainly the mixed secondary amine product over Pd/C, Pt/C, and Rh/C.[91] Hydrogenation of benzonitrile in the presence of diethylamine gave only benzylamine regardless of the catalyst or solvent. It would appear that the secondary amine was too large and steric factors precluded its condensation with the imine. However, hydrogenation of p-amino-

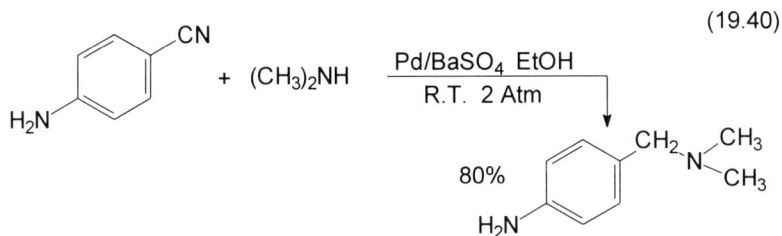

(19.40)

phenylacetonitrile in the presence of dimethylamine gave the tertiary amine in 80% yield (Eqn. 19.40).[116]

Hydrogenation of dodecylnitrile at 250°C and 50 atmospheres pressure in methanol over Cu-CrO and other supported copper catalysts gave good yields of the dimethyl amine arising from the alkylation of the product amine with the alcohol solvent at the high temperatures used.[117,118]

Reductive Hydrolysis

Advantage can be taken of the reactivity of the imine intermediate, **31** (Scheme 19.3), for the reductive hydrolytic conversion of nitriles to aldehydes. If a carbonyl reagent such as hydrazine[119] or semicarbazide[120,121] is present in the hydrogenation mixture, it can react with the imine to form the aldehyde hydrazone or semicarbazone, **39**, which are more difficult to hydrogenate than the imine. It can, therefore, be isolated and hydrolyzed to give the aldehyde, **40**. Alternately, running the hydrogenation over Raney nickel in strongly acidic media can result in the hydrolysis of the imine to produce the aldehyde, **40**, directly.[122] To avoid the problems caused by the dissolution of nickel in the acid media, supported palladium catalysts were used but the yields of aldehyde were not very good[123] unless the catalyst was first treated with a modifier such as Cu^{++} or Ag^+. Propionaldehyde was produced in near 80% yield on hydrogenation of propionitrile over a modified palladium catalyst in acidic media at 0°C and under conditions of low hydrogen availability to the catalyst.[124]

Dinitriles

The hydrogenation of a dinitrile interjects an additional degree of selectivity, that of preferentially saturating one of the nitrile functions to give an amino-nitrile product. As shown in Eqn. 19.41, the hydrogenation of 1,3-dicyanobenzene (**41**) over a Pt-Ni/Al$_2$O$_3$ catalyst at 90°C and 220 atmospheres in the presence of ammonia gave 97% of the cyanobenzylamine, **42**. Raising the reaction temperature to 120°C gave a 99% yield of the dibenzylamine, **43**. Similar results were found with the para dinitrile.[125,126] An 80% yield of the p-amino nitrile was obtained using Pd/Al$_2$O$_3$ at 75°C and 100 atmospheres pressure.[127] Hydrogenation of the o-dinitrile over Raney nickel gave the cyclization product, 2,3-dihydro-lH-isoindole (**44**), in very good yield (Eqn. 19.42).[128]

The selective hydrogenation of aliphatic dinitriles to amino nitriles appears to be best accomplished using rhodium catalysts. Succinonitrile (**45**) was

(19.41)

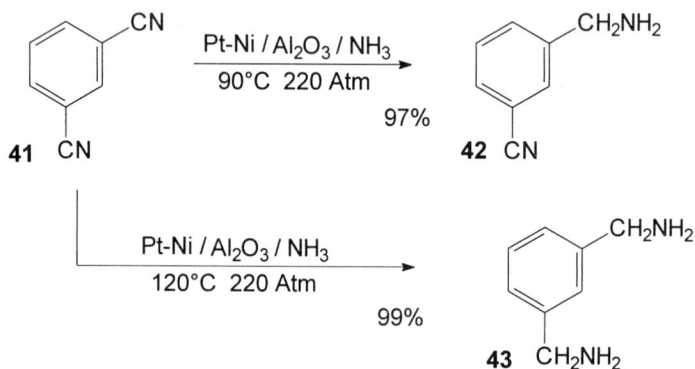

41 → 42 (97%, Pt-Ni / Al$_2$O$_3$ / NH$_3$, 90°C 220 Atm)

41 → 43 (99%, Pt-Ni / Al$_2$O$_3$ / NH$_3$, 120°C 220 Atm)

converted to the aminobutyronitrile (**46**) in 90–94% yield over 5% Rh/Al$_2$O$_3$ at 15°C and 150 atmospheres in the presence of gaseous ammonia (Eqn. 19.43).[129] The partial hydrogenation of adiponitrile (**47**) (Eqn. 19.44) was accomplished using Rh$_2$O$_3$ at 75°C and 102 atmospheres in ammoniacal MeOH/THF[130] or Rh/MgO at 100°C and 34 atmospheres.[131,132] In the latter reaction the aminonitrile, **48**, was obtained in 93% selectivity at 74% conversion. High selectivity for aminonitrile formation was also observed using Raney nickel in methanol containing sodium hydroxide or ammonia.[133]

In the hydrogenation of aliphatic dinitriles, cyclization can also be an important reaction pathway.[92] Saturation of succinonitrile, **45**, (Eqn. 19.45) over Raney nickel gave pyrrolidine, **49**, as the primary product even in the presence of ammonia.[134] Intramolecular condensation of the imine-amine intermediate is apparently a very facile reaction, taking place in preference to imine hydrogenation and the intermolecular ammonia-imine condensation. The

(19.42)

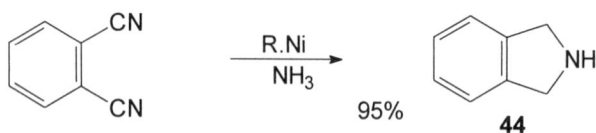

(19.43)

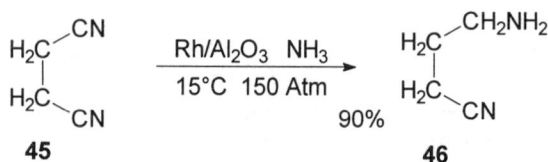

(19.44)

$$N\equiv C(CH_2)_4C\equiv N \quad \xrightarrow[\text{100°C 35 Atm}]{\text{Rh/MgO NH}_3} \quad N\equiv C(CH_2)_4CH_2NH_2$$

47   70%   **48**

(19.45)

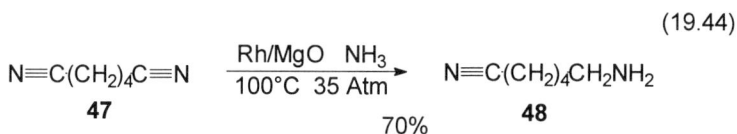

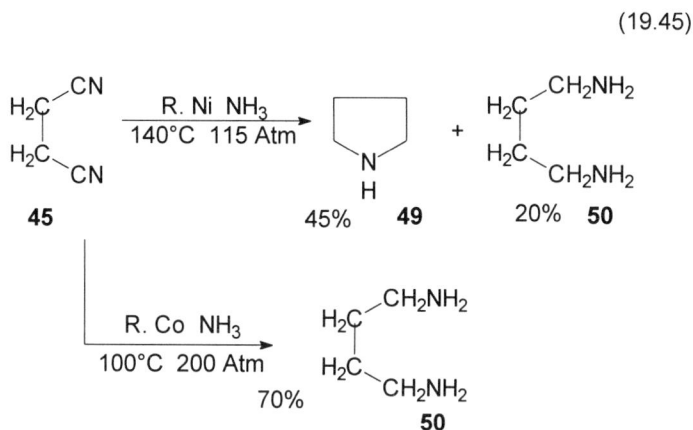

diamine, **50**, could be prepared by running the hydrogenation over Raney cobalt at 200 atmospheres and 100°C in the presence of a large amount of ammonia.[135] Selectivity can also be improved by hydrogenating very dilute solutions of the dinitrile.[136] The use of a manganese-modified Raney cobalt for this reaction gave **50** in 97% yield.[137] Hydrogenation of longer chain dinitriles is complicated by the number of intra- and inter-molecular condensation intermediates that can be present in the reaction mixture.[138-140] This is particularly true in the hydrogenation of glutaronitrile and adiponitrile where the piperidines and azacycloheptanes are relatively easily formed by condensation of the intermediate amino-imines.[141,142] The hydrogenation of adiponitrile (**47**) (Eqn. 19.46) over barium-promoted copper chromite and very finely powdered TiO_2 at 270°C and 1.4 atmospheres, with ammonia present in the reaction stream, gave caprolactam (**51**) in 91% selectivity at 85% conversion.[143]

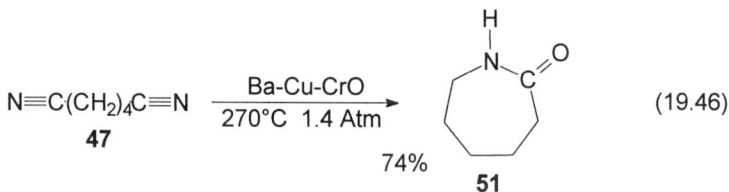

$$N\equiv C(CH_2)_4C\equiv N \quad \xrightarrow[\text{270°C 1.4 Atm}]{\text{Ba-Cu-CrO}} \quad$$

47   74%

(19.46)

51

(19.47)

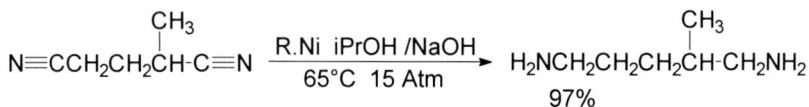

$$N\equiv CCH_2CH_2\underset{\underset{CH_3}{|}}{CH}-C\equiv N \xrightarrow[\text{65°C   15 Atm}]{\text{R.Ni iPrOH /NaOH}} H_2NCH_2CH_2CH_2\underset{\underset{CH_3}{|}}{CH}-CH_2NH_2$$

97%

(19.48)

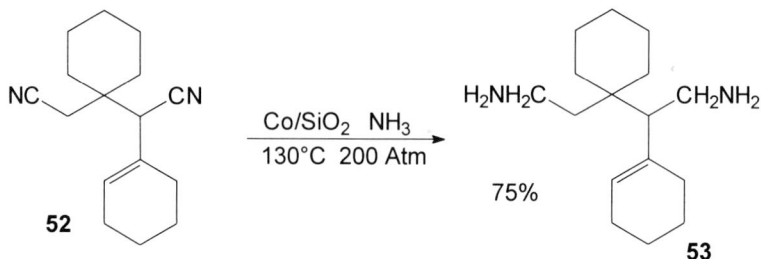

NC.   CN

$\xrightarrow[\text{130°C   200 Atm}]{\text{Co/SiO}_2 \text{ NH}_3}$

H₂NH₂C.   CH₂NH₂

75%

52

53

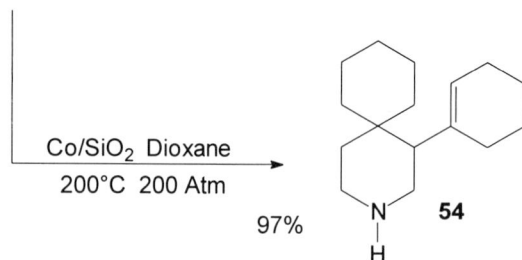

$\xrightarrow[\text{200°C   200 Atm}]{\text{Co/SiO}_2 \text{ Dioxane}}$

97%

54

Hydrogenation to the diamine is usually accomplished using a Raney nickel catalyst in alcoholic solvents at 65°–100°C and 15 atmospheres pressure (Eqn. 19.47).[141]   The presence of sodium hydroxide in the reaction mixture increases the selectivity for diamine formation.[141]   The hydrogenation of the substituted glutaronitrile, **52**, over a cobalt catalyst in ammonia gave the diamine, **53**, in 75% yield. In the absence of ammonia, the substituted piperidine, **54**, was formed (Eqn. 19.48).[144] Longer chain dinitriles are hydrogenated to the diamines in good yield using a Raney nickel catalyst in ammonaical ethanol at room temperature and 4 atmospheres.[145]

Imines and Reductive Amination

The hydrogenation of an imine or enamine generally takes place at room temperature and atmospheric pressure using palladium or Raney nickel catalysts giving good to excellent yields of the amine product. Quaternizing the nitrogen facilitates the reaction.[146,147] Hydrogenation of 4-aza-choles-4-ene (**55**) over platinum gave the 4-aza-5α-cholestane (**56**) (Eqn. 19.49)[148] rather than the 5β-isomer, which is preferentially formed on hydrogenation of cholest-4-ene.[149]

(19.49)

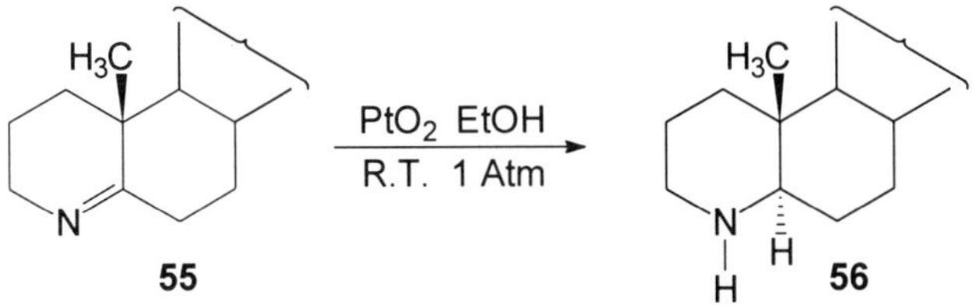

55   **56**

Imines can be saturated in preference to the hydrogenation of aldehydes, ketones or nitriles and the hydrogenolysis of benzyl ethers and amines. Alkenes, alkynes and nitro groups are, however, usually hydrogenated in preference to an imine.

(19.50)

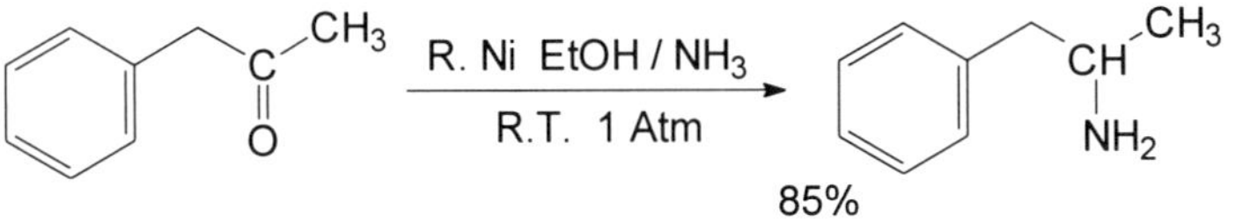

85%

Reductive alkylation of amines proceeds by the hydrogenation of the imine or enamine formed, *in situ*, by the condensation of the amine with a carbonyl compound. This reaction can give a mixture of products if the amine produced initially competes with the reactant amine in the carbonyl condensation step. The proper selection of reagent concentrations avoids this difficulty and leads to the formation of good yields of the desired product.[150] The use of a large excess of ammonia gives the primary amine as the predominant product (Eqn. 19.50).[151] An excess of a primary amine as the reactant leads to the preferential formation of the secondary amine product. An excess of the carbonyl compound gives the symmetrical secondary or tertiary amines (Eqn. 19.51).[150]

(19.51)

$$PhCHO \;+\; NH_3 \xrightarrow[\text{R.T. 1 Atm}]{\text{Pd/C EtOH}} (PhCH_2)_3N$$
(Excess)   60-80%

$$PhCHO \;+\; CH_3NH_2 \xrightarrow[\text{R.T. 1 Atm}]{\text{Pd/C EtOH}} PhCH_2NHCH_3$$
(Excess)   70-90%

$$PhCHO \;+\; CH_3NH_2 \xrightarrow[\text{R.T. 1 Atm}]{\text{Pd/C EtOH}} (PhCH_2)_2NCH_3$$
(Excess)   70-90%

(19.52)

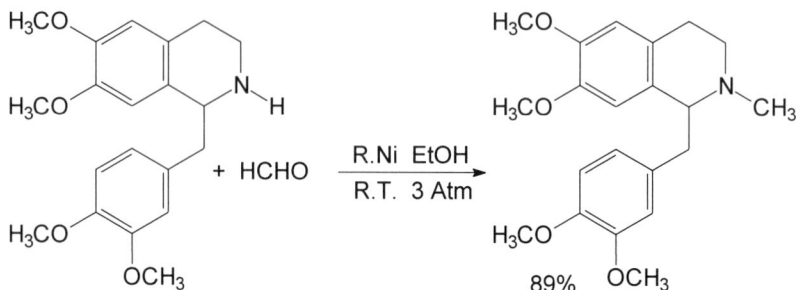

The extent to which secondary and tertiary amines are formed by reductive alkylation is controlled to some degree by the steric bulk of the amine and the carbonyl compound. The more hindered the system, the higher the temperature and hydrogen pressure needed to affect the reaction. While the reductive alkylation of secondary aliphatic amines with formaldehyde takes place under mild conditions (Eqn.19.52),[152] N,N-dimethylaniline was prepared by reductive alkylation over palladium at 120°C and 15 atmospheres pressure (Eqn. 19.53).[153,154] Reductive alkylation of aniline with acetone over palladium gave a 67% yield of the monoalkylaniline at 100°C and 40 atmospheres pressure but secondary amine formation using the more sterically accessible ketones, 2-tetralone or 2-indanone, took place at room temperature and 4 atmospheres pressure (Eqn. 19.54).[155] Palladium was the preferred catalyst in these reactions since with platinum or rhodium ring hydrogenation was also observed.[154,155]

(19.53)

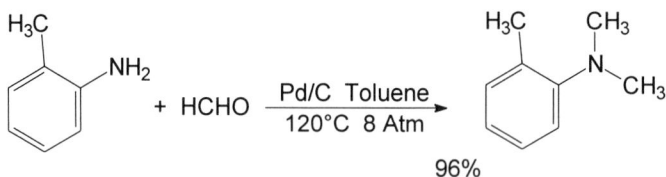

The use of a reductive alkylation for the preparation of tertiary aromatic amines having at least one secondary alkyl group is more difficult and needs higher reaction temperatures and pressures. Under these conditions, though, ring saturation can also take place with the usual metal catalysts. The sulfided platinum metals are, however, particularly useful for such reactions since they do not promote the ring saturation nor C–N or C–O hydrogenolyses, which can significantly lower the yield of the desired aromatic amines.[156,157] The reductive alkylation of aniline with acetophenone over platinum or palladium catalysts gave

(19.54)

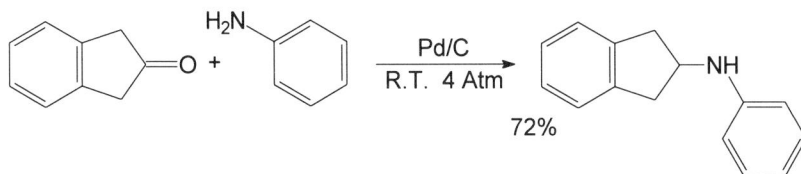

72%

only very low yields of the alkylaniline because of the concomitant hydrogenolysis of the acetophenone to ethyl benzene as well as ring saturation and hydrogenolysis of the product amine. With a sulfided platinum catalyst, however, the alkylated aniline was obtained in 94% yield (Eqn. 19.55).[157]

(19.55)

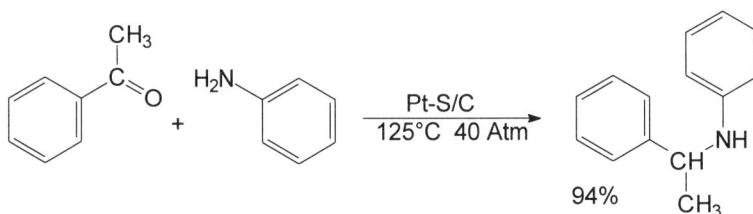

94%

Reductive alkylations with phenylglyoxal (**57**) took place exclusively on the aldehyde carbonyl group when the reaction was run over either palladium[158] or Raney nickel[159] at room temperature and atmospheric pressure (Eqn. 19.56).

(19.56)

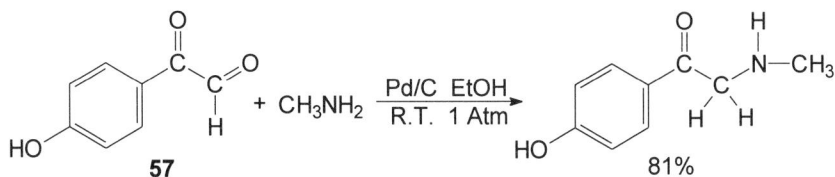

It is not essential for the reaction that the amine be present, as such, before hydrogenation. Any functional group that gives an amine on hydrogenation can be used (Eqn. 19.57).[160] Interestingly, the reductive alkylation of acid hydrazides took place in preference to the hydrogenolysis of the N–N bond when the reaction was run over either platinum or palladium at 75°–100°C and 40 atmospheres pressure (Eqn. 19.58).[161]

(19.57)

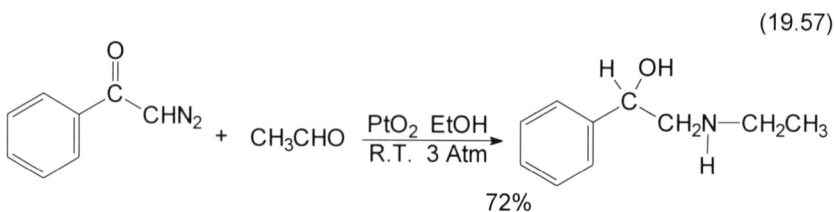

72%

(19.58)

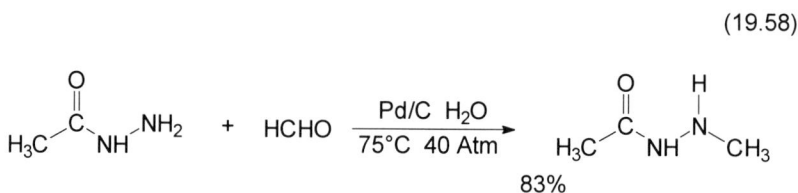

83%

Oximes

The hydrogenation of oximes to the primary amines is usually accomplished using a platinum catalyst at room temperature and 2–3 atmospheres of hydrogen[162] but palladium,[162-164] rhodium[162,165] and Raney nickel[87,166] have also been used effectively under these conditions. The platinum and palladium catalyzed reactions are usually run in acetic acid, frequently with a small amount of hydrochloric acid added to inhibit secondary amine formation. Acetylation of the oxime makes it more easily hydrogenated. If the hydrogenation is run in the presence of 2–3 equivalents of acetic anhydride, not only is the oxime acetylated prior to hydrogenation and, thus, reduced more easily, but the product amine is also isolated as the acetate, precluding any possible secondary amine formation (Eqn. 19.59).[167,168] Oximes, particularly when acetylated, are hydrogenated in preference to aldehydes, ketones, nitriles and pyridine rings. The partial hydrogenation of oximes to hydroxylamines has been accomplished using a platinum catalyst at atmospheric pressure and slightly elevated temperatures (Eqn. 19.60).[169,170]

(19.59)

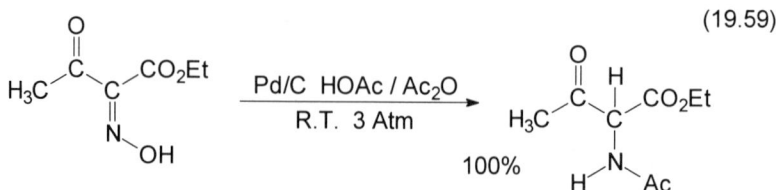

100%

The stereochemistry of the amines obtained on hydrogenation of cyclic oximes is dependent on the pH of the reaction medium in the same way as the formation of alcohols from cyclic ketones, which was discussed in Chapter 18. In

(19.60)

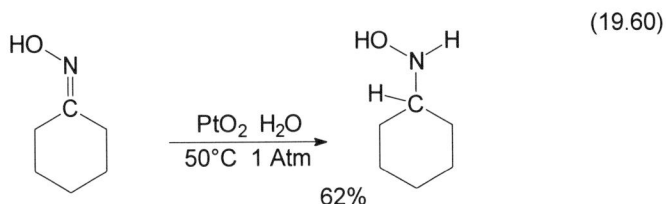

62%

neutral solution, the addition of hydrogen takes place from the least hindered side of the molecule giving mixtures of *cis* and *trans* isomers on hydrogenation of substituted cyclohexanone oximes[171,172] and cyclopentanone oximes.[174] Hydrogenation of the cyclohexanone oximes in acid leads to the primary formation of the axial amine[171,172] while in basic media the formation of the equatorial amine is favored (Eqn. 19.61).[174]

(19.61)

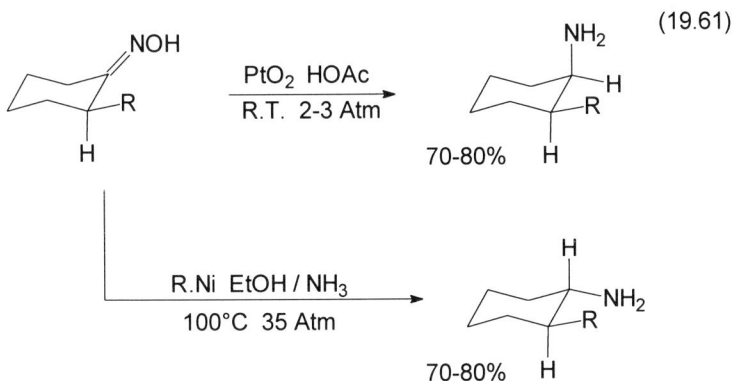

70-80%

70-80%

(19.62)

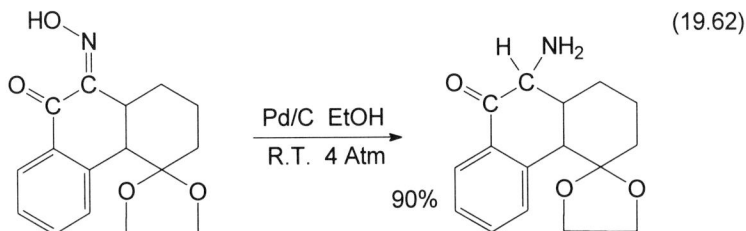

90%

A palladium catalyst is preferred for the selective hydrogenation of α-oximinoketones to the aminoketones. Even aryl carbonyl groups are unaffected in this reaction if it is stopped after the consumption of two equivalents of hydrogen (Eqn. 19.62).[175,176] A strong acid is usually present in the hydrogenation mixture to protonate the product amine and prevent bimolecular condensation to the piperazine. Further hydrogenation of the aminoketone gives the amino alcohol. In cyclic systems these groups are usually *cis* to each other and in acylic systems the *erythro* product is formed.[177,178]

The α- and β-dioximes are hydrogenated to the diamines in good yields under the conditions used for monooxime hydrogenation.[179] The amount of diamine obtained on hydrogenation of a β-dioxime is usually somewhat less that that obtained from hydrogenation of an α-dioxime because some hydrogenolysis of the β-amino group does occur. This is also observed during the hydrogenation of β-aminooximes.[180] The γ-dioximes are hydrogenated to the pyrrolidines (Eqn. 19.63).[181]

(19.63)

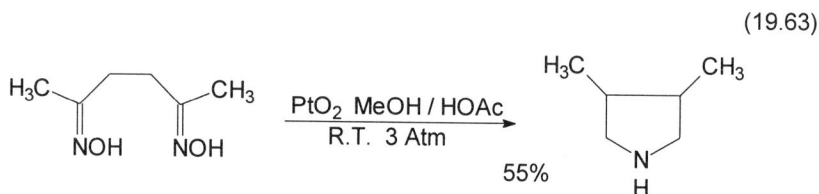

55%

Azines, Hydrazones and Semicarbazones

While azines, substituted hydrazones and semicarbazones are, respectively, more difficult to hydrogenate than oximes, the same catalysts and conditions are used for the hydrogenation of all of these functional groups. The hydrogenation of azines takes place in a stepwise fashion over platinum with the hydrazone formed on absorption of one equivalent of hydrogen (Eqn. 19.64).[182] The selective hydrogenation of semicarbazones to semicarbazides, and azines and hydrazones to hydrazines takes place over platinum and nickel, but less readily over palladium. Further hydrogenolysis to the primary amine, while occurring over all catalysts, is particularly favored with palladium. Since the primary amine is usually the

(19.64)

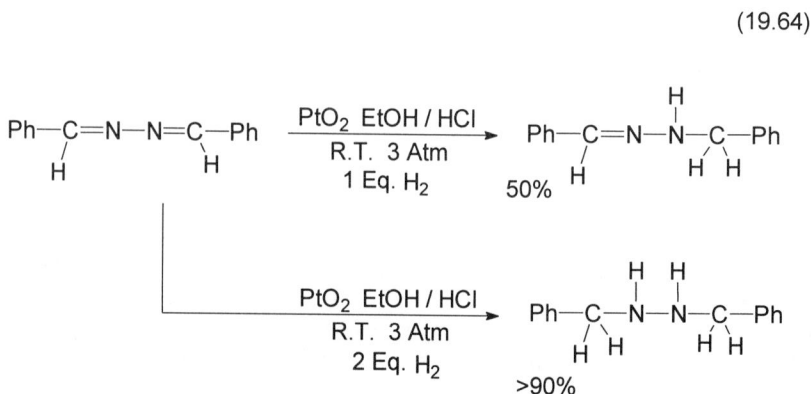

50%

>90%

(19.65)

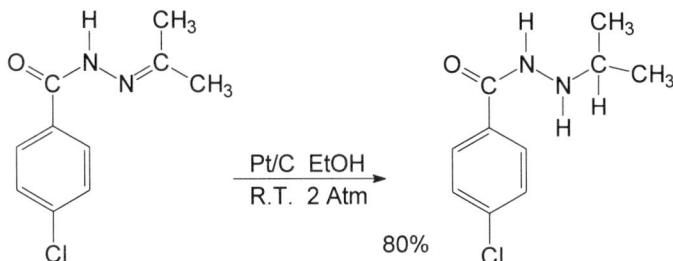

desired product, the hydrogenation is usually not stopped at the semicarbazide or hydrazine stage.

The hydrogenation of azines, hydrazones and semicarbazones takes place in preference to the reduction of aryl and alkyl carbonyl groups and nitriles but not before the saturation of alkenes, alkynes or nitro groups occurs. No hydrogenolysis of aryl fluorides or chlorides occurs during the hydrogenation of these groups, particularly if the reaction is run in acidic media. Aryl bromides are somewhat hydrogenolyzed and iodides are cleaved to a greater extent (Eqn. 19.65).[183]

References

1. J. A. Berson and T. Cohen, *J. Org. Chem.*, **20**, 1461 (1955).
2. E. Ochiai, *J. Org. Chem.*, **18**, 534 (1953).
3. T. Ishii, *J. Pharm. Soc., Japan*, **72**, 1315 (1952).
4. T. Ishii, *J. Pharm. Soc., Japan*, **72**, 1317 (1952).
5. E. V. Brown, *J. Am. Chem. Soc.*, **79**, 3565 (1957).
6. A. Onopchenko, E. T. Sabourin and C. M. Selwitz, *J. Org. Chem.*, **44**, 3671 (1979).
7. A. Onopchenko, E. T. Sabourin and C. M. Selwitz, *J. Org. Chem.*, **44**, 1233 (1979).
8. A. Onopchenko, E. T. Sabourin and C. M. Selwitz, *Chem. Ind. (Dekker)*, **18** (Catal. Org. React.), 265 (1984).
9. J. R. Kosak, *Ann. N. Y. Acad. Sci.*, **172**, 175 (1970).
10. N. P. Sokolova, A. A. Balandin, M. P. Maksimova and Z. M. Skul'skaya, *Isv. Akad. Nauk SSSR, Ser. Khim.*, 1891 (1966); *Chem. Abstr.*, **66**, 75497 (1967).
11. B. Coq, A. Tijani, R. Dutartre and F. Figueras, *J. Mol. Catal.*, **79**, 253 (1993).
12. H. S. Broadbent, *Ann. N. Y. Acad. Sci.*, **145**, 58 (1967).
13. H. Greenfield, *Ann. N. Y. Acad. Sci.*, **145**, 108 (1967).
14. C. Moreau, C. Saenz, P. Geneste, M. Breysse and M. Cacroix, *Stud. Surf. Sci. Catal.*, **59** (Heterog. Catal. Fine Chem., II), 121 (1991).
15. Y.-Z. Chen and Y.-C. Chen, *Appl. Catal. A*, **115**, 45 (1994).
16. W. Pascoe, *Chem. Ind. (Dekker)*, **33** (Catal. Org. React.), 121 (1988).
17. J. R. Kosak, *Catalysis in Organic Synthesis* (W. H. Jones, Ed.) Academic Press, 1980, p. 107.

18. W. Bohm and A. Wissner, *Brit. Pat.*, 1,191,610 (1970); *Chem. Abstr.*, **73**, 25103 (1970).
19. K. Isobe, Y. Nakano and M. Fujise, *Ger. Offen.*, 2,441,650 (1973); *Chem. Abstr.*, **83**, 113914 (1975).
20. P. Baumeister and H. U. Blaser, *Stud. Surf. Sci. Catal.*, **59** (Heterog. Catal. Fine Chem., II) 321 (1991).
21. G. Cordier, J. M. Grosselin and R. M. Bailliard, *Chem. Ind. (Dekker)*, **53** (Catal. Org. React.), 103 (1994).
22. O. M. Kut, F. Yeucelen and G. Gut, *J. Chem. Technol. Biotechnol.*, **39**, 107 (1987).
23. Nissan Chemical Industires, *Jpn., Kokai Tokkyo Koho* JP 82 80,344 (1980); *Chem. Abstr.*, **97**, 127241 (1982).
24. R. E. Lyle and J. L. LaMattina, *Synthesis*, 726 (1974).
25. W. H. Jones, W. F. Benning, P.Davis, D. M. Mulvey, P. I. Pollak, J. C. Schaeffer, R. Tull and L. M. Weinstock, *Ann. N. Y. Acad. Sci.*, **158**, 471 (1969).
26. E. S. Lazer, J. S. Anderson, J. E. Kijek and K. C. Brown, *Synth. Commun.*, **12**, 691 (1982).
27. W. H. Brunner and A. Halasz, *U. S. Pat* 3,088,978 (1963).
28. R. J. Alaimo and R. J. Storrin, *Chem. Ind. (Dekker)*, **5** (Catal. Org. React.), 473 (1981).
29. W. H. Jones, S. J. Pines and M. Sletzinger, *Ann. N. Y. Acad. Sci.*, **214**, 150 (1973).
30. P. N. Rylander, I. M.Karpenko and G. R. Pond, *Ann. N. Y. Acad. Sci.*, **172**, 266 (1970).
31. J. R. Kosak, *Chem. Ind. (Dekker)*, **33** (Catal. Org. React.), 135 (1988).
32. W. R. Tong, R. L. Seagraves and R. Weiderhorn, *Chem. Eng. Prog., Loss Prevention*, 2, **71** (1977).
33. S. L. Karwa and R. A. Rajadhyaksha, *Ind. Eng. Chem. Res.*, **26**, 1746 (1987).
34. J. LeLudec, *Ger. Offen.*, 2,455,887 (1975); *Chem. Abstr.*, **83**, 113915 (1975).
35. V. L. Mylroie, D. Allgeier, J. A. P. Diesenroth, D. S. Files, T. J. Linehan, F. T. Melia and P.-W. Tang, *Chem. Ind. (Dekker)*, **18** (Catal. Org. React.), 249 (1984).
36. K. Hata, *Urushibara Catalysts*, University of Tokyo Press, Tokyo, 1971.
37. D. A. Fidler, J. S. Logan and M. M. Boudakian, *J. Org. Chem.*, **26**, 4014 (1961).
38. V. Weinmayr, *J. Am. Chem. Soc.*, **77**, 1762 (1955).
39. A. M. Strätz, *Chem. Ind. (Dekker)*, **18** (Catal. Org. React.), 335 (1984).
40. E. L. Derrenbacker, *U. S. Pat* 4,307,249 (1981); *Chem. Abstr.*, **96**, 122409 (1982).
41. J. R. Kosak, *Chem. Ind. (Dekker)*, **5** (Catal. Org. React.), 461 (1981).
42. C. G. Coe and J. W. Brockington, *Chem. Ind. (Dekker)*, **33** (Catal. Org. React.), 101 (1988).
43. S. L. Karwa and R. A. Rajadhyaksha, *Ind. Eng. Chem. Res.*, **27**, 21 (1988).
44. V. Dubois, G. Jannes, J. L. Dallons and A. Van Gysel, *Chem. Ind. (Dekker)*, **53** (Catal. Org. React.), 1 (1994).
45. J. W. Wilson, III, C. L.Zirkle, E. L. Anderson, J. J. Stehle and G. E. Ullyot, *J. Org. Chem.*, **16**, 792 (1951).

46. D. Jerkel and H. Fisher, *Liebigs Ann. Chem.*, **574**, 85 (1951).
47. H. C. Godt, Jr. and J. F. Quinn, *J. Am. Chem. Soc.*, **78**, 1461 (1956).
48. H. J. Dauben, Jr., H. J. Ringold, R. H. Wade and A. G. Anderson, Jr., *J. Am. Chem. Soc.*, **73**, 2359 (1951).
49. H. Sasai, N. Itoh, T. Suzuki and M. Shibasaki, *Tetrahedron Lett.*, **34**, 855 (1993).
50. J. A. Robertson, *J. Org. Chem.*, **13**, 395 (1948).
51. F. G. Bordwell and R. L. Arnold, *J. Org. Chem.*, **27**, 4426 (1962).
52. P. Guyer and H. J. Merz, *Chimia*, **18**, 144 (1964).
53. E. B. Hodge, *J. Org. Chem.*, **37**, 320 (1972).
54. J. R. Bull, E. R. H. Jones and G. D. Meakins, *J. Chem. Soc.*, 2601 (1965).
55. A. Lindenmann, *Helv. Chim. Acta*, **32**, 69 (1949).
56. W. K. Siefert and Pl C. Condit, *J. Org. Chem.*, **28**, 265 (1963).
57. T. C. Meyers, R. J. Pratt, R. L. Morgan, J. O'Donnel and E. V. Jensen, *J. Am. Chem. Soc.*, **77**, 5655 (1955).
58. W. Theilacker and G. Wendtland, *Liebigs Ann. Chem.*, **570**, 33 (1950).
59. C. Grundmann and W. Ruske, *Chem. Ber.*, **86**, 939 (1953).
60. C. W. Muth, J. R. Elkins, M. L. DeMalto and S. T. Chiang, *J. Org. Chem.*, **32**, 1106 (1967).
61. G. R. Clemo and D. G. I. Felton, *J. Chem. Soc.*, 671 (1951).
62. S. B. Hendi and L. D. Basanagoudar, *Indian J. Chem., Sect. B*, **20B**, 330 (1981).
63. G. D. Bucklen and T. J. Elliot, *J. Chem. Soc.*, 1508 (1947).
64. R. C. Elderfield and H. A. Hageman, *J. Org. Chem.*, **14**, 605 (1949).
65. G. V. Smith, R. Song, M. Gasior and R. E. Malz, Jr., *Chem. Ind. (Dekker)*, **53** (Catal. Org. React.), 137 (1994).
66. E. Müller, H. Metzger and D. Fries, *Chem. Ber.*, **87**, 95 (1954).
67. P. Ruggli and G. Bartusch, *Helv. Chim. Acta*, **27**, 1371 (1944).
68. I. A. Pearl, *J. Org. Chem.*, **10**, 205 (1945).
69. C. M. McCloskey, *J. Am. Chem. Soc.*, **74**, 5922 (1952).
70. W. F. Whitmore and A. J. Revukas, *J. Am. Chem. Soc.*, **62**, 1687 (1940).
71. E. D. Nicolaides, R. D. Westland and E. L. Wittle, *J. Am. Chem. Soc.*, **76**, 2887 (1954).
72. A. K. Ghosh, S. P. McKee, T. T. Duong and W. J. Thompson, *J. Chem. Soc., Chem. Commun.*, 1308 (1992).
73. C. A. VanderWerf, R. Y. Heisler and W. E. McEwen, *J. Am. Chem. Soc.*, **76**, 1231 (1954).
74. L. Birkofer, *Chem. Ber.*, **80**, 83 (1947).
75. T. Tihanyi, K. Varga, I. Hannus, I. Kiricsi and P. Fejes, *React. Kinet. Catal. Lett.*, **18**, 449 (1981).
76. K. Matsuo and T. Arase, *Chem. Pharm. Bull.*, **42**, 715 (1994).
77. G. Reutenauer and C. Paquot, *C. R. Acad. Sci., Ser. C*, **224**, 478 (1947).
78. H. Adkins and G. M. Whitman, *J. Am. Chem. Soc.*, **64**, 150 (1942).
79. R. H. Wiley and H. Adkins, *J. Am. Chem. Soc.*, **60**, 914 (1938).
80. M. A. T. Rogers, *J. Chem. Soc.*, 590 (1943).
81. W. Barr and J. W. Cook, *J. Chem. Soc.*, 438 (1945).
82. H. Henecka, *Chem. Ber.*, **82**, 104 (1949).
83. H. Beyer and K. Leverenz, *Chem. Ber.*, **94**, 407 (1961).
84. J. Cologne and F. Guigues, *Bull. soc. chim., France*, 4308 (1967).
85. J. Stanek and J. Urban, *Coll. Czech. Chem. Commun.*, **15**, 397 (1950).

86. M. Fields, D. E. Walz and S. Rothchild, *J. Am. Chem. Soc.*, **73**, 1000 (1951).
87. W. Reeve and J. Christian, *J. Am. Chem. Soc.*, **78**, 860 (1956).
88. C. Barnett, *Ind. Eng. Chem., Prod. Res. Develop.*, **8**, 145 (1969).
89. T. W. Russell, R. C. Hoy and J. E. Cornelius, *J. Org. Chem.*, **37**, 3552 (1972).
90. J. Decombe, *C. R. Acad. Sc. Paris*, **222**, 90 (1946).
91. P. N. Rylander, L. Hasbrouck and J. Karpenko, *Ann. N. Y. Acad. Sci.*, **214**, 100 (1973).
92. C. DeBellefon and P. Fouilloux, *Catal. Revs.*, **36**, 459 (1994).
93. T. S. Work, *J. Chem. Soc.*, 426 (1942).
94. M. Freifelder and Y. H. Ng, *J. Pharm Sci*, **54**, 1204 (1965).
95. F. E. Gould, G. S. Johnson and A. F. Ferris, *J. Org. Chem.*, **25**, 1658 (1960).
96. C. F. Winans and H. Adkins, *J. Am. Chem. Soc.*, **54**, 306 (1932).
97. A. Albert and D. Magrath, *J. Chem. Soc.*, 678 (1944).
98. W. Huber, *J. Am. Chem. Soc.*, **66**, 876 (1944).
99. A. T. Fuller, I. M. Tonkin and J. Walker, *J. Chem. Soc.*, 633 (1945).
100. P. Ruggli and B. Pujs, *Helv. Chim. Acta*, **28**, 674 (1945).
101. G. Reutenauer and C. Paquot, *C. R. Acad. Sc. Paris*, **233** 578 (1946).
102. W. Reeve and W. M. Eareckson, III, *J. Am. Chem. Soc.*, **72**, 3299 (1950).
103. E. Baltazzi, *C. R. Acad. Sci., Ser. C*, **233**, 491 (1951).
104. M. Freifelder, *J. Am. Chem. Soc.*, **82**, 2386 (1960).
105. M. Besson, J. M. Bonnier and M. Joucla, *Bull. soc. chim., France*, 5 (1990).
106. M. Besson, D. Djaouadi, J. M. Bonnier, S. Hamar-Thibault and M.Joucla, *Stud. Surf. Sci. Catal.*, **59** (Heterog. Catal. Fine Chem., II), 113 (1991).
107. P.Popescu, A. Buzas, S. Serban, V. Levinta and C. Fagarasan, *Rev. Chim. (Bucharest)*, **22**, 327 (1971): *Chem. Abstr.*, **77**, 129308 (1972).
108. A. C. Cope, H. R. Nace and L. L. Estes, Jr., *J. Am. Chem. Soc.*, **72**, 1123 (1950).
109. B. Tschoubar, *Bull. soc. chim., France*, 160 (1949).
110. M. Freifelder and R. B. Hasbrouck, *J. Am. Chem. Soc.*, **82**, 696 (1960).
111. W. L. Hawkins and B. S. Biggs, *J. Am. Chem. Soc.*, **71**, 2530 (1949).
112. A. Galan, J. deMendoza, P. Prados, J. Rojo and A. M. Echavarren, *J. Org. Chem.*, **56**, 452 (1991).
113. A. Gaiffe and C. Launay, *C. R. Acad. Sc. Paris*, **266**, 1379 (1968).
114. J. Koubek, J. Pasak and J. Horyna, *Chem. Prum.*, **31**, 349 (1981); *Chem. Abstr.*, **95**, 186562 (1991).
115. J. Pasek, N. Kostova and B. Dvorak, *Coll. Czech. Chem. Commun.*, **46**, 1011 (1981).
116. K. Kindler, K. Schrader and B. Middelhoff, *Arch. Pharm.*, **283**, 184 (1950).
117. J. Barrault, N. Essayem and C. Guimon, *Appl. Catal. A*, **102**, 151 (1993).
118. J. Barrault, S. Brunet, N. Essayem, A. Piccirilli and C. Guimon, *Stud. Surf. Sci. Catal.*, **78** (Heterog. Catal. Fine Chem., III), 305 (1993).
119. W. W. Zajac and R. H. Denk, *J. Org. Chem.*, **27**, 3716 (1962).
120. H. Pleininger and G. Werst, Angew. Chem., **67**, 156 (1955).
121. A. Gaiffe and A. Padovani, *C. R. Acad. Sci., Ser. C*, **269**, 144 (1969).

122. P. Tinapp, *Chem. Ber.*, **102**, 2770 (1969).
123. T. Mallat, J.Petro, S. Mendioroz and J. A. Pajares, *Appl. Catal.*, **33**, 245 (1987).
124. T. Mallat, S. Szabo and J. Petro, *Appl. Catal.*, **66**, 91 (1990).
125. J. Baltz, J. Becker, H. Oberender, D. Timm and B. Lippert, *Ger. (East)*, 77,983 (1969); *Chem. Abstr.*, **72**, 72204 (1970).
126. D. Timm, H. Oberender, B. Lippert, K. Bekker and H. Baltz, *Z. Chem.*, **10**, 185 (1970).
127. N. I. Shcheylov and D. V. Sokol'skii, *Izv. Akad. Nauk Kaz. SSR, Ser. Khim.*, 65 (1962); *Chem. Abstr.*, **59**, 2700 (1963).
128. L. K. Freidlin and T. A. Sladkova, *Izv. Akad. Nauk SSSR Otdel Khim. Nauk*, 1859 (1959);*Chem. Abstr.*, **54**, 8836 (1960).
129. Montecatine Edison, *Ital. Pat.*, 845,999 (1969); *Chem. Abstr.*, **78**, 29279 (1973).
130. S. E. Diamond, F. Mares and A. Szalkeiwicz, *Eur. Pat. Appl. EP* 77,911 (1983); *Chem. Abstr.*, **99**, 70241 (1983).
131. J. E. Galle, F. Mares, S. E. Diamond, J. Corsi and F. Regina, *Eur. Pat. Appl. EP* 161,419 (1985); *Chem. Abstr.*, **105**, 6413 (1986).
132. F. Mares, J. E. Galle, S. E. Diamond and F. J. Regina, *J. Catal.*, **112**, 145 (1988).
133. S. B. Ziemecki, *Stud. Surf. Sci. Catal.*, **78** (Heterog. Catal. Fine Chem., III), 283 (1993).
134. H. P.Schultz, *J. Am. Chem. Soc.*, **70**, 2666 (1948).
135. E. I. duPont de Nemours, Co., *Brit. Pat*, 576,015 (1948); *Chem. Abstr.*, **42**, 591 (1948).
136. E. Strack and H. Schwaneberg, *Chem. Ber.*, **67B**, 39 (1934).
137. S.Tomita, M. Nagai, S. Kanda, A. Hasegawa and I. Shimura, *Japan. Kokai 73* 13,305 (1973);*Chem. Abstr.*, **78**, 147335 (1973).
138. P. Marion, P. Grenouillet, J. Jenck and M. Joucla, *Stud. Surf. Sci. Catal.*, **59** (Heterog. Catal. Fine Chem., II), 329 (1991).
139. P. Marion, M. Joucla, C. Taisne and J. Jenck, *Stud. Surf. Sci. Catal.*, **78** (Heterog. Catal. Fine Chem., III), 291 (1993).
140. M. Joucla, P. Marion, P. Grenouillet and J. Jenck, *Chem. Ind. (Dekker)*, **53** (Catal. Org. React.), 127 (1994).
141. G. Cordier, *Chem. Ind. (Dekker)*, **62** (Catal. Org. React.), 285 (1995).
142. F. Medina, P. Salagre, J.-E. Sueiras and J.-L. Garcia Fierro, *J. Chem. Soc., Faraday Trans.*, **90**, 1455 (1994).
143. F. Mares, R. Y. Tang, J. E. Galle and R. M. Federici, *U. S. Pat.*, 4,625,023 (1986).
144. I. Leupold and H. J. Arpe, *Angew. Chem., Inter. Ed.*, **12**, 927 (1973).
145. M. Israel, J. S. Rosenfield and E. J. Modest, *J. Med. Chem.*, **7**, 710 (1964).
146. J. M. Bobbitt and T. T. Chou, *J. Org. Chem.*, **24**, 1106 (1959).
147. F. Troxler, *Helv. Chim. Acta*, **56**, 374 (1973).
148. J. T. Edward and P. F. Morand, *Can. J. Chem.*, **38**, 1316 (1960).
149. A. Windaus, *Chem. Ber.*, **52**, 170 (1919).
150. W. S. Emerson, *Org. Reactions*, **4**, 174 (1948).
151. L. Haskelberg, *J. Am. Chem. Soc.*, **70**, 2811 (1948).
152. L. E. Graig and D. S. Tarbell, *J. Am. Chem. Soc.*, **70**, 2783 (1948).
153. A. P. Bonds and H. Greenfield, *Chem. Ind. (Dekker)*, **47** (Catal. Org. React.), 65 (1992).

154. H. Greenfield, *Chem. Ind. (Dekker)*, **53** (Catal. Org. React.), 265 (1994).
155. J. B. Campbell and E. P. Lavagnino, *Catalysis in Organic Synthesis* (W. H. Jones, Ed.), Academic Press, 1980, p 43.
156. H. Greenfield and R. E. Malz, Jr., *Chem. Ind. (Dekker)*, **18** (Catal. Org. React.), 309 (1984).
157. R. E. Malz, Jr., E. H. Jancis, M. P. Reynolds and S. T. O'Leary, *Chem. Ind. (Dekker)*, **62** (Catal. Org. React.), 132 (1995).
158. G. Fodor and O. Kovacs, *J. Am. Chem. Soc.*, **71**, 1045 (1949).
159. O. Kovacs and G. Fodor, *Chem. Ber.*, **84**, 795 (1951).
160. W. Gruber and H. Renner, *Monatsch.*, **81**, 751 (1950).
161. R. E. Malz, Jr. and H. Greenfield, *Catalysis in Organic Synthesis* (W. H. Jones, Ed.), Academic Press, 1980, p 49.
162. E. Breitner, E. Roginski and P. N. Rylander, *J. Chem. Soc.*, 2918 (1959).
163. J. Koo, *J. Am. Chem. Soc.*, **75**, 723 (1953).
164. T. Nozoe, K. Takase, N. Kawabe, T. Asao and H. Yamamoto, *Bull. Chem. Soc., Japan.*, **56**, 3099 (1983).
165. M. S. Newman and V. Lee, *J. Org. Chem.*, **40**, 381 (1975).
166. D. C. Iffland and F.-F. Yen, *J. Am. Chem. Soc.*, **76**, 4180 (1954).
167. N. F. Albertson, B. F. Tullar, J. A. King, B. B. Fishbiun and S. Archer, *J. Am. Chem. Soc.*, **70**, 1150 (1948).
168. M. Fields, D. E.Walz and S. Rothchild, *J. Am. Chem. Soc.*, **73**, 1000 (1951).
169. E. Muller, D. Fries and H. Metzger, *Chem. Ber.*, **88**, 1891 (1955).
170. BASF A.G., *Japan Kokai 75* 49,209 (1975), *Ger. Appl.*, 2,343,053 (1973);*Chem. Abstr.*, **85**, 23954 (1976).
171. P. Anziani and R. Cornubert, *C. R. Acad. Sci., Ser. C*, **221**, 103 (1945).
172. P. Anziaini and R. Cornubert, *Bull. soc. chim., France*, 857 (1948).
173. W. Huckel and R. Kupka, *Chem. Ber.*, **89**, 1694 (1956).
174. F. E. King, J. A. Barltrop and R. J. Waller, *J. Chem. Soc.*, 277 (1945).
175. D. Ginsburg and R. Pappo, *J. Chem. Soc.*, 1524 (1963).
176. R. V. Heinzelmann, H. G. Kolloff and J. H. Hunter, *J. Am. Chem. Soc.*, **70**, 1386 (1948).
177. Y. Chang and W. H. Hartung, *J. Am. Chem. Soc.*, **75**, 89 (1953).
178. T. Taguchi, M. Tomoeda and T. Ishida, *J. Pharm. Soc., Japan*, **75**, 666 (1955).
179.1S. Ser, L. Piaux and P. Freon, *C. R. Acad. Sci., Ser. C*, 229, **376** (1949).
180. N. H. Cromwell, Q. T. Wiles and O. C. Schroeder, *J. Am. Chem. Soc.*, **64**, 2432 (1942).
181. F. C. Uhle and W. A. Jacogs, *J. Biol. Chem.*, **160**, 243 (1945).
182. H. H. Fox and J. T. Gibers, *J. Org. Chem.*, **20**, 60 (1955).
183. M. Freifelder, W. B. Martin, G. R. Stone and E. L. Coffin, *J. Org. Chem.*, **26**, 383 (1961).

Hydrogenolysis

C–C Bonds

Cyclopropanes

Most hydrogenolyses take place over palladium or nickel catalysts at room temperature and atmospheric pressure but the cleavage of a C–C bond of alkyl substituted cyclopropanes is not so easily accomplished with these catalysts. The more substituted the ring, the more difficult the hydrogenolysis over palladium or nickel. 1,1-Dialkylcyclopropanes, for instance, require temperatures near 200°C and pressures of about 75 atmospheres for hydrogenolyses over Raney nickel[1] and even over palladium this reaction is quite slow.[2] It is sometimes beneficial to run these reactions without a solvent.[1] The most effective catalyst system for the cleavage of alkyl cyclopropanes is platinum oxide in acetic acid under 2–3 atmospheres of hydrogen at room temperature.[3-6] Rhodium and ruthenium catalysts are not commonly effective for this reaction and no hydrogenolysis of cyclopropanes occurs over Cu-Ba-CrO catalysts under moderately high reaction conditions.[7-9]

As illustrated in Fig. 20.1, the C–C bond that is broken in the hydrogenolysis of alkyl cyclopropanes is that one between the two least substituted carbon atoms of the ring. With unsymmetrical 1,2-dialkyl-cyclopropanes (**1**, R≠R') the bond adjacent to the smaller alkyl group is preferentially cleaved. When coupled with the facile cyclopropanation of alkenes, the hydrogenolysis of alkyl cyclopropanes over platinum oxide has served as a useful means of introducing a methyl group onto a tertiary carbon atom as illustrated by the preparation of tert-butyladamantane (**3**) (Eqn. 20.1)[4] and the 9-methyldecalins, **5** and **6** (Eqn. 20.2).[5,10]

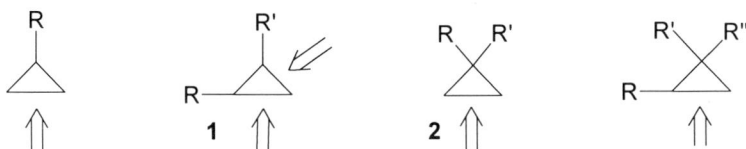

Fig. 20.1. Direction of bond cleavage for alkyl substituted cyclopropanes.

(20.1)

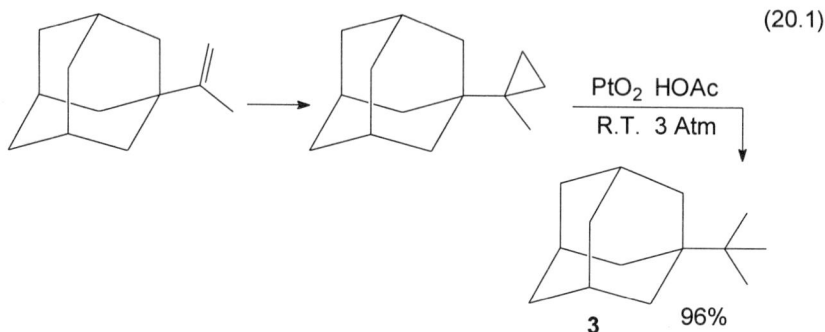

3   96%

(20.2)

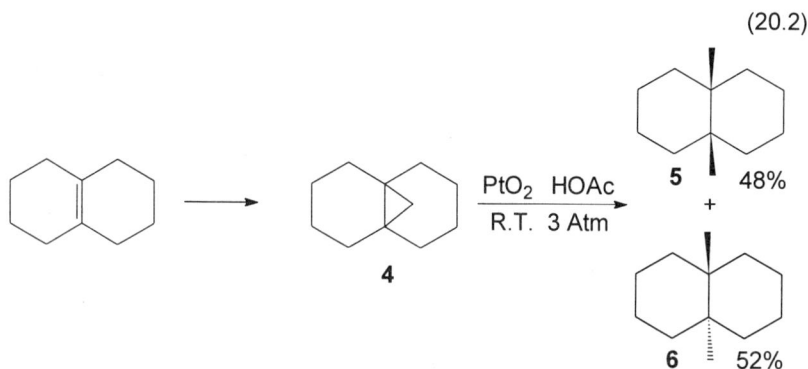

4

5   48%

+

6   52%

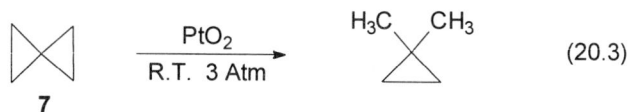

7

(20.3)

The hydrogenolysis of strained ring systems generally takes place more easily. Spiropentane (**7**) was cleaved over platinum to give 1,1-dimethyl-cyclopropane as the primary product (Eqn. 20.3).[11] All spiro-cyclopentanes (**2** in Fig. 20.1 where R and R' are a ring) are hydrogenolyzed to give the *gem* dimethyl products almost exclusively.[4,12] The steroidal bicyclobutane, **8**, was hydrogenolyzed to the unsaturated methyl substituted steroid, **9**, over palladium at room temperature and atmospheric pressure (Eqn. 20.4).[13] Bond cleavage of substituted nortricyclenes takes place primarily on that cyclopropane bond that is on the opposite side of the molecule from the substituent. Hydrogenolysis of 3-exo-acetoxy-nortricyclene (**10**) over platinum in acetic acid gave as the major product 7-acetoxynorbornane (**11**) by cleavage of the 1–6 bond. Some 2–6 bond breaking also occurred to give 2-exo-acetoxynorbornane (**12**) (Eqn. 20.5).[14,15]

(20.4)

8   Pd/C EtOAc
     R.T. 1 Atm

9   80%

(20.5)

10   PtO$_2$ HOAc/HClO$_4$
      80°C 2 Atm

11   68%

+

12   32%

13   PtO$_2$ HOAc/HClO$_4$
      80°C 2 Atm

14

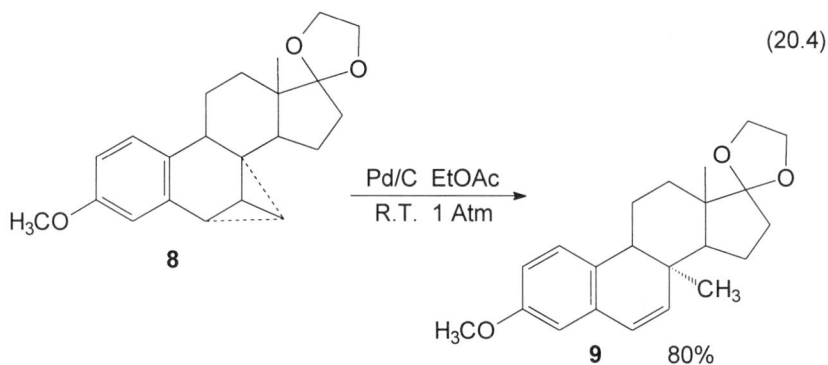

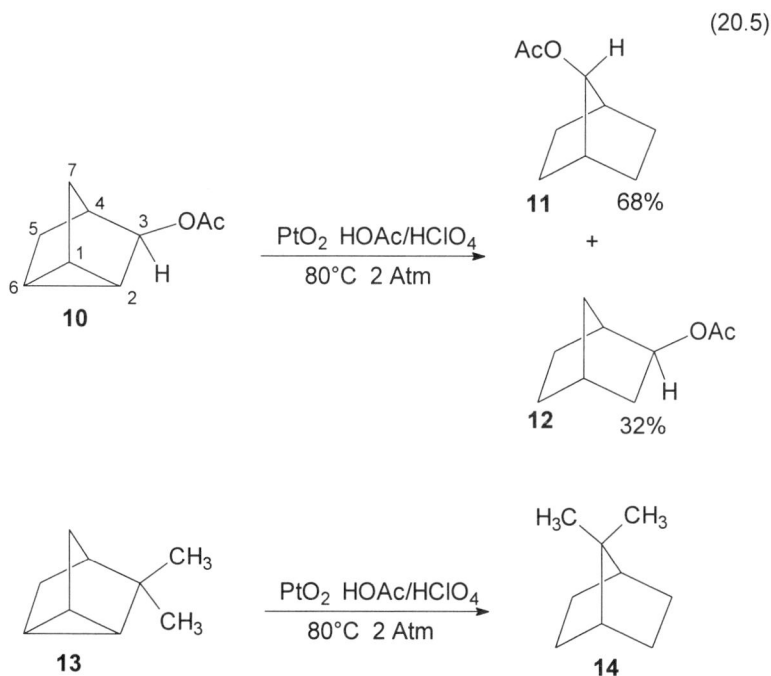

Hydrogenolysis of 3,3-dimethylnortricyclene (**13**) followed the same pattern, primary cleavage of the cyclopropane bond opposite the substituents, to give 7,7-dimethylnorbornane (**14**).[16]

Hydrogenolysis of bicyclo [n.1.0] alkanes (**15**) takes place primarily by breaking one of the exocyclic bonds to give the methyl cycloalkane product (Eqn. 20.6).[12]   When the cyclopropane ring is substituted, alkylcycloalkanes are

(20.6)

(20.7)

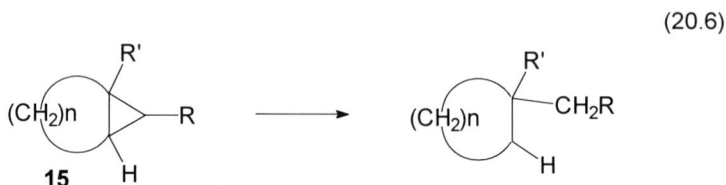

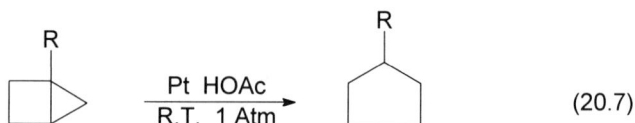

produced.[12,17]   Hydrogenolysis of bicylo [2.1.0] pentanes, however, gave the substituted cyclopentanes (Eqn. 20.7).[12,18]

Phenyl, vinyl or carbonyl substituted cyclopropanes are more easily hydrogenolyzed than are the alkyl substituted species.   Such compounds are commonly cleaved over palladium catalysts at room temperature and atmospheric pressure.   Hydrogenolyses run over platinum, rhodium or nickel catalysts frequently result in the saturation of the double bond, the benzene ring or the carbonyl group with the cyclopropane ring remaining intact or cleaved to only a slight extent.[19,20]   As illustrated in Fig. 20.2[21] the bond broken in the

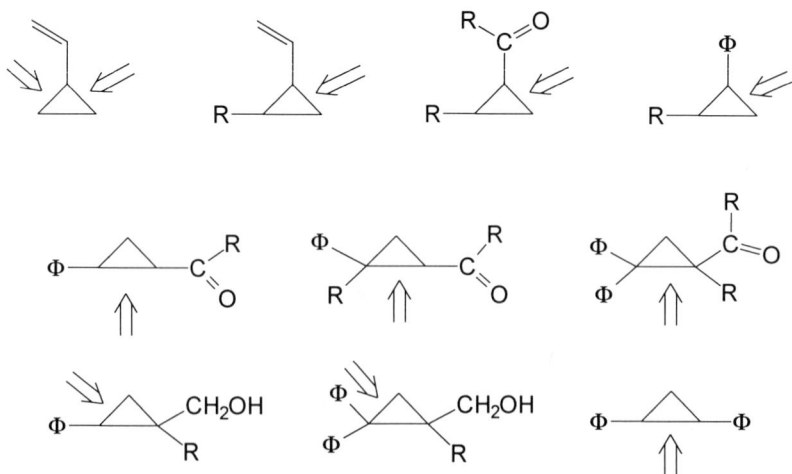

Fig. 20.2. Direction of bond cleavage for cyclopropanes having unsaturated substituents.

hydrogenolyses of these cyclopropanes is that between the unsaturated substituent and the less substituted of the other two carbons. If there are two unsaturated substituents on adjacent carbon atoms, the bond between them will be broken. Carboxylic acids, esters and nitriles promote this reaction as well as aldehydes and ketones. Aromatic rings other than benzenes are also effective. Hydrogenolysis of the phenyl substituted bicycloheptane, **16**, over palladium at room temperature and atmospheric pressure gave the *trans* 1,2 disubstituted cyclohexane, **17**, along with a small amount of phenylcycloheptane (**18**). With a Raney nickel catalyst only **17** was produced (Eqn. 20.8).[22]

(20.8)

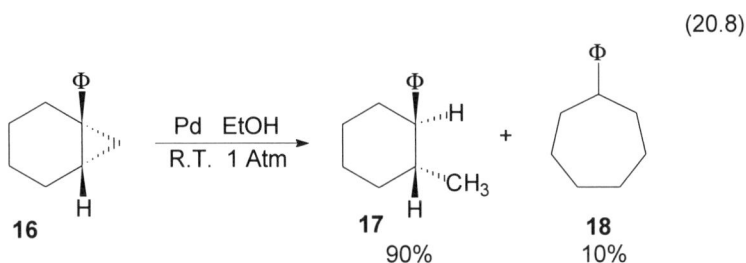

Halogen atoms on a cyclopropane ring apparently weaken the bond opposite them.[23,24] Hydrogenolysis of the phenyl fluoro bicycloheptane, **19**, over palladium gave phenylcycloheptane as the sole product (Eqn. 20.9).[24] When the halogen was on the bridgehead as in **20**, cyclopropane bond cleavage gave the substituted six membered ring system (Eqn. 20.10).[25]

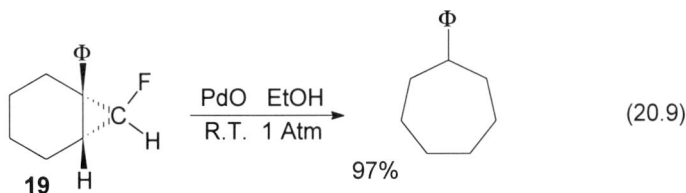

(20.9)

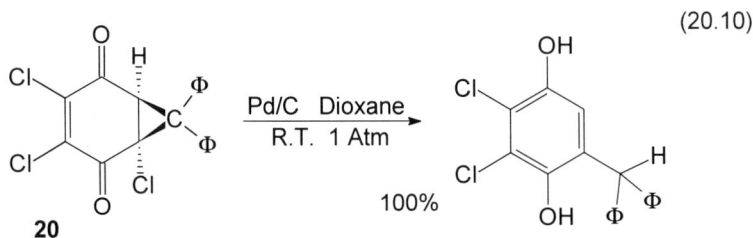

(20.10)

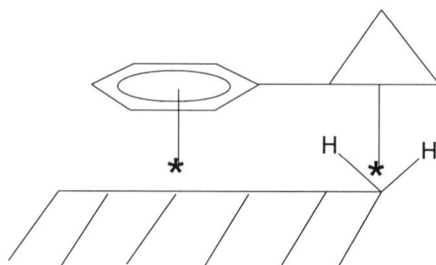

Fig. 20.3. Assisted adsorption mode for cyclopropane hydrogenolysis.

It appears that this facile bond cleavage is the result of an assisted adsorption of the cyclopropane ring promoted by the adsorption of the unsaturated species on the catalyst surface adjacent to the hydrogenolysis active site as depicted in Fig. 20.3. Unsaturated substituents are not the only species that can assist in the adsorption of the cyclopropane ring.   Cyclopropyl amines are hydrogenolyzed over palladium or Raney nickel at 80°C and 50 atmospheres by breaking one of the bonds adjacent to the amine group.[26,27]

The stereochemistry of the products obtained on hydrogenolysis of substituted cyclopropanes is not always that predicted if the reaction were to proceed by way of the analogous Horiuti-Polanyi mechanism shown in Fig. 20.4. If this were the case, the hydrogens would be added to the diadsorbed species from that side of the molecule that was initially adsorbed on the catalyst and hydrogen addition would occur from the direction of the most accessible bond of the cyclopropane. But this is not always observed. The hydrogenolysis of the α 1-5 cyclosteroid, **21**, does give the 5α product, **22**, as expected (Eqn. 20.11).[28,29] However, the epimeric β 1-5 cyclosteroid, **24**, also gives a product, **25**, which has a 5α hydrogen (Eqn. 20.12).[30] This is clearly not what would be expected from the mechanism presented in Fig. 20.4.

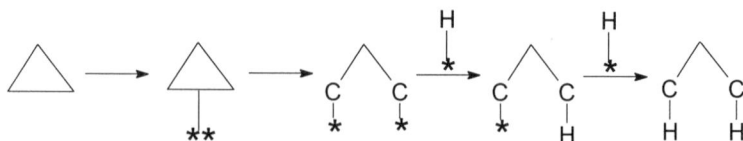

Fig. 20.4 A cyclopropane hydrogenolysis mechanism based on the
Horiuti-Polanyi alkene hydrogenation mechanism.

(20.11)

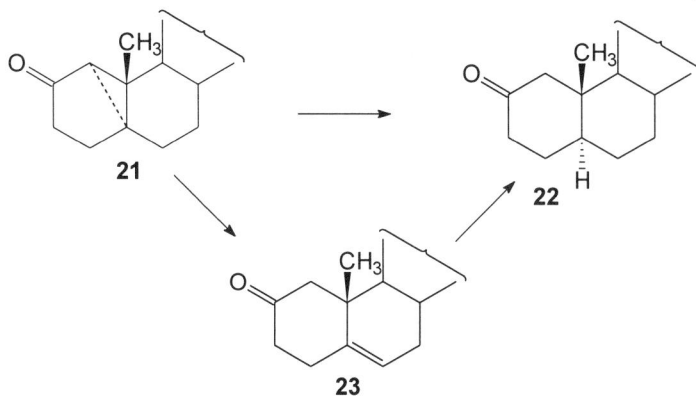

(20.12)

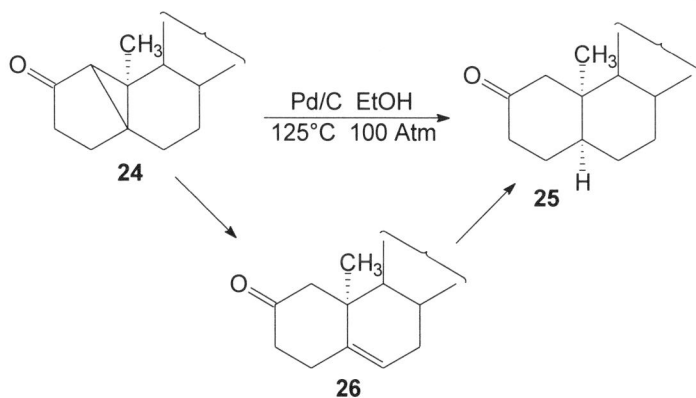

There are a number of other examples of apparent inversion of configuration taking place on hydrogenolysis of cyclopropanes. Hydrogenolysis of **27** (Eqn. 20.13) or **31** (Eqn. 20.14) over palladium at room temperature and atmospheric pressure leads to the introduction of a 4α hydrogen in each of the products, **28** and **32**.[31,32] Reaction of the diphenyl-dimethylcyclopropanes, **34** and **37**, with hydrogen over either palladium or platinum gives a mixture of the meso product, **36**, and the racemic enantiomers, **35** (Eqns. 20.15–16).[33,34] With a palladium catalyst the predominant product from either the *cis* (**37**) or *trans* (**34**) starting materials was that arising from a retention of configuration at the carbon atoms associated with the broken bond. A platinum catalyst, however, gave more inversion than retention with the *trans* material, **34**, but mainly retention with

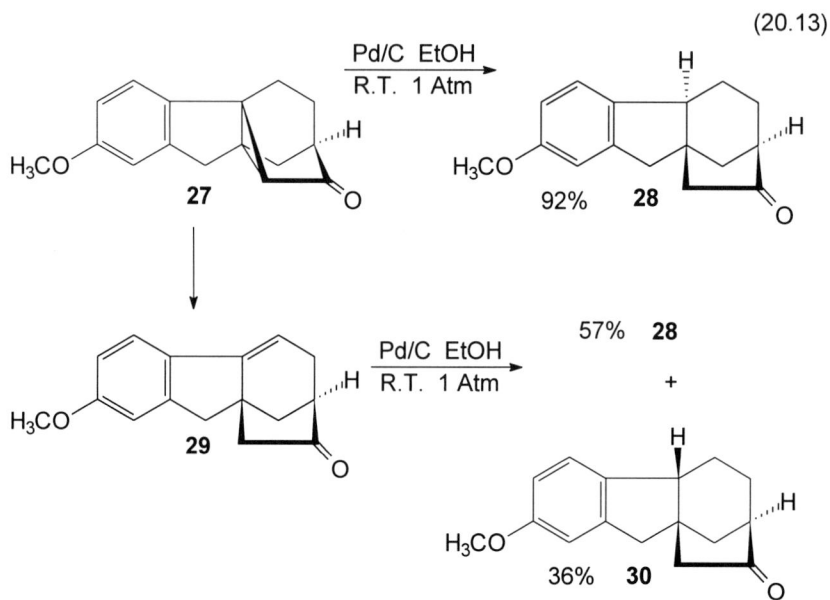

(20.13)

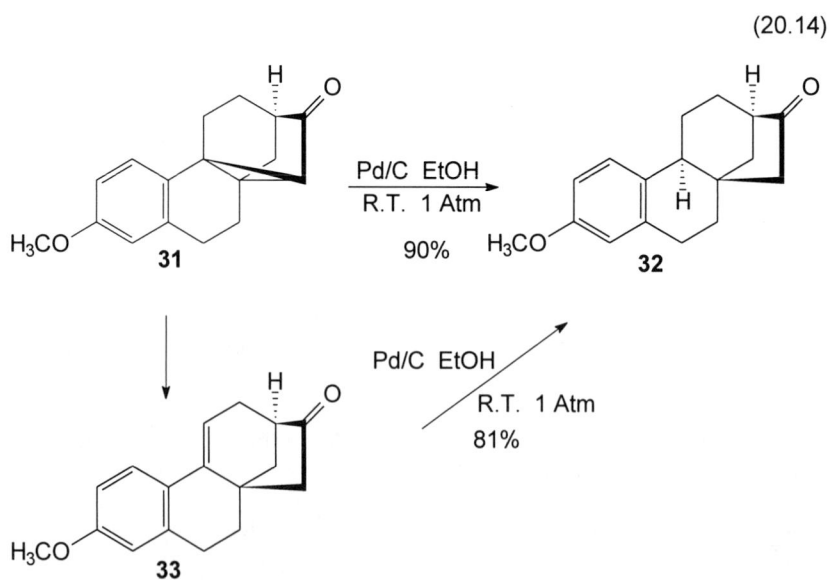

(20.14)

(20.15)

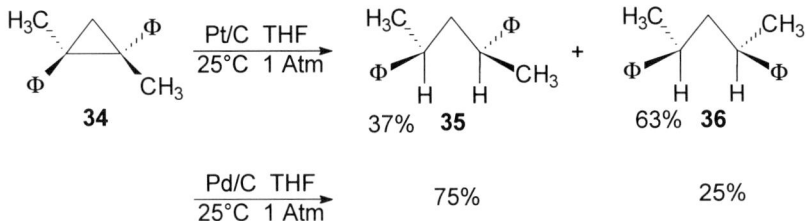

34   Pt/C THF 25°C 1 Atm   37% **35** + 63% **36**

Pd/C THF 25°C 1 Atm   75%   25%

(20.16)

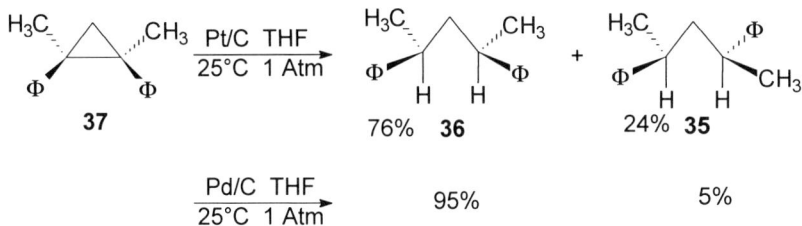

37   Pt/C THF 25°C 1 Atm   76% **36** + 24% **35**

Pd/C THF 25°C 1 Atm   95%   5%

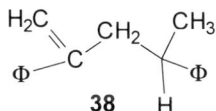

38

the *cis* species, **37**. The product formation resulting from an inversion at one of the cyclopropane carbons was not limited to the reactions of conjugated cyclopropanes. This is shown by the formation of equal amounts of both the *cis* and *trans* 9-methyl decalins, **5** and **6**, in the hydrogenolysis of **4** (Eqn. 20.2).[5]

To account for these stereochemical results it has been proposed that polyalkyl cyclopropanes are hydrogenolyzed by way of an initial olefin formation as shown in the conversion of **39** to **40** (Eqn. 20.17) and that the products formed are actually determined by the hydrogenation of the olefin.[10,30,35] When this proposal was applied to the hydrogenolysis of **4**, **21** and **24**, product formation was easily explained. From **4**, **41** would be formed. It has been shown that the hydrogenation of **41** under the reaction conditions used to hydrogenolyze **4**, gave the same product mixture as that found in the hydrogenolysis of **4** (Eqn. 20.18).[10]

The olefins, **23** and **26** are the expected intermediates in the hydrogenolysis of **21** and **24**, respectively. In both compounds the double bonds are hydrogenated give to the 5α products.[28-30] Ring opening of the substituted

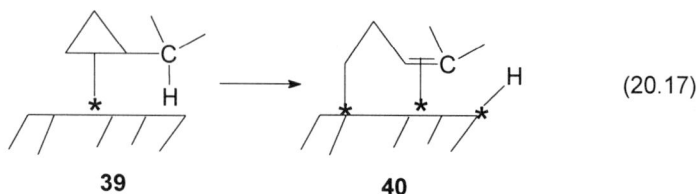

(20.17)

39   **40**

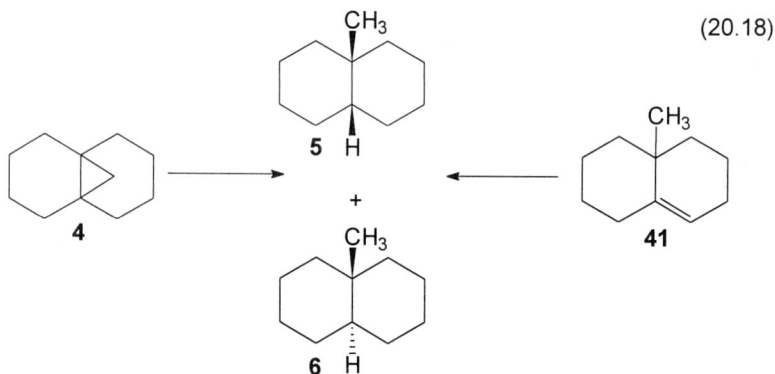

(20.18)

4

5 H

+

6 H

41

tetralin, **31**, would give the alkene, **33**, which on hydrogenation gave exclusively the 4α product as was obtained on hydrogenolysis of **31** (Eqn. 20.14).[32] However, with the substituted indane, **27**, hydrogenolysis gave only the 4α product but hydrogenation of the expected olefinic intermediate, **29**, gave a mixture of the 4α and 4β products (Eqn. 20.13).[32] The products obtained on hydrogenolysis of the substituted cyclopropanes, **34** and **37**[33,34] could be produced by hydrogenation of the substituted styrene, **38**, but if this were true then one would expect the same product mixture from both the *cis* and *trans* starting materials, and this is not the case.

It is clear, then, that the prediction of the product stereochemistry from a cyclopropane hydrogenolysis is not straightforward. In some cases such as the cleavage of the phenyl bicycloheptane, **16** (Eqn. 20.8), the product stereochemistry was that expected if the simple mechanism depicted in Fig. 20.4 were operational. In other cases it appeared that an olefinic intermediate may be formed and that the product resulted from the hydrogenation of that intermediate. There are also some instances in which neither explanation seems to apply.

C–O Bonds

Benzyl Alcohols, Ethers and Esters

The breaking of the benzyl-oxygen bond of benzyl alcohols, ethers and esters takes place most easily over a palladium catalyst at room temperature and atmospheric pressure.[36] Benzhydryl[37] and trityl[38] compounds are also hydrogenolyzed under these conditions. Activation of the C–O bond in these compounds is thought to be the result of the adsorption of the aromatic ring on the catalyst surface, which then brings the C–O bond into contact with the active site used for the bond cleavage.

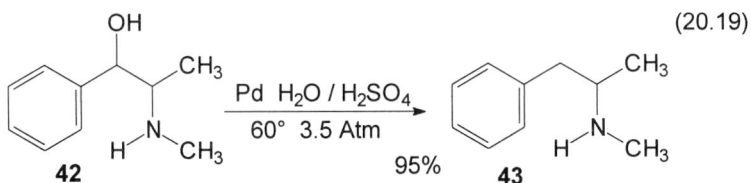

(20.19)

The addition of a small amount of concentrated hydrochloric, perchloric or sulfuric acid facilitates the reaction particularly for the cleavage of the more difficultly hydrogenolyzable compounds such as o-hydroxy- and α-amino-benzyl alcohols.[39,40]   The hydrogenolysis of the benzyl hydroxy group of the amino alcohol, **42**, over palladium black in aqueous sulfuric acid at 60°C gave the amine, **43**, in very good yield (Eqn. 20.19),[40] but the use of supported palladium catalysts for the hydrogenolysis of β-phenylserine (**44**) under a variety of acidic conditions was unsuccessful.[41]   Hydrogenolysis of the diacetate, **45**, however, took place using Pd/BaSO$_4$ in the presence of a small amount of triethyl amine (Eqn. 20.20).[41]

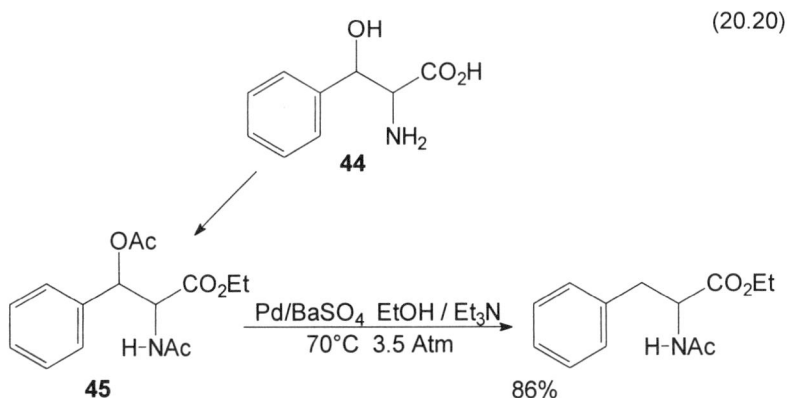

(20.20)

Ring hydrogenation occurs in preference to hydrogenolysis over ruthenium, rhodium, platinized Raney nickel and platinum oxide in the presence of a trace of acetic acid.[42-45]   With the latter system, the extent of ring hydrogenation is dependent on the nature of the alcohol used as the solvent with ring saturation increasing in the order 1° alcohols < 2° alcohols < 3° alcohols. Hydrogenation of the 1,4 dibenzyl alcohol, **46**, over platinum oxide in alcoholic acetic acid resulted in a 64% selectivity for benzene ring hydrogenation in ethanol, 71% in iso-propanol and 90% in tert-butanol (Eqn. 20.21).[44] Hydrogenation of benzyl alcohol over either platinum oxide or rhodium oxide in iso-propanol at room temperature and atmospheric pressure gave good yields of cyclohexane carboxaldehyde.[45]

(20.21)

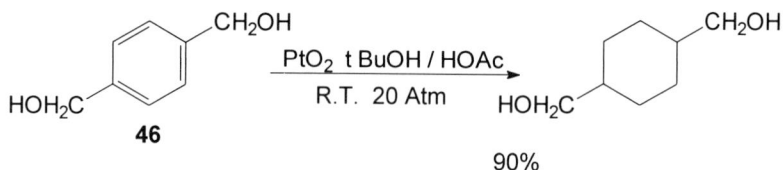

46

90%

Not surprisingly, the ease of C–O bond cleavage in the hydrogenolysis of benzyl alcohols (and ethers and esters as well) decreases with the degree of substitution at the carbinol carbon.[46] The hydrogenolysis of ring substituted benzyl alcohols over palladium is enhanced by the presence of electron-donating groups on the benzene ring.[46] With Raney nickel catalysts the reverse is true.[47]

The facile hydrogenolysis of benzyl ethers makes them a versatile blocking group for alcohols. The benzyl ether is easily prepared by the reaction of the alkoxide ion with benzyl chloride. It is not affected by the usual acids, bases, or hydride reducing agents used in synthetic reactions, and it is readily removed under mild, neutral to slightly acidic hydrogenation conditions (Eqn. 20.22).[48] The hydrogenolysis of benzyl ethers of carbohydrates takes place under conditions that permit the isolation of the unmutarotated compounds.[49]

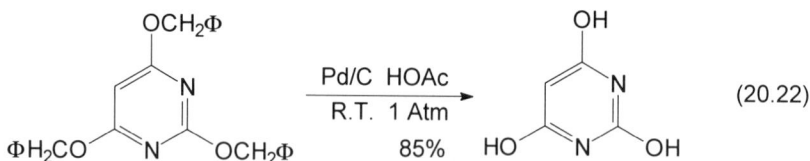

(20.22)

Only simple olefins, enol ethers, acetylenes, and nitro groups are hydrogenated in preference to the cleavage benzyl ethers. Aryl and alkyl ketones, N-oxides, and hindered double bonds are unaffected if the reaction is stopped after one molar equivalent of hydrogen has been absorbed. Benzyl[50,51] and benzhydryl[52] acetals can also be used as easily removed blocking groups for glycols (Eqn. 20.23).[51] These compounds are hydrogenolyzed with the same ease as are benzyl ethers.

(20.23)

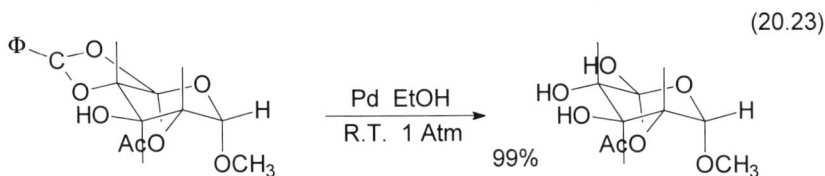

99%

Aryl benzyl ethers are cleaved more easily than are alkyl benzyl ethers. The presence of an amine hinders the hydrogenolysis of alkyl benzyl ethers but has no effect on the cleavage of aryl benzyl ethers.[53,54] Because of this the benzyl ether is selectively hydrogenolyzed in compounds containing both a benzyl amine and an aryl benzyl ether (Eqn. 20.24)[55] but the benzyl amine is selectively cleaved in compounds having both a benzyl amine and an alkyl benzyl ether (Eqn. 20.25).[53,54,56]

(20.24)

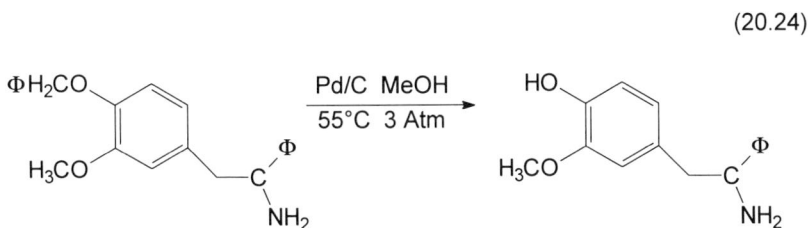

(20.25)

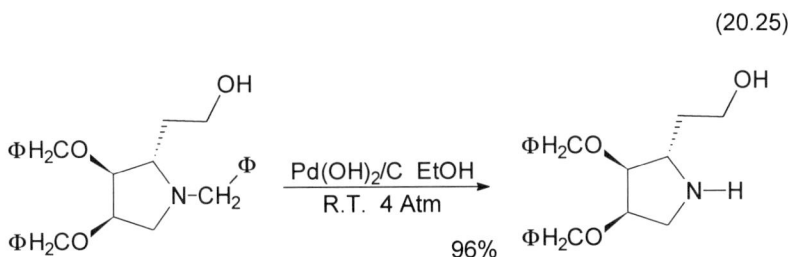

96%

The hydrogenolysis of benzyl esters provides a means for the selective conversion of the ester to the acid under mild conditions. Substituted dibenzyl malonates are easily converted to the substituted malonic acids by hydrogenolysis in neutral medium thus avoiding the acyl cleavage or decarboxylation which can result from base or acid catalyzed ester hydrolysis (Eqns. 20.26–27).[57,58] Monobenzylmalonates are selectively debenzylated and decarboxylated to give the substituted acetic esters.

The hydrogenolysis of benzyl protecting groups on poly(benzyl acrylates) and related polymers was successfully accomplished over Pd/C at room temperature and atmospheric pressure provided the palladium was present on the surface of the support and not in the pores.[59]

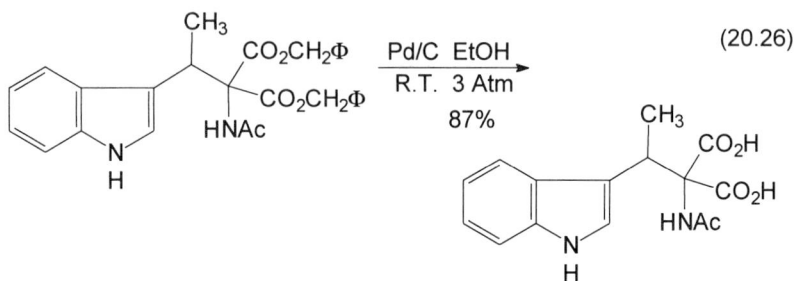

(20.26)

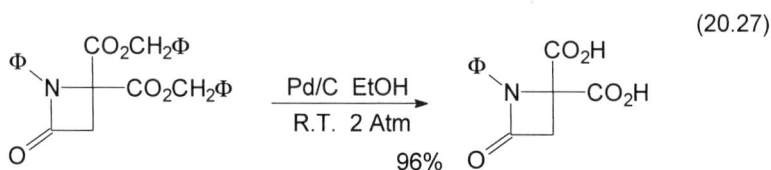

(20.27)

The removal of the carbobenzyloxy (Φ–CH2–O–CO–X) blocking group also takes place under these conditions. As depicted in Eqn. 20.28, the benzyl group is hydrogenolyzed to give the carbamate or carbonate, which loses carbon dioxide to give the amine or the alcohol.[36,60,61] Hydrogenolysis of the phenyl cyclopropylamine derivative, **47**, over palladium gave the cyclopropyl amino acid, **48**, in good yield but hydrogenolysis of the diastereomer, **49**, not only cleaved the carbobenzyloxy group but also the cyclopropane ring to give the amino acid, **50** (Eqn. 20.29).[61] Benzyl phosphates are also easily hydrogenolyzed to the phosphoric acids under these conditions (Eqn. 20.30).[62]

(20.28)

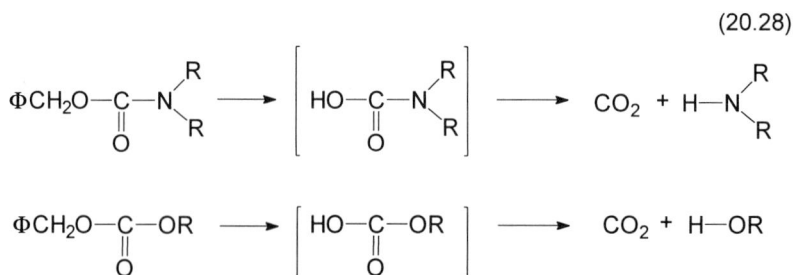

Benzyl and carbobenzyloxy protecting groups have been particularly useful in peptide syntheses but hydrogenolyses do not occur under the normal reaction conditions when sulfur containing amino acids are present. This problem was overcome, however, by running the hydrogenolyses over palladium black in liquid ammonia. In this way complete removal of the benzyl entity from benzyl

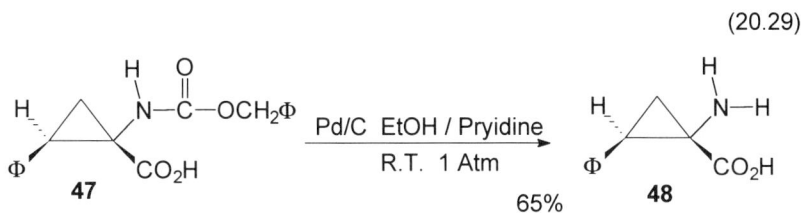

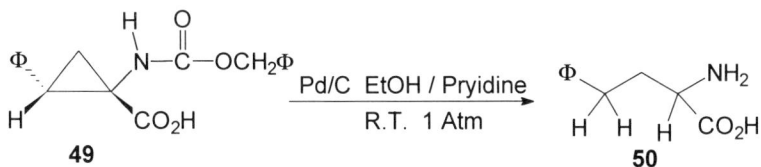

(20.29)

Pd/C EtOH / Pryidine
R.T. 1 Atm
65%

47 → **48**

Pd/C EtOH / Pryidine
R.T. 1 Atm

49 → **50**

(20.30)

Pd/C EtOH / H₂O / HCl
R.T. 1 Atm
80%

esters, ethers and carbobenzyloxy groups was accomplished in cystein or methionine containing peptides.[63]

The stereochemical changes occurring during the cleavage of benzyl oxygen bonds depends on the substituent on the oxygen and the catalyst used for the reaction. Hydrogenolysis of benzyl alcohols occurs with inversion of configuration over palladium[64-68] and retention over nickel.[64-67] Hydrogenolyses run over copper or cobalt also proceed with retention of configuration.[65,66] Cleavage of benzyl esters occurs with inversion of configuration over both palladium[68] and nickel.[69] The hydrogenolysis of alkyl benzyl ethers follows the same pattern as is observed with benzyl alcohols[64,69] but the direction of cleavage of aryl benzyl ethers is somewhat complex. Over nickel these materials are cleaved with inversion of configuration[69] while with palladium the stereochemical result depends on the nature of the aryl group attached to the benzyloxy moiety but in all cases considerable racemization is observed so the optical yields of the products obtained are only 50–69%[70] rather than the 90–95% commonly obtained in all of the other reactions.

Allyl Alcohols

Allyl alcohols are deoxygenated under the same conditions used to break a benzyl oxygen bond. Over a palladium catalyst, loss of oxygen occurs along with double bond saturation. The presence of acids promotes hydrogenolysis. With platinum, nickel, or rhodium catalysts in neutral medium, progressively less oxygen cleavage is observed and the saturated alcohol is generally the primary product obtained unless the double bond is hindered. The addition of 1% sodium nitrite to the platinum catalyzed reaction mixture further inhibits hydrogenolysis.[71,72] Cyclohexenols with ψ axial alcohol groups are cleaved more readily than those with the alcohol in the ψ equatorial conformation.[73]

(20.31)

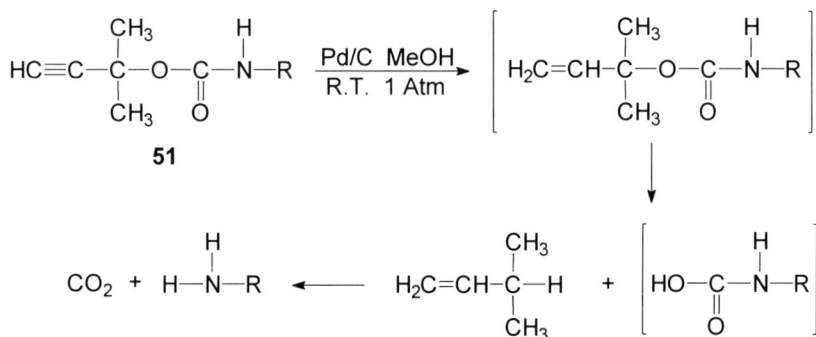

A blocking group related to the carbobenzlyoxy entity is the 1,1-dimethyl-2-propynyloxy carbonyl species shown attached to an amine group in **51**. Hydrogenation of the triple bond over Pd/C gave the allyl alcohol which on cleavage generated the carbamate. Loss of carbon dioxide then gave the amine (Eqn. 20.31).[74]

Oxiranes

The formation of alcohols by the hydrogenolytic opening of oxiranes (epoxides) takes place over palladium at room temperature and atmospheric pressure. The addition of some perchloric or hydrochloric acid not only increases the rate of the reaction, but can also influence the direction of ring opening of unsymmetric epoxides.

In acid medium the epoxide is hydrogenolyzed in such a way as to break that carbon-oxygen bond that gives the more stable carbocation. 1-Methyl-cyclohexene oxide was opened, under these conditions, to give a mixture of *cis* and *trans* 2-methylcyclohexanol. No 1-methylcyclohexanol was formed (Eqn. 20.32).[75]

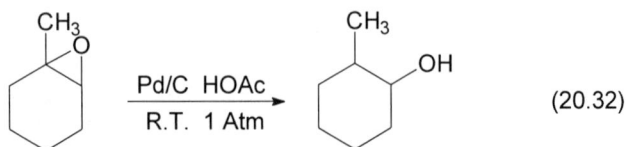

(20.32)

(20.33)

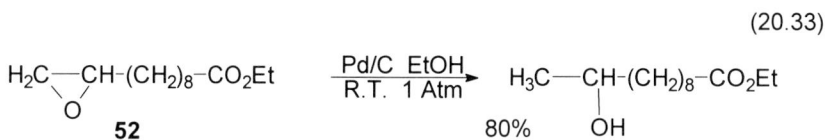

(20.34)

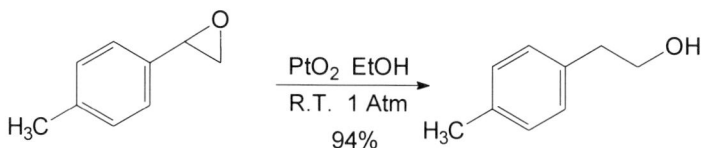

In neutral medium, ring opening generally proceeds to give the more substituted alcohol as in the hydrogenolysis of **52** (Eqn. 20.33).[76] However, hydrogenolysis of 1-decene oxide over Raney nickel in ethanol gave the primary alcohol, but in basic medium, the secondary alcohol was formed.[77] In the absence of any steric differences between the two carbons of the epoxide, both C–O bonds are cleaved with equal facility.[78,79] Most substituted styrene oxides give the β-phenyl ethanols with almost all catalysts (Eqn. 20.34).[80] The opening of α-keto-epoxides results in the formation of the α-hydroxy-ketone (Eqn. 20.35).[81,82] Cyclohexene epoxides are hydrogenolyzed to give the axial alcohols as the primary products (Eqns. 20.36-37).[83-86] Unsaturated epoxides can be hydrogenolyzed to the unsaturated alcohols by adding silver nitrate to the palladium catalyzed hydrogenation mixture.[87]

(20.35)

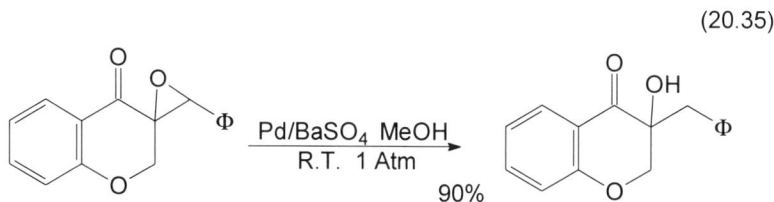

(20.36)

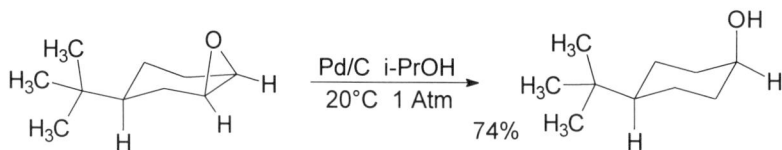

(20.37)

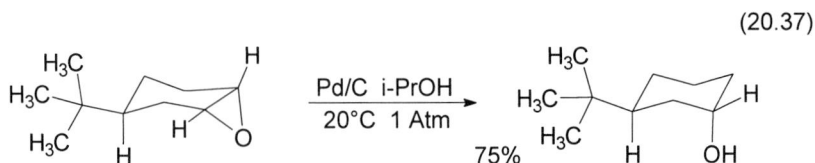

75%

Hydrogenolysis of 5,6-α-epoxy-5α-cholestan-3β-acetate (**53**) over platinum oxide at room temperature and atmospheric pressure gave a very good yield of the 5α-hydroxy product, **54**, which has the axial hydroxy group (Eqn. 20.38).[83]   Under the same conditions the 5,6-β-epoxy cholestane, **55**, gave a number of different desoxy, monohydroxy and dihydroxy products.[83,88]

(20.38)

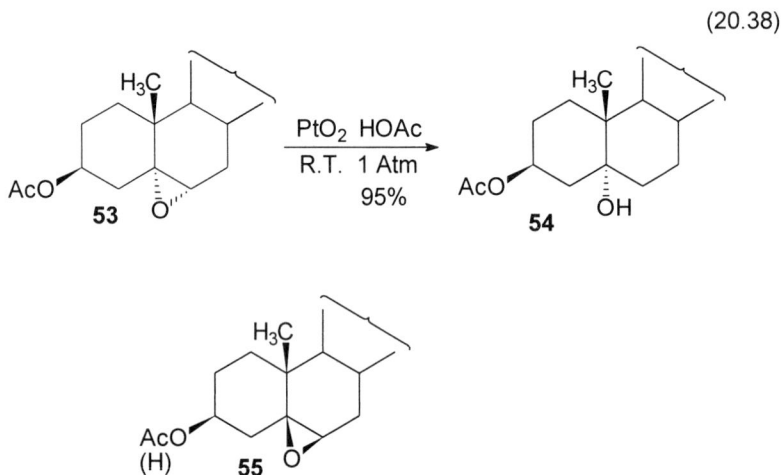

As was observed with most benzyl oxygen bond breakings, the hydrogenolysis of 1,2-dimethylcyclohexene oxide (**56**) over Raney nickel occurred primarily with retention of configuration to give the *cis* dimethyl-cyclohexanol, **57**. With platinum or palladium, however, the products from both retention, **57**, and inversion, **58**, were formed (Eqn. 20.39).[89] Cleavage of 1-phenylcyclohexene oxide over Pd/C gook place with almost exclusive inversion of configuration (Eqn. 20.40).[22]

Acetals and Ketals

The hydrogenolysis of acetals and ketals does not take place over most catalysts under normal hydrogenation conditions, even at relatively high temperatures and pressures.   About the only reasonable hydrogenolytic cleavage of these compounds takes place over rhodium at 50°–80°C under 3–4 atmospheres of hydrogen in the presence of a trace of acid.[90] The formation of ethers in this

(20.39)

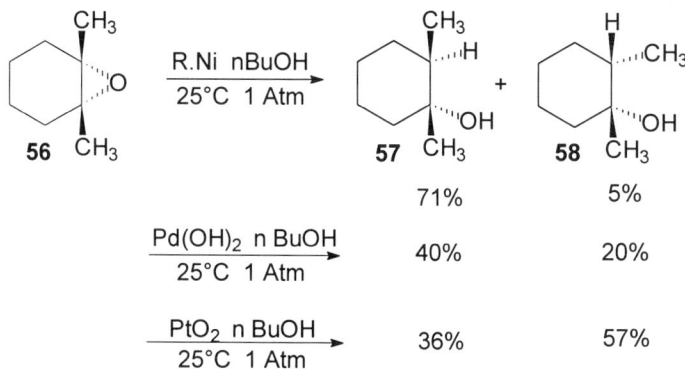

56 CH₃   R.Ni nBuOH / 25°C 1 Atm   57 + 58

	57	58
R.Ni nBuOH 25°C 1 Atm	71%	5%
Pd(OH)₂ n BuOH 25°C 1 Atm	40%	20%
PtO₂ n BuOH 25°C 1 Atm	36%	57%

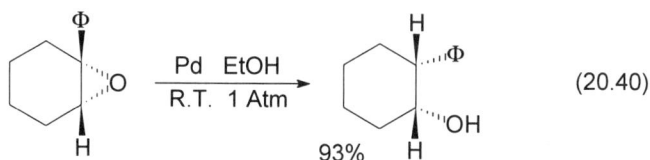

Pd EtOH
R.T. 1 Atm

93%

(20.40)

reaction is thought to proceed by the acid catalyzed loss of a molecule of alcohol to form the enol ether which is then hydrogenated.[90]

Carbonyl Groups

As mentioned in Chapter 18, aryl ketones are hydrogenolyzed to the methylene over palladium catalysts under mild conditions. The addition of a little acid to the hydrogenation mixture facilitates the reaction.[91] Since this reaction takes place under mild conditions few, if any, side reactions take place making this palladium catalyzed hydrogenolysis the preferred method for the removal of aryl carbonyl groups.[92-94]

(20.41)

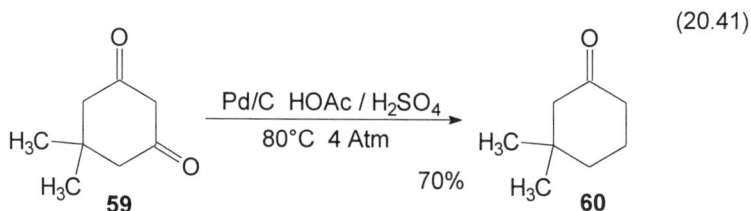

59   Pd/C HOAc / H₂SO₄ / 80°C 4 Atm   70%   60

The palladium catalyzed hydrogenolysis of 5,5-dimethyl 1,3-cyclohexane-dione (**59**) in acidic medium gave good yields of the cyclohexanone, **60** (Eqn. 20.41).[95] In the absence of acid, a platinum catalyst gave a higher selectivity for the hydrogenolysis of one of the carbonyl groups of 1,3-cyclohexanedione than did palladium.[96]

Aryl Hydroxy Groups

Aryl hydroxy groups have been removed by first converting them into an ether by reaction with a heterocycle such as 1-phenyl-5-chlorotetrazole (**61**) and then hydrogenolyzing the ether with platinum or palladium (Eqn. 20.42).[97-99] Other aryl ether groups that are readily hydrogenolyzed to give the benzene are the benzoxazole, **62**,[97,98] sulfonates, **63**,[100] and phosphates, **64**.[101]

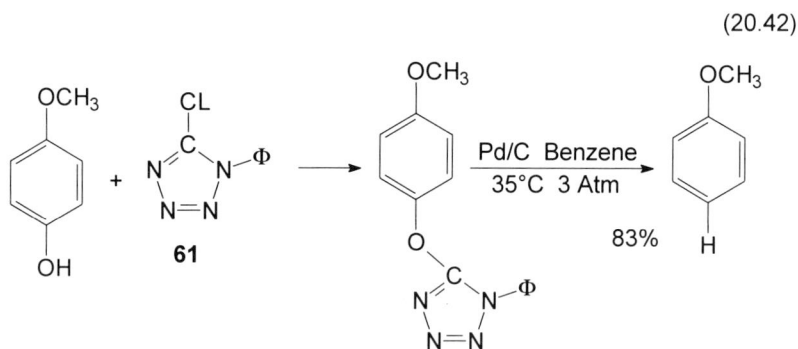

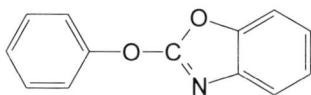

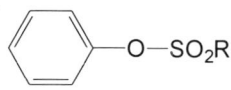

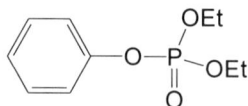

(20.42)

C–N Bonds

Benzyl Amines

The ease of hydrogenolysis of benzyl amines depends on the nature of the substituents on the nitrogen atom.[36] The debenzylation of tertiary benzyl amines takes place over palladium on charcoal at 25°–50°C and 2–4 atmospheres of hydrogen (Eqn. 20.43).[102] Platinum and rhodium catalysts generally promote

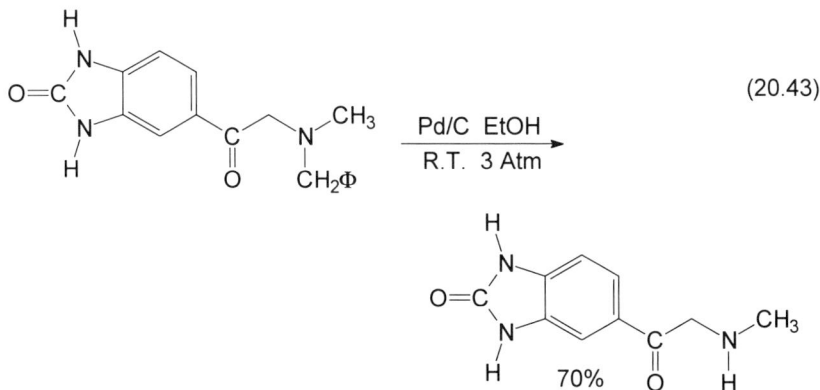

(20.43)

ring hydrogenation in preference to hydrogenolysis of the benzyl-nitrogen bond.[103] The addition of a little acid facilitates the hydrogenolysis as does the quaternization of the amine.[104] Secondary aryl benzyl amines are also debenzylated under these conditions[105] but alkyl secondary benzyl amines and benzyl amine itself are not affected and require the use of Raney nickel at 150°–160°C and 100–175 atmospheres in order for hydrogenolysis to occur.[106] Phenolic Mannich bases having a nitro group meta to the aminomethyl group are converted into amino-methylphenols over palladium under mild conditions by hydrogenation of the nitro group and hydrogenolysis of the benzyl amine (Eqn. 20.44).[107] The hydrogenation of other phenolic Mannich bases has been accomplished with a Cu-CrO catalyst at 235°C and 275 atmospheres of hydrogen (Eqn. 20.45).[108]

(20.44)

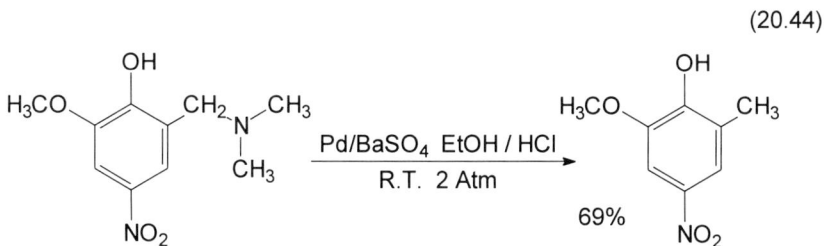

(20.45)

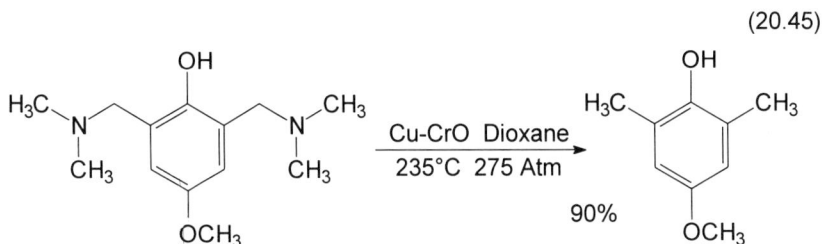

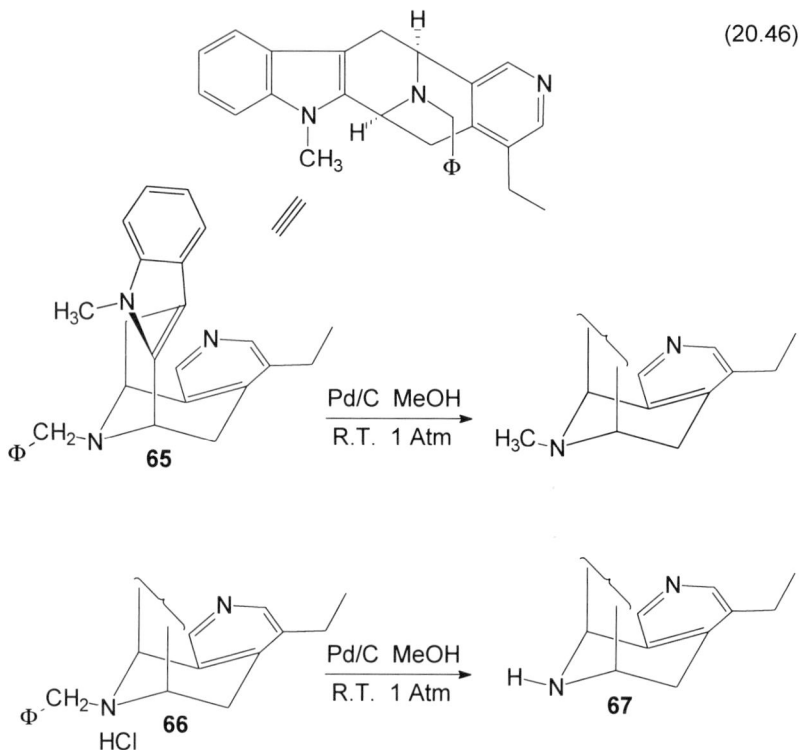

(20.46)

Tertiary benzyl amines are hydrogenolyzed in preference to the hydrogenation of aryl ketones and the hydrogenolysis of alkyl benzyl ethers[52,55] but with somewhat more difficulty than the cleavage of aryl benzyl ethers.[52,53] The presence of substituents on the aromatic ring stabilizes the compound toward hydrogenolysis. The debenzylation of dibenzyl tertiary amines with one substituted benzyl ring, invariably results in the loss of the unsubstituted benzyl group.[109] If both rings are substituted and one substituent is a para amino group, p-toluidine will be formed. In other cases, mixtures are produced.[110]

Reaction of N-benzylsuaveoline (65) with hydrogen over a large excess of Pd/C in methanol for a long period of time led to the replacement of the benzyl group by a methyl in near quantitative yield. However, hydrogenolysis of the hydrochloride, 66, gave the secondary amine, suaveoline (67) (Eqn. 20.46).[111]

Benzyl amides are more resistant to hydrogenolysis than are secondary benzyl amines (Eqn. 20.47).[112] Higher temperatures and pressures are required to debenzylate these amides.

The stereochemical course of benzyl amine hydrogenolysis is such that inversion of configuration is observed with both palladium[113-115] and nickel[115]

(20.47)

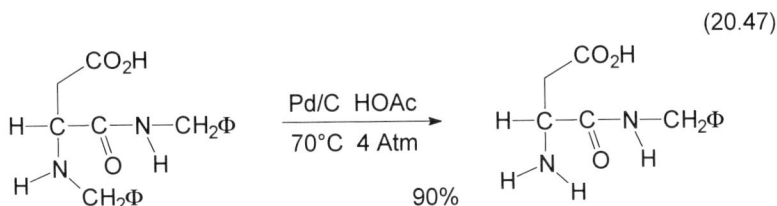

catalysts. When the reactions were run at room temperature palladium catalysts gave products with reasonably high optical yields[113,114] but for reactions run at 50°C Raney nickel catalysts gave significantly higher enantioselectivities than those observed with palladium.[115]

Aziridines

The pattern of bond cleavage in the hydrogenolysis of aziridines generally follows that observed with cyclopropanes. Alkyl substituted aziridines are cleaved at the most accessible C–N bond. This provides a good route to the production of molecules having a primary amine group on a tertiary carbon (Eqn. 20.48).[116,117] The hydrogenolysis of 1-azabicyclo[3.1.0]hexane (**68**) took place with the breaking of both C–N bonds giving 2-methylpyrrolidine and piperidine in a 2:1 ratio (Eqn. 20.49).[118] With aryl substituted aziridines the C–N bond adjacent to the aromatic ring is preferentially hydrogenolized (Eqn. 20.50).[119]

(20.48)

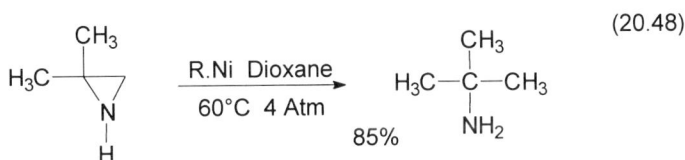

(20.49)

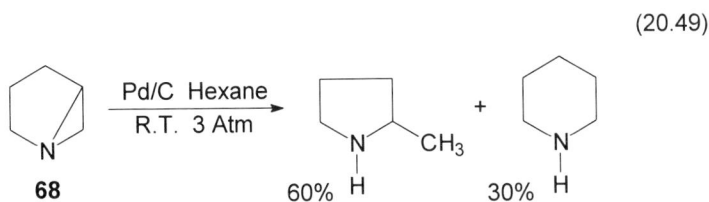

(20.50)

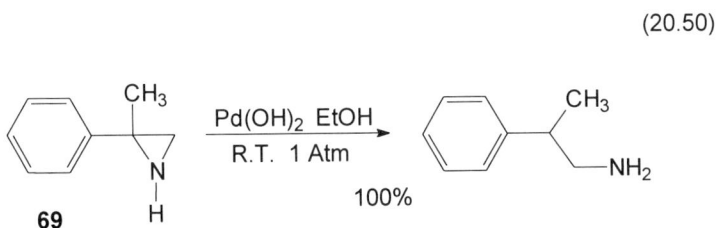

(20.51)

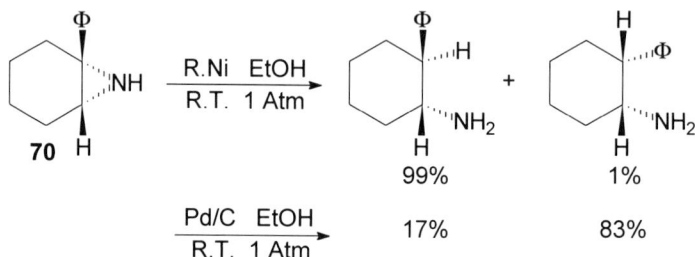

Hydrogenolysis of optically active 2-phenyl-2-methylaziridine (**69**) took place with inversion of configuration over palladium and retention over platinum, nickel and cobalt but the selectivities observed with nickel and cobalt catalysts were very low.[119-121] Hydrogenolysis of 1-phenyl-7-azabicyclo[4.1.0] heptane (**70**) occurred with essentially complete retention of configuration over Raney nickel and inversion over palladium (Eqn. 20.51).[22]

C–X Bonds

The ease with which carbon-halogen bonds undergo hydrogenolytic cleavage is dependent on the type of halogen involved, the presence of other functionality in the molecule, the catalyst and conditions used for the reaction. Alkyl halides are removed less readily than are the benzyl, aryl, allyl or vinyl halides. For all types of halogen compounds the order of reactivity is RF << RCl < RBr < RI. Palladium is superior to the other catalysts for this reaction since it not only promotes bond cleavage, but is also the least affected by the catalyst poisoning properties of the halide ions. Raney nickel is also useful but, because of this poisoning effect,[122] large amounts of nickel must be used. Platinum is also somewhat inferior to palladium for dehalogenations. The presence of base in the reaction mixture facilitates dehalogenation as does the use of polar, hydroxylic solvents.[123]

Aliphatic Halogen Compounds

In neutral or acidic medium, unactivated primary and secondary alkyl chlorides and bromides are not hydrogenolyzed over palladium, but iodides, as well as tertiary chlorides and bromides, are lost to some extent at least, on hydrogenation at room temperature and atmospheric pressure.[123,124] All alkyl fluorides, however, are inert to hydrogenolysis. Alkyl halides, except fluorides, are hydrogenolyzed in methanolic potassium hydroxide over palladium on charcoal at room temperature and atmospheric pressure. Polar solvents, such as alcohols, favor dehalogenation and the presence of base also facilitates the reaction.

Almost any type of base can be used to promote this reaction but potassium hydroxide and magnesium oxide[125] are particularly effective. Triethylamine was found to be inferior to potassium hydroxide.[126] If calcium,

(20.52)

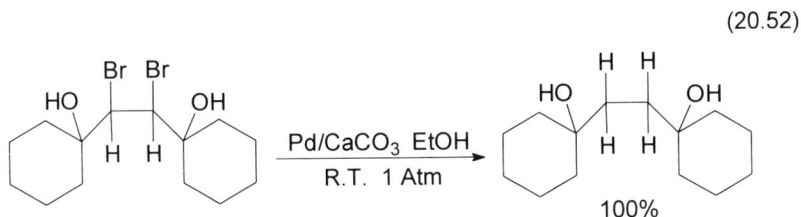

100%

strontium, or barium carbonate are used as supports for the palladium catalyst, additional base is not absolutely necessary, but these catalysts are not as effective as the palladium on charcoal:potassium hydroxide combination. They can be used, however, if the reacting molecule is sensitive to strong base (Eqn. 20.52).[127]

Raney nickel is not only less effective than palladium but also requires the use of a higher catalyst ratio. The addition of one or two equivalents of potassium hydroxide to the reaction mixture is also beneficial in nickel catalyzed dehalogenations.

(20.53)

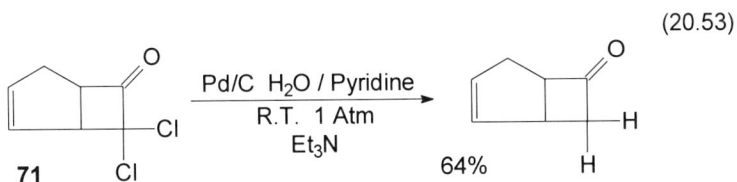

71   64%

The presence of other functional groups in the molecule can also activate the alkyl halide for hydrogenolysis. α-Alkoxy and carbonyl groups facilitate the loss of halogen but not as readily as do α-hydroxy and amine groups. The presence of more than one halogen on a carbon also promotes dehalogenation.[128] Even though the removal of these activated halides takes place quite readily, the use of the usual basic reaction conditions is still advantageous. However, palladium on a basic support material can be substituted for the palladium on charcoal:potassium hydroxide combination quite effectively for the dehalogenation of these compounds. The selective hydrogenolysis of the gem-dichlorides in **71** was affected in aqueous pyridine with added triethyl amine (Eqn. 20.53).[129] Double bond saturation took place with other bases in ethanol solution. Pre-reduction of the palladium catalyst was also beneficial.[129]

Benzyl, allyl, and vinyl halides are cleaved more readily than are alkyl halides. Vinyl halogens are dehalogenated as readily as are aryl halides. Over a palladium catalyst, saturation of the double bond takes place along with vinyl halogen cleavage (Eqn. 20.54)[130] and, unless hindered, during allyl halide hydrogenolysis as well. The use of a rhodium catalyst in cyclohexane at 100°C and 65 atmospheres, however, resulted in olefin hydrogenation without the loss of the chlorines (Eqn. 20.55).[123]

(20.54)

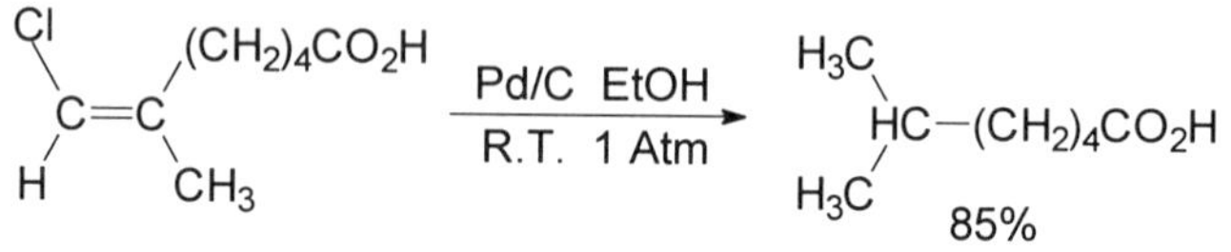

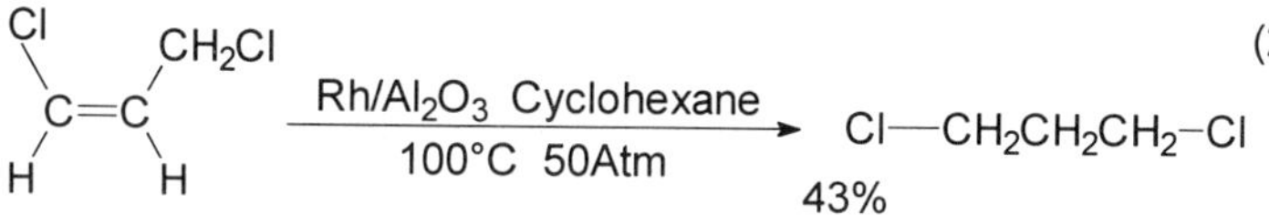

Halogens on a cyclopropane ring can be removed without hydrogenolysis of the ring itself. The selective removal of one bromine from 1,1–dibromo-cyclopropanes occurred over palladium on charcoal or platinum oxide in methanolic potassium hydroxide at room temperature and atmospheric pressure. Complete debromination resulted from the use of Raney nickel under the same conditions (Eqn. 20.56).[131]

(20.56)

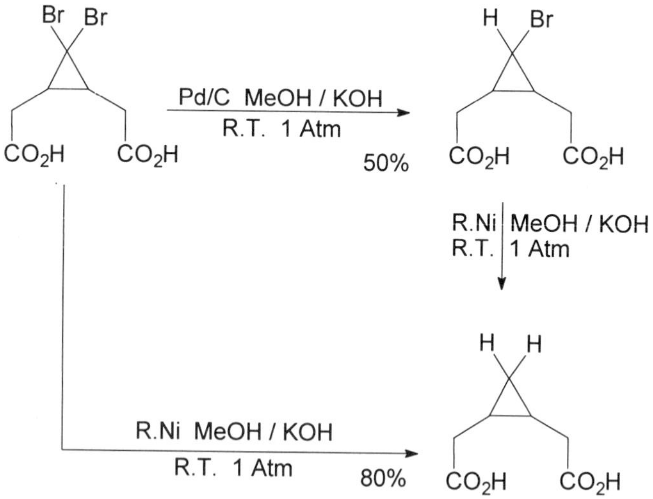

Aryl Halogen Compounds

The hydrogenolysis of aryl halides proceeds more readily than the cleavage of alkyl-halogen bonds. Aryl chlorides are rather stable to hydrogenolysis in neutral medium and bromides are dehalogenated only to a moderate extent[124,132] but iodides are readily lost. Aryl fluorides, however, are hydrogenolyzed only under

(20.57)

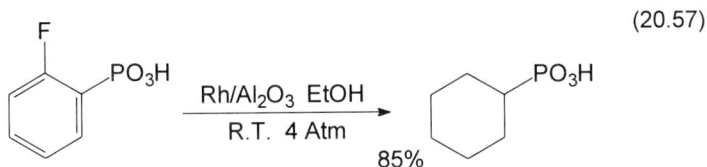

85%

conditions that also saturate the ring (Eqn. 20.57).[133] In basic medium, all of the halides, except fluorides, are dehalogenated with little difficulty. For this reason, the use of the standard alkaline, palladium-catalyzed hydrogenation conditions is also preferred for the hydrogenolysis of the phenyl-halogen bond. Palladium on basic support materials is also effective. Excellent yields of dehalogenated materials were obtained on hydrogenolysis of substituted aryl halides over palladium in hydrocarbon solutions at 50°C and atmospheric pressure in the presence of a surfactant.[134] Aryl chlorides are cleaved in preference to the hydrogenolysis of secondary benzylamines over palladium in acetic acid at room temperature and atmospheric pressure.[135]

The presence of amino, hydroxy, alkoxy, or carboxy groups on the ring facilitates the dehalogenation of a phenyl halide (Eqn. 20.58).[136,137] The relative position of the activating group to the departing halide has little effect on the amount of halogen lost.[124,136]

(20.58)

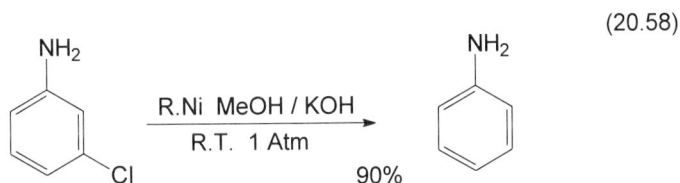

90%

This activation of an aryl halide by an amine group explains the C–X bond cleavage that commonly occurs during the hydrogenation of halo-aryl nitro, azo, and hydrazo compounds. Such hydrogenolysis, however, can be minimized by running these reactions in acidic media. This not only retards the dehalogenation reaction but also protonates the amine which destroys its activating influence.[124]

The loss of halogen can also be decreased by the use of non-polar solvents such as ethyl acetate, benzene, or cyclohexane (Eqn. 20.59).[132,138] The inactive forms of Raney nickel also catalyze the hydrogenation of an aryl nitro group at 100°C and 20 atmospheres of hydrogen without the breaking of an aryl chloride or bromide bond. Even iodides are retained to some extent.[139] Halobenzoyl-hydrazones can be hydrogenated over palladium at room temperature and 2 atmospheres to give the substituted hydrazines without loss of chlorine or bromine and only moderate loss of iodine.[140]

(20.59)

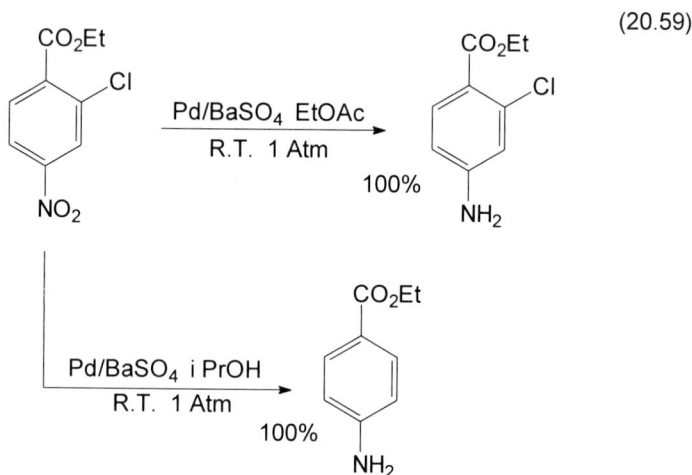

Complete dechlorination of polychlorobenzenes has been accomplished using a palladium catalyst at room temperature and 3–4 atmospheres pressure.[141] The presence of a surfactant also facilitates this reaction.[134,142] Very good yields of 3,5-dichloroaniline were obtained by the selective hydrodechlorination of poly-chloroanilines over palladium in acidic medium (Eqn. 20.60).[143,144] Dehydrochlorination of polychlorophenols under these conditions gave good yields of 3,5-dichlorophenol.[144]

(20.60)

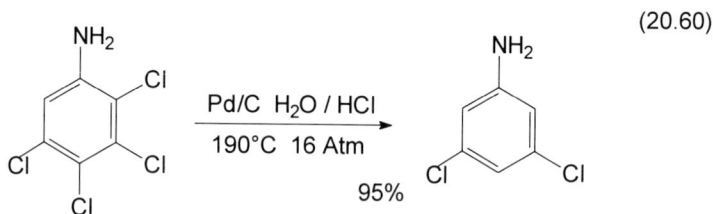

Heterocyclic Halogen Compounds

The removal of a halogen from an aromatic heterocycle takes place more readily than the hydrogenolysis of a phenyl halide of the same type (Eqn. 20.61).[145] The relative difference between the ease of dehalogenation of a phenyl and a heterocyclic halide is more pronounced over Raney nickel in basic medium than it is over palladium (Eqn. 20.62).[146]

The usual basic, palladium catalyzed dehalogenation conditions are, however, preferred over Raney nickel for nonselective halide hydrogenolysis. Dehalogenation over palladium occurs in preference to ring and cyano group

(20.61)

Pd/C EtOAc / Et₃N
R.T. 3 Atm
90%

(20.62)

R.Ni EtOH / KOH
R.T. 1 Atm
93%

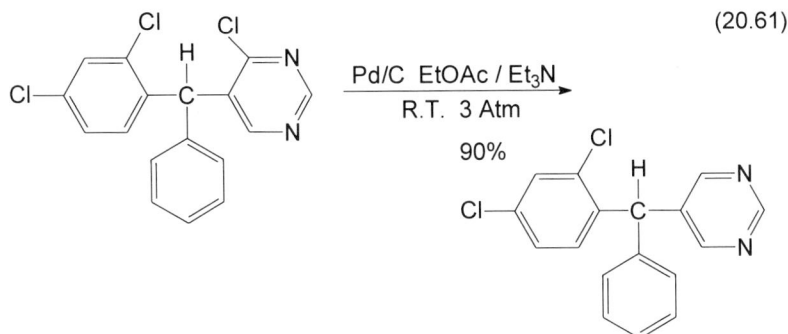

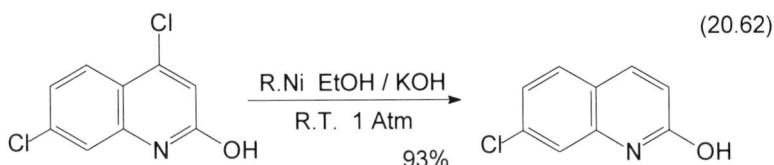

hydrogenation (Eqn. 20.63)[147] but nitro groups are hydrogenated prior to chloride hydrogenolysis in the absence of any added base or basic catalyst support material.[148]

The use of platinum catalysts in acetic acid is not recommended for the removal of halogens from heterocycles as ring hydrogenation also occurs under these conditions.

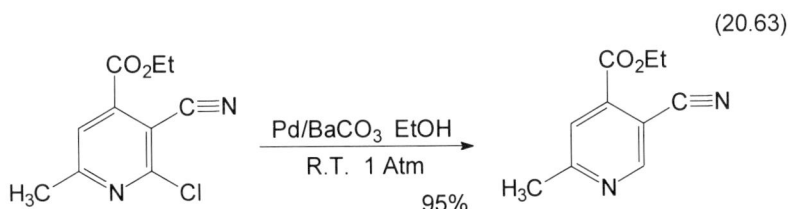

(20.63)

Pd/BaCO₃ EtOH
R.T. 1 Atm
95%

Acyl Halides

The hydrogenolysis of acyl chlorides, the Rosenmund reaction, is one of the standard methods used for the conversion of carboxylic acids to aldehydes.[149,150] The reaction is commonly run by bubbling hydrogen through a refluxing toluene or xylene solution of the acid chloride in which a Pd/BaSO₄ catalyst is suspended (Eqn. 20.64).[151] Since the rate of further hydrogenation of the carbonyl group over palladium increases rapidly with increasing temperature, keeping the reaction temperature at the minimum level required for dehalogenation decreases the chance of further hydrogenation and increases the yield of the aldehyde.

(20.64)

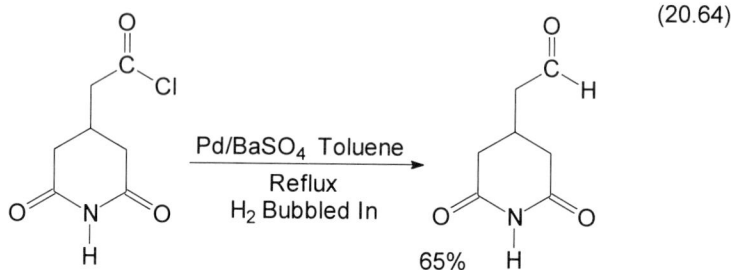

The addition of various modifiers, such as quinoline-sulfur,[149] thiourea,[152] or, especially, tetramethylthiourea[153] to the reaction mixture, is sometimes necessary in order to inhibit the further hydrogenation of the aldehyde. This is particularly true for the formation of aryl aldehydes that are hydrogenated over palladium under mild conditions.[154,155] The barium sulfate support is, itself, a moderate inhibitor and is generally sufficient for regulating the preparation of unsubstituted aliphatic aldehydes. However, if another support, such as charcoal, is used for the palladium, one of the standard inhibitors must be added to moderate the reaction.   When these added inhibitors are present, the hydrogenation of double bonds and nitro groups, as well as the hydrogenolysis of aryl chlorides, is suppressed and these groups remain intact throughout the reaction.[149]

(20.65)

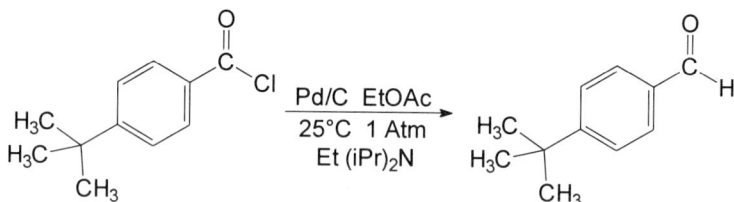

High yields of aromatic aldehydes have been obtained by the hydrogenolysis of the acid chlorides over Pd/C at room temperature and atmospheric pressure in ethyl acetate containing some ethyl di-iso-propylamine (Eqn. 20.65).[156]

C–S Bonds

The hydrogenolysis of carbon-sulfur bonds has found extensive use in both structure determination and synthesis.[157,158] A sulfur compound is desulfurized by refluxing it with a large excess of Raney nickel in a solvent, usually absolute ethanol or dioxane. Stirring is usually necessary to keep the nickel in suspension. Sulfur with oxidation states of from +2 to -6 can be removed by this procedure. Deselenization also occurs under these conditions.[159,160]

Not only is the sulfur removed in this reaction, but olefins, carbonyl and nitro groups, and, in some cases, even benzene rings are reduced. If the Raney nickel is deactivated by refluxing it in acetone before use,[161,162] the hydrogenation of these functional groups can be avoided. By the use of such a deactivated catalyst, thio enol ethers are converted to olefins (Eqn. 29.66).[163-166] Deactivation of the catalyst can also be brought about by heating it to temperatures of 200°–250°C. The use of these degassed catalysts, however, leads to the formation of side products that arise from the coupling of the desthio fragments.[157] Such products are not generally found when the acetone deactivated nickel is used.

(20.66)

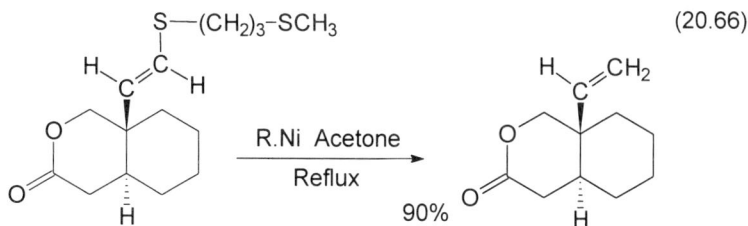

Since this procedure can be used to replace any sulfur atom in a molecule by a hydrogen, it is not surprising that it has many synthetic uses. Thioketals are converted to the hydrocarbons providing a good method for the removal of a carbonyl group in neutral medium.[157,167] Unsaturated thioketals are desulfurized over deactivated Raney nickel without saturation of the double bond (Eqn. 20.67).[168] Mercapto pyrimidines have been desulfurized in very good yields[169] and sulfonamides converted to amides.[170] Thioethers can be cleaved in preference to benzyl ethers if deactivated Raney nickel is used.

(20.67)

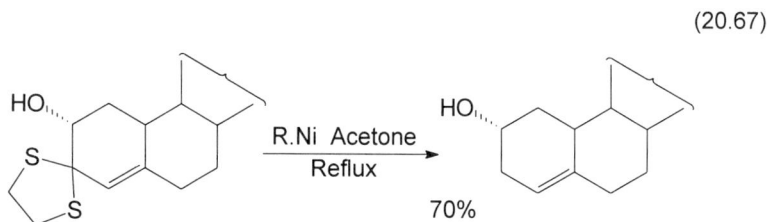

Since alkyl halides can be converted to mercaptans, thioethers, xanthates, or thiouronium compounds, the ready removal of these groups offers another

$$\text{H}_3\text{C}\underset{\text{CH}_2}{\overset{\text{O}}{\underset{}{\parallel}}}\text{C}\text{SEt} \xrightarrow[\text{Reflux}]{\text{R.Ni EtOH}} \underset{73\%}{\text{H}_3\text{C}\underset{\text{CH}_2}{\overset{\text{O}}{\parallel}}\text{C}\text{H}} \qquad (20.68)$$

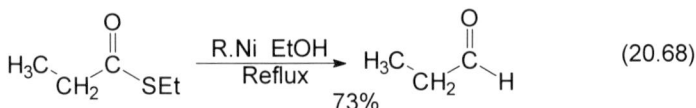

method for the dehalogenation of a molecule. The hydrogenolysis of thioesters over active Raney nickel gives the alcohol, but the use of a deactivated catalyst leads to aldehyde formation (Eqn. 20.68).[161,162,172] This provides an alternate method for the conversion of an acyl halide to an aldehyde. Acylation of thiophene followed by reductive desulfurization gives a method for the lengthening of the carbon chain by four carbons.[157] Condensation of thiophene with ketones gives di-(2-thienyl)alkanes which on desulfurization produces dialkyl-dibutylmethanes (Eqn. 20.69).[172,173]

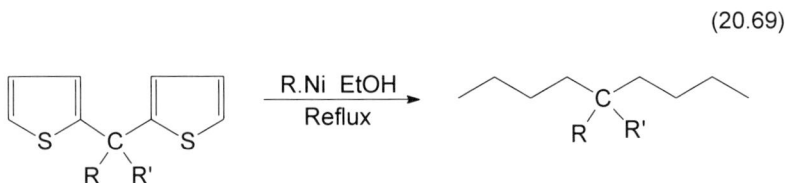

$$(20.69)$$

The major drawback to this procedure is the large amount of catalyst that is required for the reaction. This precludes the use of this method for any large scale process even though it represents an excellent way for removing a carbonyl or an alkyl halo group in neutral medium. The large amount of catalyst also tends to make product isolation somewhat difficult. It is sometimes necessary to continuously extract the catalyst or to dissolve the nickel in mineral acid in order to effect a reasonable degree of product recovery. Even with these drawbacks, the ease with which the reaction takes place and the conditions under which it is run, make it the favored method for many small scale synthetic conversions.

The reductive desulfurization of optically active sulfides[174] and sulfoxides[175] over Raney nickel gives the racemic products but desulfurization of sulfones occurs with retention of configuration.[175]

Other Bonds

S–S Bonds

The cleavage of a disulfide without the breaking of a carbon sulfur bond is possible over a palladium catalyst in aqueous acid at room temperature and atmospheric pressure. Good yields of the mercaptans have been obtained in this way.[176-178]

O–O Bonds

The cleavage of the oxygen-oxygen bond in hydroperoxides to give the alcohols takes place over platinum,[179,180] palladium[181,182] and Raney nickel[183] catalysts at room temperature and atmospheric pressure (Eqn. 20.70). The use of Lindlar's catalyst or a similarly deactivated catalyst is preferred for the cleavage of allyl hydroperoxides in order to minimize the hydrogenolysis of the resulting allyl alcohol (Eqn. 20.71).[182,183]

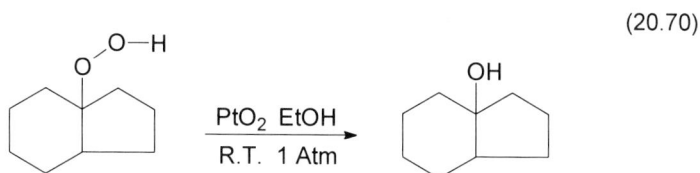

(20.70)

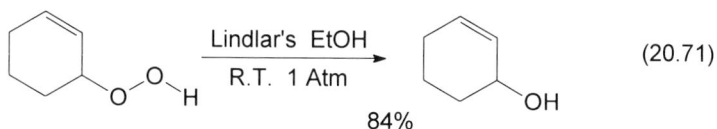

(20.71)

Hydrogenolysis of dialkylperoxides or epidioxides (cyclic peroxides) also takes place over these same catalysts under mild conditions.[184-187] If tertiary alcohols are the products, running the reaction in acid media can lead to dehydration.[184] Deactivated catalysts are also needed if the product contains an allyl alcohol that could be cleaved by the more active catalysts (Eqn. 20.72).[186,187]

(20.72)

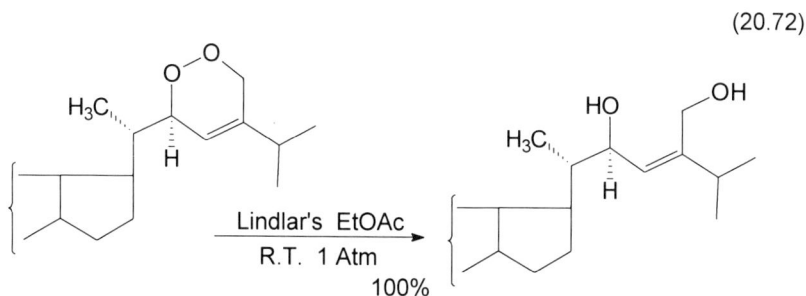

References

1. R. W. Shortridge, R. A. Craig, K. W. Greenlee, J. M. Derfer and C. E. Boord, *J. Am. Chem. Soc.*, **70**, 946 (1948).
2. C. Gröger and H. Musso, *Angew. Chem., Internat. Ed.*, **15**, 373 (1973).
3. P. v.R. Schleyer and E. Wiskott, *Tetrahedron Lett.*, 2845, (1967).
4. C. W. Woodworth, V. Buss and P. v. R. Schleyer, *J. Chem. Soc., Chem. Commun.*, 569 (1968).
5. Z. Majerski and P. v. R. Schleyer, *Tetrahedron Lett.*, 6195 (1968).
6. J. B. Hendrickson and R. K. Boeckman, Jr., *J. Am. Chem. Soc.*, **93**, 4491 (1971).
7. V. A. Slabey, P. H. Wise and L. C. Gibbons, *J. Am. Chem. Soc.*, **71**, 1518 (1949).
8. V. A. Slabey and P. H. Wise, *J. Am. Chem. Soc.*, **71**, 3252 (1949).
9. V. A. Slabey and P. H. Wise, *J. Am. Chem. Soc.*, **74**, 3887 (1952).
10. R. L. Augustine and B. A. Patel, *Tetrahedron Lett.*, 4545 (1976).
11. V. A. Slabey, *J. Am. Chem. Soc.*, **69**, 475 (1947).
12. K. J. Stahl, W. Hertzsch and H. Musso, *Liebigs Ann. Chem.*, 1474 (1985).
13. E. Galantay, N. Paolella, S. Barcza, R. V. Coombs and H. P. Weber, *J. Am. Chem. Soc.*, **92**, 5771 (1970).
14. M. N. Akhtar and W. R. Jackson, *J. Chem. Soc., Chem. Commun.*, 813 (1972).
15. M. N. Akhtar, J. J. Rooney and W. R. Jackson, *J. Chem. Soc., Perkin Trans. 2*, 595 (1976).
16. M. N. Akhtar, J. J. Rooney and W. R. Jackson, *J. Chem. Soc., Perkin Trans. 2*, 1412 (1976).
17. B. M. Mil'vitkaya, A. F. Plate, K. B. Sterin and M. P. Ivannikova, *Zh. Org. Khim.*, **4**, 271 (1968); *Chem. Abstr.*, **68**, 104356 (1968).
18. M. J. Jorgenson, *Tetrahedron Lett.*, 4577 (1968).
19. R. V. Volkenburgh, K. W. Greenlee, J. M. Derfer and C. E. Boord, *J. Am. Chem. Soc.*, **71**, 172 (1949).
20. S. R. Poulter and C. H. Heathcock, *Tetrahedron Lett.*, 5339 (1968).
21. A. L. Schultz, *J. Org. Chem.*, **36**, 383 (1971).
22. S. Mitsui, Y. Sugi, M. Fujimoto and K. Yokoo, *Tetrahedron*, **30**, 31 (1974).
23. K. Isogai, N. Nishizawa, T. Saito and J. Sakai, *Bull. Chem. Soc., Japan*, **56**, 1555 (1983).
24. K. Isogai, J. Sakai and K. Kosugi, *Bull. Chem. Soc., Japan*, **59**, 1359 (1986).
25. T. Oshima, Y. Nakajima and T. Nagai, *Chem. Lett.*, 1977 (1993).
26. K. Isogai, J. Sakai and K. Yamauchi, *Nippon Kagaku Kaishi*, 214 (1986); *Chem. Abstr.*, **105**, 171567 (1986).
27. K. Isogai, J. Sakai and K. Yamauchi, *Nippon Kagaku Kaishi*, 825 (1986); *Chem. Abstr.*, **106**, 155909 (1987).
28. J. Frei, C. Ganter, D. Kagi, K. Kocsis, M. Miljkovic, A. Siewinski, R. Wenger, K. Schaffner and O. Jeger, *Helv. Chim. Acta*, **49**, 1049 (1966).
29. S. B. Laing and P. J. Sykes, *J. Chem. Soc.,C*, 937 (1968).
30. R. L. Augustine and E. J. Reardon, Jr., *J. Org. Chem.*, **39**, 1627 (1974).
31. U. R. Ghatak, P. C. Chakraborti, B. C. Ranu and B. Sanyal, *J. Chem. Soc., Chem. Commun.*, 548 (1973).

32. P. N. Chakrabortty, R. Dasbupta, S. K.Dasgupta, S. R. Ghosh and U. R. Ghatak, *Tetrahedron,* **28**, 4653 (1972).
33. A. P. G. Kieboom, A. J. Breijer and H. van Bekkum, *Recl. Trav. Chim. Pays-Bas*, **93**, 186 (1974).
34. A. P. G. Kieboom, H. J. van Benschop and H. van Bekkum *Recl. Trav. Chim. Pays-Bas*, **95**, 231 (1975).
35. R. L. Augustine and B. A. Patel, *Catalysis in Organic Synthesis-1976,* (P. N. Rylander and H. Greenfield, Eds.) Academic Press, New York, 1976, p. 325.
36. W. H. Hartung and R. Simonoff, *Org. Reactions*, **8**, 263 (1953).
37. E. Hardegger, Z. El Heweihi and F. G. Robinet, *Helv. Chim. Acta*, **31**, 439 (1948).
38. L. J. Stegerhoek and P. E. Verkade, *Recl. Trav. Chim. Pays-Bas*, **75**, 143 (1956).
39. K. W. Rosenmund, E. Karg and F. K. Marcus, *Chem. Ber.*, **75**, 1850 (1942).
40. H. Kindler, B. Hedemann and E. Scharfe, *Liebigs Ann. Chem.*, **560**, 215 (1948).
41. J. S. Tou and B. D. Vineyard, *J. Org. Chem.*, **49**, 1135 (1984).
42. S. Nishimura, *Bull. Chem. Soc., Japan*, **32**, 1155 (1959).
43. E. Galantay, *Tetrahedron,* **19**, 319 (1963).
44. Y. Ichinohe and H. Ito, *Bull. Chem. Soc., Japan*, **37**, 887 (1964).
45. S. Nishimura and M. Hama, *Bull. Chem. Soc., Japan*, **39**, 2467 (1966).
46. A. P. G. Kieboom, J. F. de Kreuk and H. van Bekkum, *J. Catal.*, **20**, 58 (1971).
47. Y. Oikawa, T. Tanaka, K. Horia and O. Yonemitsu, *Tetrahedron Lett.*, **25**, 5397 (1984).
48. F. E. King and T. J. King, *J. Chem. Soc.,* 726 (1947).
49. C. E. Ballou, S. Roseman and K. P. Link, *J. Am. Chem. Soc.*, **73**, 1140 (1951).
50. A. S. Meyer and T. Reichstein, *Helv. Chim. Acta*, **29**, 152 (1946).
51. H. B. Wood, Jr. and H. G. Fletcher, Jr., *J. Am. Chem. Soc.*, **78**, 2849 (1956).
52. P. E. Papadakis, *J. Am. Chem. Soc.*, **58**, 665 (1936).
53. B. P. Czech and R. A. Bartsch, *J. Org. Chem.*, 49, 4076 (1984).
54. L. S. Seif, K. M. Partyka and J. E. Hengeveld, *Chem. Ind. (Dekker)*, **40** (Catal. Org. React.), 197 (1990).
55. W. D. McPhee and E. S. Erickson, Jr., *J. Am. Chem. Soc.*, **68**, 624 (1946).
56. R. D. Bernotas and R. V. Cube, *Synth. Commun.,* **20**, 1209 (1990).
57. H. R. Snyder and D. S. Matteson, *J. Am. Chem. Soc.*, **79**, 2217 (1957).
58. J. C. Sheehan and A. K. Bose, *J. Am. Chem. Soc.*, **72**, 5158 (1950).
59. A. Caron, C. Bunel, C. Braud and M. Vert, *Polymer*, **32**, 2659 (1991).
60. P. Ruggli and H. Dahn, *Helv. Chim. Acta*, **27**, 1116 (1944).
61. H. Kimura and C. H. Stammer, *J. Org. Chem.*, **48**, 2440 (1983).
62. R. H. Hall and H. G. Khorana, *J. Am. Chem. Soc.*, **76**, 5056 (1954).
63. J. Meienhofer and K. Kurojizu, *Tetrahedron Lett.*, 3259 (1974).
64. S. Mitsui, Y. Senda and K. Konno, *Chem. and Ind., (London)*, 1354 (1963).
65. S. Mitsui and Y. Kudo, *Chem. and Ind., (London)*, 381 (1965).
66. S. Mitsui, Y. Kido and M. Kobayashi, *Tetrahedron,* **25**, 1921 (1969).

67. S. Mitsui, S. Imaizumi and Y. Esashi, *Bull. Chem. Soc., Japan*, **43**, 2143 (1970).
68. A. M. Khan, F. J. McQuillin and I. Jardine, *Tetrahedron Lett.*, 2649 (1966).
69. S. Mitsui and S. Imaizumi, *Bull. Chem. Soc., Japan*, **34**, 774 (1961).
70. S. Mitsui and S. Imaizumi, *Bull. Chem. Soc., Japan*, **36**, 855 (1963).
71. M. C. Dart and H. B. Henbest, *Nature*, **183**, 817 (1959).
72. M. C. Dart and H. B. Henbest, *J. Chem. Soc.*, 3563 (1960).
73. A. Giger, M. Fetizon, J. Henniker and L. Jacque, *C. R. Acad. Sci.,* **251**, 2194 (1960).
74. G. L. Southard, B. R. Zaborowsky and J. M. Pettee, *J. Am. Chem. Soc.*, **93**, 3302 (1971).
75. F. J. McQuillin and W. O. Ord, *J. Chem. Soc.*, 3169 (1959).
76. G. V. Pigulevskii and Z. Y. Rubashko, *Chem. Abstr.*, **50**, 9291 (1956).
77. R. C. Fuson, *Reactions of Organic Compounds*, Wiley, New York, 1962, p.651.
78. S. P. Fore and W. G. Bickford, *J. Org. Chem.*, **24**, 620 (1959).
79. D. R. Howton and R. W. Kaiser, Jr., *J. Org. Chem.*, **29**, 2420 (1964).
80. S. Mitsui, S. Imaizumi, M. Hisashige and Y. Sugi, *Tetrahedron,* **29**, 4093 (1973).
81. O. Dann and H. Hofmann, *Chem. Ber.*, **96**, 320 (1963).
82. S. Mitsui, Y. Senda, T. Shimodaira and H. Ishikawa, *Bull. Chem. Soc., Japan*, **38**, 1897 (1965).
83. P. A. Plattner, T. Petrzilka and W. Lang, *Helv. Chim. Acta*, **27**, 513 (1944).
84. E. J. Agnello, R. Pinson, Jr. and G. D. Laubach, *J. Am. Chem. Soc.*, **78**, 4756 (1956).
85. G. C. Accrombessi, P. Geneste, J.-L. Olivé and A. A. Pavia, *J. Org. Chem.*, **45**, 4139 (1980).
86. J. Ishiyama, H. Yashima, H. Matsuo, Y. Senda, S. Imaizumi, K. Hanaya and T. Muramatsu, *Chem. Lett.*, 989 (1989).
87. R. Subbarao, G. Venkateswara Rao and K. T. Achaya, *Tetrahedron Lett.*, 379 (1966).
88. S. Nishimura, M. Shiota and A. Mizuno, *Bull. Chem. Soc., Japan*, **37**, 1207 (1964).
89. Y. Nagahisa, Y. Sugi and S. Mitsui, *Chem. and Ind., (London)*, **38** (1975).
90. W. L. Howard and J. H. Brown, Jr., *J. Org. Chem.*, **26**, 1026 (1961).
91. K. Kindler, H. Oelschlager and P. Henrich, *Chem. Ber.*, **86**, 501 (1953).
92. J. W. Burnham and E. J. Eisenbraun, *J. Org. Chem.*, **36**, 737 (1971).
93. E. J. Eisenbraun, J. W. Burnham, J. D. Weaver and T. E. Webb, *Ann. N. Y. Acad. Sci.*, **214**, 204, (1973).
94. F. Grass, J. M. Grosselin and C. Mercier, *Stud. Surf. Sci. Catal.*, **78** (Heterog. Catal. Fine Chem., III), 259 (1993).
95. R. A. Cormier, *Synth. Commun.*, **11**, 295 (1981).
96. T. Chihara, S. Teratani, M. Hasegawa-Ohotomo, T. Amemiya and K. Taya, *J. Catal.*, **90**, 221 (1984).
97. W. J. Musliner and J. W. Gates, Jr., *J. Am. Chem. Soc.*, **88**, 4271 (1966).
98. J. D. Weaver and E. J. Eisenbraun, *Prepr. Div. Pet. Chem., Am. Chem. Soc.*, **18**, 196 (1973).

99. R. Bognar, G. Gaal, P. Kerekes, G. Horvath and M. T. Kovacs, *Org. Prep. Proced. Int.*, **6**, 305 (1974).
100. K. Clauss and H. Jensen, *Angew. Chem., Internat. Ed.*, **12**, 918 (1973).
101. A. Jung and R. Engel, *J. Org. Chem.*, **40**, 244 (1975).
102. J. R. Vaughn, Jr. and J. Blodinger, *J. Am. Chem. Soc.*, **77**, 5757 (1955).
103. K. Ikedate, T. Harada and S. Suzuki, *Nippon Kagaku Zasshi*, **92**, 246 (1971); *Chem. Abstr.*, **76**, 24420 (1972).
104. A. P. Phillips, *J. Am. Chem. Soc.*, **76**, 2211 (1954).
105. L. Birkofer, *Chem. Ber.*, **75**, 429 (1942).
106. S. Mitsui, A. Kasahara and N. Endo, *J. Chem. Soc. Japan*, **75**, 234 (1954); *Chem. Abstr.*, **49**, 10210 (1955).
107. W. E. Solodar and M. Green, *J. Org. Chem.*, **27**, 1077 (1962).
108. W. Reeve and A. Sadle, *J. Am. Chem. Soc.*, **72**, 3252 (1950).
109. R. Baltzly and J. S. Buck, *J. Am. Chem. Soc.*, **65**, 1984 (1943).
110. R. Baltzly and P. B. Russell, *J. Am. Chem. Soc.*, **72**, 3410 (1950).
111. X. Fu and J. M. Cook, *J. Org. Chem.*, **58**, 661 (1993).
112. Y. Siwschitz and A. Zilkha, *J. Am. Chem. Soc.*, **76**, 3698 (1954).
113. C. O'Murchu, *Tetrahedron Lett.*, 3231 (1969).
114. H. Dahn, J. A. Barbarino and C. O'Murchu, *Helv. Chim. Acta*, **53**, 1370 (1970).
115. Y. Sugi and S. Mitsui, *Tetrahedron*, **29**, 2041 (1973).
116. K. N. Campell, A. H. Sommers and B. K. Campbell, *J. Am. Chem. Soc.*, **68**, 140 (1946).
117. A. H. Sommers and B. K. Campell, *Org. Synth.*, **27**, 12 (1947).
118. P. G. Gassman and A. Fentiman, *J. Org. Chem.*, **32**, 2388 (1967).
119. S. Mitsui and Y. sugi, *Tetrahedron Lett.*, 1287 (1969).
120. Y. Sugi and S. Mitsui, *Bull. Chem. Soc., Japan*, **42**, 2984 (1969).
121. Y. Sugi and S. Mitsui, *Bull. Chem. Soc., Japan*, **43**, 1489 (1970).
122. J. N. Pattison and E. F. Degering, *J. Am. Chem. Soc.*, **73**, 611 (1951).
123. G. E. Ham and W. P. Cocker, *J. Org. Chem.*, **29**, 194 (1964).
124. R. Baltzly and A. P. Phillips, *J. Am. Chem. Soc.*, **68**, 261 (1946).
125. C. Paal and C. Müller-Lobek, *Chem. Ber.*, **64B**, 2142 (1931).
126. M. G. Reinecke, *J. Org. Chem.*, **29**, 299 (1964).
127. J. D. Chanley, *J. Am. Chem. Soc.*, **71**, 829 (1949).
128. L. Horner, L. Schlafer and H. Kämmerer, *Chem. Ber.*, **92**, 1700 (1959).
129. T. Mallat and J. Petro, *React. Kinet. Catal. Lett.*, **38**, 325 (1989).
130. M. G. Reinecke, *J. Org. Chem.*, **28**, 3574 (1963).
131. P. K. Hovmann, S. F. Orochena, S. M. Sax and G. A. Jeffrey, *J. Am. Chem. Soc.*, **81**, 992 (1959).
132. K. Kindler, H. Oelschlager and P. Henrich, *Chem. Ber.*, **86**, 167 (1953).
133. L. D. Freedman, G. O. Doak and E. L. Petil, *J. Am. Chem. Soc.*, **77**, 4262 (1955).
134. C. A. Marques, M. Selva and P. Tundo, *J. Org. Chem.*, **59**, 3830 (1994).
135. M. Freifelder, *J. Org. Chem.*, **31**, 3875 (1966).
136. H. Kammerer, L. Horner and H. Beck, *Chem. Ber.*, **91**, 1376 (1958).
137. V. Ruzicka and J. Prochazka, *Coll. Czech. Chem. Commun.*, **35**, 430 (1970).
138. A. Weizmann, *J. Am. Chem. Soc.*, **71**, 4154 (1949).
139. C. F. Winans, *J. Am. Chem. Soc.*, **61**, 3564 (1939).
140. M. Freifelder, W. B. Martin, G. R. Stone and E. L. Coffin, *J. Org. Chem.*, **26**, 383 (1961).

141. E. N. Balko, E. Przybylski and F. Von Trentini, *Appl. Catal.*, *B*, **2**, 1 (1993).
142. C. A. Marques, M. Selva and P. Tundo, *J. Org. Chem.*, **58**, 5256 (1993).
143. G. Cordier and Y. Colleuille, *Chem. Ind. (Dekker)*, **18** (Catal. Org. React.), 197 (1984).
144. G. Cordier, *Stud. Surf. Sci. Catal.*, **41** (Heterog. Catal. Fine Chem.), 19 (1988).
145. J. B. Campbell, C. W. Whitehead, T. J. Kress and L. L. Moore, *Ann. N. Y. Acad. Sci.*, **214**, 216 (1973).
146. R. E. Lutz, G. Ashburn and R. J. Rowlett, Jr., *J. Am. Chem. Soc.*, **68**, 1322 (1946).
147. M. J. Reider and R. C. Elderfield, *J. Org. Chem.*, **7**, 286 (1942).
148. L. Velluz and G. Amiard, *Bull. soc. chim., France*, 136 (1947).
149. E. Mosettig and R. Mozingo, *Org. Reactions*, **4**, 362 (1948).
150. D. P. Wagner, H. Gurien and A. I. Rachlin, *Ann. N. Y. Acad. Sci.*, **172**, 186 (1970).
151. D. D. Philips, M. A. Acitelli and J. Meinwald, *J. Am. Chem. Soc.*, **79**, 3517 (1957).
152. C. Weygand and W. Meusel *Chem. Ber.*, **76**, 503 (1943).
153. S. Affrossman and S. J.Thomson, *J. Chem. Soc.*, 2024 (1962).
154. Z. Poltarzewsky, S. Galvagno, P. Staiti, P. Antonucci, A. Rositani and N. Giordano, *React. Kinet. Catal. Lett.*, **22**, 383 (1983).
155. G. Cum, R. Gallo, S. Galvagno, A. Spadaro and P. Vitarelli, *J. Chem. Technol. Biotechnol.*, **34A**, 416 (1984).
156. J. A. Peters and H. van Bekkum, *Recl. Trav. Chim. Pays-Bas*, **100**, 21 (1981).
157. G. R. Pettit and E. E. van Tamelin, *Org. Reactions*, **12**, 356 (1962).
158. H. Hauptmann and W. F. Walter, *Chem. Rev.*, **62**, 347 (1962).
159. G. E. Wiseman and E. S. Gould, *J. Am. Chem. Soc.*, **76**, 1706 (1954).
160. H. Hauptmann and W. F. Walter, *J. Am. Chem. Soc.*, **77**, 4929 (1955).
161. G. B. Spero, A. V. McIntosh, Jr. and R. H. Levin, *J. Am. Chem. Soc.*, **70**, 1907 (1948).
162. R. H. Levin, G. B. Spero, A. V. McIntosh, Jr. and D. E. Rayman, *J. Am. Chem. Soc.*, **70**, 2958 (1948).
163. G. Rosenkranz, S. Kaufmann and J. Romo, *J. Am. Chem. Soc.*, **71**, 3689 (1949).
164. J. Romo, M. Romero, C. Djerassi and G. Rosenkranz, *J. Am. Chem. Soc.*, **73**, 1528 (1951).
165. J. A. Marshall and D. E. Seitz, *Synth. Commun.*, **4**, 395 (1974).
166. J. A. Marshall and D. E. Seitz, *J. Org. Chem.*, **40**, 534 (1975).
167. F. sondhei;mer and S. Wolfe, *Can. J. Chem.*, **37**, 1870 (1959).
168. P. N. Rao and H. R. Gollberg, *Tetrahedron*, **18**, 1251 (1962).
169. H. N. Schlein, M. Israel, S. Chatterjee and E. J. Modest, *Chem. and Ind., (London)*, 418 (1964).
170. G. R. Pettit and R. E. Kadunce, *J. Org. Chem.*, **27**, 4566 (1962).
171. M. L. Wolfrom and J. V. Karabinos, *J. Am. Chem. Soc.*, **68**, 1455 (1946).
172. M. Sy and M. Maillet, *Bull. soc. chim., France*, 2253 (1966).
173. M. Sy, M. Maillet and P. David, *Bull. soc. chim., France*, 2609 (1967).
174. I. Imaizumi, *Nippon Kagaku Zasshi*, **78**, 1396 (1957); *Chem. Abstr.*, **54**, 1403 (1960).

175. W. A. Bonner, *J. Am. Chem. Soc.*, **74**, 1034, 5089 (1952).
176. M. Bergmann and J. Michalis, *Chem. Ber.*, **63**, 987 (1930).
177. K. E. Kavanagh, *J. Am. Chem. Soc.*, **64**, 2721 (1942).
178. L. Zervas and D. M. Theodoropoulos, *J. Am. Chem. Soc.*, **78**, 1359 (1956).
179. R. Creigee and H. Zogel, *Chem. Ber.*, **84,** 215 (1951**).**
180. B. Witkop and J. B. Patrick, *J. Am. Chem. Soc.*, **73**, 2188 (1951).
181. B. Lythgoe and S. Trippett, *J. Chem. Soc.*, 471 (1959).
182. M. Rebeller and G. Clement, *Bull. soc. chim., France,* 1302 (1964).
183. G. O. Schenck and O. A. Neumüller, *Liebigs Ann. Chem.*, **618**, 194 (1958).
184. P. Pladon, R. B. Clayton, C. W. Greenhalgh, H. B. Henbest, E. R. H. Jones, B. J. Lovell, G. Solverstone, G. W. Wood and G. F. Woods, *J. Chem. Soc.,* 4883, (1952).
185. R. B. Clayton, H. B. Henbest and E. R. H. Jones, *J. Chem. Soc.,* 2015 (1953).
186. G. D. Laubach, E. C. Schreiber, E. J. Agnello and K. J. Brunings, *J. Am. Chem. Soc.*, **78**, 4746 (1956).
187. M. Anastasia, P. Allevi, P. Ciuffreda, A. Fiecchi and A. Scala, *J. Org. Chem.*, **49**, 4297 (1984).

Oxidation

The oxidation of organic functional groups is an important synthetic reaction. For the synthesis of complex molecules, however, most oxidations are run using inorganic oxidants such as permanganate and chromium oxide and the disposal of their reduction products is environmentally unacceptable. In the manufacture of bulk chemicals, catalytic oxidation has become very important for the conversion of alkenes, alkanes and aromatics into the more valuable oxygen containing materials. These processes, which include transformations such as propene to acrolein,[1,2] C-4 hydrocarbons to maleic anhydride,[3,4] o-xylene to phthalic acid,[5] and ethylene to ethylene oxide[6,7] have been extensively reviewed.[8-12] Since these reactions are run in the vapor phase and the reaction conditions are usually specific for the production of a single product, they are generally not applicable to the synthesis of the more complex molecules encountered in fine chemical syntheses. Thus, these reactions will not be discussed here. Instead, the following sections will be devoted to those catalytic reactions that have found use in more classic synthetic transformations.[13-15] While most of these reactions take place in the liquid phase, some vapor phase processes will also be discussed. The primary oxidants used in these reactions are hydrogen peroxide, organic peroxides and oxygen.

Hydrogen Peroxide and Organic Peroxide Oxidants

Metal oxides such as MoO_3 and WO_3 have been used as catalysts for the epoxidation of alkenes with alkyl hydroperoxides but the activity of such catalysts is usually due to the formation of soluble peroxo complexes so these reactions are, in reality, homogeneously catalyzed.[16] Supporting the MoO_3 on silica only served to facilitate the dissolution of the molybdenum species by dispersing the molybdena over the surface of the silica and making it more accessible to the peroxide.[17] A Ti^{IV}/SiO_2 species, however, has been shown to be an active, heterogeneous catalyst for the epoxidation of double bonds with alkyl hydroperoxides.[16,18]

Titanium Silicalites

Another approach to minimizing the problem of metal peroxide dissolution was to incorporate a redox metal into a zeolite by replacing some or all of the aluminum in the framework. Of these redox zeolites, the most common catalysts for peroxide oxidations are the titanium silicalites, TS-1 and TS-2 (Chapter 10).[14,15,19-23] TS-1, the more generally used material, has a crystal structure analogous to ZSM-5 with two dimensional channels of 0.55–0.60 nm in diameter.[24,25] The catalytic sites are isolated titanium ions incorporated into the zeolite framework. Because the titanium atoms are separated from each other, they cannot be deactivated by oligomerization of the active Ti=O species as is observed with other, more conventional oxidation catalysts. The titanium atoms react with hydrogen peroxide to form titanium-peroxo species within the zeolite framework so reactions on these sites exhibit varying degrees of shape selectivity depending on the nature of the substrate.

    Molecules with a cross section diameter greater than about 0.60 nm cannot diffuse into the TS-1 channels and are, therefore, not oxidized. This size restriction limits this system essentially to the oxidation of linear molecules and monocyclic aromatic rings with at most, small substituents. Even with these limitations, TS-1, and to a lesser extent TS-2, is an effective catalyst for the selective oxidation of a number of different types of organic compounds. Thirty percent hydrogen peroxide is the most commonly used oxidant. The more bulky alkyl peroxides are not effective because of their inability to diffuse into the zeolite channels to react with the titanium sites. While the oxidation of most primary and secondary alcohols occurs with reasonable ease, methanol is sufficiently inert to oxidation under the common reaction conditions that it is the solvent of choice for most TS-1 catalyzed reactions.[19,21,26,27]

$$\text{(21.1)}$$

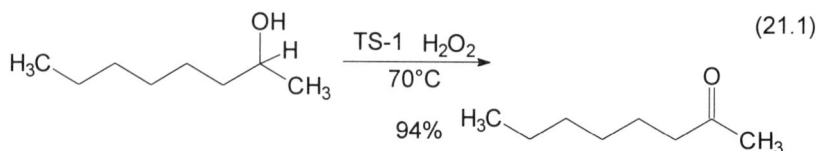

    The oxidation of secondary alcohols gives good to excellent yields of the corresponding ketones (Eqn. 21.1)[28] while primary alcohols are oxidized to the aldehydes at low conversions and the acids at high conversions.[19,27,28] Secondary alcohols on the second carbon atom (β) of a chain react more rapidly than primary alcohols while secondary alcohols on other carbon atoms of the chain (γ) react more slowly.[28,29] The general order for the oxidation of alcohols on a linear carbon chain is $\beta > \alpha > \gamma$.[28] The difference in activity between the β and γ alcohols has been ascribed to a difference in the transition state restrictions in these oxidations. These reactions are usually run in moderately dilute solutions at temperatures between 50°–70°C.

    The epoxidation of alkenes, though, is generally carried out at ambient temperature. This reaction is rapid and selective with unhindered double bonds.[19] Hindered alkenes or those having electron withdrawing groups present in the molecule require somewhat higher temperatures for epoxidation.[19] The oxidation of allyl chloride takes place at 40°C to give a 75% yield of epichlorohydrin (Eqn. 21.2).[30]

$$\text{(allyl chloride)} \quad \xrightarrow[\text{40°C}]{\text{TS-1 H}_2\text{O}_2} \quad \text{(epichlorohydrin)} \qquad (21.2)$$

75%

$$\text{(allyl alcohol)} \quad \xrightarrow[\text{40°C}]{\text{TS-1 H}_2\text{O}_2} \quad \text{(glycidol)} \qquad (21.3)$$

85%

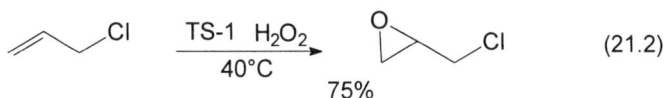

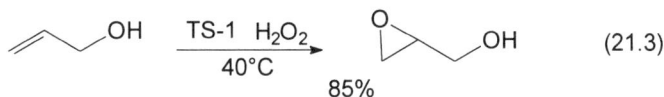

The selectivity in the reaction of unsaturated alcohols with hydrogen peroxide in the presence of TS-1 depends on the substitution on the double bond and the carbinol carbon. Reaction of allyl alcohol with hydrogen peroxide in the presence of TS-1 gave an 85% yield of the epoxide (Eqn. 21.3).[30] While the epoxidation of a terminal double bond was slower than that of internal double bonds, the oxidation of molecules having primary or secondary alcohols along with a terminal double bond gave only the epoxide with no alcohol oxidation (Eqn. 21.4),[31] but with more substituted unsaturated alcohols, oxidation of both the double bond and the alcohol took place. In these reactions the ratio of products was determined by the degree of substitution on the double bond and the carbinol carbon.[31] Increasing the steric hindrance around the double bond increases the amount of alcohol oxidation and vice versa. For example, 3-penten-2-ol (**1**) has nearly equal steric environments about both the double bond and the carbinol carbon. Oxidation of **1** over TS-1 gave both the epoxy alcohol, **2**, and the unsaturated ketone, **3**, in nearly equal amounts (Eqn. 21.5).[31]

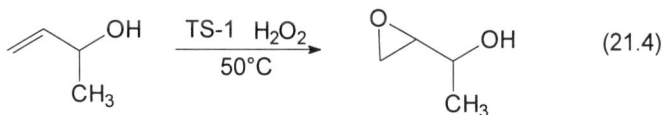

$$\xrightarrow[\text{50°C}]{\text{TS-1 H}_2\text{O}_2} \qquad (21.4)$$

$$\text{CH}_3 \qquad \qquad \text{CH}_3$$

$$(21.5)$$

$$\underset{\mathbf{1}}{\text{H}_3\text{C}\diagup\diagdown\text{CH}_3} \xrightarrow[\text{50°C}]{\text{TS-1 H}_2\text{O}_2} \underset{\mathbf{2}}{\text{H}_3\text{C} \quad \text{CH}_3} + \underset{\mathbf{3}}{\text{H}_3\text{C} \quad \text{CH}_3}$$

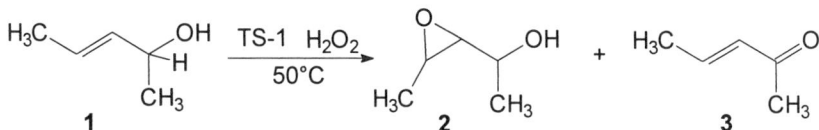

Phenol was hydroxylated with TS-1 and hydrogen peroxide to give a 1:1 mixture of catechol (**4**) and hydroquinone (**5**) (Eqn. 21.6).[26,29] Anisole and other substituted benzenes gave similar results. Secondary and tertiary aliphatic C–H bonds were oxidized to give alcohols and ketones, but primary C–H bonds were

(21.6)

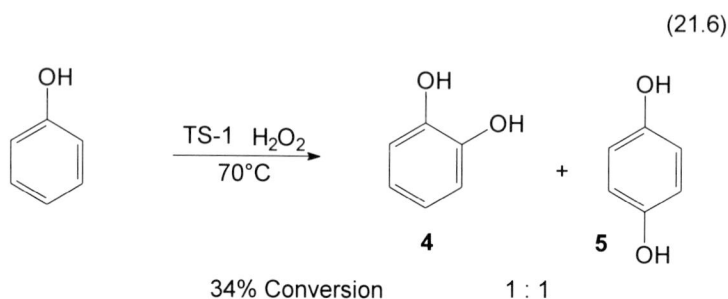

34% Conversion   1 : 1

unreactive. Methylene groups adjacent to a methyl group were preferentially oxidized to a mixture of the alcohol and ketone. A γ methylene group was more difficult to oxidize. Oxidation takes place at the linear end of a branched alkane (Eqn. 21.7).[32] The presence of an electron withdrawing group such as a halide or an ester on one end of a linear hydrocarbon forces the oxidation to take place at the other end.[26]

(21.7)

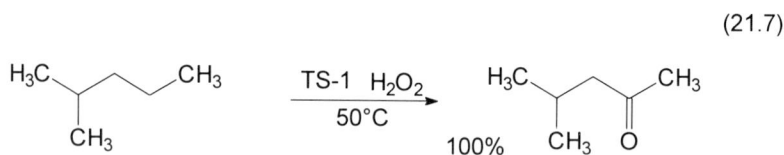

(21.8)

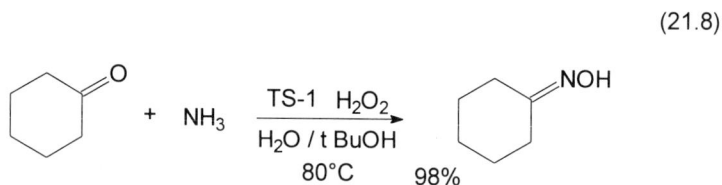

Ammonia was oxidized with hydrogen peroxide in the presence of TS-1 first to hydroxylamine and then, further to nitrogen oxides if the hydroxylamine was not trapped. When this reaction was run in the presence of a ketone, the oxime was selectively formed (Eqn. 21.8).[33] Oximes were also produced by the oxidation of primary aliphatic amines with TS-1 or TS-2 and hydrogen peroxide at ambient temperature. Linear amines such as n-propyl amine (6) gave the oxime in 73% yield at 32% conversion over the TS-1 catalyst (Eqn. 21.9).[34] The iso-propyl amine (7) was oxidized with somewhat higher selectivity over the larger pore TS-2 catalyst (Eqn. 21.10).[34] Primary aryl amines were oxidized to the symmetrical azoxybenzenes under these conditions (Eqn. 21.11).[35]

A number of other redox zeolites have peen prepared but they have not been as extensively studied as TS-1.[15,20] A titanium beta zeolite having a larger

(21.9)

$$H_3C-CH_2-CH_2-NH_2 \quad \xrightarrow[\text{MeOH } 60°C]{\text{TS-1 } H_2O_2} \quad H_3C-CH_2-CH=NOH$$

6

32% Conv. 73% Selectivity

(21.10)

$$(H_3C)_2CH-NH_2 \quad \xrightarrow[\text{MeOH } 60°C]{\text{TS-2 } H_2O_2} \quad (H_3C)_2C=NOH$$

7

31% Conv. 84% Selectivity

(21.11)

$$\text{C}_6\text{H}_5-\text{NH}_2 \quad \xrightarrow[\text{Acetone } 55°C]{\text{TS-1 } H_2O_2} \quad \text{C}_6\text{H}_5-N=N(O)-\text{C}_6\text{H}_5$$

75%

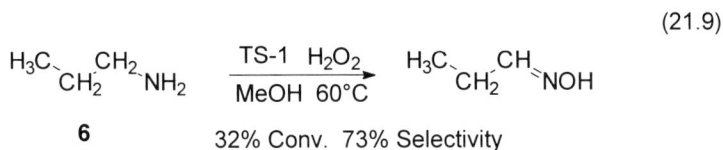

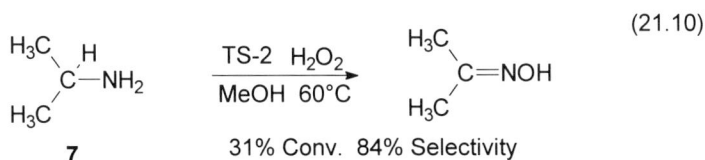

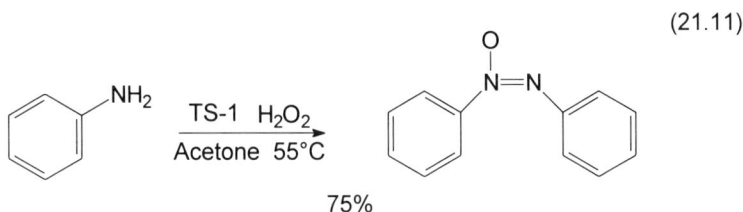

pore size than TS-1 was more reactive in the oxidation of larger molecules.[36] Selective oxidation reactions were also observed using a vanadium molecular sieve catalyst.[37,38] Zirconium-[39] and chromium-zeolites[40] have also been prepared.

Pillared Clays

Another approach to designing shape-selective heterogeneous oxidation catalysts was to use redox metal oxides as the pillaring agents in the preparation of pillared clays.[41] These redox pillared clays have been used for a number of selective oxidations. Chromium pillared montmorillonite (Cr-PILC) is an effective catalyst for the selective oxidation of alcohols with tert-butyl hydroperoxide.[42] Primary aliphatic and aromatic alcohols are oxidized to the aldehydes in very good yields. Secondary alcohols are selectively oxidized in the presence of a primary hydroxy group of a diol to give keto alcohols in excellent yields (Eqn. 21.12).[42]

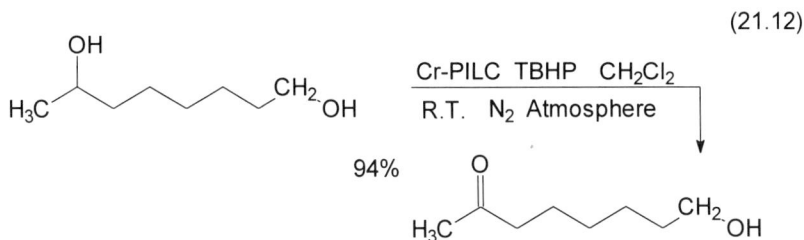

(21.12)

$$\xrightarrow[\text{R.T. } N_2 \text{ Atmosphere}]{\text{Cr-PILC TBHP } CH_2Cl_2}$$

94%

(21.13)

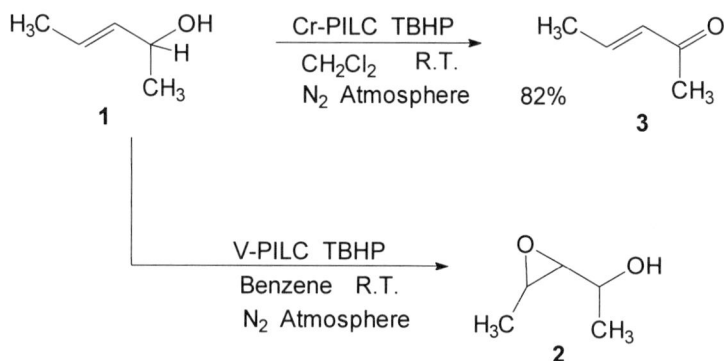

In contrast to the lack of selectivity observed in the TS-1 catalyzed oxidation of 3-penten-2-ol (**1**) (Eqn. 21.5), the oxidation of **1** with tert-butyl hydroperoxide (TBHP) over Cr-PILC gave the unsaturated ketone, **3**, in 82% yield (Eqn. 21.13)[42] while the oxidation of **1** over a vanadium pillared montmorillonite (V-PILC) gave the epoxy alcohol, **2**, in 94% yield.[43] V-PILC, however, does promote the oxidation of primary benzyl alcohols to the acids with tert-butyl hydroperoxide. This reaction exhibits shape selectivity in that para-substituted benzyl alcohols are oxidized while the ortho- and meta- substituted species are essentially inert (Eqn. 21.14).[44]

(21.14)

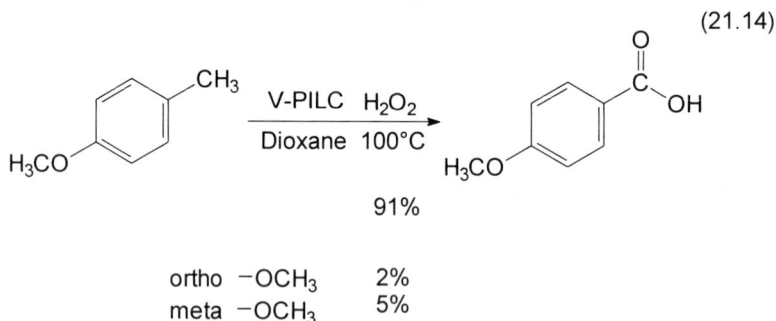

$$91\%$$

ortho $-OCH_3$ 2%
meta $-OCH_3$ 5%

Cr-PILC also catalyzes the benzylic oxidation of acylmethylene compounds to the corresponding carbonyl compounds in good to excellent yields with tert-butyl hydroperoxide as the oxidant.[45] This reaction is selective for the production of mono carbonyl compounds from substrates in which more than one benzylic methylene group is present. Tetralin (**8**) was oxidized to α-tetralone (**9**) and bibenzyl (**10**) to desoxybenzoin (**11**) in very good yields. No dicarbonyl

(21.15)

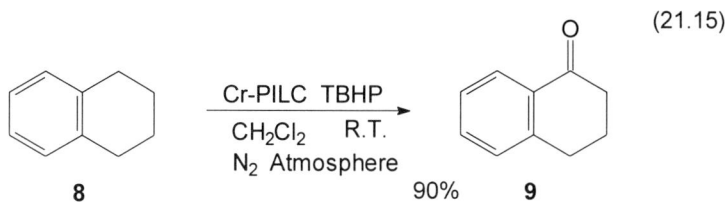

8   90%   **9**

(21.16)

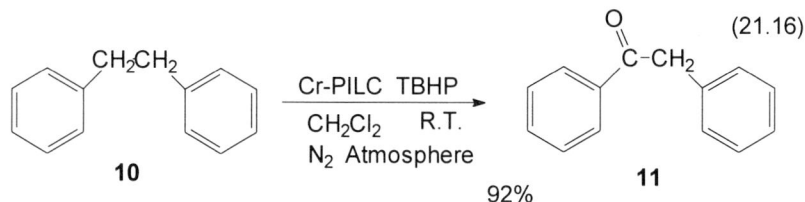

10   **11**

92%

products were formed (Eqns. 21.15-16).[45] Tertiary C–H bonds were not affected by this oxidation system. It was used to selectively oxidize para-isopropylethylbenzene (**12**) to para-isopropylacetophenone (**13**) (Eqn. 21.17).[45] Arylacetic esters have been oxidized by *tert*-butyl hydroperoxide over V-PILC to give the arylglyoxylic esters in good yields (Eqn. 21.18).[46]

(21.17)

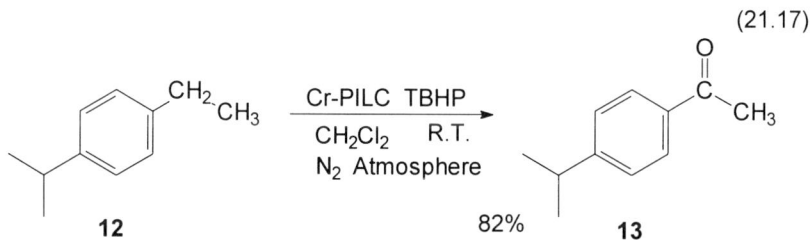

12   82%   **13**

(21.18)

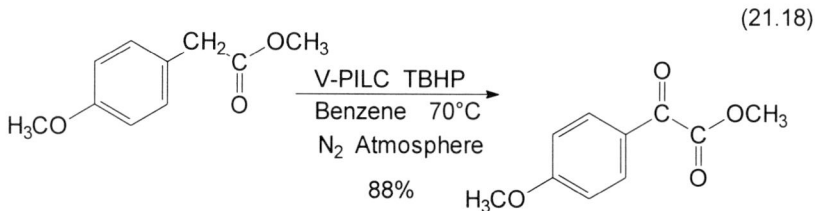

88%

Allylic oxidation is also promoted by Cr-PILC catalysts with good yields of the α,β-unsaturated ketones being produced under mild reaction conditions.[47] Again, tertiary C–H bonds were not oxidized under these conditions but the oxidation of a CH_2 group was favored over that of a CH_3 entity (Eqn. 21.19).[47]

(21.19)

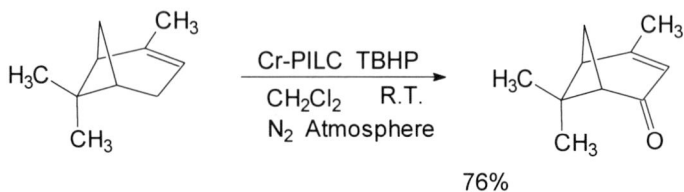

76%

(21.20)

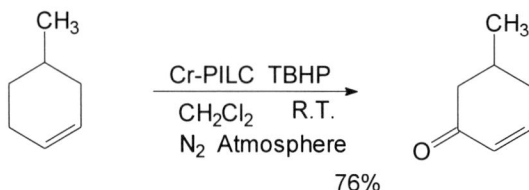

76%

With unsymmetrical alkenes the oxidation generally takes place at the least hindered methylene group with no double bond isomerization (Eqn. 21.20).[47]

The Cr-PILC catalyzed benzylic and allylic oxidations also provide a facile approach to the oxidative deprotection of allyl and benzyl ethers and amines.[45,47] Treatment of allyl or benzyl ethers with one equivalent of tert-butyl hydroperoxide in the presence of Cr-PILC at room temperature resulted in the oxidative cleavage of the allyl- or benzyl-oxygen bond to give the alcohol but when two equivalents of tert-butyl hydroperoxide (TBHP) were used, the alcohol was oxidized further to the aldehyde or ketone (Eqn. 21.21).[47] Oxidation of allyl amines resulted in the cleavage of the allyl-nitrogen bond to give the des-allyl amine.[47] Benzyl amines, however, were oxidized to the benzamides (Eqn. 21.22).[45]

(21.21)

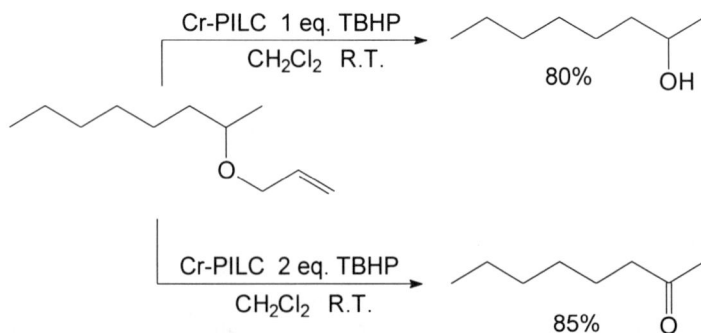

80%

85%

(21.22)

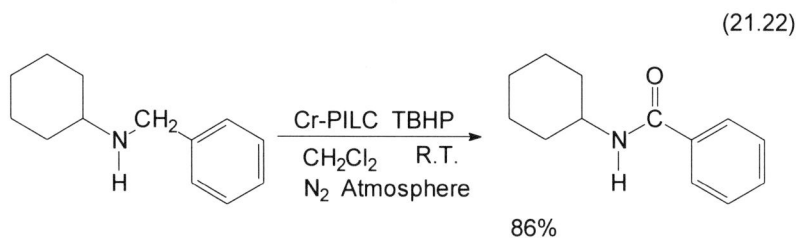

86%

   Sulfides were selectively oxidized to sulfoxides over V-PILC with tert-butyl hydroperoxide.[48] When a chiral modifier such as diethyl tartrate was present in these reactions optically active sulfoxides were obtained but the enantioselctivities were low.[49] With a titanium pillared montmorillonite (Ti-PILC) the oxidation of sulfides by tert-butyl hydroperoxide in the presence of diethyl tartrate gave the sulfoxides in good yields with good to excellent enantiomeric excesses (Eqn. 21.23).[49,50] Oxidation of unsaturated alcohols over this chirally modified Ti-PILC gave the optically active epoxides with enantiomeric excesses as good as or better than those obtained using the homogeneous Sharpless reagent (Eqn. 21.24).[51]

(21.23)

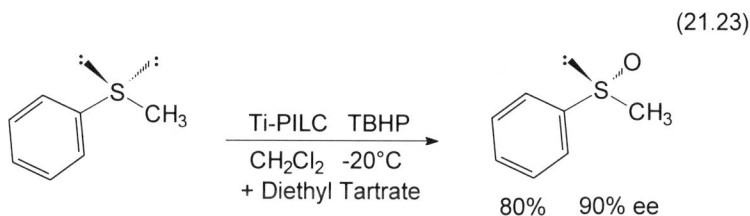

80% 90% ee

(21.24)

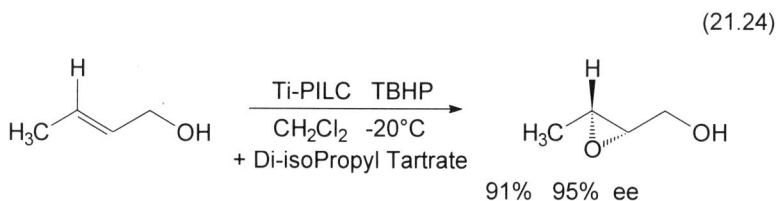

91% 95% ee

Molecular Oxygen as the Oxidant

Alcohol Oxidation

The platinum catalyzed oxidation of alcohols with molecular oxygen has been known for a long time.[52,53] Palladium and iridium are also effective for this reaction but not rhodium or ruthenium.[54] The reaction proceeds by the initial dehydrogenation of the alcohol to produce the aldehyde or ketone with the adsorbed hydrogen then reacting with the oxygen to give water.[53,55-58] These oxidations usually proceed best in aqueous medium and, thus, are particularly effective for the oxidation of water soluble substrates such as sugars.[52] A number of alcohol oxidations, however, have been run in aqueous acetone or dioxane, and, occasionally, in hydrocarbon solvents.[59] A ternary solvent system of water, heptane and methyl acetate was shown to be useful in the oxidation of water insoluble alcohols.[60] Another procedure for the oxidation of water insoluble alcohols was the use of a surfactant such as sodium dodecylbenzenesulfonate to keep the alcohol in suspension in the reaction medium. The oxidation of α-tetralol (**14**) in an aqueous suspension with this detergent gave α-tetralone (**9**) in 95% yield (Eqn. 21.25).[61]

Secondary alcohols give the respective ketones as products, regardless of the reaction conditions. Primary alcohols, however, can produce either aldehydes or carboxylic acids. Commonly, these oxidations are run in basic aqueous media. Under these conditions, the aldehydes that are formed initially react with the water to form a hydrate and this is oxidized further to the acid. Aldehydes, particularly aryl aldehydes, are the primary product only when the reaction is run in a neutral, preferably non-aqueous, medium such as a hydrocarbon solvent and usually with controlled oxygen uptake.[53,62]

One of the major reasons why this reaction is not more widely used is that the catalyst is usually deactivated before the oxidation is completed. This deactivation is thought to be caused by either the oxidation of the metal or the blocking of the metal surface by the strong adsorption of reaction by-products. A number of procedures have been employed to minimize this deactivation. Most of the early work in this area used a large amount of platinum, prepared by the hydrogenation of platinum oxide (Adam's catalyst), as the catalyst.[52] These larger metal particles are more resistant to oxidation than the smaller particles present on supported platinum catalysts.[63] In addition, the large quantity of catalyst ensures that some active species will still be available toward the end of the reaction.

Deactivated catalysts have been regenerated by stopping the reaction and either purging the system with nitrogen[64] or adding formalin to reduce the catalyst surface.[65] Because the oxygen that is deactivating the metal is the excess present after the hydrogen has been removed from the catalyst surface, the lifetime of the active catalyst can be increased by limiting the amount of oxygen in the system to that required to react with the hydrogen. This is usually accomplished by using air or even more dilute oxygen mixtures for the reaction or by bubbling air or oxygen through the reaction mixture. Such mass transport control conditions can significantly increase the catalyst lifetime.[61]

A further mode of deactivation can come from the strong adsorption of reaction by-products onto the catalyst surface. Of particular importance in this regard are the products from the self-condensation of the aldehydes or ketones produced in the oxidation of alcohols.[53,66] This condensation is enhanced in the aqueous base commonly used as the solvent in these oxidations.

Adding another metal, particularly lead or bismuth to a platinum or palladium catalyst significantly reduced the extent to which these catalysts were deactivated. The added metal also increased the catalyst activity and, in some cases, the reaction selectivity.[66-75] It is not completely clear how these added metals work, but they could be lowering the adsorption strength of the condensation products on the catalyst and/or inhibiting the oxidation of the active metal.

Most of these liquid phase catalytic oxidations have been used for the selective oxidation of carbohydrates. In the oxidation of a sugar such as glucose, four basic types of selective oxidation are possible.[14]

1) Oxidation of the C_1 aldehyde (hemiacetal) to the carboxylic acid (lactone).
2) Oxidation of the primary alcohol to the aldehyde or carboxylic acid.
3) Oxidation of a secondary alcohols to a ketone.

Scheme 21.1

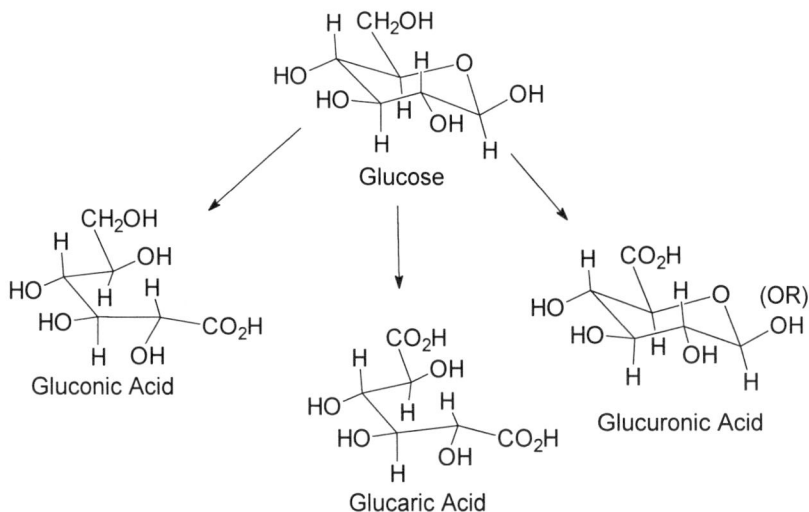

Glucose

Gluconic Acid

Glucaric Acid

Glucuronic Acid

4) Oxidative cleavage of a diol, but this is only an occasional side reaction in most oxidations.

The general order of reaction in these compounds is: $CHO > COCH_2OH > CH_2OH > CHOH_{axial} > CHOH_{equatorial}$.[52,76,77]

As depicted in Scheme 21.1 for the oxidation of glucose, three general types of acids can be produced by the oxidation of a sugar: aldonic acids in which only the aldehyde of the parent aldose has been oxidized to the acid; aldaric acids that are dicarboxylic acids with one carboxy group coming from the oxidation of the aldehyde and the other from the primary alcohol at the other end of the carbon chain; or uronic acids derived from the selective oxidation of the primary alcohol without the oxidation of the aldehyde.

The oxidation of glucose to gluconic acid was accomplished in 96% selectivity using a Pd-Pt-Bi/C catalyst at 55°C and 0.01 atmospheres of oxygen in an aqueous solution at pH 10 (Eqn. 21.26).[73,74] This oxidation successfully competes with the standard enzymatic method for the commercial preparation of gluconic acid. The platinum increases the activity of the palladium while the bismuth increases catalyst selectivity. Reactions run over unmodified Pt/C catalysts also gave good yields but not as high as those obtained with the trimetallic species.[64] Pt/C has been used for the oxidation of a number of five- and six-carbon sugars to give the corresponding aldonic acids in generally good yields.[78,79] Pentoses were oxidized more easily than were the hexoses.

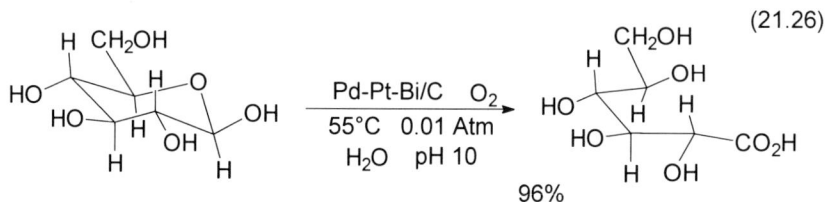

(21.26)

The continuing oxidation of gluconic acid to glucaric acid is hindered by the degradation of the oxidized species that occurs during the more prolonged oxidation reaction. The best yields of glucaric acid have been in the 55%–60% range with this procedure.[80-82]

While examining the effect of modified catalysts on aldaric acid formation it was found that Pt/C catalysts that were treated with lead salts were extremely selective for the oxidation of the hydroxy group next to the initially formed carboxylic acid to give 2-keto gluconic acid.[82,83] These catalysts also promote the selective oxidation of a number of other α-hydroxy acids. For example, lactic acid has been oxidized to pyruvic acid in greater than 95% yield (Eqn. 21.27).[83] Similar results have been observed with lead doped Pd/C catalysts.[72,75]

Uronic acids can be produced by the catalytic oxidation of the primary alcohol of an aldose, provided that the aldehyde group has first been protected, usually as a glycoside.[84,85]

(21.27)

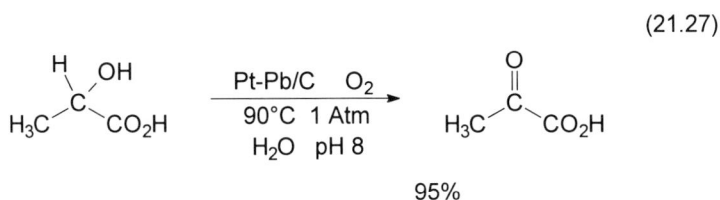

95%

(21.28)

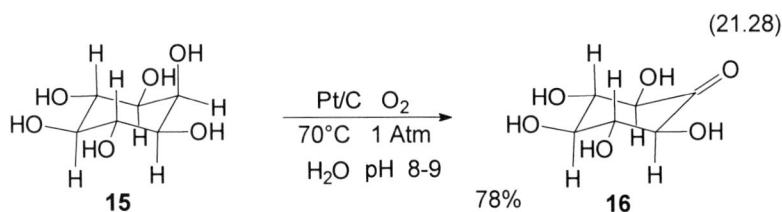

15   78%   **16**

(21.29)

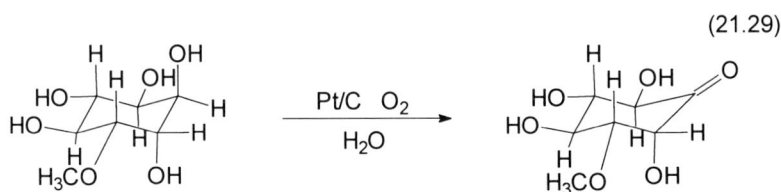

The oxidation of cyclic polyhydric alcohols shows a definite preference for the oxidation of an axial alcohol over an equatorial hydroxy group. In the oxidation of myoinositol (**15**) over Pt/C only the axial hydroxy group was oxidized to give the pentahydroxy ketone, **16** (Eqn. 21.28).[86] When two axial hydroxy groups were present, both were potentially oxidizable but only the mono keto products were formed. No further oxidation took place.[87] If one of the axial alcohols was adjacent to a methoxy group there was apparently some steric hindrance with the other axial alcohol being preferentially oxidized (Eqn. 21.29).[87] However, such steric factors do not completely repress the oxidation if a single axial alcohol is present as in the case of the isopropylidene compound, **17** (Eqn. 21.30).[87] Cyclic polyhydroxy compounds with no axial hydroxy groups appear to be inert to catalytic oxidation under these conditions.[87]

(21.30)

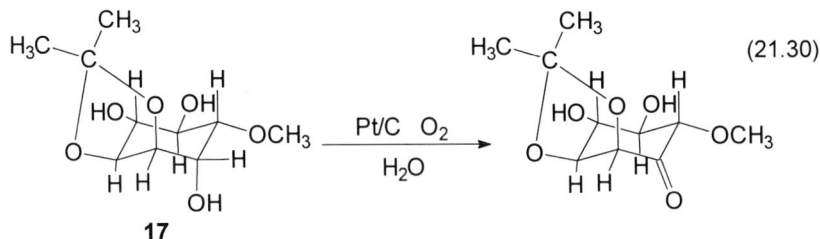

17

(21.31)

(21.32)

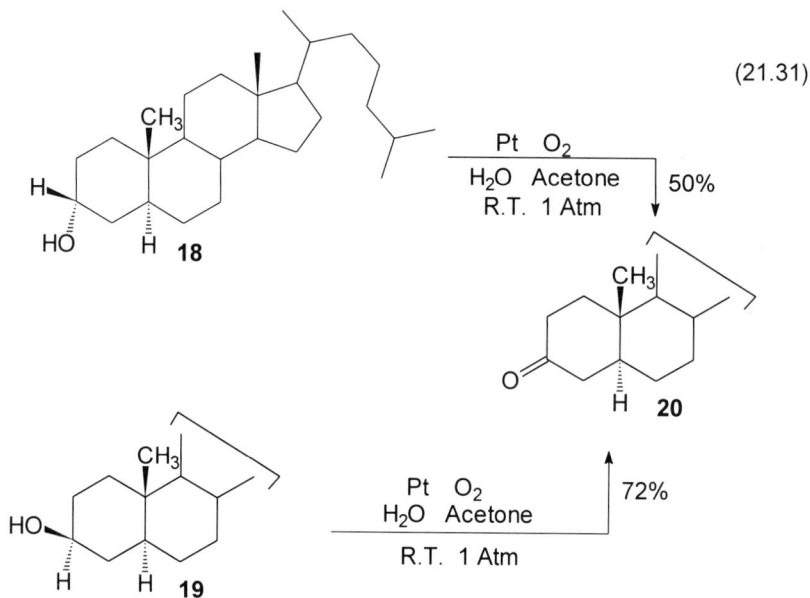

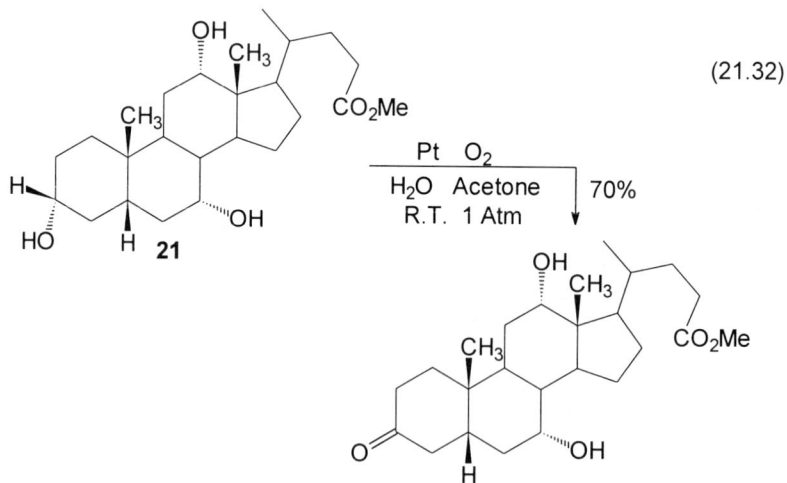

This selectivity for axial alcohol oxidation is not, apparently, universal since both the 3α- (18) and the 3β-cholestanol (19) are oxidized to cholestanone (20) with a better yield obtained on oxidation of the equatorial, 3β-alcohol (Eqn. 21.31).[88] Selectivity in the platinum catalyzed oxidation of steroid alcohols,

(21.33)

(21.34)

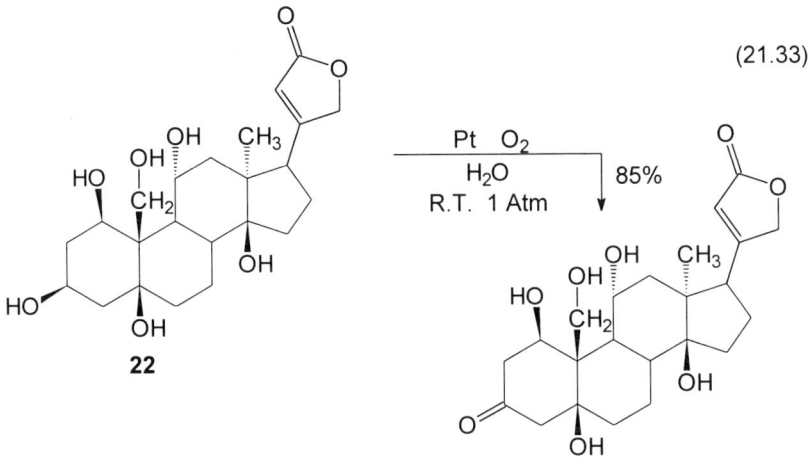

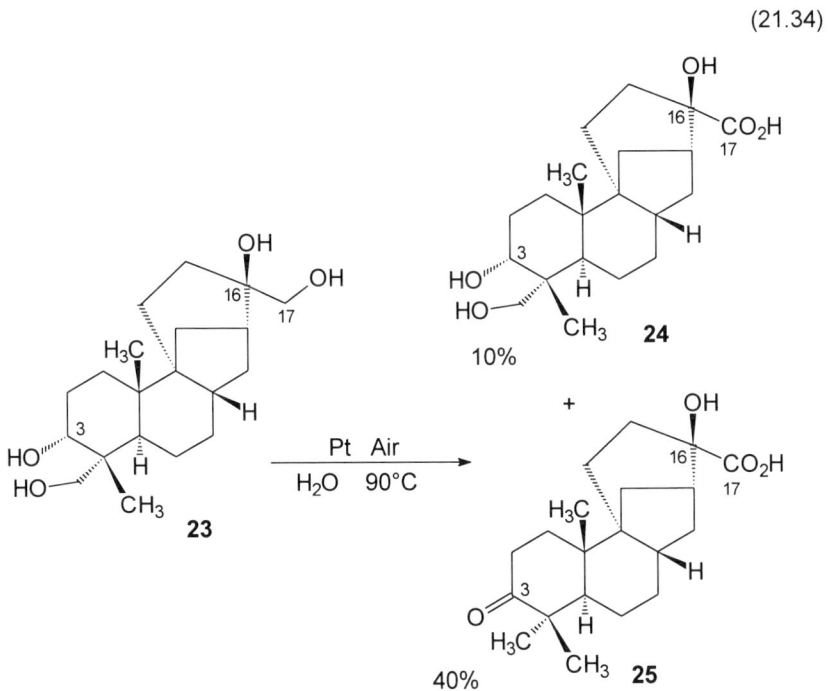

however, has been observed in that the 3-hydroxy group is converted to the ketone in preference to the oxidation of hydroxy groups on the 6, 7, 12 or 17 carbons of the steroid skeleton[88,89] as shown by the oxidation of methyl 3α, 7α, 12αtrihydroxycholanate (**21**) (Eqn. 21.32).[88] This selectivity was also observed in the oxidation of the polyoxygenated compound, ouabagenin (**22**) (Eqn. 21.33).[90] But the platinum catalyzed oxidation of aphicholin (**23**) occurs first to give the 17 carboxylic acid (**24**) and then the 3-ketone. The 3-keto-17-acid, **25**, was the predominant product (Eqn. 21.34).[91]

(21.35)

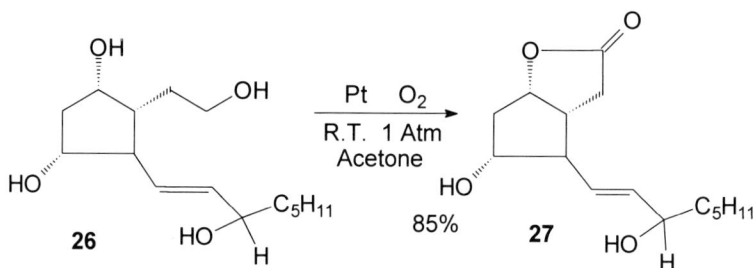

The selective oxidation of the primary alcohol in **26** gave the acid which then formed the lactone, **27**. Neither the aliphatic nor the allylic secondary alcohols were oxidized (Eqn. 21.35).[92] One aldehyde group was oxidized on treating glyoxal (**28**) with oxygen in the presence of a Pt/C catalyst. Glyoxylic acid (**29**) was obtained in about 70% yield (Eqn. 21.36).[93] In a similar manner, one hydroxy group in ethylene glycol was oxidized over Pt/C at 40°C and one atmosphere of oxygen to give glycolic acid (**30**) (Eqn. 21.37).[65,94] The catalyst was deactivated during this reaction but it was regenerated by the addition of formalin to the reaction mixture during the course of the reaction.[65]

(21.36)

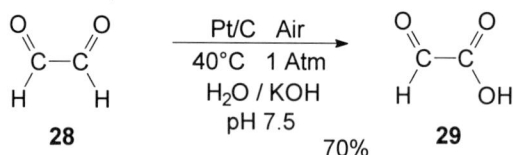

(21.37)

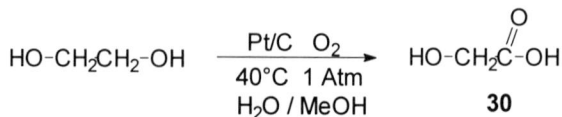

(21.38)

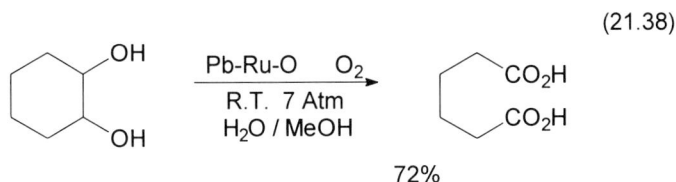

72%

Vicinal diols have been cleaved with molecular oxygen using a ruthenium pyrochlore oxide catalyst. Cyclohexane-1,2-diol was oxidized to adipic acid in 72% yield with this catalyst system (Eqn. 21.38). Cyclohexanone was also converted to adipic acid under these conditions.[95]

Other Substrates

While alcohol oxidations have been the most common metal promoted reactions involving molecular oxygen, a number of other metal catalyzed oxidations of potential synthetic interest have been reported. Supported palladium catalysts are comparable to many soluble palladium catalysts in promoting the selective oxidations of alkenes and aromatics.[96] 2-Butene was oxidized primarily to crotonic acid over Pd/C in water but methyl vinyl ketone and crotonaldehyde were also formed in significant amounts.[97] When this oxidation was run in acetic acid the allyl acetates were the major products, particularly when a Pd/Al_2O_3 catalyst was used.[96]

The Wacker oxidation of ethylene to acetaldehyde was accomplished using Cu-Pd exchanged zeolites as the catalysts. The vapor phase reaction was run at 100°C and gave acetaldehyde in 100% selectivity at 50% conversion.[98] A similar redox system involving supported palladium catalysts that were doped with a cupric ion solution were used to oxidize benzene to phenol. The reaction was run using neat benzene at 25°C with hydrogen and oxygen being bubbled through the reaction mixture. Carbon monoxide could also be used as the reducing agent required to reduce the cupric ions to the cuprous ions that are needed to convert the oxygen to hydrogen peroxide, the active hydroxylating agent.[99-101] The partial oxidation of benzene to phenol was also accomplished by passing a mixture of benzene vapor and nitrous oxide through a bed of HZSM-5 at 330°C. Phenol was produced in a near 20% yield.[102,103]

$NiO-Fe_2O_3$ catalysts were used for the vapor phase oxidation of benzoic acid to phenol. This reaction took place in a stream of air at 400°C. A catalyst having a Ni:Fe atomic ratio of one gave phenol in 88% selectivity at 100% conversion.[104]

Supported platinum catalysts have been used to promote the oxidation of alkenes in water at 180°C and two atmospheres of oxygen to give the 1,2 diols. Oxidation of ethylene gave a mixture of the diol and acetic acid in about a 2:1 ratio. Oxidation of propene gave the diol as the predominant product with some acetone also produced. When a small amount of carbon monoxide was added to the reaction mixture, catalyst activity decreased significantly but the sole products were the diols.[105]

Treating alkenes with air in the presence of iso-butraldehyde and a clay catalyst that was impregnated with nickel acetonylacetate gave good yields of the corresponding epoxides (Eqn. 21.39).[106]   The reaction of oxygen with the aldehyde gave the hydroperoxide which, in turn, was used to convert the alkene to the epoxide over this "clayniac" catalyst.

(21.39)

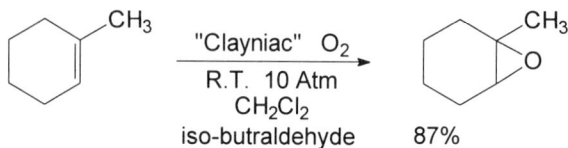

Toluene was oxidized to a mixture of benzyl acetate (**31**) and benzylidene diacetate (**32**) on reaction with oxygen in the presence of a silica supported Pd-Sn catalyst.[107-109]   A reaction run in HOAc/KOAc at 70°C under an atmosphere of oxygen gave a near 3:1 ratio of the monoacetate to the diacetate at 98% conversion (Eqn. 21.40). The two products are formed in parallel reactions as the benzylacetate does not react further under these reaction conditions. Substituted diphenylmethanes were oxidized to the diphenyl ketones by refluxing them in air in a DMF solution containing a copper powder catalyst.[110]

(21.40)

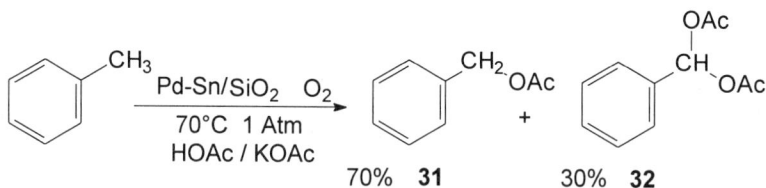

Passing a mixture of cyclohexylamine vapor and oxygen through a WO_3/Al_2O_3 catalyst bed at 160°C gave cyclohexanone oxime in 64% selectivity at 33% conversion.[111] Tertiary amines having at least one methyl group on the nitrogen have been demethylated to the secondary amines in good yields by stirring a methanol solution of the amine containing a large amount of Pd/C in air for a prolonged period of time (Eqn. 21.41). Tertiary N-ethyl and N-allyl groups were also removed by the procedure but larger alkyl groups appeared to be inert to oxidative cleavage under these conditions. Secondary amines did not react further to form the primary amines.[112]

(21.41)

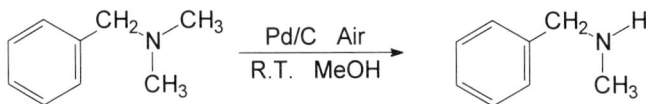

Reaction of an amine with oxygen and carbon monoxide in the presence of a metal catalyst in an aprotic solvent gives the disubstituted urea as the product. The carbamate is produced in an alcoholic solution.[113] For instance, heating a DMF solution of aniline in the presence of a Pd/C catalyst and a sodium iodide promoter at 100°C under 40 atmospheres of a 5:1 mixture of carbon monoxide and oxygen gave diphenylurea (33) in about 30% yield with 100% selectivity. In an ethanol solution the ethyl carbamate, 34, was produced along with the urea in nearly equal quantities (Eqn. 21.42).[113] When this reaction was run in methanol using a sodium iodide promoted Pd-ZSM-5 catalyst and a 55 atmosphere pressure of a 16:1 ratio of carbon monoxide and oxygen, the product formed was dependent on the temperature used. At 100°C the urea was produced in 95% selectivity at near 10% conversion; at 170°C the carbamate was formed in 95% selectivity at 99% conversion. Intermediate temperatures gave product mixtures at incomplete conversions.[114] These and other data [113] indicate that the urea is produced first and then converted to the carbamate by reaction with the alcohol used as the solvent. Almost exclusive carbamate formation was also observed when the reaction was run at 170°C over a lithium iodide promoted Rh/C catalyst.[115]

(21.42)

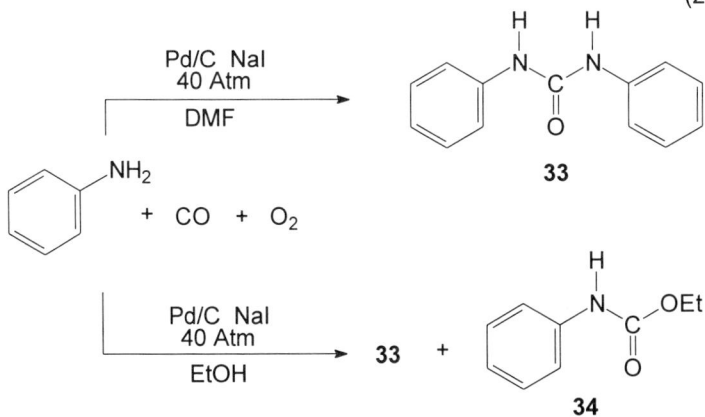

References

1. J. M. Lopez Nieto, J. L. G. Fierro, L. Gonzalez Tejuca and G. Kremenic, *J. Catal.*, **107**, 325 (1987).
2. M. Carbucicchio, G. Centi, P. Forzatti, F. Trifiro and P. L. Villa, *J. Catal.*, **107**, 307 (1987).
3. R. H. Bretton, S. W. Wan and B. F. Dodge, *Ind. Eng. Chem.*, **44**, 594 (1952).
4. J. K. Dixon and J. Longfield, in *Catalysis* (P. H. Emmett, Ed.), Vol. 7, Reinhold, New York, 1960, p183.
5. M. G. Nobbenhuis, A. Baiker, P. Parnickel and A. Wokaun, *Appl. Catal.*, *A*, **85**, 157 (1992).
6. J. V. Porcelli, *Catal. Revs.*, **23**, 151 (1981).
7. W. M. H. Sachtler, C. Backx and R. A. Van Santen, *Catal. Revs.*, **23**, 127 (1981).
8. H. H. Voge and C. R. Adams, *Adv. Catal.*, **17**, 151 (1967).
9. R. Higgins and P. Hayden, *Catalysis (London)*, **1**, 168 (1977).
10. C. F. Cullis and D. J. Hucknall, *Catalysis (London)*, **5**, 273 (1982).
11. G. I. Golodets, *Stud. Surf. Sci. Catal.*, **15** (Heterog. Catal. Reactions Involving Mol. Oxygen), Elsevier, Amsterdam, (1983).
12. R. K. Grasselli, *ACS Symp. Ser.*, **222** (Heterog. Catal.), 317 (1983).
13. R. A. Sheldon, *Stud. Surf. Sci. Catal.*, **55** (New Devel. Selective Oxidation), 1 (1990).
14. R. A. Sheldon, *Stud. Surf. Sci. Catal.*, **59** (Heterog. Catal. Fine Chem., II), 33 (1991).
15. R. A. Sheldon and J. Dakka, *Catal. Today*, **19**, 215 (1994).
16. R. A. Sheldon, in *Aspects of Homogeneous Catalysis* (R. Ugo, Ed.), Vol. 4, Reidel, Dordrecht, 1981, p 1.
17. J. Sobczak and J. J. Ziolkowski, *React. Kinet. Catal. Lett.*, **11**, 3583 (1985).
18. R. A. Sheldon, in *The Chemistry of Functional Groups-Peroxides* (S. Patai, Ed.) Wiley, New York, 1982, p.161.
19. U. Romano, A. Esposito, F. Maspero, C. Neri and M. G. Clerici, *Stud. Surf. Sci. Catal.*, **55** (New Devel. Selective Oxidation), 33 (1990).
20. C. Ferrini and H. W. Kouwenhoven, *Stud. Surf. Sci. Catal.*, **55** (New Devel. Selective Oxidation), 53 (1990).
21. M. G. Clerici, *Stud. Surf. Sci. Catal.*, **78** (Heterog. Catal. Fine Chem., III), 21 (1993).
22. C. B. Khouw, H. X. Ko, C. B. Dartt and M. E. Davis, *ACS Symp. Ser.*, **523** (Catalytic Selective Oxidation), 273 (1993).
23. B. Notari, *Catal. Today*, **18**, 163 (1993).
24. R. Millini, E. P. Massara, G. Perego and G. Bellussi, *J. Catal.*, **137**, 497 (1992).
25. M. R. Boccuti, K. M. Rao, A. Zecchina, G. Leofant and G. Petrini, *Stud. Surf. Sci. Catal.*, **48** (Structure and Reactivity of Surfaces), 133 (1989).
26. M. G. Clerici, *Appl. Catal.*, **68**, 249 (1991).
27. F. Maspero and U. Romano, *J. Catal.*, **146**, 476 (1994).
28. A. J. H. P. van der Pol and J. H. C. van Hooff, *Appl. Catal.*, A, **106**, 97 (1993).
29. U. Romano, A. Esposito, F. Maspero, C. Neri and M. G. Clerici, *Chim Ind. (Milan)*, **72**, 610 (1990); *Chem. Abstr.*, **113**, 233670 (1990).

30. C. Neri and F. Buonomo, Eur. Patent, 102,097 (1984); *Chem. Abstr.*, **100**, 209389 (1984).
31. T. Tatsumi, M. Yako, M. Nakajura, Y. Yuhara and H. Tominaga, *J. Mol. Catal.*, **78**, L41 (1993).
32. T. Tatsumi, M Nakamura, S. Negishi and H. Tominaga, *J. Chem. Soc., Chem. Commun.*, 476 (1990).
33. R. Roffia, G. Leofant, A. Cesana, M. Mantegazza, M. Padovan, G. Petrini, S. Tonti and P. Gervasutti, *Stud. Surf. Sci. Catal.*, **55** (New Devel. Selective Oxidation), 43 (1990).
34. J. S. Reddy and P. A. Jacobs, *J. Chem. Soc., Perkin Trans.,1*, 2665 (1993).
35. H. R. Sonawane, A. V. Pol, P. O. Moghe, S. S. Biswas and A. Sudalai, *J. Chem. Soc., Chem. Commun.*, 1215 (1994).
36. M. A. Camblor, A. Corma, A. Martinez, J. Perez-Pariente and J. Primo, *Stud. Surf. Sci. Catal.*, **78** (Heterog. Catal. Fine Chem., III), 393 (1993).
37. P. R. Hari Prasad Rao and A. V. Ramaswami, *J. Chem. Soc., Chem. Commun.*, 1245 (1992).
38. P. R. Hari Prasad Rao, K. R. Reddy, A. V. Ramaswami and P. Ratnasamy, *Stud. Surf. Sci. Catal.*, **78** (Heterog. Catal. Fine Chem., III), 385 (1993).
39. M. K. Dongare, P. Singh, P. P. Moghe and P. Ratnasamy, *Zeolites*, **11**, 690 (1992).
40. U. Cornaro, P. Jiru, Z. Tvaruzkova and K. Habersberger, *Stud. Surf. Sci. Catal.*, **69** (Zeolite Chem. and Catal.), 165 (1991).
41. F. Figuras, *Catal. Revs.*, **30**, 457 (1988).
42. B. M. Chaudary, A. Durga Prasad and V. L. K. Valli, *Tetrahedron Lett.*, **31**, 57885 (1990).
43. B. M. Chaudary, V. L. K. valli and A. Durga Prasad, *J. Chem. Soc., Chem. Commun.*, 721 (1990).
44. B. M. Chaudary and V. L. K. Valli, *J. Chem. Soc., Chem. Commun.*, 1115 (1990).
45. B. M. Chaudary, A. Durga Prasad, V. Bhuma and V. Swapna, *J. Org. Chem.*, **57**, 5841 (1992).
46. B. M. Chaudary, G. V. S. Reddy and K. K. Rao, *J. Chem. Soc., Chem. Commun.*, 323 (1993).
47. B. M. Chaudary, A. Durga Prasad, V. Swapna, V. L. K. Valli and V. Bhuma, *Tetrahedron*, **48**, 953 (1992).
48. B. M. Chaudary and S. S. Pani, *J. Mol. Catal.*, **75**, L7 (1992).
49. B. M. Chaudary, S. S. Rani and N. Narender, *Catal. Letters*, **19**, 299 (1993).
50. B. M. Chaudary, S. S. Rani and Y. V. Subba Rao, *Stud. Surf. Sci. Catal.*, **75** (New Frontiers in Catalysis), 1247 (1993).
51. B. M. Chaudary, V. L. K. Valli and A. Durga Prasad, *J. Chem. Soc., Chem. Commun.*, 1186 (1990).
52. K. Heynes and H. Paulsen, *Newer Methods of Preparative Organic Chemistry*, **2**, 303 (1963).
53. T. Mallat and A. Baiker, *Catal. Today*, **19**, 247 (1994).
54. H. E. Van Dam, L. J. Wisse and H. Van Bekkum, *Appl. Catal.*, **61**, 187 (1990).
55. H. Wieland, *Chem. Ber.*, **45**, 2606 (1912); **46**, 3327 (1913); **54**, 2353 (1921).
56. J. W. Nicoletti and G. M. Whitesides, *J. Phys. Chem.*, **93**, 759 (1989).

57. R. DiCosimo and G. M. Whitesides, *J. Phys. Chem.,* **93**, 768 (1989).
58. R. L. Augustine and L. K. Doyle, *J. Catal.*, **141**, 58 (1993).
59. K. Heynes and L. Blazejewicz, *Tetrahedron,* **9**, 67 (1960).
60. R. L. Augustine and L. K. Doyle, *Chem. Ind. (Dekker)*, **53** (Catal. Org. React.), 479 (1994).
61. T. Mallat, Z. Bodnar and A. Baiker, *Stud. Surf. Sci. Catal.*, **78** (Heterog. Catal. Fine Chem., III), 377 (1993).
62. M. Hronec, Z. Cvengrosova, J. Tuleja and J. Ilavsky, *Stud. Surf. Sci. Catal.*, **55** (New Devel. Selec. Oxidation), 169 (1990).
63. M. Boudart, A. Aldag, J. E. Benson, N. A. Dougharty and C. G. Harkins, *J. Catal.*, **6**, 92 (1966).
64. J. M. H. Dirkx and H. S. van der Baan, *J. Catal.*, **67**, 1 (1981).
65. M. I. A. Khan, Y. Miwa, S. Morita and J. Okada, *Chem. Phar. Bull.*, **31**, 1141 (1983).
66. T. Mallat and A. Baiker, *Appl. Catal.*, *A*, **79**, 41 (1991).
67. T. Mallat and A. Baiker, *Appl. Catal.*, *A*, **79**, 59 (1991).
68. T. Mallat and A. Baiker, *Appl. Catal.*, *A*, **86**, 147 (1992).
69. T. Mallat, T. Allmendinger and A. Baiker, *Appl. Surf. Sci.*, **52**, 189 (1991).
70. T. Mallat, Z. Bodnar, A. Baiker, O. Greis, H. Strubig and A. Reller, *J. Catal.*, **142**, 237 (1993).
71. H. Kimura, A. Kimura, I. Kokubo, T. Wakisaka and Y. Mitsuda, *Appl. Catal.*, *A*, **95**, 143 (1993).
72. H. Hayashi, S. Sugiyama, Y. Katayama, K. Kawashiro and N. Shigemoto, *J. Mol. Catal.,* **91**, 129 (1994).
73. B. M. Despeyroux, K. Deller and E. Peldszuz, *Stud. Surf. Sci. Catal.*, **55** (New Devel. Selec. Oxidation), 159 (1990).
74. K. Deller and B. Despeyroux, *Chem. Ind. (Dekker)*, **47** (Catal. Org. React.), 261 (1992).
75. T. Tsujino, S. Ohigashi, S. Sugiyama, K. Kawashiro and H. Hayashi, *J. Mol. Catal.,* **71**, 25 (1992).
76. K. Heynes and H. Paulsen, *Angew. Chem.*, **69**, 600 (1957).
77. K. Heynes and H. Paulsen, *Adv. Carbohydrate Chem.*, **17**, 169 (1962).
78. F. R. Venema, J. A. Peters and H. van Bekkum, *J. Mol. Catal.,* **77**, 75 (1992).
79. K. Heynes and R. Henemann, *Liebigs Ann. Chem.*, **558**, 187 (1947).
80. J. M. H. Dirkx and H. S. van der Baan, *J. Catal.*, **67**, 14 (1981).
81. P. J. M. Dijkgraaf, M. J. M. Rijk, J. Meuldijk and K. van der Wiele, *J. Catal.*, **112**, 329 (1988).
82. P. C. C. Smits, B. F. M. Kuster, K. van der Wiele and H. S. van der Baan, *Appl. Catal.*, **33**, 83 (1987).
83. P. C. C. Smits, B. F. M. Kuster, K. van der Wiele and H. S. van der Baan, *Carbohydrate Res.*, **153**, 227 (1986).
84. S. A. Barker, E. J. Bourne and M. Stacy, *Chem. and Ind., (London)*, 970 (1957).
85. C. L. Hehltretter, B. H. Alexander, R. L. Mellies and C. E. Rist, *J. Am. Chem. Soc.,* **73**, 2424 (1951).
86. K. Heynes and H. Paulse, *Chem. Ber.*, **86**, 833 (1953).
87. S. J. Angyal and L. Anderson, *Adv. Carbohydrate Chem.*, **14**, 135 (1959).
88. R. P. A. Sneeden and R. B. Turner, *J. Am. Chem. Soc.,* **77**, 190 (1955).

89. J. A. Edwards, H. Carpio and A. D. Cros, *Tetrahedron Lett.*, 3299 (1964).
90. R. P. A. Sneeden and R. B. Turner, *J. Am. Chem. Soc.*, **77**, 130 (1955).
91. J. F. Gordon, J. R. Hanson, A. G. Jarvis and A. H. Ratcliffe, *J. Chem. Soc., Perkin Trans., 1*, 3019 (1992).
92. J. Fried and J. C. Sih, *Tetrahedron Lett.*, 3899 (1973).
93. P. Gallezot, R. Fache, R. de Mesanstourne, Y. Christidis and G. Mattioda, *Stud. Surf. Sci. Catal.*, **75** (New Frontiers in Catalysis, Pt. A), 195 (1993).
94. M. I. A. Khan, Y. Miwa, S. Morita and J. Okada, *Chem. Pharm. Bull.*, 31, 1827 (1983).
95. T. R. Felthouse, *J. Am. Chem. Soc.*, **109**, 7566 (1987).
96. J. Lyons, G. Suld and C. Y. Hsu, *Chem. Ind. (Dekker)*, **33** (Catal. Org. React.), 1 (1988).
97. T. Seiyama, N. Yamazoe, J. Hojo and M. Hayakawa, *J. Catal.*, **24**, 173 (1972).
98. P. H. Espeel, G. DePeuter, M. C. Tielen and P. A. Jacobs, *J. Phys. Chem.*, **98**, 11588 (1994).
99. A. Kunai, K. Ishihata, S. Ito and K. Sasaki, *Chem. Lett.*, 1969 (1988).
100. A. Kunai, T. Wani, Y. Uehara, F. Iwasaki, Y. Kuroda, S. Ito and K. Sasaki, *Bull. Chem. Soc., Japan*, **62**, 2613 (1989).
101. K. Sasaki, S. Ito and A. Kunai, *Stud. Surf. Sci. Catal.*, **55** (New Devel. Selec. Oxidation), 125 (1990).
102. R. Burch and C. Howitt, *Appl. Catal.*, A, **86**, 139 (1992).
103. R. Burch and C. Howitt, *Appl. Catal.*, A, **103**, 135 (1993).
104. J. Miki, M. Asanuma, Y. Tachibana and T. Shikada, *Appl. Catal.*, A, **115**, L1 (1994).
105. M. A. Benvenuto and A. Sen., *J. Chem. Soc., Chem. Commun.*, 970 (1993).
106. E. Bouhlel, P. Laszlo, M. Levart, M.-T. Mmontaufier and G. P. Singh, *Tetrahedron Lett.*, **34**, 1123 (1993).
107. S. K. Tanielyan and R. L. Augustine, *J. Mol. Catal.*, **87**, 311 (1994).
108. S. K. Tanielyan and R. L. Augustine, *J. Mol. Catal.*, **90**, 267 (1994).
109. S. K. Tanielyan and R. L. Augustine, *Chem. Ind. (Dekker)*, **53** *(Catal. Org. React.)*, 553 (1994).
110. F. G. Farrell, D. Moskowitz and F. Terrier, *Synth. Commun.*, **23**, 231 (1993).
111. J. N. Armor, E. J. Carlson, R. Riggitano, J. Yamanis and P. M. Zambri, *J. Catal.*, **83**, 487 (1983).
112. N. A. Chaudhuri, P. Servando, B. Markus, I. Galynker and M.-S. Sung., *J. Indian Chem. Soc.*, **62**, 899 (1985).
113. S. P. Gupte and R. V. Chaudhari, *J. Catal.*, **114**, 246 (1988).
114. S. Kanagasabapathy, A. Thangaraj, S. P. Gupte and R. V. Chaudauri, *Catal. Letters*, **25**, 361 (1994).
115. K. V. Prasad and R. V. Chaudhari, *J. Catal.*, **145**, 204 (1994).

22

Other Reactions

Reactions used in organic synthesis fall into two general categories. They either modify a functional group on a given molecule or they change the carbon skeleton of the starting material by making or breaking C–C bonds. With the exception of the C–C bond breaking in the hydrogenolysis of cyclopropanes and the oxidative cleavage of vicinal diols, all of the reactions discussed in the previous eight chapters were used to modify functional groups. This chapter, however, will cover a number of heterogeneously catalyzed reactions that are used to make C–C bonds. These reactions are promoted by solid acids and solid bases as well as by some metals and oxides. Some of these catalysts, particularly the solid acids, can also affect functional group modification as in the nitration of aromatic rings. Because of the varied nature of these reactions, the following material is organized according to the nature of the catalyst. Acid catalyzed reactions are considered first, followed by solid base promoted reactions, metal catalyzed reactions other than hydrogenations and oxidations, and, finally, alkene metathesis.

Solid Acid Catalyzed Reactions

Acid catalyzed reactions have played a significant role in the synthesis of a variety of compounds. While most such reactions were initially run with simple protic or Lewis acids, the use of such materials now presents serious environmental problems concerning the disposal of the unused acids and their salts. To remedy this situation, solid acids have been developed to promote a number of different reactions. Not only are these solid acids easily separated from the product but they are also usually not destroyed in the reaction and can frequently be recycled for prolonged use. Generally, these solid acids are easy to handle and are non-toxic, non-volatile and non-corrosive.

Not only do these acids present an economic advantage over conventional acid technology, but there are also frequent selectivity enhancements and changes in reactivity associated with the use of these heterogeneous catalysts. A number of different types of solid acids have been involved in synthetically useful reactions. They range from the hydrogen forms of various ion exchange resins[1] and the perflourinated resin sulfonic acid, Nafion-H,[2,3] to the amorphous acidic oxides, silica and aluminum silicate,[4,5] the crystalline zeolites[6,7] and the natural clays.[8,9]

Friedel-Crafts Alkylations

The sulfonic acid resins such as Dowex-50 and Amberlyst-15 have been used to promote the alkylation of the more active aromatic rings but attempts to increase their acidity generally resulted in the degradation of the solid.[1,2] The more strongly acidic perfluorinated resin sulfonic acid, Nafion-H,[2,3] has, however, been used to promote the alkylation of benzene and other aromatic compounds. Nafion-H catalyzed the vapor phase reaction between toluene and methanol. When run at 185°C a 12% yield of the isomeric xylenes was obtained with the ortho isomer the major product.[10] Methylation of phenol at 205°C over this catalyst gave, at 63% conversion, 37% anisole and 10% of a mixture of the ortho and para cresols in a 2:1 ratio. Reaction of anisole with methanol under these conditions resulted in a 14% selectivity to the methyl anisoles at 40% conversion, with the ortho and para isomers formed in nearly equal amounts.[11]

The alkylation of toluene with methanol at 400°C over H-ZSM-5 gave, at 36% conversion, a 69% selectivity for xylene formation, of which 27% was the para isomer.[12] Aluminum phosphate based molecular sieve catalysts such as CoAPO gave a lower conversion but higher selectivities for p-xylene formation.[12] Metallosilicates such as As-silicate having a ZSM-5 structure produced an 82% selectivity for p-xylene at 21% conversion.[13]

(22.1)

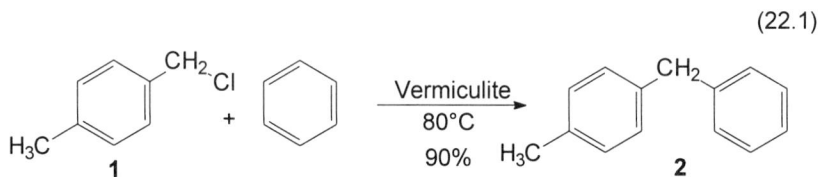

With a vermiculite catalyst, the alkylation of benzene with p-methyl benzyl chloride (**1**) at 80°C gave the mono-alkylated product, **2**, in greater than 90% yield.   No dialkylated product was formed (Eqn. 22.1).[14]   With a montmorillonite catalyst the reaction proceeded to give a 50% yield of this product and then stopped.[14]

Passing a mixture of benzene vapor and ethylene through a bed of Nafion-H at 180°C gave a 36% yield of ethyl benzene. With propylene a 30% yield of cumene was obtained.[15] The H-ZSM-5 catalyzed reaction between toluene and iso-propanol gave, at 7% conversion, both the cymenes, **3** and **4**, and the n-propyl toluenes, **5** and **6**, (Eqn.22.2).[16] Running this reaction over H-Y or H-mordenite (H-M) also gave primarily the meta cymene (**3**) but no n-propyl toluene was formed. Using an iron doped H-ZSM-5 catalyst gave the para cymene (**4**) in 75% selectivity.[16]

When alkyl chloroformates were used as the alkylating agents in Nafion catalyzed reactions, better yields of the alkylated benzene were obtained. The liquid phase reaction of toluene with iso-propyl chloroformate at 100°C gave a mixture of the ortho, meta and para cymenes (**7**, **3**, and **4**) in a 42:21:37 ratio at

(22.2)

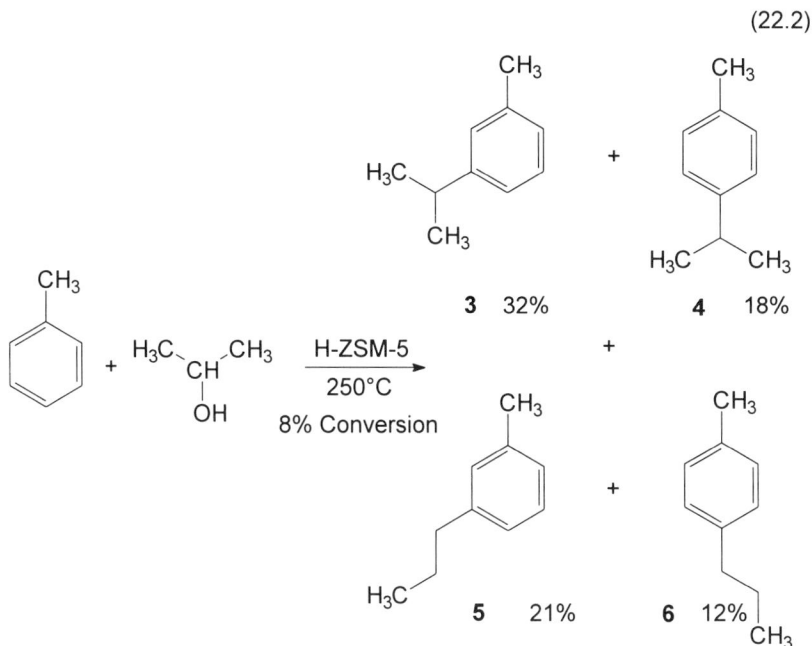

3   32%

4   18%

5   21%

6   12%

80% conversion. The corresponding vapor phase reaction run at 190°C gave **7**, **3**, and **4** in a 4:66:30 ratio at 76% conversion (Eqn. 22.3).[17] Dialkyl oxalates could also be used as alkylating agents in this reaction.[17]

(22.3)

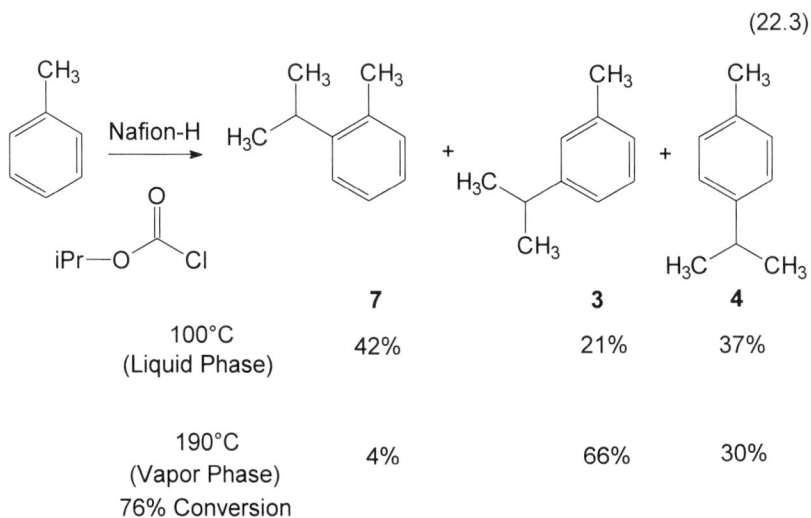

	7	3	4
100°C (Liquid Phase)	42%	21%	37%
190°C (Vapor Phase) 76% Conversion	4%	66%	30%

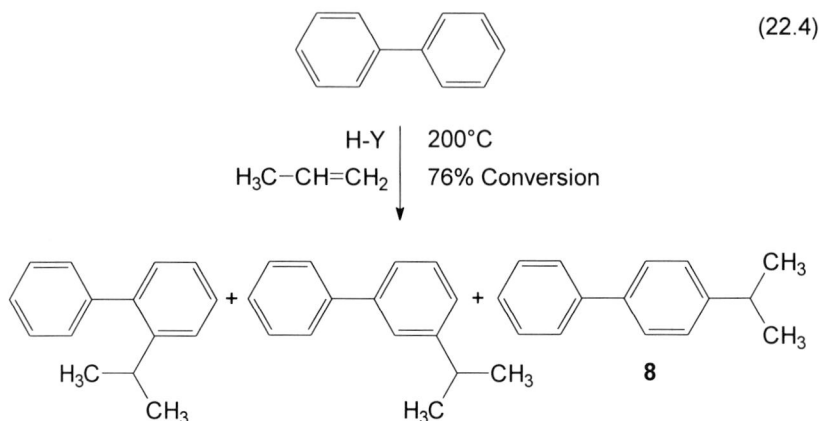

(22.4)

Reaction of biphenyl with propylene over H-Y gave a 41% selectivity to 4-isopropyl biphenyl (**8**) at 76% conversion (Eqn. 22.4).[18] The use of H-L zeolite gave comparable results but with H-M and H-ZSM-5 considerably lower conversions were obtained. When amorphous aluminum silicate was used as the catalyst a product mixture similar to those obtained with the crystalline zeolites was also obtained, suggesting that these reactions may be taking place on the exterior surface of the zeolites and not in the pores or cages.[18]

(22.5)

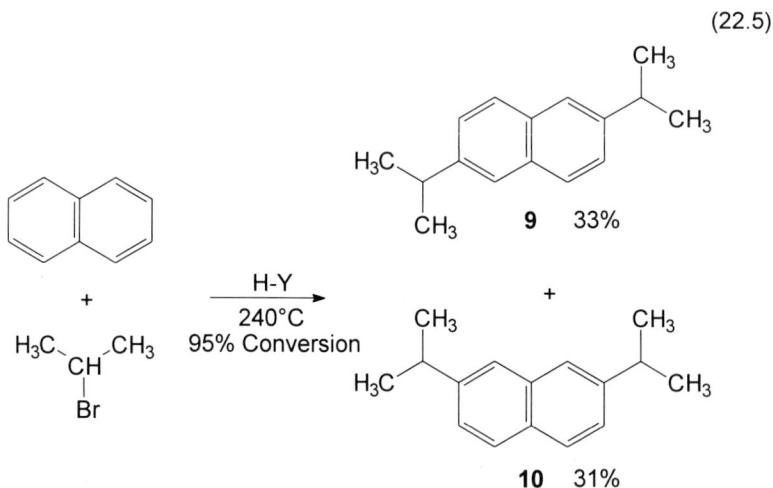

Alkylation of naphthalene with propylene over H-Y and H-L zeolites gave primarily the di-isopropyl naphthalenes with the 2,6- (**9**) and 2,7- (**10**) isomers formed in nearly equal amounts (Eqn. 22.5).[18] Similar results were observed on reaction of naphthalene with iso-propyl bromide and cyclohexyl bromide over an H-Y catalyst.[19]

The montmorillonite promoted alkylation of benzene with long chain primary and secondary alcohols gave mixtures of the 1- 2- and 3-phenyl alkanes. No 1-phenyl products were obtained when secondary alcohols were used. Reaction with tertiary alcohols gave a single alkylated product in better than 95% yield after about a five minute reaction at 80°C (Eqn. 22.6).[20]

(22.6)

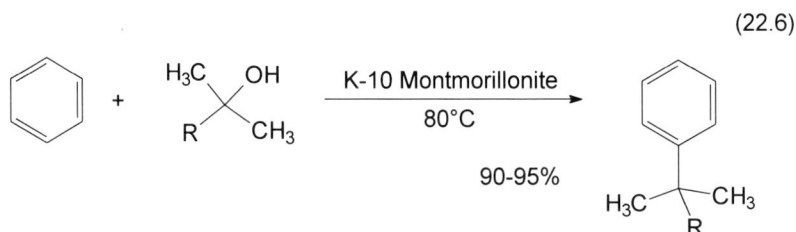

While alkylations using terminal alkenes or primary alcohols led to predominant, if not exclusive, formation of the 2-arylalkanes, alkylations using allyl alcohols give the terminally substituted aryl alkenes. The reaction of toluene with a substituted allyl alcohol over montmorillonite gave a near quantitative yield of the terminally substituted aryl alkenes (Eqn. 22.7).[5] The isomeric 3-hydroxy-1-alkenes were only slightly less effective as alkylating agents but they also gave the same terminally substituted products.[5]

(22.7)

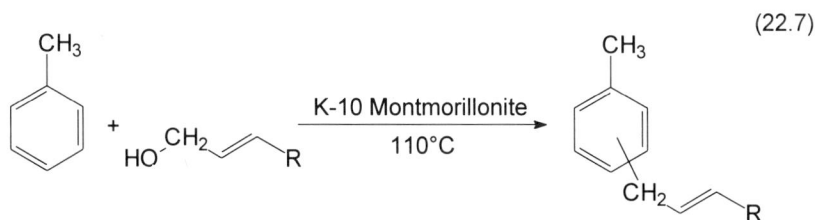

Friedel-Crafts Acylations

Nafion-H is an effective catalyst for the liquid phase acylation of aromatic hydrocarbons with aroyl chlorides and anhydrides but reactions with acetyl chloride are complicated by concomitant ketene formation.[21] Aliphatic anhydrides, however, can be used. Reaction of thiophene with aliphatic anhydrides over Nafion-H gave the 2-acylthiophenes in very good yields (Eqn. 22.8).[22]

H-ZSM-5 has been used to catalyze the liquid phase benzoylation of activated aromatic compounds such as substituted and unsubstituted phenols,

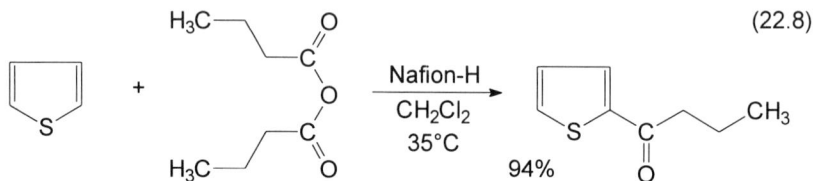

(22.8)

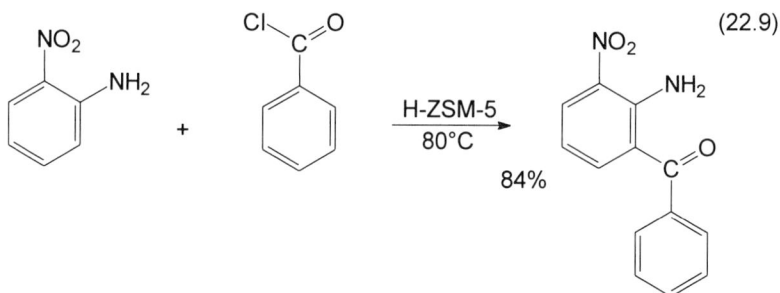

(22.9)

anisoles and anilines (Eqn. 22.9)[23] while dealuminated zeolites were found to be particularly effective catalysts for the acylation of benzofuran (Eqn. 22.10).[24] Lanthanum exchanged Y zeolites promoted the liquid phase acylation of toluene with aliphatic and aromatic acid chlorides giving almost exclusive para isomer formation. Little reaction was observed using an H-Y catalyst under these conditions.[25]

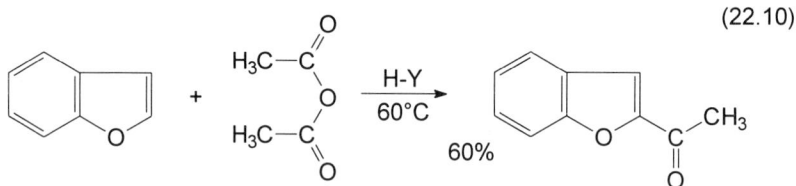

(22.10)

Aldol and Related Reactions

The self-condensation of acetone using a Nafion-H catalyst gave mesityl oxide in about 20% yield. When this reaction was run in a hydrogen atmosphere in the presence of Pd/C, the unsaturated product was hydrogenated giving methyl-isobutyl ketone. Low reaction temperatures were used to minimize the hydrogenation of acetone.[26] Running this condensation-hydrogenation in the vapor phase with a Pt-H-ZSM-5 catalyst gave methyl-isobutyl ketone in about 10% yield at 15% conversion.[27] The vapor phase condensation of formaldehyde with acetone at 200°C gave methyl vinyl ketone in a 52% yield over a V_2O_5-P_2O_5

(22.11)

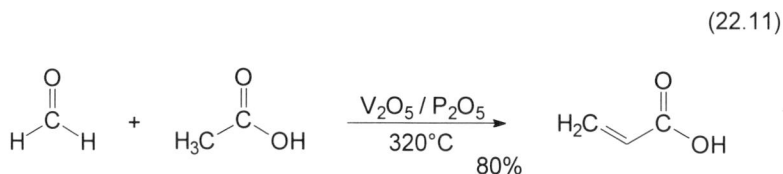

catalyst.[28]   An 86% yield of acrylic acid and a 70% yield of acrolein were obtained on reaction of formaldehyde with acetic acid and acetaldehyde, respectively, over this catalyst at 320°C (Eqn. 22.11).[28]   Condensation of methanol and n-propanol over a VO_x-TiO_2 catalyst at 350°C gave isobutryaldehyde in 60% yield at 95% conversion.[29]   Condensation of acetone, formaldehyde and ammonia over H-ZSM-5 at 420°C gave a mixture of 2-picoline (**11**) and lutidine (**12**) (Eqn. 22.12).[30]   With ethanol, formaldehyde and ammonia 2- and 3-picoline and lutidine were formed but in lower yields.[31]

(22.12)

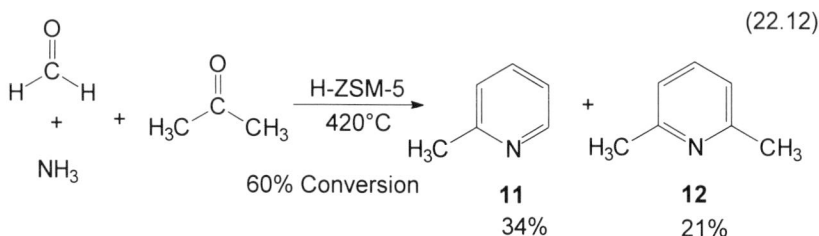

Aldol reactions between silyl enol ethers and aldehydes or ketones are versitile C–C bond forming reactions that are generally run in homogeneous systems containing an acid. This reaction, however, is effectively promoted by aluminum exchanged monmorillonite, usually at sub-zero temperatures.   The reaction between trimethylsiloxy cyclohexene (**13**) and benzaldehyde took place in methylene chloride at -95°C to give a 91% yield of the product after two hours (Eqn. 22.13).[9]   The threo (**14**) and erythro (**15**) diastereomers were produced in a 63:37 ratio.   When the reaction was run in toluene at -50°C the overall yield decreased to 82% but the amount of the threo diasteromer, **14**, increased to 71%. Running the reaction in dimethoxyethane, however, gave predominantly the erythro isomer, **15**.[9]   Similar results were observed with a number of aliphatic and aromatic aldehydes and silyl enol ethers.[9,32,33]

Diels-Alder Reaction

While Diels-Alder reactions can proceed in the absence of a catalyst, higher temperatures are usually needed and this can present a problem with heat sensitive molecules.[34]   Lewis acid catalysts have been used to lower the temperatures needed for Diels-Alder reactions[35,36] but these reactions are frequently accompanied by diene polymerizations.   Further, an excess of the catalyst is

(22.13)

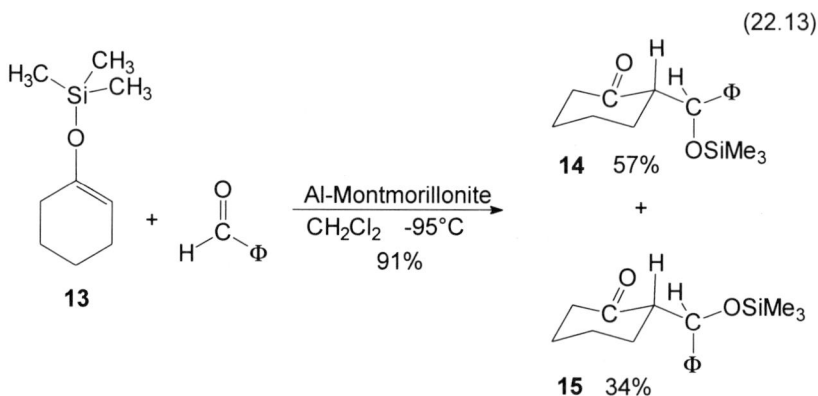

13

14   57%

+

15   34%

usually needed, especially when the dienophile contains a carbonyl or other oxygen containing group. These disadvantages can be overcome by the use of a solid acid catalyst. The uncatalyzed reaction of 1,3-cyclohexadiene with acrolein gave only a 25% yield of the adduct, **16**, after heating at 100°C for 3.5 hours but running this reaction in the presence of a Nafion-H catalyst at 25°C for 40 hours gave **16** in 88% yield (Eqn. 22.14).[37]

(22.14)

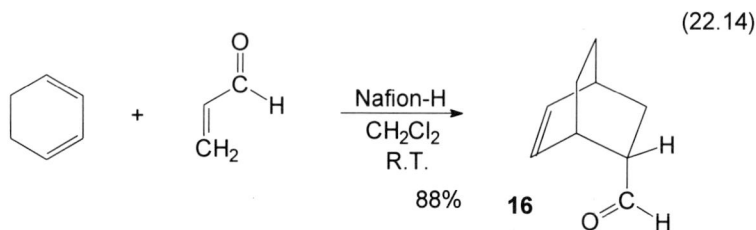

Cation exchanged montmorillonites are also effective catalysts for the Diels-Alder reaction. The uncatalyzed reaction of cyclopentadiene with methyl acrylate gave, after 24 hours, a 54% yield of the adducts having an endo:exo (**17:18**) ratio of 3.7. With a Ca^{++}-montmorilonite catalyst a 98% yield of the product having an endo:exo ratio of 14 was obtained. When a Zr^{IV} exchanged montmorillonite was used a 97% yield of the adduct having an endo:exo ratio of 14.4 was obtained after only 1.5 hours (Eqn. 22.15).[38]   Diethyl ether and dichloromethane were the best solvents for these reactions.[39] Reactions run under these conditions with acrylate esters of optically active alcohols gave chiral products with ee's as high as 50%.[38-42]   K-10 montmorillonite, which had been exchanged with Fe^{+3} or $AlCl_3$ was also an effective catalyst for Diels-Alder reactions.   In most cases very good to excellent yields were obtained at temperatures between 0°–20°C.[43-46]   This catalyst also promoted the reaction

(22.15)

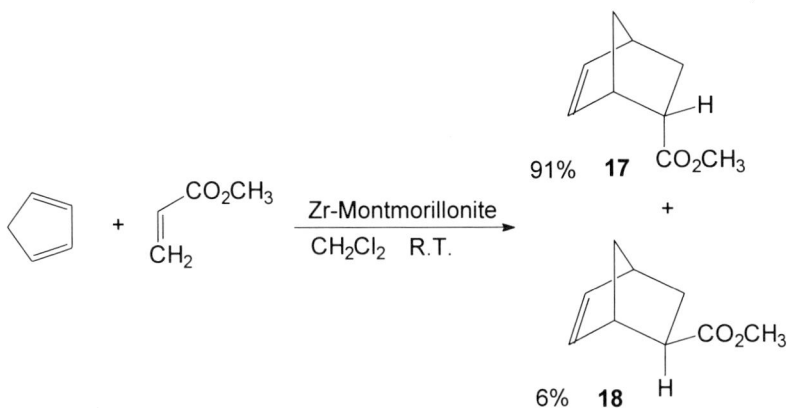

$$\text{cyclopentadiene} + \underset{CH_2}{\overset{CO_2CH_3}{\diagup}} \xrightarrow[\text{CH}_2\text{Cl}_2 \quad \text{R.T.}]{\text{Zr-Montmorillonite}}$$

91% **17** CO_2CH_3

+

6% **18** H

(22.16)

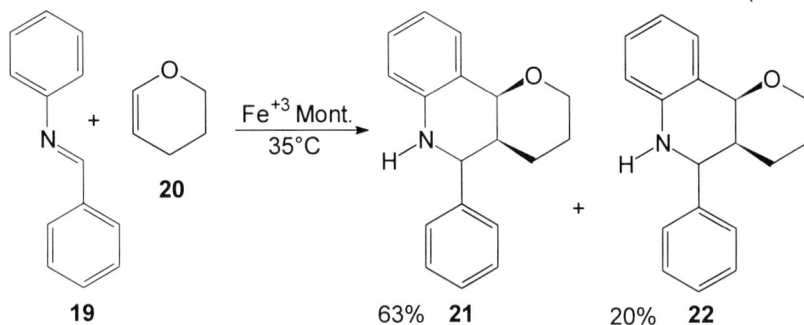

19 + **20** $\xrightarrow[35°C]{Fe^{+3} \text{ Mont.}}$ 63% **21** + 20% **22**

between N-benzylidene aniline (**19**) and enol ethers such as dihydropyran (**20**) to give the substituted tetrahydroquinolines, **21** and **22** (Eqn.22.16).[47]

The Diels-Alder reaction between dihydropyran (**20**) and acrolein gave only a 5% yield of *cis* 1,8-dioxaoctahydronaphthlene (**23**) after 15 hours at 150°C. When the reaction was run at 0°C in the presence of a dealuminated Y zeolite, **23** was obtained in 62% yield after 10 minutes (Eqn. 22.17).[48] The Diels-Alder reaction between cyclopentadiene and methyl acrylate took place at 0°C with a

(22.17)

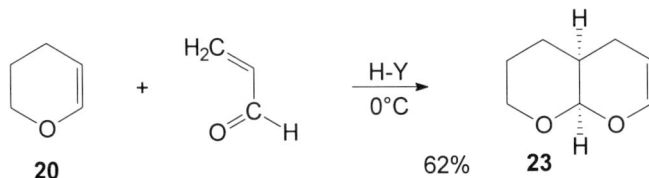

20 + $\underset{O}{\overset{H_2C}{\diagdown}}\overset{}{\underset{H}{C}}$ $\xrightarrow[0°C]{\text{H-Y}}$ 62% **23**

ZnBr$_2$ doped Ce exchanged Y zeolite. After one hour a 96% yield of the endo product, **17**, was obtained. No exo isomer was formed.[49]

Rearrangements

Pinacol Rearrangement

Heating 2,3-dimethyl-2,3-diol, pinacol (**24**), at 80°C in the presence of Nafion-H gave a 92% yield of pinacolone (methyl-tert-butyl ketone) (**25**) (Eqn. 22.18).[50] Similar yields were obtained with other ditertiary vicinal diols. Reaction of pinacol with either La-HY or H-ZSM-5[51] or aluminum exchanged montmorillonite[52] gave both pinacolone (**25**) and 2,3-dimethylbutadiene (**26**), resulting from the dehydration of the diol, in nearly equal quantities. Only a small amount of dehydration was observed when the rearrangement was run over Amberlyst-15.[52]

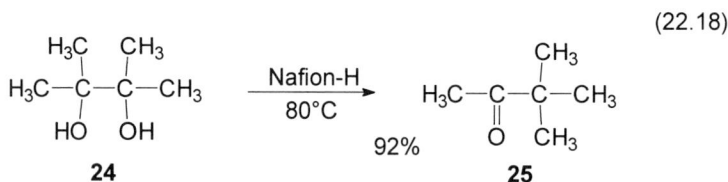

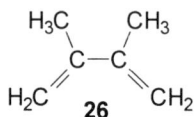

(22.18)

Fries Rearrangement

Heating a phenol ester in the presence of Nafion-H gave good yields of the hydroxy phenyl ketone.[53] This Fries rearrangement gave a mixture of the ortho and para hydroxy compounds in a 1:2-2.5 ratio (Eqn. 22.19). A study of the rearrangement of phenyl benzoate over a number of solid acid catalysts indicated that Nafion-H was more effective than montmorillonite or amorphous aluminum silicate for this reaction.[54]

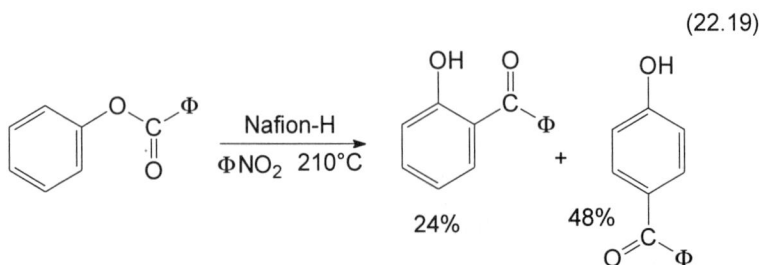

(22.19)

(22.20)

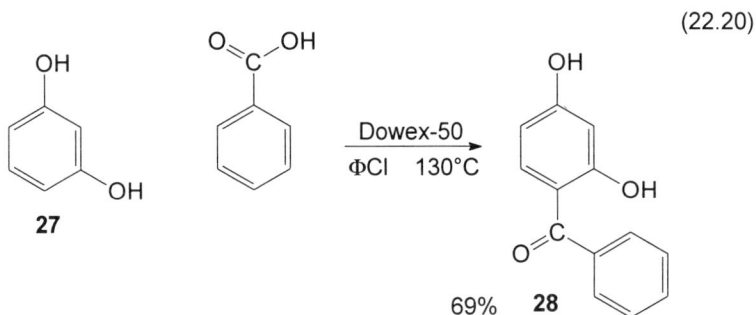

69% **28**

An indirect Fries reaction involving first the *in situ* formation of the ester from resorcinol (**27**) and benzoic acid followed by rearrangement to the 2,4-dihydroxybenzophenone (**28**) was run over a number of different solid acids.[55] When the reaction was run at 130°C, the zeolite, H-beta, gave a 42% yield of the benzophenone, **28**, as did a Nafion catalyst but both Dowex-50 and Amberlyst-15 gave 65-70% yields. Raising the reaction temperature to 160°C gave the substituted benzophenone, **28**, in 70% yield over H-beta (Eqn. 22.20).[55] Comparable yields were obtained using substituted benzoic acids in this reaction. It was concluded that H-beta was preferred for this reaction since no by-products were formed with this catalyst and it was stable at the elevated temperatures needed to obtain good product yields.[55] However, the yields obtained with Amberlyst-15 and Dowex-50 at 130°C were comparable to those obtained with H-beta at 160°C.

Beckmann Rearrangement

Reaction of ketoximes with acids generally leads to a rearrangement to the amide or lactam while aldoximes are usually dehydrated to the nitriles. Reaction of cyclohexanone oxime (**29**) with various zeolites at temperatures between 250°–380°C gave varying amounts of caprolactam (**30**) (Eqn. 22.21).[56] The small pore size zeolite, H-A, produced caprolactam in only 4% selectivity at 14% conversion while the medium pore sized H-ZSM-5 and H-M zeolites gave caprolactam in about 50% selectivity at 100% conversion. With the large pore, H-Y, caprolactam was formed in 62% selectivity at 84% conversion.[57] Running the reaction over a silica supported tantalum oxide catalyst at 300°C gave caprolactam in 98%

(22.21)

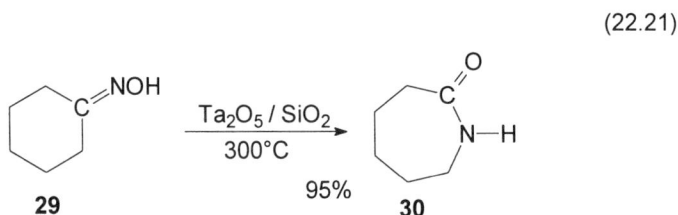

95%

(22.22)

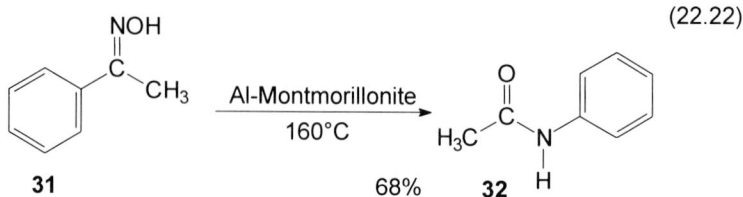

31   68%   32

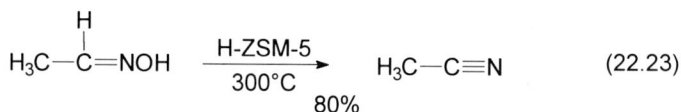

(22.23)

selectivity at 97% conversion.[56] Aluminum exchanged montmorillonite catalyzed the rearrangement of acetophenone oxime (**31**) to acetanilide (**32**) in 68% yield (Eqn. 22.22).[58]

Acetaldoxime was converted to acetonitrile in about 80% yield over an H-ZSM-5 catalyst at 300°C (Eqn. 22.23).[57]   Heating aldoximes with montmorillonite KSF in refluxing toluene gave the nitriles in 65–85% yields.[59]

Rupe Reaction

The acid catalyzed rearrangement of tertiary acetylenic alcohols, the Rupe reaction, is a viable procedure for the preparation of α,β-unsaturated carbonyl compounds.   Under standard acidic conditions, however, the products can polymerize and various by-products can be produced.[60] With solid acids these problems are minimized.   The reaction of an acetylenic tertiary alcohol with Nafion-H gave the α,β-unsaturated ketone in good to very good yields (Eqns. 22.24–25).[60]   When this reaction was run using a vanadium pillared montmorillonite as the catalyst, the products were the α, β-unsaturated aldehydes instead of the ketones, which are obtained under most other conditions. These aldehydes were formed in very good yields (Eqn. 22.25).[61] Cyclic alcohols, such as 1-ethynylcyclohexanol (**33**), however, were inert under the vanadium catalyzed reaction conditions.[61]

(22.24)

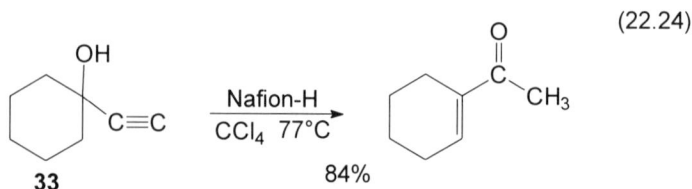

33   84%

(22.25)

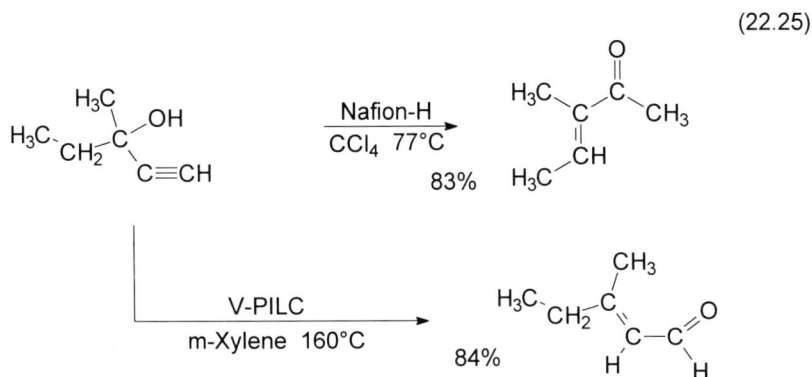

Fisher Indole Synthesis

Running the Fisher indole synthesis on an unsymmetrical phenyl hydrazone gives a mixture of 2,3-disubstituted indoles. For example, reaction of the phenyl hydrazone, **34**, with acid can give both **35** and **36** (Eqn. 22.26). Soluble acids and Amberlyst-15 give these two products in a 75:25 ratio at 100% conversion. With an H-M catalyst they are formed in a 65:35 ratio but over a dealuminated H-beta zeolite, the selectivity is reversed and **36** is produced in an 82% yield at 100% conversion.[62] It was proposed that the preferential formation of **36** over the H-beta catalyst was the result of a restricted transition state selectivity.[62]

(22.26)

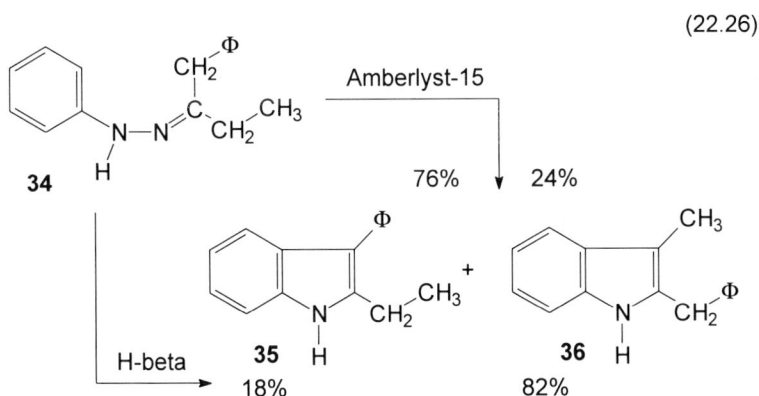

Aromatic Nitrations

Benzene was converted to nitrobenzene in 80% selectivity at 40% conversion by passing benzene and concentrated nitric acid through a reactor containing an H-Y catalyst at 170°C.[63] The liquid phase nitration of benzene has been accomplished by heating a mixture of benzene and n-butyl nitrate at 80°C in the presence of

Nafion-H. Nitrobenzene was produced in 77% yield.[64,65] The nitration of toluene by this procedure gave a 96% yield of the nitrotoluenes having an ortho:meta:para ratio of 50:3:47.[64,65] Using acetone cyanohydrin nitrate as the nitrating agent with a Nafion catalyst gave slightly lower yields of the nitrotoluenes having about the same ortho:meta:para ratio.[65] Reaction of toluene with n-propyl nitrate in the presence of H-ZSM-5 gave a 67% yield of the nitrotoluenes of which 93% was the para isomer (Eqn. 22.27).[66] Reactions with acyl nitrates promoted by H-ZSM-11 that had been treated with tributylamine to block the external acid sites, gave 95% selectivity for p-nitrotoluene formation.[67]

(22.27)

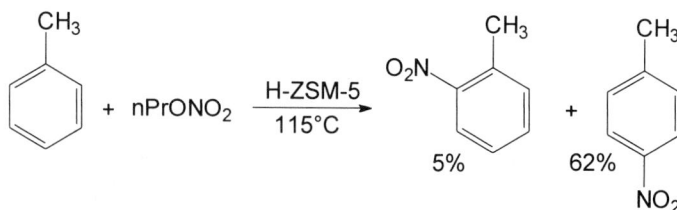

Nitration of chlorobenzene with n-butyl nitrate and a Nafion-H catalyst gave only a 15% yield of the chloronitrobenzenes but with acetone cyanohydrin nitrate a 49% yield was obtained with the para isomer produced with 70% selectivity.[65] Using an iron exchanged montmorillonite to promote the nitration of chlorobenzene with nitric acid and acetic anhydride gave a 90% yield of the nitro chlorobenzenes in 15 minutes at 80°C. The para isomer was produced in 92% selectivity (Eqn. 22.28).[68]

(22.28)

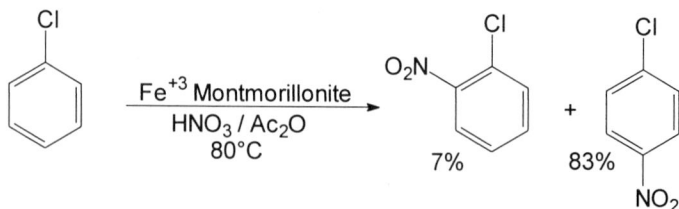

Aromatic Halogenations

Chlorination of toluene with tert-butyl hypochlorite in the presence of silica gel gave a mixture of chlorotoluenes in which the ortho isomer predominated.[4,69] Running this reaction using a large pore zeolite, H-X, gave p-chlorotoluene in

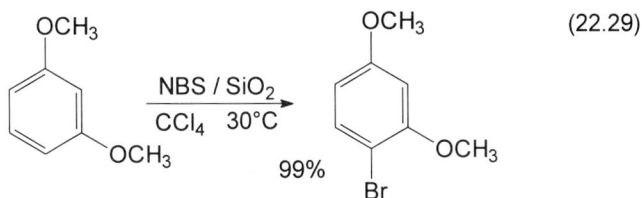

(22.29)

99%

82% selectivity.[70] The chlorination of other substituted benzenes also proceeded with the predominant formation of the p-chloro product.[70]

Brominations run using tert-butyl hypobromite and an H-X catalyst gave predominantly p-bromo isomer formation.[5] N-Bromosuccinimide (NBS) with silica gel is an effective brominating agent for a number of aromatic ring systems. It has been used as a general method for preparing p-bromo aromatic ethers (Eqn. 22.29).[71] Bromination of indoles by the classic procedure does not always lead to the formation of a brominated indole,[72] but the NBS-silica procedure provided a rapid and quantitative method for preparing either the mono- or di-bromo indole or benzimidazole, depending on the amount of NBS used (Eqns. 22.30–31).[73]

(22.30)

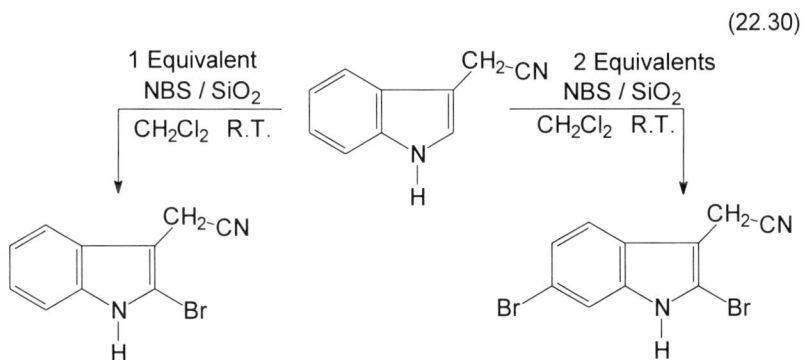

Carbon–Oxygen Bond Formation

Esters have been prepared in good to excellent yields by passing a mixture of the carboxylic acid and the alcohol through a bed of Nafion-H at 125°C at such a rate that the time of contact with the catalyst is in the order of 5–7 seconds.[74] Esters can also be prepared in very good yields by stirring a mixture of the acid, alcohol and Amberlyst-15 at room temperature for times generally ranging from 5 to 15 hours.[75]

Stirring a mixture of a carbonyl compound, triethyl orthoformate and Amberlyst-15 at 0°–5°C gave the acetal or enol ether in good to excellent

(22.31)

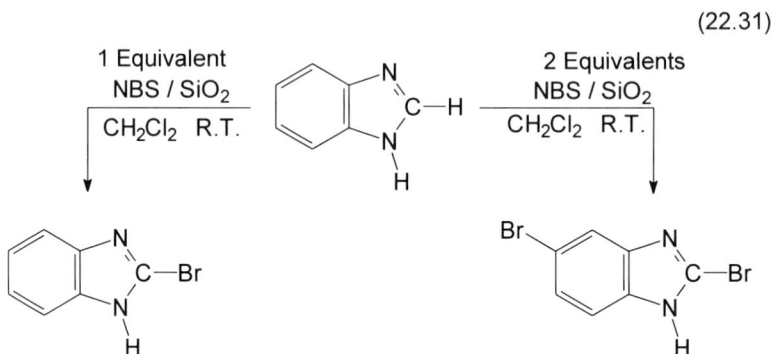

yields.[76] Similar results were also obtained using a Nafion-H catalyst at room temperature. Acetals were also produced in very good yields by stirring a mixture of an aldehyde, methanol and H-ZSM-5 at room temperature.[77] A Nafion-H catalyst was particularly effective for the preparation of ethylene dithioacetals by refluxing a mixture of the carbonyl compounds and 1,2-ethanedithiol with the catalyst in benzene with the azeotropic removal of the water formed in the reaction (Eqn. 22.32).[78]

(22.32)

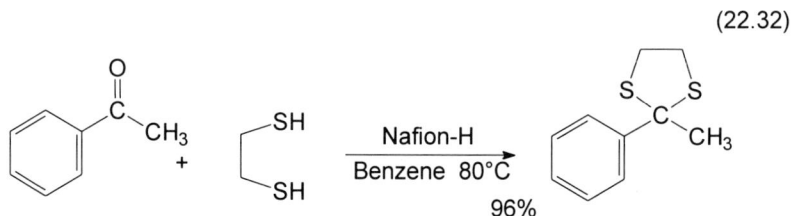

Amberlyst-15 was also used as a catalyst for the reaction of alcohols and phenols with tetrahydropyran (Eqn. 22.33).[79] Refluxing a mixture of an alcohol and dimethoxymethane in the presence of a Nafion-H catalyst gave the methoxy methyl ethers in very good yields (Eqn. 22.34).[80] Nafion-H was also used to catalyze the conversion of diols to cyclic ethers.[81]

(22.33)

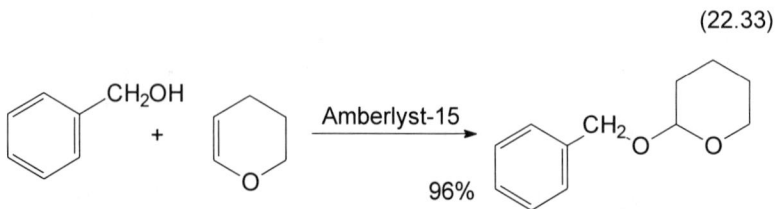

(22.34)

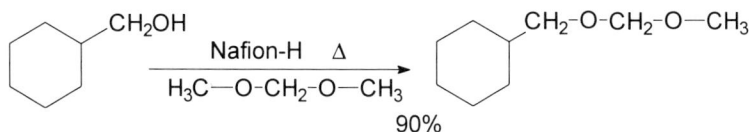

90%

Epoxide Ring Opening

Solid acids also catalyze the reaction of epoxides with nucleophiles. Nafion-H promotes the hydrolysis of epoxides giving good yields of the diols at 20°C.[82] Nafion-H catalyzed reactions run in methanol give the methoxy alcohol with the methoxy group on the more substituted carbon atoms as expected from a reaction that proceeds by way of nucleophilic attack on the more stable carbocation (Eqn. 22.35).[82] Oxiranes are also cleaved by nucleophiles in the presence of montmorillonite. Reaction of epoxides with thiophenol in the presence of montmorillonite gave the hydroxy thioethers in good yields (Eqn. 22.36).[83] With sodium p-toluene sulfinate, hydroxy sulfones are obtained in 50–80% yields.[84]

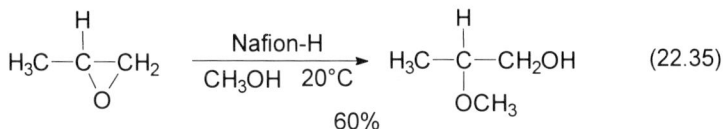

(22.35)

60%

(22.36)

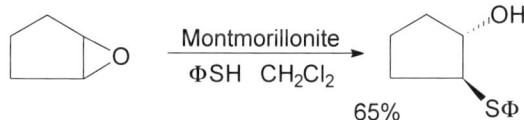

65%

(22.37)

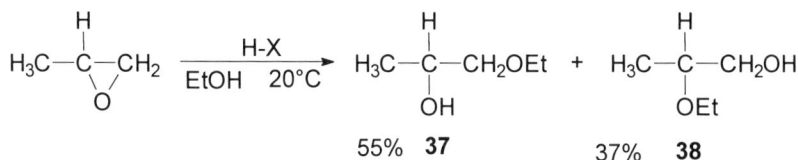

55% **37**   37% **38**

Reaction of propylene oxide with nucleophiles in the presence of the zeolite, H-X, gave a mixture of products, **37** and **38**. In contrast to the ring opening by methanol that was promoted by Nafion-H (Eqn. 22.35), when H-X was used as the catalyst the major product was that in which the nucleophile was found on the terminal carbon atom. Reaction of propylene oxide with ethanol

(22.38)

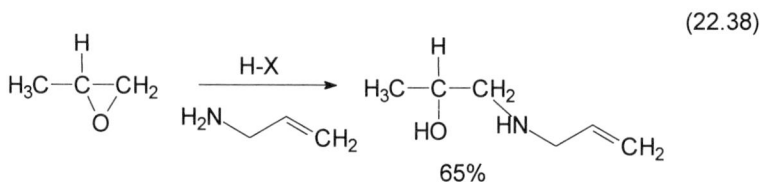

65%

gave a mixture of 1-ethoxy-2-propanol (**37**) and 2-ethoxy-1-propanol (**38**) in 92% yield with a **37**:**38** ratio of 1.5 (Eqn. 22.37).[85] Reaction with butyl sulfide gave a 61% yield of hydroxy thioethers in which the terminal thioether was present in a 40:1 ratio. Reaction with allyl sulfide or allyl amine gave only that product having the nucleophile on the primary carbon atom (Eqn. 22.38).[86]

Ring opening of 3-cyclohexyl-2,3-epoxypropan-1-ol (**39**) with sodium azide over H-X gave an 85% yield of the isomeric azido diols, **40** and **41**, with the vicinol diol, **40**, the almost exclusive product. With other zeolites and clays, the yields were lower but the regioselectivity still favored the formation of **40** (Eqn. 22.39).[86]

(22.39)

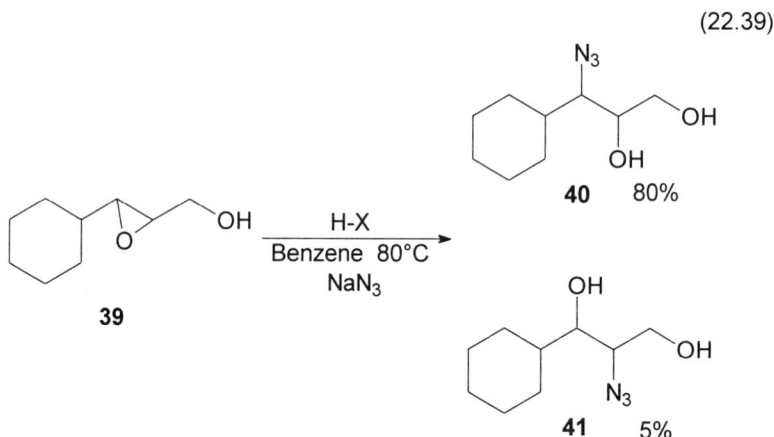

40   80%

39

41   5%

Hydrolysis Reactions

Stirring an ester in an aqueous suspension of Dowex-50 is an efficient method for hydrolyzing the ester to the acid.[87] Similar results have been obtained by refluxing an ester in an aqueous suspension of H-beta zeolite.[88] Stirring acetals in an acetone suspension of wet silica gel[89] or moistened Amberlyst-15[90] is a convenient method for converting an acetal to the aldehyde or ketone without having side reactions such as double bond isomerizations also take place (Eqn. 22.40).[89]

Oximes, hydrazones and semicarbazones have been hydrolyzed in very good yields by heating in an aqueous suspension of Dowex-50[91] or an aqueous acetone suspension of Amberlyst-15.[92]

$$(22.40)$$

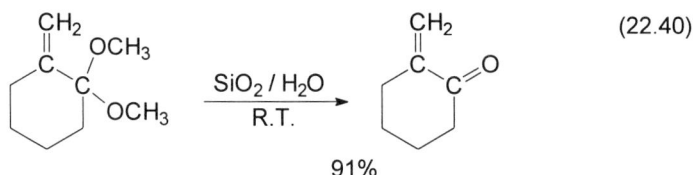

91%

Carbon–Nitrogen Bond Formation

The vapor phase N–alkylation of aniline with ethanol took place over an H-ZSM-5 catalyst at temperatures between 250° and 500°C. The maximum selectivity for N-monoethylation occurred at the lower temperatures with increasing amounts of diethyl aniline produced at higher reaction temperatures.[93]   With a montmorillonite catalyst, reaction at 400°C gave a 64% selectivity for N-ethyl aniline formation at 77% conversion.   A vanadium exchanged montmorillonite was more active but less selective giving N-ethyl aniline in 48% selectivity and N, N-diethyl aniline in 37% selectivity at 97% conversion.[94]

$$(22.41)$$

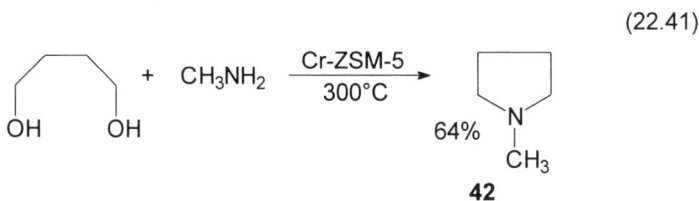

The vapor phase reaction of butane diol with methyl amine at 300°C in the presence of a Cr-ZSM-5 catalyst gave a 65% yield of N-methyl pyrrolidine, **42** (Eqn. 22.41). Tetrahydrofuran was also converted to **42** in 98% yield under these conditions.[95]

Reaction of sulfur dioxide with the imine, **43**, over a ZSM-5 catalyst gave 4-methylthiazine (**44**) in yields as high as 65% (Eqn. 22.42).[96]

$$(22.42)$$

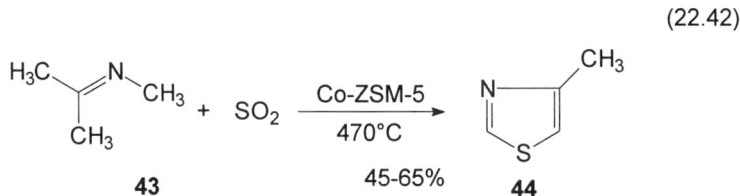

Solid Base Catalyzed Reactions

Solid base catalyzed reactions are not as commonly used in synthetic processes as are those reactions promoted by solid acids. While several so-called super bases have been developed, they have not been generally used for the types of reactions involved in the synthesis of even somewhat complex organic molecules.[97-100] The most commonly used solid bases are alkali metal ion exchanged zeolites. Since the acidity of a zeolite decreases with decreasing aluminum content and the basicity generally increases with increasing size of the exchanged cation, Cs-X is the most basic of these exchanged zeolites.[100] This catalyst has been used for the vapor phase alkylation of toluene with methanol to give a mixture of ethyl benzene and styrene.[100-102]

Condensations

Of more synthetic interest is the Cs-X catalyzed liquid phase condensation of benzaldehyde with active methylene compounds such as ethyl cyanoacetate, ethyl malonate and ethyl acetoacetate[103-105] but the yields in these reactions were only in the 40%–70% region. Higher yields were obtained using a germanium substituted faujasite[1-2,106] or a nitrided aluminophosphate[107] as the basic catalyst.

(22.43)

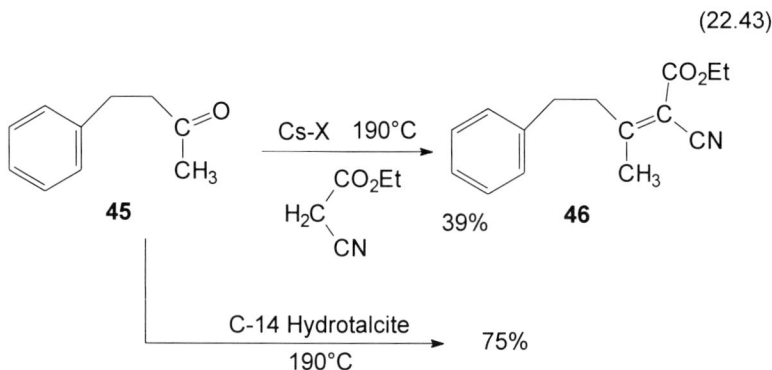

The condensation of 4-phenyl-2-butanone (**45**) with ethyl cyanoacetate took place over Cs-X to give the product, **46**, in only a 39% yield but when a magnesium oxide containing hydrotalcite clay was used a 75% yield of **46** was obtained (Eqn. 22.43).[108] Formaldehyde was condensed with acetaldehyde at 425°C over a ZSM-5 zeolite containing MgO. Acrolein was obtained in almost 70% yield.[109]

The condensation of benzaldehyde with different ring-sized cyclic ketones was run in the liquid phase over Cs-X, Cs-M, Cs-ZSM-5 and Cs-A (Eqn. 22.44).[110] The highest conversions were obtained using the large pore, Cs-X catalyst. The mono-benzylidine product, **47**, was the only one obtained with cyclooctanone but with cyclohexanone and cyclopentanone dibenzylidine, **48**, formation predominated over all catalysts.[110] The fact that the small pore Cs-A also promoted the condensation with cyclooctanone and the finding that blocking the surface sites resulted in a significant reduction in the rates of the condensations led to the conclusion that these reactions were probably taking place on the exterior surface of the zeolites.[111] Interesingly, condensation of

(22.44)

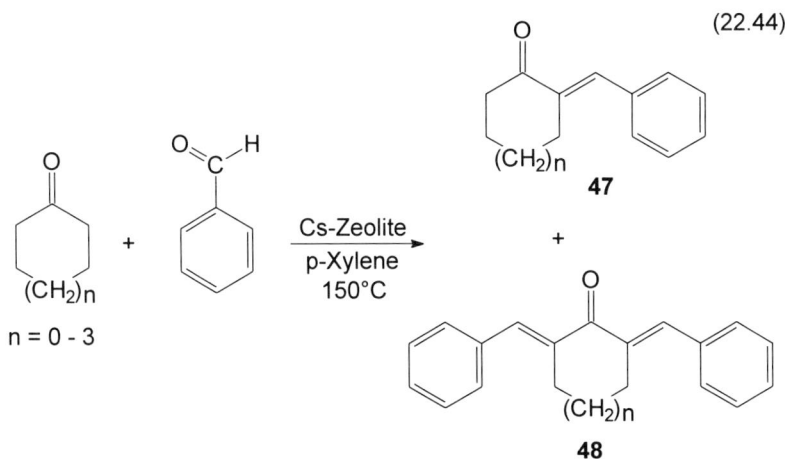

Cs-Zeolite
p-Xylene
150°C

n = 0 - 3

(22.45)

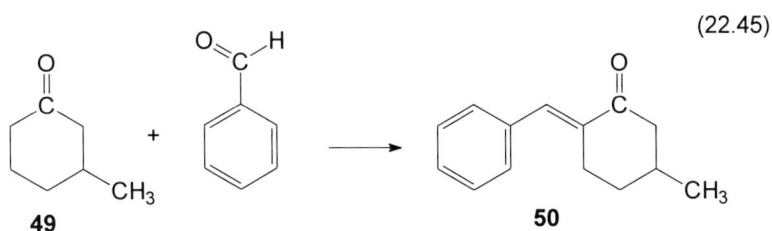

3-methyl-cyclohexanone (49) with benzaldehyde gave almost exclusively the monobenzylidine compound, 50, regardless of the zeolite used (Eqn. 22.45).[110]

Alkylations

Cs-X has also been used to promote the selective O-methylation of phenol with dimethyl carbonate.[112] N-Monomethylaniline was obtained in 93% yield by reacting aniline with dimethyl carbonate over a K-Y catalyst at 180°C (Eqn. 22.46).[113] The reaction of alcohols with chloromethyl ether over Na-Y gave the resulting methoxymethyl ethers in 70%–90% yields (Eqn. 22.47).[114]

(22.46)

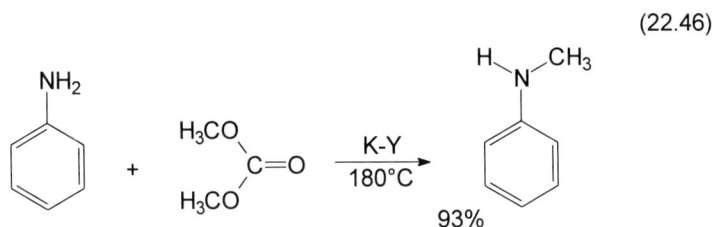

$$HC\equiv C-CH_2OH$$

+ $\xrightarrow[CH_2Cl_2 \quad 35°C]{Na-Y}$ $HC\equiv C-CH_2-O-CH_2-O-CH_3$ (22.47)

$$Cl-CH_2-O-CH_3$$ 89%

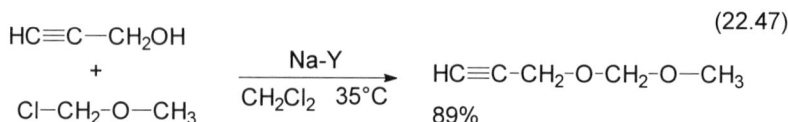

Metal Catalyzed Reactions

Hydroformylations

The hydroformylation of alkenes is commonly run using soluble metal carbonyl complexes as catalysts but there are some reports of heterogeneously catalyzed reactions of olefins with hydrogen and carbon monoxide. Almost all of these are vapor phase reactions of ethylene or propylene with hydrogen and carbon monoxide catalyzed by rhodium,[115-120] ruthenium,[121] nickel,[122,123] cobalt,[123,124] and cobalt-molybdenum[123] catalysts as well as various sulfided metal catalysts.[123,125,126]

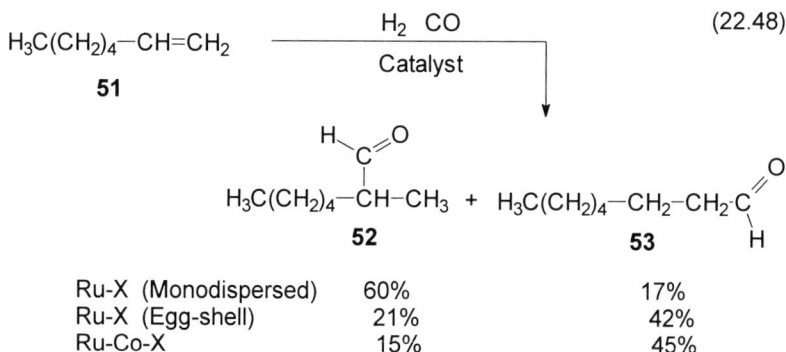

$$H_3C(CH_2)_4-CH=CH_2$$ $\xrightarrow[Catalyst]{H_2 \quad CO}$ (22.48)

51

H   O
 \\ //
  C
  |
$$H_3C(CH_2)_4-CH-CH_3$$ + $H_3C(CH_2)_4-CH_2-CH_2-C$ (=O, H)

52   **53**

	52	53
Ru-X (Monodispersed)	60%	17%
Ru-X (Egg-shell)	21%	42%
Ru-Co-X	15%	45%

The vapor-phase hydroformylation of 1-octene (**51**) over Ru-X catalysts gave primarily the product, **52**, having the aldehyde group on the secondary carbon and a small amount of over-hydrogenation to the alcohol when the ruthenium was dispersed evely throughout the support. A catalyst that had an egg-shell distribution of the ruthenium was somewhat more active and promoted less over-hydrogenation. It also gave the straight chain, **53**, and branched aldehydes, **52**, in about a 2:1 ratio (Eqn. 22.48).[127] A Ru-Co-X catalyst promoted a little more over-hydrogenation to the alcohols but the straight chain aldehyde was formed in a 3:1 ratio.[127] Hydroformylation of 1-hexene over Ir-ZSM-5 resulted in extensive hydrogenation of the alkene. Only about 30% of the hexene was converted to the aldehydes. Adding formaldehyde to the reaction stream significantly reduced the extent of alkene hydrogenation.[128]

(22.49)

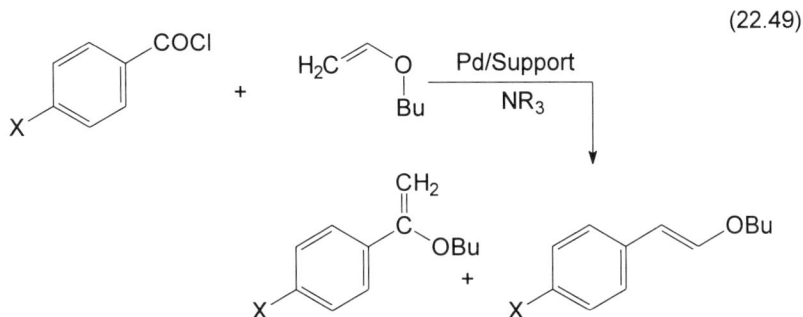

X	Support	**54**	**55**
NO$_2$	SiO$_2$	5%	46%
NO$_2$	Al$_2$O$_3$	6%	49%
NO$_2$	MgO	2%	14%
H	SiO$_2$	12%	70%
H	Al$_2$O$_3$	11%	55%
H	MgO	10%	3%
CH$_3$	SiO$_2$	20%	41%
CH$_3$	Al$_2$O$_3$	32%	33%
CH$_3$	MgO	12%	1%

Organometallic Reactions

Soluble transition metal complexes have been used to catalyze a wide variety of synthetically useful C–C bond forming reactions.[129] There are, however, only a few reports on the use of supported metals as catalysts for such reactions.[130-134] Supported palladium catalysts have been used to promote the Heck arylation of butyl vinyl ether (Eqn. 22.49).[130-133] In this reaction the selectivity for either α- (**54**) or β-product (**55**) formation is regulated by the electronic character of the substitutent on the benzene ring and the nature of the support used in the catalyst. As the data show electron withdrawing groups on the benzene ring enhance β-product, **55**, formation as does the use of more acidic supports such as silica. When electron donating groups are present or basic supports such as magnesia are used, α-product, **54**, formation is favored. The basic support also decreases catalyst activity.[133]

Similar support effects have been observed in the supported metal catalyzed allylation reaction shown in Eqn. 22.50.[134] When silica was used as the support, alkyaltion at the more stable, secondary, allyl cation was favored while with magnesia, allylation took place predominantly at the primary carbon atom.[134]

$$\text{(22.50)}$$

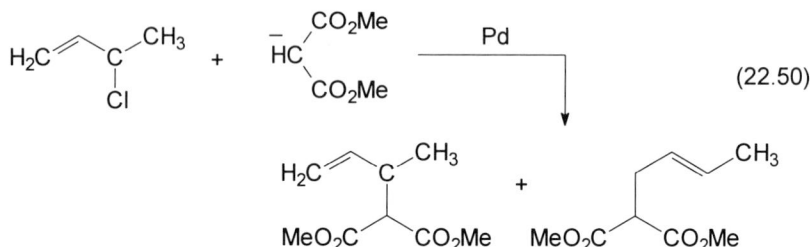

Nitrile Hydrolysis

Nitriles have been hydrolyzed to the amides by heating them in aqueous medium over copper catalysts.[135,136] Refluxing the nitrile, **56**, in an aqueous suspension of Raney copper gave the amide, **57**, in 50–60% yield (Eqn. 22.51).[136]

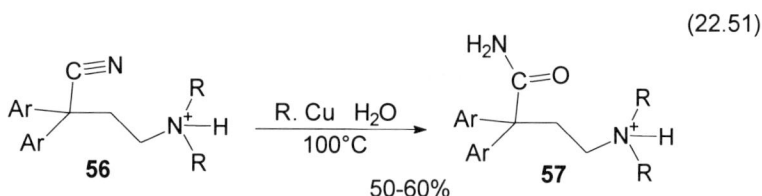

$$\text{(22.51)}$$

50-60%

Alkene Metathesis

Olefin metathesis is a reaction in which there is a net breaking and re-formation of two olefininc C–C double bonds as depicted in Eqn. 22.52.[137-140] This reaction is catalyzed by a variety of heterogeneous catalysts, primarily the oxides of molybdenum, tungsten and rhenium usually supported on silica or alumina. The metathesis of alkenes can be placed into two categories: 1) self metathesis of a single alkene and 2) cross metathesis of two alkenes. Double bond isomerization occurring during the reaction can lead to complex product mixtures.

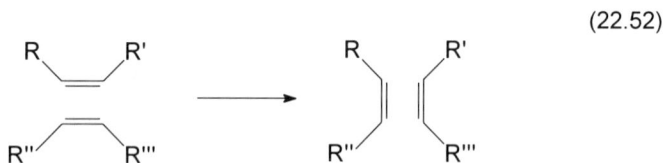

$$\text{(22.52)}$$

As an example of the first category, this reaction has been used to convert linear unsaturated nitriles having a terminal double bond into unsaturated di-nitriles (Eqn. 22.53).[141] This reaction works well when there are at least two carbon atoms between the double bond and the nitrile. Shorter carbon chains do

(22.53)

$$H_2C=CH(CH_2)_3CN$$

$$H_2C=CH(CH_2)_3CN$$

$\xrightarrow[\text{Et}_4\text{Sn} \quad \text{R.T.}]{\text{Re}_2\text{O}_7 / \text{Al}_2\text{O}_3}$

98%

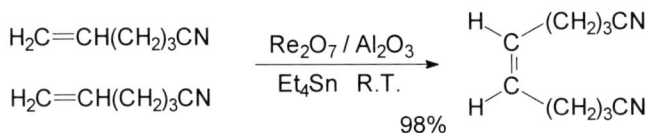

(22.54)

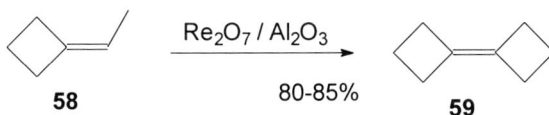

58 $\xrightarrow[\text{80-85\%}]{\text{Re}_2\text{O}_7 / \text{Al}_2\text{O}_3}$ **59**

not react.[141] Reaction of ethylenecyclobutane (**58**) gave bicyclobutylidene (**59**) in 80–86% yield (Eqn.22.54).[140]

An example of the second category is the cross metathesis of cyclooctene and ethylene giving 1, 9-decadiene (**60**) in about 75% yield (Eqn. 22.55).[142] This reaction between cyclic olefins and ethylene provids an excellent method for the preparation of α,ω-diolefins.[138]

(22.55)

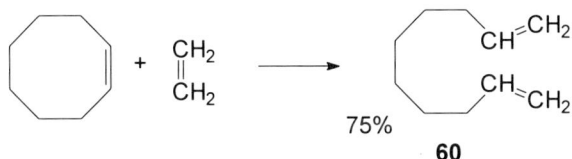

75%

60

References

1. G. A. Olah, *Friedel-Crafts Chemistry*, Wiley, New York, 1973, pp 356-358.
2. G. A. Olah, P. S. Iyer and G. K. S. Prakash, *Synthesis*, 513 (1986).
3. G. K. S. Prakash and G. A. Olah, in *Acid-Base Catalysis* (K. Tanabe, H. Hattori, T. Yamaguchi and T. Tanaka, Eds.) VCH, New York, 1989, p 59.
4. K. Smith *Stud. Surf. Sci. Catal.*, **59** (Heterog. Catal. Fine Chem., II), 55 (1991).
5. K. Smith, *Chem. Ind. (Dekker)*, **62** (Catal. Org. React.), 91 (1995).
6. C. B. Dartt and M. E. Davis, *Catal. Today*, **19**, 151 (1994).
7. W. F. Hölderich, in *Acid-Base Catalysis* (K. Tanabe, H. Hattori, T. Yamaguchi and T. Tanaka, Eds.) VCH, New York, 1989, p1.
8. J. H. Purnell, *Catal. Letters*, **5**, 203 (1990).
9. Y. Izumi, M. Kawai, H. Sakurai, M. Onaka and K. Urabe, in *Acid-Base Catalysis* (K. Tanabe, H. Hattori, T. Yamaguchi and T. Tanaka, Eds.) VCH, New York, 1989, p 21.

10. J. Kaspi, D. D. Montgomery and G. A. Olah, *J. Org. Chem.*, **43**, 3147 (1978).
11. J. Kaspi and G. A. Olah, *J. Org. Chem.*, **43**, 3142 (1978).
12. S. H. Oh and W. Y. Lee, *Stud. Surf. Sci. Catal.*, **75** (New Frontiers in Catalysis, Pt. B), 1693 (1993).
13. S. Namba, H. Ohta, J.-H. Kim and T. Yashima, *Stud. Surf. Sci. Catal.*, **75** (New Frontiers in Catalysis, Pt. B), 1685 (1993).
14. S. Okada, K. Tanaka, Y. Nakadaira and N. Nakagawa, *Bull. Chem. Soc., Japan*, **65**, 2833 (1992).
15. G. A. Olah, J. Kaspi and J. Bukala, *J. Org. Chem.*, **42**, 4187 (1977).
16. J. Cejka, G. A. Kapustin and B. Wichtelova, *Appl. Catal. A*, **108**, 187 (1994).
17. G. A. Olah, D. Meidar, R. Malhotra, J. A. Olah and S. C. Narang, *J. Catal.*, **61**, 96 (1980).
18. Y. Sugi and M. Toba, *Catal. Today*, **19**, 187 (1994).
19. P. Moreau, A. Finiels, P. Geneste, F. Moreau and J. Solofo, *Stud. Surf. Sci. Catal.*, **78** (Heterog. Catal. Fine Chem., III), 575 (1993).
20. O. Sieskind and P. Albrecht, *Tetrahedron Lett.*, **34**, 1197 (1993).
21. G. A. Olah, R. Malhotra, S. C. Narang and J. A. Olah, *Synthesis*, 672 (1978).
22. H. Konishi, K. Suetsugu, T. Okano and J. Kiji, *Bull. Chem. Soc., Japan*, **55**, 957 (1982).
23. V. Paul, A. Sudalai, T. Daniel and K. V. Srinivasan, *Tetrahedron Lett.*, **35**, 2601 (1994).
24. F. Richard, J. Drouillard, H. Carreyre, J. L. Lemberton and G. Perot, *Stud. Surf. Sci. Catal.*, **78** (Heterog. Catal. Fine Chem., III), 601 (1993).
25. D. E. Akporiaye, K. Daasvatn, J. Solberg and M. Stocker, *Stud. Surf. Sci. Catal.*, **78** (Heterog. Catal. Fine Chem., III), 521 (1993).
26. C. U. Pittman, Jr., and Y. F. Liang, *J. Org. Chem.*, **45**, 5048 (1980).
27. L. Melo, E. Rombi, J. M. Dominguez, P. Magnoux and M. Guisnet, *Stud. Surf. Sci. Catal.*, **78** (Heterog. Catal. Fine Chem., III), 701 (1993).
28. M. Ai, *Stud. Surf. Sci. Catal.*, **75** (New Frontiers in Catalysis, Pt. B), 1199 (1993).
29. F. Wang, W. Lee, Y. Liou and L. Chen, *Ind. Eng. Chem. Res.*, **32**, 30 (1993).
30. A. V. Rama Rao, S. J. Kulkarni, R. R. Rao and M. Subrahmanyam, *Appl. Catal. A*, **111**, L101 (1994).
31. S. J. Kulkarni, R. R. Rao, M. Subrahmanyam and A. V. Rama Rao, *Appl. Catal. A*, **111**, 1 (1994).
32. J. Kawai, M. Onaka and Y. Izumi, *Bull. Chem. Soc., Japan*, **61**, 2157 (1988).
33. J. Kawai, M. Onaka and Y. Izumi, *Chem. Lett.*, 381 (1986).
34. L. W. Butz and A. W. Rytina, *Org. React.*, **5**, 136 (1949).
35. P. Yates and P. Eaton, *J. Am. Chem. Soc.*, **82**, 4436 (1960).
36. G. I. Fray and R. Robinson, *J. Am. Chem. Soc.*, **83**, 849 (1961).
37. G. A. Olah, D. Meidar and A. P. Fung, *Synthesis*, 270 (1970).
38. C. Cativiela, F. Figueras, J. M. Fraile, J. I. Garcia, M. Gil, J. A. Mayoral, L. C. de Menorval and E. Pires, *Stud. Surf. Sci. Catal.*, **78** (Heterog. Catal. Fine Chem., III), 495 (1993).
39. C. Cativiela, J. M. Fraile, J. I. Garcia, J. A. Mayoral and F. Figueras, *J. Mol. Catal.*, **68**, L31 (1991).

40. C. Cativiela, F. Figueras, J. M. Fraile, J. I. Garcia and J. A. Mayoral, *Tetrahedron: Asymmetry*, **4**, 223 (1993).
41. C. Cativiela, J. M. Fraile, J. I. Garcia, J. A. Mayoral, J. M. Campelo, D. Luna and J. M. Marinas, *Tetrahedron: Asymmetry*, **4**, 2507 (1993).
42. C. Cativiela, J. M. Fraile, J. I. Garcia, J. A. Mayoral E. Pires, *Tetrahedron,* **48**, 6467 (1992).
43. P. Laszlo and J. Lucchetti, *Tetrahedron Lett.*, **25**, 1567 (1984).
44. P. Laszlo and J. Lucchetti, *Tetrahedron Lett.*, **25**, 2147 (1984).
45. P. Laszlo and J. Lucchetti, *Tetrahedron Lett.*, **25**, 4387 (1984).
46. P. Laszlo and H. Moison, *Chem. Lett.*, 1031 (1989).
47. J. Cabral and P. Laszlo, *Tetrahedron Lett.*, **30**, 7237 (1989).
48. R. Durand, P. Geneste, J. Joffre and C. Moreau, *Stud. Surf. Sci. Catal.*, **78** (Heterog. Catal. Fine Chem., III), 647 (1993).
49. Y. V. S. Narayana Murthy and C. N. Pillai, *Synth. Commun.,* 783 (1991).
50. G. A. Olah and D. Meidar, *Synthesis*, 358 (1978).
51. C. P. Bezouhanova and F. A. Jabur, *J. Mol. Catal.,* **87**, 39 (1994).
52. E. Gutierrez, A. J. Aznar and E. Ruiz-Hitzky, *Stud. Surf. Sci. Catal.*, **41** (Heterog. Catal. Fine Chem.), 211 (1988).
53. G. A. Olah, M. Arvanaghi and V. V. Krishnamurthy, *J. Org. Chem.,* **48**, 3359 (1983).
54. K. R. Lassila and M. E. Ford, *Chem. Ind. (Dekker)*, **47** (Catal. Org. React.), 169 (1992).
55. A. J. Hoefnagel and H. van Bekkum, *Appl. Catal. A*, **97**, 87 (1993).
56. T. Ushikubo and K. Wada, *J. Catal.*, **148**, 138 (1994).
57. T. Curtin and B. K. Hodnett, *Stud. Surf. Sci. Catal.*, **78** (Heterog. Catal. Fine Chem., III), 535 (1993).
58. E. Guiterriz, A. J. Aznar and E. Ruiz-Hitzky, T. Curtin and B. K. Hodnett, *Stud. Surf. Sci. Catal.*, **59** (Heterog. Catal. Fine Chem., II), 539 (1991).
59. H. M. Meshram, *Synthesis*, 943 (1992).
60. G. A. Olah and A. P. Fung, *Synthesis*, 473 (1981).
61. B. M. Choudary, A. D. Prasad and V. L. K. Valli, *Tetrahedron Lett.*, **31**, 7521 (1990).
62. M. S. Rigutto, H. J. A. de Vries, S. R. Magill, A. J. Hoefnagel and H. van Bekkum, T. Curtin and B. K. Hodnett, *Stud. Surf. Sci. Catal.*, **78** (Heterog. Catal. Fine Chem., III), 661 (1993).
63. L. E. Bertea, H. W. Kouwenhoven and R. Prins, T. Curtin and B. K. Hodnett, *Stud. Surf. Sci. Catal.*, **78** (Heterog. Catal. Fine Chem., III), 607 (1993).
64. G. A. Olah and S. C. Narang, *Synthesis*, 690 (1978).
65. G. A. Olah, R. Malhotra and S. C. Narang, *J. Org. Chem.*, **43**, 4628 (1978).
66. T. J. Kwok and K. Jayasuriya, *J. Org. Chem.*, **59**, 4939 (1994).
67. S. M. Nagy, K. A. Yarovoy, L. A. Vostrikova, K. G. Ione and V. G. Shubin, *Stud. Surf. Sci. Catal.*, **75** (New Frontiers in Catalysis, Pt. B), 1669 (1993).
68. B. M. Choudary, M. R. Sarma and K. V. Kumar, *J. Mol. Catal.,* **87**, 33 (1994).
69. K. Smith, M. Butters and W. E. Paget, *Synthesis*, 1155 (1985).
70. K. Smith and M. Butters, *Synthesis*, 1157 (1985).

71. H. Konishi, K. Aritomi, T. Okano and J. Kiji, *Bull. Chem. Soc., Japan,* **62**, 591 (1989).

72. J. Parrick, A. Yahya and Y. Jin, *Tetrahedron Lett.*, **25**, 3099 (1984).

73. A. G. Mistry, K. Smith and M. R. Bye, *Tetrahedron Lett.*, **27**, 1051 (1986).

74. G. A. Olah, T. Keumi and D. Meidar, *Synthesis*, 929 (1978).

75. M. Petrini, R. Ballini, E. Marcantoni and G. Rosini, *Synth. Commun.,* **18**, 847 (1988).

76. S. A. Patwardhan and S. Dev, *Synthesis*, 348 (1974).

77. M. V. Joshi and C, S, Narasimhan, *J. Catal.*, **128**, 63 91991).

78. G. A. Olah, S. C. Narang, D. Meidar and G. F. Salem, *Synthesis*, 282 (1981).

79. A. Bongini, G. Cardillo, M. Orena and S. Sandri, *Synthesis*, 618 (1979).

80. G. A. Olah, A. Husain, B. G. B. Gupta and S. C. Narang, *Synthesis*, 471 (1981).

81. G. A. Olah, A. P. Fung and R. Malhotra, *Synthesis*, 474, (1981).

82. G. A. Olah, A. P. Fung and D. Meidar, *Synthesis*, 280 (1981).

83. A. K. Maiti, G. K. Biswas and P. Bhattacharyya, *J. Chem. Res., Synop.*, 325 (1993).

84. G. K. Biswas and P. Bhattacharyya, *Synth. Commun.*, **21**, 569 (1991).

85. H. Takeuchi, K.Kitajima, Y. Yamamoto and K. Mizuno, *J. Chem. Soc., Perkin Trans.2,* 199 (1993).

86. N. Onaka, K. Sugita and Y. Izumi, in *Acid-Base Catalysis* (K. Tanabe, H. Hattori, T. Yamaguchi and T. Tanaka, Eds.) VCH, New York, 1989, p 33.

87. M. K. Basu, D. C. Sarkar and B. C. Ranu, *Synth. Commun.*, **19**, 627 (1989).

88. M. J. Climent, A. Corma, H. Garcia, S. Iborra and J. Primo, *Stud. Surf. Sci. Catal.*, **59** (Heterog. Catal. Fine Chem., II), 557 (1991).

89. E. Huet, A. Lechevallier, M. Pellet and J. M. Conia, *Synthesis*, 63 (1978).

90. G. M. Coppola, *Synthesis*, 1021 (1984).

91. B. C. Ranu and D. C. Sarkar, *J. Org. Chem.*, **53**, 878 (1988).

92. R. Ballini and M. Petrini, *J. Chem. Soc., Perkin Trans.2,* 2563 (1988).

93. S. Narayanan, V. D. Kumari and A. S. Rao, *Appl. Catal. A*, **111**, 133 (199994).

94. S. Narayanan, K. Deshpande and B. P. Prasad, *J. Mol. Catal.*, **88**, L271 (1994).

95. Y. V. Subba Rao, S. J. Kulkarni, M. Subrahmanyam and A. V. Rama Rao, *J. Org. Chem.*, **59**, 3998 (1994).

96. F. P. Gortsema, B. Beshty, J. J. Friedman, D. Matsumoto, J. J. Sharkey, G. Wildman, T. J. Blacklock and S. H. Pan, *Chem. Ind. (Dekker)*, **53** (Catal. Org. React.), 445 (1994).

97. S. Malinowski and J. Kijeriski, *Catalysis (London)*, **4**, 130 (1981).

98. H. Hattori, Mater. Chem. Phys., **18**, 533 (1988).

99. S. Malinowski and M. Marczewski, *Catalysis (London)*, **8**, 107 (1989).

100. H. Hattori, *Stud. Surf. Sci. Catal.*, **78** (Heterog. Catal. Fine Chem., III), 35 (1993).

101. P. Beltrame, P. Funagalli and G. Zuretti, *Ind. Eng. Chem. Res.*, **32**, 26 (1993).

102. A. N. Vasiliev and A. A. Galinsky, *React. Kinet. Catal. Lett.*, **51**, 253 (1993).

103. A. Corma, V. Fornes, R. M. Martin-Aranda, H. Garcia and J. Primo, *Appl. Catal.*, **59**, 237 (1990).
104. A. Corma, R. M. Martin-Aranda and F. Sanchez, *Stud. Surf. Sci. Catal.*, **59** (Heterog. Catal. Fine Chem., II), 503 (1991).
105. I. Rodriguez, H. Cambon, D. Brunel, M. Lasperas and P. Geneste, *Stud. Surf. Sci. Catal.*, 78 (Heterog. Catal. Fine Chem., III), 623 (1993).
106. A. Corma, R. M. Martin-Arands and F. Sanchez, *J. Catal.*, **126**, 192 (1990).
107. P. Grange, P. Bastians, R. Conanec, R. Marchand and Y. Laurent, *Appl. Catal. A*, **114**, L191 (1994).
108. A. Corma, S. Iborra, J. Primo and F. Rey, *Appl. Catal. A*, **114**, 215 (1994).
109. E. Dumitriu, V. Hulea, N. Bilba, G. Carja and A. Azzouz, *J. Mol. Catal.*, **79**, 175 (1993).
110. L. S. Posner and R. L. Augustine, *Chem. Ind. (Dekker)*, **62** (Catal. Org. React.), 531 (1995).
111. L. S. Posner, Ph. D. Dissertation, Seton Hall University, 1995.
112. Z. Fu and Y. Ono, *Catal. Letters*, **21**, 43 (1993).
113. Z. Fu and Y. Ono, *Catal. Letters*, **18**, 59 (1993).
114. P. Kumar, S. V. N. Raju, R. S. Reddy and B. Pandey, *Tetrahedron Lett.*, **35**, 1289 (1994).
115. S. Natio and M. Tanimoto, *J. Catal.*, **130**, 106 (1991).
116. S. S. C. Shuang and S. Pien, *J. Catal.*, **138**, 536 (1992).
117. N. Takahashi, Y. sato, Y. Uchiumi and K. Ogawa, *Bull. Chem. Soc., Japan*, **66**, 1273 (1993).
118. M. Lenarda, R. Ganzerla, S. Enzo, L. Storaro and R. Zanoni, *J. Mol. Catal.*, **80**. 105 (1993).
119. M. Lenarda, R. Ganzerla, L. Storaro and R. Zanoni, *J. Mol. Catal.*, **79**, 243 (1993).
120. G. Srinivas and S. S. C. Chuang, *J. Catal.*, **144**, 131 (1993).
121. S. Pien and S. S. C. Chuang, *J. Mol. Catal.*, **68**,313 (1991).
122. S. S. C. Chuang and S. Pien, *Catal. Letters*, **6**, 389 (1990).
123. Z. Vit, J. L. Portefaix and M. Breysse, *Appl. Catal. A*, **116**, 259 (1994).
124. K. Takeuchi, T. Hanaoka, T. Matsuzaki, Y. Sugi, S. Ogasawara, Y. Abe and T. Misono, *Catal. Today*, **20**, 423 (1994).
125. S. S. C. Chuang, *Appl. Catal.*, **66**, L1 (1990).
126. S. S. C. Chuang, S. Pien and C. Sze, *J. Catal.*, **126**, 187 (1990).
127. W. Huang, L.-H. Yin and C.-Y. Wang, *Stud. Surf. Sci. Catal.*, **75** (New Frontiers in Catalysis, Pt. C), 2359 (1993).
128. J.-Z. Zhang, Z. Li and C.-Y. Wang, *Stud. Surf. Sci. Catal.*, **75** (New Frontiers in Catalysis, Pt. A), 919 (1993).
129. J. P. Colllmann, L. S. Hegedus, J. R. Norton and R. G. Finke, *Principles and Applications of Organotransition Metal Chemistry*, University Science Books, Mill Valley, CA, 1987.
130. R. L. Augustine, S. T. O'Leary, K. M. Lahanas and Y.-M. Lay, *Stud. Surf. Sci. Catal.*, **59** (Heterog. Catal. Fine Chem., II), 129 (1991).
131. R. L. Augustine and S. T. O'Leary, *J. Mol. Catal.*, **72**, 229 (1992).
132.. S. T. O'Leary and R. L. Augustine, *Chem. Ind. (Dekker)*, **47** (Catal. Org. React.), 351 (1992).
133. R. L. Augustine and S. T. O'Leary, *J. Mol. Catal.*, **95**, 277 (1995).
134. S. V. Malhotra and R. L. Augustine, *Chem. Ind. (Dekker)*, **62** (Catal. Org. React.), 553 (1995).

135. M. A. Kohler, J. C. Lee, M. S. Wainwright, D. L. Trimm and N. W. Cant, *Appl. Catal.*, **35**, 237 (1987).
136. M. G. Scaros, J. P. Westrich, O. J. Goodmonson and M. L. Prunier, *Chem. Ind. (Dekker)*, **47** (Catal. Org. React.), 373 (1992).
137. R. L. Banks, *Catalysis (London)*, **4**, 100 (1981).
138. R. L. Banks, *Applied Industrial Catalysis*, **3**, 215 (1984).
139. M. Sibeijn, E. K. Poels, A. Bliek and J. A. Moulijn, *J. Am. Oil Chem. Soc.*, **71**, 553 (1994).
140. E. S. Finkel'shtein, V. I. Bykov and E. B. Portnykh, *J. Mol. Catal.*, **76**, 33 (1992).
141. G. C. N. van den Aardweg, R. H. A. Bosma and J. C. Mol, *J. Chem. Soc., Chem. Commun.*, 262 (1983).
142. R. L. Banks, D. S. Banasiak, P. S. Hudson and J. R. Norell, *J. Mol. Catal.*, **15**, 21 (1982).

Index